IFMBE Proceedings

Volume 56

Series editor

Ratko Magjarevic

Deputy Editors

Fatimah Binti Ibrahim
Igor Lacković
Piotr Ładyżyński
Emilio Sacristan Rock

The International Federation for Medical and Biological Engineering, IFMBE, is a federation of national and transnational organizations representing internationally the interests of medical and biological engineering and sciences. The IFMBE is a non-profit organization fostering the creation, dissemination and application of medical and biological engineering knowledge and the management of technology for improved health and quality of life. Its activities include participation in the formulation of public policy and the dissemination of information through publications and forums. Within the field of medical, clinical, and biological engineering, IFMBE's aims are to encourage research and the application of knowledge, and to disseminate information and promote collaboration. The objectives of the IFMBE are scientific, technological, literary, and educational.

The IFMBE is a WHO accredited NGO covering the full range of biomedical and clinical engineering, healthcare, healthcare technology and management. It is representing through its 60 member societies some 120.000 professionals involved in the various issues of improved health and health care delivery.

More information about this series at http://www.springer.com/series/7403

Fatimah Ibrahim · Juliana Usman
Mas Sahidayana Mohktar · Mohd Yazed Ahmad (Eds.)

International Conference for Innovation in Biomedical Engineering and Life Sciences

ICIBEL2015, 6–8 December 2015, Putrajaya, Malaysia

Springer

Editors
Fatimah Ibrahim
Centre For Innovation in Medical Engineering
Department of Biomedical Engineering
Faculty of Engineering
Unviversity of Malaya
Kuala Lumpur
Malaysia

Juliana Usman
Department of Biomedical Engineering
Faculty of Engineering
University of Malaya
Kuala Lumpur
Malaysia

Mas Sahidayana Mohktar
Centre For Innovation in Medical Engineering
Department of Biomedical Engineering
Faculty of Engineering
University of Malaya
Kuala Lumpur
Malaysia

Mohd Yazed Ahmad
Centre For Innovation in Medical Engineering
Department of Biomedical Engineering
Faculty of Engineering
University of Malaya
Kuala Lumpur
Malaysia

ISSN 1680-0737 ISSN 1433-9277 (electronic)
IFMBE Proceedings
ISBN 978-981-10-0265-6 ISBN 978-981-10-0266-3 (eBook)
DOI 10.1007/978-981-10-0266-3

Library of Congress Control Number: 2015958318

The IFMBE Proceedings is an Official Publication of the International Federation for Medical and Biological Engineering (IFMBE)

Printed on acid-free paper

This Springer imprint is published by SpringerNature
The registered company is Springer Science+Business Media Singapore Pte Ltd.

About IFMBE

The International Federation for Medical and Biological Engineering (IFMBE) is primarily a federation of national and transnational societies. These professional organizations represent interests in medical and biological engineering. The IFMBE is also a Non-Governmental Organization (NGO) for the United Nations and the World Health Organization (WHO), where we are uniquely positioned to influence the delivery of health care to the world through Biomedical and Clinical Engineering. The IFMBE's objectives are scientific and technological as well as educational and literary. Within the field of medical, biological and clinical engineering IFMBE's aims are to encourage research and application of knowledge, and to disseminate information and promote collaboration. The ways in which we disseminate information include: organizing World Congresses and Regional Conferences, publishing our flagship journal Medical & Biological Engineering & Computing (MBEC), our web-based newsletter – IFMBE News, our Congress and Conference Proceedings, and books. The ways in which we promote collaborations is through networking programs, workshops and partnerships with other professional groups, e.g., Engineering World Health.

Mission

The mission of the IFMBE is to encourage, support, represent and unify the world-wide Medical and Biological Engineering community in order to promote health and quality of life through advancement of research, development, application and management of technology.

Objectives

The objectives of the International Federation for Medical and Biological Engineering shall be scientific, technological, literary, and educational. Within the field of medical, clinical and biological engineering its aims shall be to encourage research and the application of knowledge, and to disseminate information and promote collaboration.
In pursuit of these aims the Federation may, in relation to its specific field of interest, engage in any of the following activities: sponsorship of national and international meetings, publication of official journals, co-operation with other societies and organizations, appointment of commissions on special problems, awarding of prizes and distinctions, establishment of professional standards and ethics within the field, or in any other activities which in the opinion of the General Assembly or the Administrative Council would further the cause of medical, clinical or biological engineering. It may promote the formation of regional, national, international or specialized societies, groups or boards, the coordination of bibliographic or informational services and the improvement of standards in terminology, equipment, methods and safety practices, and the delivery of health care.
In general the Federation shall work to promote improved communication and understanding in the world community of engineering, medicine and biology.

Foreword

The International Conference for Innovation in Biomedical Engineering and Life Sciences (ICIBEL 2015) is organized by the Center of Innovation in Medical Engineering (CIME), Department of Biomedical Engineering, Faculty of Engineering, under the Innovative Technology Research Cluster, University of Malaya. This conference is also co-organized by the Society of Medical and Biological Engineering, Malaysia (MSMBE), and endorsed by the International Federation for Medical and Biological Engineering (IFMBE), United States of America.

We are honored to present to you the collection of papers in the area of Biomedical Engineering and Life Sciences. The papers are contributed by presenters from Australia, China, Egypt, Germany, India, Malaysia, Mexico, New Zealand, Pakistan, Singapore, Japan, Hong Kong, Canada, Bangladesh and United States of America. We hope the content of the proceedings will stimulate curiosity, increase your understanding in these fields, and beneficial to your current and future research activities.

On behalf of the organizing committee of ICIBEL2015, it is our pleasure to take this opportunity to convey our highest appreciation to our reviewers, editors, keynote and invited speakers, paper presenters, members of the organizing committee and advisory panels for ICIBEL2015. Without the contributions, commitment and support of everyone, the compilation of this proceedings would not be possible. Thank you.

Prof. Ir. Dr. Fatimah Ibrahim
Chairperson,
ICIBEL 2015 Organizing Committee
President
Society of Medical and Biological Engineering, Malaysia (MSMBE)

Conference Details

Name
International Conference for Innovation in Biomedical Engineering and Life Sciences

Short Name
ICIBEL 2015

Venue
6th, 7th and 8th December 2015
Pullman Hotel, Putrajaya, Malaysia

Proceedings Editors
Fatimah Ibrahim
Juliana Usman
Mas Sahidayana Mohktar
Mohd Yazed Bin Ahmad

Organized by
Centre for Innovation in Medical Engineering,
Department of Biomedical Engineering,
Faculty of Engineering,
University of Malaya,
Malaysia

Co-Organized by
Society of Medical and Biological Engineering, Malaysia (MSMBE)

Endorsed by
International Federation for Medical and Biological Engineering (IFMBE)

Supported by
IEEE University of Malaya Student Branch
IEEE Engineering in Medicine and Biology Society (EMBS)
Biomedical Engineering Society (Singapore)

Organizing Committee

Chairperson

Fatimah Ibrahim

Co-chair

Tan Maw Pin

Publication

Fatimah Ibrahim
Juliana Usman

Secretary

Noraisyah Mohamed Shah

Treasurer

Mas Sahidayana Mohktar

Publicity

Norhayati Soin

Technical

Mas Sahidayana Mohktar
Mohd Yazed Ahmad

Sponsorship/Exhibition

Siti Zawiah Binti Md Dawal

Workshop Committee

Bashar Yafouz
Mohammad Mahdi Aeinehvand

Secretariat Committee

Bashar Yafouz
Faiz Zulkeflee
Karunan Joseph
Mazita Binti Mohamad
Mohammad Mahdi Aeinehvand
Nalisa Shamyra Binti Johari
Nurhaslina binti Abd Rahman
Reena Sri Selvarajan
Ridgwan Welfred
Shah Mukim Uddin
Yuslialif Yusup

International Advisory Board

A.S. Md. Abdul Haseeb, Malaysia
Abdul Aziz Bin Abdul Raman, Malaysia
Ahmad Khairi Abdul Wahab, Malaysia
Alan Murray, United Kingdom
Awg Bulgiba Bin Awg Mahmud, Malaysia
Chin Ai-Vyrn, Malaysia
James Goh Cho Hong, Singapore
Khairul Azmy Bin Kamaluddin, Malaysia
Lim Kheng Seang, Malaysia
Marc Madou, United States
Mohd Nasir Taib, Malaysia
Nigel Lovell, Australia
Nitish Thakor, Singapore
Noor Azuan Bin Abu Osman, Malaysia
Noorsaadah Binti Abd Rahman, Malaysia
Ramesh Singh A/L Kuldip Singh, Malaysia
Shahrul Bahyah Binti Kamaruzzaman, Malaysia
Sharifah Bee, Malaysia
Sulaiman Wadi Harun, Malaysia
Tauran Zaidi Ahmad Zaidi, Malaysia
Toshiyo Tamura, Japan
Wahidah Mansor, Malaysia
Wan Abu Bakar Bin Wan Abas, Malaysia
Yong Ping Zheng, Hong Kong
Zamane Abdul Rahman, Malaysia
Zhong Liang, Singapore

Contents

Electrode Considerations, Excitation Methods and Measurement Techniques for Electrical Impedance Tomography

C. Venkatratnam[1,2] and N. Farrukh[2]

[1]Asia Pacific University, School of Engineering, Bukit Jalil, Malaysia
[2]Universiti Tenaga Nasional, Faculty of Mechanical Engineering, Kajang, Malaysia

Abstract— **Careful selection of electrode material, configuration, dimensions, number and the inter electrode distance can improve the spatial resolution of the reconstructed images in electrical impedance tomography (EIT) applications. Considering the best of these features, a wet, non-polarizable circular, single type, Ag/AgCl electrode of 10mm diameter is chosen for the study. The inter electrode distance is maintained as per the standard of 20 mm. An EIT system with eight electrodes was developed and studied for different excitation methods and measurement techniques. The statistical analysis suggest that the adjacent method with a standard deviation as low as 0.06 is still the most suitable excitation method for mono polar voltage measurements.**

Keywords— **Electrodes, Adjacent method, Multiplexer, Standard deviation, Electrical impedance tomography.**

I. INTRODUCTION

The spatial resolution of an electrical impedance tomography (EIT) system does not only depend on the hardware components but also on the accuracy of the measurements. Electrodes, excitation methods and measurement patterns play an important role towards the spatial resolution of the image. Electrodes are the sensors used for measuring the two dimensional (2D) electrical field on the surface of the object under study. Majority of the EIT systems work on input-adjacent current method and output-differential voltage measurement technique.

II. LITERATURE REVIEW

This section is divided into two parts, the first part discusses on the electrode considerations of the EIT systems and the second part highlights on the excitation methods and measurement techniques studied by other researchers.

A. EIT Electrodes

Maximum contact stability and minimum contact impedance are the two desired features expected form any electrodes that are used for the EIT applications. A point electrode would meet the above requirements in an ideal situation. Rather, all the electrodes have some contact area and the above two requirements must be met within this scope. The electrode material, configuration, dimensions, number and the inter electrode distance (IED) contribute towards the quality of the reconstructed image(s).

a) Electrode Material

Electrodes could be either wet or dry type. The wet type comes with a conductive gel between the electrode itself and the test object for the better flow of current while the dry type do not require the conductive gel. Hence the former has better contact stability while the latter has less of it [1]. Wet type electrodes are most suitable for the EIT measurements [2]. The electrode materials can also be classified as either polarizable or non-polarizable electrodes. The polarizable electrodes are characterized by a capacitive behavior i.e. when a potential source is applied, there is no current flowing between the electrode and the test object. Examples include electrodes made up of Gold and Platinum. The non-polarizable electrodes are characterized by a resistive behavior i.e. when a voltage is applied, there is a free flow of current across the electrode – test object interface. The non-polarizable electrodes are highly stable and have lower noise levels compared to the polarized electrodes [2]. Silver/Silver Chloride (Ag/AgCl) electrode is the widely used wet type, non-polarizable material for the EIT applications in this category [3].

b) Electrode Configuration

The EIT systems can be either mono polar or of differential configurations. The mono polar configurations refer to the information collected from the individual electrodes. This approach is exclusive for research applications. The differential configurations on the other hand pick up the information between two electrodes. This approach gives the electrical field distribution as a function of time as well as space [1].

c) Electrode Dimensions

Most of the EIT systems either use a circular or a rectangular shaped electrodes. These can be of single type or compound type [4]. In the single type, the current is injected and voltage is measured from the same electrode. The compound electrode consist of two individual electrodes,

F. Ibrahim et al. (eds.), *International Conference for Innovation in Biomedical Engineering and Life Sciences*,
IFMBE Proceedings 56,
DOI: 10.1007/978-981-10-0266-3_1

where the larger one is used for the current injection and the smaller one is used for the voltage measurements.

The compound electrodes help in minimizing the contact impedance but however tend to limit the inter electrode distance [5], does not actually help in improving the image resolution [6] and tend to reduce the current distribution to the interior of the test tank when compared with a single electrode [7].

Some more electrode designs can be seen in the literature. An inflatable belt with carbon electrodes was used for ease of use as compared to the sticky electrodes. Instead, they gave rise to more noise [8]. Similarly, a flexible belt with conductive nano-fiber electrodes was used to provide low contact impedance during EIT measurements. However the hardware had cumbersome wires [9]. To avoid this an active electrode was designed with less wiring by bringing all the electronic components closer to the test object. However, the system was bulky and uncomfortable [10].

d) Electrode Number

It is a common belief that the increasing the number of electrodes in EIT applications will result in better image resolution. But, there is a limit for increasing the number of electrodes beyond which the image resolution is no longer improved [7]. More electrode also mean more hardware and hence more processing time. Moreover, if an EIT system uses a number of electrodes, then all the electrodes must be driven by the input currents simultaneously for better results [11].

e) Inter Electrode Distance

The center-to-center distance between two consecutive electrodes is the inter electrode distance (IED) and should not be more than 20 mm for EIT applications.

In view with the above discussion, a wet, non-polarizable circular, single type Ag/AgCl electrode of 10 mm diameter is chosen for the study. The inter electrode distance is maintained as per the standard of 20 mm. Only 8 electrodes are used for hardware simplicity.

B. *Excitation Methods and Measurement Patterns.*

AC current is safer than the DC current in EIT studies. A common practice is to apply the currents and measure the voltages rather than applying the voltages and measuring the currents. This is mainly done to reduce the noise due to the spatial variations when the currents are applied. Also floating currents are not suitable for the EIT applications. Instead, grounded current sources are used [12]. Individual voltage measurements is carried out to get complete information on the test object rather than the differential voltage measurements which is a function of both time and space [1].

The EIT systems could be either applied current type (ACT) or applied potential type (APT). In the ACT type, the input current is applied to the electrodes. Further, ACT can be single current source type or multi-current type [13]. In the single current source method, the input current is being applied to each single electrode one at a time in a cyclic manner while in the multi-current source method all the electrodes are injected with independent input current simultaneously using independent multiple current sources. The ACT with multiple current sources has the advantage of delivering maximum power to the electrodes and hence better resolution of the images is expected from the EIT systems [14]. On the other hand, these systems become more expensive because of the independent multiple current sources which equal to the number of electrodes are thereby making the EIT systems more complex [13]. In the ACT with single current source, a single current source is used and the current is applied to a pair of electrodes one at a time and repeated for all possible electrode pairs. The ACT with single current source are generally used in the EIT systems rather than the APT systems as they mainly overcome the contact impedance errors.

The electrode pair combination for the single current source can be neighboring, opposite or cross [15]. The neighboring method or also known as the adjacent method is proposed as the best excitation method especially for circular test objects [16], for a single current source [6], for static images [17] and for the boundary measurements [18]. But [19] did not support this method saying that the adjacent method is usually chosen to support the back propagation algorithm.

III. MATERIALS AND METHODS

A wet, non-polarizable circular, single type Ag/AgCl electrode of 10 mm diameter was used for the following study. In this type, Fig 1, the metal stud allows even the smallest voltages to be measured, the backing foam material provides the necessary stability to the electrode, Ag/AgCl ensures the accuracy of the readings, the adhesive is suitable for the long term monitoring at patients comfort and the gel acts as a conductive medium between the electrode and the human body.

A test tank made up of polypropylene material is used. The dimensions of the test tank are 1760 mm^2 × 120 mm. Eight electrodes are placed on the outer surface of the test tank which is filled with 10% saline water. The inter electrode distance of 20mm is maintained.

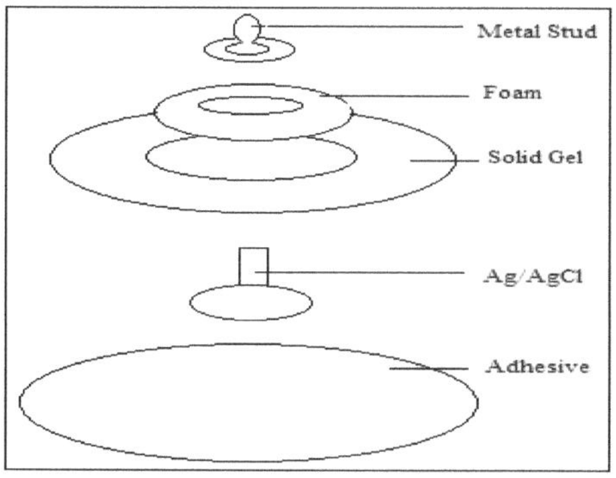

Fig. 1 Anatomy of Ag/AgCl electrode

A voltage-controlled oscillator (VCO) using AD9850 generates an alternating voltage of 1 V_{PP} at 100 KHz. This is converted into a current of 1 mA considering a load of 1 KΩ. This current of is then applied to a pair of electrodes using neighboring, opposite and cross methods as shown in Fig 2.

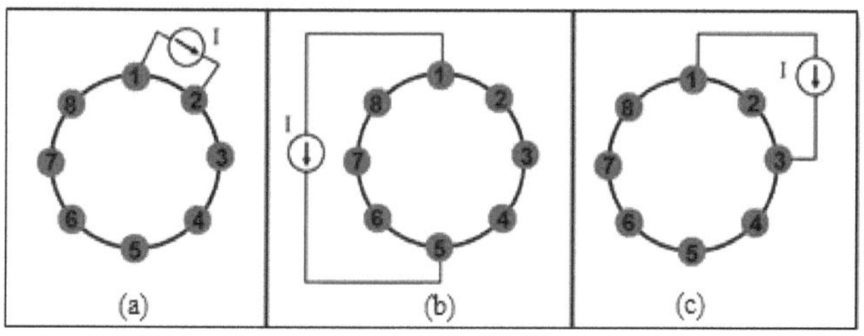

Fig. 2 (a) Neighboring (b) Opposite (c) Cross methods

The mono-polar voltages are finally measured from the remaining non-current carrying electrodes. The voltages of the current carrying electrodes are not measured due to a large voltage drop across them. Later all the measured voltages are summed up for better readings [20]. The entire experimental set up is shown in Fig 3.

A LabVIEW program is developed for shifting the currents according to a sequence using multiplexer circuit. The switching order of the multiplexers for the three different methods is according to [21]. Each electrode will get eight analog voltage readings for each cycle. However, two of these readings are zeros, because each electrode will inject current once, and once will be grounded. A VI is developed to get the sum of the eight readings from each electrode.

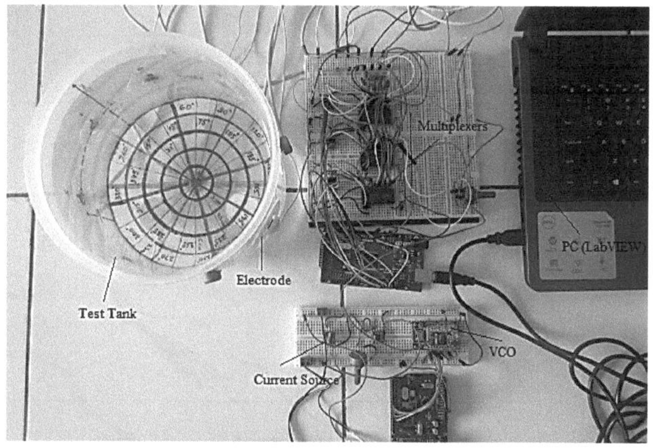

Fig. 3 Complete experimental set up

IV. Results and Discussion

The mono-polar voltages are tabulated for the three different methods of current injection. The measurements were carried out five times for each excitation pattern. The mono-polar voltages acquired by the neighboring, opposite and cross methods are shown in Table 1, Table 2 and Table 3 respectively.

Table 1 Measured voltages by the neighboring method

Test	Mono polar voltages of each electrode							
No	1	2	3	4	5	6	7	8
1	0.99	0.95	0.96	1.02	0.64	0.77	0.87	0.82
2	0.99	0.82	0.83	1.31	0.88	0.85	0.89	0.77
3	0.80	1.02	0.96	0.95	1.12	0.87	0.78	0.52
4	0.95	0.66	0.94	0.96	0.89	0.94	0.81	0.91
5	0.75	1.18	0.65	0.60	1.26	0.94	0.74	0.97
Avg	0.89	0.93	0.87	0.97	0.96	0.87	0.82	0.80

Table 2 Measured voltages by the opposite method

Test	Mono polar voltages of each electrode							
No	1	2	3	4	5	6	7	8
1	0.98	0.99	0.98	0.93	0.54	1.68	0.48	0.67
2	0.87	0.95	0.96	0.89	0.75	0.95	0.91	0.96
3	0.84	0.93	0.93	0.91	0.70	0.99	1.29	0.80
4	0.80	0.76	0.97	0.80	0.52	1.40	1.22	0.92
5	1.06	0.87	1.17	0.98	0.96	1.14	0.94	0.64
Avg	0.91	0.90	1.00	0.90	0.69	1.23	0.97	0.80

Table 3 Measured voltages by the cross method

Test	Mono polar voltages of each electrode							
No	1	2	3	4	5	6	7	8
1	0.96	0.85	1.32	0.62	0.71	0.81	0.70	0.73
2	0.67	1.10	0.81	1.48	0.93	1.07	0.88	0.96
3	0.79	1.12	1.14	0.99	1.09	0.35	0.92	0.64
4	0.76	0.70	0.95	0.75	0.81	0.76	0.76	0.68
5	0.74	1.31	1.45	0.80	1.14	1.32	0.74	0.64
Avg	0.79	1.02	1.13	0.93	0.94	0.86	0.80	0.73

The average voltages are then plotted w.r.t. the electrodes for the above three methods as shown in Fig 4.

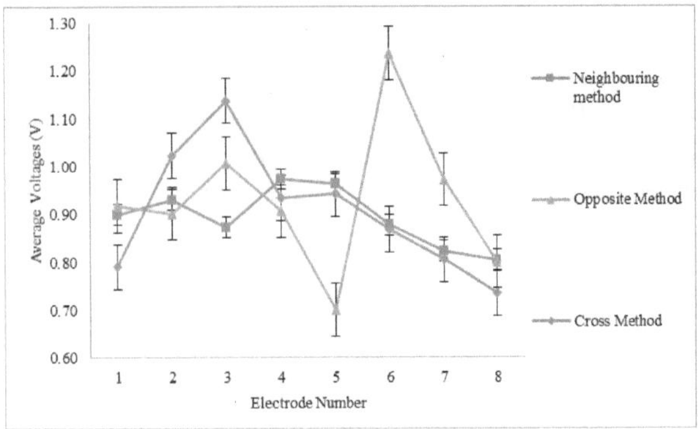

Fig. 4 Comparative graphs of the three excitation methods

The variation of the average mono polar voltage readings among the electrodes is more in the opposite methods and the cross methods as compared with neighboring method of current injection. In order to quantify this amount of variation or dispersion of a set of data values, the standard deviation (σ) is calculated for the three methods as,

$$\sigma = \sqrt{\frac{\Sigma(x-\bar{x})^2}{n-1}} \quad (1)$$

where, x is the individual mono-polar voltages, $\bar{x}$ is the mean value of the voltages and n is the number of voltage values. The calculated values are shown in Table 4.

Table 4 Standard deviation of the three excitation methods

Method	$\bar{x}$	σ
Neighboring	0.89	0.06
Opposite	0.93	0.15
Cross	0.90	0.13

Unlike most of the researchers, the mono polar voltages are measured rather than the differential voltages in order to get more spatial information of the test object. The statistical analysis suggest that the adjacent method with a standard deviation as low as 0.06 is still the most suitable excitation method for EIT applications as compared to the opposite and cross methods of current excitation.

V. CONCLUSION

A wet, non-polarizable circular, single type Ag/AgCl electrode of 10 mm diameter has been found suitable for the electrical impedance tomography (EIT) studies. The inter electrode distance must not be greater than 20 mm. Apparently the adjacent method of excitation shows more stabilized mono polar voltage readings, compared with the opposite and cross methods, as the standard deviation of the adjacent method is less than that of the opposite and cross methods. Hence, the adjacent method has the lowest dispersion or variation in the mono-polar voltage values and is still suggested as the most suitable excitation method for the EIT applications.

CONFLICT OF INTERSET

The authors declare that they have no conflict of interest.

REFERENCES

1. Alberto Botter et.al. (2013) "Surface Electromyogram Detection" Chapter 6, Introduction to Neural Engineering for Motor Rehabilitation, First Edition, The Institute of Electrical and Electronics Engineers, Inc., John Wiley & Sons, Inc.
2. W Wang et.al. (2007) "Study into the repeatability of the electrode-skin interface utilizing electrodes commonly used in Electrical Impedance Tomography" ICEBI, IFMBE Proceedings 17, pp. 336–339.
3. R. Merletti and H. J. Hermens (2004) "Detection and Conditioning of the Surface EMG Signal" Chapter 5, Electromyography: Physiology, Engineering, and Noninvasive Applications, Institute for Electrical and Electronics Engineers, Inc.
4. Wang Yan et.al. (2005) "An Optimizing Design Method for Electrode Structure Parameter of Electrical Impedance Tomography" Proceedings of the IEEE, Engineering in Medicine and Biology.
5. Ping Hua et.al. (1993) "Using Compound Electrodes in Electrical Impedance Tomography" IEEE Transactions on Biomedical Engineering, Vol. 40, No.1.
6. Thomas F. Schuessler and Jason H. T. Bates (1998) "Current Patterns and Electrode Types for Single-Source Electrical Impedance Tomography of the Thorax" Annals of Biomedical Engineering, Vol. 26, pp. 253–259.
7. Cheng-Ning Huang et.al. (2007)"Rotational electrical impedance tomography" Measurement Science and Technology, IOP Publishing.
8. R.J. Sadleira et.al. (2000) "Inflatable belt for the application of electrode arrays" Review of Scientific Instruments Volume 71, Number 2, American Institute of Physics.
9. Tong In Oh et.al. (2012) "Flexible electrode belt for EIT using nanofiber web dry electrodes" Measurement Science And Technology,IOP Publishing.

10. Pascal Olivier Gaggero et.al. (2012) "Electrical impedance tomography system based on active electrodes" Measurement Science and Technology, IOP Publishing.
11. Margaret Cheney et.al. (1999) "Electrical Impedance Tomography" SIAM Review, Vol. 41, No. 1, pp. 85-101.
12. LI Xiaoke et.al. (2012) "Analysis of Constant-Current Characteristics for Current Sources" 24th Chinese Control and Decision Conference (CCDC), IEEE.
13. Ping Hua et.al. (1992) "Improved Methods to Determine Optimal Currents in Electrical Impedance Tomography" IEEE Transactions on Medical Imaging. Vol. 1, No. 4.
14. G.J.Saulnier et.al. (2011) "Electrical Impedance Tomography" IEEE Signal Processing Magazine.
15. Jaakko Malmivuo & Robert Plonsey (1995) "Impedance Tomography" Chapter 26, Bioelectromagnetism book.
16. M.J. Booth and I. Basarab-Horwath (1996) "Comparing electrode configurations for electrical impedance tomography" Electronics Letters, Vol. 32 No. 7.
17. Canhua Xu et.al. (2008) "Comparison of Drive Patterns for Single Current Source EIT in Computational Phantom" 978-1-4244-1748-3/08 © IEEE.
18. Shu Zhao et.al. (2013) "The Impact of the Measurement Accuracy and the Excitation Pattern on EIT Image Reconstruction" 6th International Conference on Biomedical Engineering and Informatics.
19. Andy Adler et.al. (2011) "Adjacent stimulation and measurement patterns considered harmful" Physiological Measurement, IOP Publishing.
20. CC Barber et.al. (1983) "Imaging Spatial Distributions of Resistivity using Applied Potential Tomography" Electronics Letters, Vol. 19 No. 22.
21. Venkatratnam Chitturi et.al (2014) "A Low Cost Electrical Impedance Tomography (EIT) for Pulmonary Disease. Modelling and Diagnosis" Page 83-87, The Second International Conference on Technological Advances in Electrical, Electronics and Computer Engineering.

Address of the corresponding author:

Author:	Venkatratnam Chitturi
Institute:	Universiti Tenaga Nasional
Street:	Kajang
City:	Selangor
Country:	Malaysia
Email:	chitturi@apu.edu.my

Developing an Optimized Single-Trial P300-Based Brain Computer Interface System

F.E. Motlagh[1] and F. Ibrahim[1]

[1]Department of Biomedical Engineering, Faculty of Engineering, University of Malaya, Kuala Lumpur, Malaysia

Abstract— **Brain event related potentials (ERP) have been used in developing brain computer interface (BCI) systems. P300 as a robust ERP has been utilized in BCI and clinical researches. A common P300-based BCI system consist of brain signal recording, pre-processing, P300 features extraction, and classification units. Achieving a high accuracy in detection of single-trial P300, using fast computational algorithms is the main challenge of designing these systems. However, there is trade-off between accuracy and computational time. In this study, various well-developed algorithms controlled by a rule-based platform to optimize the detection algorithm. P300 feature extraction algorithms has been developed by using wavelet transform techniques, while SVM with linear/Gaussian kernels and logistic regression applied as alternative supervised learning classifiers. Principle component analysis also was used for feature selection in order to speed up the classification procedure. This optimization system make decision on selecting the proper P300 detection method via selecting the group of channels, feature extraction algorithm, number of selected principle components, and type of classifier. Controller used cross validation data set to calculate the accuracy and ratio of computational time for each possible combination, and the optimized method was assessed using test data set. The results suggest that designing a P300-BCI system with the ability to select the proper method of detection can be utilized in different applications to benefit the user with a better performance.**

Keywords— **Brain computer interface, Event related potentials, Wavelet transform, Support vector machine, Supervised learning.**

I. INTRODUCTION

BCI systems are designed to transform brain electrophysiological signals into commands for computers. Brain computer interface (BCI) systems has been used excessively for research purposes and clinical diagnostics in the past decade [1].There have been various BCI systems based on different attributes of brain signals. However, event-related potential (ERP)-based BCI systems are known as effective systems in this field. P300 is a robust positive ERP which has been utilized in BCI systems and shows promising results in terms of accuracy and robustness [2]. P300 occurs as a response to rare task-relevant stimuli in a series of task irrelevant stimuli around 300 ms after stimuli [3]. Farwell and Donchin described P300-BCI system to communicate with computers without utilizing voluntary muscle activity [4], using oddball paradigm to evoke P300 component [5]. Oddball paradigm demonstrate a random sequence of desired and undesired events, which is supposed to probe P300 during the novel desired stimuli. The aim of BCI detection algorithm is to detect the target signals (carry P300) among non-targets.

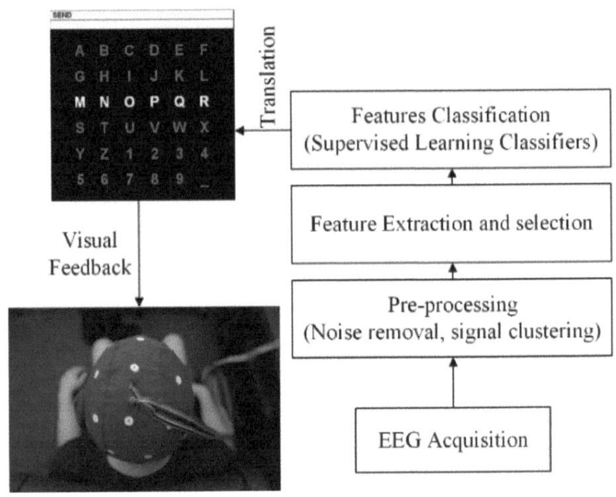

Table 1 Fig. 1 BCI system components, including data acquisition, signal processing and BCI application units. Brain signals are recorded using an EEG machine. Signal processing using is consist of Pre-prosessing, feature extraction, and classification blocks. BCI application includes translated command of brain and a paradigm for stimulating the brain.

Typical P300-BCI system consists of three major blocks: (1) signal acquisition and pre-processing, (2) P300 features extraction, and (3) classification (Figure 1). There are challenges in developing components of this system; each unit of the system should be developed in order to optimize the performance of whole system. Designing an algorithm for providing noise and artifact free signals, accurate P300 feature extraction and selection, and finally efficient classification of these features are the main criteria of a robust P300 detection method. However, when it comes to real time and online applications, processing time, number of channels, accuracy of single-trial detection, and complexity of the classifier are significance issues to be considered.

This paper aims to utilize a rule-based controller for optimizing the single-trial P300 detection accuracy and

F. Ibrahim et al. (eds.), *International Conference for Innovation in Biomedical Engineering and Life Sciences*,
IFMBE Proceedings 56,
DOI: 10.1007/978-981-10-0266-3_2

enhancement of processing time in real-time P300-BCI systems. This optimization platform evaluates the EEG recording channels and employs the proper algorithm for pre-processing, feature extraction, and classification units.

II. P300-BCI SYSTEM

A. Signal Acquisition

Although various types of current scalp electroencephalogram (EEG) equipment record efficient signals using convenient and user-friendly electrodes; there are two significant considerations about the recording signals and number of recording channels in a P300-BCI system. Developing a P300-BCI system using large number of electrodes cause user discomfort as well as longer processing time. Therefore, employing smaller subset of electrodes was suggested to reduce the processing time and increasing user comfort, while providing enough information for an accurate detection of P300. Some studies aimed to define an optimal subset of electrodes applicable in P300-BCI systems, e.g. Krusienski et al, suggested 8-channel electrodes set (Fz, Cz, P3, Pz, P4, PO7, PO8, Oz) [6], and Motlagh et al, suggested five channels (C1, Cz, Cpz, Pz, Fcz) [7].

In this study, dataset were obtained based on 10–20 system using 19 EEG Channels ($Fp_{1/2}$, $F_{7/8}$, F_z, $F_{3/4}$, $T_{3/4}$, $C_{3/4}$, C_z, $T_{5/6}$, $P_{3/4}$, P_z, $O_{1/2}$) with average of $A_{1/2}$ as reference during the performance of an oddball paradigm. Nicolet EEG diagnostics system (Care Fusion Corporation, 3750 Torrey View Court, San Diego, CA 92130) was used to capture the EEG activities within the frequency band of 0.5-70Hz (with a sampling rate of 256 Hz). Before data collection, the impedances of all the electrodes were monitored during the EEG recording, to verify its value to be under 5 kΩ, and the paradigm timing system and the EEG recorder were synchronized.

B. Pre-Processing

EEG is highly susceptible to various forms and sources of noises, which present significant difficulties and challenges in analysis and interpretation of EEG data. Preprocessing the data as the most essential step in development of a reliable BCI system should be accurate and efficient. In this study, an automated standard pre-processing steps was applied on the signals in the initial phase.

Each channel's signal band-pass filtered between 0.1-45 Hz using *slepian multitaper* spectrum (MATLAB *"pmtm"* function) by applying four orthogonal tapers, (a combination of a high pass and low pass filter), in order to remove power line, high frequency noises and DC biases.

Signal mean, standard deviation, skewness, kurtosis and median (five first cumulates of distribution) were calculated and stored. Signals data-points distribution from each channel shows an estimation of quality of EEG recording from that channel. Kolmogrov-smirnov test applied to estimate the distribution of the signal of each channel subsequently. The result of this test at a significant level of $P \leq 0.05$ shows whether the data distribution of signal is Gaussian or not; thus, each channel would be labeled based on the equation (1) criteria.

$$Channel\ Number \begin{cases} 1 & P\ value \leq 0.05 \\ 0 & P\ value > 0.05 \end{cases} \quad (1)$$

Channels with label "0" should be eliminated from the rest of the procedure. In order to have a uniform and standard procedure for detecting the EEG artifacts, all Gaussian signals were divided to epochs with duration of one-second period, and following steps were applied:

Removing linear trend: During acquisition of EEG, recording-induced current drifts and electrode movements cause occurrence of linear drifts in EEG trials. This type of artifacts was removed by fitting a straight line to the data-points of the signal, and if the slope of the calculated line is more than 60 μv, then the epoch's linear trend was removed.

ICA decomposition: Independent component analysis (ICA) algorithm was used for removing the artifacts, eye-movements and blinking using *"runica infomax"* in MATLAB. ICA decomposed the channels' signals into independent signals (based on orthogonality using singular value decomposition algorithm) as a common method for artifact removal and source localization of EEG signals (this method described in [8, 9]).

ICA Reconstruction: After removing the artifactual independent components (IC), remaining signals was used for reconstruction of channels' EEG. Then, ICA decomposition was applied again in order to guarantee the noise removal procedure.

Windowing: Each signal should be synchronized with the stimuli onset timing of the paradigm; therefore, 256 samples (1second) from the onset of the stimuli was selected as a single trial.

Training set labeling: Single-trial P300 detection algorithm is based on a supervised learning classification. therefore 60% of the dataset were selected randomly as training set, and each trial of training set were labeled as target or non-target.

C. Feature Extraction

The major challenge in optimizing the performance of the P300-BCI is to enhance the real-time detection of P300. The process of real-time detection consists of an optimal P300 features extraction in order to employ a simpler classification algorithm to increase the processing speed in real-time applications. P300 like other event related potentials has a very low amplitude (μv) compared to baseline EEG signals (mv); this per se compromise the detection accuracy and cause a trade-off between detection speed and accuracy. Additionally, high signal-to-noise ratio (SNR) of EEG signals make this very challenging [10]. Various studies have been employed orthogonal linear transformation [11], blind source separation, wavelet transform [12, 7, 2, 13-15] and other advanced techniques to overcome these challenges. The key point of established feature extraction techniques is to utilize most distinctive features to reduce the computational time.

In this study, two feature extraction algorithms were designed based on wavelet transform (WT) properties:

CWT: In the first approach, continuous WT (CWT) were applied on the trials using *Mexican-hat* wavelet (scales 30-100). It was shown that this wavelet has a good similarity on scales of 30-100 with P300 component (this range of scale is associated with 0.6-2 Hz in frequency domain) [7]. In this method, each trial was swept by different scales of wavelet and their correlation was calculated for each time-scale as similarity coefficients. Wavelet coefficients of a signal x (t) at time point p are defined as:

$$\int_0^{256} x(t)\varphi(s,p,t)dt \qquad (2)$$

Where s is the wavelet scale, t is trial data-points, and φ is the chosen wavelet (*Mexican-hat*).

Then, CWT coefficients were averaged over different scales and extermum values of obtained vector was stored. It is assumed that the maximum of averaged curve has the amplitude of A_0 that happens at time T_0. The goal is to find the two local minimums, i.e., one just after A_0 and another just before A_0 with amplitudes of B_1 and B_2 respectively. Using "equations (3) and (4)", and the obtained extermum properties, two heuristic features over averaged scales can be defined. "A" as the similarity amplitude and "T" as ratio of latency.

$$A = (A_0 - B_1) + (A_0 - B_2) \qquad (3)$$

$$T = {|T_0 - 300|}/{300} \qquad (4)$$

For detecting P300 wave the amplitude of the peak feature should have "large" value and time difference feature should be as "small" as possible (zero is considered ideal). Therefor, *A* and *T* are two features to be extracted by applying CWT. This method was confirmed to provide robust features for single-trial P300 detection, although calculation of correlation and sweeping the signals for all scales increase the processing time [7, 13].

(DC)WT: In the second approach, a combination of discrete WT (DWT) and CWT was applied for providing more robust features. In this method, discrete wavelet transform was used for multi-resolution decomposition of signal into 'details' and 'approximation' (high frequency and low frequency) components. B-Splines wavelets were chosen as mother function in this study due to their high resemblance with brain evoked responses. Five levels of DWT transformed the signals into 64–128 Hz, 32–64 Hz (gamma), 16–32 Hz (beta), 8–16 Hz (alpha), 4–8 Hz (theta) and the last approximation giving the activity in the 0.1–4 Hz (delta). Since, each step of DWT decomposition divide the signal into two components by down-sampling, delta and theta band (0.1-8 Hz) contains 16 data points. These 16 data points were stored as DWT features.

Thanks to low frequency of evoked potentials, delta and theta decompositions were used for reconstructing the signal and up-sampling. The reconstructed signal from delta and theta band results in a smooth signal 0-8 Hz. CWT as it explained earlier was applied on reconstructed signal and CWT and DWT features was stored (18 features).

Final number of features in this method is these features multiple by the number of selected channels. Since, dealing with large number of features leads more computational time for classification; selected features should be reduced into lower dimensions using principle component analysis (PCA). PCA reduced the features dimensionality into lower orthogonal dimensions using Eigen vectors of features. The number of principle component should be chosen based on the percentage of variance that retained. The optimal percentage of retained variance is supposed to be 99%. The whole procedure of this approach is depicted in Figure 2.

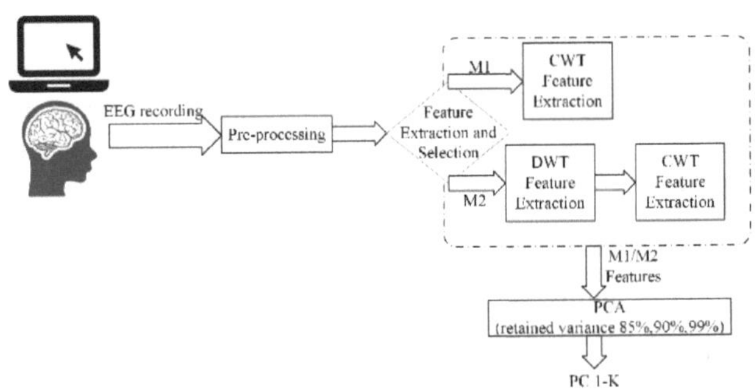

Fig. 2 Feature extraction methods designed based on two approaches. CWT and DWT features of each channel should transformed to a lower dimensions using PCA algorithm. The controller decide on the number of principle components (K) to be selected as the inputs of classifier.

D. Classification

Single-trial P300 detection requires accurate classification of extracted features. Numerous studies have attempted to enhance the classification algorithm by utilizing linear and nonlinear methods [16-18]. However, avoiding complex but reliable classifiers benefits the enhancement of computational time. Logistic regression and support vector machine (SVM) (both linear and Gaussian kernels) were used as fast binary classifiers in this study.

Selected principle components and trials label of training dataset would be used in these three supervised classification algorithms (logistic regression, SVM with linear kernel, and SVM with Gaussian kernel).

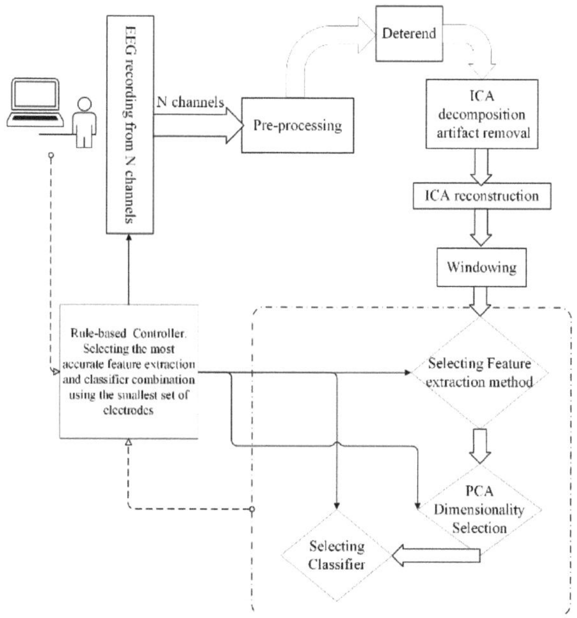

Fig. 3 Flowchart of rule-based controller of P300-BCI system. The user will choose the criteria for accuracy and computational time based on the application and system would be modified.

E. Controller

The aim of controlling the single-trial P300 detection algorithm is to choose the best procedure for detecting P300 accurately in minimum time. In this study, 60% of the data was selected randomly for training, 20 % for cross validation, the rest of data (20 %) as test dataset, and a rule-based controller was chosen for selecting the best combination of the algorithms in order to fulfill the accuracy and computational time conditions. This controller can decide on number of selected channels (NC), feature extraction algorithm (FE), number of selected features based on PCA output (FS), and selecting the type of classifier (CS). In this step, training and cross-validation set were used to evaluate each possible combination of parameters. However, there are huge number of different possibility for NC and calculation of all possible combinations is impossible. Therefore, a certain set of well-studied channels was used (Table 1 shows some of the outputs of this part).

There are five different level for each method, namely very low, low, normal, high, and very high. Computational time was normalized between 0-1, and 0.2 as threshold of each level (0-0.2 very low, 0.2-0.4 low, 0.4-0.6 normal, 0.6-0.8 high, and 0.8-1 very high). Accuracy percentage of cross validation set was divided to 50-65% very low, 65-75% low, 75-85% normal, 85-90% high, and 90-100% as very high. The controller aims to maintain the system to perform in a condition which user can define (e.g. very low computational time and high or standard accuracy percentage), shown as user condition (UC).

The system start with using Cz channel signals, CWT method of feature extraction and using just one principle component and logistic regression (the fastest method). The rule-based controller works as follow:

Step 1. If the accuracy is lower than UC, then other classifiers would be evaluated, and the one with highest accuracy is selected.

Step 2. If the accuracy is lower than UC, increase the number of K (number of principle components) that 99% variance retained.

Step 3. If the accuracy is lower than UC, evaluate both feature extraction methods and select the most accurate algorithm.

Step 4. If the accuracy is lower than UC, evaluate the second set of channels.

Step 5. Repeat steps 1-4 until the UC condition for accuracy percentage is provided.

In this system, accuracy percentage condition has priority over the computational time; therefore, after fulfilling the accuracy criterion, the computational time would be evaluated using all possible methods without changing the number of channels and the fastest method would be chosen; then, the UC condition for accuracy percentage should be reassessed. By defining the method, system use test dataset to evaluate the system again and results would be shown to the user and waiting for confirmation or new set of criteria. The flowchart of this system is depicted in Figure 3.

Table 2 Accuracy (A) and normalized computational time (NCT) for some of the possible combinations.

NC	FE	FS	CS	A%	NCT
Cz	2	3	LR	68.23	0.21
Cz, Pz	1	3	SVML	66.12	0.16
Pz	2	3	SVMG	73.29	0.19
C1	1	3	SVMG	61.19	0.18
Cpz	1	3	SVML	54.49	0.11
Fcz	2	3	LR	60.10	0.16
C1,Cz	2	2	SVMG	65.12	0.21
Cz, Pz, Fcz	2	2	SVMG	75.45	0.25
5	1	2	SVML	86.28	0.35
8	1	2	SVML	89.36	0.54
19	2	3	LR	96.38	0.96

III. CONCLUSIONS

The P300-BCI appears to be the most commonly used BCI system. Despite its popularity among researchers, it is apparent that many P300-BCI systems must be improved before they can be considered as an alternative communication device for individuals. In this paper, a rule-based controller system was applied to optimize the accuracy and processing time of single-trial P300-BCI system.

The accuracy of each set of data would be compared to select the smallest set of channels providing fastest computation and highest possible accuracy. Once the sub-group of channels selected, the system can be work in test condition using less number of channels. This controlling system can provide a better performance of a typical BCI system in various applications. Future work is focus on testing this system in various applications and compare it with current systems using fuzzy controllers and other soft computing methods for a better modification.

ACKNOWLEDGMENT

We would like to acknowledge and thank the University of Malaya and Ministry of Higher Education for providing High Impact Research Grant, account codes E000007-20001 to fund this project. This research was also supported by Special Prime Minister's project (Project No: 66-02-03-0061/oracle 8150061).

CONFLICT OF INTEREST

The authors declare that they have no conflict of interest.

REFERENCES

1. Burns A, Adeli H, Buford JA. Brain–Computer Interface after Nervous System Injury. The Neuroscientist. 2014:1073858414549015.
2. Motlagh FE. P300 detection of brain signals using a combination of wavelet transform techniques: Universiti Putra Malaysia; 2012.
3. Sanei S, Chambers JA. EEG signal processing. John Wiley & Sons; 2013.
4. Farwell LA, Donchin E. Talking off the top of your head: toward a mental prosthesis utilizing event-related brain potentials. Electroencephalography and clinical Neurophysiology. 1988;70(6):510-23.
5. Donchin E. Surprise!... surprise? Psychophysiology. 1981;18(5): 493-513.
6. Krusienski DJ, Sellers EW, McFarland DJ, Vaughan TM, Wolpaw JR. Toward enhanced P300 speller performance. Journal of neuroscience methods. 2008;167(1):15-21.
7. Motlagh F, Tang S, Motlagh O, editors. Optimal accuracy and runtime trade-off in wavelet based single-trial P300 detection. Signal and Image Processing Applications (ICSIPA), 2013 IEEE International Conference on; 2013: IEEE.
8. Schlögl A, Keinrath C, Zimmermann D, Scherer R, Leeb R, Pfurtscheller G. A fully automated correction method of EOG artifacts in EEG recordings. Clinical neurophysiology. 2007;118(1):98-104.
9. Li Y, Ma Z, Lu W, Li Y. Automatic removal of the eye blink artifact from EEG using an ICA-based template matching approach. Physiological measurement. 2006;27(4):425.
10. Li K, Raju VN, Sankar R, Arbel Y, Donchin E. Advances and challenges in signal analysis for single trial P300-BCI. Foundations of Augmented Cognition. Directing the Future of Adaptive Systems. Springer; 2011. p. 87-94.
11. Dien J, Spencer KM, Donchin E. Localization of the event-related potential novelty response as defined by principal components analysis. Cognitive Brain Research. 2003;17(3):637-50.
12. Xu N, Gao X, Hong B, Miao X, Gao S, Yang F. BCI competition 2003-data set IIb: enhancing P 300 wave detection using ICA-based subspace projections for BCI applications. IEEE transactions on biomedical engineering. 2004;51(6):1067-72.
13. Motlagh F, Tang S, Motlagh O, editors. Combination of continuous and discrete wavelet coefficients in single-trial P300 detection. Biomedical Engineering and Sciences (IECBES), 2012 IEEE EMBS Conference on; 2012: IEEE.
14. Quiroga RQ, Garcia H. Single-trial event-related potentials with wavelet denoising. Clinical neurophysiology. 2003;114(2):376-90.
15. Bostanov V, Kotchoubey B. The t-CWT: a new ERP detection and quantification method based on the continuous wavelet transform and Student's t-statistics. Clinical neurophysiology. 2006;117(12):2627-44.
16. Kaper M, Meinicke P, Grossekathoefer U, Lingner T, Ritter H. BCI competition 2003-data set IIb: support vector machines for the P300 speller paradigm. Biomedical Engineering, IEEE Transactions on. 2004;51(6):1073-6.
17. Krusienski DJ, Sellers EW, Cabestaing F, Bayoudh S, McFarland DJ, Vaughan TM et al. A comparison of classification techniques for the P300 Speller. Journal of neural engineering. 2006;3(4):299.
18. Cecotti H, Gräser A. Convolutional neural networks for P300 detection with application to brain-computer interfaces. Pattern Analysis and Machine Intelligence, IEEE Transactions on. 2011;33(3):433-45.

Author: Farid Esmaeili Motlagh
Institute: Centre for Innovation in Medical Engineering (CIME)
City: Kuala lumpur
Country: MALAYSIA
Email: farid_motlaq@hotmail.com

Neural-Response-Based Text-Dependent Speaker Identification Under Noisy Conditions

M.A. Islam, M.S.A. Zilany, and A.J. Wissam

Department of Biomedical Engineering, Faculty of Engineering, University of Malaya, Kuala Lumpur, Malaysia

Abstract— **Speaker identification is a technique of determining an unknown speaker's identity and is very essential for security, crime investigation, and telephoning. In this study, a text-dependent speaker identification technique using the neural responses of a physiologically-based computational model of the auditory periphery is proposed. Neurograms were constructed from the responses of the auditory-nerve model to sentences of different speakers. The proposed features were then used to train and test the recognition system using the support vector machine and Gaussian mixture model classification techniques. The proposed method was tested on a text-dependent database in quiet and under noisy conditions for a range of signal-to-noise ratios. Although the performance of the proposed method in quiet was comparable to the performance of a traditional Mel frequency cepstral coefficients-based method and also to the result of a very recent Gammatone frequency cepstral coefficient-based system, the neural-response-based method showed a substantially better classification accuracy under noisy conditions. The proposed method could be extended to design a text-independent speaker identification and verification system.**

Keywords— **Speaker Identification, Auditory periphery, Neurogram, MFCC, SVM.**

I. INTRODUCTION

The traditional features of speech such as frequency, formants, energy, power, and pitch vary from speaker to speaker due to variation of the vocal fold position and length of the vocal tract. Every speech signal conveys some information to the listener like what being said (message), gender, speaker identity, emotions, and the location of the source. Automatic speaker identification system is a biometric modality that uses an individual speaker's voice for identification purposes [1]. Automatic speaker identification system is related to extract, characterize, and recognize information in the speech signal of speaker.

Substantial research and development on automatic speaker identification methods have been undertaken for decades. Generally, the most common traditional speaker identification approaches use features such as Mel frequency cepstral coefficients (MFCC) [2] and linear prediction cepstral coefficients (LPCCs)[3]. In a very recent study, the auditory-based Gamma-tone frequency cepstral coefficient (GFCC) feature have been used to propose a robust speaker identification system [4]. Speaker identification techniques are often subject to additive noise and input speech distortion which degrade identification performance substantially. The challenge is to increase the identification accuracy under noisy conditions, i.e., to make the identification system more robust.

Two types of automatic speaker identification system exist: text-independent and text-dependent.This paper proposes a new robust text-dependent approach to classify speakers based on the neural responses simulated using a physiologically-based computational model of the auditory nerve (AN)[5]. This AN model captures most of the nonlinear behaviors of the peripheral auditory system, such as nonlinear tuning, two-tone suppression, and adaptation in the inner-hair-cell-AN synapse. The motivation of this approach hinges on the fact that the neural responses are robust against different types of noises[6].

A neurogram is a 2D representation (time-frequency) and can be constructed by simulating the responses of AN fibers over a range of characteristic frequency in response to a speech signal. The neurogram contains important information about the identity of the speaker that is extracted and trained using common classifiers such as support vector machine (SVM) and Gaussian mixture model (GMM) to identify the speaker. Figure 1 shows the block diagram of the proposed method for a robust speaker identification system.

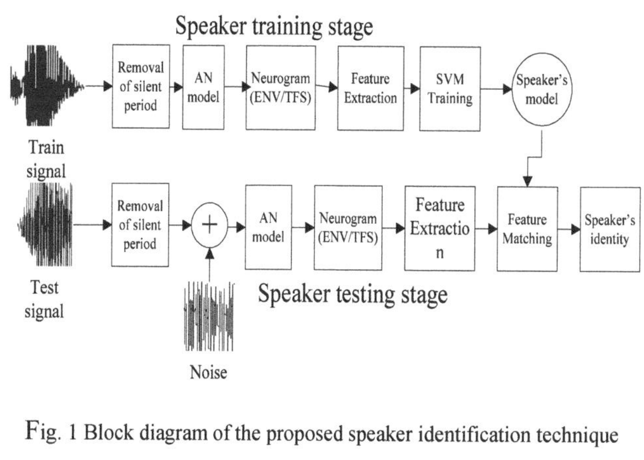

Fig. 1 Block diagram of the proposed speaker identification technique

F. Ibrahim et al. (eds.), *International Conference for Innovation in Biomedical Engineering and Life Sciences*,
IFMBE Proceedings 56,
DOI: 10.1007/978-981-10-0266-3_3

II. METHODOLOGY

In this study, a speech database with 39 speakers (25 males and 14 females) from the auditory neuroscience lab in the University of Malaya has been used. Every speaker was asked to utter 'Universiti Malaya' for 10 times. The signals were recorded in a sound proof booth with a sampling frequency of 8 kHz. The silent periods from recorded signal were removed using the voice-activity-detector (VAD) algorithm provided in the voice box toolbox[7]. Seventy percent of speech signals from each speaker was randomly chosen to train the SVM/GMM classifier, and the rest of the signals was used to test the performance of the proposed system. In order to determine the robustness of the method, the test speech signals were distorted by a white Gaussian noise over a range of signal to noise ratios (0 to 15 dB in steps of 5 dB).

A. AN Model and Feature Extraction

In recent years, the responses of AN model has been used quite extensively in the field of speech processing applications. The neural responses are very robust against noise due to the property of phase locking [8,9]. The maximum frequency limit for phase locking lies at about 4~5 kHz. In this study, the computational AN model developed by Zilany and colleagues [5] was employed to simulate the neural responses. The input to the AN model is a speech signal which was up sampled to 100 kHz required by the AN model, and the output is the spike times for a certain value of characteristic frequency (CF).The neural responses of each CF were averaged with a bin (time window) size of 100 μs, and then the responses were divided into frames using a Hamming window (50% overlap between adjacent frames) of length 128 samples. The combination of binning to 100 μs and smoothing with the 128-sample accounted for spike synchronization to frequencies up to ~160 Hz $[1/(100\times10^{-6}\times128\times0.5)]$. Thus the resulting neurogram excludes spike timing information aboutthe temporal fine structure (which might go up to ~4 kHz); however, it includes information about the envelope (mostly low frequency) of the signal.

Neurogram is basically a 2D representation in which neural responses of a wide range of CFs are displayed as a function of time [10]. In this paper, the responses were simulated for 12 CFs logarithmically spaced from 250 Hz to 1 kHz. It was observed that the responses of AN fibers above 1 kHz were not quite consistent for the same speaker in response to different instances of the same speech utterance, and thus only the responses of AN fibers with CFs up to 1 kHz were considered in this study. The responses of three different types (high, medium, and low spontaneous rate) of AN fibers were simulated and weighted by the reported distribution of spontaneous rates (SR) of AN fibers (60% high, 20% medium, and 20% low) [11].

B. MFCC

Mel-frequency cepstral coefficient (MFCC) is a representation of linear cosine transform of log power spectrum of a sound signal. The MFCC features have been widely used in automatic speaker and speech identification systems. In this study, the VAD algorithm was used to remove the silent periods from the acoustic signals. The rastamat toolbox [12] was then employed to calculate the MFCC coefficients from speech signals. The coefficients were calculated using a frame size of 25 ms with a Hanning window and an overlap of 50% between consecutive frames. Each frame resulted 39 MFCC features consisting of three groups: Ceps (cepstral coefficients), Del (derivatives of ceps) and the Ddel (derivatives of Del) with 13 features for each group.

C. GFCC

GFCC is an auditory-based feature which was derived from the responses of Gammatone filter bank. According to physiological observation, Gammatone filter bank resembles more to the cochlear filter bank [13]. In order to compute GFCC, the same window size and overlapping between frames were used, as employed in the computation of MFCC. The basic difference between MFCC and GFCC is that Gammatone filter bank and cubic root operation are used in GFCC derivation whereas triangular filter bank and log operation are applied in MFCC extraction[14]. In this study, the lowest 23 GFCCs (among 128 coefficients) from each frame have been used, since these coefficients retain most information of Gammatone feature (GF) due to energy compaction property of DCT [15].

D. SVM

The Matlablibsvm toolbox [16] was used to generate 39 SVM speaker models for speaker identification. In this study, the one-against-all classification technique was employed to train the SVM models. Note that the size of the feature vector determines the dimensional space of the classifier. In this study, 7 samples were used to train the SVM models, and the remaining 3 samples were used to evaluate the performance of the system. In this study, the feature size of the proposed method was m × 12, where m was the number of envelope points in the neurogram. For the MFCC- and GFCC-based methods, there were $n \times 39$ and $n \times 23$ features for each speech sample; where n is the number of frames of the speech signal.

In the proposed method, the features were normalized in such a way that training data array was bounded to a mean value of zero and a standard deviation of one (1). The default type of kernel function (Radial basis function, RBF) was used in this study. Using a cross validation algorithm [17], the best C and γ (associated to the RBF kernel) were chosen in such a way that it resulted best accuracy. Once the training model was obtained, it was saved and tested against test samples for range of SNRs. In this study, the following parameters: cost function (c), gamma (g), SVM type (s), shrinking parameter (h) were set to 8, 8, 0, and 0, respectively. The speaker identity of the unknown test speech was determined by the model that resulted a maximum decision value.

E. GMM

The use of GMM classification technique in speaker identification is very common and well established. The application of expectation maximization (EM) algorithm [18] makes the GMM a successful classifier for speaker identification. Mean vectors, covariance matrices, and mixture weights from all component densities parameterized speaker GMM model.

In this study, all neurograms of the training set from each speaker were combined together to form an input array for training. Thirty nine (39) GMM models were generated using EM algorithm. Diagonal covariance matrices were used to reduce the computational overhead and thus make the system faster. Ten (10) mixture components were used to obtain 12 dimensional feature vectors, and it was noticed that the proposed system performance dropped substantially when the number of mixture component was increased above 10. Speaker identification process was accomplished by comparing each unknown test utterance to all GMM speaker models and calculating the maximum likelihood values using the probability density function (pdf).

III. Results and Discussions

This section illustrates the performance of the proposed neural-response-based speaker identification system for both in quiet and under noisy environment. The estimated accuracy of the proposed system was also compared with the accuracy of the MFCC-and GFCC-based speaker identification system. To check the stability of the proposed system, the performance of the proposed method was evaluated for five times independently for each SNR, and the average results are reported in Figs. 2 and 3.

Figure 2 shows the accuracy performance of the proposed system (solid line) along with the identification performance of the baseline-feature-based methods using GMM as a classifier. The performance is shown as a function of SNR. It is clear that the performance of all the systems in quiet was near 100% and very comparable to each other. As more noise was added to the speech signal, the identification performance of all systems degraded accordingly. However, the neural response-based proposed system outperformed the traditional acoustic-feature-based systems at all SNRs studied. Also, GFCC-based system showed better performance compared to the results using MFCC as features under noisy conditions, consistent with the observation in[15,19].

Figure 3 shows the performance of the identification systems using SVM as a classifier. In general, the results are similar compared to the results described in Fig. 2 (using GMM) except that the performance of the proposed and MFCC-based methods showed better identification results at all SNRs studied, and GFCC-based system exhibited a substantially poorer performance compared to the identification result of the MFCC-based system under noisy conditions (due to less number of support vector).Based on these results, it has been observed that the performance of the proposed system was almost same irrespective of classifier, whereas the GFCC- and MFCC-feature-based system performances were dependent on the classifier.

To make resemble with the human auditory system, a very simple feature has been proposed in this study. The

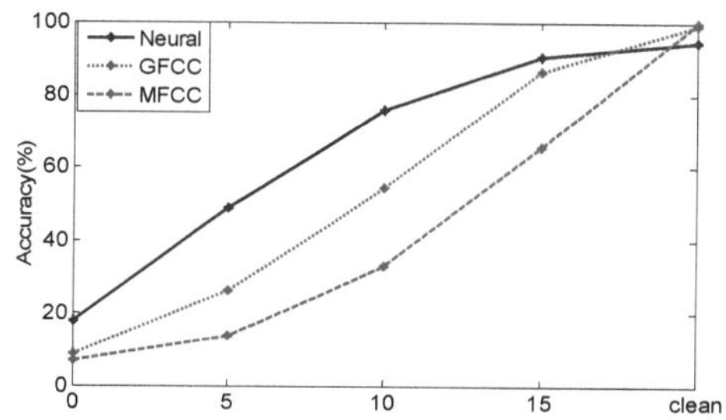

Fig. 2 Illustration of speaker identification rate for the proposed, GFCC, and MFCC features using GMM as a classifier (for 39 speakers).

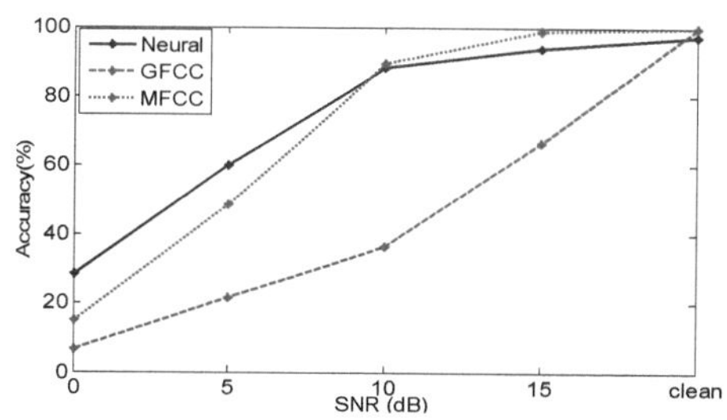

Fig. 3 Illustration of speaker identification rate for the proposed, GFCC, and MFCC features using SVM as a classifier (for 39 speakers).

3. To investigate the reason behind the robustness of speaker identification, the proposed neural feature was calculated (using responses of the AN model) from several speech samples of different speakers. It has been observed that the responses at lower frequencies (below 1 KHz) were

comparatively less affected by noises than the responses at higher frequencies, consistent with the phase-locking property of the auditory neurons at low CFs (<~4 kHz).

IV. CONCLUSION

In this paper, a neural-response-based feature was proposed for the text-dependent speech-based speaker identification system. This novel feature captures essential information about the speaker and was derived from the responses of a well-established physiological-based model of the auditory periphery. The performance of the proposed system was evaluated both in quiet and under noisy conditions and compared to the performance of two traditional acoustic-feature-based identification system, GFCC and MFCC. In clean condition, the speaker identification rate of the proposed system was comparable to the result from the acoustic-feature-based systems, however, the proposed system outperformed under noisy conditions. It was also observed that the performance of the GFCC- and MFCC-based systems was dependent on the classifier, whereas the identification rate of the proposed method was almost independent of the classifier. Since the same front-end is used for both automatic speech recognition and speaker recognition, the proposed feature could also be used for speaker verification, text-independent speaker identification, and speech recognition.

ACKNOWLEDGEMENT

This study was supported by the high impact research (HIR) in the University of Malaya under the grant UM.C/625/1/HIR/152.

CONFLICT OF INTEREST

The authors declare that they have no conflict of interest.

REFERENCES

1. D. A. Reynolds, "An overview of automatic speaker recognition," in *Proceedings of the International Conference on Acoustics, Speech and Signal Processing (ICASSP)(S. 4072-4075)*, 2002.
2. M. R. Hasan, M. Jamil, M. Rabbani, and M. Rahman, "Speaker identification using Mel frequency cepstral coefficients," variations, vol. 1, p. 4, 2004.
3. J. Makhoul, "Linear prediction: A tutorial review," Proceedings of the IEEE, vol. 63, pp. 561-580, 1975.
4. Q. Li and Y. Huang, "An auditory-based feature extraction algorithm for robust speaker identification under mismatched conditions," Audio, Speech, and Language Processing, IEEE Transactions on, vol. 19, pp. 1791-1801, 2011.
5. M. S. Zilany and I. C. Bruce, "Modeling auditory-nerve responses for high sound pressure levels in the normal and impaired auditory periphery," The Journal of the Acoustical Society of America, vol. 120, pp. 1446-1466, 2006.
6. J. E. Rose, J. F. Brugge, D. J. Anderson, and J. E. Hind, "Phase-locked response to low-frequency tones in single auditory nerve fibers of the squirrel monkey," Journal of neurophysiology, vol. 30, pp. 769-793, 1967.
7. M. Brookes, "Voicebox: Speech processing toolbox for matlab," Software, available [Mar. 2011] from www. ee. ic. ac. uk/hp/staff/dmb/voicebox/voicebox. html, 1997.
8. M. I. Miller, P. E. Barta, and M. B. Sachs, "Strategies for the representation of a tone in background noise in the temporal aspects of the discharge patterns of auditory-nerve fibers," The Journal of the Acoustical Society of America, vol. 81, pp. 665-679, 1987.
9. A. Palmer and I. Russell, "Phase-locking in the cochlear nerve of the guinea-pig and its relation to the receptor potential of inner hair-cells," Hearing research, vol. 24, pp. 1-15, 1986.
10. N. Mamun, W. Jassim, and M. S. Zilany, "Prediction of Speech Intelligibility Using a Neurogram Orthogonal Polynomial Measure (NOPM)," Audio, Speech, and Language Processing, IEEE/ACM Transactions on, vol. 23, pp. 760-773, 2015.
11. M. C. Liberman, "Auditory-nerve response from cats raised in a low-noise chamber," The Journal of the Acoustical Society of America, vol. 63, pp. 442-455, 1978.
12. D. Ellis, "Rastamat Toolbox."
13. R. Patterson, I. Nimmo-Smith, J. Holdsworth, and P. Rice, "An efficient auditory filterbank based on the gammatone function," in a meeting of the IOC Speech Group on Auditory Modelling at RSRE, 1987.
14. X. Zhao, Y. Shao, and D. Wang, "CASA-based robust speaker identification," Audio, Speech, and Language Processing, IEEE Transactions on, vol. 20, pp. 1608-1616, 2012.
15. Y. Shao, S. Srinivasan, and D. Wang, "Incorporating auditory feature uncertainties in robust speaker identification," in *Acoustics, Speech and Signal Processing, 2007. ICASSP 2007. IEEE International Conference on*, 2007, pp. IV-277-IV-280.
16. C.-C. Chang and C.-J. Lin, "LIBSVM: A Library for Support Vector Machines. ACM Transactions on Intelligent Systems and Technology, 2: 27: 1–27: 27, 2011," *Software available at http://www. csie. ntu. edu. tw/cjlin/libsvm*, 2011.
17. J. A. Suykens, T. Van Gestel, J. De Brabanter, B. De Moor, J. Vandewalle, J. Suykens, and T. Van Gestel, *Least squares support vector machines* vol. 4: World Scientific, 2002.
18. J. Bilmes, "A gentle tutorial of the EM algorithm and its application to parameter estimation for Gaussian mixture and hidden Markov models," " vol. 60 IEEE CIRCUITS AND SYSTEMS MAGAZINE SECOND QUARTER 2011, 1998.
19. A. KROBBA, M. DEBYECHE, and S.-A. SELOUANI, "Comparison of Auditory Feature Based GFCCs and MFCCs for Robust Speaker Identification in Noisy Environment applied to Arabic Speech."

Author:	Md. Atiqul Islam, M. S. A. Zilany and Wissam A. Jassim
Institute:	University of Malaysia
Street:	Jalan university
City:	WP Kuala Lumpur
Country:	Malaysia
Email:	atiq.atrai@gmail.com, zilany@um.edu.my, binaye2001@yahoo.com

Mental Stress Quantification Using EEG Signals

F.M. Al-shargie[1], T.B. Tang[1], N. Badruddin[1], and M. Kiguchi[2]

[1] Universiti Teknologi PETRONAS
Centre of Intelligent Signal and Imaging Research, Department of Electrical and Electronics Engineering,
32610 Bandar Seri Iskandar, Perak, Malaysia
[2] Hitachi Ltd., Research & Development Group, 2520, Hatoyama, Saitama, 350-0395, Japan

Abstract— **Mental stress has been identified as one of the major contributing factors that leads to various diseases such as heart attack and stroke. To avoid this, stress quantification is very important for clinical intervention and disease prevention. In this study, we investigate the feasibility of exploiting Electroencephalography (EEG) signals to discriminate stress from rest state in mental arithmetic tasks. The experimental results showed that there were significant differences between the rest state and under stress at three levels of arithmetic task levels with p-values of 0.03, 0.042 and 0.05, respectively. We thus confirm the feasibility of EEG signals in detecting mental stress levels. Using support vector machine (SVM) we could detect mental stress with an accuracy of 94%, 85%, and 80% at level one, level two and level three of arithmetic problem difficulty respectively.**

Keywords— **Stress, EEG, Wavelet transform, SVM.**

I. INTRODUCTION

People suffer from stress in their daily life. Stress has been defined as "the non-specific response of the body to any demand for change" [1]. Stress can change the responsiveness of central-peripheral regulatory systems rendering them less efficient in supporting health. It has been recognized as one of the major factors contributing to chronic disorders and productivity loses. It influences the desire to work, performance at work and attitude toward life. Chronic stress has been linked to a range of health problems [2]. Previous studies have shown a correlation between long-term exposure to stress and risk factors such as cardiovascular diseases [3, 4].

Stress response can be evaluated from perceptual, behavioural and physical responses to mental stress task. Evaluation of perceptual responses to stress involves subjective estimations and perceptions. Self-report questionnaires are one of the most commonly used methods to measure an individual's level of stress [5]. However, evaluating the stress using questionnaires is subjective method [6]. Therefore, clinicians evaluated the stress by measuring cortisol and α-amylase levels [7]. Stress response involves the activation of hypothalamus-pituitary-adrenocortical axis (HPA) and sympathetic nervous system (SNS) causing an increase in the glucocorticoid/cortisol secretion in the adrenal cortex.

Beside the release of cortisol, stress can be quantified from human bio-signals [8]. Studies have found a relationship between salivary cortisol levels and physiological variables changes such as heart rate variability (HRV), skin temperature (ST) and blood pressure (BP) [9]. Heart rate variability refers to the beat-to-beat alternations in heartbeat intervals. Stress causes a decrease in the high frequency components of the heart beat interval and an increase in the low frequency components of that heartbeat interval signals respectively. Thus, heart rate variability analysis has been established as an instantaneous quantitative measure of ANS activity associated with mental stress. Skin conductivity on the other hand varies with the changes in skin moisture level revealing the changes in sympathetic nervous system. Skin conductivity has been reported to increase with stressful task and can be acquired simultaneously using galvanic skin response (GSR) [10].

Furthermore, the changes in ANS can be effectively represented by electroencephalography (EEG) signals [11]. Electroencephalogram (EEG) is one of the most common sources of information used to study brain functions and conditions. It is a very complex signal and can be recorded non-invasively using surface electrodes from the scalp. EEG is the most studied non-invasive brain imaging device due to its excellent temporal resolution, ease of use, and low set-up cost. Additionally, EEG benefits from its high temporal resolution, enable it to measure the changes in cognitive activity within millisecond scale [12]. EEG signals are categorized by frequency bands; Delta (0.5-4 Hz), Theta (4-8 Hz), Alpha (8-13 Hz) and Beta (14-30 Hz). Each of the frequency band represents a state of the person. An increase of EEG power spectra in the Beta frequency band associated with increase in the alertness and arousal; Alpha increased with relaxation and Theta occurred during the sleep state [13].

EEG signals have been previously used in the assessment of variation in the state of the subjects during cognitive tasks. Compared to HRV and blood pressure, EEG gives more information about relaxation and alertness condition [12, 14]. In literature, both decreased and increased in alpha and beta power have been found as a sign of mental stress [15]. EEG has successfully classified stress from the rest state in [16] with an average accuracy of 85.55% using Yule

F. Ibrahim et al. (eds.), *International Conference for Innovation in Biomedical Engineering and Life Sciences*,
IFMBE Proceedings 56,
DOI: 10.1007/978-981-10-0266-3_4

Walker and in [17] with average classification rate of 90%. Another studies have combined EEG signal with several physiological signals to study mental stress. Skin conductance, heart rate variability and EEG signals combined to classify stress in [18] resulted in average accuracy of 84.1% using psychological signals and 82.7% using the EEG signals. Blood pressure, skin conductance, heart rate variability and EEG signals in [19] model the stress with high classification rate of 95% using all physiological signals and 91% using EEG signals.

According to previous studies, no stress levels has been studied yet. Stroop colour word test [20], mental arithmetic task[21], public speaking [22], cold pressor [23], computer work [24] and videos [25] have been successfully used in previous studies as stress stimuli. In this study, we developed a stress stimuli to elucidate three levels of stress using arithmetic task. The aim of this study is to discriminate between stress levels and rest state based on EEG signals collected while performing mental arithmetic tasks. We simulated the brain using mental arithmetic task with three levels of difficulty as proposed by [26]. In this study, we proposed wavelet transform (WT) for feature extraction. According to previous study, [27] WT has the ability to deal with stationary and non-stationary signals. As EEG signals are non-stationary, WT may give good features that highly correlate with mental stress levels.

II. Methodology

A. Subjects

Twelve healthy male right-handed adults with an age ranges from 20-24 years participated to this study. All subjects were informed prior to the experiment and they gave written consent, in accordance with the declaration of Helsinki and ethical approval granted by local ethics committee at Unversity Teknologi PETRONAS. None of these participants had a history of psychiatric, neurological illness or psychotropic drug use. The participants were asked to minimize their head movements and to keep calm during the entire experiment.

B. Experimental Set-up and Protocol Design

We measured EEG signals from the frontal cortex using BrainMaster 24E system with seven active electrodes [FP1, F3, F7, Fz, FP2, F4, and F8] and one reference A1 attached to the earlobes. All the electrodes placed on the surface scalp based on the international 10-20 system of electrode placement. The sampling frequency for EEG was set to 256 Hz. The impedance of EEG was minimized using small amount of gel directly to the scalp.

The mental stress experiment was designed based on Montreal Imaging Stress Task (MIST) [26]. The experiment protocol were performed in four steps. First, brief introduction was given to the participants to be familiar with the proposed tasks. Second, the participants were trained for five minutes at each level of difficulty in the mental arithmetic (MA) task to estimate time taken to answer each question. Third (i.e. control phase), the participants had their EEG signals recorded for total duration of 15 minutes while solving arithmetic problems at three levels of difficulty *without any time limit per question.* After the EEG recording, a questionnaire was filled by the participants self-reporting about task loading according to NASA-TLX rating scale. Fourth (i.e. stress phase), the average time recorded during the training phase was reduced by 10% to induce stress on the participants. Similar as in the control phase, the EEG was recorded for 15 minutes and the participants completed another questionnaires about the task loading. The task in level one (L1) involved 3-one digit integer (ranging from 0 to 9) and used the operands of + or – (example 7+2-4). In level two (L2), the task involved 3 integers (ranging from 0 to 99) with at least 2 two-digit integers using the operands of +, –, and × (example 14×3-39). In level three (L3), the task involved 4 integer numbers (ranging from 0 to 99) and the operands include +, –, ×, / and ÷ (example 7-99/3+35). In this experiment, we developed the control technique by sending marker via channels 23-24 of EEG BrainMaster as '1' to mark the start of the task and '0' for the end of the task for each block. The entire record which had a total of nearly 1 hour, consisted of four blocks. Figure 1, gives an overview of the experimental protocol and the block design. Each block consisted of 40 seconds of mental arithmetic task and 30 seconds of rest.

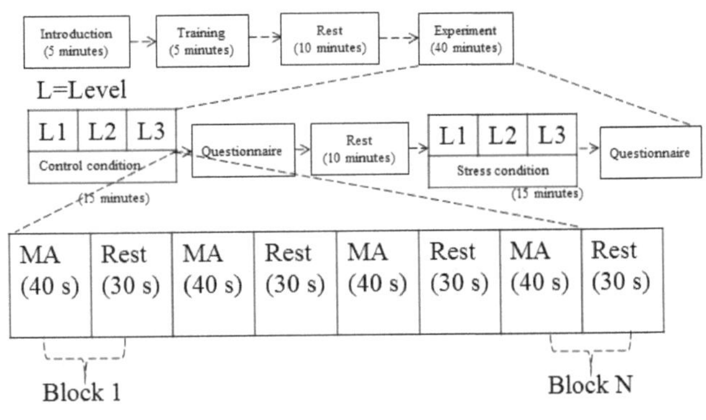

Fig. 1 Experimental protocol of mental stress study. The labels L1, L2 and L3 represent the levels of mental arithmetic task and MA stands for mental arithmetic. Six recordings were performed in this experiment; three for control case and three for stress case. In each record, there were four blocks. In each block, mental arithmetic was allocated for 40 s followed by 30 s rest.

C. EEG Analysis

The Pre-processing of EEG data was performed offline. EEG data was Pre-processed with EEGLAB 2013a toolbox. First, the EEG data was bandpass filtered in the range of 0.5 Hz to 30 Hz using 3rd order Butterworth filter. Second, the artefacts were removed using independent components analysis technique (ICA).

EEG signal is a non-stationary signals which have different frequency elements at different time intervals. In this study, EEG signals was analysed using wavelet transform (WT) [28]. WT is suitable method for multiresolution time-frequency analysis. In this work, WT decomposed the EEG into four sub-frequency bands; Delta (1-4Hz), Theta (4-8 Hz), Alpha (8-12.5 Hz) and Beta (12.5-30 Hz). The wavelet decomposition level was set to 5-levels and one final approximation since we are interested in the frequency range of 0-30 Hz only. Table 1 gives a summary of the frequency distribution with wavelet decomposition levels.

Table 1 EEG frequency band decomposition levels.

Decomposition level	Frequency bandwidth	Frequency band
DL1	64 Hz -128 Hz	Noisy signal
DL2	32 Hz -64 Hz	Noisy Gamma
DL3	16 Hz -32 Hz	Beta
DL4	8 Hz -16 Hz	Alpha
DL5	4 Hz -8 Hz	Theta
AL5	0-4 Hz	Delta

From the wavelet coefficients we extracted the mean of the absolute values of the wavelet coefficient in each sub-band and the average power and energy. The power spectral density was calculated by:

$$P = \frac{1}{N}\sum_{n=k}^{k+N-1}|x(n)|^2 , \quad (1)$$

where *x(n)* represents the segmented EEG signal and *N* is the length of the recorded EEG signal. The energy of EEG frequency bands was calculated based on:

$$E = \frac{1}{N}\sum_{n=-\infty}^{\infty}|x(n)|^2 , \quad (2)$$

In this work, we used a segmentation of s time interval rectangular window to calculate the features of EEG signals.

D. Features Normalization

The features extracted from EEG signals were normalized to interval [0, 1] before feeding them into the classifier by calculating:

$$x' = \frac{x - min\,(x)}{max(x) - min\,(x)} , \quad (3)$$

where x is the entire feature set, *min(x)* is the minimum value in the feature set and *max(x)* is the maximum value in the feature set respectively.

E. Classification

In this work, we used support vector machine (SVM) with 10-fold cross validation to classify stress from rest state. The SVM algorithm is a nonprobabilistic binary linear classifier that build a model to predict which category the new case belongs to. In this phase, only power values extracted from alpha band were used for classification. There was a total of 840 features for each subject in each recording phase (120 power values calculated from alpha frequency band multiply by 7 EEG measured electrodes). Features of one trial were used for testing and the other features (from other trials) were used for training the classifier. The formulation for obtaining SVM can be found in [29].

III. RESULT AND DISCUSSION

The goal of this study was to discriminate between stress and rest state based on EEG signals collected while solving mental arithmetic task with three levels of difficulty. Stress levels in this experiment were based on time pressure and negative feedback. The developed mental stress stimulus in this paper induced variations in the brain cortical activities as captured by EEG signals. We investigated on the effects of mental stress levels induced by arithmetic tasks by calculating the alpha and beta rhythm power values in all electrodes for all the subjects. The obtained results of EEG records found that, subjects failed to relax and appeared less attentive when facing high level of stress. This indicates that, the cortical activation increased with low level of stress and decreased with high level of stress and time pressure. Our results demonstrated a decrease in alpha rhythm power from level one to level two and failed to drop in level three of mental stress. Beta rhythm power, on the other hand increased with response speed of mental arithmetic, suggesting that subjects needed to pay more attention to finish the task under time pressure but as they faced high level of stress their attention would reduce.

In particular, by studying each level of mental stress separately, we found a very significant difference between the control/rest and the stress condition. We employed two-sample t-test analysis between every pair of the tasks to study their significant differences. In the analysis part, we focused on alpha rhythm due to its negative correlation with mental stress and its significant variation with mental workload [16]. There was a significant decrease in alpha rhythm power from control condition to stress condition in level one of mental stress with mean *p-value* of 0.03. Level two however, showed a significant reduce in alpha rhythm power with mean *p-value* of *0.042*. In level three we found that, the mean difference between control and stress condition was less significant as compared to level one and level two, mean *p-value* of 0.05. By comparing the three levels of mental stress induced by arithmetic tasks, there was a great decline in alpha power from level-one-to-level-two and increased again from level-two-to-level-three. This indicates that the cortical activation increased from task level-one to task level-two and failed at task level-three, verified with questionnaires. Figure 2 showed the mean differences between each level of the control task and stress task.

NASA-TLX rating scales showed no significant differences in the three mental stress levels. The results suggested that, subjective assessment using questionnaire was not sensitive for measuring mental stress levels and revealed the effectiveness of physiological measurement using EEG signals.

Based on the repititive of the stress tasks (three times recording for three levels of mental stress), we studied the dominant brain regions to mental stress. The dominant was obtained based on the significant responses from the right and left asymmetry as calculated by two-sample t-test. The study revealed the dominant of the right prefrontal cortex (PFC) to mental stress (PFC reduce with stress) in all the three levels of mental stress. This finding is consistent with previous study [30, 31]. Furthermore, we classified the stress levels using support vector machine classifier. The mean classification accuracy obtained by the classifier was 94%, 85% and 80% for level one, level two and level three of mental stress respectively.

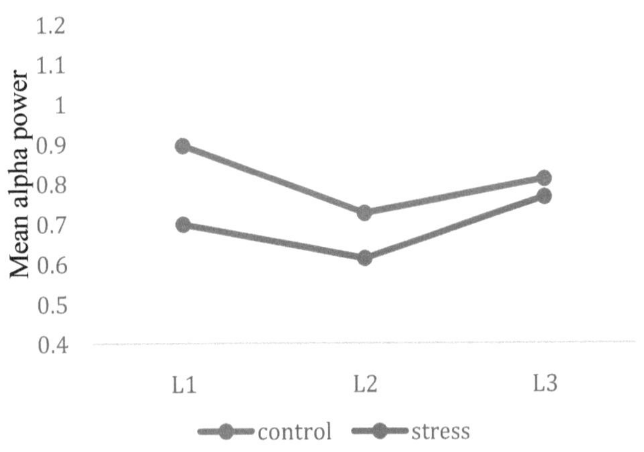

Fig. 2 Mean Alpha rhythm power distribution in two mental states; control and stress. The L1-represents level one of mental arithmatic task, L2-represents level two of mental arithmatic task and L3- represents the level three of arithmatic task.

IV. CONCLUSION

In a study with 12 subjects, we have shown that EEG signals can be used to reliably identify mental stress levels from rest state. The study reported a significant differences between the tasks (control and stress) as measured by two-sample t-test with mean *p-values* of *0.03, 0.042, and 0.05* for level one , level two and level three of arithmetic task respectively. Furthermore, the study revealed the dominant of the right prefrontal cortex to mental stress. The questionnaire about task loading indicated that with increasing level of difficulty especially level three, the engagement of participants reduced significantly. The experimental results supported the suggestion of using EEG to detect mental stress, and reported level one of mental arithmetic task as the most suitable stress stimuli.

ACKNOWLEDGMENT

This research is funded by the Ministry of Education, Malaysia under the HiCOE grant at Centre for Intelligent Signal and Imaging Research.

CONFLICT OF INTEREST

The authors have declared that no competing interests exist.

REFERENCES

1. H. Selye, "The stress syndrome," The American Journal of Nursing, vol. 65, pp. 97-99, 1965.
2. B. S. McEwen, "Central effects of stress hormones in health and disease: Understanding the protective and damaging effects of stress and stress mediators," European Journal of Pharmacology, vol. 583, pp. 174-185, 2008.
3. S. Cohen, D. Janicki-Deverts, and G. E. Miller, "Psychological stress and disease," Journal of the American Medical Association, vol. 298, pp. 1685-1687, 2007.
4. A. Steptoe and M. Kivimäki, "Stress and cardiovascular disease," Nature Reviews Cardiology, vol. 9, pp. 360-370, 2012.
5. T. H. Holmes and R. H. Rahe, "The social readjustment rating scale," Journal of Psychosomatic Research, vol. 11, pp. 213-218, 1967.
6. T.-K. Liu, Y.-P. Chen, Z.-Y. Hou, C.-C. Wang, and J.-H. Chou, "Non-invasive evaluation of mental stress using by a refined rough set technique based on biomedical signals," Artificial Intelligence in Medicine, vol. 61, pp. 97-103, 2014.
7. M. Yamaguchi, T. Kanemori, M. Kanemaru, N. Takai, Y. Mizuno, and H. Yoshida, "Performance evaluation of salivary amylase activity monitor," Biosensors and Bioelectronics, vol. 20, pp. 491-497, 2004.
8. R. Zheng, S. Yamabe, K. Nakano, and Y. Suda, "Biosignal analysis to assess mental stress in automatic driving of trucks: Palmar perspiration and masseter electromyography," Sensors (Switzerland), vol. 15, pp. 5136-5150, 2015.
9. H. Ashton, R. D. Savage, J. W. Thompson, and D. W. Watson, "A method for measuring human behavioural and physiological responses at different stress levels in a driving simulator," British Journal of Pharmacology, vol. 45, pp. 532-545, 1972.

10. G. Chanel, K. Ansari-Asl, and T. Pun, "Valence-arousal evaluation using physiological signals in an emotion recall paradigm," in Conference Proceedings - IEEE International Conference on Systems, Man and Cybernetics, 2007, pp. 2662-2667.
11. S.H. Seo and J.T. Lee, "Stress and EEG," Convergence and Hybrid Information Technologies, Marius Crisan, InTech, pp. 413-426, 2010
12. C. Berka, D. J. Levendowski, M. M. Cvetinovic, M. M. Petrovic, G. Davis, M. N. Lumicao, et al., "Real-time analysis of EEG indexes of alertness, cognition, and memory acquired with a wireless EEG headset," International Journal of Human-Computer Interaction, vol. 17, pp. 151-170, 2004.
13. X.-W. Wang, D. Nie, and B.-L. Lu, "Emotional state classification from EEG data using machine learning approach," Neurocomputing, vol. 129, pp. 94-106, 4/10/ 2014.
14. G. Chanel, J. J. M. Kierkels, M. Soleymani, and T. Pun, "Short-term emotion assessment in a recall paradigm," International Journal of Human Computer Studies, vol. 67, pp. 607-627, 2009.
15. E. Puterman, A. O'Donovan, N. E. Adler, A. J. Tomiyama, M. Kemeny, O. M. Wolkowitz, et al., "Physical activity moderates effects of stressor-induced rumination on cortisol reactivity," Psychosomatic Medicine, vol. 73, pp. 604-611, 2011.
16. A. Saidatul, M. P. Paulraj, S. Yaacob, and M. A. Yusnita, "Analysis of EEG signals during relaxation and mental stress condition using AR modeling techniques," in Proceedings - 2011 IEEE International Conference on Control System, Computing and Engineering, ICCSCE 2011, 2011, pp. 477-481.
17. R. Khosrowabadi, C. Quek, K. K. Ang, S. W. Tung, and M. Heijnen, "A Brain-computer interface for classifying EEG correlates of chronic mental stress," in Proceedings of the International Joint Conference on Neural Networks, 2011, pp. 757-762.
18. S. A. Hosseini and M. A. Khalilzadeh, "Emotional stress recognition system using EEG and psychophysiological signals: Using new labelling process of EEG signals in emotional stress state," in 2010 International Conference on Biomedical Engineering and Computer Science, ICBECS 2010, 2010.
19. N. Sharma and T. Gedeon, "Modeling stress recognition in typical virtual environments," in Proceedings of the 7th International Conference on Pervasive Computing Technologies for Healthcare, 2013, pp. 17-24.
20. M. Svetlak, P. Bob, M. Cernik, and M. Kukleta, "Electrodermal complexity during the Stroop Colour Word Test," Autonomic Neuroscience: Basic and Clinical, vol. 152, pp. 101–107, 2010.
21. K. Ushiyama, T. Ogawa, M. Ishii, R. Ajisaka, Y. Sugishita, and I. Ito, "Mental Physiologic Neuroendocrine Arousal by Mental Arithmetic Stress Test in Healthy Subjects," The American Journal of Cardiology, vol. 67, pp. 101-103, 1991.
22. P. F. Miller, K. C. Light, E. E. Bragdon, M. N. Ballenger, M. C. Herbst, W. Maixner, et al., "Beta-endorphin response to exercise and mental stress in patients with ischemic heart disease," Journal of Psychosomatic Research, vol. 37, pp. 455-465, 1993.
23. S. S. Hassellund, A. Flaa, L. Sandvik, S. E. Kjeldsen, and M. Rostrup, "Long Term Stability of Cardiovascular and Catecholamine Responses to Stress Tests an 18-Year Follow-Up Study," Journal of American Heart Association, Vol. 55, Pp. 131-136, 2010.
24. N. Hjortskov, D. Rissen, A. K. Blangsted, N. Fallentin, U. Lundberg, and K. Sogaard, "The Effect of Mental Stress on Heart Rate Variability and Blood Pressure during Computer Work," Eur J Appl Physiol, vol. 92, pp. 84-89, 2004.
25. A. Choppin, "EEG-Based Human Interface for Disabled Individuals: Emotion Expression with Neural Networks", Master thesis, Information processing, Tokyo institute of technology, Yokohama, Japan, 2000.
26. K. Dedovic, R. Renwick, N. K. Mahani, V. Engert, S. J. Lupien, and J. C. Pruessner, "The Montreal Imaging Stress Task: using functional imaging to investigate the effects of perceiving and processing psychosocial stress in the human brain," Journal of Psychiatry and Neuroscience, vol. 30, p. 319, 2005.
27. M. Akay, "Wavelet applications in medicine," Spectrum, IEEE, vol. 34, pp. 50-56, 1997.
28. R. N. Khushaba, S. Kodagoda, S. Lal, and G. Dissanayake, "Driver drowsiness classification using fuzzy wavelet-packet-based feature-extraction algorithm," Biomedical Engineering, IEEE Transactions on, vol. 58, pp. 121-131, 2011.
29. C. Cortes and V. Vapnik. Support-vector networks. Machine Learning, 20:273–297, 1995.
30. R. S. Lewis, N. Y. Weekes, and T. H. Wang, "The effect of a naturalistic stressor on frontal EEG asymmetry, stress, and health," Biological Psychology, vol. 75, pp. 239-247, 7// 2007.
31. A. F. Arnsten, M. A. Raskind, F. B. Taylor, and D. F. Connor, "The effects of stress exposure on prefrontal cortex: translating basic research into successful treatments for post-traumatic stress disorder," Neurobiology of stress, vol. 1, pp. 89-99, 2015.

Author:	Fares Al-shargie
Institute:	Universiti Teknologi PETRONAS,
	Centre of Intelligent Signal and Imaging Research,
	Department of Electrical and Electronic Engineering
Street:	Bandar Seri Iskandar,
City:	Perak
Country:	Malaysia
Email:	fares_g02737@utp.edu.my

Mean Glandular Dose and Figure of Merit of 2 Dimensional and 3 Dimensional Imaging in Digital Breast Tomosynthesis

K. Izdihar[1], K.C. Kanaga[2], M. Nawal[3], K. Vijaylakshimi[4], and T. Sulaiman[5]

[1,3] Medical Imaging, KPJ Healthtacare University College,Nilai Negeri Sembilan Malaysia
[2] Diagnostic Imaging and Radiotherapy Programme, Faculty of Health Sciences, University Kebangsaan Malaysia, Kuala Lumpur, Malaysia
[4] Diagnostic Imaging department, Hospital Tuanku Ampuan Rahimah, Klang, Selangor Malaysia
[5] The National Cancer Society of Malaysia, Women's Cancer Detection and Breast Clinic, 66 Jalan Raja Muda Abd Aziz, Kuala Lumpur, Malaysia

Abstract— **Studies of Mean Glandular dose (MGD) and Image quality for 3D imaging are still relatively new compared to 2D imaging. Therefore the objective of this research is to compare mean glandular dose (MGD) and figure of merit (FOM) obtained using ionization chamber between 2D and 3D breast imaging. This experimental study was conducted at the National Cancer Society using the Hologic Selenia Digital Breast Tomosynthesis. Three types of CIRS phantom with 50/50, 30/70 and 20/80 glandularity were exposed with an Auto - Filter mode in the Cranio-caudal projection. The ionization chamber (IC) was used to measure the Entrance Surface Air Kerma (ESAK) and Half Value Layer (HVL). The MGD value was calculated using Euref protocol and finally the image quality was assessed qualitatively by two radiologists and quantitatively by figure of merit (FOM). The results of the experiment showed that the MGD value obtained from the experiment increased with thickness and the MGD value for 3D projection is consistently higher than 2D projection. The spectrum W/Rh has a higher FOM value compared to spectrum W/Al for all phantoms. For 3D imaging the FOM value decreased when the depth of the object increased and the object is clearer in the center compared with other slices. In summary, it is estimated the MGD value for combined 2D and 3D increased by 1.5 - 2.5 times than single 2D mammography. The combined 2D and 3D imaging MGD value for 50% was 2.27 mGy, for 30% was 3.07 mGy and for 20% was 4.09mGy. *The combined 2D and 3D imaging for 50% glandularity is less than FDA limit for single projection and resulted in superior image quality compared to 2D imaging alone.**

Keywords— **Mean Glandular Dose (MGD), Figure of Merit (FOM), Image quality, Digital breast tomosynthesis (DBT).**

I. INTRODUCTION

Breast tomosynthesis is a three-dimensional imaging technology that involves acquiring images of a stationary compressed breast at multiple angles (x-ray tube rotate from -7.5 degrees to + 7.5 degrees) during a short scan [1]. The individual images are then reconstructed into a series of thin high-resolution slices that can be displayed individually or in a dynamic ciné mode.

Acquisition of DBT image can be done in both 2D and 3D imaging or maybe with 3D imaging is sufficient to show breast abnormality. However by additional projection like 3D imaging the radiation dose received by the patient will be higher. Therefore, this study is designed to compare the dosage and image quality for 2D and 3D imaging.

The objective of this research was to compare Mean Glandular dose (MGD) and figure of merit (FOM) value obtained using an ionization chamber between 2D and 3D breast imaging.

II. MATERIAL AND METHOD

This experimental study was conducted at the National Cancer Society using Digital Breast Tomosynthesis (Selenia Dimension breast tomosynthesis, Hologic, Bedford, United States). Selenia Dimensions with AWS 5000 is equipped with a Tungsten (W) anode x-ray tube and with a choice for Rhodium (Rh) or Silver (Ag) filtration (0.050 mm thick, respectively) for 2D mammography whilst for 3D the filter used is Aluminium with 0.70 mm thickness.

There were three types of phantom used for this experiment. Phantoms are breast-shaped and made of epoxy resin, with standard thickness (4 cm, 5 cm and 6cm), equivalent to 50%, 30% and 20% glandular breast tissue, respectively [2]. The first phantom composed of 20% glandular and 80% adipose tissue of tissue equivalent breast phantom (Breast phantom, Computerized Imaging Reference Systems (CIRS), Norfolk, Virginia) with 6 cm thickness, the second phantom 30/70 with 5cm thickness and the third phantom is 50/50 with 4cm thickness respectively [3]. As this study was a phantom study, ethical clearance was not obtained.

The Entrance Surface Air Kerma was measured using the calibrated ionization chamber (Radcal 9095, Radcal Corporation, Monrovia, California, USA). The ionization chamber was calibrated by Malaysia Institution Nuclear Technology with correction factor provided, so that it can be used to correct the obtained ESAK values. The CIRS

F. Ibrahim et al. (eds.), *International Conference for Innovation in Biomedical Engineering and Life Sciences*,
IFMBE Proceedings 56,
DOI: 10.1007/978-981-10-0266-3_5

breast phantom was placed on the surface of the image receptor in craniocaudal (CC) position. The exposure was made in Auto-filter mode, which enables the automatic selection of the target, filter, kVp and mAs during patient screening.

To obtain an ESAK value for each phantom the method described in International Atomic Energy Agency (IAEA) TRS 457 for the assessment of dose was followed [4]. The method consists of two parts: the determination of the tube loading for correct exposure of the phantom and the measurement of the incident air kerma at the mammographic reference point for this tube loading [4]. Firstly the machine was set-up in Craniocaudal position, the phantom was placed on the breast table with its longest edge aligned with the chest wall edge of the table and centred laterally [4]. The compression plate was brought down onto the phantom. Exposures were made according to the tube output chose by the system. The resulted tube loading (mAs), tube voltage and target/filter combination used were recorded. The experiment was repeated for three times to find the mean value for each phantom. Later the phantom was removed and the ionization chamber was placed on the reference point of the radiation detector at the mammographic reference point, 45 mm above the cassette table, and 40 mm from the chest wall edge and centred with respect to the lateral direction. The compression plate was in contact with the detector. The exposures were made using the tube loading obtained from the previous experiment [4]. The entire experiment was repeated for three times to check for consistency.

In order to calculate the MGD, half value layer (HVL) for each tube output (kVp) was also measured using the same method describe in TRS 457 report [4]. HVL measurements were carried out for obtaining suitable conversion factors to estimate the MGD.

The MGD was calculated for each combination parameter of those settings (target/filter combination, kVp and mAs values) chosen by the system, by applying published conversion factors to the ESAK previously measured using the European protocol [3,4].

Factor g is the conversion factor of ESAK to MGD, the factor **c** corrects for any differences in breast composition from 50% glandularity and the factor **s** corrects for any differences from the original tabulation due to the use of a different x-ray spectrum. Formalism is proposed to estimate the MGD for breast tomosynthesis, which it introduces *t*-factors for the calculation of breast dose from a single projection and *T*-factors for a complete exposure series [5].

$$MGD=K.g.c.s.T \quad (1)$$

Contrast-to-noise ratio (CNR) value was recorded for each exposure parameter from the screen of the DBT system and followed the same method as described by Kanaga et al [6].

$$CNR = \frac{MPV_{SIGNAL} - MPV_{BACKGROUND}}{\sqrt[2]{\frac{SD^2_{SIGNAL}+SD^2_{BACKGROUND}}{2}}} \quad (2)$$

MPVsignal and MPV background is the mean pixel value measured in an area of 100% glandularity of the stepwedge (center of the stepwedge) and in the reference zone respectively. Whilst the SDsignal and SDbackground are the standard deviations of each region of interest (ROI) respectively. Then, Figure of Merit (FOM) is determined by using the formula below.

$$FOM=\frac{CNR^2}{MGD} \quad (3)$$

Image quality assessments were done on CIRS 11A breast phantom for line pair (resolution), speck, nylon fiber and masses for each exposure. In evaluating the image, two radiologists with twenty years working experience were asked to rate the visibility of each phantom test object by using validated scoring form. More than one radiologist is needed to avoid bias.

III. RESULTS

Table 1 showed the normalized MGD for 2D imaging and 3D imaging. The results were divided into three phantoms 20/80 (6cm), 30/70 (5cm) and 50/50 (4cm). The MGD value obtained from the experiment increased with thickness meanwhile, MGD value for 3D projection is consistently higher than 2D projection. This results in an MGD of less than two times compared to a single-view of 2D imaging.

The spectrum of W/Rh has a higher value of FOM for all phantoms compared to spectrum of W/Al. For 3D imaging the FOM value decreased when the depth of the object increased as shown in Table 2. The object is clearer in the center compared with other slices.

The score of the image quality was between 23 to 32 marks. All of the score of image quality had an acceptable value as recommended by the manufacturer (the detectability must be 15-16 line pair/mm identifiable. All specks larger than 0.196 mm must be visible, for nylon fibers, four largest sizes must be visible, and for masses, four largest sizes must be visible). Figure 1 show the distribution of marks for each phantom 20/80, 30/70 and 50/50 glandularity.

Table 1 Parameters chosen under fully Auto-Filer mode2D, two-dimension; 3D, three-dimension; mAs, mili-ampere second; MGD, mean glandular dose; CNR, contrast to noise ratio; FOM, figure-of merit.

Image	kVp	Phantom	Target/ filter	MGD	CNR	Figure Merit
2D	31	20/80	W/Rh	1.90±0.11	7.38±0.12	30.3±0.95
3D	33	20/80	W/Al	2.19±0.09	4.58± 0.23	8.4±1.03
2D	29	30/70	W/Rh	1.27±0.01	7.70± 0.32	33.1±2.68
3D	31	30/70	W/Al	1.80±0.01	4.70± 0.02	8.98±0.10
2D	28	50/50	W/Rh	0.96±0.01	6.56±0.27	38.9±1.16
3D	29	50/50	W/Al	1.31±0.02	4.70±0.11	9.6±0.46

Table 2 The CNR and FOM values for 3D imaging 2D, two-dimension; 3D, three-dimension

Image	Phantom	Thickness	Slice depth	CNR	Figure of Merit
3D	20/80	6cm	50mm	3.4±0.06	4.70±0.18
3D	20/80	6cm	30mm	4.6±0.23	8.4±1.03
3D	30/70	5cm	40mm	3.3±0.08	7.9±0.40
3D	30/70	5cm	25mm	4.7±0.02	8.9±0.10
3D	50/50	4cm	30mm	4.4±0.33	8.5±1.26
3D	50/50	4cm	20mm	4.7±0.11	9.6±0.46

IV. DISCUSSION

Table 1 shows the mean MGD and FOM of each phantom. An IC was used to measure the MGD since the source of uncertainty for TLD 100H could reach ±12.98%. In particular to the radiation dose of 2D and 3D imaging, the MGD for the 20% breast glandularity for 2D imaging was 1.9 mGy. Meanwhile, the MGD for 3D imaging was 2.19 mGy, which was 15.2% higher than 2D. Thus, a combination of 2D and 3D imaging resulted in an MGD of 4.09 mGy, 2.1 times than the single-view 2D imaging. On the other hand, the MGD for 30% glandularity with 5cm thickness for 2D imaging was 1.27 mGy. However, the MGD for 3D imaging was 1.8 mGy, which was 41.7% higher than 2D. Thus, a combination of 2D and 3D imaging resulted in an MGD of 3.07 mGy, 2.4 times that of a single-view 2D imaging. Finally for 50% glandularity with 4cm thickness, the resulted MGD for 2D imaging was 0.96 mGy and for 3D imaging was 1.31 mGy, 36.4% higher than 2D, thus the combination of 2D and 3D imaging resulted in an MGD of 2.27 mGy.

The CNR values increased with the increasing phantom thickness due to higher absorbed dose. In this study, the FOM values for 4cm breast phantom is higher than 6cm breast phantom. Besides, for the 3D image the CNR value

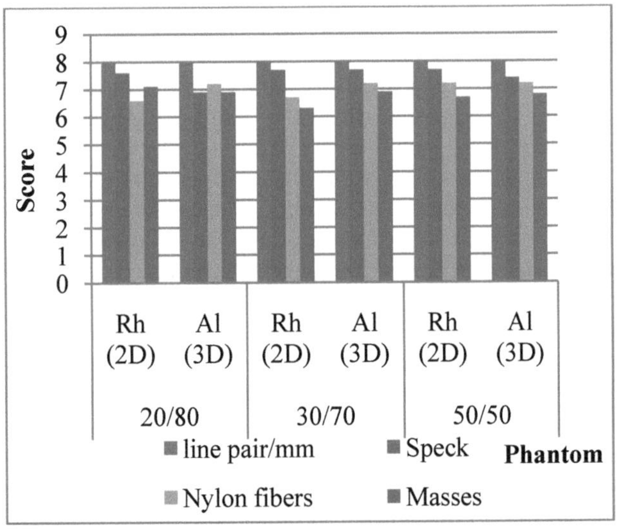

Fig. 1 Diagnostic value for each phantom and filter. 2D, two-dimension; 3D, three-dimension

for the center slice (half of the phantom thickness) is higher than the slice at 10mm above the detector. The deeper image has lower CNR compared with center image as the speck, masses, nylon fibres and other anatomical structures were located in central phantom.

In addition to above image quality assessment, a human observer was used to evaluate the image as the interpretation of mammogram is done by a radiologist using image quality criteria [7]. However, it may be subjective as different observers have their own evaluation, thus Kappa analysis is used to find the agreement between observers [8]. In this study both observers had shown an excellent agreement with k=0.88. The specks were better visualized compared with all features (line pair, masses, fibers) and it is similar to Myung-Su K et al's finding [9]. Meanwhile, the line pair assessment 3D imaging showed better image quality than 2D imaging. Due to tube movement from -7.5 degrees to +7.5 degrees, the image was produced at different angles and separate structures at differing heights. The increased number of image as compared to single 2D image may increase visibility of the object.

Radiation doses from digital breast tomosynthesis was the primary interest in this study. Earlier research results showed the MGD value for breast thickness 4.7cm with Molybdenum target resulted in 1.97 ± 0.95 mGy that was higher than single 3D imaging (Table 1) [10]. Meanwhile other research reported that 'when comparing doses with the "average" breast (compressed thickness of 5 cm, 50 percent glandular fraction), a DBT acquisition resulted in 1.30 mGy, only an 8 percent higher mean glandular dose than the FFDM acquisition of 1.20 mGy which is comparable to our study in which a combined 2D and 3D imaging resulted in MGD of 2.27 mGy [11].

Our results may be compared and are in good agreement with those of other work. Steve and Sechopoulos [20] based their work on the characterize the dosimetric properties of a clinical digital breast tomosynthesis (DBT) system for the acquisition of mammographic and tomosynthesis images. They reported that for a combined 2D and 3D imaging for 50% glandularity resulted in 2.5 mGy which is less than Mammography Quality Standards Act limit for a twoviewcscreening mammography study [11].

Meanwhile for 20% breast glandularity the mean MGD estimated in this study was also in good agreement with those estimated by Olgar et al [21] the mean MGD values in 2D imaging mode per exposure for the standard breast were 1.57 mGy, while the mean MGD values after correction for real breast composition were 1.82 mGy for Craniocaudal view. The mean MGD values in 3D imaging mode per exposure for the standard breast were 2.19 mGy while the mean MGD values after correction for the real breast composition were 2.53 mGy for Craniocaudal view. Whilst in this study the MGD for 20% breast glandularity is 2.19 mGy.

On the other hand, it can be estimate if a patient with normal breast 50% undergoing two procedure 2D and 3D imaging the total MGD received is less than ±3mGy that meets the FDA approval limit for each projection. Meanwhile for 20% the MGD value for single 2D and 3D imaging was less than 3mGy for each projection. This study suggests that the combined 2D and 3D imaging can be used as a breast screening tool as the combine 2D and 3D the MGD is still low and its show the benefit is far outweigh the dose

V. Conclusion

In summary, it is estimated the MGD value for combined 2D and 3D increase by 1.5-2.5 times than single 2D mammography. The combined 2D and 3D imaging MGD value for 50% was 2.27 mGy, for 30% was 3.07 mGy and for 20% was 4.09mGy. The combined 2D and 3D imaging for 50% glandularity is less than FDA limit for single projection and resulted in superior image quality compared to 2D imaging alone. The CNR value for 2D is higher than 3D since the measurement done without scatter reduction and the method is not specifically for 3D image analysis. Nevertheless the radiologist observation revealed that the 3D image is a complementary projection for the 2D image since it produce image at different slice and level.

Conflict of Interest

The author declare that they have no conflict of interest.

References

1. S. Andrew, Fundamentals of breast tomosynthesis, white paper, Hologic Inc. Bedford, U.S.A.2008 at www.hologic.com/data/WP-00007_Tomo_08-08.pdf
2. Baldelli P, Phelan N, Egan G (2010) Investigation of the effect of target/filter materials on the dose and image quality of a digital mammography system based on an amorphous selenium flat panel detector. Brit J Radiol. 83(988): 290–295
3. Hammerstein GH, Miller DW, White DR, Masterson ME, Woodard HQ, Laughlin JS (1979) Absorbed radiation dose in Mammography. Eur Radiol. 130: 485-491
4. IAEA. Dosimetry In Diagnostic Radiology: An International Code Of Practice at www.pub.iaea.org/MTCD/publications/PDF/TRS457_web.pdf
5. Dance DR, Young KC, Van Engen R (2011) Estimation of mean glandular dose for breast tomosynthesis: factors for use with the UK, European and IAEA breast dosimetry protocols. Phys. Med. Biol. 56:453–471
6. Kanaga KC, Yap HH, Laila SE, Sulaiman T, Zaharah M, Shantini A (2010) A critical comparison of three full field digital mammography systems using figure of merit. Med J Malaysia. 65(2):119–122
7. Fischmann A, Siegmann KC, Wersebe A, Claussen CD et al.(2005) Comparison of full-field digital mammography and film–screen mammography: image quality and lesion detection. Br J Radiol. 78: 312-315
8. Kazuyuki K and Kazuhiro S (2006) Intra- and interobserver agreement and performance score of breast phantom image interpretation: influence of ambient room lighting levels. Nagoya J. Med. Sci., 68,147- 153
9. Myung-Su K, Hak HK ,Joo HC et. al (2013) Dose Reduction in Automatic Optimization Parameter of Full Field Digital Mammography: Breast Phantom Study. J Breast Cancer. 16(1): 90–96
10. Sookepeng S and Ketted P (2006) Mean glandular dose from routine mammography. Nare Uni. J. 14(3):19-26
11. Steve SJF and Sechopoulos BS (2012) Clinical Digital BreastTomosynthesis system: Dosimetric Characterization. Radiology. 263(1): 35-42

Corresponding author: Izdihar Binti Kamal
Institution: Medical Imaging,
KPJ Healthtacare University College,
Nilai, Negeri Sembilan Malaysia
Telephone : 0166366383
Email: izdiharkamal@yahoo.com

The Effect of Hearing Augmentation on Cognitive Assessment Scores: A Pilot Crossover Randomized Controlled Trial

I. Amirah Fatin[1,2], L.A. Khatijah[2], M.S.A. Zilany[3], A.Z. Zuheir[3], S.H. Ong[3], and M.P. Tan[4,5,6]

[1] Kulliyyah (Faculty) of Nursing, International Islamic University Malaysia (IIUM), PO Box 141, 25710 Kuantan, Pahang, Malaysia
[2] Department of Nursing Sciences, Faculty of Medicine, University of Malaya, Kuala Lumpur 50603, Malaysia
[3] Department of Biomedical Engineering, Faculty of Engineering, University of Malaya, Kuala Lumpur 50603, Malaysia
[4] Division of Geriatric Medicine, Department of Medicine, Faculty of Medicine, University of Malaya, Kuala Lumpur 50603, Malaysia
[5] Ageing and Age-Associated Disorders Research Group, Faculty of Medicine, University of Malaya, Kuala Lumpur 50603, Malaysia
[6] Centre for Innovation in Medical Engineering, University of Malaya, Kuala Lumpur 50603, Malaysia

Abstract— **This randomized cross-over pilot study aimed to evaluate the effect of hearing augmentation on cognitive assessment scores and duration to complete cognitive assessment among the elderly in-patients in a teaching hospital. A hearing amplifier was used for hearing augmentation and the Montreal Cognitive Assessment (MoCA) test was used to assess cognition. Seventy one patients were allocated into Group A (n=33) or Group B (n=38) using block randomization. There was no significant difference in total MoCA scores with and without hearing augmentation (p = 0.622). There was a significant improvement in the total scores on the second test that suggests a learning effect ($p < 0.05$). There was also no significant difference in time taken to complete cognitive assessment with and without hearing augmentation (p = 0.879). Similar statistical tests performed on a subgroup of patients with hearing impairment did not reveal significant results. The results of this study will now inform a larger randomized controlled study evaluating the use of hearing amplifiers as cost-effective solutions to hearing impairment in our older population.**

Keywords— **Dementia, Cognition, Aged, Hearing augmentation, Hearing impairment.**

I. INTRODUCTION

Individuals aged 60 years and above currently represents 11.7% of the world population, and this is projected to be doubled by the year 2050 to 21.1% [1], with Asia witnessing the largest number (225 million) in growth of their older population in the past 20 years. As a result, age-associated conditions such as hearing loss and dementia are expected to increase accordingly [2-4]. Hearing loss affects one-third of individuals aged over 65 years, and its prevalence is usually greatest among those living in low and middle-income countries such as countries in the Asia Pacific region [5]. However, only a small portion of older people in these countries currently use hearing aids because they accept hearing loss as a part of old age, poor visual acuity, decreasing manual dexterity or the cost of devices [6; 7]. Being custom made, personalized hearing aids are costly. The use of alternative, less costly hearing devices to augment hearing may be a cost effective solution in societies where affordability issues are common.

There is evidence of an association between lower hearing ability with lower cognitive function among the elderly [3; 8; 9]. Hearing loss may also occur as a result of loss of neurons and slow brain processing time resulting from an underlying degenerative pathology within the brain [10]. Internationally, there has been several studies on the effects of hearing on cognition [2; 3; 8; 9; 11; 12]. A study on in an Australias population found an association between hearing loss and lower cognitive test scores after adjusting for age, sex, education and a history of stroke [3]. Nevertheless, all of the above evidence were obtained from observational studies whereby the conclusions need to be drawn with caution due to the absence of true cause and effect.

A study on hearing augmentation using a hearing amplifier had been conducted in a prospective randomized study among older patients at a teaching hospital in Scotland. The study reported the effects of hearing augmentation using hearing amplifiers on two cognitive performance measures and found that the Mini Mental State Examination (MMSE) scores improved significantly, while no improvements was observed in the Abbreviated Mental Test [13]. The samples in the study were not blinded which may lead to a response bias which can affect the conclusion. There was also uneven sample allocation, with far fewer individuals in the control group compared to the intervention group, which may bias the statistical conclusion. Therefore, further studies were suggested to verify the benefit of hearing augmentation during routine cognitive screening.

In view of the above available evidence, we conducted a pilot study using a randomized crossed-over design for the effects of hearing augmentation on cognitive assessment scores among older in-patients at a teaching hospital in Kuala Lumpur, Malaysia.

F. Ibrahim et al. (eds.), *International Conference for Innovation in Biomedical Engineering and Life Sciences*,
IFMBE Proceedings 56,
DOI: 10.1007/978-981-10-0266-3_6

II. METHODS

A. Design

This was a randomized cross-over design. This design was suggested to ensure that the subjects had the highest possible equivalence in its background compared to the treatment given[14]. In this study, the patients were randomly assigned to Group A or Group B (Figure 1). Group A received cognitive assessments with hearing augmentation first, while Group B had their cognitive assessment without hearing augmentation first. The sequence was reversed in the second cognitive assessment after a wash-out period. A brief wash-out period of 24 to 48 hours was observed before the second cognitive assessment. This study was approved by the Medical Ethics Committee, University of Malaya Medical Centre (MEC ID. No: 201410-713).

B. Participants

The study was conducted among in-patients aged 60 years and above from medical, obstetrics & gynecology, cardiology, geriatric, rehabilitative, surgical and orthopedic wards in a teaching hospital in Malaysia.

C. Inclusion and Exclusion Criteria

Informed consent was obtained from patients prior to participation in this study, and potential participants were excluded if they declined to participate or were unable to provide informed consent. Patients with any language disorder such as dysphasia were excluded. Patients with visual impairment or severe hand deformity who were unable to participate in reading, writing and drawing tasks were also excluded.

D. Randomization

Block randomization was generated using a computerized random sequence allocation, and the treatment allocation was stored in sealed, opaque envelopes. The envelopes were arranged in sequence, and the researcher used the envelopes in the correct sequence. A block size of 10 was used to ensure that out of each 10 patients enrolled, 5 will come from the intervention group and 5 from the control group. The envelopes were opened once the participants agreed to participate.

E. Procedure

Eligible participants were approached by the researcher. Demographic data and background information were first obtained from consenting participants. Hearing augmentation was performed using a hearing amplifier. The hearing amplifier is a type of assistive listening device (ALDs), consisting of a microphone, amplifier and headphone assembled together. The researcher would first of all demonstrate the use of the device. Participants were then assisted with appropriate use of the device. The volume was adjusted according to the patients' needs.

During both cognitive assessments, with and without hearing augmentation, the time duration required to complete the test was measured in minutes using a calibrated stopwatch to ensure accurate measurement of the time duration. The researcher started timing the patient when the first question of the Montreal Cognitive Assessement (MoCA) test, "Please draw a line …" was administered and stopped timing when participants answered the final question (orientation to city).

F. Outcomes

a) Cognitive performance

The Montreal Cognitive Assessment scale was used to assess cognitive performance, either with or without hearing augmentation according to the treatment allocation. It assessed nine cognitive domains (attention, concentration, executive function, memory, language, visuoconstructional skills, conceptual thinking, calculations, and orientation to time and place). The MoCA test has a combination of dichotomous and ordinal scales of measurements with the highest possible score of 30 points. Both English and Malay versions of the MoCA were used. The Malay version of the MoCA has been validated in the Malaysian population [15].

b) Additional measures

The presence of hearing impairment was determined with self-perceived hearing assessment using the screening version of the Hearing Handicap Inventory for the Elderly (HHIE-S) assessment scale [16]. The HHIE-S contains 10 items with a range of scores from 0 to 40. The cut-off used for determining the presence of hearing impairment was a total score of more than eight.

G. Statistical Methods

Data analysis was performed using the IBM® SPSS® Version 21 statistical software. Categorical data were presented as frequencies with percentages. Normally distributed data were presented as mean with standard deviation, and non-normally distributed data were presented as median with interquartile range. The main outcome, MoCA scores with and without hearing amplifier were compared using the Mann-Whitney U test because of the high rate of dropouts. A subset of 20 patients who reported having self-perceived hearing impairment were evaluated with an additional subgroup analysis comparing cognitive performance with and without hearing augmentation using an independent sample t-test. We adjusted for potential learning effects between the first and second test using linear regression methods.

A p value of less than .05 was considered statistically significant. As this was designed as a pilot study to inform a larger randomized-controlled study for hearing amplification, we were not adequately powered to detect any significant differences in cognitive outcomes. Potential differences observed would be used to inform power calculations for a future larger study.

III. RESULTS

A. Response and attrition rates

One hundred and fifty seven patients were assessed for eligibility. However, 86 patients were excluded because i) they did not met the inclusion criteria (n = 35), ii) refused to give consent (n = 51) and iii) were discharged (n = 13). Seventy one patients were randomized into Group A (n=33) and Group B (n=38). A total of 37 (52%) patients out of the initial 71 recruited completed the study, 34 (48%) patients did not complete the second cognitive assessment. One patient refused to have any cognitive assessment after hearing assessment. In Group A, 17 (52%) out of 33 patients completed both cognitive assessments, and 16 patients dropped out of the study. While in Group B, 20 (53%) out of 38 patients completed both cognitive assessments. Figure 1 shows the flow of participants and the reasons recorded for discontinued participation, with the highest recorded reasons being refusal (n = 15), followed by patients being discharged (n = 13), patients too ill to continue with cognitive assessments (n = 4), and patients transferred out to another ward (n = 2). Only 37 (52%) patients completed both cognitive assessments.

Table 1 Demographic data and background information

Items	Total (n=71)	Group A (n=33)	Group B (n=38)	p-value
Age, Mean (SD)	70.97 [a] (8.171)	71.81 (8.101)	69.43 (7.178)	.229 [b]
Gender				
Male, n (%)	33 (47.1%)	16 (48.5%)	17 (44.7%)	.832 [c]
Female, n (%)	37 (52.9%)	17 (51.5%)	20 (52.6%)	
Hearing impaired, n (%)	20 (28.2%)	6 (18.2%)	14 (36.8%)	.081 [c]
Hearing score, median (IQR)	2.00 (12)	2.00 (5)	3.00 (16)	.521 [b]
Use of hearing aid	0	0	0	-
Completed second test, n (%)	37 (52.1%)	17 (51.5%)	20 (52.6%)	-
Existing medical condition				
Cardiovascular disease, n (%)	20 (29%)	12 (38%)	8 (22%)	.147 [c]
Hypertension, n (%)	45 (65%)	22 (69%)	23 (62%)	
COPD, n (%)	10 (14%)	2 (6%)	8 (22%)	
Diabetes, n (%)	30 (43%)	12 (38%)	18 (49%)	
Thyroid dysfunction, n (%)	0	0	0	

[a] mean count for 68 patients
[b] Calculated with Mann-Whitney test
[c] Calculated with Chi-square test

B. Baseline Demographic data and HHIE-S scores

Table 1 summarizes the demographic data, existing medical conditions, use of hearing aid and hearing scores. The mean age for both groups was 70.97± 8.1, and there were more females than males in this study. Six (18%) patients in Group A and 14 (37%) patients in Group B have a HHIE-S score greater than eight. The median HHIE-S score was two. No patient reported use of a hearing aid. Twenty-eight (39.4%) patients scored less than two while 43 (60.6%) patients scored more than two. Twenty seven (38%) individuals reported difficulty hearing people whisper. Only six (9%) individuals reported that their personal or social life was hampered by hearing difficulties.

C. Cognitive performance

Table 2 summarizes the cognitive performance scores for individual items presented in median and interquartile range. The p value obtained using Mann Whitney-U showed no significant difference in each item of cognitive assessment scores with and without the use of a hearing amplifier. The total mean cognitive performance scores for both groups with hearing amplifier was 18.84 ± 6.5 and that without hearing amplifier was 17.55 ± 7.7, (t (101) = -0.083, p = 0.622).

Table 2 Itemized cognitive performance scores

Item	With hearing amplifier (n=52)	Without hearing amplifier (n=55)	z score	p [a]
Visuospatial/ executive, median (IQR)	3.00 (3)	3.00 (3)	-.177	.859
Naming, median (IQR)	3.00 (0)	3.00 (0)	-.728	.466
Attention reading digits, median (IQR)	2.00 (1)	2.00 (1)	-.381	.703
Attention reading letters, median (IQR)	1.00 (0)	1.00 (1)	-.982	.326
Attention serial subtraction, median (IQR)	1.00 (1)	1.00 (1)	-1.22	.223
Language repeat sentence, median (IQR)	.00 (1)	.00 (1)	-843	.399
Language fluency, median (IQR)	.00 (0)	.00 (0)	-1.26	.208
Abstraction, median (IQR)	1.00 (1)	1.00 (1)	-.387	.699
Delayed recall, median (IQR)	1.00 (3)	1.00 (3)	-.268	.789
Orientation, median (IQR)	5.00 (2)	5.00 (3)	-.195	.845

[a] p value calculated using Mann Whitney Test

The cognitive performance scores obtained with a hearing amplifier was higher (19.70 ± 6.4) than the scores without a hearing amplifier (16.85 ± 6.8) in Group B. However, in Group A, the cognitive performance scores without a

Table 3 Comparison between groups

Item	Group A n=32	Group B n=36	p-value
MoCA at T1[a] Mean (SD)	17.72 (6.5)	17.75 (6.9)	.985
Completed both tests	n=17	n=20	
MoCA at T1 Mean (SD)	16.47 (7.3)	16.85 (6.8)	.871
MoCA at T2 Mean (SD)	19.00 (7.7)	19.70 (6.4)	.765
Difference with and without HA Mean (SD)	2.38 (1.7)	3.35 (2.2)	.159
Time taken T1 (mins), mean (SD)	12.51 (4.5)	13.13 (3.5)	.660
Time taken T2 (mins), mean (SD)	10.74 (4.5)	12.21 (3.4)	.287
Groups Combined	n= 37		
MoCA With HA Mean (SD)	18.22 (6.9)		.819
MoCA Without HA Mean (SD)	17.84 (7.2)		
Time with HA (mins), mean (SD)	12.35 (3.9)		.816
Time without HA (mins), mean (SD)	12.11 (4.2)		

Group A had hearing amplification at T1
Group B had hearing amplification at T2
[a]Total MoCA scores including T1 scores of drop outs

hearing amplifier were higher (18.88 ± 8) than the scores with a hearing amplifier (16.44 ± 7.5). The overall increase in scores from the first test to the second test regardless of the use of a hearing amplifier suggests a learning effect despite the washout period ($F = 22.88, p < 0.0005$). Table 3 summarizes the cognitive assessment scores and duration to complete the assessment between groups.

Table 4 compares the MoCA scores and the time taken to complete each test for individuals with HHIES scores ≥ 2 vs HHIES scores <2.

When we then considered the 20 individuals who had a HHIE-S of above 8, which was the recommended cut-off to define hearing impairment, the mean scores of cognitive performance with a hearing amplifier were 17.17±7.98 and without a hearing amplifier were 15.5±7.97 (t (28) = 0.261, p = 0.796).

D. Duration

The mean duration required to complete cognitive assessment with hearing augmentation was 12.22±3.65 and without hearing augmentation was 12.35±4.25 (t (88) = -0.153, p = 0.879).

Table 4 Comparison of MoCA scores and duration

Item	HHIES >=2 Group A n= 20	HHIES >=2 Group B n= 20	p-value	HHIES <2 Group A n= 12	HHIES <2 Group B n= 14	*p*-value
MoCA at T1[a] Mean (SD)	18.30 (6.7)	17.90 (6.4)	.848	16.75 (6.4)	19.36 (6.2)	.304
	n= 9	n= 12		n= 8	n= 8	
MoCA at T1 Mean (SD)	16.11 (8.0)	16.83 (7.0)	.828	16.88 (7.0)	16.88 (7.0)	1.00
MoCA at T2 Mean (SD)	19.22 (9.0)	19.42 (6.6)	.832	18.75 (6.6)	20.13 (6.5)	.681
Difference with and without HA Mean (SD)	2.89 (2.6)	3.42 (2.3)	.627	2.63 (2.0)	3.25 (2.3)	.578
Time taken T1 (mins), mean (SD)	12.36 (5.2)	13.34 (4.3)	.676	12.65 (4.3)	13.51 (2.5)	.652
Time taken T2 (mins), mean (SD)	10.38 (4.5)	12.34 (3.7)	.302	11.10 (4.8)	11.93 (2.7)	.642
Groups Combined	n= 21			n= 16		
With HA Mean (SD)	18.29 (7.1)		.630	18.50 (6.7)		.462
Without HA Mean (SD)	17.86 (7.8)			17.81 (6.7)		
Time with HA (mins), mean (SD)	12.45 (4.3) [b]		.749	12.34 (5.6) [c]		.832
Time without HA (mins), mean (SD)	12.07 (4.7) [b]			12.16 (4.1) [c]		

Group A had hearing amplification at T1
Group B had hearing amplification at T2
[a]Total MoCA scores including T1 scores of drop outs
[b]n=17
[c] n=14

IV. Discussion

In this pilot study, we have evaluated and highlighted various issues with a cross-over design for the universal use of hearing augmentation for cognitive assessments in older inpatients in a tertiary hospital in Kuala Lumpur. Dropout rates were high following the washout period and there appeared to be a learning effect for the MoCA from the first to the second test. There appeared to be a small but statistically non-significant difference between the use of hearing amplifier for cognitive assessments, and this difference appeared larger, but remained non-significant, when only individuals with hearing impairment were considered. There appeared to be little difference in the time taken to perform the test with and without hearing amplification.

Previous studies evaluating the effect of hearing augmentation to improve cognitive performance among the elderly have yielded inconclusive results [13; 17-19]. The study populations were different across these studies where most of the studies were from Western developed countries. Older people in industrialized nations could have been exposed to greater noise pollution, resulting in higher rates of hearing problems among the older population. This could have explained variations in outcomes in the previous studies. In this pilot study, a lower rate of hearing problems was detected.

The improvement in cognitive assessment scores with time confirmed the concern of the author of the MoCA test [20] that a repeated MoCA test should be done with a different set of questions, although no studies have reported any learning effect on cognitive assessment using the same set of assessment tools [13; 21]. This, therefore, limited our ability to perform the cross-over design despite a brief washout period. A longer washout period was also not feasible as ours was an acute hospital. Even with the washout period of 24-48 hours, 15 of our recruited participants had already been discharged prior to the second test. Individuals were also reluctant to be tested again, presumably because they did not perceive a repeat test as valuable, further limiting the cross-over design. This finding also has important implications for other studies with uncontrolled pre and post-test designs and for clinical settings where repeated testing is used over time to assess treatment response and disease progression [22].

The lack of observed benefit with hearing augmentation may reflect the small sample size, however, the observed trend also suggests there to be no clinically significant benefits to the universal use of hearing augmentation. We had postulated that the universal use of hearing augmentation in busy, noisy hospital settings may be useful as hearing impairment is common, and at least some mild difficulties exist in older individuals. Universal use will reduce the need to screen for hearing problems prior to cognitive testing, and in self-reported screening tools, older individuals may not willingly admit to hearing problems due to embarrassment or denial. In addition, in a multilingual society that makes up Malaysia where multiple Malay, Chinese and Indian dialects as well as variable command of the English language exists, and it is not easily discernible whether difficulties in communication are due to hearing or language issues. However, problems may exist with the use of hearing augmentation. None of our participants had previously used hearing aids. Hearing amplification will accentuate background noise as well as speech, and while the volume of speech is amplified, it may not be of adequate clarity to allow comprehension. Furthermore, previous studies have also suggested an association between hearing impairment and cognitive decline. However, as accurate assessment of cognition among individuals with sensory impairment remains a major challenge, this relationship has yet to be accurately characterized. This may, however, account for our unremarkable findings as correction of hearing impairment by amplification is unlikely to improve cognitive performance should the central processes required to interpret sound stimuli remain afflicted.

V. Limitations and Recommendations

Due to the limitation in resources and reluctance to additionally burden inpatients, actual audiological tests were not performed. As the HHIE-S is a self-reported assessment of hearing, it is almost invariably subject to bias. Our study also had not enquired about the patients' perception of the use of hearing augmentation. Our pilot study has highlighted that the use of cross-over design is not appropriate for our study aimed at determining the value of hearing augmentation during cognitive testing in an in-patient setting. The future larger study should be limited to a randomized-controlled design with less stringent exclusions, as cognitive assessments should be universally performed and incorporated into the routine assessment performed for all older hospital in-patients. We feel that the hearing amplifier should still be applied universally, however, this should be complemented by formal audiological testing and self-reported hearing assessment tools to determine how many individuals with hearing impairments will be missed without universally offering hearing augmentation. Hearing amplifiers could also be offered as a low cost alternative to hearing aids which are unaffordable to majority of older individuals living in developing nations. However, this will require additional evaluation in a separate randomized controlled study with personalized hearing aids or placebo, assessing the effect of a hearing amplifier on quality of life, activity of daily living and cognition over a longer period.

VI. CONCLUSION

Our pilot randomized-controlled study has revealed that a cross-over design to evaluate the benefits of hearing amplification for cognitive testing in older inpatients is not appropriate due to a rapid patient turnover, loss of patient cooperation and learning effect associated with the cognitive test chosen. Our findings also have additional implications for the serial use of cognitive tests to assess progress or treatment response. The results of our pilot will now inform a future larger, randomized controlled study evaluating the use of hearing amplifiers in developing country settings during cognitive testing as a short term outcome, and also to improve independence and quality of life as a long term outcome.

CONFLICT OF INTEREST

The authors declare no conflict of interest in publishing this research article.

FUNDING

This research project was partially funded by the Postgraduate Research Grant (PPP), project number: PO017-2014B. The researcher's salary was paid by the Malaysian Elders Longitudinal Research (MELoR) study High Impact Research grant while conducting this study (UM.C/625/1/HIR-MOHE/ASH/02).

REFERENCES

1. United Nations. (2013). World Population Ageing. New York: United Nations, Department of Economic and Social Affairs, Population Division.
2. Tun, P., Williams, V. A., Small, B. J., & Hafter, E. R. (2012). The Effects of Aging on Auditory Processing and Cognition. American Journal of Audiology, 21, 344-350.
3. Tien, T., Jie Jin, W., Kifley, A., Lindley, R., Newall, P., & Mitchell, P. (2006). Sensory and Cognitive Association in Older Persons: Findings from an Older Australian Population. Gerontology, 52(6), 386-394.
4. Cruickshanks, K. J., Wiley, T. L., Tweed, T. S., Klein, B. E. K., Klein, R., Mares-Perlman, J. A., & Nondahl, D. M. (1998). Prevalence of hearing loss in older adults in Beaver Dam, Wisconsin. The Epidemiology of Hearing Loss Study. American Journal of Epidemiology, 148(9).
5. World Health Organization. (2014). Deafness and hearing loss, Media Centre.
6. Lemone, P., & Burke, K. (2004). Medical-surgical nursing Critical Thinking in Client Care (3rd ed.). United States: Pearson Education International.
7. Hartley, D., Rochtchina, E., Newall, P., Golding, M., & Mitchell, P. (2010). Use of Hearing Aids and Assistive Listening Devices in an Older Australian Population. Journal of the American Academy of Audiology, 21(10), 642-653.
8. Lin, Ferrucci, L., Metter, E. J., An, Y., Zonderman, A. B., & Resnick, S. M. (2011). Hearing loss and cognition in the Baltimore Longitudinal Study of Aging. Neuropsychology, 25(6), 763-770.
9. Flatau, E., Syndulko, K., Flatau, Y., & Osterweil, D. (1998). Hearing impairment in residents of a long-term care facility: prevalence and relationship to cognitive dysfunction. Annals of Long Term Care, 6(13), 410-413.
10. Kozier, B., Erb, G., Burke, K., & Berman, A. J. (2000). Fundamentals of Nursing: Concepts, Process, and Practice, 6th Edition. New Jersey: Prentice Hall Health.
11. Gates, G., Cobb, J., Linn, R., Rees, T., Wolf, P., & D'Agostino, R. (1996). Central auditory dysfunction, cognitive dysfunction, and dementia in older people. Arch Otolaryngol Head Neck Surg, 122(2), 161-167.
12. Beck, D. (2010). Cognition and Audition: introductory concepts, Continuing Education: International Hearing Society.
13. MacDonald, A. A., Joyson, A., Lee, R., Seymour, D. G., & Soiza, R. L. (2012). The Effect of Hearing Augmentation on Cognitive Assessment Scales at Admission to Hospital. American Journal of Geriatric Psychiatry, 20(4), 355-361.
14. Polit, D. F., & Beck, C. T. (2014). Essentials of Nursing research. Appraising evidence for nursing practice. China: Wolters Kluwer health Lippincott Williams & Wilkins.
15. Razali, R., Jean-Li, L., Jaffar, A., Ahmad, M., Shah, S. A., Ibrahim, N., Din, N. C., Jaafar, N. R. N., Midin, M., Sidi, H., & Ahmad, S. (2014). Is the Bahasa Malaysia version of the Montreal Cognitive Assessment (MoCA-BM) a better instrument than the Malay version of the Mini Mental State Examination (M-MMSE) in screening for mild cognitive impairment (MCI) in the elderly? Comprehensive Psychiatry, 55(Supplement 1), S70-S75.
16. Ventry, I., & Weinstein, B. (1982). The Hearing Handicap Inventory for the Elderly: A new tool. Ear and Hearing, 3, 128-134.
17. Allen, N. H., Burns, A., Newton, V., Hickson, F., Ramsden, R., Rogers, J., Butler, S., Thistlewaite, G., & Morris, J. (2003). The effects of improving hearing in dementia. Age and Ageing, 32(2), 189-193.
18. Lin, Yaffe, K., Xia, J., Xue, Q.-L., Harris, T. B., Purchase-Helzner, E., Satterfield, S., Ayonayon, H. N., Ferrucci, L., & Simonsick, E. M. (2013). Hearing loss and cognitive decline in older adults. JAMA Internal Medicine, 173(4), 293-299.
19. van Hooren, S. A. H., Anteunis, L. J. C., Valentijn, S. A. M., Bosma, H., Ponds, R. W. H. M., Jolles, J., & van Boxtel, M. P. J. (2005). Does cognitive function in older adults with hearing impairment improve by hearing aid use? ¿La función cognitiva en adultos mayores hipoacúsicos mejora con el uso de auxiliares auditivos?, 44(5), 265-271.
20. Nasreddine, Z. S., Phillips, N. A., Bédirian, V., Charbonneau, S., Whitehead, V., Collin, I., Cummings, J. L., & Chertkow, H. (2005). The Montreal Cognitive Assessment, MoCA: a brief screening tool for mild cognitive impairment. Journal of the American Geriatrics Society, 53(4), 695-699.
21. Choi, A. Y., Shim, H. J., Lee, S. H., Yoon, S. W., & Joon, E. J. (2011). Is Cognitive Function in Adults with Hearing Impairment Improved by the Use of Hearing Aids? Clinical and Experimental Otorhinolaryngology, 4(2), 72-76.
22. Connelly, L. M. (2014). Understanding crossover design, 267.

Corresponding author:

Author: Amirah Fatin Ibrahim
Institute: Kulliyyah (Faculty) of Nursing,
International Islamic University Malaysia (IIUM),
PO Box 141, 25710 Kuantan, Pahang, Malaysia.
Email: ami.fatin@gmail.com

Development and Characterization of Polypyrrole-Based Nanocomposite Adsorbent and Its Applications in Removal of Radioactive Materials

M.A. Olatunji[1], M.U. Khandaker[1], Y.M. Amin[1], and H.N.M. Ekramul Mahmud[2]

[1] Department of Physics, University of Malaya, 50603 Kuala Lumpur, Malaysia
[2] Department of Chemistry, University of Malaya, 50603 Kuala Lumpur, Malaysia

Abstract— **Development of environmental friendly materials is desirable in the fields of science and engineering for various purposes. In environmental science, they find applications commonly in environmental remediation and the recovery of valuable metals. This aspect is of particular concern to health sciences due to the toxic effects and non-biodegradable nature of many substances that have found their way into the human body via environmental media. As for instance, though cobalt possesses some importance in diet it has been linked to health problems like cardiomyopathy, asthma etc. via overexposure. As a result, removal of this substance from the aqueous media of our environment is crucial. In this study, polypyrrole conducting polymer incorporated with biomass waste (sawdust) was prepared by simple chemical oxidative polymerization as a composite adsorbent for ^{57}Co radionuclide removal from aqueous solution. The as-prepared composite material was characterized by FESEM, XRD and BET surface analysis. In the ^{57}Co adsorption studies, the effects of several factors such as adsorbent dosage, contact time and competitive metal ions were investigated for real wastewater treatment applications. The results indicated that the removal process was rapid and reached saturation within 3 hours of contact of adsorbent and adsorbate at ambient temperature. Both Na^+ and K^+ coexisting ions were found to affect ^{57}Co adsorption and the uptake percentage was reduced from 54.9 to ~ 20% at 0.1M of both Na^+ and K^+ ions. Langmuir isotherm model was used to fit the maximum sorption capacity.**

Keywords— **Adsorption, ^{57}Co radioisotope, sawdust, polypyrrole nanocomposite.**

I. Introduction

Conducting polymers have attracted much attention in scientific community since their discovery in the last century. This is due to the physico-chemical properties exhibited by the conducting polymers which have made them applicable for different purposes. Conducting polymers are available in polymeric batteries, wire, microactuators, electronic devices, functional membranes and as bio- and gas- sensors among others [1]. However, in the field of environmental science, the flexibility and strong adhesive forces of conducting polymers have been explored in decontamination of heavy metals and radioactive materials from water and wastewater [2].

Concerning liquid radioactive wastes treatment, among the radionuclides that need urgent attention to be removed is the radioisotopes of cobalt. The reason is that these radioisotopes are commonly released from different extractive industries such as nuclear power plants, medicine, mining, metallurgy, painting, pigments and electronics [3,4]. From various applications, cobalt is released as part of the liquid wastes into the environment and transfer through soil-to-foodstuff route into human and animals, leading to internal exposure. Once ingested, cobalt can cause serious health effects in the human body some of which are diarrhea, paralysis, low blood pressure, neurotoxicological symptoms such as headaches and changes in reflexes [5]. Cardiomyopathy outbreak in beer was once linked to ingestion of cobalt (Lauwerys and Lison, 1994). The effects of radioisotopes of cobalt are more severe because they can induce many a lifelong disease among which is cancer [5]. All these have expedited increasing worldwide concern to decontaminate cobalt and its isotopes from water and wastewater.

In the present study, we have investigated the possibility of polypyrrole/ saw dust composite adsorbent for ^{57}Co removal from its aqueous solution by adsorption. The choice of biomass material (saw dust) with polypyrrole conducting polymer in this study was due to its cheap accessibility, pressence of many functional groups which could support adsorption and possibly increase sorption capacity when coated with polypyrrole. Previous studies have revealed that enhancement of sorption capacity is possible through incorporation of conducting polymers into the inorganic and biomass adsorbents [2,7]. Apart from this, organic-inorganic composites have proven stable to decomposition or to the release of adsorbed radionuclides and are easily separated from aqueous solutions due to insoluble nature [8,9]. Hence, polypyrrole nanocomposite adsorbent based on sawdust is expected to be safer for final disposal after adsorption of the target radionuclide.

To the best of our knowledge, no study has been conducted for ^{57}Co removal from waste effluents by using polypyrrole/ sawdust composite. As a result, adsorption studies were conducted to obtain the effects of common environmental conditions on the adsorption process.

F. Ibrahim et al. (eds.), *International Conference for Innovation in Biomedical Engineering and Life Sciences*,
IFMBE Proceedings 56,
DOI: 10.1007/978-981-10-0266-3_7

II. Materials and Methods

A. Materials

Sawdust used in this study was obtained freely from the Mechanical workshop, Physics department and separated from the dirt before washing and drying to obtain in clean form. The dried sawdust was sieved to small particle sizes (0.45-0.60mm) before use.

B. Reagents

All chemicals used in this work were of analytical reagent grade. Iron (III) chloride ($FeCl_3.6H_2O$) and pyrrole monomer (98%) were obtained from Merck. The pyrrole was distilled for high purity before use. Distilled water obtained from Milli-Q system was used throughout the experiment. The radiotracer ^{57}Co (traceable in 4M HCl solution) was purchased from Eckert and Ziegler Analytics, supplied in a flame sealed vial. All chemicals and materials were used as obtained from the supplier without further purification.

C. Synthesis of Polypyrrole and Composite

The synthesis and modification of polypyrrole was performed by the following steps: 1.680g of FeCl3.6H2O was dissolved in 250ml of distilled water and after; 250ml of 5.705g of previously distilled pyrrole monomer was added to the oxidant solution and left for about 30mins in a static mode for the formation of polypyrrole at room temperature.

The modification of polypyrrole with sawdust was conducted by first adding a requisite amount of sawdust (about 2.5 g) to 250ml of 1.680g aqueous solution of pyrrole monomers and stirred at 300rpm for about 30 mins. After stirring by magnetic stirrer, 250ml of 5.705g aqueous solution of FeCl3.6H2O as oxidant was added to the mixture and stirring continued at 300rpm for the next 30mins for the formation of the polypyrrole / sawdust composite. The solid was then separated from the solution using Whatman #2 filter paper under vacuum and thoroughly washed with distilled water. The remaining solid was dried at 50±0.1°C under vacuum for 48 h. The as-prepared composites and polypyrrole adsorbents were crushed with mortar and pestle and sieved before use.

D. Adsorbent Characterization

PPy/SD composites were characterized by Field emission scanning electron microscopy (FESEM), X-ray diffractometry (XRD) and Brunauer-Emmett-Teller (BET) instrumentations. The FESEM (Hitachi Technology) photograph was taken to show the surface morphology of the sorbents at 2000 volt accelerating potential and 11300 mA emission current. The X-ray diffraction (XRD) data (on powders) were collected using PANalytical EMPYREAN diffractometer system in a 2θ configuration employing the Cu kα radiation (λ = 1.54 Å) consisting of soller slits (0.04rad), a fixed divergence slit (0.5°), a fixed anti-scatter slit (1°) and a rotating sample stage (rotating time of 8s). The sample was scanned between 2θ = 10° and 70° with the PIXcel-3D detector. PANalytical High Score Plus software, version 3.0d (3.0.4) was employed for qualitative analysis. The Brunauer, Emmett and Teller (BET) surface area and porosity were measured from the N_2 gas-adsorption/ desorption isotherm at 77.4K using Micromeritics (ASAP 2020 and Tritar II 3020 Kr). The total pore volume (Vpore) was determined with the manufacturer's software (SAP 2020 V4.01). Moreover, pore size distribution was inferred using the BJH theory [10]. The t-plot method was used to calculate the micropore volume (Vmicro) and exterior surface area [11].

E. Batch Experiment

The adsorption of ^{57}Co onto PPy/SD composites was carried out at room temperature (25°C) and pH3.65 in batch mode. Solutions of different concentrations were prepared by diluting an appropriate portion of the ^{57}Co radiotracer stock solution in 250ml polyethylene beaker.

Several batch experiments were carried out on mechanical platform shaker at 250 rpm with 50ml of solution to identify the optimum values of contact time, adsorbent dosage and the effect of interfering ions. The isotherm studies were carried out in different initial concentrations (2.0 to 7.55 Bq/L) at the optimum time and adsorbent dosage.

The contents of the polyethylene beakers were filtered through #2 whatman filter paper. The initial solution and the filtrate were analyzed using high-purity germanium detector (HPGe) coupled with multi-channel analyzer for data acquisition at γ-ray energy = 122.1 KeV, Iγ = 85.6%. The removal percentage of the radionuclide and the amount sorbed per unit mass of the adsorbent (q_e) were calculated using the following equations (1 and 2):

$$Removal\ (\%) = \frac{C_i - C_f}{C_i} \times 100 \quad (1)$$

$$q_e = \frac{C_i \times V_i - C_f \times V_f}{m} \quad (2)$$

Where C_i= initial concentration, C_f= filtrate concentration, V_i= initial volume, V_f= filtrate volume m= adsorbent dosage amount.

III. Results and Discussion

A. Adsorbent Characterization

The surface area measurement of the samples in dried powder form was carried out to investigate the adsorption efficiency of the sorbents by BET technique. The BET surface areas of the polypyrrole and composites are 10.57 and

7.06 m^2/g, respectively whereas, the total pore volume was 0.021 cm^3/g (single point adsorption volume of pores less than 3549.22 Å with p/p_0 = 0.995) and 0.017 cm^3/g (single point adsorption volume of pores less than 3253.07 Å with p/p_0 = 0.994) for both the polypyrrole and composite before ^{57}Co adsorption. The difference in the BET surface area and the pore volume for polypyrrole and its composite confirmed the formation of polypyrrole/sawdust. The nitrogen gas adsorption-desorption isotherm curve (Type II curve) is obtained as shown in Fig. 1.

The surface morphological structures of the adsorbents were obtained by FESEM images. The sorbent images suggest that polypyrrole has been coated on the sawdust (Fig. 2). The incorporation of sawdust does not seem to impair the typical PPy pore network as there was no much change in composite image except for some PPy precipitate formed on the composite. These observations have well been reported by previous works on polypyrrole and its composites [12].

The XRD patterns of synthesized PPy and PPy coated sawdust are shown in Fig. 3. Since there is no diffraction peak except for the broad peak at position 2θ: 22–29 in the XRD pattern of polypyrrole hence, indicates its amorphous structure [13]. However, the PPy coated sawdust shows a backward shift in the PPy broad peak to position 2θ: 10-20 and a sharp peak at position 2θ: 22.7 related to the crystalline structure of the sawdust.

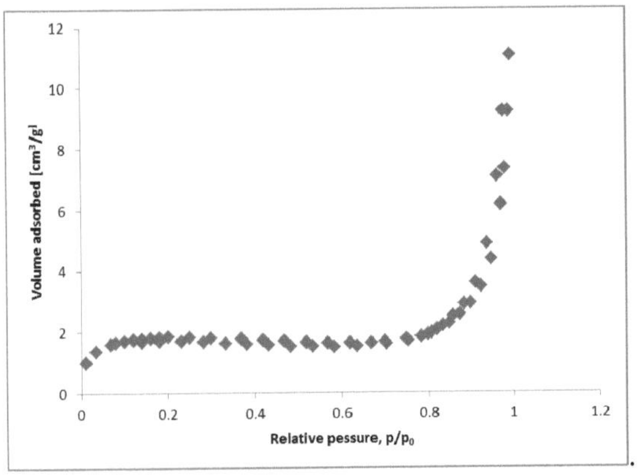

Fig. 1 N2 gas adsorption and desorption isotherms of the polypyrrole/ sawdust composite

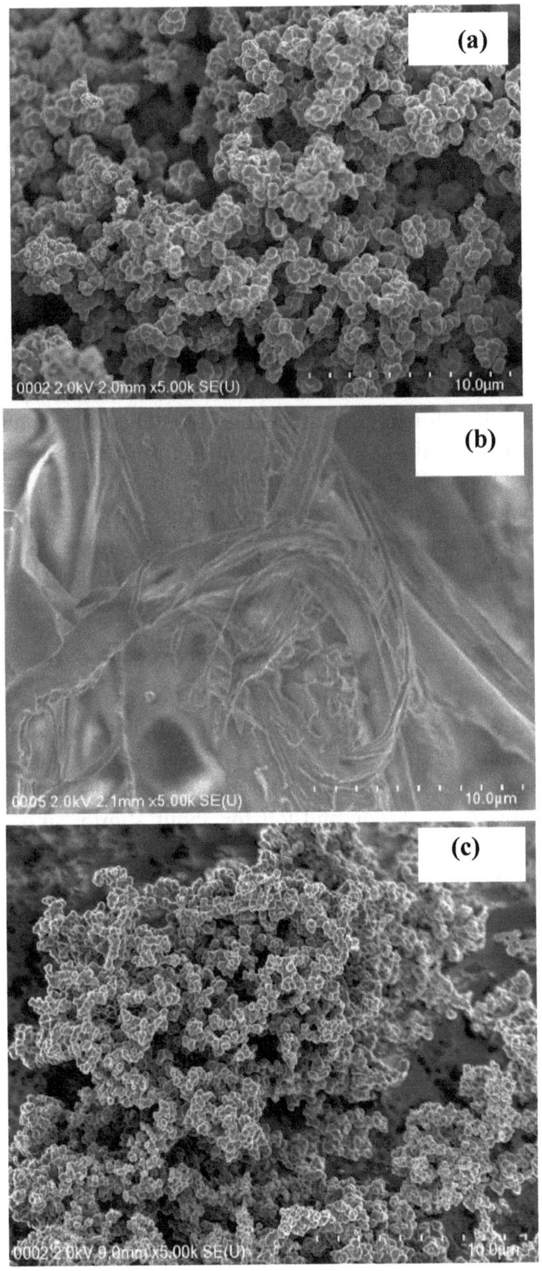

Fig. 2 Field Emission Scanning Electron Micrograph of the (a) PPy, (b) Sawdust and (c) PPy/SD sorbents

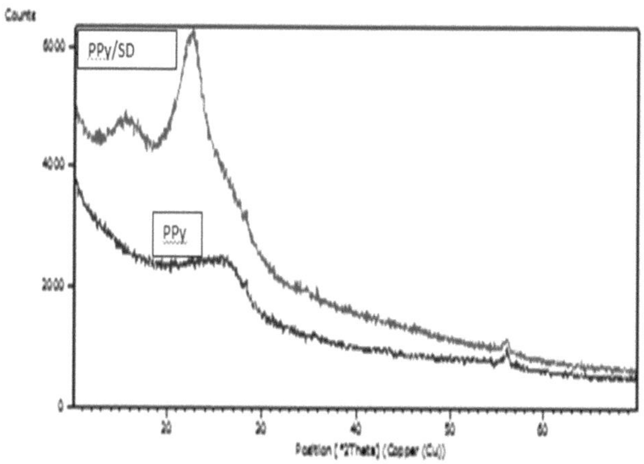

Fig. 3 XRD pattern for polypyrrole and polypyrrole coated sawdust

B. *Effects of Adsorption Parameters on ^{57}Co Uptake*

Effect of Contact Time

Adsorption time is crucial in justifying if an appropriate technology has been designed. If the treatment technologies implemented for separation of any adsorbates from their solutions are appropriate and considered economically viable then, optimum equilibrium and adsorption time must be short with higher adsorption rate [14]. So, 50ml of aqueous solution (2.0Bq ^{57}Co/L) was contacted with 0.3g of PPy and PPy/SD separately at different time up to 17h at pH3.65 and 25±0.1°C. As shown in Fig. 4, adsorption rate of ^{57}Co onto PPy/SD was rapid in the first hour and becomes steady at 2h until equilibrium is reached in 3h. At this time, there was no much increase in the uptake amount for the next 2h and started declining in 6h until 17h. The maximum removal efficiency at 3h was 54.9% which gradually reduced to ~2% in 17h. In case of PPy, the process was rather very slow and no significant amount was removed for the duration of contact time (figure not shown). However, higher adsorption capacity and faster adsorption rate of PPy is possible through coating of the polymer on other materials. This observation has been reported by previous authors using polypyrrole/sawdust to remove Zn (II) from aqueous solution [15] and polypyrrole–impregnated porous carbon for lead, silver and mercury removal [16]. As a result, PPy/SD was used for subsequent studies at 3h optimum contact time.

Effect of PPy/SD Dosage

The effect of adsorbent dosage on removal of the radionuclide was assessed at 0.1 to 1.0g dosage on 50ml of 2.0 Bq/L initial ^{57}Co concentrations, at room temperature (25°C) and at 3h optimum contact time. As shown (Fig. 5), adsorption percentage of ^{57}Co increased as sorbent amount increased from 0.1 to 0.3g. This observation is expected as increase in sorbent amount corresponds to increase sorption surface area and hence, more sorption sites are made

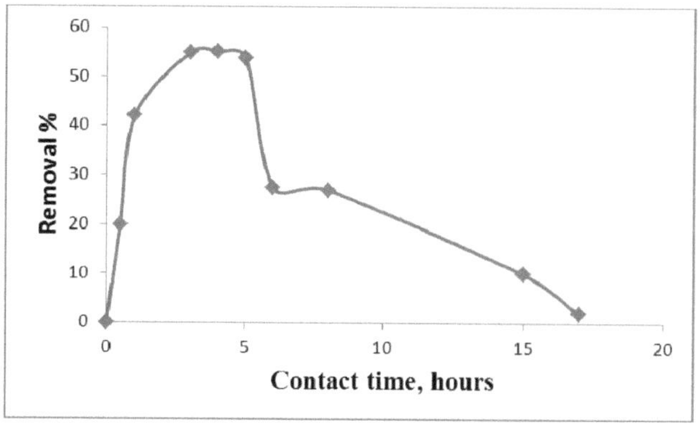

Fig. 4 The effect of contact time on ^{57}Co removal onto polypyrrole coated sawdust at 25°C and pH3.65

available for adsorption [15]. However, as dosage further increased from 0.3 to 0.8g, there was no significant increase in adsorption percentage as a result of less adsorbate concentration compared to available active sites and as dose increased to 1.0g, there was a decline in the amount of adsorbed ^{57}Co. So, further experiments were performed using 0.3g of PPy/SD at optimum time of 3h.

Effect of Interfering Ions

Selectivity is a vital factor to consider for an adsorbent to be used in real wastewater treatment technology. This is because, industrial waste effluents like liquid nuclear waste (the common source of cobalt radionuclide) containing several ions that may impede adsorption of a specific ion. As a result, the effect of co–existing ions on ^{57}Co adsorption was studied using sodium and potassium ions (0.1 to 0.3M concentrations). In this study, the presence of both alkali metal ions had much negative influence on ^{57}Co adsorption within the concentration range studied. Adsorption efficiency decreased to 20% in the presence of 0.1M of Na^+ and K^+ as against the 54.9% adsorption obtained in the absence of the metal ions. This shows that the presence of these metal ions is not favourable for cobalt adsorption from aqueous solution.

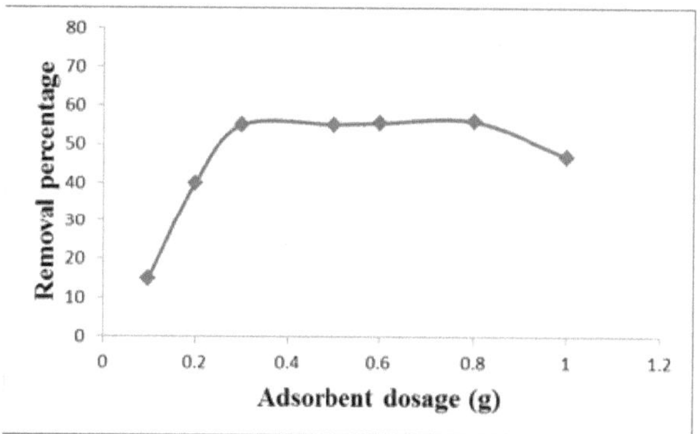

Fig. 5 The effect of adsorbent dosage on ^{57}Co removal onto PPy/SD at 25°C, pH3.65 and 3h

C. Adsorption Isotherm Modelling

Assuming that the adsorption sites are equivalent and the ability of a particle to bind there is independent of whether or not adjacent sites are occupied [17]. Thus, we can estimate the existing relationship between the adsorbate concentrations in solution and the amount sorbed by the adsorbent at a particular temperature using isotherm models. The most common isotherm models (Langmuir and Freundlich isotherm models) have been used in the present study.

Langmuir isotherm model [18]

$$\frac{1}{q_e} = \frac{1}{q_m} + \frac{1}{q_m b C_e} \quad (3)$$

Where q_e, q_m and b represent adsorption capacity at equilibrium concentration C_e (Bq/L), maximum monolayer adsorption capacity and Langmuir constant, respectively. Adsorption capacity at equilibrium (q_e) is calculated from the equation 2. Both q_m and b are estimated by plotting $1/q_e$ versus $1/C_e$ as shown in the Fig. 6. The Langmuir isotherm plot suggests the model could be applicable to the cobalt adsorption with the correlation coefficient R^2= 0.974. Both q_m and b estimated from the plot are compiled in Table 1.

The R_L value which indicates the type of isotherm was estimated by using equation 4 [19]. The values of R_L between 0 and 1 suggest favourable adsorption of the adsorbate. As shown in the table (Table 2), the values of R_L are below 1 for all the initial concentrations of the cobalt radionuclide, indicating favourable adsorption.

$$R_L = \frac{1}{1+bC_i} \quad (4)$$

Meanwhile, Freundlich isotherm model was also used to analyze the adsorption data using equation 5:

$$log q_e = log k_f + \frac{1}{n} log C_e \quad (5)$$

Where, k_f and n are Freundlich constants representing adsorption capacity and adsorption intensity, respectively. The plot of $logq_e$ against $logC_e$ as shown in Fig. 6 suggests the model may not be applicable to the adsorption system in this study as the correlation coefficient, R^2= 0.957 obtained is less than the value obtained by Langmuir model. Hence, adsorption data followed more to Langmuir isotherm model based on the correlation coefficient obtained compared to Freundlich isotherm model (see Table 1). However, a good fit of experimental data to an isotherm model might not be sufficient for its validity and furthermore interpretation of the model for liquid–solid system is expected to be difficult compared to gas–solid system [20]. But still, isotherm model which represents equilibrium data might give some insight into the nature of sorption. Based on the Langmuir plot, it can be assumed that the sorbent surface carries a limited number of sites that are characterized by equal energy of adsorption, indicating monolayer coverage of adsorbate onto a homogeneous adsorbent surface [21] and that the sorption of a molecule at a particular site does not depend on the occupation of the neighboring sites [22].

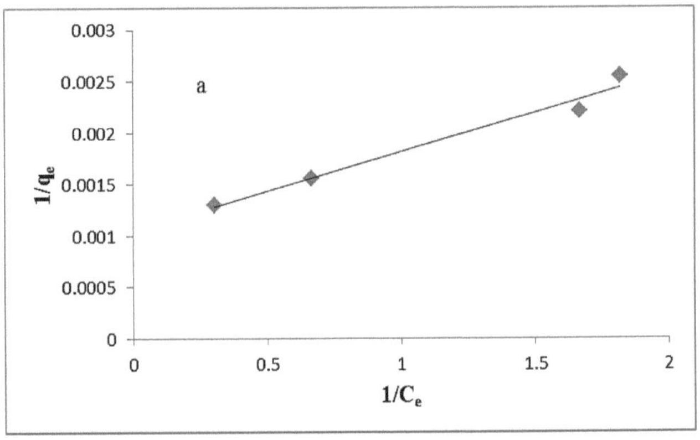

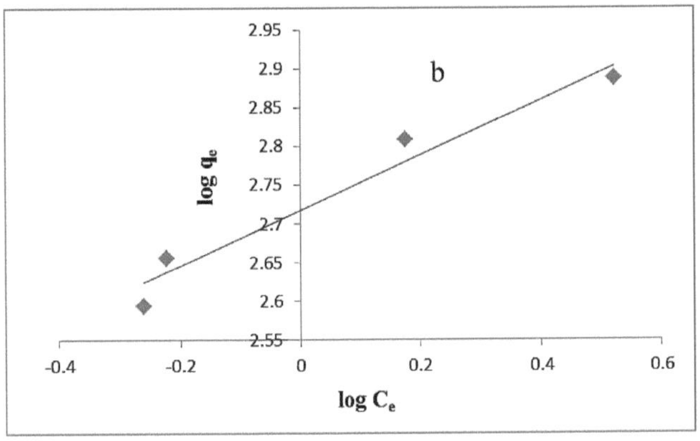

Fig. 6 Adsorption isotherms of cobalt adsorption onto polypyrrole coated sawdust (a) Langmuir isotherm model, (b) Freundlich isotherm model

Table 1 Isotherm constants for cobalt radionuclide adsorption onto polypyrrole coated sawdust

Parameter	Value	R^2
Langmuir isotherm q_m (mBq/g)	909.1	0.974
b (L/mBq)	1.38	
Freundlich isotherm k_f (mBq/g)	521.6	0.957
n	2.79	

Table 2 RL values for adsorprtion of ^{57}Co onto polypyrrole coated sawdust at 25°C

Intial Co–57 concentration (Bq/L)	2.84	3.25	5.18	7.55
$R_L \times 10^{-3}$	0.203	0.182	0.123	0.088

IV. CONCLUSION

The toxicity, easy migration within the environmental media and subsequent health implications of cobalt metal and its radioisotopes warrant their removal from aqueous solutions in the environment. The results presented in this study show that polypyrrole coated sawdust can be used as an adsorbent alternative to high cost materials for the removal of cobalt radioisotope (^{57}Co) from its aqueous solution. Several environmental factors have been investigated to optimize the use of the adsorbent for maximum adsorption in actual wastewater treatment applications. The study revealed that polypyrrole coated sawdust could be applied for real wastewater treatment at ambient temperature. Langmuir isotherm model could favourably be used to describe the adsorption of cobalt radionuclide onto the polypyrrole coated sawdust.

ACKNOWLEDGEMENT

The work has been carried out under the University of Malaya Grant no. PG027–2014A.

CONFLICT OF INTEREST

The authors declare that there is no conflict of financial interest or any other.

REFERENCES

1. Campbell TE, Hodgson AJ, Wallace GG (1999) Incorporation of erythrocytes into polypyrrole to form the basis of a biosensor to screen for Rhesus (D) blood groups and Rhesus (D) antibodies. *Electroanalytical 11, 215–222*
2. Saberi R, Nilchi A, Garmarodi SR et al (2010) Adsorption characteristic of ^{137}Cs from aqueous solution using PAN-based sodium titanosilicate composite. *J Radioanal Nucl Chem. 284:461–469*
3. Baun DL, Christensen TH (2004) Speciation of heavy metals in landfill leachate: A review. *Waste Management & Research, 22(1): 3-23*
4. Babel S, Kurniawan TA (2003) Low-cost adsorbents for heavy metals uptake from contaminated water: a review. *J. Hazard. Mater. B97: 219–243*
5. Parab H, Joshi S, Sudersanan M et al (2010) Removal and recovery of cobalt from aqueous solutions by adsorption using low cost lignocellulosic biomass—coir pith. *J. Environ. Sci. Health, Part A: Toxic/Hazardous Substances and Environmental Engineering, 45:5, 603-611*, DOI: 10.1080/10934521003595662
6. Lauwerys R, Lison D (1994) Health risks associated with cobalt exposure. Sci. Total Environ. 150 (1-3): 1–6
7. Wang YQ, Zou BF, Gao T et al (2012) Synthesis of orange-like Fe_3O_4/PPy composite microspheres and their excellent Cr(VI) ion removal properties. *J Mater Chem 22: 9034–9040*
8. Molt M, John J, Šebesta F (1997) Composite absorbers of inorganic ion exchangers and polyacrylonitrile binding matrix. *J. Radioanal. Nucl. Chem. 222: 205-207*
9. Narbbutt J, Bilewicz A, Baratos B (1989) Composite ion exchangers for radiocesium removal from nuclear reactor wastes. In: Proceedings of an international symposium on management of low and intermediate level radioactive wastes. IAEA, Vienna
10. Barrett EP, Joyner LG, Halenda PP (1951) The determination of pore volume and area distributions in porous substances. I. Computations from nitrogen isotherms. J. Am. Chem. Soc. 73: 373–380
11. Wu FC, Tseng RL, Juang RS (2009) Initial behavior of intraparticle diffusion model used in the description of adsorption kinetics. *Chem. Engr. J. 153: 1-8*
12. Mahore RP, Burghate DK, Kondawar SB (2014) Development of nanocomposites based on polypyrrole and carbon nanotubes for supercapacitors. *Adv. Mater. Lett. 5(X): XXX-XXX*
13. Javadian H, Taghani M (2014) Application of novel Polypyrrole/thiol-functionalized zeoliteBeta/MCM-41 type mesoporous silica nanocomposite for adsorptionof Hg^{2+} from aqueous solution and industrial wastewater: Kinetic, isotherm and thermodynamic studies. *Appl. Surf. Sci. 289: 487– 494*
14. Sheha RR, El-Khouly SH (2013) Adsorption and diffusion of cesium ions in zirconium (IV) iodomolybdate exchanger. *Chem. Engr. Res. Des. 91: 942–954*
15. Omraei M, Esfandian H, Katal R et al (2011) Study of removal of Zn (II) from aqueous solution using polypyrrole nanocomposite. *Desa. 271: 248-256*
16. Choi M, Jang J (2008) Heavy metal ion adsorption onto polypyrrole-impregnated porous carbon. *J. Colloid and Interface Science, 325: 287-289*
17. Upendra K, Manas B (2006) Sorption of cadmium from aqueous solution using pretreated rice husk. *Bioresour. Technol. 97: 104-109*
18. Bhatnagar A, Minocha AK, Sillanpää M (2010) Adsorptive removal of cobalt from aqueous solution by utilizing lemon peel as biosorbent. *Biochemical Engr. J. 48:181-186*
19. Ahmadpour A, Tahmasbi M, Bastami TR et al (2009) Rapid removal of cobalt ion from aqueous solutions by almond green hull. *J. Hazard. Mater. 166:925-930*
20. Abusafa A, Yücel H (2002) Removal of ^{137}Cs from aqueous solutions using different cationic forms of a natural zeolite: clinoptilolite. *Sep. Purificat. Tech. 28: 103–116*
21. Wei J, Zhang X, Liu Q et al (2014) Magnetic separation of uranium by $CoFe_2O_4$ hollow spheres. *Chem. Engr. J. 241: 228-234*
22. Cornell RM (1993) Adsorption of cesium on minerals: a review. *J. Radioanal. Nucl. Chem., 171: 483–500*

Author:	M.U. Khandaker
Institute:	Department of Physics, University of Malaya
Street:	50603 Kuala Lumpur
City:	Kuala Lumpur
Country:	Malaysia
E-mail:	mu_khandaker@um.edu.my

Grey Matter Volume Differences of Textual Memorization: A Voxel Based Morphometry Study

A.H. Sapuan[1], N.S. Mustofa[1], M.Z. Che Azemin[2], Z.A. Abdul Majid[1], and I. Jamaludin[1]

[1] Department of Diagnostic Imaging and Radiotherapy, Kulliyyah of Allied Health Sciences, International Islamic University Malaysia, Kuantan, Pahang, Malaysia
[2] Department of Optometry and Visual Science, Kulliyyah of Allied Health Sciences, International Islamic University Malaysia, Kuantan, Pahang, Malaysia

Abstract— **Textual memorization involves intensive memorization techniques and processes, including repetition, rehearsal, and retrieval. Yet, little is known about the possible effects of this robust memorization training on brain structures. Therefore, this study was designed to investigate the presence of structural plasticity in the brain, which may develop after long-term practice of textual memorization techniques. The subjects are a group of huffaz who underwent one year of textual memory training. Twenty-eight volunteers (14 males) were enrolled in this study. Brain images of the subjects were obtained by employing 1.5 Tesla MRI and 3D-MPRAGE imaging sequence. Voxel-based morphometry was applied to assess the gray matter volume differences in the memorizers and the control subjects. The current study found that gray matter volume in the anterior cingulate gyrus, orbitofrontal cortex, left inferior temporal gyrus, right occipitotemporal gyrus, left inferior parietal gyrus, right perirhenal cortex, superior parietal cortex, posterior cingulate cortex and anterior cingulate cortex of the memorizers were significantly increased as compared with the control subjects. The results suggest textual memorization involves with extensive connection across the brain.**

Keywords— **voxel-based morphometry, grey matter volume, textual memorization, huffaz, Quran.**

I. INTRODUCTION

Numerous studies have been carried out on the relation between specific types of memory and structural brain variation. Most studies documented the structural plasticity in hippocampus and its gray matter volume associated with spatial memory [1]–[3]. Apart from the studies conducted on the hippocampal contribution to spatial memory, there is a growing need to determine whether textual memorization could directly be associated with any changes in the brain structures. Textual memorization involves comprehensive memory construction and retrieval processes which may leave permanent changes in the brain structures.

To our knowledge there is no scientific study that has been performed on the effect of textual memorization on regional grey matter volume. A study has found a positive correlation between gray matter volume with extensive learning [4] but it is still undetermined specifically for textual memorization. Another study concluded that changes in gray matter induced by learning new task is more vital for the brain plasticity than a constant learning of previously trained activity [5]. The study found only certain areas in gray matter in the occipito-temporal cortex responsive towards the training. The investigation was done on visio-motor skill after three months of training.

Results from a study among semantic dementia patients reported atrophy in left anterior temporal region is positively associated with the degree of memory impairment, but not in the hippocampus compared to age-matched control subjects [6]. They also located areas that are significantly affected by the disease. However, the effect of semantic memory training to the gray matter volume has not yet been investigated.

Long-term memory (LTM) has enormous capacity in holding information. The acquired information needs to be consolidated first before it can be stored in LTM. Formation of LTM occurs gradually following learning when new memory becomes consolidated into a relatively long lasting state by reverberating neural circuits. This will induce lasting changes in synaptic connections that provide the basis for LTM. This is evident in a study of the effects of musical training on brain function during LTM retrieval in a functional study [7]. On the other hand, the study has been extended to reveal that there is also an increase in gray matter volume in hippocampal region of musicians compared to non-musicians. Study in the memory-related regions with regard to textual memorization, however, is still not well explored.

The aim of the current study is to investigate whether grey matter volume changes associated with verbatim and extensive textual memorization regime can be detected anywhere in healthy individuals.

II. MATERIAL AND METHOD

A total of 28 healthy volunteers with ages ranging from 21-25 years participated in the study. The participants comprised of 12 subjects (6 males) who had undergone extensive textual memory training for 1 year. This group of participants is called the *huffaz*. The Quran

F. Ibrahim et al. (eds.), *International Conference for Innovation in Biomedical Engineering and Life Sciences*,
IFMBE Proceedings 56,
DOI: 10.1007/978-981-10-0266-3_8

text has been used at the memorization material for the subjects. It is written in Arabic with 114 chapters and each chapter consists of varying length of verses. The total number of verses is 6236 verses with 77437 total words. The control group consisted of 16 subjects (8 males) matched for age. Participants were strongly right-handed, reported no past head injury, no significant medical illness, no known diagnosis of psychiatric, neurological and endocrine disorder and no contraindication for MRI scans (e. g. floating metallic bodies, claustrophobia for small spaces). The study protocol has been approved by local ethics committees. The study is carried out in accordance with the Declaration of Helsinki. This study had been registered in Malaysian National Medical Research Registration (NMRR) with research identification number is NMRR-14-107-19576. All participants gave their written informed consent to participate in the research after deliberate explanation about the objectives and methods of the research.

A. MRI Acquisition

High resolution structural MRI brain scans were obtained with 1.5 Tesla Siemens Magnetom Avanto scanner (Siemens Medical Solutions, Erlangen, Germany). Participants underwent a 10 minute resting–state scan in which they were given no specific instructions except to keep their eyes closed and hold still during the image acquisition. The MRI protocol scan was T1-weighted three-dimensional magnetization-prepared rapid gradient-echo (T1W-3D-MPRAGE) sequence. This sequence is the most common sequence used in the department for producing a high resolution image. Sequence parameters were: TR/TE/TI = 1890/2.93/1100 ms with 15° flip angle. A field-of-view (FOV) of 280mm was used with 0.5 x 0.5 x 0.7mm voxel size producing 208 contiguous T1-weighted sagittal slices with slice thickness of 0.7mm. The generated images were checked for any presence of motion artifact before considered as approved and continued to the next subjects. All brain images were then transferred from the host computer (Syngo acquisition workplace) into compact discs in the digital imaging and communication in medicine (DICOM) format.

B. Voxel-Based Morphometry (VBM) Analysis

Voxel-based morphometric analysis of data was performed using Statistical Parametric Mapping 8 (SPM 8; version 4290) software (developed by the Wellcome Trust Centre for Neuroimaging, in the Institute of Neurology at University College London, UK) running under MATLAB, version 7.12.0 (The MathWorks Inc., Natick, MA, USA). The data processing steps are illustrated in Figure 2.1.

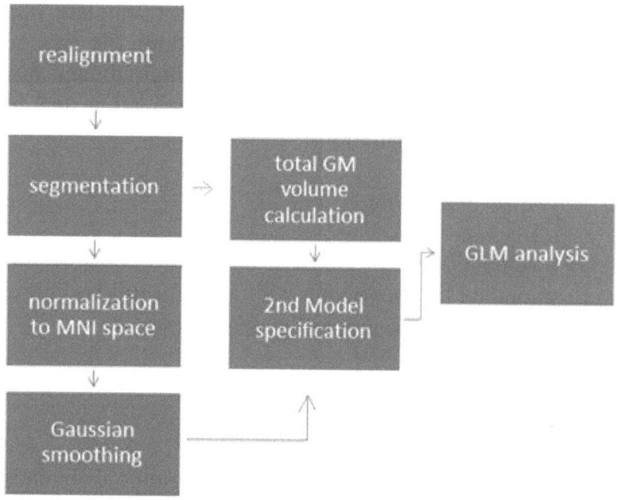

Fig. 1 Voxel based morphometry (VBM) analysis steps performed using SPM software.

C. Realignment of Structural Brain Images

Before pre-processing and analysing the images, the DICOM format images were converted into the NIfTI format so that it can be read by SPM 8. Then, all of the structural images were visually checked for artifacts and manually realigned the crosshair so that the origin of the coordinate system was located at the anterior commissure.

D. Segmentation

Using the segmentation procedure implemented in SPM8, the images were segmented into gray matter (GM), white matter (WM) and cerebrospinal fluid (CSF). For each subject, this resulted in a set of 3 images in the same space as the original T1-weighted image, in which each voxel was assigned a probability of it being GM, WM and CSF respectively. The original images also formed the tissue probability maps of GM, WM and CSF and inverse deformation field for each image. For this experiment, GM images were utilized for analysis later.

E. Total Gray Matter Volume (GMV) Calculation

Total gray matter volume (GMV) for each participant was calculated using a customized MATLAB script. The resultant total gray matter volume was used later in the 2nd model specification stage as a covariate.

F. Normalization to Montreal Neurological Institute (MNI) Space

This step involved two processes; first creating an average GM template for all participants and then normalizing this template to each participant. Both processes were executed using Diffeomorphic Anatomical Registration Through Exponentiated Lie (DARTEL) tools in the SPM.

The first process produced 7 template files (Template_0.nii until Template_6.nii) and the Template_6.nii file was selected for the normalization step. This high dimensional warping process results in improved localization and increased sensitivity in analyses.

G. Gaussian Smoothing

The optimally processed and normalized images then were smoothed by convolving with an 8mm full width-half maximum (FWHM) isotropic Gaussian kernel.

H. Second Model Specification and General Linear Model (GLM) Analysis

The investigation of the gray matter volume differences between *huffaz* and control subjects was done by using voxel-by-voxel analysis of variance test. Because the groups did not show a significant difference in age and sex distribution, hence these variables were not inserted as covariates. The relationship between the textual memorization status and gray matter volume changes was investigated by regression analysis. During conducting the analysis, great care was taken at each step to ensure that the matrix vector for the group of participants and the grey matter volume for each participant were always in exactly the same order.

For the statistical analysis, all of the voxels with a gray matter value below 0.1 (of a maximum value of 1) were excluded to avoid possible edge effects around the border between gray and white matter and to include only voxels with sufficient gray matter, respectively. *T-contrast* was set to find positive and negative differences between grey matter volume and textual memorization. Positive differences mean textual memorizer participants have greater grey matter volume than control participants and the negative differences is the opposite. The findings were considered significant at a voxel level of p=0.005, Family Wise-Error (FWE) corrected for multiple comparisons, with an extended threshold looking for clusters with at least 250 contiguous voxels.

III. RESULTS

There were no significant group differences in age, gender and total gray matter volume as shown in Table 1.

Table 1 Characteristics of participants.

Characteristics	Huffaz	Control	P-Value
Mean Age ± Stdev (Year)	22.83 ± 0.718	23.19 ± 1.109	0.344
Male (%)	50	50	-
Total Gray Matter Volume ± Stdev (mm3)	0.703 ± 0.031	0.677 ± 0.031	0.185

The VBM analysis in comparing these two groups (*huffaz* and non-*huffaz*) was threshold at the cluster-wise significance level at p=0.005 (FWE-corrected). At this threshold, there were nine gray matter regions that were found positively correlated with textual memorization status as shown in Table 2.

The involved brain structures are in the medial areas of the anterior cingulate gyrus, orbitofrontal cortex, posterior parietal cortex, posterior cingulate cortex and anterior parietal cortex. Additional positive correlations with textual memorization status were found in the right occipitotemporal cortex, right perirhenal cortex, left inferior parietal lobe and left inferior temporal gyrus. Figure 2, Figure 3, and Figure 4 indicate the representative peaks for each lobe on axial, sagittal and coronal view of a transparent canonical brain and labeled for reference.

The involved brain structures are in the medial areas of the anterior cingulate gyrus, orbitofrontal cortex, posterior parietal cortex, posterior cingulate cortex and anterior parietal cortex. Additional positive correlations with textual memorization status were found in the right occipitotemporal cortex, right perirhenal cortex, left inferior parietal lobe and left inferior temporal gyrus. Figure 2, Figure 3, and Figure 4 indicate the representative peaks for each lobe on axial, sagittal and coronal view of a transparent canonical brain and labeled for reference.

For the analysis to look for the negative relationship between textual memorization status and structural brain changes, there are very few peaks in the relatively small area of parietal cortex for the effect to reach significance (cluster size: 46mm^3, cluster p-value=1.00 FWE-corrected). Thus, it can be concluded that no area in the control group that was significantly larger than the *huffaz*.

Table 2 Brain regions with positive correlations between total gray matter volume and textual memorization

Cluster size (mm^3)	MNI coordinates (x, y, z)	Cluster *p*-value (corrected)	Voxel *t*-statistic	Z- value	Brain region	BA
5383	3, 42, 21	<0.001	11.87	6.82	Anterior cingulate gyrus	32
	-2, 41, -17	<0.001	11.34	6.68	Orbitofrontal cortex	11
815	-23, -11, -17	<0.001	8.38	5.73	Inferior temporal gyrus_L	20
482	50,-54, -20	<0.001	9.94	6.27	Occipitotemporal gyrus_R	37
456	-63, -50, 36	<0.001	9.11	6.00	Inferior parietal gyrus_L	40
406	14, -7, -18	<0.001	8.56	5.80	Perirhenal cortex_R	35
374	-6, -62, 36	<0.001	9.87	6.25	Superior parietal cortex	7
255	2, -6, 42	<0.001	9.31	6.06	Posterior cingulate cortex	23
	3, 5, 43	<0.001	8.23	5.68	Anterior cingulate cortex	24

Area	Axial	Sagittal	Coronal
Anterior cingulate gyrus (BA32)	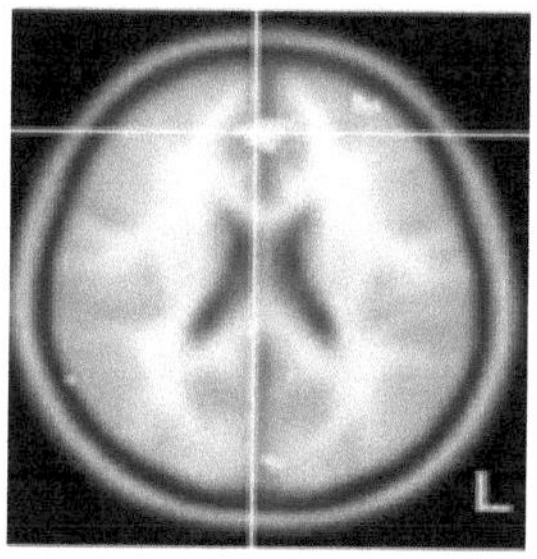	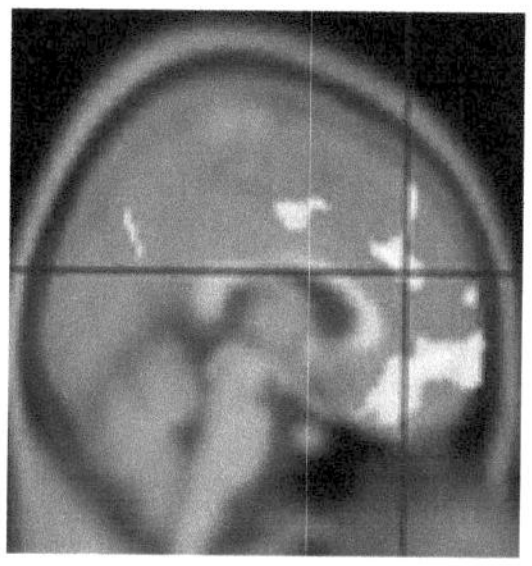	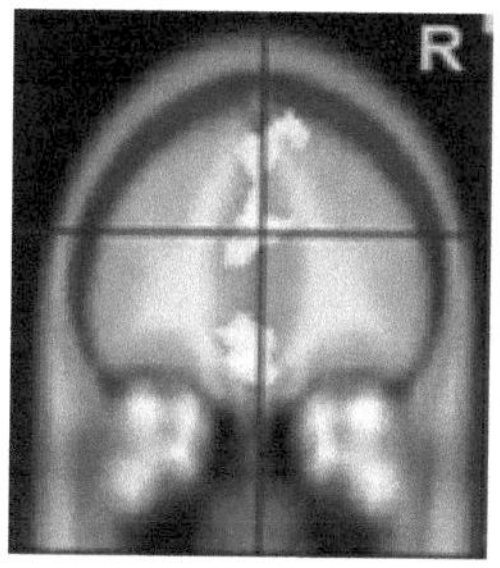
Orbitofrontal cortex (BA11)	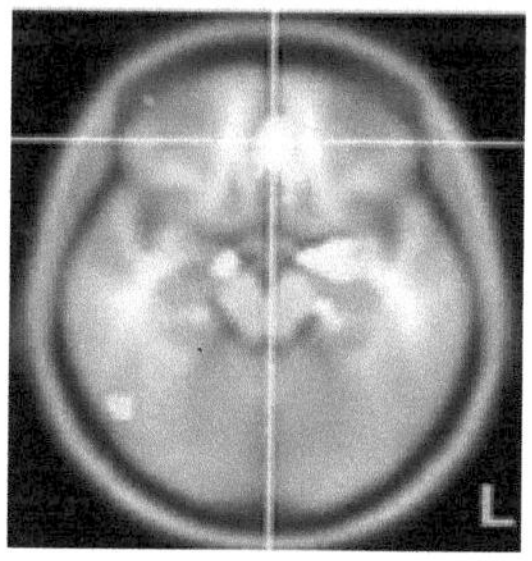	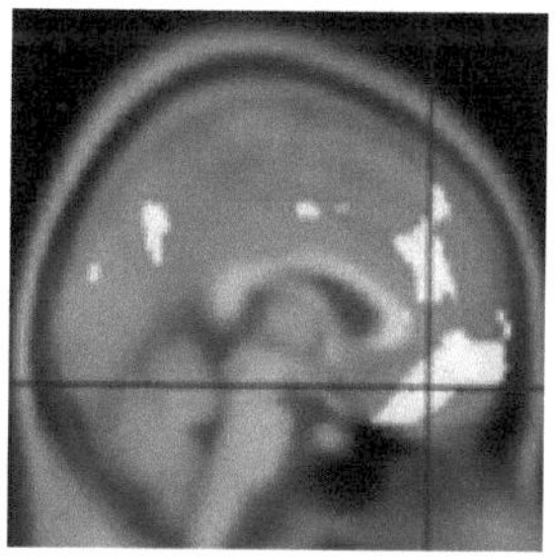	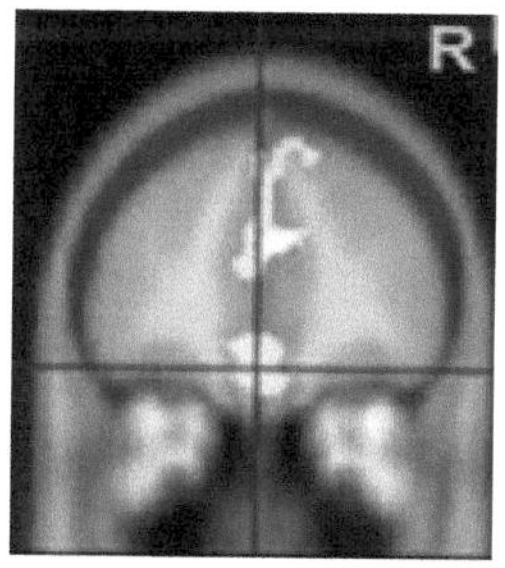
Anterior cingulate cortex (BA24)	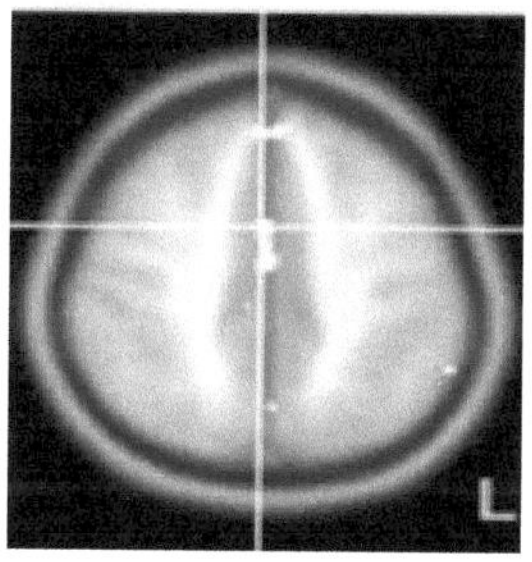	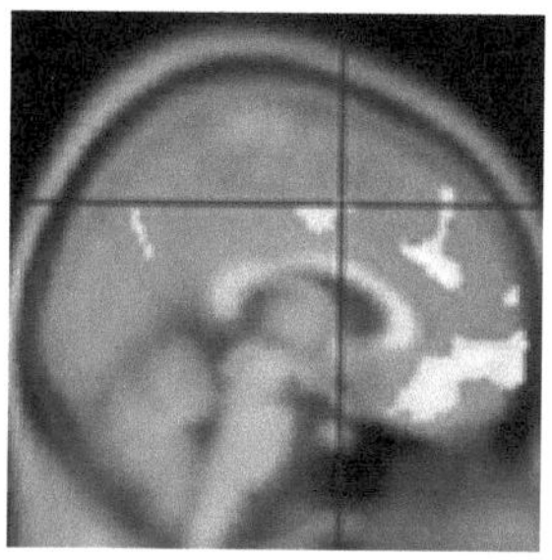	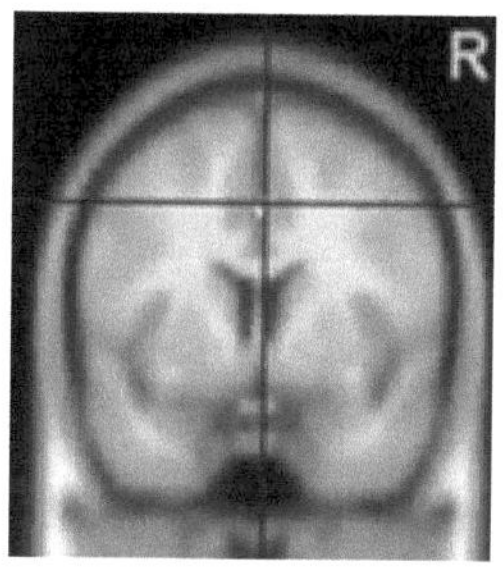

Fig. 2 Frontal lobes areas with increased gray matter volume in *huffaz* group.

Area	Axial	Sagittal	Coronal
Occipito-temporal cortex (BA37)	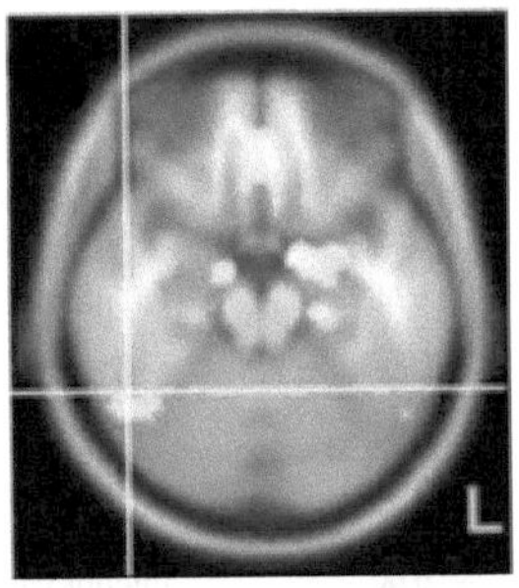	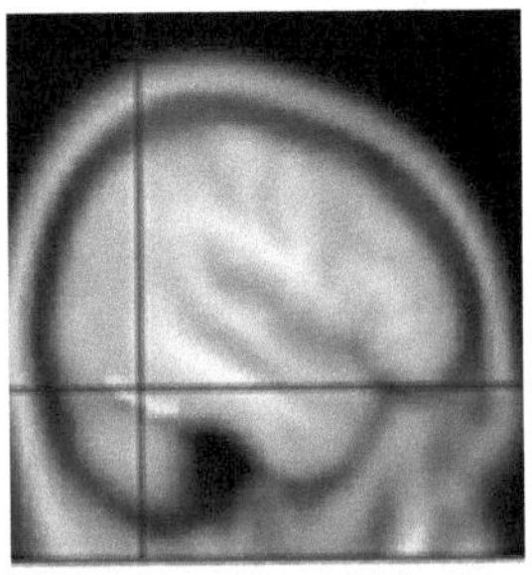	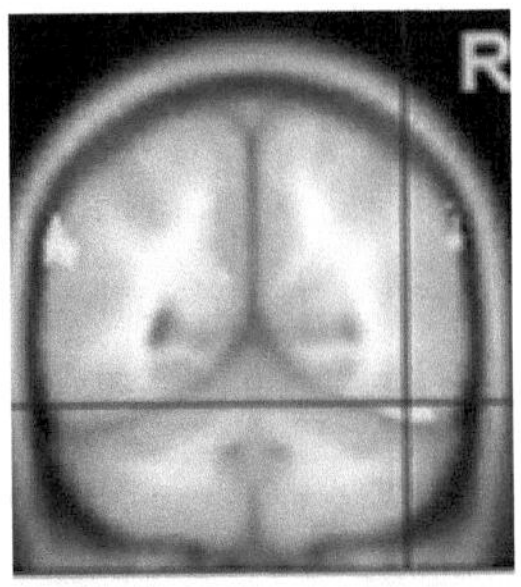
Perirhenalcortex (BA35)	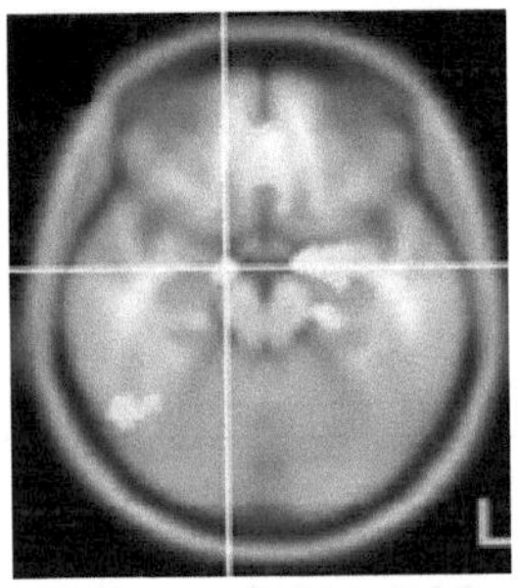	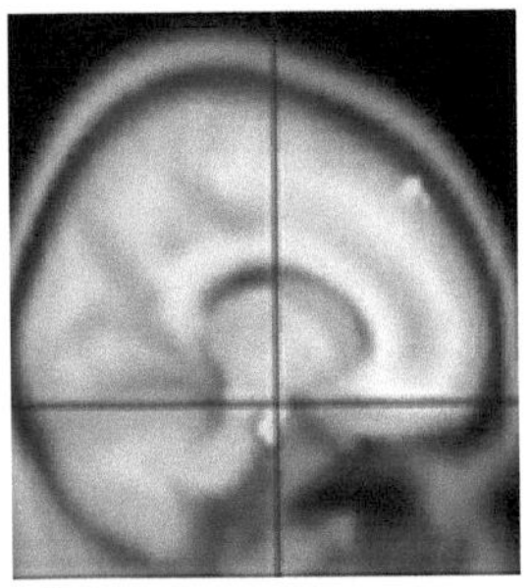	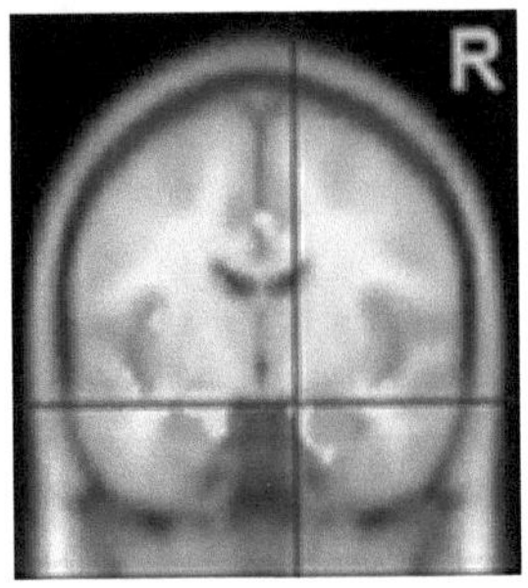
Inferior temporal gyrus (BA20)	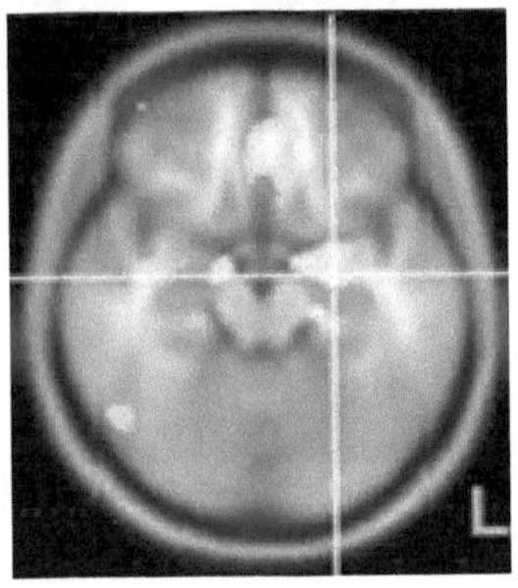	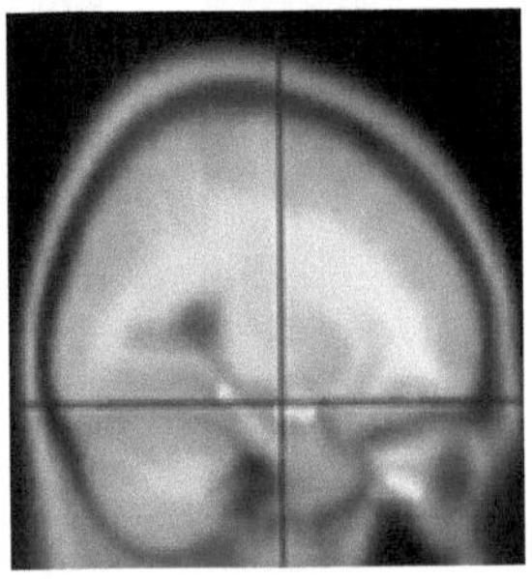	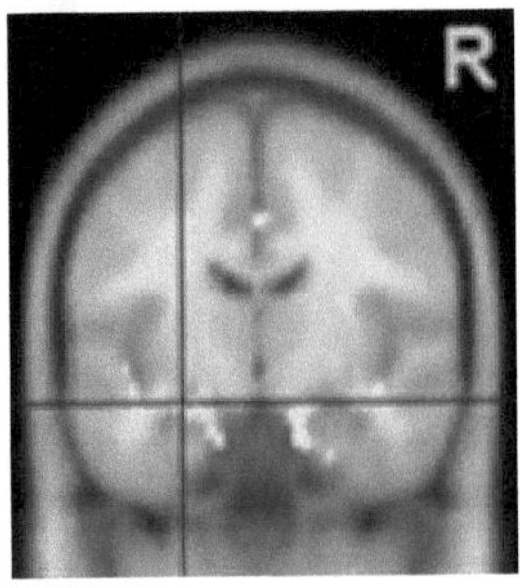

Fig. 3 Areas with increased gray matter volume in *huffaz* group in temporal lobes

Area	Axial	Sagittal	Coronal
Superior parietal cortex	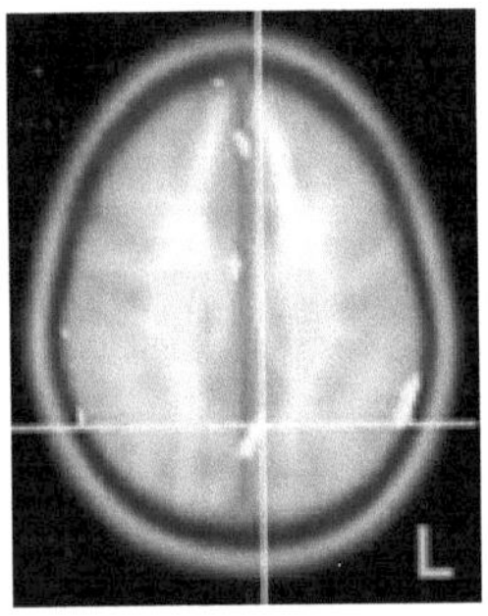	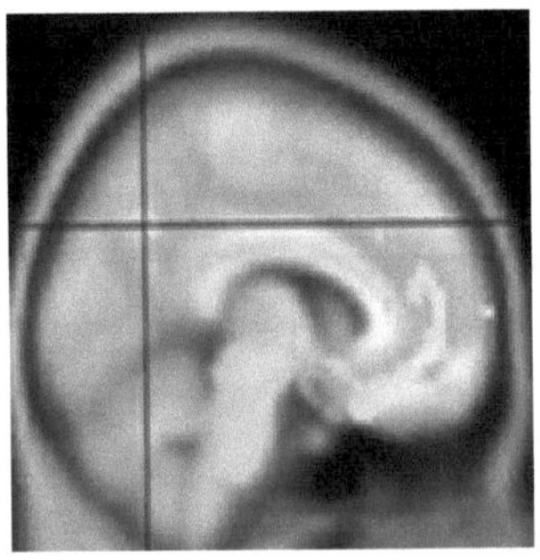	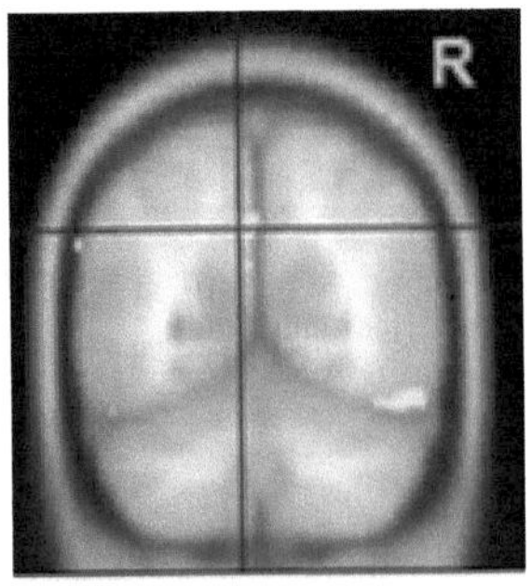

Fig. 4 Parietal lobes areas with increased gray matter volume in *huffaz* group

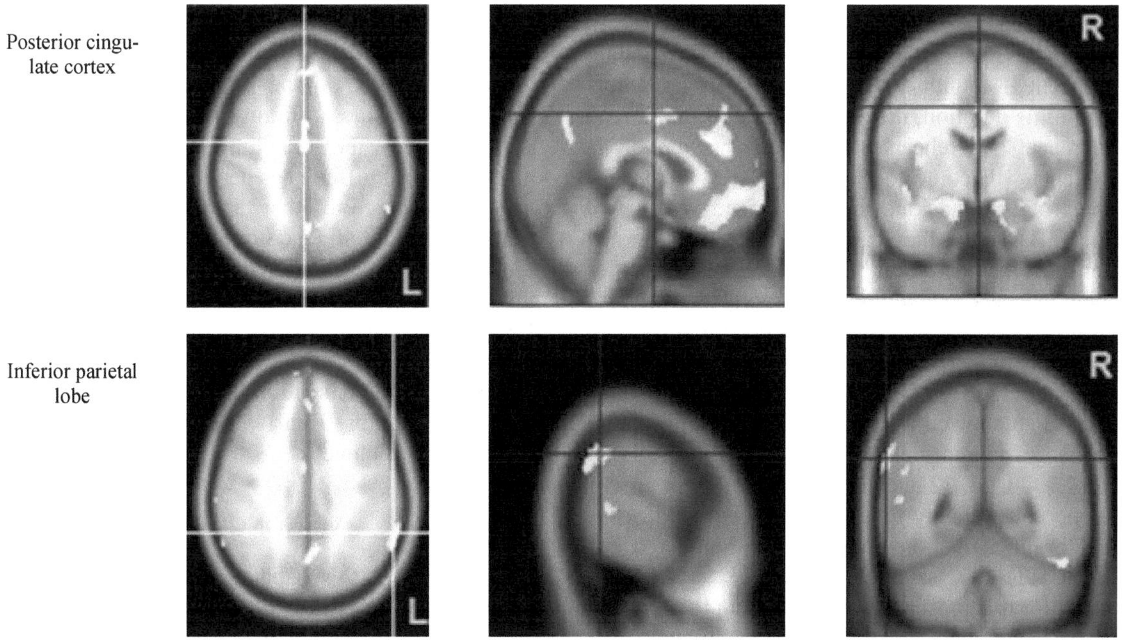

Fig. 4 (*Continued*)

IV. DISCUSSION

The present study aimed to address two questions. First, does brain morphology correlate with the textual memorization status? Second, if so, where are the areas that contribute to the difference as compared to the control group? The results provided positive answers to both questions, which reveal the areas that show an increase in the gray matter volume and subsequently clarify their contribution to textual memorization process.

The VBM analysis that explored the overall relationship between gray matter volume and textual memorization reports several main findings: 1) There are no significant difference in terms of total gray matter volume between *huffaz* and non-*huffaz* group and 2) Increased regional gray matter volume of the *huffaz* group was observed in several brain areas namely anterior cingulate gyrus, orbitofrontal cortex, left inferior temporal gyrus, right occipitotemporal gyrus, left inferior parietal gyrus, right perirhenal cortex, superior parietal cortex, posterior cingulate cortex and anterior cingulate cortex. Together, these results clearly indicate that the involved brain areas are critical for the memory maintenance, manipulation, and retrieval processes in the textual memorization activity. Hence, these areas can be anticipated to have functional roles in textual memorization process.

In order to understand the findings of the study, it is important to know the components involved during textual memorization process. Since lack of study has been done in this area, it is quite difficult to demarcate type of memory and storage area associated with textual memorization. However, the results suggest a pattern of differences in the gray matter distribution between *huffaz* and non-*huffaz* that include working memory, visual, attention and language regions. Thus, it can be assumed that textual memorization involves with extensive neural connections and processes across the brain.

Results showed that *huffaz* group has more gray matter volume in the occipitotemporal cortex, perirhenal cortex and inferior temporal gyrus. The key role of temporal lobe in memory has extensively been studied before [8]. Perirhenal is a part of hippocampal complex that strongly involves in receiving input from frontal, temporal and parietal as well as projecting the processed information back to the neocortices for storage [9], which contributes in memory consolidation process.

Besides, inferotemporal gyrus and occipitotemporal cortex are considered as memory storage areas. The finding of an increased gray matter volume of the occipitotemporal cortex, perirhenal cortex and inferior temporal gyrus are consistent with previous studies that reported changes in brain activation or size in this area to memory tests [3].

During Quran memorization, intense repetition of the verses is exercised to ensure it is strongly and correctly being stored in the memory which may explain this finding. *Huffaz* had spent 18 months to complete the Quran memorization. Along this duration, they devoted their time to read, memorize, repeat and rehearse their memorization of the Quran. Students will come to class every day to rehearse several times before checking the memorized verses with their teachers. Basically, they can only proceed to the next verses after their memorization has been verified to be perfect by their teachers. Therefore, it supports the proposal that larger gray matter volume seen in *huffaz* might actually cause by adaptations to long-term robust memorization training.

This study also showed a strong increase in gray matter volume related to Quran memorization status in the Brodmann area of 7, 20, 24, 32 and 40. The role of these areas is thereby supported by many functional imaging studies that have found brain activation associated with the retrieval success [10], [11].

The reason for this present result may relate to the Quran memorization technique. Aforementioned technique to memorize the Quran has highlighted the need for a constant revision of the memorization, which requires *huffaz* to recall the memory in front of the teachers or friends repeatedly, just to ensure everything has been memorized flawlessly. Besides, this finding seemed to agree with Kozlovskiy (2012) findings which show increased in the size of the anterior and posterior cingulate cortex correlates with less number of memory errors.

The findings showed the relationship between the *huffaz* and gray matter volume of the orbitofrontal cortex. Previous researches show the brain activity in the orbitofrontal region (BA11) in relation with reward-related stimuli and motivation (Rolls, 2000; Hikosaka & Watanabe, 2000; Barbey, Koenings, & Grafman, 2011). It is therefore likely that such connection exist between motivation and Quran memorization. Memorizing the whole Quran is not an easy task, especially if the Arabic language is not a primary language of the *huffaz*.

Several studies have generally discovered a conventional laterality effect in regard to brain structures, whereby left-sided structures are more important for the verbally-coded material, while right-sided structures are more crucial for nonverbal and visuospatial material. When referring to Quran memorization activity, it is linked with memory that established by verbally-coded material since *huffaz* have to recite the verses during the memorization process. However, the findings of the current study demonstrated the increased gray matter volume in both right and left brain structures. This suggest that *huffaz* not only memorized the Quran by merely reading the verses, but they also incorporate other techniques such as giving careful attention to the each verse symbol and location in the page during the recitation. The Quranic verses are verbally, visually and spatially encoded in the brain, which leave a significant effect on both sides of brain structures.

The present study was limited in several ways. The brain morphological differences between *huffaz* and control group may be attributed to other factors such as innate predisposition, gender, intelligence quotient (IQ) level, demographic factors and others, rather than to Quran memorization status. Thus, it would be better to perform IQ test and consider IQ scores as a cofactor in the statistical analyses of voxel-based morphometry. However, there was no significant difference in the education levels between the study groups, so the individuals from both groups would be expected to have similar IQ levels.

V. CONCLUSIONS

In summary, this paper used voxel-based morphometry (VBM) techniques to explore whether brain morphology is associated with textual memorization status. Primary data were collected by scanning the brain of *huffaz* and the control group. The study found that total gray matter volume of *huffaz* was not significantly differed from the control group. However, *huffaz* showed regionally increased gray matter volume in the anterior cingulate gyrus, orbitofrontal cortex, left inferior temporal gyrus, right occipitotemporal gyrus, left inferior parietal gyrus, right perirhenal cortex, superior parietal cortex, posterior cingulate cortex and anterior cingulate cortex. While differences were found in relevant brain areas demonstrating the reliability of the VBM technique, the present results should be interpreted with caution. Considering the lack of literature concerning textual memorization and structural brain variation, it is believed that the results of the present study establish a basis for future studies probing more direct relationship between long-term textual memorization activity and related structural changes in specific brain regions.

CONFLICT OF INTEREST

The authors declare that they have no conflict of interest.

REFERENCES

1. E. A. Maguire, D. G. Gadian, I. S. Johnsrude, C. D. Good, J. Ashburner, R. S. Frackowiak, and C. D. Frith, "Navigation-related structural change in the hippocampi of taxi drivers.," Proc. Natl. Acad. Sci. U. S. A., vol. 97, pp. 4398–4403, 2000.
2. E. A. Maguire, K. Woollett, and H. J. Spiers, "London taxi drivers and bus drivers: A structural MRI and neuropsychological analysis," Hippocampus, vol. 16, pp. 1091–1101, 2006.

3. E. A. Maguire, R. S. Frackowiak, and C. D. Frith, “Recalling routes around london: activation of the right hippocampus in taxi drivers.,” J. Neurosci., vol. 17, pp. 7103–7110, 1997.
4. B. Draganski, C. Gaser, G. Kempermann, H. G. Kuhn, J. Winkler, C. Buchel, and A. May, “Temporal and spatial dynamics of brain structure changes during extensive learning,” J. Neurosci., vol. 26, pp. 6314–6317, 2006.
5. J. Driemeyer, J. Boyke, C. Gaser, C. Büchel, and A. May, “Changes in gray matter induced by learning--revisited.,” PLoS One, vol. 3, p. e2669, 2008.
6. C. J. Mummery, K. Patterson, C. J. Price, J. Ashburner, R. S. J. Frackowiak, and J. R. Hodges, “A voxel-based morphometry study of semantic dementia: Relationship between temporal lobe atrophy and semantic memory,” Ann. Neurol., vol. 47, pp. 36–45, 2000.
7. M. Groussard, R. La Joie, G. Rauchs, B. Landeau, G. Chételat, F. Viader, B. Desgranges, F. Eustache, and H. Platel, “When music and long-term memory interact: Effects of musical expertise on functional and structural plasticity in the hippocampus,” PLoS One, vol. 5, 2010.
8. B. C. Dickerson and R. A. Sperling, “Functional abnormalities of the medial temporal lobe memory system in mild cognitive impairment and Alzheimer’s disease: Insights from functional MRI studies,” Neuropsychologia, vol. 46, pp. 1624–1635, 2008.
9. C. Mühl, J. Griego, and U. Friese, “Memory consolidation and the hippocampal complex,” Cognitive Science. p. 75, 2004.
10. M. D. Rugg, P. C. Fletcher, C. D. Frith, R. S. J. Frackowiak, and R. J. Dolan, “Differential activation of the prefrontal cortex in successful and unsuccessful memory retrieval.,” Brain, vol. 119, pp. 2073–83, 1996.
11. N. M. Hall, A. Gjedde, and R. Kupers, “Neural mechanisms of voluntary and involuntary recall: A PET study,” Behav. Brain Res., vol. 186, pp. 261–272, 2008.

Author: Abdul Halim Sapuan
Institute: International Islamic University Malaysia
Street: Kuantan Campus
City: Kuantan
Country: Malaysia
Email: abd_halim@iium.edu.my

Beat-to-Beat Blood Pressure Variations While Standing are Associated with Postural Changes in Blood Pressure

C.H. Goh[1,2], S.C. Ng[1], and M.P. Tan[2]

[1] Department of Biomedical Engineering, Faculty of Engineering, University of Malaya, Kuala Lumpur, Malaysia
[2] Department of Medicine, Faculty of Medicine, University of Malaya, Kuala Lumpur, Malaysia

Abstract— **While blood pressure variability (BPV) and orthostatic hypotension (OH) are independently related to an increase in prevalence of cardiovascular disease, the relationship between BPV and OH remains poorly understood. We evaluated the association between postural changes in blood pressure with beat-to-beat BPV during supine rest and standing. Continuous beat-to-beat blood pressure recordings were obtained from 832 individuals, aged 55 years and above, during ten minutes' supine rest and three minutes' standing. The maximal drop in systolic blood pressure (ΔSBP) and diastolic blood pressure (ΔDBP) with standing from the supine position, BPV in the supine and standing positions and standing to lying BPV ratio were computed. Standing systolic BPV and SLR for systolic BPV were correlated significantly with ΔSBP (r=0.497, p=<0.001; r=0.476, p=<0.001). Standing diastolic BPV and SLR of diastolic BPV showed a moderate correlation ΔDBP (r=0.397, p=<0.001; r=0.432, p=<0.001). Standing BPV and SLR for BPV were significantly higher for subjects with documented OH. Standing systolic BPV, height, weight and age were accounted 27% of the variability in ΔSBP while the SLR of diastolic BPV accounted for 19% of the variability in ΔDBP. The SLR for diastolic BPV, body weight, and cerebrovascular disease were independent predictors for OH (R^2=0.157). The significance of beat-to-beat BPV remains unclear. Our study suggests that supine and erect BPV are influenced by different mechanisms. The impaired compensatory mechanism leading to OH, which are sympathetic hyporesponsiveness, are also associated with increased BPV while standing.**

Keywords— **Blood pressure variability, orthostatic hypotension, aged, autonomic nervous system, baroreflex.**

I. INTRODUCTION

Blood pressure variability (BPV) is the measure of fluctuations in blood pressure but can be measured in relation to differences in office visit to office visit measurements, changes over a 24-hour period and beat-to-beat variations [1, 2]. Daily BPV has been found to have prognostic significance in terms of target-organ damage [3-7], incidence of cardiovascular event [1, 7] and mortality [7, 8]. Visit-to-visit variability has been found to independently contribute to stroke event [8, 9], target-organ damage, cardiovascular event and all-cause mortality [1, 2, 8]. Recent studies have also found that the visit-to-visit BPV is associated with cognitive function in the elderly [10, 11].

Orthostatic hypotension (OH) is one of the most common disorders of blood pressure regulation [12]. The prevalence of OH using existing definitions has been reported as 5% to 30%, depending on the population studied and the methods of measurement [13, 14]. The American Autonomic Society and the American Academy of Neurology consensus committee defined OH as a drop in systolic blood pressure (SBP) of 20 mmHg or greater and/or a drop in diastolic blood pressure (DBP) of 10 mmHg or greater [13]. This definition has remained unchanged from the original consensus definition published in 1996. Many researchers have, however, acknowledged that with the increasing use of continuous non-invasive BP monitoring, this definition may need to be reconsidered [12, 15].

Greater changes in postural BP is related to silent cerebral infarcts and cardiac overload in hypertensive patients [16, 17]. Moreover, OH has been associated with falls [18], and increased risk of myocardial infarction, coronary heart disease, transient ischaemic attack [17, 19], fracture and increased mortality [19, 20]. Orthostatic hypotension is also associated with increasing age [21], isolated systolic hypertension [22], depression [23] and reduced body weight [22].

Increased mean BP and BPV are independently related to an increase in prevalence and severity of target-organ damage [3, 6]. Previous research findings have, however, been conducted primarily on office BP measurements [24]. Regardless of mean BP, the severity of target-organ damage is also higher in patients with greater visit-to-visit and 24-hour BPV [17].

While BPV and OH have been found to be independently associated with similar conditions, the relationship between BPV and OH remains poorly understood. This study therefore evaluated the relationship between BPV with the reduction in blood pressure with posture change as well as OH, using current consensus definition. Our specific hypothesis was that OH and changes in blood pressure with posture change is associated with standing BPV and standing to lying BPV ratio.

II. METHODS

A. *Study Population*

This study included 832 subjects from the Malaysian Elders Longitudinal Research (MELoR) study. The MELoR study is a longitudinal cohort study based in greater Kuala

F. Ibrahim et al. (eds.), *International Conference for Innovation in Biomedical Engineering and Life Sciences*,
IFMBE Proceedings 56,
DOI: 10.1007/978-981-10-0266-3_9

Lumpur. Individuals aged 55 years and above were selecting through simple random sampling from the electoral rolls of the Parliamentary constituencies of Petaling Jaya Utara, Petaling Jaya Selatan and Lembah Pantai. One hundred and thirty six subjects were excluded due to inability to stand for 3 minutes, non-availability of continuous blood pressure measurements, or less than 128 beats of adequate quality blood pressure signals available. The study had obtained a favorable opinion from the University of Malaya Medical Ethics Committee (MEC Ref No: 925.4) and informed consent was obtained from all subjects prior to their inclusion.

B. Baseline and Beat-to-Beat Blood Pressure Assessment

Baseline characteristics including age, gender and medical history were obtained from all subjects. Body weight and body height were measured for each subject. All subjects were assessed using continuous non-invasive beat-to-beat blood pressure monitoring using the vascular unloading technique (Task Force ®, CNSysem Austria) [25]. The active stand test was performed on all eligible subjects. Continuous blood pressure was recorded during 10 minutes of supine rest and 3 minutes of standing [26]. Assistance was provided whenever required to ensure that the transition from lying to standing occurred smoothly. Subjects were instructed not to move their hands fitted with the finger cuff during data collection to reduce artefacts. The presence of any symptoms of dizziness during standing was recorded for all subjects.

C. Signal Processing

The continuous blood pressure signals for individual participants were exported to MATLAB and analyzed with a software custom-written using MATLAB (version R2009a, the MathWorks Inc., Natick, Massachusetts, United States). Orthostatic blood pressure change in SBP and DBP was determined by calculating the difference between the mean blood pressure for 20 beats immediately before standing to the minimum blood pressure during 3 minutes of standing [27]. Blood pressure variability was computed separately for supine and standing blood pressure segments using the coefficient of variation (standard deviation/mean).

The ratio of standing BPV to lying BPV (SLR) was then computed for each subject. This derived measure represents the changes in variability from the lying position to the standing position.

D. Statistical Analysis

Individual data on maximal systolic and diastolic drop with standing and the presence of OH were considered Continuous variables were expressed as mean ± standard deviation while discrete variables were expressed as frequencies with percentages in parenthesis. The independent t-test was applied to obtain the difference between groups for continuous parametric variables and the Chi-square test for categorical variables. The Pearson's r correlation coefficient was calculated for univariate comparisons between continuous variables such as SBPV with ΔSBP. Multivariate analyses were conducted using linear and logistic regression methods to determine predictor equations for postural changes in SBP and DBP and the presence of OH. A p-value of <.05 was considered statistically significant.

III. RESULTS

A. Baseline Assessment

Table 1 shows the characteristics of included particicipants. There was no significant difference in the maximal drop of in systolic blood pressure (ΔSBP) for all listed variables. Subjects with cerebrovascular disease (stroke and/or transient ischemic attack) had a significantly lower drop of diastolic blood pressure with standing (ΔDBP) (p=0.025). Subjects who reported dizziness with standing had significantly greater ΔDBP (p=0.016). Subjects with cerebrovascular disease were significantly less likely to have documented OH.

B. Postural Changes in Blood Pressure and Blood Pressure Variability

From Table 2, in the supine position, only systolic BPV had a weak but statistically significant negative correlation with ΔSBP (r=-0.101, p=0.008). In the erect position, there was a significant strong correlation between standing systolic BPV and ΔSBP (r-0.497, p=<0.001) and a moderately strong correlation between standing systolic BPV with ΔDBP (r=0.357, p=<0.001). There was a moderately strong correlation between diastolic BPV with ΔSBP and ΔDBP (r=0.388, p=<0.001; r=0.397, p=<0.001). Standing systolic BPV and diastolic BPV were significantly higher among subjects with OH (p=<0.001). A similar relationship was observed between the ratio of standing to lying (SLR) systolic and diastolic BPV with ΔSBP, ΔDBP and OH. There were moderate significant correlations between systolic and diastolic SLR with ΔSBP and ΔDBP, while SLR for systolic and diastolic BPV for subjects with OH were also significantly higher than subjects without OH.

C. Predictor Equations for Postural Changes in Blood Pressure and Orthostatic Hypotension.

Table 3 and Table 4 contains the best predictor models for ΔSBP and ΔDBP respectively using multivariate linear regression analyses. Standing systolic BPV, height, weight and age accounted for 27% of the variability in ΔSBP (Table 3). In Table 4, the SLR for diastolic BPV accounted for 19% of the variability in ΔDBP. The standing to lying ratio

Table 1 Basic Demographic of Maximum Drop of Blood Pressure due to Postural Changes

Variable	ΔSBP (mmHg), mean (SD)					ΔDBP (mmHg), mean (SD)			OH, n(%)		
	n	Yes	n	No	p-value	Yes	No	p-value	Yes	No	p-value†
Age (>70ys)	269	28 (16)	427	28 (18)	0.923	15 (11)	16 (12)	0.157	212 (39.7%)	57 (35.2%)	0.301
Gender (Female)	385	27 (17)	311	29 (16)	0.350	15 (11)	17 (12)	0.087	292 (54.7%)	93 (57.4%)	0.541
Hypertension	347	27 (17)	331	29 (17)	0.306	12 (11)	17 (12)	0.172	260 (49.9%)	87 (55.4%)	0.226
Diabetes mellitus	177	28 (17)	486	28 (17)	0.928	16 (11)	16 (12)	0.706	138 (27.0%)	39 (25.7%)	0.742
Falls history	143	26 (17)	534	28 (17)	0.182	15 (11)	16 (12)	0.424	107 (20.6%)	36 (22.9%)	0.527
Cardiovascular disease	70	28 (16)	605	28 (17)	0.877	16 (10)	16 (12)	0.727	57 (11.0%)	13 (8.3%)	0.341
Cerebrovascular disease	26	23 (18)	645	28 (17)	0.155	11 (12)	16 (12)	0.025†	14 (2.7%)	12 (7.7%)	0.005*
Smoking	145	28 (17)	531	28 (17)	0.608	17 (12)	16 (12)	0.412	112 (21.6%)	33 (21.0%)	0.881
Dizziness	135	29 (19)	561	28 (16)	0.231	18 (13)	15 (11)	0.016†	109 (20.4%)	26 (16.0%)	0.219

*p<0.05, ΔSBP=maximal drop in SBP with standing, ΔDBP= maximal drop in SBP with standing, OH=orthostatic hypotension, †Pearson's Chi-squared test

Table 2 Correlation of Drop of Blood Pressure and Orthostatic Hypotension with Blood Pressure Variability

Variables	ΔSBP (mmHg)		ΔDBP (mmHg)		OH		p-value
	r	p-value	r	p-value	Yes (n=534)	No (n=162)	
Resting BPV							
Systolic	-0.101	0.008*	0.002	0.965	0.07 (0.05)	0.07 (0.05)	0.082
Diastolic	0.018	0.632	0.023	0.549	0.08 (0.05)	0.08 (0.05)	0.333
Standing BPV							
Systolic	0.497	<0.001**	0.357x	<0.001**	0.08 (0.05)	0.06 (0.03)	<0.001**
Diastolic	0.388	<0.001**	0.397**	<0.001**	0.09 (0.06)	0.06 (0.03)	<0.001**
SLR							
Systolic	0.476**	<0.001**	0.333**	<0.001**	1.48 (0.87)	1.26 (0.72)	<0.001**
Diastolic	0.423**	<0.001**	0.432**	<0.001**	1.03 (0.56)	0.85 (0.38)	<0.001**

*p<0.01, **p<0.001, ΔSBP=maximal drop in SBP with standing, ΔDBP= maximal drop in SBP with standing, OH=orthostatic hypotension, BPV=blood pressure variability, SLR= standing to lying ratio of blood pressure variability

for diastolic BPV, body weight, and cerebrovascular disease remained independent predictors for the presence of OH (R^2=0.157) using a multivariate logistic regression analysis (Table 5).

Table 3 Independent Predictors of Maximum Drop of Systolic Blood Pressure

Variables	Standard Coefficient	95% Confidence Interval	p-value	R-square
Standing systolic BPV	0.503	171.96 to 222.84	<0.001*	0.268
Height	0.121	0.10 to 0.37	0.001*	
Weight	-0.108	-0.24 to -0.047	0.003*	
Age	-0.097	-0.39 to -0.08	0.004*	
Constant		-21.50 to 22.67	0.959	

* p<0.05, BPV=blood pressure variability

Table 4 Independent Predictors of Maximum Drop of Diastolic Blood Pressure

Variables	Standard Coefficient	95% Confidence Interval	p-value	R-square
Diastolic SLR	0.432	6.24 to 8.54	<0.001*	0.186
Constant		5.76 to 8.87	<0.001*	

* p<0.05, SLR= standing to lying ratio of blood pressure variability

Table 5 Independent Predictors of Orthostatic Hypotension

Variable	B	SE	Adjusted OR (95% CI)	p-value	R-square
Diastolic SLR†	1.586	0.238	4.884 (3.062-7.789)	<0.001*	0.157
Weight†	0.017	0.007	0.983 (0.970-0.997)	0.019*	
Cerebrovascular disease	1.040	0.435	0.354 (0.151-0.830)	0.017*	
Constant	0.699	0.505	2.012	0.166	

* indicate p<0.05, †per unit increase, SLR=standing to lying ratio of blood pressure variability

IV. Discussion

Our study has found that postural changes in blood pressure were significantly associated with standing BPV and not resting BPV in the supine position. The standing to lying ratio for BPV was also significantly associated with postural blood pressure changes. Using the current consensus criteria for OH, OH was significantly associated with standing BPV and SLR for BPV. Our multivariate analysis, however, revealed that our best predictor models only accounted for 16-27% of the variability in OH and the degree of blood pressure reduction with posture change.

The prevalence of OH reported in population studies involving older people is 6-33% [13, 18]. A higher prevalence of OH of 55% was, however, reported in a study involving 342 older individuals [20]. In our study, 72% of all individuals evaluated fulfilled the consensus criteria for OH. Our population consisted of older individuals from different ethnic groups (Chinese, Indian and Malay) compared to previously published studies consisting of mainly Caucasian populations. Thirty six percent of our overall population had a pre-existing medical diagnosis of diabetes mellitus, which was much higher than that reported by other study populations, due to the increased susceptibility of Asian ethnic groups to diabetes. While the increased prevalence of diabetes may account for the increased risk of OH, another likely explanation is that ours was the only study so far to have used continuous beat-to-beat assessments of blood pressure in a large study population. Previous diagnosis of OH had been based on manual or oscillometric blood pressure measurements. As mentioned earlier, beat-to-beat measurements are far more likely to detect the true minimum blood pressure with standing than oscillometric measurement [12]. Many researchers have therefore proposed a revision of the definition OH for occasions when measurements are made with continuous blood pressure monitoring [12].

The actual significance of very short term blood pressure variability detected with non-invasive methods of beat-to-beat blood pressure measurement remains unclear. It is expected that fluctuations in blood pressure are subject to the complex interplay between different cardiovascular control systems such as the baroreceptor reflex, the renin-angiotensin system (RAS), the vascular myogenic response and the release of nitric oxide (NO) from endothelium [7]. While the mechanisms listed above are likely to account for daily or longer term variations in blood pressure, some of the above named mechanisms may not have adequate temporal sensitivity to influence beat-to-beat variations in blood pressure. It is likely that baroreflex responsiveness and vasomotor reactivity will have a larger influence on very short term blood pressure variability. As suggested by our findings, beat-to-beat BPV changes from the supine to erect posture. In the supine posture, the parasympathetic nervous system is expected to be activated. Beat-to-beat variations in BPV may therefore correspond to changes in heart rate associated with regular breathing, and may therefore be associated with baroreflex sensitivity. Conversely, when individual assumes the erect posture the sympathetic nervous system is activated in order to maintain systemic blood pressure against gravitational forces. The degree of fluctuation of in blood pressure in the erect posture may therefore be dependent both on the baroreflex response and the myogenic response. Larger variations in blood pressure in the erect posture may therefore be present if the function of inhibitory systems is impaired.

In the erect posture, gravitational forces will lead to pooling of the intravascular volume in the lower extremities, which then leads to a reduction in blood pressure. The baroreceptors are then expected to activate the sympathetic nervous system, leading to a rise in heart rate as well as an increased peripheral vascular tone, in order to maintain systemic blood pressure [29]. A failure in this response system is therefore expected to lead to OH. Underlying causes of OH in individuals, however, do differ, and common causes include depletion in intravascular volume, autonomic failure and medications [13, 14]. In our study, some of the postural changes in blood pressure is accounted for by the BPV in the erect posture, as well as the ratio between erect and supine BPV. This suggests that the same impaired compensatory mechanisms which are expected to be under sympathetic control are also linked to increased BPV while standing. The presence of cerebrovascular disease, however, appears to be protective against OH. As cerebrovascular disease is associated with increased arterial stiffness, it would appear that it paradoxically blunts the strong relationship between diastolic blood pressure drop with standing and the standing to lying ratio in diastolic BPV.

As the data available is currently only cross-sectional in nature, we are only able to attribute associations and not causation in our findings. In addition, much of the variability in postural changes in blood pressure remains unaccounted for, particularly as we have not evaluated medication consumption in this study. However, our findings, have contributed to the current limited knowledge on very short term or beat-to-beat blood pressure variability. In addition, it has confirmed previous unconfirmed reports that beat-to-beat blood pressure variability with standing accounts for some of the drop in blood pressure observed with posture change. Future studies should now seek to determine other factors that could influence very short term blood pressure variability in order to elucidate the mechanisms that underlie beat-to-beat blood pressure variability which remains poorly understood.

V. CONCLUSIONS

Standing BPV is significantly correlated with postural changes in blood pressure, with the correlation between supine BPV and postural changes in blood pressure being much weaker. This suggests that supine and erect BPV are influenced by different factors, but some common mechanisms do exist between orthostatic blood pressure changes and BPV while standing. While the significance of beat-to-beat BPV remains unclear, our study has demonstrated that it is possible to evaluate this in population studies. Further evaluation of the medication use and prospective evaluation of disease outcomes associated with beat-to-beat BPV is now indicated, and may help improve secondary prevention measures in non-communicable disorders such as hypertension and stroke.

ACKNOWLEDGMENT

The authors would like to thanks High Impact Research-Ministry of Education Grant (UM.C/625/HIR-MOHE/ASH/02) for providing the funds for Malaysian Elders Longitudinal Study (MELoR). The authors also would like to thank the members from MELoR for helping with subject recruitment and data collection.

CONFLICT OF INTEREST

The authors declare that they have no conflict of interest.

REFERENCES

1. Mancia, G.: 'Short-and long-term blood pressure variability present and future', Hypertension, 2012, 60, (2), pp. 512-517
2. Parati, G., Ochoa, J.E., Lombardi, C., and Bilo, G.: 'Assessment and management of blood-pressure variability', Nature Reviews Cardiology, 2013, 10, (3), pp. 143-155
3. Parati, G., Pomidossi, G., Albini, F., Malaspina, D., and Mancia, G.: 'Relationship of 24-hour blood pressure mean and variability to severity of target-organ damage in hypertension', Journal of hypertension, 1987, 5, (1), pp. 93-98
4. Kario, K., and Pickering, T.G.: 'Blood pressure variability in elderly patients', The Lancet, 2000, 355, (9215), pp. 1645-1646
5. Mancia, G., and Grassi, G.: 'Mechanisms and clinical implications of blood pressure variability', Journal of cardiovascular pharmacology, 2000, 35, pp. S15-S19
6. Parati, G.: 'Blood pressure variability: its measurement and significance in hypertension', Journal of hypertension, 2005, 23, pp. S19-S25
7. Höcht, C.: 'Blood Pressure Variability: Prognostic Value and Therapeutic Implications', ISRN Hypertension, 2013
8. Poortvliet, R.K., Ford, I., Lloyd, S.M., Sattar, N., Mooijaart, S.P., de Craen, A.J., Westendorp, R.G., Jukema, J.W., Packard, C.J., and Gussekloo, J.: 'Blood pressure variability and cardiovascular risk in the PROspective Study of Pravastatin in the Elderly at Risk (PROSPER)', PLOS one, 2012, 7, (12), pp. e52438
9. Rothwell, P.M., Howard, S.C., Dolan, E., O'Brien, E., Dobson, J.E., Dahlöf, B., Sever, P.S., and Poulter, N.R.: 'Prognostic significance of visit-to-visit variability, maximum systolic blood pressure, and episodic hypertension', The Lancet, 2010, 375, (9718), pp. 895-905
10. 1Lattanzi, S., Luzzi, S., Provinciali, L., and Silvestrini, M.: 'Blood pressure variability predicts cognitive decline in Alzheimer's disease patients', Neurobiology of aging, 2014, 35, (10), pp. 2282-2287
11. Sabayan, B., Wijsman, L.W., Foster-Dingley, J.C., Stott, D.J., Ford, I., Buckley, B.M., Sattar, N., Jukema, J.W., van Osch, M.J., and van der Grond, J.: 'Association of visit-to-visit variability in blood pressure with cognitive function in old age: prospective cohort study', BMJ, 2013, e347
12. Romero-Ortuno, R., Cogan, L., Foran, T., Kenny, R.A., and Fan, C.W.: 'Continuous noninvasive orthostatic blood pressure measurements and their relationship with orthostatic intolerance, falls, and frailty in older people', Journal of the American Geriatrics Society, 2011, 59, (4), pp. 655-665
13. Lanier, J.B., Mote, M.B., and Clay, E.C.: 'Evaluation and management of orthostatic hypotension', American family physician, 2011, 84, (5), pp. 527-536
14. Ward, C., and Kenny, R.A.: 'Reproducibility of orthostatic hypotension in symptomatic elderly', The American journal of medicine, 1996, 100, (4), pp. 418-422
15. Wieling, W., Krediet, C., Van Dijk, N., Linzer, M., and Tschakovsky, M.: 'Initial orthostatic hypotension: review of a forgotten condition', Clinical science, 2007, 112, pp. 157-165
16. Eguchi, K., Kario, K., Hoshide, S., Hoshide, Y., Ishikawa, J., Morinari, M., Hashimoto, T., and Shimada, K.: 'Greater change of orthostatic blood pressure is related to silent cerebral infarct and cardiac overload in hypertensive subjects', Hypertension Research, 2004, 27, (4), pp. 235-241
17. Kario, K.: 'Orthostatic Hypertension A Measure of Blood Pressure Variation for Predicting Cardiovascular Risk', Circulation Journal, 2009, 73, (6), pp. 1002-1007
18. Heitterachi, E., Lord, S.R., Meyerkort, P., McCloskey, I., and Fitzpatrick, R.: 'Blood pressure changes on upright tilting predict falls in older people', Age and ageing, 2002, 31, (3), pp. 181-186
19. Ejaz, A.A., Kazory, A., and Heinig, M.E.: '24-Hour Blood Pressure Monitoring in the Evaluation of Supine Hypertension and Orthostatic Hypotension', The Journal of Clinical Hypertension, 2007, 9, (12), pp. 952-955
20. Poon, I., and Braun, U.: 'High prevalence of orthostatic hypotension and its correlation with potentially causative medications among elderly veterans', Journal of clinical pharmacy and therapeutics, 2005, 30, (2), pp. 173-178
21. Kartheek, B.R., Kumar, G., Ameerunnisabegum, S., and Venkateswaraiah, S.: 'Postural changes in blood pressure associated with Ageing', International Journal of Life Science and Pharma Research, 2011, 1, pp. 88-93
22. Rutan, G.H., Hermanson, B., Bild, D.E., Kittner, S.J., LaBaw, F., and Tell, G.S.: 'Orthostatic hypotension in older adults. The Cardiovascular Health Study. CHS Collaborative Research Group', Hypertension, 1992, 19, (6 Pt 1), pp. 508-519
23. Vasudev, A., O'Brien, J.T., Tan, M.P., Parry, S.W., and Thomas, A.J.: 'A study of orthostatic hypotension, heart rate variability and baroreflex sensitivity in late-life depression', Journal of affective disorders, 2011, 131, (1), pp. 374-378
24. Rothwell, P.M.: 'Limitations of the usual blood-pressure hypothesis and importance of variability, instability, and episodic hypertension', The Lancet, 2010, 375, (9718), pp. 938-948
25. Fortin, J., Marte, W., Grüllenberger, R., Hacker, A., Habenbacher, W., Heller, A., Wagner, C., Wach, P., and Skrabal, F.: 'Continuous non-invasive blood pressure monitoring using concentrically interlocking control loops', Computers in biology and medicine, 2006, 36, (9), pp. 941-957
26. Tan, M.P., Newton, J.L., Chadwick, T.J., and Parry, S.W.: 'The relationship between carotid sinus hypersensitivity, orthostatic hypotension, and vasovagal syncope: a case–control study', Europace, 2008, 10, (12), pp. 1400-1405
27. McLaren, A., Kerr, S., Allan, L., Steen, I.N., Ballard, C., Allen, J., Murray, A., and Kenny, R.A.: 'Autonomic function is impaired in elderly stroke survivors', Stroke, 2005, 36, (5), pp. 1026-1030
28. Murata, J., Murata, S., Horie, J., Ohtao, H., and Miyazaki, J.: 'Relationship Between Orthostatic Blood Pressure Changes and Postural Sway When Standing up from a Chair in Older Adult Females', International Journal of Gerontology, 2012, 6, (3), pp. 182-186

Corresponding author:

Author: Maw Pin, Tan
Institute: University of Malaya
Street: Kuala Lumpur
City: Kuala Lumpur
Country: Malaysia
Email: mptan@ummc.edu.my

Curvature Analysis of Multifractal Spectra for Time Series of RR Intervals for Patients with Congestive Heart Failure

A. Muñoz-Diosdado[1] and A.M. Aguilar-Molina[2]

[1] Instituto Politécnico Nacional, UPIBI, Basic Sciences Department, Mexico City, Mexico
[2] Instituto Politécnico Nacional, ESFM, Physics Department, Mexico City, Mexico

Abstract— **Multifractal analysis have been used to study time series of physiological systems such as the series of heartbeat intervals. Using the multifractal formalism we analyze RR interval time series from 54 healthy subjects and 44 patients with congestive heart failure (CHF). Their multifractal spectra were calculated for each time series and it was observed that around the maximum of the multifractal spectra of healthy subjects, the spectra are smooth but the multifractal spectra of CHF patients tended to be pointed, therefore it is proposed to measure the curvature of all multifractal spectra. We conclude that this parameter provides information about the health state of the persons in the sample.**

Keywords— **Fractal, multifractal spectrum, curvature, congestive heart failure, NYHA index.**

I. INTRODUCTION

Heartbeat intervals are the result of the interaction of many physiological components that operate at different time scales. The mathematical techniques used to study these long series described them as non-stationary, non-homogeneous and with multifractal features in healthy persons, and there is a multifractality loss in patients with chronic heart disease [1-3].

Multifractal analysis has recently been used to study time series of physiological systems such as the heart [1-3]. Today, heart diseases constitute a risk because they can lead to serious complications difficult to treat. One of these complications is CHF that occurs when the heart is weakened by diseases or conditions that injure the heart muscle and as a result, the heart cannot pump the blood the organs need to carry out its functions. In the majority of cases, CHF can be treated, but not cured. The NYHA classification (New York Heart Association) is commonly used as a functional method for classifying the severity of patients with heart failure [4]. CHF is one of the most important problem of public health in Mexico, it is well known that heart diseases are the first cause of global mortality and within these, the CHF is emerging as one of the direct causes [5].

The proposal of this article consists of measuring the curvature of the multifractal spectra of RR interval time series, since we noticed that these spectra tended to be sharp for patients with CHF, and our hypothesis is that this method provides a correct appreciation of the health state of subjects in our sample.

II. METHODOLOGY

A. Multifractal Analysis

Physiological signals are generated by auto regulated complex systems processing entries with a wide range of features. Monofractal signals are homogeneous in the sense that they have the same properties of scaling, which locally can be characterized by a fractal dimension through the entire signal. On the other hand, multifractal signals can be decomposed into many subsets (possibly an infinite number) characterized by different fractal dimensions, which quantify the singular local behavior and therefore refer to local time-series scaling. Multifractal signals require several dimensions to fully characterize its scaling properties and are inherently more complex and non-homogeneous than monofractal signals [1, 2].

To obtain the multifractal spectrum the method proposed by Chhabra and Jensen was used, which is an accurate method for the direct calculation of the multifractal spectrum [6], for brevity we do not describe here the method, the multifractal spectrum has the appearance of the graph with downward concavity which is shown in Fig. 1. In such graphic we plot the fractal dimension $f(\alpha)$ versus α, the average power of the singularity or Hölder exponent in terms of a parameter q. Usually the applications of multifractal spectrum to the analysis of physiological signals are focused on the spectrum width, $\Delta\alpha = \alpha_{max} - \alpha_{min}$ [1, 2], which is known to decrease with age or disease , and in the asymmetry of the spectrum, in which case we use $\Delta\alpha_{right} = \alpha_{max} - \alpha_0$ and $\Delta\alpha_{left} = \alpha_0 - \alpha_{min}$ where α_0 is the value of α for which the spectrum is maximum [7]. In this paper we propose the curvature of the spectrum around its peak as an additional criterion to characterize these spectra.

We show in Fig. 1 two multifractal spectra and we can see from this figure that around the maximum the multifractal spectrum of patients with CHF tends to be more pointed than healthy subjects.

F. Ibrahim et al. (eds.), *International Conference for Innovation in Biomedical Engineering and Life Sciences*,
IFMBE Proceedings 56,
DOI: 10.1007/978-981-10-0266-3_10

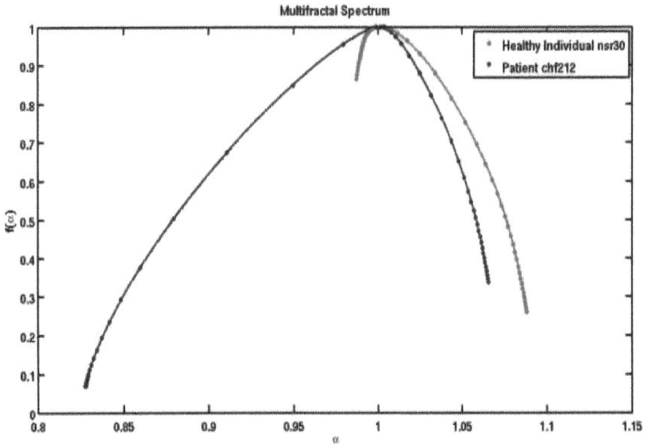

Fig. 1 Multifractal spectra of heartbeat intervals time series of a 70 years old healthy subject and a CHF patient with NYHA II

B. Curvature

If $T(t)$ is the unit tangent vector to a curve C at a point P (Fig. 2), s is the arc length measured from a point P of C chosen arbitrarily, and s grows as t increases, then the curvature of C at P, denoted by $K(t)$, is defined as

$$K(t) = \left|\frac{d\varphi}{ds}\right| \quad (1)$$

it is the absolute value of the rate of change of φ with respect to the measure of the arc length along the curve [20].

To compute the curvature of a planar curve $y = G(x)$, that it is an equation of the curve C, we have

$$K = \frac{\left|\frac{d^2y}{dx^2}\right|}{\left[1+\left(\frac{dy}{dx}\right)^2\right]^{3/2}} \quad (2)$$

And we calculate the curvature by previously evaluating the first and second numerical derivatives.

C. Databases

The heartbeat intervals records were obtained from the Physionet databases [8]. The Normal Sinus Rhythm RR Interval Database contains records of 54 healthy individuals (30 men, ages 28 to 76 years, and 24 women, aged between 58 and 73). The data of the CHF patients were obtained from two databases with the intention to have a greater number of patients. The first database is called Congestive Heart Failure RR Interval Database and contains 29 patients (34–79 years for both sexes), with congestive heart failure (NYHA classes I, II and III). The second is the BIDMC Congestive Heart Failure Database and contains 15 patients (11 male from 22 to 71 years, and women aged 54 to 63), with severe congestive heart failure (NYHA class III–IV).

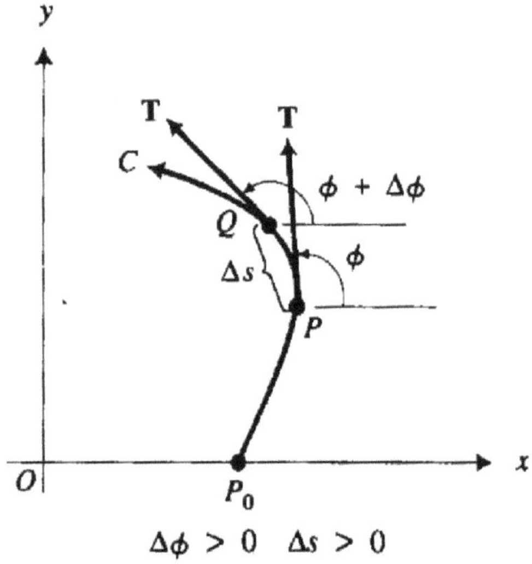

Fig. 2 Curvature

In these series we eliminated artifacts and we separated two segments of 6 hours each one, the first one when CHF patients and healthy people are asleep, and the second one when they are awake. So we analyzed total series (24 hours) and awake and asleep series (6 hours).

III. RESULTS

For each one of the heartbeat interval time series we obtained the multifractal spectrum and calculated α_{min}, α_o and α_{max} which were used to measure the width of the spectrum $\Delta\alpha$, and their asymmetry with $\Delta\alpha_{right}$ and $\Delta\alpha_{left}$. The calculations were done from $q = -30$ to $q = 30$. We observe that the multifractal spectra of patients with CHF tend to be more pointed around the maximum, which means that the curvature in these spectra is higher than the curvature of the spectra of healthy persons. For this reason, we calculated the curvature for each of the multifractal spectra of the 54 healthy subjects and 44 CHF patients for the total series, 6 hours series in the asleep stage and 6 hours series in the awake stage, in order to gain a better description of them.

We show in Fig. 3 two graphics of the first numerical derivative of the multifractal spectrum of a healthy person and a patient with CHF. Both curves are smooth but the curve of patients with CHF presents greater variation. The second derivative of the graphics in Fig. 3, shows that the CHF graphic is more irregular. Moreover, in our opinion, the curvature is also the best way to quantify differences in both the first and second derivatives, since it depends on both derivatives.

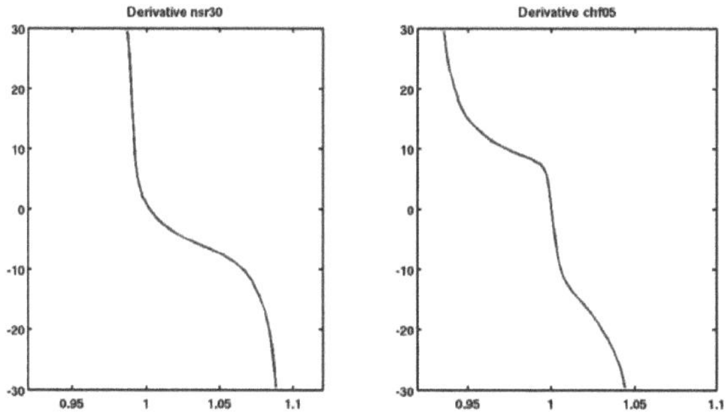

Fig. 3 Graphs of the first numerical derivative of multifractal spectrum of a 70 years old healthy person, and a 59 years old CHF patient with NYHA index III–IV

The curvature was calculated for each of the 98 multifractal spectra (54 healthy individuals and 44 CHF patients) and the average values were obtained, the same calculations were done for the 6 hours time series (awake and asleep). In Fig. 4 we plot the average curvature of the multifractal spectra of the total series for CHF patients (left) and healthy individuals (right). There is a significant difference between both graphs, since the maximum value of the average curvature in patients is 1723.68 and 505.59 in healthy subjects, i.e. nearly three times the average curvature of healthy persons. Also it can be seen in both cases that the pronounced peak occurs around the maximum of the multifractal spectra, so we can confirm that spectra of the patients tend to be sharper that the spectra of healthy persons around the maxima.

Fig. 5 shows the multifractal spectra average curvature of the 6 hours time series when they are awake and when they are asleep for CHF patients (left), and healthy persons (right). There is a big difference between both average curves when they are asleep and when they are awake. The maximum value of the average curvature in patients in the asleep and awake stage are 1970.61 and 2013.57 respectively, while for healthy people are 1176.93 and 586.55 respectively. Comparing the values of the two phases (asleep-awake) in the case of patients it is observed that most pointy multifractal spectra corresponds to the awake though the difference between the maxima does not seem to be significant.

In the case of healthy people the average curvature value in the awake stage is less than the average curvature value at the asleep stage, which means that the spectra of the asleep stage in healthy people tends to be more pointed, in this case we can also appreciate that the difference between the maxima in both stages seems to be double in the asleep phase than when they are awake.

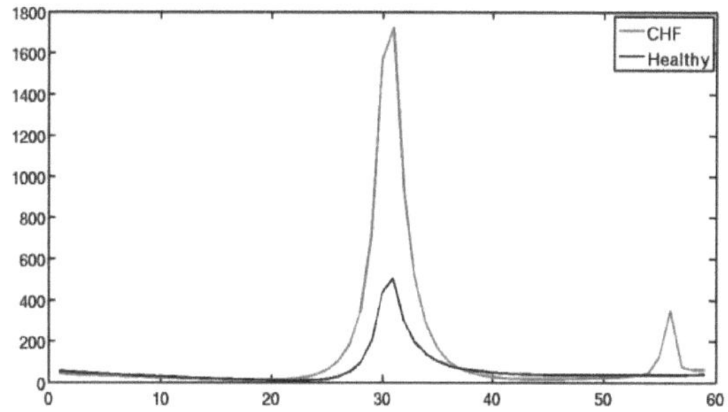

Fig. 4 Comparison of the average curvature of healthy persons with the CHF patients for total series (24 hours)

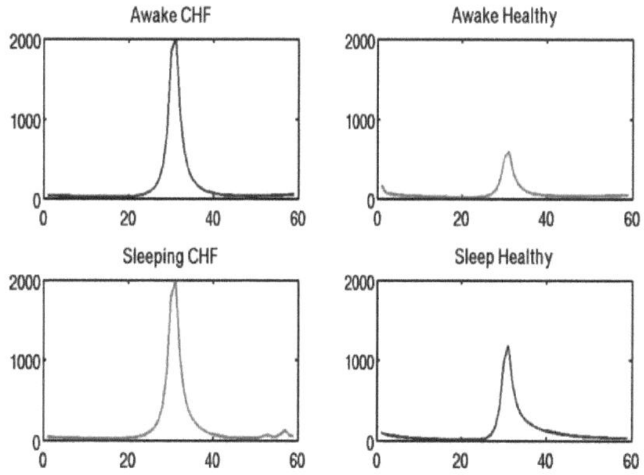

Fig. 5 The average curvature of the 6 hours series are shown in the awake and asleep stages in healthy persons and patients with CHF

It can be seen that there is a difference on the average curvature of patients and healthy people, but now in these two stages. The multifractal spectra average curvature of patients is about 3 times the average curvature of healthy people in the awake stage and 2 times the average curvature of the multifractal spectra of healthy people during asleep stage (Fig. 6).

Fig. 7 shows the comparison of the total series against the series of 6 hours in the wake and sleep stages. As can be seen, in the case of healthy people the average curvature value of the total series is approximately equal to the average curvature series of 6 hours in the awake state, this behavior is expected since healthy people is awake more time than asleep. The average curvature of the 6 hours series in the sleep phase is larger in the case of the healthy subjects. Conversely the average curvature of the 6 hours series of CHF patients when they are awake and when they are asleep are almost equal, this is probably related to the fact that patients with CHF are aggravated when they are sleeping; depending on the symptoms and the evolution of the disease, in fact some of them suffer from shortness of breath, so it is difficult to sleep well.

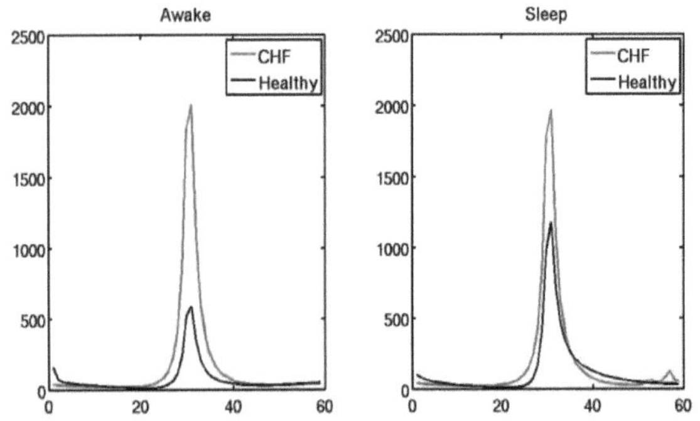

Fig. 6 Comparison of the multifractal spectra average curvature of 6 hours heartbeat time series of healthy persons and CHF patients, the heartbeat intervals time series have 6 hours in the awake and asleep stages

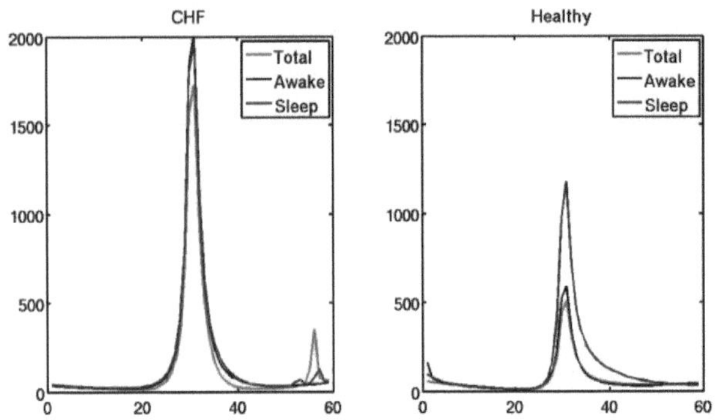

Fig. 7 Comparison of the average curvature of the total series and 6 hours series in the wake and sleep stages in CHF patients and healthy subjects

IV. CONCLUSIONS

We have found that the curvature of the multifractal spectra is another parameter that helps us to determine the health state of subjects. We have confirmed our hypothesis that the multifractal spectra of the patients with CHF tend to be more pointed than the spectra of healthy persons. The same behavior was observed in the series of 6 hours in the wake and sleep stages. When we made the comparison between the two stages, we noticed that both values tend to be approximately the same in the case of patients, and on the other hand, in the case of healthy people the value of the average curvature during the awake is double than when they are sleeping.

ACKNOWLEDGMENT

We acknowledge SIP, EDI, COFAA IPN and CONACyT Mexico for partial support.

CONFLICT OF INTEREST

The authors declare that they have no conflict of interest.

REFERENCES

1. Ivanov P Ch, Goldberger A L, Stanley H E (2002) Fractal and Multifractal Approaches in Physiology, in *The Science of Disasters*, Bunde A, Knopp J, Schellnhuber H J eds., Springer Verlag, Germany.
2. Ivanov P Ch, Nunez Amaral L A, Goldberger A L, Havlin S, Rosemblum M G, Struzik Z R and Stanley H E (1999) Multifractality in human heartbeat dynamics. *Nature* (London) 399: 461–465.
3. Goldberger AL, Amaral L A N et al. (2002) Fractals in physiology: Alterations with disease and aging. Proceedings of the National Academy of Sciences of the United States of America 99(Suppl 1): 2466 – 2472. doi:10.1073/pnas.012579499.
4. Raphael C, Briscoe C, Davies J, Whinnett ZI, Manisty C, Sutton R, Mayet J and Francis D P (2007) Limitations of the New York Heart Association functional classification system and self-reported walking distances in chronic heart failure. HEART. 93(4):(476–482).
5. Argüero Sánchez R., Magaña Serrano J A (2008) Congestive Heart Failure, www.facmed.unam.mx/sms/seam2k1.
6. Chhabra A B and Jensen R V (1989) Direct determination of the f (a) singularity spectrum. Physical Review Letters 62:1327–1330.
7. Muñoz-Diosdado A. and Del Río-Correa J L (2006) Further Study of the Asymmetry for Multifractal Spectra of Heartbeat Time Series. Proceedings IEEE EMBS Annual International Conference 28: 1450-1453.
8. Goldberger A L, Amaral, L A N et al. (2000) PhysioBank, PhysioToolkit, and PhysioNet: Components of a New Research Resource for Complex Physiologic Signals. Circulation 23:e215–e220.

Corresponding author:

Author: A. Muñoz-Diosdado
Institute: Instituto Politécnico Nacional, UPIBI
Street: Avenida Acueducto S/N, CP 07340
City: Mexico City
Country: Mexico
Email: amunozdiosdado@gmail.com

Elemental Analysis of Nigerian and Nigerien Foods Using Neutron Activation and Estimation of Daily Intake

A.R. Usman[1,4], M.U. Khandaker[1], N.F. Isa[2], and Y.A. Ahmed[3]

[1] Department of Physics, University of Malaya, 50603 Kuala Lumpur, Malaysia
[2] Department of Physics, Bayero University, Kano, Nigeria
[3] Center for Energy Research and Training, Ahmadu Bello University, Zaria, Nigeria
[4] Department of Physics, Umaru Musa Yar'adua University, Katsina, Nigeria

Abstract— **Realizing the importance of individual elements to human body, 14 different food samples (mainly cereals: beans, millet, sorghum and rice) collected from two different countries; Nigeria and Republic of Niger, were analyzed using Instrumental Neutron Activation Analysis (INAA) for the determination of the elemental content. Despite the commercial activity of uranium in the Republic of Niger, the concentration of Uranium, Arsenic, Chromium etc. in the studied samples show below detection limit. Elevated concentrations of K, Ca and Mg were found from the sampled beans of this work. Daily intake per person of all assessed elements were estimated and compared with Recommended Daily Intake (RDI) to establish the accumulation information of mineral, toxic and radioactive elements in the body of people of Nigeria and Niger Republic. Assessed elemental concentrations were compared with the available literature data on relevant cereals, and found reasonable agreement. Analytical quality control service (AQCS) reference material (lichen) from International Atomic Energy Agency (IAEA) was also irradiated and analyzed in this work for ensuring quality control of the method and facility (nuclear reactor) used.**

Keywords— **Cereals, Katsina, Maradi, NAA, Mineral, Toxic and Radioactive elements.**

I. INTRODUCTION

The need to have a detailed and accurate information on the elemental concentrations of our foods cannot be over emphasized. This is because, trace elements form an important component of the biotic environment for their roles in the health of human being, animals and plants. Taking of right food with the correct proportion of nutrients is the ideal requirement of the body. Most of our dietary compositions are now understood to be either essential to, or complimenting the proper usage of nutrients in human body [1, 2]. Cereals, one of the major food category, is the staple food of the sub-Sahara region of northern Nigeria and Niger republic. This is largely produce due to soil type and amount of annual rainfall in the regions. The main crops being cultivated in Niger Republic are millet, sorghum (guinea corn), and beans. Rice is mainly cultivated through irrigation farming around the River Niger and River Komadougou of the country [3]. On the other hand, maize, millet, sorghum and beans are the main cultivated edible crops in Nigeria.

In order to determine elemental content of the selected crops, this study employed (thermal) neutron activation analysis for its advantages over several others techniques that includes ability to detect many elements in a wider range of matrix, freedom from reagent blank, high accuracy and improved sensitivity [4, 5, 6,]. The study was carried out using Nigerian Research Reactor (NIRR-1) situated at the Center for Energy Research and Training (CERT), Ahmadu Bello University (ABU), Zaria – Nigeria. NIRR-1 was designed as a Miniature Neutron Source Reactor (MNSR) such that it has a configuration of tank-in-pool structure, a fuel of highly enriched uranium (90.2 %), light water serving as both moderator and coolant, metallic beryllium (Be) surrounding core of the reactor as reflector, rated power of 31kW and a corresponding thermal neutron flux 1.0×10^{12} $cm^{-2}s^{-1}$. Details of the reactor structural and configurations can further be found elsewhere [1]. Present study broadly bears the followings as some of its major goals. First is to determine and compare all the detectable elements in the selected food items of Nigeria and Republic of Niger, including the risk elements since these regions are major producers of the itemized food products. The second goal arose from considering that Republic of Niger has uranium exploration commercial activities and thus aimed to find if those activities have (through leaching, human activities and wind) led to sufficient migration of related parent and daughter elements to the selected regions, and consequently contaminated the distant soil and crops. The third is to use the obtained data to calculate elemental contribution of dietary intake to the people of these communities. This would ultimately increase more available data on this subject from these regions, especially Republic of Niger.

II. METHODOLOGY

Through careful selection, a total of 14 food samples comprising two each of sorghum (guinea corn), millet, beans and rice were collected from cities of Bakori, Funtua, Katsina, and Charanci Local Governments of Katsina state

F. Ibrahim et al. (eds.), *International Conference for Innovation in Biomedical Engineering and Life Sciences*,
IFMBE Proceedings 56,
DOI: 10.1007/978-981-10-0266-3_11

in Nigeria and Some parts of Maradi in Republic of Niger after the best harvesting period of the samples. In order to ascertain the locality of the samples, the purchase was mostly directly from the farmers. The dried and grinded samples were then packed into separates polythene bags. A small portion of each powdered sample was prepared, weighed using high precision balance (METTLER AE 240) and packed using 8 ml polyethylene vials. An air drier was used to heat seal the vials. Each sample was prepared in twice for short and long irradiations. With the help of rabbit carriers, the samples and standard were sent into the reactor through the pneumatic transfer system which uses pneumatic pressure. NIRR – 1 irradiated the samples with a thermal neutron flux of $2.5x10^{11}$ $ncm^{-2}s^{-1}$for 5 minutes and $5.0x10^{11}$ $ncm^{-2}s^{-1}$ for 6 hours for short and long irradiations, respectively. The whole system was equipped with electronic timers which help in monitoring the exact irradiation and decay times. The samples were then taken to a measurement room consisting of a high purity germanium (HPGe) detector, connected to a PC-based Multi-Channel Analyzer (MCA).

A. Calculation of Dietary Intake for People of Nigeria and Niger Republic

The Estimated Daily Intake (EDI) were based on mg per person per day. We used the published data of [8]2009 for per capita average consumption of millet, sorghum, rice and beans as 95, 101, 58, and 24 g/day, respectively by Nigerians. The same source shows that people of Niger Republic consume 425, 124, and 88 g/day of millet, sorghum and beans, respectively. The EDI is compared with the Recommended Daily Intake (RDI) to have an insight of the daily contribution and loading of the studied elements in the bodies of people of these regions.

III. RESULTS AND DISCUSSIONS

Quantitative information of this work is presented in the Tables 1 and 2. Table 2 represents the EDI and RDI of

Table 1 Analytical Results for Various Elements from 7 Biological Samples in This Research.

Elements (ppm)*	Nigeria (average Values)				Niger Republic (average Values)		
	S1(Sorghum)	R (Rice)	B1 (Beans)	M1 (Millet)	M2 (Millet)	B2 (Beans)	S2 (Sorghum)
^{27}Mg	884 ± 89	305 ± 67	1297 ± 101	556 ± 72	893 ± 82	1167 ± 98	757 ± 72
^{28}A1 (ppm)	43 ± 5	54 ± 5	23 ± 2	66 ± 6	29 ± 3	29 ± 3	39 ± 4
^{49}Ca (ppm)	291 ± 75	280 ± 73	1489 ± 249	448 ± 93	455 ± 94	1333 ± 223	310 ± 73
^{56}Mn (ppm)	15 ± 1	6.1 ± 0.3	14.4 ± 0.6	11.6 ± 0.5	12.8 ± 0.6	27.3 ± 1.2	21 ± 1
^{24}Na (ppm)	8 ± 1	56 ± 3	41 ± 2	12 ± 1	13 ± 1	33 ± 2	5.9 ± 0.4
^{42}K (ppm)	4297 ± 305	1842 ± 138	15420 ± 1079	5198 ± 3698	4510 ± 325	16310 ± 1142	3866 ± 278
^{82}Br (ppm)	4.3 ± 0.2	0.24 ± 0.02	0.54 ± 0.04	1.27 ± 0.07	11.7 ±1.1	2.3 ± 0.1	7.5 ± 1.2
^{140}La (ppm)	0.06 ± 0.01	BDL	0.11 ± 0.02	0.11 ± 0.01	0.08 ± 0.01	0.09 ± 0.01	BDL
^{65}Zn (ppm)	60 ± 6	18 ± 3	44 ± 6	40 ± 5	32 ± 6	57 ± 7	31 ± 6
^{86}Rb (ppm)	6 ± 1	15 ± 2	20 ± 2	13 ± 2	9 ±2	23 ± 3	7 ± 1
^{232}Th(ppm)	0.4 ± 0.1	0.3 ± 0.1	0.4 ± 0.1	0.6 ± 0.1	0.4 ±0.1	0.3 ± 0.1	0.5 ± 0.1

*ppm – parts per million (i.e., PPM = $\mu g\ g^{-1}$=$mg\ kg^{-1}$), BDL – below detection limits, NA – not analyzed.

Table 2 Nigerian and Nigerien (Republic of Niger) foods; Estimation of Daily Intake and its comparison to Recommended Daily Intake

Elements	Estimated daily intake for Nigerian food (mg/day)				Estimated daily intake for Niger Republic food (mg/day)			Recommended Daily Intake (mg/day)
	Millet	Beans	Sorghum		Millet	Beans	Sorghum	
^{27}Mg	123.215	31.128	89.284	17.69	379.525	102.696	93.868	420(320)*
^{28}A1	2.185	0.552	4.343	3.132	12.325	2.552	4.836	1
^{49}Ca	141.455	35.736	29.391	16.24	193.375	117.304	38.44	1000
^{56}Mn	1.368	0.3456	1.515	0.3538	5.44	2.4024	2.604	2.3(1.8)*
^{24}Na	3.895	0.984	0.808	3.248	5.525	2.904	0.7316	2300
^{42}K	1464.9	370.08	433.997	106.836	1916.75	1435.28	479.384	4700
^{82}Br	0.0513	0.01296	0.4343	0.01392	4.9725	0.2024	0.93	8
^{140}La	0.01045	0.00264	0.00606	-	0.034	0.00792	_	375
^{65}Zn	4.18	1.056	6.06	1.044	13.6	5.016	3.844	11(8)*
^{86}Rb	1.9	0.48	0.606	0.87	3.825	2.024	0.868	400
^{232}Th	0.038	0.0096	0.0404	0.0174	0.17	0.0264	0.062	760

* *Where there is difference between male and female requirement, the bracket term is for female.*

various elements for the people of Nigeria and Republic of Niger, respectively via the consumption of studied foods. A short briefing about the determined elements are given under the following sections:

A. Toxic Elements

Thorium (Th): Thorium has been detected in all the samples, with the Nigerian millet showing the highest value. When ^{232}Th is ingested as a food element, it predominantly escapes from the body via urine and feces within few days. Usually only little amount of ^{232}Th is left in the body which subsequently enters the bloodstream and finally deposited on bones for years. As ^{232}Th undergoes radioactive decay to ^{228}Ra, concurrently it emits ionizing radiation such as alpha particle with accompanying gamma radiation. Increase risk of cancer is the major concern of low to moderate exposure level of ionizing radiation. Studies have indicated that inhaling thorium dust causes an increased chance of lung and pancreatic cancers. Due to likelyhood of thorium storage in bone, the risk of bone cancer is also increased [9]. A single thorium administration of 760 mg/kg body weight per day (84nCi/kg/day = 3100 Bq/kg/day) resulted to no mortality of mice whereas a single gavage dosage of 1000 mg thorium per kg body weight per day (110 nCi/kg/day = 4070 Bq/kg/day) in form of thorium nitrate lead to the death of 4/20 mice [10].

Aluminum (Al): This study reports the presence of aluminum in virtually all the samples. Among the studied samples, Nigerian cereals show the highest mean concentrations of Al. There has not been any report of any significant role of Al in human. However, increase in evidence of its toxicity relating to its accumulation in brain and effects on nervous system has consequently led WHO to revise its earlier stand of 7 mg/kg of body weight Provisional Tolerable Weekly Intake in 2006 to 1 mg/kg of body weight [11]. Millet and several other cereals demonstrate higher level of this toxic element, above the recommended daily intake.

B. Mineral Elements

Manganese (Mn): Manganese shows highest concentration of 27.3 μg g^{-1} in Nigerien beans, 14.4 μg g^{-1} for Nigerian beans while (Nigerian) rice has the least content of 6.1 μg g^{-1} of all the samples. Mn is an essential element for metabolism of carbohydrates and lipid and also for bone growth of human beings. It also serves as an enzyme activator for some enzymes. Toxicity of manganese has also been documented, but the danger is very unlikely through dietary daily exposure except under distinct situations [11].

Calcium (Ca): This study reported highest concentrations of Ca (1489 μg g^{-1}) from Nigerian beans while the least concentration of the element is seen in rice as 280 μg g^{-1}. Calcium balance in the body is very crucial, it is not only useful to the skeleton but also plays important roles in regulating nerve excitability, muscle contraction and blood coagulation.

Magnesium (Mg): Mg is reported to have the highest concentration of 1297 μg g^{-1} from Nigerian beans and least concentration of 305 μg g^{-1} from rice. The variation is highly pronounced from one food item to the other, signifying importance of individual element for particular mineral complementation. Literatures have shown a usual reported deficiency of Mg in human and indicated nausea, growth retardation and muscle weakness as some of its symptoms and also suggested cardiac malfunction due to deficiency of the element [11].

Sodium (Na): Varying concentrations of sodium has also been reported here from one food item to the other with the highest (56 μg g^{-1}) in rice and least (5.9 μg g^{-1}) in Nigerien sorghum. Natural level of Na is usually sufficient in most foods but is still further supplemented especially in processed foods. Dietary excess of sodium results to hypertension, with the most adverse consequences seen in people of historic renal disease, congestive heart failure, or cirrhosis.

Zinc (Zn): With the exception of rice that shows lowest amount of 18 μg g^{-1} in zinc, the other food items have moderately high concentration of this element. Zn is an activator of several enzymes and stabilizes the structure of DNA, RNA and Ribosomes. Additionally, it plays a crucial role in hormonal metabolism. Diarrhea, growth retardation, skin lesions etc. have been reported as some of its deficiency syndromes.

Potassium (K): Potassium (^{42}K) was detected in high concentration in all the samples, with the most elevated value seen in Nigerien beans. Potassium is essential for human life and is considered as one of the most abundant element in our body. It serves as a regulator in the acid–base and the osmotic balance in the cell, plays a significant role in muscle contraction and transmission of nerve impulse, and also as a many enzymes' co-factor involved in the storage of carbohydrates and synthesis of protein. Despite the above advantages, potassium is also known to be radioactive and thus can have detrimental effect to health when at very high level.

Bromine (Br): Nigerian rice shows 0.24 μg g^{-1} as the least Br content. This suggest that for higher Br requirement of the body, Nigerien millet and sorghum (11.7 μg g^{-1} and 7.5 μg g^{-1}) are good choice while rice is very poor in that context. The major food sources of Br are grains, nuts, sea foods, sea salt and bread. Eusinophilic leukocytes of blood uses Br for immune defense [12]. A study suggests intake of 8 mg per day as adequate to the body and similarly, the study further cited insomnia and retarded growth as the deficiency symptoms of Br [12].

Generally, this study reports a lower concentration for the elements K, Ca, Zn and Mn, compared to the earlier data reported in [1]. Maihara et al. [13] however reported a much lower concentration of Zn in Brazilian cereals than those reported here. Arogunjo et al. [14] reported activity concentrations of ^{40}K, ^{238}U and ^{232}Th as 297.87 Bq/Kg, Nil and 5.95 Bq/Kg, respectively in Nigerian white beans of Sokoto region, and 42.29 Bq/Kg, Nil and 4.01 Bq/Kg in local rice from Ekiti state of Nigeria. The level of radioactive concentration reported in this study is below the adverse level suggested as tolerable daily intake (TDI) by [10] based on conducted studies on mice and dog. This study also indicated that beans is very rich in Mg, Ca and K as compared to other sampled cereals.

IV. Conclusions

This study detected the presence of Mg, Al, Ca, Mn, Na, K, Br, Zn, La, Rb and Th in the most staple food items of Northern Nigeria and Republic of Niger. The Estimated Daily Intake for all of the detected elements was obtained for adult population, and compared with the Recommended Daily Intake to have an insight of the daily contribution of these elements into their body. Despite the commercial activity of uranium in Republic of Niger, excessive leaching may not have yet occurred to the extent of contaminating distant soil and crops of the selected regions. This study also provides new data on cereals to the selected regions, and suggest more studies especially in Republic of Niger as only very little data was found from the region. Due to the important roles of these nutrients in biochemical processes within the body, it is thus vital that we should consider them in unison rather than in isolation for collective improvement of our health.

Acknowledgment

This work was partly supported by University of Malaya Research Grant, Malaysia and Umaru Musa Yar'adua University, Katsina, Nigeria via Education Trust Fund (ETF).

Concflict of Interest

The Authors declared that there is no conflict of interest.

Refrences

1. Ahmed Y A, Landsberger S., O'Kelly et al (2010) Compton suppression method and epithermal NAA in the determination of nutrients and heavy metals in Nigerian food and beverages. *Applied Radiation and Isotopes, 68*(10), 1909-1914.
2. Chukwuma Sr C, (1995) Evaluating baseline data for copper, manganese, nickel and zinc in rice, yam, cassava and guinea grass from cultivated soils in Nigeria. *Agriculture, Ecosystems & Environment, 53*(1), 47-61
3. FAO (2006) Country posture/forage Resource Profile - Niger Republic. In). <http://www.fao.org/ag/AGP/AGPC/doc/counprof/niger/niger.htm>: DIETER GEESING and HASSANE DJIBO
4. IAEA (1990) *Practical Aspects of Operating a Neutron Activation Laboratory* (IAEA TECDOC – 564 ed.) Vienna, Austria: International atomic energy agency
5. Kapsimalis R, Landsberger S, Ahmed Y A (2009) Determination of uranium in food samples by Compton suppression epithermal neutron activation analysis. *Applied Radiation and Isotopes, 67*(12), 2097-2099
6. Nyarko B J B, Akaho E H K, Fletcher, J J, et al (2008) Neutron Activation analysis for Dy, Hf, Rb, Sc and Se in some Ghanaian cereals and vegetables using short-lived nuclides and Compton suppression spectrometry. *Applied Radiation and Isotopes, 66*(8), 1067-1072
7. Ahmed Y A, Mansir I B, Dewu B B M (2013) Installation of permanent cadmium-lined channel as a means for increasing epithermal NAA capabilities of miniature neutron source reactors. *Nuclear Engineering and Design, 263*(0), 70-76
8. FAOstat F, (2009) Agriculture Organization of the United Nations. Statistical Database. http://faostat3.fao.org/home/E
9. US EPA (2014) Radiation Protection: Thorium at http://www.epa.gov/radiation/radionuclides/thorium.html
10. A.T.S.D.R, U.S.P.H.S., Environmental Protection Agency, US, (1990) Toxicological Profile For Thorium. Agency for Toxic Substances and Disease Registry U.S. Public Health Service in collaboration with: U.S. Environmental Protection Agency
11. Antoine J M R, Hoo Fung L A, Grant C N, et al (2012) Dietary intake of minerals and trace elements in rice on the Jamaican market. *Journal of Food Composition and Analysis, 26*(1–2), 111-121
12. Kohlmeier M (2003). *Nutrient Metabolism: Structures, Functions, and Genetics*. Academic Press, 2003
13. Maihara V A, Avegliano R P, Santos P S, et al. (2009) Neutron Activation Analysis Applied To Nutritional And Foodstuff STUDIES. In *International Nuclear Atlantic Conference - INAC*, (pp. ISBN: 978-985-99141-99103-99148). Rio de Janeiro,RJ, Brazil
14. Arogunjo A M, Ofuga E E, Afolabi M A (2005) Levels of natural radionuclides in some Nigerian cereals and tubers. *Journal of Environmental Radioactivity, 82*(1), 1-6

Corresponding Author: M.U. Khandaker
Institute: University of Malaya
Street: 50603 Jalan University
City: Kuala Lumpur
Country: Malaysia
Email: mu_khandaker@yahoo.com

Optimization of CNT Based MEMS Piezoresistive Pressure Sensor for Intracranial Pressure Monitoring

S.H.A. Rahman[1], N. Soin[1], and F. Ibrahim[2]

[1] Department of Electrical Engineering, Faculty of Engineering, University of Malaya, Kuala Lumpur, Malaysia
[2] Department of Biomedical Engineering, Faculty of Engineering, University of Malaya, Kuala Lumpur, Malaysia

Abstract— **Optimum design considerations for modified square diaphragm of intracranial pressure sensors are carried out using Taguchi method. The optimum slot sizes and diaphragm thickness are analyzes by taking into account the effects of maximum displacement and maximum stress of diaphragm. A new slots are introduced to replace typical square diaphragm for the maximization of displacement and stress on the diaphragm. It is found that the optimum design is as follows: thickness =2 µm, slot width =5µm, and slot length =8µm to achieve the result of increment of almost two folds in maximum displacement and more than 10 folds of maximum stress as compared to previous typical square diaphragm design.**

Keywords— **MEMS pressure sensor, Optimization, Taguchi method, Square diaphragm.**

I. INTRODUCTION

Piezoresistive Micro electro mechanical system (MEMS) pressure sensor is a device that measure change in pressure by means of a change in resistivity of a piezoresistor placed on the maximum strain area of a diaphragm that is proportional to applied stress and, subsequently, to applied pressure. In order to design a pressure sensor for biomedical applications, it is important to maximize the performance of the sensor which can be quantified by the product of signal to noise (S/N) ratio and sensitivity to the temperature coefficient of piezoresistance to meet the desired design specification.

The excellent strain of CNTs produces a highly piezoresistive network, which benefits pressure sensors and microscale/nanoscale strains with fine resolution. The effects of stress are far more significant on crystalline materials' with nanometer thick planes than they are on bulk materials [1]. Many studies have examined the fabrication of highly sensitive pressure sensors by depositing piezoresistive CNTs onto the fixed silicon substrate in which single-walled and multi-walled carbon nanotubes (SWNTs and MWCNTs, respectively) are utilized as active sensing elements [2]–[4].

Most of the researches has shown that the most influential parameter for the piezoresistive pressure sensor diaphragm design in the thickness of the diaphragm. Lin et al. (1999) found out, the output voltage proportionally increases or decreases, with the increase of the resistance of the sensing resistor (by increasing the sheet resistance) or decreased (by increasing the width). Increasing the diaphragm thickness will also drastically reduce the sensor output as the sensitivity drops [5]. The same goes to Tsai et al., (1999) who has confirmed that the control factors such as the of sensor membrane thickness, number of electrode finger pairs and the electrodes are essential design parameters in that they significantly influence the precision and sensitivity of the sensor [6]. Currently, a study also proved that the optimum design parameters are contributed by the sensitive layer including thickness and coverage area [7].

MEMS technology has paved a new solution for the invasive intracranial pressure monitoring (ICP) devices that is currently being used to monitor the patient with brain injury [8], [9].Besides having the specialty on its small size, the high sensitivity of the pressure sensor is another important characteristic to be achieved in designing the ICP MEMS pressure sensor. However, problems such as signal loss and design complexity need to be considered as they become the main quality trade off to achieve high sensitivity pressure sensors.

Optimization of MEMS pressure sensor design parameters to achieve a set of quality attributes is important in bridging up the sensitivity and ease of fabrication especially in designing the pressure transducer which in this case the pressure sensor diaphragm. The quality attributes considered in this study are maximum displacement and maximum stress on pressure sensor diaphragm. In this paper we present a pressure sensor based on piezoresistive effect of carbon nanotubes (CNT). The sensor is 2, 8 and 10 µm thick, 50µm wide modified square diaphragm with CNT piezoresistor located on the maximum strain area. Modification is done by forming two slots at the edge of square diaphragm. Taguchi method was widely adopted in order to determine the optimum parameters in the design stages [6], [10] and as well as during the manufacturing processes [11], [12] and has been proven to significantly improved the quality [13].

II. PRESSURE SENSOR DIAPHRAGM DESIGN

In order to achieve more sensitive device and reducing the effect of residual stress and stiffness of the diaphragm,

F. Ibrahim et al. (eds.), *International Conference for Innovation in Biomedical Engineering and Life Sciences*,
IFMBE Proceedings 56,
DOI: 10.1007/978-981-10-0266-3_12

slotted diaphragm is proposed. Eight slots with the same dimension and geometry was formed on the square diaphragm. The cross sectional view of the slotted square pressure sensor is shown in Figure 1 below. The pressure sensor is designed for a maximum pressure range of 0.001 Mpa where this is typical of intracranial pressure range [14], [15]. Finite Element Analysis was used to see the effect of different pressure ranges on the diaphragm. These diaphragms were subjected to pressure range of 0.0001 to 0.001 of pressure in Termoelectromechanical (TEM) module of Intellisuite software. Prior to this, the design model of diaphragms were developed using 3DBuilder module in Inteliusite software and is shown in Fig. 2.

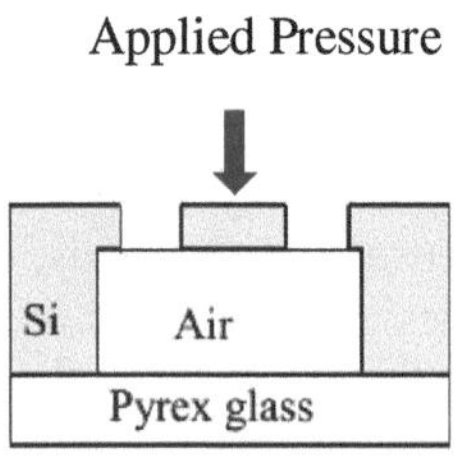

Fig. 1 Cross sectional view of slotted square diaphragm ICP pressure sensor

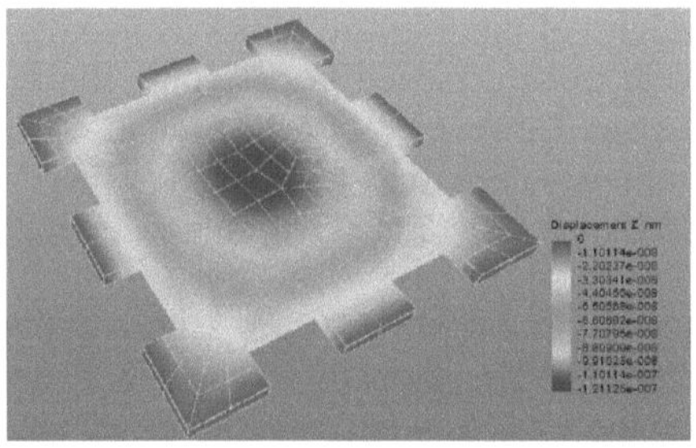

Fig. 2 FEM diaphragm model for designed pressure sensor

III. Taguchi Method

Optimizing one quality attribute may lead to loss of other quality attribute. Hence in order to simultaneously satisfy all the quality requirements a multi objectives optimization is required. In this study, Taguchi method was adopted to find the optimum slot dimension and diaphragm thickness to achieve maximum displacement and stress on the slotted square diaphragm. Three levels of each parameter was selected and the value for every level is shown in Table 1 below.

Table 1 Three levels of parameter for diaphragm design optimization.

Factor Name	Factor Letter	level 1	level 2	level 3
thickness (μm)	A	2	5	10
slot width (μm)	B	4	5	6
slot length (μm)	C	6	8	10

Referring to the OA selector [16] and few other study that using the same three level factors, OA of L9 is selected in this optimization [12], [17]. Table 2 below tabulate the complete experimental layout for maximum pressure applied within the ICP pressure range and it is worth to note that a good diaphragm design need to consider the burst pressure of 5x and 10x of maximum pressure applied [18]–[20]. The factors considered for the diaphragm optimization is shown in Figure 3 below.

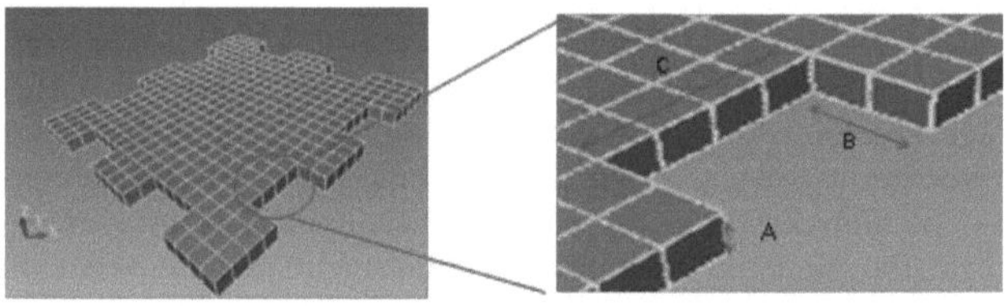

Fig. 3 Three level factors considered for slotted square diaphragm's Taguchi optimization are thickness, slot width and slot length labeled as A, B and C respectively.

Taguchi method identify three categories of S/N ratios which are:

The lower the better quality characteristic:

$$S/N = -10\log_{10}\left(\frac{y_1^2 + y_2^2 + y_3^2 + \ldots + y_n^2}{n}\right) \quad (1)$$

Nominal the best quality characteristic:

$$S/N = -10\log_{10}\left[\frac{(y_1 - y_0)^2 + (y_2 - y_0)^2 + (y_3 - y_0)^2 + \ldots + (y_n - y_0)^2}{n}\right] \quad (2)$$

The higher the better quality characteristic:

$$S/N = -10\log_{10}\left[\frac{\frac{1}{y_1^2} + \frac{1}{y_2^2} + \frac{1}{y_3^2} + \ldots + \frac{1}{y_n^2}}{n}\right] \quad (3)$$

where y1, y2, etc., are the simulation results and n is the number of observations with the same values of factors (in this current study, n=1). For this case, the larger the better characteristic is chosen so the S/N is calculated as shown in Table 2 below.

Figures 4 and 5 indicate the magnitudes of the average response effects of the various control factors for the maximum stress and for the maximum displacement, respectively. From figure 3, to obtain the maximum stress area for piezoresistor placement on slotted square diaphragm, the three parameters should be set as follows A=A1=2 μm, B=B1=4μm, C=C2=8μm. Figure 4 illustrates that to obtain a diaphragm with highest value of maximum displacement, the parameters should be A=A1=2μm, B1=B=4 μm, and C=C2=8 μm. These optimization results illustrate slot dimension and diaphragm thickness settings for achieving both maximum displacement and maximum stress simultaneously coincide.

Table 2 Signal to noise (S/N) response table for slotted square diaphragm performance

Factors	A	B	C	Simulation results		S/N Ratio	
experimental number				max disp	max stress	max disp	max stress
1	2	4	6	8.13E-05	6.83E-02	-81.80	-23.31
2	2	5	8	8.63E-05	7.45E-02	-81.28	-22.56
3	2	6	8	8.68E-05	7.36E-02	-81.23	-22.66
4	5	4	8	6.22E-06	1.18E-02	-104.12	-38.54
5	5	5	8	6.24E-06	1.18E-02	-104.09	-38.54
6	5	6	6	5.90E-06	1.07E-02	-104.59	-39.38
7	10	4	10	1.22E-06	3.44E-03	-118.27	-49.27
8	10	5	6	1.00E-06	2.61E-03	-119.97	-51.65
9	10	6	8	1.08E-06	2.86E-03	-119.35	-50.86

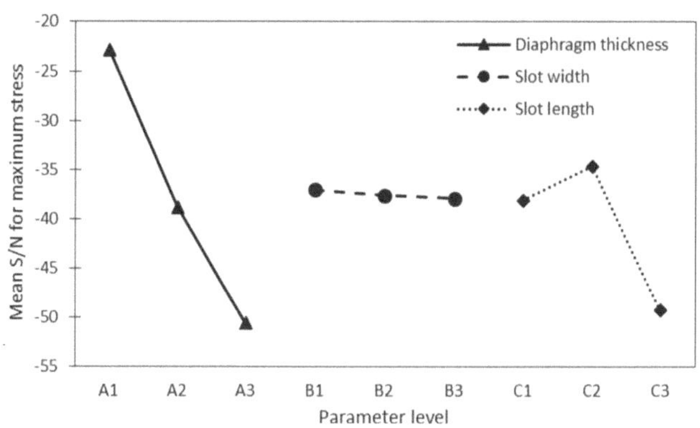

Fig. 4 Mean S/N ratio for maximum stress at maximum applied pressure (0.001 Mpa) Optimum condition: (the larger the better) A1, B2, C2

IV. SIMULATION OF OPTIMIZED SQUARE DIAPHRAGM

After the modification of square diaphragm was found to increase the effect of pressure on the diaphragm, and the best parameter condition for slot size and diaphragm thickness was obtained, a comparison of the verification run between the initial and optimal conditions was carried out and the result indicate that almost two folds of increment on maximum displacement and more than 10 folds of maximum stress on the optimized slotted square diaphragm as compared to the typical square one.

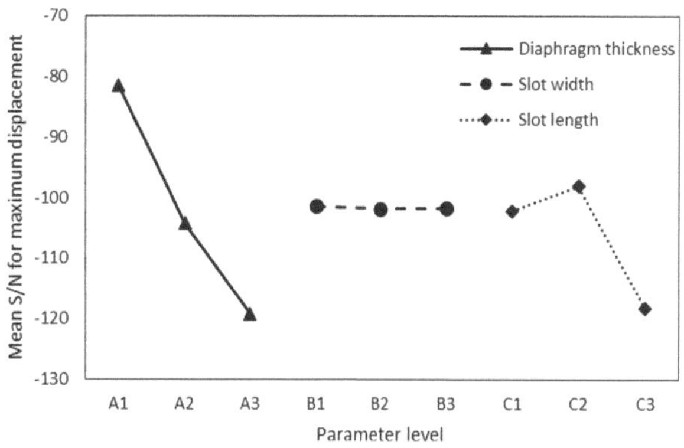

Fig. 5 Mean S/N ratio for maximum displacement at maximum applied pressure (0.001 Mpa). Optimum condition: (the larger the better) A1, B2, C2

Table 3 Comparison of the verification run between the initial and optimal conditions

Performance characteristics	Initial condition (typical 50x50μm square diaphragm)	Optimal condition A1, B1, C2	Gain
Maximum displacement	8.49e-7	8.63E-05	8.55E-05
Maximum stress	1.90e-3	7.45E-02	7.26E-02

A piezoresitive study was also carried out. The slotted square diaphragm is integrated with CNT piezoresistor and the change in resistance due to the strain was obtained through simulation using CNT features in Intellisuite TEM module. It is assumed that the CNT experience the same stretching as the modified square diaphragm, similar to the doped-silicon strain gauges used in state-of-the-art MEMS pressure sensors. Four CNT piezoresistor is attached to electrodes placed at the maximum stress area on center edge of diaphragm as shown in Figure 6 below. It is crucial to spread the piezoresistors like the traditional strain gauges and placed in fashion so that the resistivity of two piezoresistors increased and that of two decreased with applied pressure in Wheatstone bridge configuration [21], [22]

Unlike the conventional way, where the simulations are done to determine the most sensitive part on the diaphragm, this simulation is run in order to determine the effect of different size of slot and thickness that effect the maximum value of stress and displacement at the most sensitive part of the diaphragm. This is due to the fact that this the most sensitive part on the square diaphragm is already well known by the sensor community and there were already studies that validate the correctness of the analytical models [5].

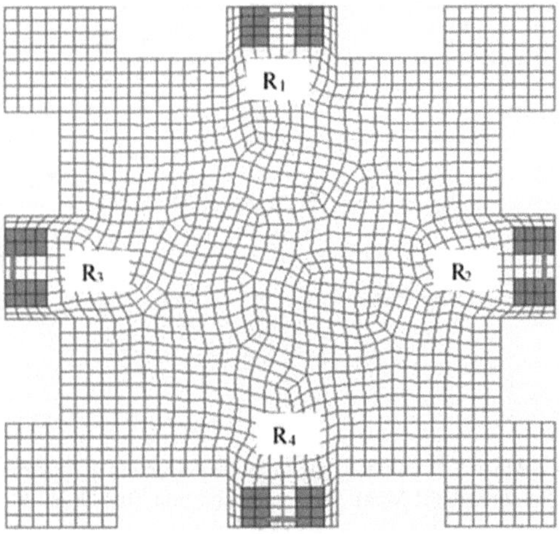

Fig. 6 Slotted square diaphragm with four CNT piezoresistors placed at maximum stress area.

The plot of the change in resistance of the piezoresistive CNT on the optimized diaphragm design for intracranial pressure range of 0 to 1 1Kpa is given in the graph below (Figure 7). From the plotted graph, the linear change in in resistance of the piezoresistive CNT can be observed which is implored for a sensor.

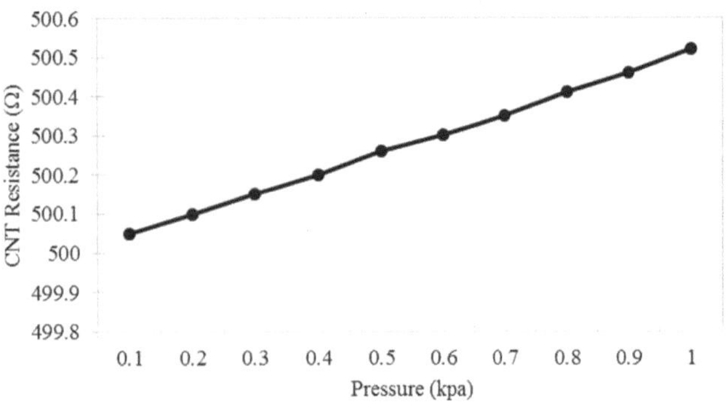

Fig. 7 CNT resistance change on optimized slotted square diaphragm

V. CONCLUSIONS

It is suggested the optimum condition of the diaphragm should have the thickness of 2µm, slot width 4 µm and slot length of 8 µm respectively. The maximum displacement has almost two fold (175%) improvement while maximum stress shows more than 10 (1048%) folds improvement. Results of this study indicate that a huge improvement in the value of maximum displacement and maximum stress value was obtained. The results also show that applying Taguchi's method for the purpose of product improvement is feasible.

ACKNOWLEDGMENT

This research is funded by the University Malaya Research Grant (UMRG)-AET (Grant Number RP009B-13AET). The authors would also like to express their gratitude to the Micro Electronic laboratory, Faculty of Engineering for allowing the facilities to be used.

CONFLICT OF INTEREST

The authors declare that they have no conflict of interest.

REFERENCES

1. K. Z. Kaourosh and F. Benjamin, Nanotechnology-Enabled Sensor. Melbourne: Springer, 2008.
2. M. A. S. Mohammad Haniff, H. W. Lee, D. C. S. Bien, A. S. Teh, and I. A. Azid, "Highly sensitive integrated pressure sensor with horizontally oriented carbon nanotube network.," Nanoscale Res. Lett., vol. 9, no. 1, p. 49, Jan. 2014.
3. C. Stampfer, T. Helbling, D. Obergfell, B. Schöberle, M. K. Tripp, a Jungen, S. Roth, V. M. Bright, and C. Hierold, "Fabrication of single-walled carbon-nanotube-based pressure sensors.," Nano Lett., vol. 6, no. 2, pp. 233–7, Feb. 2006.
4. G. Y. Li, P. M. Wang, and X. Zhao, "Pressure-sensitive properties and microstructure of carbon nanotube reinforced cement composites," Cem. Concr. Compos., vol. 29, no. 5, pp. 377–382, May 2007.
5. L. Lin, H.-C. Chu, and Y.-W. Lu, "A simulation program for the sensitivity and linearity of piezoresistive pressure sensors," J. Microelectromechanical Syst., vol. 8, no. 4, pp. 514–522, 1999.
6. H. H. Tsai, D. H. Wu, T. L. Chiang, and H. H. Chen, "Robust design of SAW gas sensors by Taguchi dynamic method," Sensors, vol. 9, no. 3, pp. 1394–1408, 2009.
7. H. Guo, X. Chen, and Z. Wu, "A feasible simulation method for vapor sensor based on polymer-coated NEMS diaphragm," Measurement, vol. 68, pp. 219–224, 2015.
8. li-Anne Liew and V. M.Bright, "Disposable CMOS catheter-tip pressure sensor for intracranial pressure measurement - Microtechnologies in Medicine and Biology, 1st Annual International, Conference On. 2000," no. l, pp. 10–15, 2000.
9. Y. Zhang, Z. Zhang, B. Pang, L. Yuan, and T. Ren, "Tiny MEMS-Based Pressure Sensors in the Measurement of Intracranial Pressure," Tsinghua Sci. Technol., vol. 19, no. 2, pp. 161–167, 2014.
10. M. L. Ya and N. Soin, "Design and Optimization of a Low-Voltage Shunt Capacitive RF-MEMS Switch," no. April, 2014.
11. P. Ananthakumar and M. Ramesh, "Optimization of Turning Process Parameters Using Multivariate Statistical Method-PCA Coupled with Taguchi Method," Int. J. Scintific Eng. Technol., vol. 2, no. 4, pp. 263–267, 2013.
12. I. Asiltürk and H. Akkuş, "Determining the effect of cutting parameters on surface roughness in hard turning using the Taguchi method," Meas. J. Int. Meas. Confed., vol. 44, no. 9, pp. 1697–1704, 2011.
13. H.-C. Liao, "Multi-response optimization using weighted principal component," Int. J. Adv. Manuf. Technol., vol. 27, no. 7–8, pp. 720–725, 2006.
14. Y. Tian, Z. Wang, Y. Jia, S. Li, B. Wang, S. Wang, L. Sun, J. Zhang, J. Chen, and R. Jiang, "Intracranial pressure variability predicts short-term outcome after intracerebral hemorrhage: a retrospective study.," J. Neurol. Sci., vol. 330, no. 1–2, pp. 38–44, Jul. 2013.

15. J. Bellner, B. Romner, P. Reinstrup, K.-A. Kristiansson, E. Ryding, and L. Brandt, "Transcranial Doppler sonography pulsatility index (PI) reflects intracranial pressure (ICP).," Surg. Neurol., vol. 62, no. 1, pp. 45–51; discussion 51, Jul. 2004.
16. Stephanie Fraley, M. Oom, B. Terrien, and J. Zalewski, "Design of experiments via taguchi methods: orthogonal arrays," 2006. [Online]. Available: https://controls.engin.umich.edu/wiki/index.php/Design_of_experiments_via_taguchi_methods:_orthogonal_arrays. [Accessed: 09-May-2015].
17. M. S. Ranganath, "Optimization of Process Parameters in Turning Operation Using Taguchi Method and Anova: A Review," vol. 1, no. 1, pp. 31–45, 2013.
18. M. P. Orthner, S. Buetefisch, J. Magda, L. W. Rieth, and F. Solzbacher, "Development, fabrication, and characterization of hydrogel based piezoresistive pressure sensors with perforated diaphragms," Sensors Actuators, A Phys., vol. 161, pp. 29–38, 2010.
19. K. J. Suja, S. Mathew, and R. Komaragiri, "Optimized Design of a Stacked Diaphragm MEMS Pressure Sensor for Tsunami Warning System," pp. 346–351, 2013.
20. G. Bistué, J. G. Elizalde, S. García-Alonso, E. Castaño, F. J. Gracia, and a. García-Alonso, "A design tool for pressure microsensors based on FEM simulations," Sensors Actuators A Phys., vol. 62, no. 1–3, pp. 591–594, 1997.
21. K. Singh, R. Joyce, S. Varghese, and J. Akhtar, "Physical Fabrication of electron beam physical vapor deposited polysilicon piezoresistive MEMS pressure sensor," Sensors Actuators A. Phys., vol. 223, pp. 151–158, 2015.
22. T. Jeong, "Design and modeling of sensor behavior for improving sensitivity and performance," Measurement, vol. 62, pp. 230–236, 2015.

Author: Siti Hajar Abd Rahman
Institute: University of Malaya
Street: Jalan Universiti
City: Wilayah Persekutuan Kuala Lumpur
Country: Malaysia
Email: sitiahajar.abdrahman@gmail.com

Removal of Ocular Artifacts in EEG Signals Using Adapted Wavelet and Adaptive Filtering

M. Aniket[1], L. Arpit[1], and B.N. Krupa[2]

[1] Department of Telecommunication Engineering, PES Institute of Technology, Bengaluru, India
[2] Department of Electrical and Electronics Engineering, PES Institute of Technology, Bengaluru, India

Abstract— **EEG is the recording of electrical activity along the scalp. The raw EEG signals generally contain a large amount of unwanted artifacts. The movement of eyeballs and eye blinks introduce artifacts, also known as electrooculogram (EOG) signal. In this study, the EEG signals were recorded from 24 subjects while they were performing a particular task in a noise free environment. For the removal of these ocular artifacts (OAs), pattern matching using a continuous adapted wavelet followed by adaptive filtering technique has been proposed to obtain a refined EEG signal. This new approach is able to identify multiple eye blinks at different instants within the specified interval of time. An existing method has also been applied to all the recorded signals. The proposed method is compared with this existing approach. The performance evaluation of these methods is done using Mean Square Error (MSE) and Signal-to-Noise ratio (SNR). This showed that the average SNR of the proposed method is 48.43% higher than the existing method.**

Keywords— **EEG, Pattern matching, Adapted Wavelet, Adaptive filtering, Wavelet Transform, Ocular Artifacts (OAs), SNR, MSE.**

I. Introduction

EEG measures the voltage fluctuations within the neurons of the brain. EEG signals are susceptible to various forms of undesired signals, which make it difficult for the researchers to analyze and interpret the raw data. EEG signals are non-stationary and non-linear and have very low SNR ratios [1]. Based on their origin, artifacts can be of two types. Biological artifacts caused by cardiac activities referred to as electrocardiogram (ECG), movements caused by muscle contraction commonly known as electromyogram (EMG), and ocular activities caused by eyeball movement and blinking of eyes called electrooculogram (EOG) and External artifacts caused by line-interference, leads, and electrodes [2]. ECG signal is unpreventable, but also has the lowest effect on the recorded EEG signal while EMG noise can be avoided or reduced by asking the subject to find a comfortable position and relax before the start of a recording session, and by avoiding tasks that require verbal responses or large movements [3]. Eye blinks are difficult to avoid; one possibility is to ask subjects not to blink during critical periods of the task and then provide cues for periods when they can blink freely. While such strategies can effectively reduce occurrence of blinks and eye movements in critical task periods, they also have significant drawbacks [3]. As both blinking and spontaneous eye movements are involuntary in nature, withholding either of them might hinder with task performance of the subject while recording the EEG signal. Researchers have used a number of techniques to remove or attenuate such ocular artifacts (OAs) from recorded EEG either in the time domain [4] or in the frequency domain [5]. EOG signals are characterized by amplitudes which are much larger than the EEG signals, and hence, pose a serious problem for further analysis [6].

A method involving the direct subtraction of the EOG from the EEG signal is not advisable as useful information may be lost [7]. Early methods use a simple filtering concept ignoring low frequencies [8]. This is impractical as EEG is dominated by frontal slow waves in the delta band [9]. Principal Component Analysis (PCA) cannot differentiate between OAs from the EEG signal, when they possess comparable amplitudes [10]. Independent Component Analysis (ICA), developed with respect to blind source separation (BSS) is not automated and requires visual classification of the components [11]. Stationary Wavelet Transform (SWT) is a method which neither relies upon the reference OAs nor visual inspection, but as the OAs have an overlapping spectrum; it cannot be removed by SWT [12]. Adaptive filtering and Wavelet Transform (WT) are two techniques that are extensively used for denoising EEG signals [13], [14]. WT helps in transforming a signal in time domain into time and frequency localization which provides more information about the behavior of the signal [15]. The adaptive filter adjusts its coefficients to minimize the squared error between its output and a primary signal [12]. In this paper, a novel method is proposed to enhance EEG signals by removing OAs using adapted wavelet followed by adaptive filtering. Adaptive filtering is performed using a simple yet effective, Least Mean Square (LMS) algorithm. The proposed method is also compared with another approach presented in [12] for removing ocular artifacts. The performance evaluation is done by calculating MSE and SNR.

F. Ibrahim et al. (eds.), *International Conference for Innovation in Biomedical Engineering and Life Sciences*,
IFMBE Proceedings 56,
DOI: 10.1007/978-981-10-0266-3_13

II. METHODOLOGY

A. Data Acquisition

A total of 24 subjects, 14 female and 10 male, in the age group of 18-23 years were considered. The device used for recording EEG signals is NeuroSky MindWave, with a sampling frequency of 512 Hz. The subject was informed about the research, its implications, the role and tasks involved in volunteering for the same. According to Declaration of Helsinki [16], an informed consent was undertaken from all the volunteers at the time of data acquisition. The stimulus used while recording was an interactive game which is very intuitive in nature. The game was played on a Google Nexus 4 smartphone with sufficient screen resolution and brightness with pop-up notifications disabled. Initially, the subject was asked to play the game for a few minutes to get familiarized with the game and the interface of the device, before performing the actual task. The subject was then instructed to play the game for a minute regardless of the score and level. The task was performed in a room with minimum background noise, sufficient lighting and atmosphere. The subjects were instructed to keep their body movements, facial expressions to a minimum and be in a relaxed state. During the whole process, the NeuroSky Mindwave Device was mounted on the head with the electrode being placed on the frontal lobe of the subject. All the subjects were treated to the same procedure as stated above.

B. Data Enhancement Approach

Ocular artifacts are generally represented by sudden unusual spikes in the EEG signals; these are eliminated to obtain a clean EEG signal, which can be used for further processing. Another aspect that needs to be considered while removing such artifacts from the recorded signals is that the useful information is not lost from the signal. We have compared two approaches for the enhancement of EEG signals. One is the proposed method and the other is the existing method. In these two approaches we have used three important techniques; adaptive wavelets, discrete wavelet transform (DWT) and adaptive filtering. These three techniques are briefly discussed below.

a) Proposed adapted wavelet:

As shown in Fig. 1, a pattern matching technique using Continuous Wavelet Transform (CWT) is described here. CWT uses a group of wavelets by "continuously" scaling and translating a localized function called the mother wavelet [17]. For this method, an eye blink signal is recorded. From the recording, the duration of a single eye blink is found to be 310 milliseconds. This eye blink pattern is used to design an admissible wavelet. This wavelet was designed by approximating the eye blink pattern by least squares fitting using different polynomial order values starting from two. A polynomial order of 6 provided the best approximation to the eye blink pattern as shown in Fig. 2. It satisfies the existing criterion for wavelets [18], as in equation (1) and (2):

$$\int_{-\infty}^{+\infty} |\psi(t)|^2 dt < \infty \tag{1}$$

$$c_\psi = 2\pi \int_{-\infty}^{+\infty} \frac{|\Psi(\omega)|^2}{|\omega|} d\omega < \infty \tag{2}$$

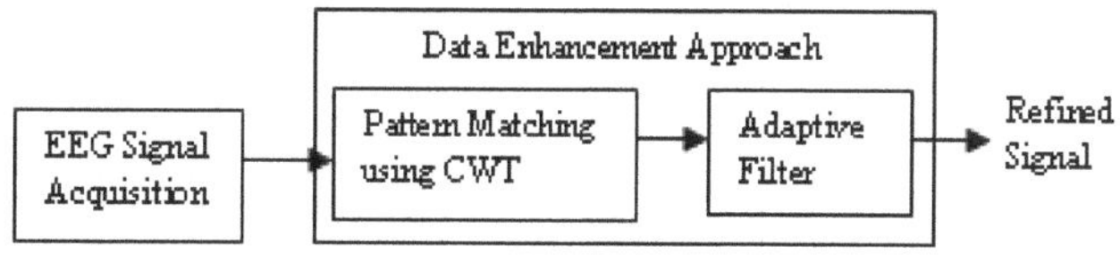

Fig. 1 Block Diagram of the Proposed Method

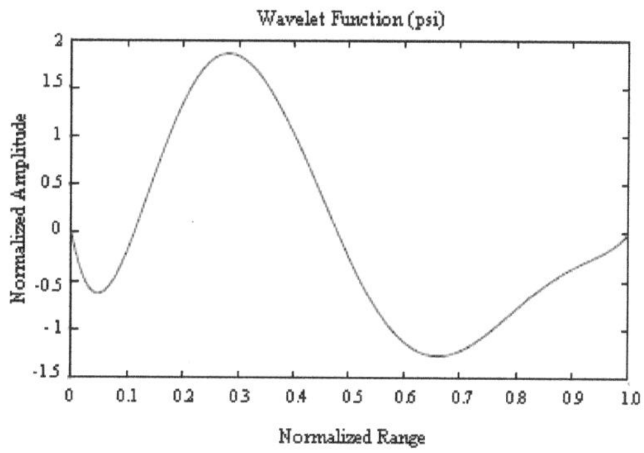

Fig. 2 Block Diagram of the Proposed Method

Where, Ψ is the Fourier transform of ψ. The first condition implies finite energy of the function, and the second condition, the admissibility condition, implies that if $\Psi(w)$ is smooth then $\Psi(0) = 0$. The function ψ is the mother wavelet. The mother wavelet is scaled (or dilated) by a factor of 'a' and translated (or shifted) by a factor of 'b' to give (equation (3))

$$\psi_{a,b}(t) = \frac{1}{\sqrt{a}}\psi\left(\frac{t-b}{a}\right) \tag{3}$$

The new wavelet has a time domain representation as the wavelet function $\psi(t)$ and it is rescaled in such a way that the pattern lies on the closed interval between 0 and 1 [19].

The CWT of the EEG signal is found using the adapted wavelet. A total of 320 levels of continuous wavelet coefficients are computed. For a particular sample, the percentage of energy for each coefficient is determined. Based on this energy calculation of each coefficient, the position and duration of the artifact is determined as shown in the scalogram (Fig. 3). A threshold is set at 25% of the maximum

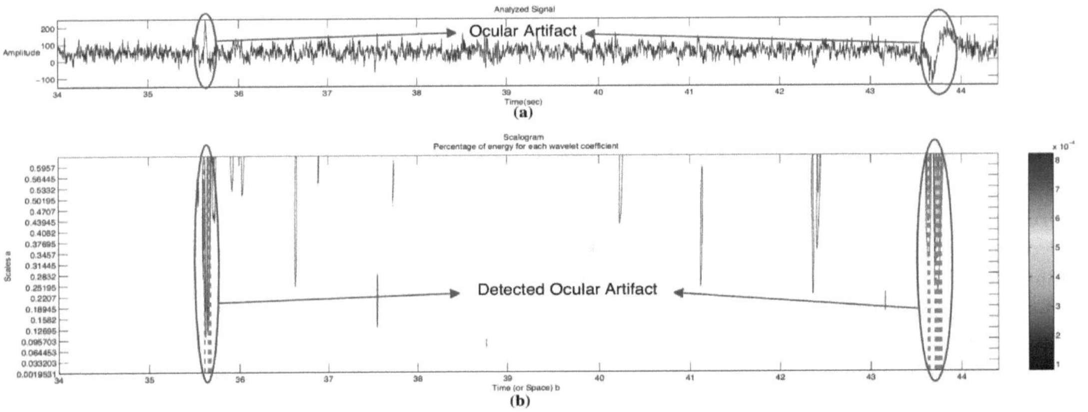

Fig. 3 (a) A portion of the recorded EEG signal (b) Detection of multiple ocular artifacts

energy value, which means any energy value for a particular coefficient above this threshold value will be detected and considered an OA. Thus, OAs' are identified using the proposed method.

b) DWT for Generating the Noise Signal (OAs):

As shown in Fig. 4, the second approach is the existing approach based on the method discussed in [12] where the EOG signal is extracted from the raw EEG signal using DWT. The signal is decomposed to seven levels (seven detailed coefficients and one approximate coefficient).The mother wavelet used is Daubechies 3, because of its simple coefficients. The coefficients of the last three levels contain the maximum information about the EOG signal. The OAs are mainly present in the lower frequency bands. A threshold value is computed on the detailed coefficients of the last three levels. The three conditions discussed in [12] are used for this purpose. Based on the threshold values, the new wavelet coefficients are determined. Then, wavelet reconstruction is applied on these new coefficients resulting in a signal which contains the information about the OAs.

c) Adaptive Filtering:

Both the approaches shown in Fig. 1 and Fig. 4 employ an adaptive filtering technique to remove the OAs and to enhance the EEG signals. The signal containing OAs is the reference signal and the recorded EEG signal is the desired signal plus noise. In the proposed method (Fig. 1) the reference OAs signal obtained with the help of adapted wavelets as explained in section *B (a)* is fed to the adaptive filter, using the LMS algorithm, along with the raw EEG signal containing OAs to obtain the desired EEG signal free of OAs. In the existing method (Fig. 4) the OAs signal obtained using DWT as explained in section *B (b)* is considered as a secondary signal for adaptive filtering. This resulted in EEG signals free of OAs. Adaptive filters do not have constant filter coefficients and no prior information of the signal and noise characteristics.

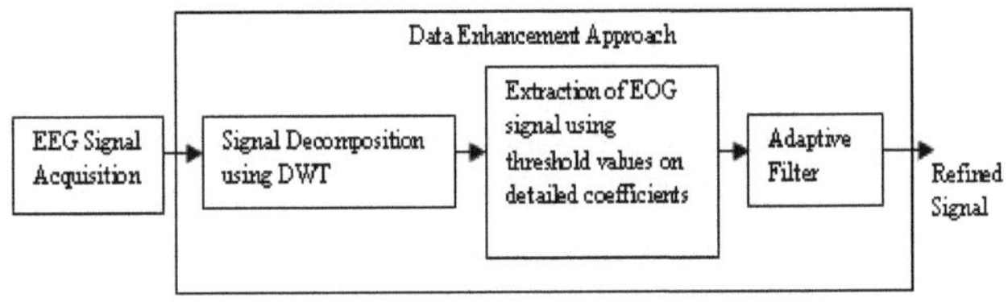

Fig. 4 Block diagram of the existing method

These filters are useful when the filter coefficients are variable, adaptable to changing conditions, and there is spectral overlap between the signal and the noise. It exploits the error signal *e(n)*, as in equation (4), produced at every instant to obtain the adaptive filter coefficient vector *w(n)* in each successive iterations. The LMS algorithm is outlined in the following set of equations [20].

$$e(n) = d(n) - y(n) \quad (4)$$

where, *d(n)* is the primary signal containing desired signal and noise components, the *x(n)* is the input secondary signal, $y(n) = x^T(n)w(n) = w^T(n)x(n)$ is the output of adaptive filter, *w(n)* is the filters weight coefficients. Using the steepest descent method, we get the update recursive relation of filter weight coefficients, equation (5):

$$w(n+1) = w(n) + 2\mu(n)[d(n) - x^T(n)w(n)]^T \quad (5)$$

where, μ is the convergence factor. For our analysis, the filter order was 20 and the μ value was 0.08 calculated using equations (5) and (6).

$$w(n+1) = w(n) + 2\mu(n)e(n) \quad (6)$$

Outcome of both the approaches, the proposed method and the existing method, are compared using MSE and SNR as defined by equations (7) and (8), respectively.

$$\sigma^2 = \frac{1}{N}\sum_{n=1}^{N}(x_n - y_n)^2 \quad (7)$$

where, x_n is the input data sequence, y_n is the reconstructed data sequence and N is the length of data sequence.

$$SNR = 10log_{10}\frac{\sigma_x^2}{\sigma_d^2} \quad (8)$$

where, σ_x^2 is the average square value of original data sequence and σ_d^2 is the MSE.

III. RESULTS AND DISCUSSION

A total of 24 subjects were considered for this research. While the EEG signals were recorded, each subject had to perform the same task in an environment with minimum background noise for a span of one minute. The sampling frequency was 512 Hz. After the adaptive filtering stage, the refined EEG signal is obtained. Fig. 5(a) shows the waveform of the recorded EEG signal for a portion of the sample and the Fig. 5(b) shows the waveform of the refined signal obtained after passing through the algorithm of the proposed method i.e., the first approach (Fig. 1). Enhanced EEG signal obtained from the second approach i.e., the existing method (Fig. 4), is shown in Fig. 5(c).

Finally, the MSE and SNR for each refined signal are calculated. For the proposed method, the MSE values calculated using (7) fell in the range of 40.43 and 14595.86 and the average MSE was computed to be 2205.20. On the other hand, the SNR determined using (8) had values that ranged from 3.03 to 19.91 and the average SNR was found to be 9.36 (Table 1). The average number of blinks recorded for 24 subjects is 6 blinks per minute.

Similarly, for the existing method, the MSE was determined to be in the range 212.57 to 307097.85 with the average value being 32173.02 and the SNR was between 2.30 and 12.67 with an average value of 6.30 (Table 1). Both the methods are able to detect multiple OAs which may be present at different intervals in the recorded EEG signal. On comparing the average MSE value obtained between the two methods, it is observed that the proposed method as a value which is 93.15% lesser compared to the existing method.

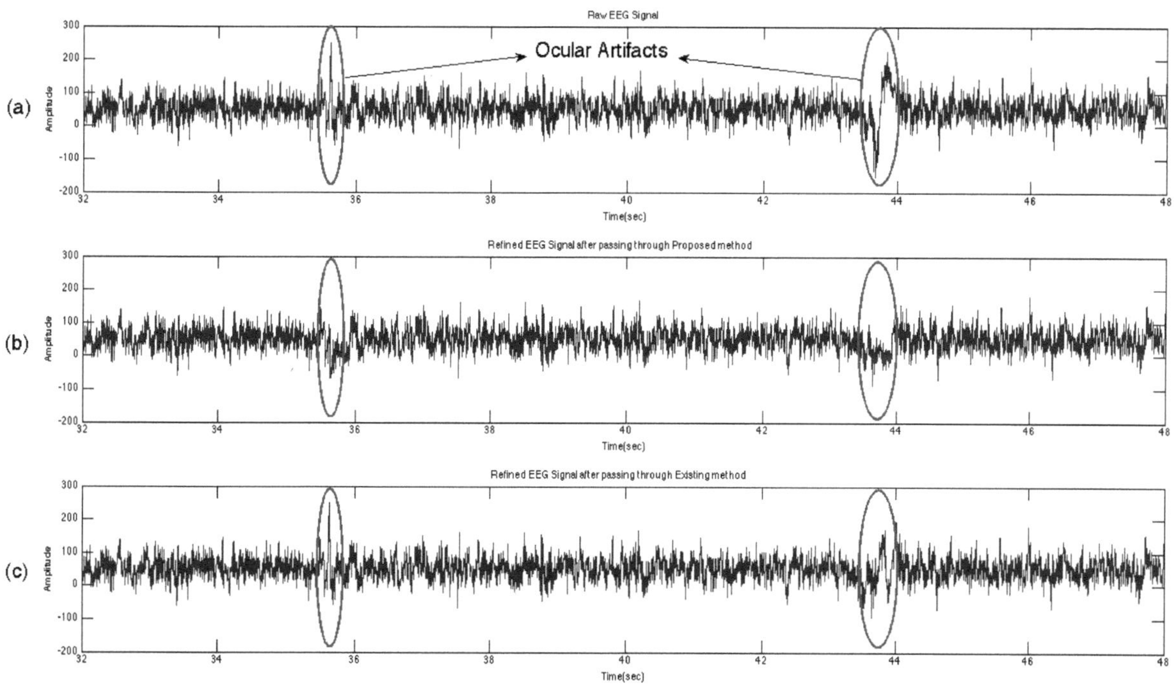

Fig. 5 5 (a) A portion of the recorded EEG signal with OAs marked (b) The refined signal after passing through the proposed method at the same instants of time (c) The refined signal after passing through the existing method at the same instants of time.

Table 1 A comparison of the proposed and existing methods

	Proposed Method	Existing Method
Average MSE	2205.2	32173.02
Average SNR	9.36	6.30

This in turn affects the average SNR value. It is seen that the proposed method has an average SNR value which is 48.43% higher than the existing method. Equation (9) is used for calculating the percentage difference. Higher the SNR for any signal better is the quality of the signal. In this aspect, the proposed method is preferred for removing ocular artifacts from the EEG signal

$$\% \, diff. = \frac{proposed\ method - existing\ method}{existingmethod} \times 100 \quad (9)$$

IV. CONCLUSIONS

In this study, two methods have been used to identify and remove OAs keeping the useful information from the raw EEG signal, intact. All the subjects had to perform the interactive task, the recording was done using Neurosky Mindwave device, and the performance evaluation was based on MSE and SNR as discussed in the above sections. The proposed method gives better results compared to the existing method in obtaining a refined EEG signal. Future work involves testing for the accuracy of the proposed method on more number of EEG signals. Another avenue that needs to be explored is that the proposed method should be applied on signals recorded using different devices when the subject is exposed to a different set of stimuli.

ACKNOWLEDGMENT

This study was supported by the Center for Intelligent Systems, PES Institute of Technology, India. We express our sincere thanks to Anil B. Murthy who helped in the Acquisition of the EEG signals and all the student volunteers who gave their consent for recording the signals.

COMPLIANCE WITH ETHICAL STANDARDS

Ethical approval: All procedures performed in studies involving human participants were in accordance with the ethical standards of the institutional and/or national research committee and with the 1964 Helsinki declaration and its later amendments or comparable ethical standards.

Informed consent: Informed consent was obtained from all individual participants included in the study.

CONFLICT OF INTEREST

The authors declare that they have no conflict of interest.

REFERENCES

1. Romo-Vazquez R, Ranta R, Louis-Dorr V, Maquin D (2007) EEG ocular artefacts and noise removal, IEEE 29th Annual International Conference, EMBS, Lyon, Aug 2007, pp. 5445-5448. DOI:10.1109/IEMBS.2007.4353577
2. Jafarifarmand A, Badamchizadeh MA (2013) Artifacts removal in EEG signal using a new neural network enhanced adaptive filter. Neurocomputing 103:222-231 DOI:10.1016/j.neucom.2012.09.024
3. Repov˘s G (2010) Dealing with noise in EEG recording and data analysis. Informatica Medica Slovenica 15:18-25
4. Gratton G, Coles MGH, Donchin E (1983) A new method for offline removal of ocular artifact. Electroencephalography and Clinical Neurophysiology 55:468-484 DOI:10.1016/0013-4694(83) 90135-9
5. Woestengurg JC, Verbaten MN, Slangen JL (1982) The removal of the eye movement artifact from the EEG by regression analysis in the frequency domain. Biological Physiology 16:127-147. DOI:10.1016/0301-0511(83)90059-5
6. Flexer A, Bauer H, Pripfl J, Dorffner G (2005) Using ICA for removal of ocular artifacts in EEG recorded from blind subjects. Neural Networks 18:998-1005 DOI:10.1016/j.neunet.2005.03.012
7. Jung TP et al. (2000) Removing electroencephalographic artifacts by blind source separation. Society for Psychophysiological Research, Psychophysiology 37:163-178.
8. Gotman J et al. (1973) Clinical applications of spectral analysis and extraction of features from electroencephalograms with slow waves in adult patients. Electroencephalography and Clinical Neurophysiology 35:225-235 DOI:10.1016/0013-4694(73)90233-2
9. Zhou W, Gotman J (2009) Automatic removal of eye movement artifacts from the EEG using ICA and the dipole model. Progress in Natural Science 19:1165-1170 DOI:10.1016/j.pnsc.2008.11.013
10. Lagerlund TD, Sharbrough FW, Busacker NE (1997) Spatial filtering of multichannel electroencephalographic recordings through principal component analysis by singular value decomposition. Journal of Clinical Neurophysiology 14:73-82
11. Wallstrom GL, Kass RE, Miller A, Cohn JF, Fox NA (2004) Automatic correction of ocular artifacts in the EEG: a comparison of regression-based and component-based methods. International Journal of Psychophysiology 53:105-119
12. Huang R et al. (2014) Artifacts reduction method in EEG signals with wavelet transform and adaptive filter. Brain Informatics and Health 8609:122-131 DOI: 10.1007/978-3-319-09891-3_12
13. Yu L (2009) EEG de-noising based on wavelet transformation, ICBBE, 3rd International Conference, Beijing, China, 2009, pp 1-4
14. Correa MAG, Leber EC (2011) Noise removal from EEG signals in polisomnographic records applying adaptive filters in cascade. Adaptive Filtering Applications, Dr Lino Garcia edn, INTECH, pp 173-196 DOI:10.5772/17219
15. Salwani M.D, Jasmy Y (2005) Comparison of few wavelets to filter ocular artifacts in EEG using lifting wavelet transform, TENCON, IEEE Region 10, Melbourne, Nov 2005, pp 1-6 DOI:10.1109/TENCON.2005.301044
16. WMA Declaration of Helsinki - Ethical principles for medical research involving human subjects (1964-2014). http://www.wma.net/en/30publications/10policies/b3/
17. Kirkove M, François C, Verly J (2014) Comparative evaluation of existing and new methods for correcting ocular artifacts in electroencephalographic recordings. Signal Processing 98:102-120 DOI:10.1016/j.sigpro.2013.11.015
18. Lee DTL, Yamamoto A (1994) Wavelet analysis: Theory and applications. Hewlett-Packard Journal, 44-54

19. Mesa H (2005) Adapted wavelets for pattern detection, Proc. 10th Iberoamerican Congress on Pattern Recognition (CIARP 2005), Havana, Cuba, Nov 2005, pp 933-944
DOI:10.1007/11578079_96
20. Dhankhar P, Khaleri S (2014) Eye blink artifact removal in EEG using adaptive FIR filter-A review. International Journal of Emerging Technology and Advanced Engineering 4:700-703

Corresponding author:

Author: Prof. Dr. B. Niranjana Krupa
Institute: PES Institute of Technology
Street: 100 Feet Ring Road
City: Bengaluru
Country: India
Email: bnkrupa@pes.edu

Real Time Eye Blink Artifacts Removal in Electroencephalogram Using Savitzky-Golay Referenced Adaptive Filtering

F. Abd Rahman[1,2] and M.F. Othman[1]

[1] Malaysia-Japan International Institute of Technology (MJIIT) & Centre for Artificial Intelligence and Robotics (CAIRO), Universiti Teknologi Malaysia (UTM), Kuala Lumpur, Malaysia

[2] Department of Electrical and Computer Engineering, Faculty of Engineering, International Islamic University of Malaysia, Kuala Lumpur, Malaysia

Abstract— **Eye blink artifacts cause a major problem to electroencephalograph (EEG) signals since they introduce serious distortion to the signals. The implementation of algorithm to eliminate those artifacts automatically and at real time becomes highly important when considering the recent focus on portable EEG applications. In this paper, we propose a real-time eye blinks artifact removal using adaptive filtering without an additional reference electrode. The proposed method utilizes a Savitzky-Golay (SG) filter to estimate a blink component from the noisy EEG signals, and then is used as a reference in adaptive noise cancellation system. We demonstrate the reliability of the proposed method by using both simulated and real EEG datasets. The estimated blink artifacts by the proposed method are compared to the original blink signals measured by electrooculograph (EOG). The performance of the filters in terms of removing the blink components from the noisy signals are also compared between the proposed SG referenced and a conventional EOG referenced adaptive filtering. The results show that the proposed method is able to estimate the blink artifacts with high correlation to the original blink signals and remove the artifact successfully.**

Keywords— **Eye blink artifact, adaptive filtering without reference, Savitzky-Golay filter, Electroencephalogram (EEG).**

I. INTRODUCTION

The contamination of physiological artifacts such as pulse, muscle and ocular artifacts in electroencephalogram (EEG) signal is inevitable. These artifacts reduce the quality of the brain signals which could bias the analysis results. Electrooculograph (EOG) artifact, particularly, is a major problem that causes distortion in EEG signal generated by blinks and eye movements. The magnitude of this artifact can goes up to ten and hundred times larger than the original EEG signals [1]. In many cases, artifacts are handled by discarding the contaminated segments based on a visual judgment from the expert. However, this could bring a great loss of valuable data and it is a long process, thus, is not suitable for long term recording such as sleep stage study. Hence, the process to remove or suppress the artifacts while leaving the relevant EEG signals become highly important.

Blind source separation (BSS) is a popular approach used in separating the artifacts from a noisy signal. Among the BSS approaches, independent component analysis (ICA) is one of the method used for EEG artifacts removal which was first proposed by Makeig *et. al* [2]. It has become a widely accepted method, especially for blink removal owing to its good performance in separating the artifact component from noisy EEG signals. However, ICA requires visual inspection in order to choose and to remove the artifact components. Thus, several approaches which are a combination of ICA with other techniques have been proposed to overcome those limitations [3]–[5]. Besides, ICA entails a data from multi channel EEG to perform the separation of independent components, making it difficult to be implemented in single-channel EEG. In recent years, the focus is on the wearable EEG device for real-time applications. Thus, single-channel EEG is more reliable for its simplicity, light weight and suits the real environment.

Adaptive noise cancellation system is an approach for noise removal using adaptive filtering [6] which can be operated in real time and available for both single and multi channels EEG. It is suitable for non-stationary signals through its ability to adjust itself according to the changing environment. The filter suppresses the noise by subtracting a reference from the measured EEG signal based on linear regression process. A reference is a signal which is correlated with the artifact and carries the properties of the artifact. Usually, this reference is taken by placing an additional electrode to record the artifact signal, and delivers it as an input reference to the filter. For example, a method of using a cascade of three adaptive filters was proposed [7]. This method uses three different additional channels to cancel out the ocular, muscle and pulse artifacts. Recently, a method of using a camera based or eye tracker based reference to eliminate blinks and movement artifacts have been proposed [8], [9]. The filtering results of using a camera based reference outperform the EOG electrode based reference. Nevertheless, the requirement for those additional sensors or channel in adaptive filter brings both the hardware size and cost problems. This is unfavorable for recent applications that demand on compactness and portability in the real environment. On top of that, adaptive filter requires different reference channel for different type of artifacts.

F. Ibrahim et al. (eds.), *International Conference for Innovation in Biomedical Engineering and Life Sciences*,
IFMBE Proceedings 56,
DOI: 10.1007/978-981-10-0266-3_14

In this paper, we present a method of removing the eye blink artifacts on real time using adaptive filtering without an additional reference channel. We propose a method to estimate eye blink from single-channel EEG by adopting a Savitzky-Golay (SG) filter. The output from the SG filter is then used as a reference input in adaptive noise cancellation system. We demonstrate the potential of this method by using a simulation and a real EEG dataset. The proposed method is then compared with the conventional method that use EOG channel for reference. This paper is organized as follows. Section II discusses the overview of the adaptive noise canceling system and the implementation of SG filter in the system, followed by the description on how to prepare the dataset. The result is then presented in Section III. Finally, Section IV concludes the findings.

II. METHODOLOGY

A. Savitzky-Golay Filter

A Savitzky-Golay filter is a method of data smoothing based on local least squares polynomial approximation, proposed by Savitzky and Golay in 1964 [10]. This filter was originally developed for noise removal in analytical chemistry, but then has been found to be attractive in the area of signal processing and image processing. The filter output is computed as shown in the equation below.

$$Y_j = \sum_{n=-n_L}^{n_R} C_i y_{j+i} \qquad (1)$$

Here, n_L is the number points used to the left of a data point i, and n_R is the number of points to the right after point i. C_i is a fixed value taken from a set of weighting factors developed by Savitzky and Golay. n describes the window size where $n=n_R+n_L$. The center value is replaced with the output from the model. The major advantage of this method is that it has a property to preserve the peak shape, such as peak height and width, which are often affected when adopting other smoothing techniques. This feature is advantageous in estimating the magnitude and frequency of eye blink signal. The measured EEG signal will be filtered using the above equation, and the filtered value will be set as the input reference for adaptive noise cancellation system.

B. Adaptive Filtering

The use of conventional fixed filtering methods, such as low-pass filter and high-pass filter to attenuate the artifacts is not really recommended since there is a possibility of overlapping frequency spectra between EEG signal and the artifacts [11]. Adaptive filtering is one of the best choices to overcome a time-invariant problem of those non-adaptive approaches. An adaptive algorithm updates the filter coefficient by a feedback process to make the output as close as possible to the desired response. In this study, we use ANFIS as the adaptive algorithm. Figure 1 illustrates a schematic of adaptive filtering for adaptive noise canceller system. From the diagram, the signal source is the original recorded EEG signal with blink artifacts, d(n). The adaptive filter aims to estimate clean EEG signal e(n), by predicting the artifacts (blinks) components, y(n) from the measured EEG signal, d(n). For this, we require a reference input, x(n) which correlates with the blinks artifact tend to be removed. Thus, here we use the smoothed values from SG filter as an artifact reference to estimate the eye blinks.

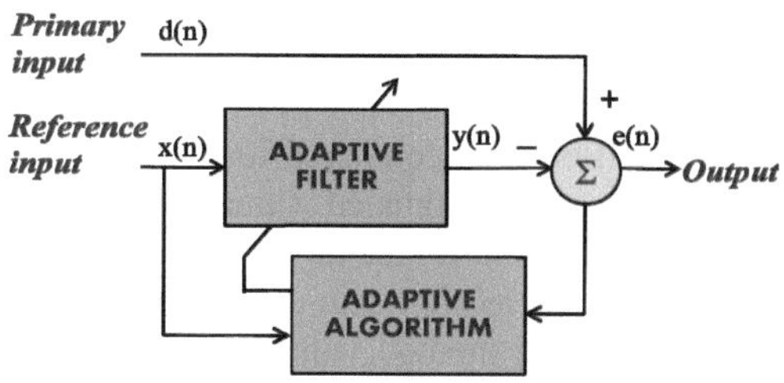

Fig. 1 Adaptive Noise Cancellation

C. Simulation Dataset

The algorithm is tested using simulation dataset and real EEG dataset with eye blink artifacts. For simulation dataset, we first simulate the clean EEG signal using a MATLAB function proposed by Yeung *et.* al [12]. The blink artifacts are simulated by a sinusoid with random amplitudes and shapes, added with 20 dB of white noises. The simulated blink-contaminated EEG signals were then obtained by combining the simulated clean EEG signal with the simulated eye blink. For a real EEG dataset, the recordings were performed using a cap type multi-channel EEG, g.tec, with 256 Hz of sampling rate. The device measured 14 locations of EEG channel plus 1 vertical EOG signal. For dataset preparation, we only used a measured signal at Fp1, since this channel is considered as a severely affected channel by blinks. The subjects were asked to blink according to metronome sound that will be played every 5 seconds. The total length of the experiment is 55 seconds.

D. Performance Analysis

For validation purpose, we compare the results of the proposed method with the conventional method which uses the measured EOG signals as artifact reference. We measure the similarity between the corrected signal of the proposed method and the reference (estimated by EOG referenced) using signal to noise ratio (SNR), mean squared error (MSE) and correlation coefficient. SNR and MSE are calculated as follows:

$$SNR = 10\, log \frac{\sum_{i=1}^{N} x(i)^2}{\sum_{i=1}^{N}(x(i) - \bar{x}(i))^2} \quad (2)$$

$$MSE = \frac{1}{N}\sum_{i=1}^{N}(x(i) - \bar{x}(i))^2 \quad (3)$$

where $x(i)$ is the target EEG signal (corrected EEG by EOG reference) and $\bar{x}(i)$ is the corrected EEG signal by the proposed method. N represents the length of the signal.

III. RESULTS

The filtering results of the simulated signals are shown in Figure 2. The estimated eye blink shown in figure indicates no significant differences compared to the simulated blink signals. Similarly, the estimated signal by the proposed method successfully follows the target signal and eliminates the blink components as shown in the figure. The SNR and MSE of the proposed method with reference to the target signal are 20.23 and 4.60×10^{-6}, respectively.

The result for the real dataset is shown in Figure 3. We compared the estimated blinks and estimated FP1 signal after being corrected by the proposed and conventional method. From the figure, it can be seen that the performance of both filters are comparable in terms of estimating the eye blinks and correcting the measured EEG signals. The figure indicates that the estimated blink using the SG filter is very similar to the one that use the conventional approach. The SNR and MSE value between EOG referenced filter and SG referenced filter is 16.98 and 3.39×10^{-5} respectively. Next, we analyzed the result using correlation coefficient at zero lag to measure the similarity between the reference signal and the estimated signals from both filters. The correlation coefficient indicates the similarity degree between two signals where one is the highest correlation. The coefficient values between the measured EEG signal and estimated EEG obtained using SG reference and EOG reference filtering are presented in Table 1. From the table, the estimated FP1 from both methods present low correlation with measured FP1 showing that the filter largely removed the blink component from the noisy signal. Meanwhile, high correlation can be seen in estimated eye blinks prove that the estimated eye blinks are very similar to the original blinks signal measured by EOG electrode. The proposed method presents close values to the values presented by conventional method which confirm its reliability in removing the artifact even though without an additional artifact channel reference.

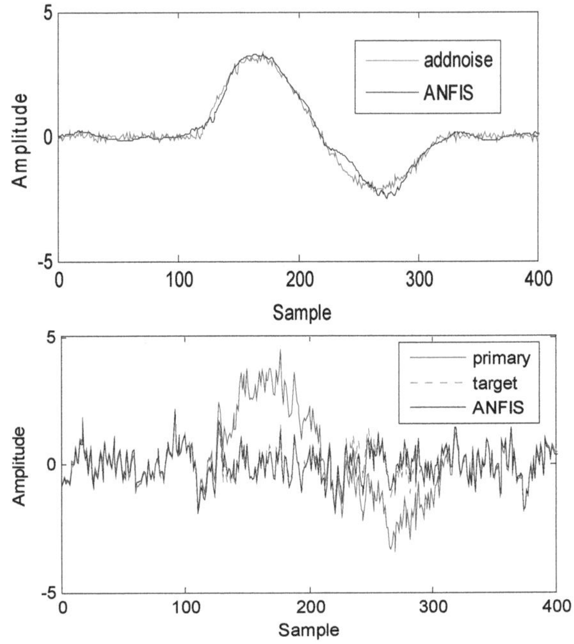

Fig. 2 Performance of the proposed method on simulated signals; simulated eye blink and estimated eye blink by ANFIS (top), simulated EEG signal and estimated signal by ANFIS (bottom).

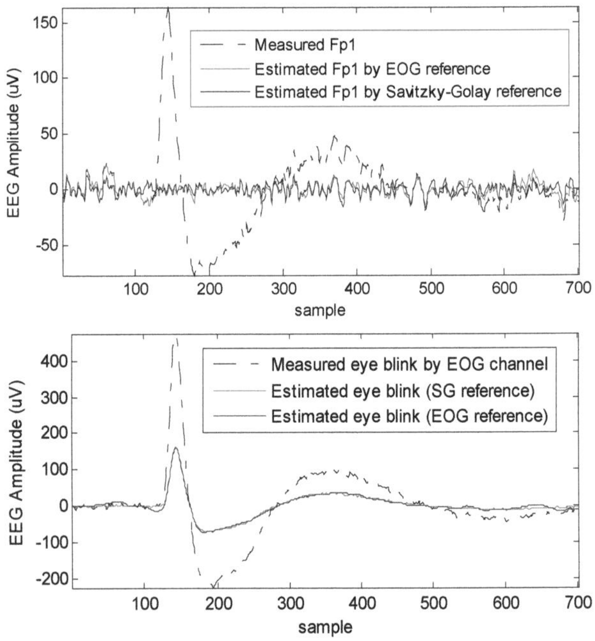

Fig. 3 Performance comparison of adaptive filters on real EEG dataset; measured and estimated EEG signal (top), measured and estimated eye blink signal (bottom).

Table 1 Correlation coefficients on both methods

Compared Signals	*Correlation Coefficient*	
	Savitzky-Golay referenced	EOG referenced
Measured FP1 / Estimated FP1	0.1478	0.1892
Measured EOG / Estimated eye blink	0.9899	0.9984

IV. CONCLUSIONS

The major advantage of adaptive filter is its availability for online processing and single-channel data. Both conditions are vital for recent applications on portable EEG. The requirement for reference channel in adaptive filter however, is something need to be considered. Hence, the filter which can be operated without an additional reference channel is very promising to tackle both of the size and cost problem for portable EEG. The general concept of SG filter is to filter out high frequency components while leaving the lower frequency components in the output. It is a type of low-pass filter, thus we found it is suitable to be used in estimating the blink component which is concentrated on lower frequencies. Nevertheless, there are some limitations in this study. Since the proposed SG filter is a type of low-pass filter, a verification on lower frequency in the corrected EEG signals need to be conducted to ensure minimal loss of relevant data on low frequency. The second limitation is the validation analysis is performed by assuming that the corrected signal using EOG reference is a clean signal in order to measure the value of SNR and MSE. In real environment, the real value of EEG signal is unknown, thus a systematic artificial model is required to measure the percentage of real EEG signal retained in the corrected signal.

In this paper, we proposed an eye blinking artifact removal method using Savitky-Golay filter referencing in adaptive filtering. We performed the proposed method for blink artifact removal using single-channel EEG signal (FP1). The proposed method has successfully estimated the blink signal with high correlation to the original blink signal recorded by EOG electrode. We then adopted ANFIS in adaptive noise cancellation system to remove the blink component from measured EEG signal, by applying the estimated blink signal generated by Savitzky-Golay filter as artifact reference. The filter performance using the proposed referencing method is then compared with the ANFIS filter using EOG reference. We verified the effectiveness of the proposed method on both simulation dataset and real dataset.

ACKNOWLEDGMENT

This research is supported by Ministry of Education Malaysia (MOE), Universiti Teknologi Malaysia (UTM) and Malaysia-Japan International Institute of Technology (MJIIT) under grant vot. no. R.K430000.7743.4J011. We would like to thank Associate Professor Dr. Yasue Mitsukura from Graduate School of Science and Technology, Keio University and all her laboratory members for the provision of EEG data.

CONFLICT OF INTEREST

The authors declare that they have no conflict of interest.

REFERENCES

1. Croft R J, Barry R J (2002) Issues relating to the subtraction phase in EOG artefact correction of the EEG. Int J Psychophysiol 44(3):187–195
2. Jung T P, Humphries C, Lee T W et al. (1998) Extended ICA removes artifacts from electroencephalographic recordings. Adv in Neural Information Processing Systems 10: 894–900
3. Daly I, Billinger M, Scherer R et al. (2013) On the Automated Removal of Artifacts Related to Head Movement From the EEG. IEEE Trans on Neural Sys and Rehab Eng 21(3):427–434
4. Li Yandong, Ma Z, Lu W et al. (2006) Automatic removal of the eye blink artifact from EEG using an ICA-based template matching approach. Physiol Meas 27: 425-436
5. Hamaneh M B, Chitravas N, Kaiboriboon N et al. (2014) Automated removal of EKG artifact from EEG data using independent component analysis and continuous wavelet transformation. IEEE Trans Biomed Eng 61(6):1634–1641
6. Haykin S (1996) Adaptive filter theory, Third. New Jersey, USA.
7. Correa A G, Laciar E, patino H D et al. (2007) Artifact removal from EEG signals using adaptive filters in cascade. J Phy:Conf Series 9:1-10
8. Noureddin B, Lawrence P D, Birch G E (2012) Online Removal of Eye Movement and Blink EEG Artifacts Using a High-Speed Eye Tracker. IEEE Trans Biomed Eng 59(8):2103–2110
9. Kierkels J J M, Riani J, Bergmans W M et al. (2007) Using an eye tracker for accurate eye movement artifact correction. IEEE Trans. Biomed. Eng 54(7):1256–1267
10. Savitzky A, Golay M J E (1964) Smoothing and Differentiation of Data by Simplified Least Squares Procedures. Anal Chem 36(8): 1627–1639
11. Gao J, Sultan H, Hu J et al. (2010) Denoising nonlinear time series by adaptive filtering and wavelet shrinkage: A comparison. IEEE Signal Process. Lett 17: 237–240
12. Yeung, Bogacz R, Holroyd C B et al.(2004) Detection of synchronized oscillations in the electroencephalogram: An evaluation of methods. Psychophysiol 41(6): 822–832

Author: Mohd Fauzi Othman
Institute: Malaysia-Japan International Institute of Technology (MJIIT), Universiti Teknologi Malaysia
Street: Jalan Sultan Yahya Petra, 54100
City: Kuala Lumpur
Country: Malaysia
Email: mdfauzi@utm.my

Classification of Image Processing Software Tools for Cardiovascular Image Analysis

K.A. Rasoul Banaeeyan[1], Y.K. Chiam[1], Z.H. Azizul[2], T.K. Chiew[1], S.H. Ab Hamid[1], and T. Thasaratharajah[1]

[1] Department of Software Engineering, Faculty of Computer Science and Information Technology, University of Malaya, 50603 Kuala Lumpur, Malaysia
[2] Department of Artificial Intelligence, Faculty of Computer Science and Information Technology, University of Malaya, 50603 Kuala Lumpur, Malaysia

Abstract— **Cardiovascular disease (CVD) is one of the leading causes of death in the modern world. Cardiac image analysis data is crucial in biomedical modelling and simulation to predict and diagnose CVD. Despite the importance of cardiovascular image analysis (CVIA) software tools for prediction, diagnosis, and therapy of CVD, to the best of our knowledge, there is no comprehensive review and a classification framework for the CVIA tools. In this paper, we review the literature related to the application of software tools for CVIA. In order to determine how image processing software tools are used for the CVIA diagnosis and prediction, this paper reviews the applications and features of these tools, through a survey of literature and the classification of articles, from January 2005 to December 2014. Keyword indices and article abstracts were used to identify 86 articles concerning image processing software tools for the CVIA. In this research, we review and classify 66 identified software tools for CVIA, with respect to the following four areas: (1) supported cardiovascular application, (2) cardiovascular imaging dimensionality, (3) cardiovascular imaging modality, and (4) post-processing ability. The results of our review and classification provide guidelines for utilization of CVIA software tool and future research on them.**

Keywords— **medical image processing, software tool, cardiovascular image analysis, cardiovascular disease.**

I. INTRODUCTION

CVD is reported as the one major cause of death globally [5]. Applied medical research highly depends on the diagnosis and outcome prediction of cardiac images which contributed to the growing number of CVIA tools in the market. Researchers often sought these tools to enlighten the tasks from eliciting useful diagnostics information investigate the gist of a cardiovascular (CV) image to constructing patient-specific simulation models. Examples of CVIA software tools are Segment [1-2] and Slicer or 3D Slicer [3].

Over the last decade, biomedical researchers and experts have employed and developed various CVIA software tools to meet their requirements. Although there are many CVIA tools available, the lack of classification scheme for these tools somewhat hinders its wide usage. As each tool comes with different functionality and applicability, biomedical researchers and experts find it hard to choose an appropriate one among the various tools which fits the purpose of their use. Inappropriate selection of CVIA tools can cause major effect on the analysis and prediction of CVD.

Apart from lacking classification, the process of comparing and selecting a tool from the abundant number of tools itself, is a difficult process. Biomedical researchers may not possess sufficient technical knowledge to study the technicality of every CVIA tools. One needs to either try using the tool or study the online documentation/tutorials. This consumes time as there are CVIA tools available.

In this study, we reviewed and classified academic journal articles on CVIA software tools that were published for the past decade, in order to gain insight on CVIA software tools. CVIA software tools are generally categorized based on (1) applications, (2) supported image dimensionality, (3) supported image modality, and (4) post-processing.

This paper is organized as follows. Section II discusses the research methodology used in this study. Section III presents the analysis of CVIA software tools and their classifications. The implications of this study are discussed in Section IV. Section V presents conclusions and future work.

II. RESEARCH METHODOLOGY

This research aims at reviewing the existing CVIA software tools by inquiring the published journal articles. Time span starts from 1st January 2005 until 31st December 2014.

A. Search Strategy

We searched the following digital databases to identify a list of papers for this review: IEEExplore, Science Direct (SD), ACM, Web of Science (WoS), PubMed and SpringerLink (SL). The next step is identifying keywords for electronic search. The literature search was based on thir-

F. Ibrahim et al. (eds.), *International Conference for Innovation in Biomedical Engineering and Life Sciences*,
IFMBE Proceedings 56,
DOI: 10.1007/978-981-10-0266-3_15

teen keywords, constructing one query string, combined by "AND" and "OR" logical operators as follows:

Query string: ("Image Processing" OR "Image Analysis" OR "Medical Imaging") AND (Software OR Tool OR Framework) AND ("Myocardial Infarction" OR Cardiac OR "Coronary Artery Disease" OR "Heart Attack" OR "Cardiac Ischemia" OR Angina OR "Cardiovascular Disease").

The search initially produced, approximately, 28,000 articles across different digital libraries as shown in Table 1. The number of papers was reduced to 86 after considering the inclusion and exclusion criteria.

Table 1 Search results categorized by publication date and digital library

Year	IEEE	SD	ACM	WoS	PubMed	SL	Total
2005	96	949	34	15	97	339	1,530
2006	112	1,199	42	27	94	410	1,884
2007	113	1,257	95	31	115	451	2,062
2008	152	1,344	92	35	120	550	2,293
2009	186	1,505	115	55	134	660	2,655
2010	181	1,652	123	30	147	724	2,857
2011	213	1,768	133	36	162	883	3,195
2012	243	2,022	131	38	148	968	3,550
2013	298	2,096	148	37	151	1,051	3,781
2014	197	2,444	132	43	100	1,233	4,149
Total	1,791	16,236	1,045	347	1,268	7,269	27,956

B. *Selection Criteria*

The inclusion and exclusion criteria are as follows:

- Include articles published in computer science or medicine, which mentioned CVIA software tool, and written in English.
- Include only the tools that are still available.
- Exclude conference papers, dissertations, textbooks and unpublished working papers.

Each article was carefully reviewed to identify the image processing software tools for CVIA. Fig. 1 shows the selection process. Although this search was not exhaustive, it serves as relevant studies for understanding CVIA image processing tools. Initially, 109 CVIA software tools were identified. After checking the availability of the tools (either as a research prototype, open source or a commercial product), only 66 tools were shortlisted for this review. The shortlisted CVIA tools are categorized based on a classification framework described in the next section.

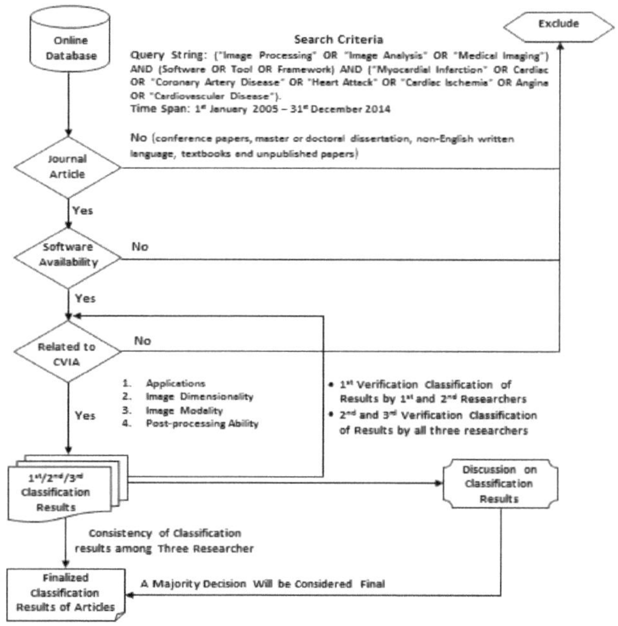

Fig. 1 Selection of CVIA tools

III. CLASSIFICATION FRAMEWORK

In this section, we provide a classification framework to study the CVIA software tools (see Fig. 2). This classification scheme consists of four main categories: (1) applications, (2) supported image dimensionality, (3) supported image modality, and (4) post-processing. We then apply this framework to review CVIA software tools. The following subsections discuss each classification categories together with the shortlisted tools in detail.

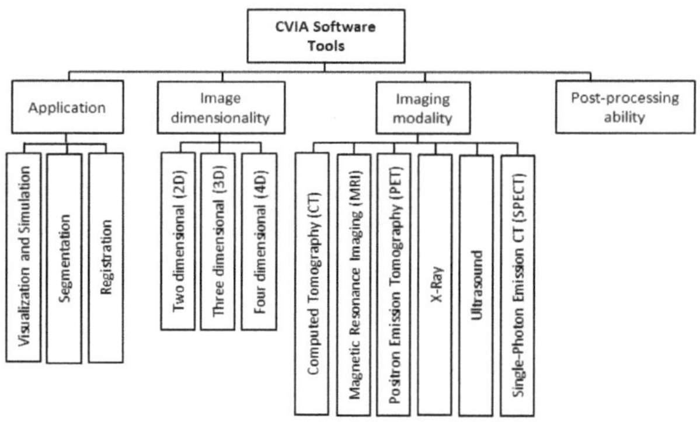

Fig. 2 Classification Framework

A. Application

Software tools are indispensable for the analysis of data in CV images, which require complex computation, or extraction of quantitative information. The following applications of CVIA software tools were analyzed.

Visualization and simulation: Visualization of CV images is used to determine the quantitative information about the properties of the heart and vessels and their functions that relate to and are affected by diseases. CV images can be simulated from digital models of the human heart for a variety of applications in research and industry. Our review recorded 32 citations for CVIA tools applied for visualization and simulation. The tools are NCAT, INRIA GHS3D, Argus, Proteus, ITK, VTK, MASS, Matlab, Syngo, MunichHeart, CTA, Patran, EchoPac, MeVisLab, ANSYS CFX, ANGIO Mentor, VMTK, SMARTVis, LabView, GE, Chaste, SOFA, SimVascular, GIMIAS, ADINA, CIM, 4D Cardio-View, CARP, Cubit, CHeart, 3mensio Structural Heart, Heart Score View.

Segmentation: CV image segmentation is the process of partitioning a cardiac into a few segments that are simpler and easier to analyze. This application focuses on analyzing the anatomical structure of heart, identifying region of interest (RoI), abnormalities, measure heart volume, and do treatment planning before therapy. Our review recorded 26 citations for CVIA tools used for segmentation. The tools are Vessel View, ITK, VTK, Analyze, BioImage, HeAT, Matlab, TrakEM2, CASS-MRV, CTA, MeVisLab, VVI, QMass, Live-vessel, Vitrea, ITK-SNAP, 3D Slicer, Seg3D, SimVascular, Mimics, SEGMENT, CardioViz3D, MITK, 3mensio Structural Heart, Myometrix, Heart Score View.

Registration: CV image registration is converting different sets of data into one coordinate system and is used to align multiple scenes into a single integrated image. Registration is considered successful once the corresponding points from different input views are successfully mapped together. In order to make the registration beneficial in medical diagnosis or treatment, the mapping must be clinically meaningful. This is normally achieved by a system that includes registration as a sub-system. However, only 7 citations of CVIA tools used for registration were recorded in our review and the tools cited are NCAT, Proteus, ITK, MASS, Patran, IRTK, and BestPhase.

B. Image Dimensionality

Different CVIA software tools produce different image dimensionalities (see Table 2). Three types of image dimensionalities reported for specific software tools include two dimensional (2D), three dimensional (3D), and four dimensional (4D). 4D type refers to those 3D images which are taken over time as another extra dimension (3D+T).

Table 2 List of identified software tools with different image dimensionalities.

Dimension	Software Tools
2D	All the tools.
3D	NCAT, Vessel View, Proteus, ITK, VTK, MASS, BioImage, Matlab, Syngo, MunichHeart, MeVisLab, Mentor, VMTK, ITK-SNAP, 3D Slicer, Seg3D, SMARTVis, ADINA, SEGMENT, CIM, Bouguet, MVTa, 4D Cardio-View, MITK, Cubit, CHeart, Myometrix, SPECT, CoroEval
4D	NCAT, ITK, VTK, HeAT, 4D Cardio-View

C. Image Modality

Visual representations of the interior of the heart and vessels are crucial for CV imaging to reveal internal structures of the heart hidden by the skin and bones, as well as to diagnose and treat CVD. Different image modalities are developed and utilized by experts in CVIA, and among them, six imaging modalities are covered which include the MRI, CT, SPECT, Ultrasound, X-Ray, and PET. We analysed the usage of the tools in terms of image modalities and found out that the MRI and CT are given higher priority with the availability of 30 and 26 software tools respectively. The Ultrasound and SPECT are made available in 7 and 8 software tools respectively whereas only 1 software tool is available for each X-Ray and PET modalities. The full list of the CVIA tools is presented in Table 3.

Table 3 List of identified software tools supporting different image modalities.

Modality	Software Tools
MRI	INRIA GHS3D, Argus, Vessel View, Proteus, ITK, VTK, MASS, Analyze, BioImage, HeAT, Matlab, CASS-MRV, Patran, MeVisLab, QMass, ITK-SNAP, 3D Slicer, Seg3D, SMARTVis, Diagnosoft, IRTK, SOFA, SimVascular, Mimics, Functool, GIMIAS, ADINA, SEGMENT, CIM, CoroEval
CT	Argus, Vessel View, Proteus, ITK, VTK, Matlab, Syngo, CTA, MeVisLab, ANGIO Mentor, VMTK, Vitrea, ITK-SNAP, 3D Slicer, Seg3D, SMARTVis, GE, COR Analyzer, SimVascular, Mimics, BestPhase, SEGMENT, 3mensio Structural Heart, Aquarius, HeartBeat, CoroEval,
SPECT	NCAT, Matlab, Myometrix, Cedars-Sinai, SPECT, CoroEval, Heart Score View, QPS
Ultrasound	EchoPac, VVI, ITK-SNAP, 4D Cardio-View, MITK, Cubit, CHeart,
X-Ray	LabView
PET	MunichHeart

D. Post-processing Ability

The goal of post-processing ability in CVIA is to alter an image to improve the interpretation of diagnosis. For example, images can be post-processed to perform the quantitative image analysis. An input image can be transformed into an output image to suit the diagnosis needs of the observer [4]. In general, post-processing ability is employed to perform some tasks like overall visual analysis, fluid-structure quantification, specific ventricle and vessels assessment. Our review recorded 20 citations of CVIA tools which offered post-processing and the tools are MASS, Analyze, Syngo, CTA, Patran, EchoPac, VVI, QMass, Vitrea, Cedars-Sinai, SMARTVis, Diagnosoft, GE, COR Analyzer, SOFA, Functool, GIMIAS, VesselMap, Bouguet, MVTa, Stereology, 3mensio Structural Heart, Myometrix, SPECT, Aquarius, HeartBeat, CoroEval, QPS.

IV. DISCUSSION

A. Research Implications

Based on the distribution of CVIA software tools analyzed by year and citations (see Fig. 3), interest in CVIA software tools will grow significantly in the future. More research is required on software tools for image registration and supporting 4D images because of rapid development of imaging devices and, resultantly, varying image modalities. Results in this research can be served for researchers and experts to find the most suitable CVIA software tools matched to their objectives. Our classification method can help biomedical researchers to select their CVIA software tools based on the application, image dimensionalities, and image modalities supported by the CVIA software tools. An appropriate software tool which is multi-purpose, multi-dimensionality, and multi-modality can save time and cost.

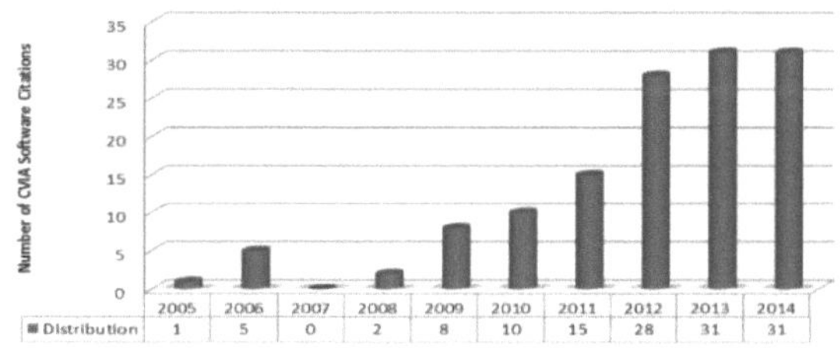

Fig. 3 Distribution of citations of CVIA software tools by year

B. Threats to Validity

This section discusses some threats that may affect validity of findings of this review. Firstly, we only searched papers that were published from January, 2005 to December, 2014. A total of 86 studies were shortlisted from data sources that are highly recognized. Therefore, if the search had been extended to cover other data sources, the validity of the results can be improved by expanding the data sources. Secondly, our findings are based on academic journal articles. If articles from conferences were included, the analysis results may be more diverse. Thirdly, our study was performed based on a search query only. This may limit the search results.

V. CONCLUSION AND FUTURE WORK

CVIA software tools are crucial for diagnosis and prediction of CVD. We have identified 86 studies that have used software tools on CVIA, published from January 2005 to December 2014. Our review recorded 66 CVIA software tools which were categorized based on the classification framework described in Section III. This review shows the current trend of CVIA-related research/application and our classification framework is useful to guide CVD researchers to select the appropriate tools to match their objectives. For future work, a more detailed feature analysis can be conducted to propose a selection method to compare specific features supported by the CVIA software tools.

ACKNOWLEDGMENT

This work is supported by University of Malaya Research Grant (UMRG), project code RP028C-14HTM.

CONFLICT OF INTEREST

The authors declare that they have no conflict of interest.

REFERENCES

1. Medviso at http://medviso.com/products/
2. Heiberg, E, Sjögren, J, Ugander, M, Carlsson, M, Engblom, H, Arheden, H (2010) Design and validation of segment – freely available software for cardiovascular image analysis. BMC Medical Imaging, 10(1):1471-2342
3. 3D Slicer at http://www.slicer.org/
4. Seeram, E, Seeram, D (2008) Image Postprocessing in Digital Radiology-A Primer for Technologists. Journal of Medical Imaging and Radiation Sciences, 39:23–41
5. Säring, D., Relan, J., Ehrhardt, J., Müllerleile, K., Bahrmeyer, a, Groth, M., & Handels, H. (2009a). Reproducible extraction of local and global parameters for functional analysis of the left ventricle in 4D MR image data. Methods of Information in Medicine, 48(2), 216–24. doi:10.3414/ME9210

Author: Yin Kia CHIAM
Institute: Department of Software Engineering, Faculty of Computer Science and Information Technology, University of Malaya
Street: Lembah Pantai
City: Kuala Lumpur
Country: Malaysia
Email: yinkia@um.edu.my

Global Based Thermal Image Registration for Diagnosis of Morphoea

L.I. Izhar[1], T. Stathaki[2], and K. Howell[3]

[1] Department of Electrical and Electronic Engineering, Universiti Teknologi PETRONAS, Tronoh, Perak, Malaysia
[2] Department of Electrical and Electronic Engineering, Imperial College London, London, UK
[3] Institute of Immunity and Transplantation, University College London, London, UK

Abstract— **Analyzing and interpreting of thermograms have been increasingly employed in the diagnosis and monitoring of diseases thanks to its non-invasive, non-harmful nature and low cost. This paper presents a thermal image analysis system based on image registration for morphoea disease diagnosis. A novel system is proposed to improve the diagnosis and monitoring of morphoea based on integration with the published lines of Blaschko. In this application, image registration based on global and local registration methods are found inevitable. A modified normalized gradient cross-correlation (NGC) method to reduce large geometrical differences between two multimodal images of different subjects that are represented by smooth gray edge maps is proposed for the global registration approach. It is shown in this paper that the NGC method outperforms phase correlation (PC) method by a lower rate of misregistration. This demonstrates that by using the gradients of the gray edge maps, the performance of the PC based image registration method can be greatly improved.**

Keywords— **Thermal imaging, Image registration, Morphoea, Blaschko's lines.**

I. INTRODUCTION

In recent years, there is an increasing number of applications which involve image analysis tasks that require integration between images from different types of sensors. In general, the integration between multimodal images can enhance the amount of information for a given image scene. For example, in medicine, combining data from different imaging modalities can be very important to better understand patient's condition, such as in monitoring of tumor growth, treatment verification, and comparison of a patient's data with anatomical atlases [6, 9, 13]. Extracted information from any given multimodal images can be utilized to complement one source image to another and thus integration between them can be very useful for further analysis [6, 8, 10, 13].

The integration of these multimodality data leads to an important need of a process to transform the different sets of data into one coordinate system which is generally referred to as image registration [7, 12]. Some of the most popular methods are correlation based methods [1], feature based methods [3, 8], and Fourier transform based methods [5, 12, 14]. In image registration, given a set of two images, one of the images is usually referred to as the reference or source image, while the other is called the target or sensed image [6]. Correlation based methods work by calculating a correlation matrix between the source image and the target image [1]. This approach is highly dependent on the brightness/magnitude of the pixels. On the other hand, feature based methods identify distinct patterns/features such as lines, contours and shapes which occur in both images to align between the images [3]. However, in the case of large image size, the computational complexity of the features in the correlation based methods becomes too high. In the Fourier transform based methods, the frequency components in images are utilized. A fast technique developed for implementation of the Fourier transform algorithms known as the FFT algorithm has contributed to the wide use of the Fourier transform based methods [5, 12, 14]. Moreover, the FFT based methods are found to be more resilient to differences in illumination as well as random noise [16]. They are also more favored than the other methods owing to their ability in solving for the best correlation of the frequency domain features. It has been the main motivation of this work to utilize the information provided by the Blaschko's lines in not only the diagnosis of morphoea [15], but also in monitoring of the disease progress and its treatment efficacy. In this paper, we present a rigid/global based image registration using modified normalized gradient cross-correlation (NGC) method for integration between face thermograms and a face sketch (with and without the Blaschko's lines) (see Figure 1). By integrating the information from both the images, the ambiguity in the thermogram can be reduced greatly and the access to its

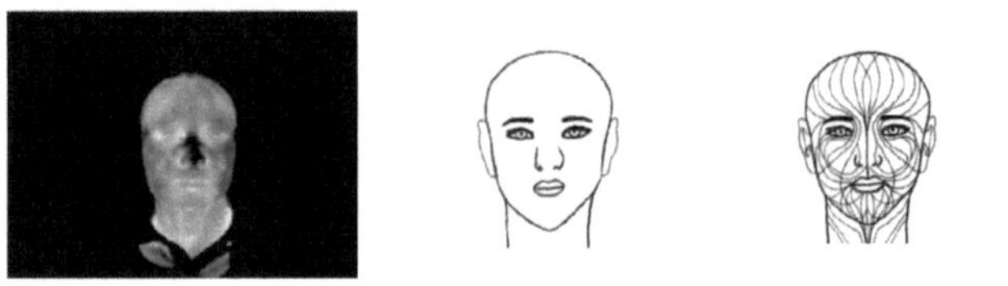

(a) (b) (c)

Fig. 1 (a) Face thermogram as a reference image, (b) a drawing of the face or a face sketch as a target image and (c) a face sketch with the Blaschko's lines that represent lesion patterns of linear morphoea (Reproduced from [15].)

F. Ibrahim et al. (eds.), *International Conference for Innovation in Biomedical Engineering and Life Sciences*,
IFMBE Proceedings 56,
DOI: 10.1007/978-981-10-0266-3_16

information can be enhanced. Hence, the current diagnosis andmonitoring system of morphoea and other medical problems based on thermal imaging can be improved. A modified normalized gradient cross-correlation (NGC) method to reduce large geometrical differences between two multimodal images of different subjects that are represented by smooth gray edge maps is proposed for a global registration approach towards registration between face thermogram and a drawing of a face with and without the Blaschko's lines.

II. METHODOLGY

A. *Pre-Processing*

In this work, to register a thermal image with a drawing, edge detection is firstly performed [5, 10, 17] using Sobel edge detector based on image gradient as it gives good performance in detecting edges in low noise images [11]. For the drawing of the face, as the drawing consists of outlines in black, the negative of the drawing image is taken so that the outlines in the image becomes white. Next, thickening of the edge/outline pixels to a 3 pixel width in size is carried out followed by Gaussian smoothing [11] in order to have smoother, more distributed and spread edge map and drawing outlines to enhance the performance of the image registration based on phase correlation.

Prior to phase correlation, spectral leakage effect that can be caused by the FFT needs to be eliminated or reduced. A cosine tapered window, known as Tukey window [2] is used as it sufficiently suppresses leakage error upon FFT processing.

B. *Normalized Gradient Cross-Correlation (NGC)*

The Phase correlation (PC) technique is based on the Fourier shift property [1, 4, 10, 12]. This PC method has recently been modified to work with the gradient field of an image instead of the gray values and in this way it has been made more robust [14]. In this normalized gradient based cross-correlation (NGC) algorithm, image gradients from both the gray edge maps for both the reference and the target images are derived along the x-axis and the y-axis to obtain gradients along the vertical direction, g_x, and the horizontal direction, g_y, respectively. Using these directional gradients, a complex gradient, g of an image f is given by

$$g(x,y) = \frac{\delta}{\delta x} f(x,y) + j\frac{\delta}{\delta y} f(x,y) = g_x + j\, g_y \quad (1)$$

where j= $\sqrt{-1}$. The gradient cross-correlation, GC, between two images can be defined as $\mathrm{GC(u,v)} = \iint g_1(x,y) g_2^*(x+u, y+v)dxdy$, where * shows the complex conjugate operation. In frequency domain, $GC(u,v)$ takes the form $GC(u,v) = F^{-1}\{G_1(w_{1,}w_2)G_2^*(w_{1,}w_2)\}$, where G_1 and G_2 are the 2D Fourier transfofrm of g_1 and g_2 respectively and F^{-1} is the inverse FFT function. Now the normalized gradient correlation (NGC) can be given as

$$NGC(u,v) = \frac{GC}{F^{-1}\{|G_1(w_1 w_2)||G_2^*(w_1 w_2)|\}} \quad (2)$$

It can be seen that the NGC definition [equation 2] is similar to the inverse Fourier transform of the normalized cross power spectrum $\frac{F_1(u,v)F_2^*(u,v)}{|F_1(u,v)F_2^*(u,v)|}$ in the PC method except that the NGC scheme is based on image gradients. In this work, for the FFT computation in Cartesian coordinate (the first FFT), instead of using the complex gradients, the directional image gradients are combined by taking their magnitudes for both the reference and the target image. However, the complex gradients of both images as shown in [equation 1] are used as the image functions for the FFT computation in log-polar coordinate (the second FFT). In recovering for translational differences between the two images once the scale and/or rotation differences are removed from the target image, again the magnitudes of the gradients are used instead of their complex representation as edge magnitude gives pixel locations strong edge responses and suppress the contribution of areas of constant intensity level which do not provide any reference points for motion estimation [14].

The results of image registration based on the NGC method between a face thermogram (reference image) and the face drawing (target image) that differs not only in scaling and translation, but also in rotation are shown in Figures 2 and 3 respectively.

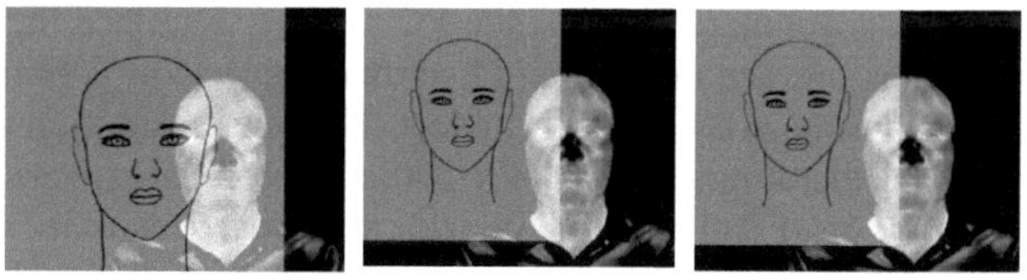

(a) (b) (c)

Fig. 2 Image overlay (a) before image registration is performed, (b) after removing of the scaling of the target image and (c) after translation is corrected (fully registered image) using the NGC method, between the face sketch and a face thermogram of Subject S4.

As mentioned earlier, the drawing used in this registration with thermograms is originally extracted from a sketch of a face with the Blaschko's lines as shown in Figure 4(a). Hence the sketches with and without the Blaschko's lines are similar in all aspects except for the Blaschko's line patterns. This enables the same parameters obtained to remove the translational, scaling and rotational differences from the face sketch without the Blaschko's lines to be applied to the face sketch with the Blaschko's lines, in order to register between the face sketch with the Blaschko's lines and its reference image (i.e., face thermogram) as can be seen in Figure 4(b).

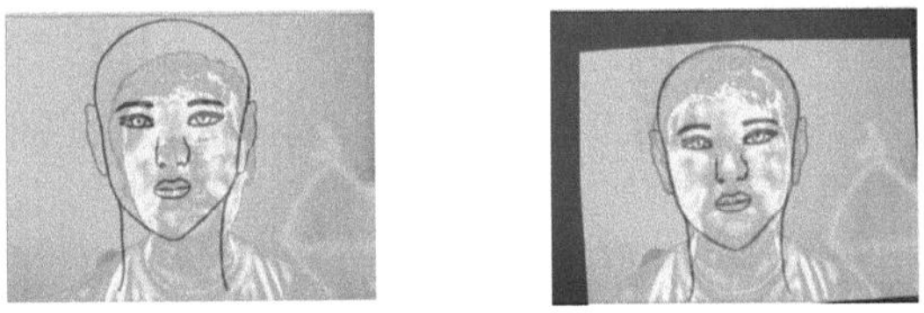

(a) (b)

Fig. 3 Image overlay (a) before image registration is performed, and (b) after scaling, rotation and translation are corrected (fully registered image) using the NGC method, between the face sketch and a face thermogram of Subject S2.

(a) (b)

Fig. 4 (a) The face sketch with the Blaschko's lines (b) Overlay between registered image (a) and its reference image.

III. Comparison Analysis

In this section, the results and analysis of registration between face thermograms and the face sketch using the NGC method is compared against the PC method. The face thermograms are acquired from 11 normal subjects at 7 timepoints during a day for each subject. The pixel resolution of the face thermograms is 386 x 290, acquired using VarioCAM hr head 700 by InfraTec with a thermal sensitivity of up to 30mK or 0.03°C. On the other hand, the pixel resolution of the face sketch is 348 x 306. From the 77 thermograms, 21 of them (total from 3 subjects) are used as training data for the development of the NGC algorithm and in fine tuning of its parameters while the rest of the 56 thermograms (from the remaining 8 subjects) are used as test data. The dataset consists mostly thermograms that differ only in scaling and translation with the sketch. However, it is observed that some data are slightly differed in rotation as well. Fusion of the registered images presented in this chapter is performed by image overlay based on alpha blending by utilizing the alpha data property of any one of the registered images. Two types of quantitative analysis were performed for this comparison analysis; based on (1) under-detection error, e_u, and over-detection error, e_o and (2) scaling factor error. For the first quantitative analysis, these measures are obtained between the registered face sketch and its reference image based on their edge pixels. The edge pixels of the reference image are extracted manually for the quantitative comparison purpose between the NGC and the PC methods so that unwanted edge pixels from features that are not present in the face sketch are not included in the computation of the performance parameters as shown in Figure 5.

A. Quantitative Analysis I

In this section, the registrations between the face sketch and the thermogram data are compared between the two discussed methods; the PC method (see Figure 6) and the NGC method (see Figures 2 and 3). For this comparison purpose, a quantitative analysis is carried out based on the results obtained by both methods using performance parameters such as under-detection error, e_u, and over-detection error, e_o. These measures are obtained between the registered face sketch and its reference image based on their edge pixels. The edge pixels of the reference image are extracted manually so that unwanted edge pixels from features that are not present in the face sketch are not included in the computation of the performance parameters. The only features that are manually extracted are such as the face outline, eyes, nose, mouth, neck and head as shown in Figure 5. The images with the manually extracted edge pixels are then binarized for the analysis purpose.

(a) (b)

Fig. 5 (a) Manually traced outlines from a thermogram (Subject S9) and (b) its extracted traced lines; (c) another example of manually traced outlines from a thermogram (Subject S8) and, (d) its extracted traced lines

Based on the manually extracted edge pixels for each thermogram, $R(x,y)$, and the registered face sketch (intensity image), $S(x,y)$, once both images are in registration, to compute e_u and e_o, the followings are performed.

1. For every black pixel (edge pixel (EP)) in $S(x,y)$, with coordinates (i,j) , check whether there is an edge pixel with the same coordinates in $R(x,y)$. If yes, count it with a counter A; $A = \Sigma_{x=i,y=j}[S(x,y) = R(x,y) = \text{EP}]$, else, count it

with a counter B; $B = \Sigma_{x=i,y=j}$ $[(S(x,y) = \mathrm{EP}) \cap (R(x,y) = \mathrm{Non\ EP})]$. Then, compute the under-detection error e_u as $B/(A+B)$.

2. For every edge pixel in $R(x,y)$, with coordinates (i,j), check whether there is a black pixel (edge pixel) in $S(x,y)$, with the same coordinates. If yes, count it with a counter C; $C = {}_{x=i,y=j}[R(x,y) = S(x,y) = EP]$, else, count it with a counter D; $D = {}_{x=i,y=j}[(R(x,y) = EP) \cap (S(x,y) == Non\ EP)]$. Then, compute the over-detection measure e_o as D/(C+D).

The computing of these parameters is performed based on point (one-pixel) and area (8-pixel neighborhood) processings. Unlike the point processing, in the 8-pixel neighborhood processing, when we look for a match in the other binary map for pixel (i,j), we do not just check for position (i,j), but also the neighbouring 8 positions. The first is referred to as case A while the latter is referred to as Case B. The parameters (e_u and e_o) for both cases A and B for all the 11 subjects, each with 7 face thermograms taken at different time, are presented based on minimum, maximum and standard deviation (std). Based on the overall values, the NGC method outperforms the PC method in terms of lower range of e_u and e_o obtained. It is found that for case A, in average, the range of values obtained for e_u and e_o for the NGC method are 0.76-0.87 and 0.46-0.70 respectively while for the PC method, the range of values are higher, 0.84-0.95 for its e_u and 0.59-0.85 for its e_o respectively. On the other hand, for case B, as predicted, the range of values for both the e_u and the e_o are reduced/improved to a good extent; 0.57-0.77 and 0.32-0.57 respectively for the NGC method, 0.71-0.89 and 0.48-0.77 respectively for the PC method. However, the average errors found in case B for the PC method are still higher than that achieved by the NGC method. As predicted, by considering 8 neighbouring pixels, the overall error parameters for both the NGC and the PC methods are reduced. The average stds computed for the results obtained for each subjects for all cases using both the methods are small, from 0.011 to 0.32 and comparable.

From the mean errors obtained based on the 77 images, it is observed that both the NGC and the PC methods suffer from under-detection more than over-detection. Using case B, the gap between the two error measures is more significant in the NGC method since the value of over-detection error achieved by this method is reduced greatly. The average standard deviations computed for the results obtained for the 77 images for both cases A and B are small, in the range of 0.004-0.012 for the NGC method and in the range of 0.11-0.16 for the PC method. Overall, as expected, the NGC method outperforms the PC methods by having less under-detection and over-detection errors in registration between two multimodal images based on face thermogram and the face sketch.

B. Quantitative Analysis II

Apart from the comparisons performed based on the e_u and the e_o, a comparison based on the recovered scaling factor is also carried out in this study. For this comparison purpose, a ground truth of the scaling parameter needed for the sketch to be aligned to the thermograms are obtained from a semi-automated registration approach using some of the MATLAB built-in functions which involves human intervention in the selection of the control points for registration. As this approach involves human intervention, the transformation parameters (i.e., scaling factor, rotation angle and translation) obtained using this semi-automated method to register between the sketch and the thermograms are considered as the ground truth. This process is performed for all the 77 thermograms. From this analysis, it is found that the NGC method achieved a scaling error of as small as ± 0.02 to as high as ± 0.10 while the PC method achieved a scaling error of as small as ± 0.03 to as high as ± 0.60 as compared to the semi-automated method. The average and standard deviation (std) of the errors calculated for all 11 subjects are ~0.06 and ~0.029 respectively for the NGC method while a higher value is obtained by the PC method; ~0.27 and ~0.148 respectively. Furthermore, the scaling factor that can be recovered by the PC method for the registration between the sketch and the face thermograms ranges from 1 to 1.4939 unlike the NGC method where a scaling factor of up to 1.7713 is achieved.

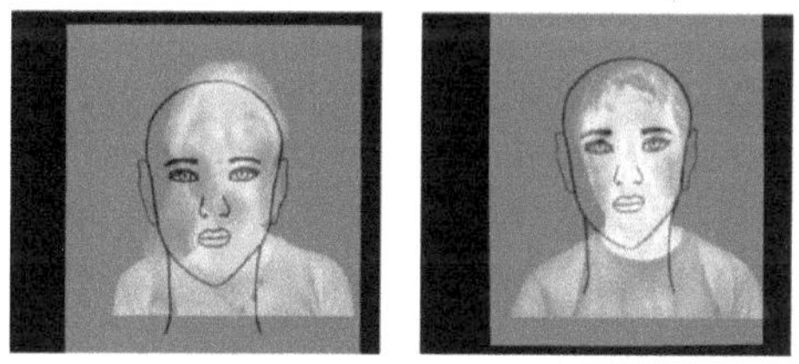

(a) (b)

Fig. 6 Registration results using the PC method for (a) Subject S1 (at 9 am) and (b) Subject S2 (at 9 am)

In the PC based registration method, most of the time, due to the inability of the method to recover the scale difference between the images to be registered accurately, the method suffers a poor retrieval of the displacement parameters for translational differences between the images to be registered. In terms of rotational angle, both the registration methods are able to retrieve a sufficient amount of angle to reduce the rotational difference between the images. Futhermore, the difference between the recovered angles by both the methods is small; from 0.13 to 10.87 in degrees.

IV. Discussion and Conclusion

Overall, based on the first quantitative analysis conducted using the under-detection and over-detection errors, it is shown that the NGC method outperforms the PC method by lower range of values achieved for both the error measures using both case A and case B. This demonstrates that by using the gradients of the gray edge maps, the performance of the PC based image registration method can be improved. This is due to the property of the gradients that consists of pixels that strongly correlates with the amount of information that each pixel in both images carries. However, it is found that windowing is still needed in the registration based on the NGC method to remove or suppress the edge pixels near the border at the bottom part of the images especially for the reference image. The overall registration results for the NGC method are fairly good even though the average under-detection range for case A is found to be quite high; above 0.7. It is observed that a value of e_u of up to as high as 0.8 for the under-detection error still display an acceptable registration result while a value of above 0.9 indicates a poor registration result. On the other hand, for the over-detection measure, a value of below 0.4 shows fairly good registration results while a value of over 0:7 indicates poor registration results. Hence, the measure based on the value of e_o is more reliable in demonstrating the performance of the NGC method. These outcomes may be caused by the fact that the target image (the sketch) and the reference image are not taken from the same image source or subject. Even though the features such as the face outline, the nose and the eyes between the face sketch and the thermograms are already in good registration, as the shape or line patterns of the features are not identical or exactly the same, this has considerably affected the accuracy of the computed error measures. Therefore, it is found that the error measures computed for case B display the performance of the registration method more agreeably than the values obtained for case A as it took into consideration this issue by considering more neighbour pixels. This problem may be minimized by replacing the face sketch with an atlas image of a face generated using optical images of the same 11 subjects integrated by group wise based registration as this atlas image may resemble a face feature that is much similar to the face feature in thermograms than the face sketch.

However, as the NGC method is a global affine transformation based registration approach, there are still some limitations in achieving an accurate registration between the face sketch and a thermogram that can be observed. Since the target and the reference images involved in this work are not of the same nature (i.e., between an artificially created data (face sketch) and a real data (face thermogram)), the registration between these images can be quite challenging. Thus, a need for the registration between these types of images to be approached not only by global or rigid transformation based method alone but also by a local or non-rigid transformation based method is found inevitable.

Conflict of Interest

The authors declare that they have no conflict of interest.

References

1. D. I. Barnea and H. F. Harvey, "A Class of Algorithms for Fast Digital Image Registration", IEEE Transactions on Computers, C-21(2):179-186, 1972
2. P. Bloomfield, Fourier Analysis of Time Series: An Introduction, Wiley-Interscience, New York, 6:69, 2000.
3. L.G. Brown, "A survey of image registration techniques", ACM, Computing Surveys, 24:326-376.
4. D. Casasent and D. Psaltis, "Position, rotation, and scale invariant optical correlation", Applied Optics, 15:1795-1799, 1976.
5. C. E. Costa and M. Petrou, "Automatic registration of ceramic tiles for the purpose of fault detection", Machine Vision and Applications, 11: 225-230, 1999.
6. A. Goshtaby, "2-D and 3-D Image Registration for Medical, Remote Sensing, and Industrial Applications", Wiley Press, 2005.
7. J. V. Hajnal, D. L. G. Hill and D. J. Hawkes. Medical Image Registration, The Biomedical Engineering Series, CRC Press, 2001.
8. A. Jarc, J. Pers, P. Rogelj, M. Perse and S. Kovacic, "Texture features for affine registration of thermal (FLIR) and visible images", Computer Vision Winter Workshop, Graz Technical University, St. Lambrecht, Austria, February 6-8, 2007.
9. L. I. Izhar, M. Petrou. Thermal Images in Medicine. Advances in Imaging and Electron Physics, 171(2): 41-114, 2012.
10. M. Petrou, "Image Registration: an overview", Advances in Imaging and Electron Physics, 130:243-291, 2004.
11. M. Petrou and C. Petrou, Image Processing: the fundamentals, 2nd Edition, John Wiley, 2010.
12. B. S. Reddy and B. N. Chatterji, "An FFT-Based technique for translation, rotation and scale-invariant image registration", IEEE Transactions on Image Processing, 5(8): 1266-1271, 1996.
13. G. Schaefer, R. Tait, K. Howell, A. Hopgood, P. Woo and J. Harper, "Automated overlay of infrared and visual medical images", User Centered Design for Medical Visualization, 8:174- 183, 2008.
14. G. Tzimiroupoulos, V. Argyriou, S. Zafeiriou and T. Stathaki, "Robust FFT-based scale-invariant image registration with image gradients", IEEE Transactions on Pattern Analysis and Machine Intelligence, 32(10):1809-1908, 2010.

15. L.Weibel and J.I. Harper, "Linear morphoea follows Blaschko's lines", British Journal of Dermatology, 159:175-181, 2008.
16. B. Zitova and J. Flusser, "Image registration methods: a survey", Image and Vision Computing, 21:977-100, 2003.
17. B. Zitova, J. Flusser and F. Sroubek, "Image Registration: a survey and recent advances", ICIP Tutorial, Genoa, Italy, 2005.

AUTHORS CONTACT DETAILS

Author: L. I. Izhar
Institute: Universiti Teknologi PETRONAS
Street: Bandar Seri Iskandar
City: Tronoh, Perak
Country: Malaysia
Email: lila.izhar@petronas.com.my

Author: T. Stathaki
Institute: Imperial College London
Street: South Kensington
City: London
Country: United Kingdom
Email: t.stathaki@ic.ac.uk

Author: K. Howell
Institute: University College London
Street: Gower Street
City: London
Country: United Kingdom
Email: rmhakh0@ucl.ac.uk

EEG Wavelet Spectral Analysis During a Working Memory Tasks in Stroke-Related Mild Cognitive Impairment Patients

N.K. Al-Qazzaz[1,5], S.H.M. Ali[1], S. Islam[2], S.A. Ahmad[3], and J. Escudero[4]

[1] Department of Electrical, Electronic and Systems Engineering, Faculty of Engineering and Built Environment, Universiti Kebangsaan Malaysia, UKM, 43600, Bangi, Selangor, Malaysia
[2] Institute of Microengineering and Nanoelectronics (IMEN), Universiti Kebangsaan Malaysia, UKM, 43600, Bangi, Selangor, Malaysia
[3] Department of Electrical and Electronic Engineering, Faculty of Engineering, Universiti Putra Malaysia, UPM, 43400, Serdang, Selangor, Malaysia
[4] Institute for Digital Communications, School of Engineering, The University of Edinburgh, Edinburgh EH9 3JL, UK
[5] Department of Biomedical Engineering, Al-Khwarizmi College of Engineering, Baghdad University, 47146, Baghdad, Iraq

Abstract— **The aim of this study was to analyse the electroencephalography (EEG) background activity of 10 stroke-related patients with mild cognitive impairment (MCI) using spectral entropy (*SpecEn*) and spectral analysis. These spectral features were used to test the hypothesis that the EEG dominant frequencies slowdown in MCI in comparison with 10 age-match control subjects. Nineteen channels were recorded during working memory and were grouped into 5 recording regions corresponding to scalp areas of the cerebral cortex. EEG artifacts were removed using wavelet analysis (WT). The *SpecEn* analysis of the EEG data suggested a broad and flat spectrum in the normal EEG. The relative powers (*RP*) in delta (δ*RP*), theta (θ*RP*), alpha (α*RP*), beta (β*RP*), and gamma (γ*RP*) were calculated. *SpecEn* was significantly lower in stroke-related MCI patients at parietal, occipital and central regions (*p*-value < 0.05, Student's *t*-test). Moreover, the other significant differences can be observed in increasing the δ*RP*, θ*RP* and γ*RP* and decreasing the α*RP* and β*RP* of the stroke-related MCI group in all regions (*p*-value < 0.05, Student's *t*-test). It can be concluded that the *SpecEn* and spectral analysis are useful tool to inspect the slowing in the EEG signals in post-stroke MCI patients' and the healthy controls' EEG.**

Keywords— **Electroencephalography, Relative power, Spectral entropy, Wavelet, Mild cognitive impairment.**

I. INTRODUCTION

Cognitive and working memory impairment are common after stroke. 30% of stroke patients are prone to develop vascular dementia (VaD) within the first year of stroke onset . VaD is the second case of dementia after Alzheimer's disease (AD), between 1% and 4% of elderly people age of 65 years are suffer from VaD and the prevalence will be double every 5-10 years after this age [1]. Clinically, mild cognitive impairment (MCI) is defined as a decline in cognitive function greater than expected with respect to the individual's age and education level, but that does not interfere notably with the activities of daily life [2]. Traditionally, it is considered as a stage between early normal brain cognition and late severe dementia. Attention and executive function are the most affected domains due to vascular lesion that results from ischemic and hemorrhagic stroke [3, 4]. For several decades, EEG has been considered an effective physiological technique which reflects the hidden cortical abnormalities by providing a quantitative insight to diagnose or evaluate potential predictors of dementia severity [5] . In the last two decades, several attempts have been made to quantify the EEG activity using computerized signal processing and analysis techniques in order to interpret the degree of EEG abnormality and dementia [6]. Typically, the clinical EEG wave forms have an amplitude around 10-100 µv and frequency range of 1 to 100 Hz . EEG can be classified into five frequency bands: Delta waves (δ), Theta waves (θ), Alpha waves (α), Beta waves (β), and Gamma waves (γ) [7]. However, the EEG is affected by non-cerebral sources called artifacts that may mimic the brain pathological activity and therefore influence the analysis. Many artifacts can have a physiological origin, like muscle activity, pulse and eye blinking. Others are non-physiological, such as power line interference. Numerous methods have been used to deal with artifacts that affect the EEG recordings. Wavelet (WT) is a time-frequency analysis that used to denoise the non-stationary bio-signals such as EEG [8]. Researchers have used WT in different ways. For instance, WT has been used to detect epileptic spike signals and to predict the changes in patients with epilepsy and to separate burst in ECG waves. Moreover, WT has been used to remove ocular artifacts and tonic components for electromyography (EMG) signals. Furthermore, WT can be an efficient technique to extracted features from the EEG sub-bands as wavelet decomposition [9].

In this paper, WT has been used as a pre-processing step to denoise the EEG datasets. The spectral entropy (*SpecEn*) and spectral analysis were extracted to examine the EEG background activity in MCI patients and control healthy subjects.

F. Ibrahim et al. (eds.), *International Conference for Innovation in Biomedical Engineering and Life Sciences*,
IFMBE Proceedings 56,
DOI: 10.1007/978-981-10-0266-3_17

II. METHODS AND MATERIALS

A. Subjects and EEG Recording Procedure

EEG datasets were recorded for ten healthy control subjects, aged 57.9 ± 5.7 years, and ten MCI patients, aged 58.2 ± 8.7 years (mean ± standard deviation, SD). The patients were recruited from the stroke ward in Pusat Perubatan Universiti Kebangsaan Malaysia (PPUKM), the Medical center of National University of Malaysia, Malaysia. The post-stroke patient satisfied the National Institutes of Health Stroke Scale (NIHSS) [10]. The healthy control had no previous history of mental and neurological abnormalities. Both groups were underwent cognitive evaluation including Mini-Mental State Examination (MMSE), [11] and Montreal Cognitive Assessment (MoCA) [12], (stroke patients MMSE 20.5±6.3, MoCA 16.35±6.9; normal control MMSE 29.6±0.7, MoCA 29.06±0.8, mean ± SD). All experiment protocols were approved by the Human Ethics Committee of the National University of Malaysia. An information consent forms were also signed by the participants. The EEG activity was recorded using the NicoletOne systems (V32), VIASYS Healthcare Inc., USA. A total of 19 electrodes, plus the ground and system reference electrodes, were positioned according to the 10-20 international system (Fp1, Fp2, F7, F3, Fz, F4, F8, T3, T5, T4, T6, P3, Pz, P4, C3, Cz, C4, O1, and O2). NicoLetOne EEG system is sampled at 256 Hz sampling frequency, impedance of electrode/skin was below 10 kilo ohms, sensitivity of 100 µv/cm, low cut of 0.5 Hz and high cut of 70 Hz using referential montage. The EEG was recorded for 60 seconds during working memory task. Patients were asked to memorize five words for 10 seconds [3]. Then, each patient was asked to remember the five words while the EEG was recorded with the eyes closed. After 1 min they were asked to open the eyes and enumerate the five words they could remember. The 19 channels from the EEG datasets of the 10 healthy and the 10 stroke patients were grouped into 5 recording regions corresponding to the scalp area of the cerebral cortex. These are the frontal region (seven channels: Fp1, Fp2, F3, F4, F7, F8 and Fz), the temporal region (four channels: T3, T4, T5 and T6), the parietal region (three channels: (P3, P4 and Pz), the occipital region (two channels: O1 and O2), and central region (three channels: C3, C4 and Cz). Conventional filtering were used to process the 19 channels EEG data. Notch filter at (50 Hz) has been used to remove the power line interference noise and a band pass filter of (0.5-64 Hz) frequency range has been used to limit the band of the recorded EEG signals. The general block diagram of our proposed system is shown in Figure 1.

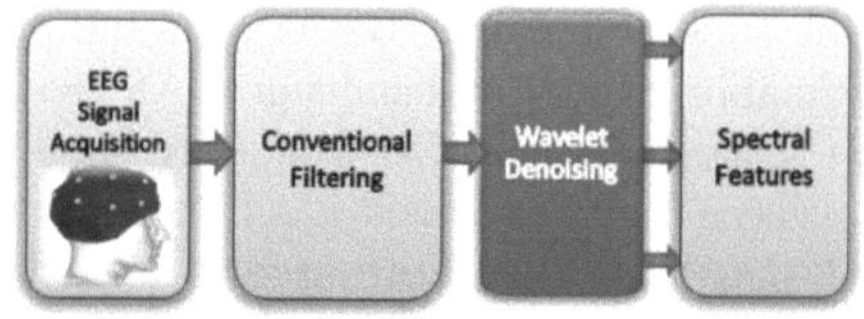

Fig. 1 The block diagram of the proposed method

Table 1 The EEG signal decomposition into five frequency bands

Decomposition levels	Decomposed Signals	EEG bands	Frequency bands (Hz)
1	D1	Noises	64-128
2	D2	Gamma	32-64
3	D3	Beta	16-32
4	D4	Alpha	8-16
5	D5	Theta	4-8
5	A5	Delta	0-4

B. Wavelet Analysis

Wavelet transform is a popular denoising technique. It has been introduced to process the non-stationary signals, as EEG and EMG. The mathematical equation of the discrete wavelet transform (DWT) can be processed by obtaining the discrete value of the parameters *a* and *b* as in (1).

$$DWT_{m,n}(f) = a_0^{-m/2} \int f(t)\, \psi(a_0^{-m}t - nb_0)dt \qquad (1)$$

Where a_0 and b_0 values are set to 2 and 1, respectively. Where $\psi(t)$ is the mother wavelet (MWT) function which is shifted by the location parameter (b) and dilated or contracted by scaling parameter (a), as in 2

$$\psi_{a,b}(t) = \frac{1}{\sqrt{a}}\psi\left(\frac{t-b}{a}\right), a\epsilon R^+, b\epsilon R \qquad (2)$$

Discrete wavelet transform (DWT) was used for denoising purposes, the symlets orthogonal MWT family of order 9 'sym9' was used due to its high compatibility with the recorded EEG datasets which produce the best denoising results [13]. In this study, five decomposition levels were chosen to decompose the EEG signals, since the sampling frequency used in this study was 256 Hz. The decomposition coefficients represented the frequency content from the EEG signal. The SURE threshold, is an adaptive soft thresholding method, which is finding the threshold limit for each level based on Stein's unbiased risk estimation [14] and commonly used value in [13, 15, 16]. Once the thresholded coefficients have been extracted from each level, the effects of the noises on the EEG signals are removed. Finally, the signals at each level have to be reconstructed using inverse discrete wavelet transform (IDWT). The first reconstructed details D1 is considered to be mainly the noise components of the EEG signal, the four reconstruction details of the sub-band signals D2–D5 and the reconstruction approximation of the sub-band signal A5 yielded signal information related to each EEG frequency band as shown in Table 1. These bands provided a compact representation of the EEG signal and they were used to extract the EEG spectral features.

C. Feature Extraction

Spectral analysis has been used extensively to detect abnormalities in the spectra of dementia patients' EEGs. Spectral entropy (*SpecEn*) measures the flatness of the signal spectrum and it considered as a convenient way that suitable in quantify slowing in frequency due to dementia. In the present work, to quantify EEG changes, *SpecEn* and the relative power (*RP*) in delta (δ*RP*), theta (θ*RP*), alpha (α*RP*), beta (β*RP*), and gamma (γ*RP*) were calculated to the WT decomposed signals to distinguish stroke-related MCI patients EEGs' from the normal age-match healthy subjects. In order to estimate the *SpecEn*, the PSD was normalized to a scale from 0 to 1 to get normalized PSD (PSD_n) so that $\sum PSD_n(f) = 1$, afterwards, *SpecEn* is computed applying the Shannon's entropy to the PSD_n as shown in 3 [17]:

$$SpecEn = \frac{-1}{\log(N)} \sum_{f=0.5Hz}^{64\,Hz} \log[PSD_n(f)] \qquad (3)$$

The *RP* for each selected frequency band δ, θ, α, β, and γ can be calculated using equation 4

$$RP(\%) = \frac{\sum Selected\ frequency\ range}{\sum Total\ range\ (0.5 - 64\ Hz)} \qquad (4)$$

D. Statistical Analysis

Normality was assessed with Kolmogrov-Smirnov test, whereas homoscedasticity was verified with Levene's test. Therefore, the student's *t*-test was applied to compare between the features of the two groups of MCI patients and control subjects using SPSS 22. These comparisons were done for each feature according to regions separately between the two groups. First of all, the statistical difference between the spectral features of the stroke-related MCI and healthy subjects were evaluated. Second, the EEG bands *RP* for the two groups were assessed for each band separately. Differences were considered statistically significant if the *p* value was lower than (0.05).

III. Results and Discussion

In Table 2, the SpecEn and the *RP* values for the MCI patients and age-match control subjects in the five scalp regions are given. It is evident that lower SpecEn values in the MCI patients than the control subjects at parietal, occipital and central regions are found, achieving significant differences (p-value < 0.05). The spectral analysis of the *RP* showed significant increases in δ*RP*, θ*RP* and γ*RP* activities for the MCI patients in all regions (p-value < 0.05). It can also be observed the decrease in both α*RP* and β*RP* activities in MCI patients significantly in all regions of the MCI patients (p-value < 0.05). Our findings agreed other studies. For instance, Klimesch described the changes in the brain activity which are strongly associated with cognitive and attentional working memory performance as decreasing in both alpha and beta but increasing in both delta and theta in [18]. Gevins et al. attributed the changes during working memory task to alpha and theta. Finally, Lundqvist et al. correlated the changes in brain activity to encoding one or more items in WM and these changes have associated with increase in theta and gamma and decrease in alpha and beta power [19, 20].

Table 2 The average values (Mean ± SD) of EEGs for the MCI patients and the control subjects for all the five scalp regions. Significant group differences are marked with an asterisk

Features	Regions	MCI (Mean ± SD)	Control (Mean ± SD)	p-value
	Frontal	0.75±0.09	0.778±0.082	0.203
	Temporal	0.767±0.088	0.798±0.065	0.101
SpecEn	Parietal	0.764±0.083	0.789±0.035	0.002*
	Occipital	0.74±0.089	0.759±0.047	0.046*
	Central	0.787±0.084	0.807±0.038	0.001*
	Frontal	0.51±0.218	0.438±0.185	0.045*
	Temporal	0.424±0.232	0.359±0.17	0.024*
δ*RP*	Parietal	0.42±0.216	0.341±0.113	0.001*
	Occipital	0.426±0.249	0.274±0.145	0.02*
	Central	0.378±0.229	0.325±0.126	0.001*
	Frontal	0.125±0.067	0.102±0.035	0.05*
	Temporal	0.147±0.088	0.11±0.03	0.05*
θ*RP*	Parietal	0.14±0.077	0.114±0.037	0.05*
	Occipital	0.172±0.093	0.106±0.044	0.05*
	Central	0.148±0.083	0.124±0.034	0.05*
	Frontal	0.145±0.098	0.201±0.108	0.408
	Temporal	0.193±0.121	0.252±0.128	0.779
α*RP*	Parietal	0.216±0.139	0.306±0.109	0.179
	Occipital	0.221±0.151	0.417±0.174	0.324
	Central	0.198±0.103	0.27±0.102	0.847
	Frontal	0.107±0.068	0.137±0.055	0.034*
	Temporal	0.118±0.074	0.157±0.044	0.006*
β*RP*	Parietal	0.117±0.075	0.141±0.028	0.001*
	Occipital	0.095±0.055	0.124±0.033	0.03*
	Central	0.135±0.084	0.165±0.04	0.002*
	Frontal	0.122±0.066	0.114±0.093	0.035*
	Temporal	0.123±0.059	0.118±0.108	0.026*
γ*RP*	Parietal	0.107±0.102	0.099±0.05	0.031*
	Occipital	0.085±0.067	0.08±0.048	0.102
	Central	0.141±0.13	0.116±0.053	0.011*

IV. Conclusions

In the current study 'sym9' MWT basis function has been used as a pre-processing to denoise the EEG datasets of both control subject and the post-stroke patient using SURE thresholding method. Nineteen channels from different regions on the scalp were recorded during working memory and were grouped into 5 recording regions corresponding to the scalp area of the cerebral cortex. Spectral analysis has been used to detect abnormalities in the spectra of stroke-related MCI patients EEGs'. The *SpecEn* and the relative powers of both low and high frequencies reflected the slowing in the electrical brain activity in MCI patients which results in shifting their power spectrum profiles. It can be noticed an increase in δRP, θRP and γRP activities for the MCI patients in all regions and decrease in both αRP and βRP activities in MCI patients in all regions. As the EEG has been widely used as a potential screening technique in clinical practice due to its low cost and portability, it could become a reference in planning and customizing an optimal therapeutic program to address the changes associated with MCI and dementia. This study suggests that the spectral analysis of EEG background activity in stroke-related MCI patients using *SpecEn* and the relative powers might be helpful in providing useful diagnoses indexes using EEG.

Conflict of Interest

The authors report no conflicts of interest in this work.

References

1. McVeigh, C. and P. Passmore, *Vascular dementia: prevention and treatment.* Clinical interventions in aging, 2006. **1**(3): p. 229.
2. Korczyn, A.D., V. Vakhapova, and L.T. Grinberg, *Vascular dementia.* J Neurol Sci, 2012. **322**(1-2): p. 2-10.
3. Al-Qazzaz, N.K., et al., *Cognitive impairment and memory dysfunction after a stroke diagnosis: a post-stroke memory assessment.* Neuropsychiatric disease and treatment, 2014. **10**: p. 1677.
4. Al-Qazzaz, N.K., et al., *Cognitive assessments for the early diagnosis of dementia after stroke.* Neuropsychiatric disease and treatment, 2014. **10**: p. 1743.
5. Al-Qazzaz, N.K., et al., *Role of EEG as Biomarker in the Early Detection and Classification of Dementia.* The Scientific World Journal, 2014. **2014**.
6. Poil, S.-S., et al., *Integrative EEG biomarkers predict progression to Alzheimer's disease at the MCI stage.* Frontiers in Aging Neuroscience, 2013. **5**.
7. Al-Kadi, M.I., M.B.I. Reaz, and M.A.M. Ali, *Evolution of Electroencephalogram Signal Analysis Techniques during Anesthesia.* Sensors, 2013. **13**(5): p. 6605-6635.
8. Zikov, T., et al. *A wavelet based de-noising technique for ocular artifact correction of the electroencephalogram.* in *Engineering in Medicine and Biology, 2002. 24th Annual Conference and the Annual Fall Meeting of the Biomedical Engineering Society EMBS/BMES Conference, 2002. Proceedings of the Second Joint.* 2002. IEEE.
9. Güler, I. and E.D. Übeyli, *Adaptive neuro-fuzzy inference system for classification of EEG signals using wavelet coefficients.* Journal of neuroscience methods, 2005. **148**(2): p. 113-121.
10. Brott, T., et al., *Measurements of acute cerebral infarction: a clinical examination scale.* Stroke, 1989. **20**(7): p. 864-870.
11. Folstein, M.F., S.E. Folstein, and P.R. McHugh, *Mini-mental state.* A prac-32, 1998.
12. Smith, T., N. Gildeh, and C. Holmes, *The Montreal Cognitive Assessment: validity and utility in a memory clinic setting.* Canadian Journal of Psychiatry, 2007. **52**(5): p. 329.
13. Al-Qazzaz, N.K., et al. *Selection of mother wavelets thresholding methods in denoising multi-channel EEG signals during working memory task.* in *Biomedical Engineering and Sciences (IECBES), 2014 IEEE Conference on.* 2014. IEEE.
14. Stein, C.M., *Estimation of the mean of a multivariate normal distribution.* The annals of Statistics, 1981: p. 1135-1151.
15. Romo-Vazquez, R., et al. *EEG ocular artefacts and noise removal.* in *Engineering in Medicine and Biology Society, 2007. EMBS 2007. 29th Annual International Conference of the IEEE.* 2007. IEEE.
16. Estrada, E., et al. *Wavelet-based EEG denoising for automatic sleep stage classification.* in *Electrical Communications and Computers (CONIELECOMP), 2011 21st International Conference on.* 2011. IEEE.
17. Escudero, J., et al., *Blind source separation to enhance spectral and non-linear features of magnetoencephalogram recordings. Application to Alzheimer's disease.* Medical engineering & physics, 2009. **31**(7): p. 872-879.
18. Klimesch, W., *EEG alpha and theta oscillations reflect cognitive and memory performance: a review and analysis.* Brain research reviews, 1999. **29**(2): p. 169-195.
19. Lundqvist, M., P. Herman, and A. Lansner, *Theta and gamma power increases and alpha/beta power decreases with memory load in an attractor network model.* Journal of cognitive neuroscience, 2011. **23**(10): p. 3008-3020.
20. Jeong, J., *EEG dynamics in patients with Alzheimer's disease.* Clinical neurophysiology, 2004. **115**(7): p. 1490-1505.

Author: Noor Kamal Al-Qazzaz
Institute: Department of Electrical, Electronic and Systems Engineering, Faculty of Engineering and Built Environment, Universiti Kebangsaan Malaysia
Street: Bangi
City: Selangor 43600
Country: Malaysia
Email: noorbmemsc@gmail.com

Ion-selective Electrode Biochip for Applications in a Liquid Environment

W.W.A. Wan Salim[1,2,5], A.C. Hermann[1,2], M.A. Zietchek[1,2], J.E. Pfluger[3], J.H. Park[1,2], A. ul Haque[4], F. Sanober[5], and D.M. Porterfield[1,2]

[1] Department of Agricultural and Biological Engineering
[2] Birck Nanotechnology Center and Bindley Bioscience Center Physiological Sensing Facility
[3] School of Chemical Engineering, Purdue University, West Lafayette, IN, USA
[4] Cooley LLP, Boston, MA, USA
[5] Department of Biotechnology Engineering, Kulliyyah of Engineering, International Islamic University Malaysia (IIUM), Gombak, Kuala Lumpur, Malaysia

Abstract— **Physiological sensing conducted in a liquid environment requires electrodes with long lifetime. The development of a robust ion-selective electrode–based biochip in a lab-on-a-chip platform is described. To compare electrode lifetime, which is driven by the transducer layer, electrochemical measurements were performed in a custom-made flow-cell chamber. The results of potentiometric measurement of cationic analytes demonstrate the electrodes to have a near-Nernstian slope profile even after they are stored for almost a month in liquid medium. The electrodes also achieved H_2O_2 amperometric sensitivity (1.25 and 3.32 $\mu AmM^{-1}cm^{-2}$ for PEDOT:PSS and PEDOT:$CaSO_4$ respectively) and lower detection limit (2.21 µM, 8.4 µM, 3.44 µM, for H^+, NH_4^+, Ca^{2+} respectively) comparable to that of wire-type electrodes. Furthermore, the lifetime is dependent on the electrodeposition method of the conductive polymer, and the transducer layer must be modified to fit the analyte types. These results indicate that extended lifetime of microfabricated ion-selective electrodes in a multiplex format can be realized by optimizing the microfabricated electrode surface functionalization.**

Keywords— **electrochemical sensor, all-solid-state ion-selective electrode, PEDOT, lab-on-a-chip.**

I. Introduction

Over the past decades, interest in polymeric-based ion-selective electrodes (ISEs) has been extensive, simply owing to their being one of the most promising chemical sensors that can detect and measure specific chemical species in natural liquid samples with simple instrumentation [1-4]. The requirement for an ion-selective electrode (ISE) to have continuous improvement in response time, sensitivity, selectivity and detection limit is mainly fuelled by the recent ISE applications in environmental and biological sensors [5-7]. These applications have additional need for real-time, on-site, and point-of-care dynamic measurement capability in complex natural samples. The challenge of ISE-based sensors is clearly demonstrated in biological, biomedical, and environmental research, where measurements with ISEs have to be made in liquid media over several days [8-13]. Furthermore, some applications require the initial measurement to be made only after several weeks' storage of the ISE in a liquid-medium environment. In spaceflight application, ISEs are to be integrated in a flight-compatible packaging unit with a controlled environment. These needs for robust measurement requirements have been clearly demonstrated in microgravity C9 flight experiments utilizing ISEs [14].

Miniaturization of ISEs for a lab-on-a-chip platform, while preserving their sensing capabilities, has been the key step in the improvement and innovation in the aforementioned applications. Furthermore, miniaturization is low cost and the resulting ISE requires small sample volumes, thus enabling close proximity sensing to biological samples [15, 16]. The importance for miniaturization becomes more apparent in spaceflight astrobiology research, where measurements are conducted in minimally equipped and compact laboratory conditions, with limited crew time in a relatively high-stress environment [17]. For miniaturization purposes, it is advantageous to replace traditional liquid-filled ion-selective electrodes [18] and coated-wire electrodes [19] with an all-solid-state ISE (ASISE) [20]. In an ASISE, the reference solution is replaced with a solid material, mainly conductive polymer (CP), with both ionic and electronic conductivity. This enables the CP to have a transducing property. CP is also an attractive material for such purposes thanks to the tailored electrochemical property [21]. This trait is especially important in sensor development, owing to the complex measurement requirement of natural samples. The ISE needs to perform within a certain linear range, detection limit, sensitivity, and selectivity [22, 23]. The need for tailored transducing properties can be seen from the ISE fabrication point of view – modifying the transducer property is more time efficient and cost effective than modifying the sensor fabrication process. CPs as the transducer material in ASISEs also demonstrate potentiometric stability [24-26] and are less susceptible to interference from dissolved gas [27]. Thus, an ASISE can be miniaturized while preserving measurement requirements.

The CP-based ASISE sensors employed in such systems also require extended lifetime capability. For instance, in

F. Ibrahim et al. (eds.), *International Conference for Innovation in Biomedical Engineering and Life Sciences*,
IFMBE Proceedings 56,
DOI: 10.1007/978-981-10-0266-3_18

spaceflight nanosatellite astrobiology research, the ISE needs to survive prolonged storage in solution before a measurement is conducted [17]. The ISE electrode calibration measurements may be conducted weeks prior to integration of the ISE-based device into a biological measurement system. Drifts in the calibration slope need to be within an acceptable tolerance value. The ASISE lifetime is determined by the condition of the ion-selective membrane (ISM) and the conductive polymer transducer layer. The ISM lifetime is limited by the leaching of membrane components into the bulk solution [28], and the CP lifetime is dependent on its mechanical property [29]. Conductive polymer transducer layers can become brittle when they experience prolonged storage in a liquid-medium environment [30]. Hence, in summary, ASISE lifetime is dependent on the chemical and mechanical properties of the ISM and the transducer layer, respectively.

Here we describe the development of robust multiple ASISEs in a lab-on-a-chip platform, named the multi-analyte biochip (MAB) (Figure 1), for physiological sensing in a liquid environment.

Fig. 1 Multi-analyte biochip (MAB) consists of multiple working and reference electrodes.

The lifetimes of the biochip's ASISEs in liquid medium are tested in a microfluidic chamber. To achieve lifetime capability, we utilized PEDOT conductive polymer conjugates [i.e. poly(3,4-ethylenedioxythiophene)-poly-(styrenesulfonate) (PEDOT:PSS)] as the transducer material, mainly because it is one of the most thermally and chemically stable forms of CP [31-34]. A PEDOT-based transducer is superior in its conductive property in both the ionic and electronic state [31, 32]. Furthermore, it can be electrodeposited on the MAB's multiple microfabricated electrodes in an aqueous medium and hence is compatible with batch processing [33]. The MAB's ASISEs demonstrated lifetime capability, as well as potentiometric and amperometric measurement stability in a biological liquid-medium environment [i.e. algal medium, Moorashige Skoog (MS) medium, phosphate buffer saline (PBS)]. The overall effort shows the potential to realize an ASISE-based lab-on-a-chip system for physiological sensing in liquid-medium environmental conditions. In addition, the system could help advance the technology for spaceflight astrobiology research and other measurements that require longer sensor lifetime.

II. MAB Biochip Design and Microfabrication

The design schematic of the MAB is shown in Figure 2. The MAB is a 10 x 11-mm biochip that consists of 9 Pt electrodes ($\varnothing$ = 240 μm) intended as ISE working electrodes (WEs) and 5 Pt ($\varnothing$ = 480 μm) electrodes intended as reference electrodes (REs). Two sets consisting of three electrodes share a reference electrode (RE), while the remaining 3 have their own RE. The sets of three WEs with shared REs are intended for potentiometric measurement, while the rest are intended for amperometric measurement. The WEs and their corresponding REs are equally spaced at 1.4 mm apart. These electrodes are connected to Pt contact pads (0.5 x 0.5 mm) located at one end of the biochip. The contact pads are further interfaced with an amplifier circuit.

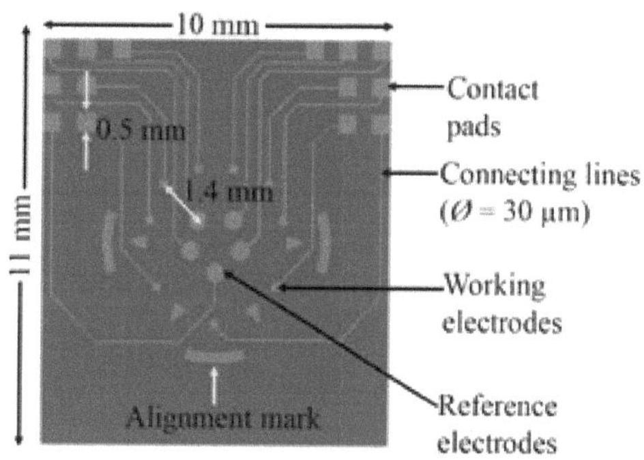

Fig. 2 Multi-analyte biochip (MAB) design schematic.

Based on the layout, a simple fabrication process was developed (Figure 3). Electrode definition via lift-off: A 4-inch silicon wafer with 1-μm thermal oxide was purchased from Noel Technology, Campbell, CA. Positive photoresist (AZ 9260) was spin-coated onto the oxide wafers at 2500 rpm to yield a ~10 μm–thick film. Photolithography was performed in a MA-6 mask aligner (Karl Suss MicroTec, Germany) using a dark field mask. Next a 30-nm titanium layer followed by a 150-nm platinum layer was deposited on the wafer using e-beam lithography (Varian e-beam evaporator). A lift-off process was performed to pattern microelectrodes arrays (Figure 3 a). The optical image of the electrodes defined by lift-off is shown in Figure 3 b, top right.

Electrode passivation: A 500 nm–thick silicon oxide was deposited on the wafer through plasma-enhanced chemical vapor deposition (STS PECVD, Germany) using 5% SiH_4 and N_2O. AZ-9260 photoresist was again spin coated and patterned on the wafer as described before. This process opens up windows in the photoresist corresponding to the electrode sensing areas and the electronic interface pads. This is followed by a reactive ion etching (RIE) process performed in SF6 plasma (Plasmatech, Suss MicroTec, Germany). The plasma etches the oxide, exposing the underlying metal. The oxide layer covers the perimeter of the electrodes to avoid exposure of the reactive seed metal layer. Removal of the photoresist post-etching in acetone leaves behind electrodes with passivated electrode leads (Figure 3 c). The optical image of the passivated electrodes with exposed active electrodes' area is shown in figure 3d.

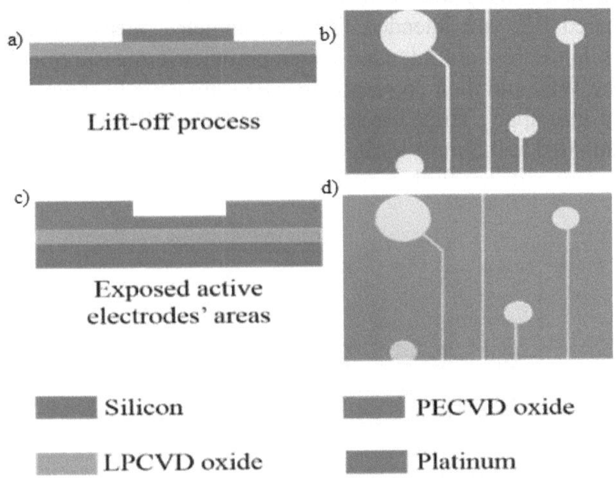

Fig. 3 Microfabrication process of the planar microelectrodes: a) cross section of the electrodes after lift-off process; b) optical image of the lift-off patterned electrodes; c) cross-section of the electrodes after passivation; d) optical image of passivated microelectrodes.

III. MATERIALS AND METHODS

A. Reagents and Chemicals

(i) Ion-selective-membrane: Three types of ion-selective membrane (ISM) have been used to determine the lifetime of ASISEs: pH ion-selective membrane, calcium-selective membrane, and ammonium-selective membrane. The composition of these ISMs is mentioned in percentage by weight/volume. a) pH ion-selective membrane consists of H^+ ionophore [1 wt. %], polyvinyl chloride (PVC) [9.9%], potassium tetrakis(4-chlorophenyl)borate (KTpClB) [0.5%], bis (2-ethylhexyl) sebacate [65.5%], cyclohexanone [10% wt/vol], and silicon tetrachloride ($SiCl_4$) [4.3 mg/100 mg without solvent]. b) The composition of calcium-selective membrane is Ca^{2+} ionophore [1 wt. %], polyvinyl chloride (PVC) [19.0%], polyurethane (PU) [10.0%], potassium tetrakis(4-chlorophenyl)borate (KTpClB) [0.7%], bis (2-ethylhexyl) sebacate [69.3%], cyclohexanone [10% wt/vol], and silicon tetrachloride ($SiCl_4$) [4.3 mg/100 mg without solvent]. c) Ammonium-selective membrane is made up of NH_4^+ ionophore I [1 wt. %], polyvinyl chloride (PVC) [22.5%], polyurethane (PU) [9.7%], potassium tetrakis(4-chlorophenyl)borate (KTpClB) [0.7%], bis (2-ethylhexyl) sebacate [66.1%], cyclohexanone [10% wt/vol], and silicon tetrachloride ($SiCl_4$) [4.3 mg/100 mg without solvent].

(ii) Surface functionalization of electrodes: a) Ag/AgCl REs were purchased from Ercon, Inc. (Wareham, MA). The active WEs were treated with three types of solution prepared as follows: a) PEDOT:PSS electropolymerization solution was prepared by combining 3,4-ethylenedioxythiophene (EDOT) monomer solution [10 mM] with sodium polystyrene sulfonate solution (NaPSS) [0.1 mM]. Both solutions were prepared in DI water. The solution was bubbled for 20 min before and during the electropolymerization process. b) To prepare PEDOT:$CaSO_4$ electropolymerization solution, NaPSS was replaced with 0.1 M $CaSO_4$ in 10 mL DI water. c) PEDOT:$(NH_4)_2SO_4$ electropolymerization solution: was prepared by replacing NaPSS with 0.1 M $(NH_4)_2SO_4$ in 10 mL DI water.

(iii) WEs conditioning solution: a) Three types of conditioning solution were used to prolong the WEs lifetime. Firstly pH membrane solution was mixed in 1:1:1 proportion by volume of potassium chloride (KCl) [5mM], sodium bicarbonate ($NaHCO_3$) [10 μM], and Algal buffer adjusted to pH 4.0. All aqueous solutions were prepared in DI water. b) Secondly Ca^{2+} membrane: solution was prepared by mixing 1:1 by volume of 0.1 M $CaCl_2$ and 10 μM $NaNO_3$ in 0.5x MS media. c) NH_4^+ membrane: 0.1 M NH_4Cl and PBS (pH 7.4) were mixed in 1:1 proportion. All materials were purchased from Sigma-Aldrich, St. Louis, MO, USA.

(iv) Measurement solution: a) The redox solution for CV characterization was a mixture of 2 mM potassium ferricyanide ($K_3Fe(CN)_6$) and 1 M potassium nitrate (KNO_3). b) 847 Algal medium (ATCC medium) was prepared by combining 1.0 L Bristol's solution and 1.0 g protease peptone (BD Diagnostic Systems, Sparks, MD, USA). The composition of Bristol's solution is sodium nitrate ($NaNO_3$) [10.0 g], calcium chloride dihydrate ($CaCl_2.2H_2O$) [1.32 g], magnesium sulphate heptahydrate ($MgSO_4.7H_2O$) [3.0 g], dipotassium phosphate (K_2HPO_4) [3.0 g], potassium dihydrogen phosphate (KH_2PO_4) [7.0 g], sodium chloride (NaCl) [1.0 g]. All Bristol's solution materials were purchased from Sigma-Aldrich. Six stock solutions (400 ml) were prepared for each salt. To prepare 1 L of Bristol's solution, 10 ml from the 400 ml stock solution was added to 940 ml DI water. At the end a drop of 1.0% $FeCl_3$ solution was added to the Bristol's solution. c) pH testing solution is an algal medium adjusted to the pH of interest with HCl or NaOH. d) Moreover, Ca^{2+} testing solution was prepared by pouring 0.01 M $CaCl_2$ in 0.5x MS medium titrated with 100 mM $CaCl_2$ in MS medium. e) NH_4^+ testing solution was prepared by dissolving NH_4Cl in PBS with pH=7.6 to yield a 0.1 M NH_4Cl solution.

B. Characterization Apparatus

Scanning electron microscopy was performed using a JEOL model 840 microscope. Optical microscopy was performed using a Nikon Eclipse model LV150 microscope. Electrochemical measurements was conducted using a three electrode potentiostat system (Bioanalytical Systems, Inc. (BASI), West Lafayette, IN, USA).

C. Surface Functionalization

An ion-selective electrode (ISE) consists of three layers: noble metal (e.g., Pt) electrodes, a conductive polymer conjugate transducer layer (e.g., PEDOT:PSS for H^+ ions; PEDOT:$CaSO_4$ for Ca^{2+} ions; PEDOT:$(NH_4)_2SO_4$ for NH_4^+ ions), and the ISM for each ion. The transducer layer was deposited via electropolymerization. Before the electropolymerization process, unwanted working electrodes areas are masked with photoresist (MicroChem AZ9260, hard-bake at 110°C). The electropolymerization deposition methods for PEDOT:PSS are double step chronoamperometry (DSCA) and cyclic voltammetry (CV), and are described. (i) DSCA: The first step was applying 2 mA for 0.5 s, followed by 0 mA for 0.5 s. These steps were repeated 999 for a total of 500 s at a 2 mA constant current. This process creates a PEDOT:PSS conductive polymer conjugate, designated PEDOT:PSS (DSCA). (ii) CV: The potential was performed at 0 to 1.1 V, with a 20-mV scan rate. This process creates a PEDOT:PSS conductive polymer conjugate, designated PEDOT:PSS (CV). The PEDOT:$CaSO_4$ and PEDOT:$(NH_4)_2SO_4$ deposition method is chronoamperometry at 814 nA for 30 min (0.2 mA/cm2). After the electropolymerization process, the MABs were dried under nitrogen and heated at 70°C on a hotplate for 30 min. The MABs were stored in a nitrogen-purged box before ISM deposition.

The surface functionalization process flow of the MAB biochip is shown in Figure 4. It starts with the electropolymerized MAB WEs (Figure 4 a). The darker color is indicative of the PEDOT polymer conjugate layer. The ISMs were deposited on the electropolymerized working electrodes by spin coating the MABs at 1500 rpm for 30 sec. This resulted in an ISM thickness of ~1 μm. The spin-coated biochips were vacuumed immediately after ISM deposition to avoid bubble entrapment in the membrane overnight. Then, the Pt contact pads were exposed by removing the membrane with cyclohexanone (Figure 4 b). Next, the reference electrodes (REs) were drop coated with silver chloride (Ag/AgCl) ink. After that, the MAB was baked in an oven at 70° C for 4 hours to activate the silanization process and dry the Ag/AgCl ink. The functionalized MAB is shown in Figure 4 c.

The MABs were stored in a nitrogen-purged box overnight in a closed container covered with aluminium foil, which prevents the effect of light on the electropolymerized conductive polymer conjugate layer. Before measurement, the MABs were bathed in conditioning solution for a minimum of 12 hrs. The conditioning solution was heated to 25°C before it was applied to the MAB surface.

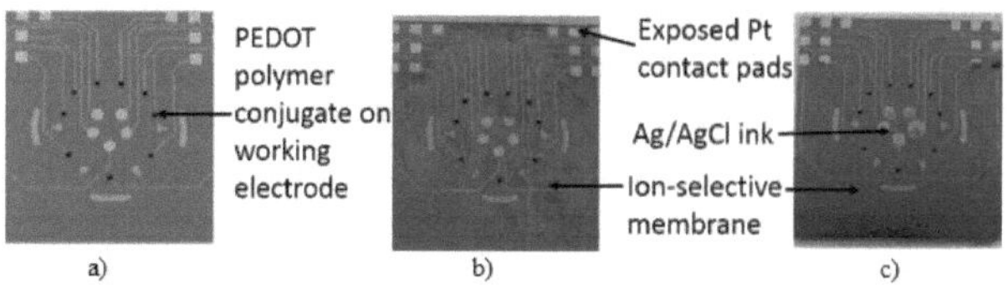

Fig. 4 Optical micrograph images of MAB electrode surface functionalization process, starting with a) electropolymerization of conductive polymer conjugate and b) spin-casting of ion-selective membrane onto Pt working electrodes; c) drop-coating of Ag/AgCl ink on reference electrode.

D. Electrochemical Measurement Setup

The setup for electrochemical measurement is shown in Figure 5. A rubber gasket was placed on the MAB and 150 μL measurement solution was pipetted within the gasket area (Figure 5 a); the MAB was positioned in a microfluidic chamber with two inlets and two outlets, with two syringes filled with 1 ml solution and two others left empty.

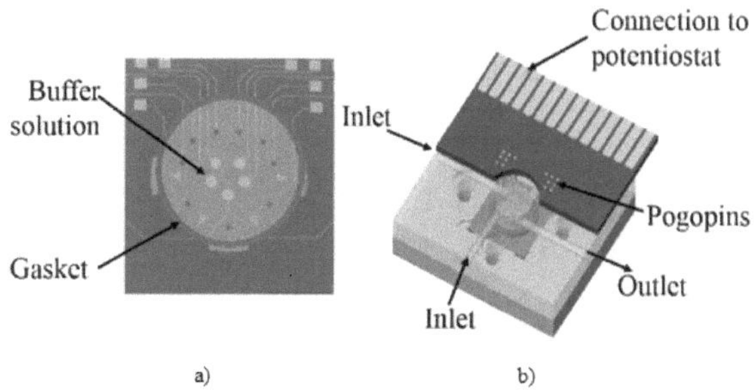

Fig. 5 Setup for electrochemical calibration measurement showing a) the MAB biochip with contained solution within gasket, b) the microfluidic chamber.

Before measurement, the measurement solution was passed back and forth between the two pairs of syringes to remove unwanted bubbles in the flow-cell chamber illustrated in Figure 5 b; pogo pins touched the MAB contact pads and were connected to the PCB board for electrical connection to the BASI cell stand; electrochemical measurement was performed using a three-electrode potentiostat. Electrochemical measurements were all conducted in a class-100 clean-room environment at 19 °C.

IV. Result and Discussion

A. Electrode Electrical Characterization

Cyclic voltammograms of MAB electropolymerized WEs were performed in redox-active solution (2 mM potassium ferricyanide ($K_3Fe(CN)_6$) and 1 M potassium nitrate (KNO_3)). The CV profiles (Figure 6) have a pair of redox peaks separated by 31, 44, and 50 mV at a lower scan rate (25 mV/s) for ISEs with PEDOT:PSS-DSCA, PEDOT:PSS-CV, and PEDOT:$CaSO_4$ transducers, respectively. The bare Pt electrode has redox peaks separated by 50 mV. The existence of small peak separation suggests a reversible electron-transfer reaction for the unmodified and modified WEs. Randles-Sevcik analysis was used to find the effective surface areas of the modified WEs [35, 36]. Based on Randles-Sevcik, the effective surface areas for PEDOT:PSS (DSCA), PEDOT:PSS (CV) and PEDOT:$CaSO_4$ WEs were found to be on the same order of magnitude: 4.37, 7.29 and 5.83 x 10^{-11} cm^2 respectively. This indicates that the ISE potentiometric and amperometric responses are dependent only on the types of counter-ions used in the synthesis of the conductive polymer conjugates of the transducer layer.

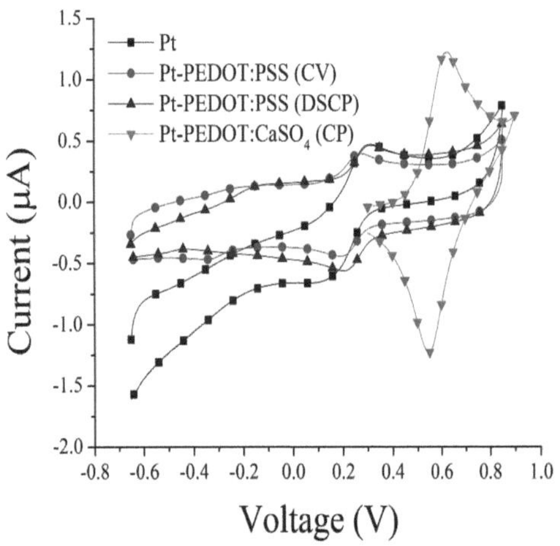

Fig. 6 Cyclic voltammetry profile of electropolymerized MAB and bare Pt working electrodes in redox solution.

B. Potentiometric Measurements

Potentiometric measurements were conducted with the ISEs for both monovalent (e.g., H^+ and NH_4^+) and divalent (e.g., Ca^{2+}) cations. Figures 7 a, b show the pH potentiometric profiles for ISEs with PEDOT:PSS (DSCA) and PEDOT:PSS (CV) transducers, respectively.

As can be seen, a linear cationic response at ~ 40 mV per pH change from 4 to 9 was achieved for both types. The result makes sense owing to the increase in OH^- concentration. An ISE with PEDOT:$(NH_4)_2SO_4$ transducer was used for NH_4^+ measurement. The pH was adjusted between 7.0 and 7.6 by PBS for all NH_4^+ ISE tests. Figure 7 c shows a sub-Nernstian positive slope of 55 mV per decade change in NH_4^+ concentration. For Ca^{2+} measurement, an ISE with PEDOT:$CaSO_4$ transducer was used; Figure 7 d shows the potentiometric profile with a positive slope at 30 mV per decade change in Ca^{2+}. The slope is slightly higher than the Nernstian slope value of 29 mV. The Nernstian slope profile is also dependent on the polymer conjugate counter-ion type. The slope profile for cations NH_4^+ and Ca^{2+} measurements was found to be negative when PEDOT:PSS was used instead of PEDOT:$(NH_4)_2SO_4$ and PEDOT:$CaSO_4$ respectively. We believe this could be due to the binding of NH_4^+ and Ca^{2+} to the SO_4^{2-} group of the PEDOT backbone or to the negative charges of the counter-ions of NaPSS. Further study is needed to verify this claim.

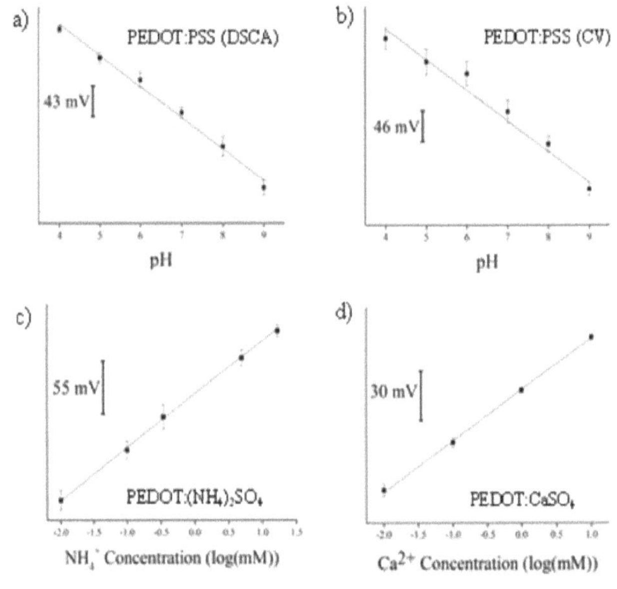

Fig. 7 ISE potentiometric profile for (a) PEDOT:PSS (DSCA) and (b) PEDOT:PSS (CV) by titration of algal buffer adjusted to pH 4-9 with 0.1 M HCl and 0.1M NaOH; (c) PEDOT:$(NH_4)_2SO_4$-based ISE potentiometric profile by titration of 0.1 M $(NH_4)_2SO_4$ in PBS adjusted to pH 7.4-7.6; (d) PEDOT: $CaSO_4$-based ISE potentiometric profile by titration of with 0.1 M $CaCl_2$ in MS medium.

C. Amperometric Measurements

An amperometric H_2O_2 test was conducted for ISEs with PEDOT:PSS (CV) and PEDOT:$CaSO_4$ transducers at 500 mV. Figures 8 a, b show the amperometric calibration plots, with the inset image showing the linear regression. The ISEs exhibit sensitivity of 3.32 and 1.25 $\mu AmM^{-1}cm^{-2}$, with a detection limit of 2.21 µM and 3.44 µM for PEDOT:PSS (CV) and PEDOT:$CaSO_4$, respectively. In comparison to previous biosensor work utilizing electrodes modified with carbon nanotubes and platinum nanoparticles, the sensitivity to H_2O_2 is in the same order of magnitude [35, 37]. These results demonstrate the potential of the MAB planar micro-fabricated ISEs to be utilized as amperometric bio-sensors.

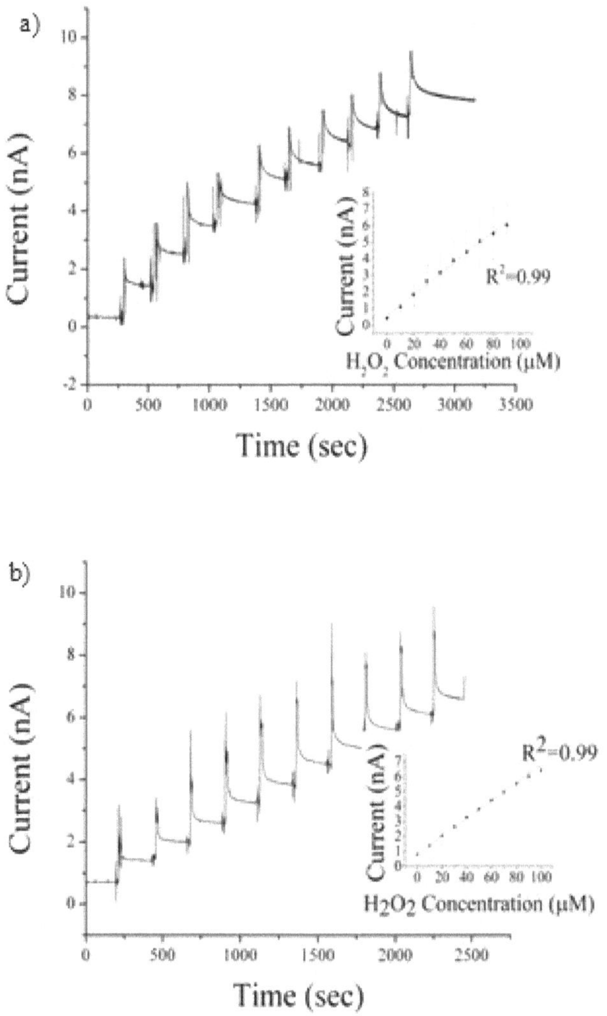

Fig. 8 Current response to the addition of 10 µM aliquots of H_2O_2 for ISE transducer based on (a) PEDOT:PSS (CV) and (b) PEDOT:$CaSO_4$ at a linear potential of 350 mV.

D. Lifetime Measurements

Application of ISE-based sensors in many fields (e.g., space flight, environment, and biomedicine) requires them to have long lifetime and be functional after several months' storage in a liquid-medium environment [38]. Solid-contact ISE lifetime is dependent on the ion-selective membrane (ISM) layer and the PEDOT polymer conjugate transducer layer.

For the ISM layer, various studies have demonstrated how lifetime is reduced by the leaching of ionophores into the external liquid medium [39] and the loss of membrane adhesion to the electrode solid surface [40]. These effects are further enhanced when ISEs are stored in the liquid medium for a prolonged period. To help reduce the effect of leaching of positive-charge ionophores (i.e., H^+, NH_4^+, Ca^{2+}) into the bulk solution, the ISE conditioning solution was prepared to have a high concentration of potassium chloride (KCl) [5mM], which could reduce the loss of the primary cationic ionophores [1]. In addition, to increase adhesion of the polymeric-based ISM to the electrode solid surface, $SiCl_4$ was used to create a silanized surface for ISM adhesion. The ISM silanol (SiO_3^-) groups are covalently bonded to the ISE PEDOT transducer layer [40].

The lifetime of an ISE is also dependent on the quality of the transducer layer. In order to demonstrate the lifetime capability of the MAB ISEs, pH potentiometric measurements were conducted over several days, for ISEs with PEDOT:PSS (DSCA) and PEDOT:PSS (CV) transducer layers respectively. The pH ISEs were stored in conditioning solution overnight before measurements were made. For pH measurements, the buffer medium was set at pH 4 and titrated with NaOH to adjust the pH from 4 to 9.

On the first day, both pH ISEs demonstrated a near-Nernstian slope value, with the PEDOT:PSS (DSCA) having a sub-Nernstian slope of -43.17 ± 3.03 $mVdec^{-1}$ and the PEDOT:PSS (CV) having a supra-Nernstian slope of -69.35 ± 7.97 $mVdec^{-1}$. However, after 3 days' storage in solution, the slope for the ISE with a PEDOT:PSS (DSCA) transducer decreased to -8.08 ± 2.07 $mVdec^{-1}$, and eventually died – no change in voltage was observed with pH change. This was not the case with the ISE based on a PEDOT:PSS (CV) transducer.

Figure 9 shows the change in slope for the ISE with PEDOT:PSS (CV) transducer over several days, when titrated with NaOH from pH 4 to pH 9. As mentioned earlier, the slope starts as supra-Nernstian and eventually goes down to 45.48 $mVdec^{-1}$ at day 10. The slope is lowest after two weeks and eventually stabilizes at 42.69 $mVdec^{-1}$ at day 28. Since the slope takes time to reach a steady-state value, we could conclude that meaningful measurements can be made only after sufficient storage in the conditioning solution. This makes the ISE based on a PEDOT:PSS (CV) transducer useful for applications that require prolonged storage in a liquid-medium environment before initial measurements are made. It is important to point out here that the conditioning solution used is medium with added KCl at 5 mM, promising its use as a robust biosensor. The short lifetime of the ISE with PEDOT:PSS (DSCA) transducer could be due to the following factors: the ISM detached from the transducer layer; the transducer layer became brittle and was removed from the substrate; the Cl^- ions from the liquid medium bound effectively to the PEDOT; the primary cations bound to the sulphur groups at the PEDOT backbone [41, 42]. These results suggest that when ISM deposition parameters are the same for both ISEs, the conductive polymer electrodeposition method could affect the ISE lifetime.

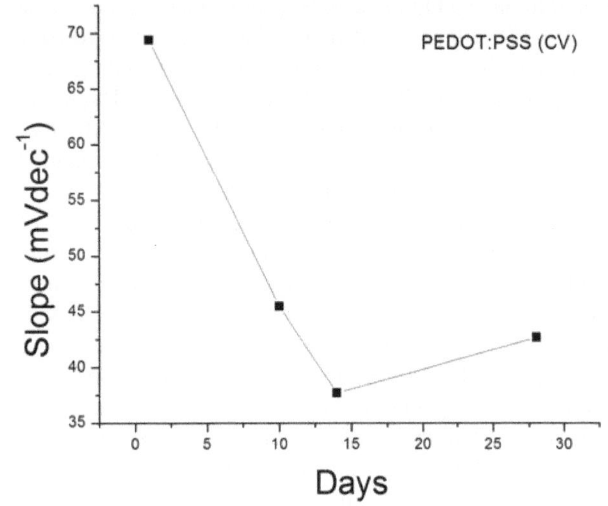

Fig. 9 Slope profile over several days for pH ISE based on a PEDOT:PSS (CV) transducer, from pH 4 to pH 9 for a 28-day period.

We performed SEM to investigate the surface morphology of both PEDOT:PSS transducer types electrodeposited specifically on MAB WEs. The surface morphology could provide an insight into the lifetime of the transducer layer. Figure 10 shows that the PEDOT:PSS (DSCA) has a film-like surface with embedded smaller grains, in contrast to PEDOT:PSS (CV), which has a film-like surface with embedded larger grains and bright agglomerates. Both transducer types have a granular morphology, indicative of amorphous structure in which grains are embedded in a polymer matrix [43]. Robust PEDOT polymer conjugates indicate effective immobilization of bulky dopants within the polymer structure and are denser in their structure substrate. After prolonged storage of PEDOT polymer conjugates in solution, a less dense film can become brittle and can have poor adhesion to the solid substrate [44]. Since the PEDOT:PSS (DSCA) is more film-like with fewer agglomerates, this could imply a less dense structure; hence the NaPSS binding to the PEDOT chain is loose. This allows Cl^- ions to bind effectively to the PEDOT, and the primary cations to bind effectively to the NaPSS anions. Furthermore, a less dense film could mean a weak mechanical property, allowing the film to quickly become brittle and detach itself from the solid substrate. More research is needed to verify these claims.

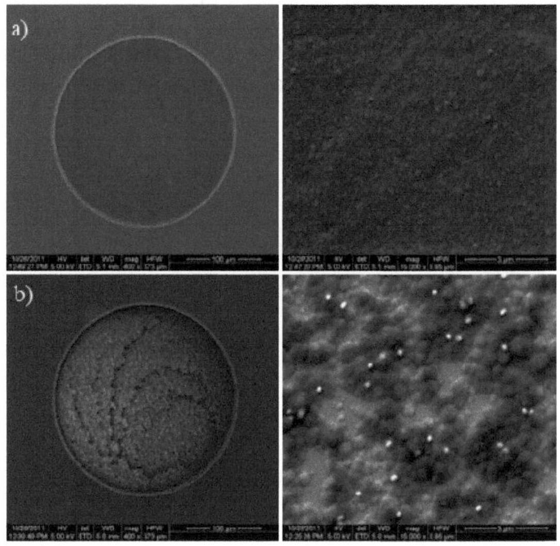

Fig. 10 Scanning electron micrographs images of MAB WEs electrodeposited with (a) PEDOT:PSS (DSCA) and (b) PEDOT:PSS (CV).

V. CONCLUSIONS

We have demonstrated the design of a biochip with multiple ASISEs that achieves long lifetime in a liquid-medium environment. The PEDOT:PSS electrodeposition methods were found to affect the ASISE lifetime, due to the different PEDOT polymer conjugate morphologies obtained via different electrodeposition methods. The lifetimes of the electrodes were tested for a month and show a stable Nernstian response after two weeks when PEDOT:PSS (CV) was utilized as the transducer layer. Although the ISM layer is thinner in comparison to that of conventional ISEs, it demonstrates stability and repeatability of ion measurements. These results suggest the possibility of utilizing the MAB biochip as a biosensor. The longer lifetime capability makes it suitable for utilization in applications that require prolonged sensor storage in a liquid-medium environment before initial measurements are made. Future work will focus on modifying the MAB electrode design to achieve higher sensitivity and faster response time for fundamental space biology research and high-throughput water-quality screening.

ACKNOWLEDGMENT

We would like to thank NASA Astrobiology Science Technology Instrumentation Development (ASTID) and EPA-STAR for funding support, Gale Lockwood of Birck Nanotechnology Center for the wire bonding, and Guy Telesnicki of the Birck Nanotechnoology Center for the dicing of the MAB devices.

CONFLICT OF INTEREST

The authors declare that they have no conflict of interest.

REFERENCES

1. P. Bühlmann, L. D. Chen, 2012, In *Supramolecular Chemistry: From Molecules to Nanomaterials,* ed. P. A. Gale, J. W. Steed, John Wiley & Sons, **5**, 2539.
2. E. Bakker, P. Bühlmann, E. Pretsch, 1997, *Chem. Review,* **97**, 3083.
3. E. Pretsch, 2002, *Anal. Chem.*, **74**, 420A.
4. A. Radomska, S. Singhal, H. Ye, M. Lim, A. Mantalaris, X. Yue, E. M. Drakakis, C. Toumazou, A. E.G. Cass, 2008, *Biosen Bioelectron,* **24**, 435.
5. R. De Marco, G. Clarke, B. Pejcic, *Electroanalysis,* 2007, **19**, 1987.
6. E. Bakker, M. E. Meyerhoff, 2007. Ion-Selective Electrodes for Measurements in Biological Fluids. *Encyclopedia of Electrochemistry.*
7. D. G. Pijanowska, W. Torbicz, 2008, *Biocybernetics and Biomedical Engineering,* **28**, 11.
8. E. Lindner, V. V. Cosofret, S. Ufer, R. P. Buck, W. J. Kao, M. R. Neuman, J. M. Anderson, 1994, *J. Biomed. Mater. Res.*, **28**, 591.
9. V. V. Cosofret, M. Erdosy, R. P. Buck, W. J. Kao, J. M. Anderson, E. Lindner, M. R. Neuman, 1994, *Analyst,* **119**, 2283.
10. H. Xue, W. G. Sunda, 1997, *Environ. Sci. Technol,* **31**, 1902.
11. E. Fogt, 1990, *Clin. Chem.*, **36**, 1573
12. H. M. Elqudaby, E. Y. Z. Frag, G. G. Mohamed, M. A. E. Mohamed, 2011, *Anal. Bioanal. Electrochem.*, **3**, 420.
13. E. Kress-Rogers, 1986, *J. Phys. E.: Sci. Instrum.*, **19**, 13.
14. A. ul Haque, M. Rokkam, A. R. De Carlo, S. T. Wereley, S. J. Roux, P. P. Irazoqui, D. M. Porterfield, 2007, *Sens. Actuator B,* **123**, 391.
15. D. Amman, 1986, *Ion-selective Microelectrodes,* Springer Verlag, Berlin.
16. M. A. Messerli, I. Kurtz, P.J. S. Smith, 2008, *Anal. Bioanal. Chem.*, **390**, 1355.
17. S. L. Bonting, 1992, *Adv. Space Bio. Med.,* **2**, 263.
18. W. Oelβner, S, Hermann and H. Kaden, *1997, Aerospace Science and Technology, 1,* 291.
19. J. W. Hines, S. Arnaud, M. Madou, J. Joseph, A. Jina, 2001, *NASA Washington, Technology: The Second National Technology Transfer Conference and Exposition,* **1**, 515.
20. F. Song, J. Ha, B. Park, T. H. Kwak, I. T. Kim, H. Nam, G. S. Cha, 2002, *Talanta,* **57**, 263.
21. J. Bobacka, 2006, *Electroanalysis,* **18**, 7.
22. A. Michalska, J Dumanska, K. Maksymiuk, 2003, *Anal. Chem.,* **75**, 4964.
23. N. A. Alarfaj, M. F. El-Tohamy, 2012, *Int. J. Phys. Sci.,* **7**, 1403.
24. A. Michalska, K. Maksymiuk, *Anal. Chim. Acta.,* **523**, 97.
25. J. Bobacka, 1999, *Anal. Chem.,* **71**, 4932.
26. S. Anastova-Ivanova, U. Mattinen, A. Radu, J. Bobacka, A. Lewenstam, J. Migdalski, M. Danielewski, D. Diamond, 2010, *Sensor Actuat. B-Chem.,* **146**, 199.
27. M. Vázquez, J. Bobacka, A. Ivaska, A. Lewenstam, 2002, *Sensor Actuat. B-Chem,* **82**, 7.
28. B. Paczosa-Bator, R. Piech, A. Lewenstam, 2010, *Talanta,* **81**, 1003.
29. T. Satchwill, D. J. Harrison, 1986, *J. Electroanal. Chem.,* **202**, 75.
30. R. De Marco, J. P. Veder, G. Clarke, A. Nelson, K. Prince, E. Pretsch, E. Bakker, 2008, Phys. Chem. Chem. Phys., 10, 73.
31. J. Bobacka, 2006, *Electroanalysis,* **1**, 7.
32. A. Elschner, S. Kirchmeyer, W. Lövenich, U. Merket, K. Reuter, 2011, In *PEDOT: Principles and Applications of an Intridically Conductive Polymer*, CRC Press, Taylor and Francis Group LLC, Boca Raton, FL, Chapter 9.
33. B. Somboonsub, M. A. Invernale, S. Thongyai, P. Praserthdam, D. A. Scola, G. A. Sotzing, 2010, *Polymers,* **51**, 1231.
34. J. Bobacka, A. Ivaska, and M. Grzeszczuk, 1991, *Synth. Met.,* **44**, 9.
35. J. Shi, J. C. Claussen, E. S. McLamore, A. ul Haque, D. Jaroch, A. R. Diggs, P. Calvo-Marzal, J. L. Rickus, D. M. Porterfield, 2011, *Nanotechnology,* **22**, 1.
36. A. J. Bard, L. R. Faulkner, 2000, *Electrochemical Methods: Fundamentals and Applications,* Wiley, New York.
37. J. C. Claussen, S. S. Kim, A. ul Haque, M. S. Artiles, D. M. Porterfield, T. S. Fisher, 2010, *J. of Diab. Sci. Tech.,* **4**, 312.
38. J. Bobacka, M. McCarrick, A. Lewenstam, 1994, *Analyst,* **119,** 1985.
39. P. Bühlmann, Y. Umezawa, *Anal. Chem.,* 2000, **72**, 1843.
40. G. S. Cha, D. Liu, M. E. Meyerhoff, H. C. Cantor, A. R. Midgley, H. D. Goldberg, R. B. Brown, *Anal. Chem.,* 1991, **63**, 1666.
41. M. Vázquez, J. Bobacka, M. Luostarinen, K. Rissanen, A. Lewenstam, A. Ivaska, 2005, *J. Solid State Electrochem,* **9**, 312.
42. A. Michalska, A. Galuskiewicz, M. Ogonowska, M. Ocypa, K. Maksymiuk, 2004, *J. Solid State Electrochem,* **8**, 381.
43. U. Lang, E. Müller, N. Naujoks, J. Dual, 2009, *Adv. Funct. Matr.,* **19**, 1215.
44. S. Ashizawa, R. Horikawa, H. Okuzaki, 2005, *Synth. Met.,* **153,** 5.

Corresponding author:

Dr. Wan Wardatul Amani Wan Salim
Department of Biotechnology Engineering
Kulliyyah (Faculty) of Engineering
International Islamic University Malaysia (IIUM)
Jalan Gombak, 50728 Kuala Lumpur, Malaysia
Email: asalim@iium.edu.my

Tuning Methods for Wireless Power Transfer in Biomedical Device

E.S. Lai, N.A. Ita, M.R. Besar, and M.Y. Ahmad

Department of Biomedical Engineering, Faculty of Engineering, University of Malaya (UM), 50603, Kuala Lumpur, Malaysia

Abstract— **This paper presents tuning methods on a Class-E power amplifier for a wireless power transfer (WPT) system based on resonant inductive coupling. In this paper, the scope of research is focusing on providing a stable power for Wireless Capsule Endoscopy (WCE). Shifting of the resonance frequency of a WPT system is unavoidable mainly due to various factors including variations in its working environments. This reduces the performance of WPT system leading to inconsistent voltage at the receiver. To overcome this problem, two tuning approaches namely frequency tuning and voltage tuning are proposed. The Frequency tuning method works to tune the power amplifier circuit so that it operates at the frequency close the resonant frequency of the transmitter which is also ensured the optimal switching condition in the driving circuit. The voltage tuning helps to regulate the receiver's voltage output to make it stable and constant. Our experimental results indicate that the proposed approach has improved the performance of the WPT.**

Keywords— **Wireless Power Transfer (WPT), Wireless Capsule Endoscopy (WCE), Tuning circuit.**

I. INTRODUCTION

Wireless power transfer (WPT) system is a potential solution to address power deficiency problems typically faced by conventional implantable medical devices [1]. WPT system is advantageous because the power at the receiver can be adjusted, can sustain longer operating time and more importantly it is safe to be used within living organism [2] [3].

Conventional WCEs are battery-powered but its power capacity is unable to meet up the power requirement of future WCEs. Advanced features such as inclusion of actuators, higher image resolution system, longer operating time, etc are always desired for future WCE. Thus these demand higher power which can range up to 400mW [4]. Therefore, WPT system is important to be investigated to support future WCEs.

However, in WPT system, the voltage output at the receiver coil is not always consistent due to special operating environment of WCE. Factors such as the displacement, angle and lateral misalignment (between the transmitter and the receiver), transmitter coil deformation, loss in transfer medium, variation of load impedance are among the main reasons for low efficiency of WPT system which cause drop in voltage at the receiver [5].

There are many works have been reported to improve the efficiency of WPT system. The scopes of researches however are mainly focused on the coil design [6], the resonant topology [7], tuning circuit such as [14], etc.

In this paper, a simple and novel method is proposed to tune the driving circuit so that it operates at a frequency close to transmitter coil's resonant frequency and also operates at optimum switching condition. The optimum switching is a condition of zero-voltage switching (ZVS) and zero-voltage derivative switching (ZVDS) and it is important for the high efficiency of the driving circuit, [8]. The aim of this paper is to achieve a constant voltage output at the receiver side even in various displacements between the transmitter and the receiver.

This paper is organized as follows. Section II provides an overview on the resonant inductive link and power amplifier circuit. Section III discusses our proposed methods to improve voltage output at the receiver which involve frequency tuning and voltage tuning methods. Section IV describes the experimental setup and discusses the validation of the proposed methods. Finally, Section V provides conclusion and future work.

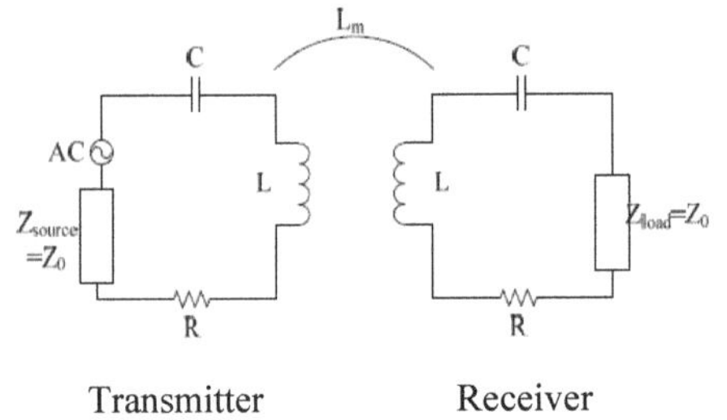

Fig. 1 The Simple inductive link circuit model [2]

II. INDUCTIVE COUPLING MODEL

A. Resonant Inductive Coupling

This section provides a brief overview on the inductive coupling links. Inductive coupling is considered as most common technique in the WPT technology [5]. It can be classified into: loose–coupling (LC) mechanism [9] and strong coupling (SC) mechanism [10][11]. The LC mechanism is commonly consists of two coils which are the transmitting coil and the receiving coil. The LC mechanism is

F. Ibrahim et al. (eds.), *International Conference for Innovation in Biomedical Engineering and Life Sciences*,
IFMBE Proceedings 56,
DOI: 10.1007/978-981-10-0266-3_19

able to maintain efficiency power transfer even in relatively far separation distance (tens of millimeters) and low coupling factors between the transmitter and the receiver. The SC mechanism involves multiple-coils typically three to four coils. This paper focuses on the tuning system on LC mechanism because its simplicity design is more suitable and practical to be used for WCE system.

In the LC mechanism, the transmitting coil is driven by a driving circuit, also known as power amplifier circuit which operates at the LC resonant frequency. Ideally, the resonant frequency of the transmitting coil and the receiving coil must be the same to ensure maximum power transfer from the transmitter to the receiver.

The resonant frequency can be tuned by the additional capacitance whether in the series or parallel topology which is depending on the load value, mutual inductance and resonant frequency [12]. So the ultimate tuning parameter for the WPT system is the capacitance tuning. There are two approaches which are being investigated and they are an additional capacitor bank and a variable capacitor. First approach may have problems with the mechanical switching due to high voltage condition which possibly results in sparks, whereas later approach is very expensive and bulky due to variable capacitor itself and also the involvement of stepper motor [13][14]. Therefore, a more feasible alternative is needed to improve the efficiency of WPT system. Figure 2 and 3 show the design of the transmitter and receiver model used in the paper to study the efficiency of the WPT system.

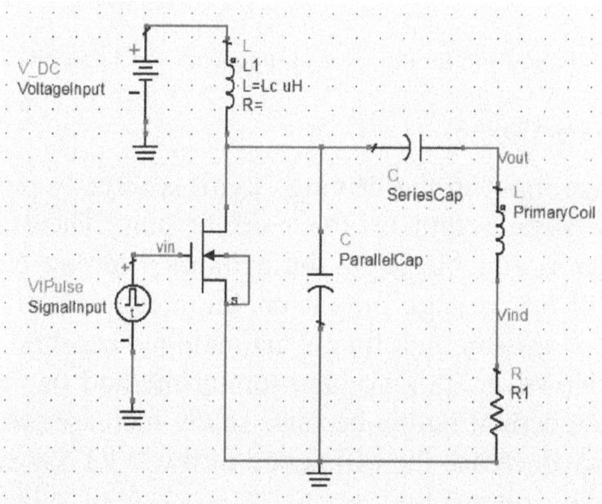

Fig. 2 The transmitter model with Class-E driving circuit

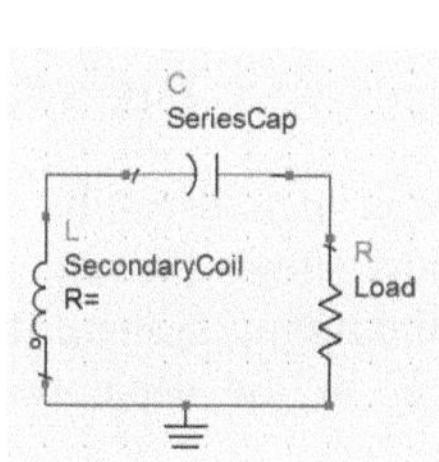

Fig. 3 The series-resonant topology receiver model

The main reason for the resonance is to improve the efficiency of the WPT system by maximize the reflected impedance seen by a power source in the primary coil. [15]

Equation below shows the calculation of resonance frequency of the primary coil based on the series capacitor and inductor.

$$F = \frac{1}{2\pi\sqrt{LpCs}} \tag{1}$$

Where,
L_p = Inductance of primary coil
Cs = Series capacitance

B. Class-E Power Amplifier

A power Amplifier is a driving circuit for the primary coil. It has different topologies and each with different characteristic to meet up the requirement of the inductive coupling link. In WPT system, alternating magnetic field is needed and the magnetic field's intensity is proportional to the current's amplitude in the primary coil. Therefore, power amplifier is needed to amplify the signal and drive the transmitter coil.

The class-E driving circuit, first introduced by Sokal and Sokal, is a high efficient driver and it is capable to drive large power at high frequency [16]. It also presents simplicity and low component count in the circuit design. Thus, it is suitable to be used as primary coil driver in the WPT system for the WCE application. Generally, it consists of a MOSFET, Choke inductor, series capacitor, parallel capacitor and primary coil, which are showed in the figure 2 on its circuit design.

III. THE POPOSED TUNING METHODS

The disadvantage of the Class-E driving circuit is that it is sensitive to the variation such as the change of distance and also loads impedance. A small deviation caused by variable conditions may significantly degrade the efficiency of the class-E driving circuit. This is mainly due to the shifting of the resonant frequency of transmitter part away from the operating frequency which leads to non-optimal switching in the driving circuit. The non-optimal switching condition may damage the components inside the driving circuit, especially the MOSFET due to the sudden current spike.

In the application of the WCE, its capsule is not fixed but freely moving and rotating in the digestive tract, thus there are many variable conditions, such as the change of displacement (between capsule and transmitter) and the deformation of the transmitting coil. These factors may cause the change of reflected impedance seen by the primary coil. Consequently, the voltage output will be affected and becoming fluctuating. Those variations are accounted for the low performance and non-consistent voltage output of the existing WPT system.

Several types of class E tuning method have been introduced in previous researches to solve the efficiency problems such as Impedance Network Switching [17], Phase control [18], etc. However, these methods may be too complex to be applied on the WPT system for WCE application.

In this paper, three parameters on the driving circuit have been identified to be effective in control of voltage output at receiver side. They are the frequency tuning, voltage tuning and lastly capacitance tuning. However, only frequency and voltage tuning methods are discussed due to its simplicity. Furthermore, the concern of this paper is mainly focused on the tuning system for the transmitter part, instead of receiver part. This is due to the size of the receiver is strictly restricted to swallowable size for the WCE application.

A. Frequency Tuning

For a resonant condition to occur in the WPT system, there are three important frequencies need to be considered for each coupling coefficient. They are the resonant frequency of the transmitting and receiving coils and the operating frequency. Due to the restricted size of the receiver, this paper investigated the operating frequency of the class E driving circuit as a frequency tuning method for the WPT system.

In the WCE application, the difference of the size between the transmitter and receiver is very large and thus the coupling coefficient between the coils is very low. Therefore, the shifting of the resonance frequency is not very significant at the receiver coil. Transmitter coil, on the other hand, may get additional resonance frequency shifting due to its coil flexibility in a portable WCE application which results in the coil deformation. The frequency deviation between the operating frequency and resonant frequency of transmitting coil may degrade the maximum power transfer to the transmitting coil from the source which leads to low efficiency of driving circuit. This causes the receiving power output to be inconsistent. Therefore, a reasonable tuning on the operation frequency will be effectively increasing the efficiency of the driving circuit and also maintaining optimum switching condition in the driving circuit. These are the main concern in this paper in achieving the high efficiency of driving circuit and the stable voltage output.

Figure 4 shows that resonance frequency of the both coils in the ideal condition where both of them resonant at the same frequency. It is important for the high efficiency of WPT system. Therefore, the operating frequency tuning method in this paper is designed with the range of tuning frequency in the bandwidth of the fixed resonant frequency of receiver, as shown in the figure 5. The large bandwidth of receiving coil, due to the lower Q factor allows larger operating frequency tuning range and prevents the high Q factor transmitting coil to be shifted far away from the receiver's resonant frequency. Overall, the limited tuning frequency tuning range is to prevent over tuning of operating frequency away from receiver's resonance frequency which may cause significant loss of efficiency of the WPT system.

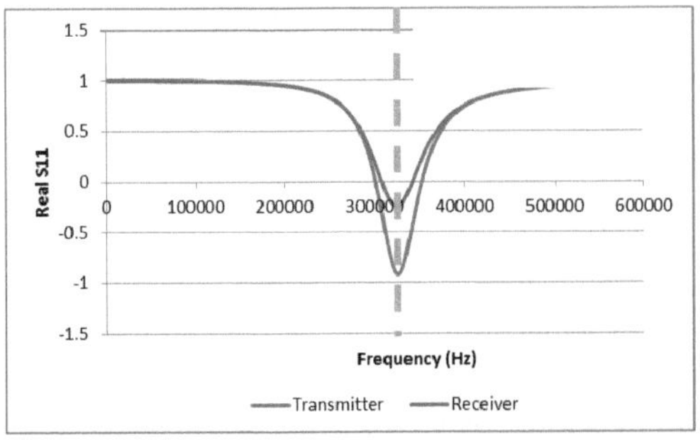

Fig. 4 Ideal resonant frequency of transmitter's and receiver's coil. The Green line indicate the resonant frequency of both coils.

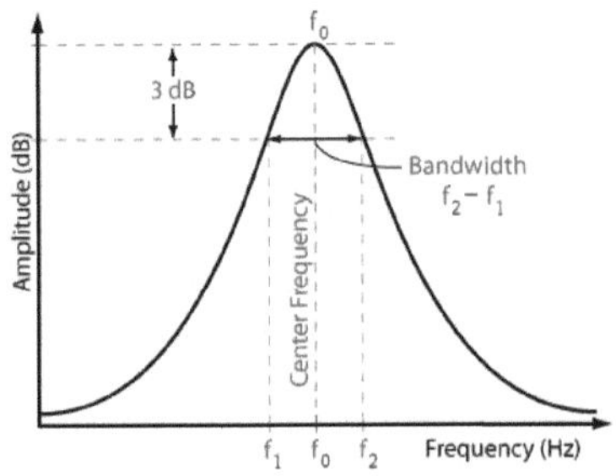

Fig. 5 Bandwidth for the resonant frequency for the receiver part

B. Voltage Tuning

Voltage input in the driving circuit is directly proportion to the voltage output on the receiver part. Therefore, this parameter is suitable to be the additional tuning parameter to control the voltage output on the receiver part. This is because it is very helpful in maintaining constant voltage output. However, this voltage tuning method has to be limited to a certain range, because solely increase the voltage input may decrease the efficiency of the WPT system. As a result, the voltage tuning has to work with other tuning parameters in order for the high efficient WPT system.

IV. RESULT AND DISCUSSION

In this section, experimental results are showed and discussed. Figure 6 below shows the experimental setup for a series of studies on the tuning system. It consists of a transmitter and a receiver in the separation distance varying between 1- 20 cm.

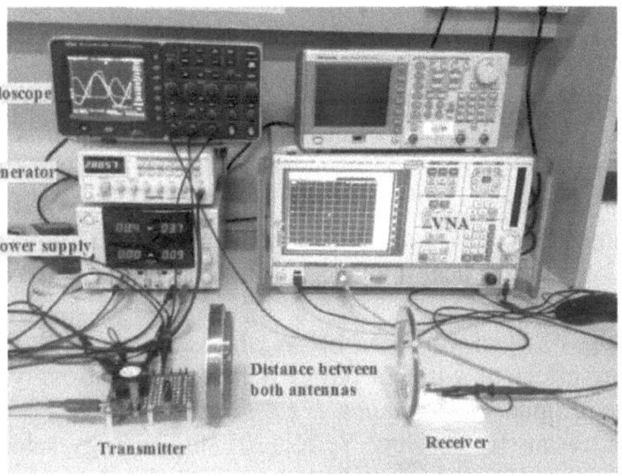

Fig. 6 The experiment set-up to study the tuning method in the WPT system

First experiment is to study the efficiency of the driving circuit corresponds to the operating frequency tuning. Its results showed that the power transfer to the transmitting coil was the maximum when the operating frequency was tuned to the resonant frequency of the transmitting coil (344 kHz), shown in figure 7. Furthermore, when the distance between both coils varied in the range 1-20 cm, the efficiency of the driving circuit improved after the frequency tuning method is applied, shown in figure 8. This concept is also applicable for the deformation of transmitter coil, where deformation of coil alters the resonant frequency of transmitting coil.

Figure 9 shows the comparison of voltage output at receiver side, for the cases of without tuning method, with frequency tuning method and with both frequency and voltage tuning methods. Without the tuning method, the voltage output at receiver side was inconsistent. When frequency tuning is implemented, the voltage output result is improved. However, merely frequency tuning method is not able to sustain the constant voltage output for large displacement. Therefore, additional voltage tuning is able to achieve the ultimate aim of constant voltage output at receiver side.

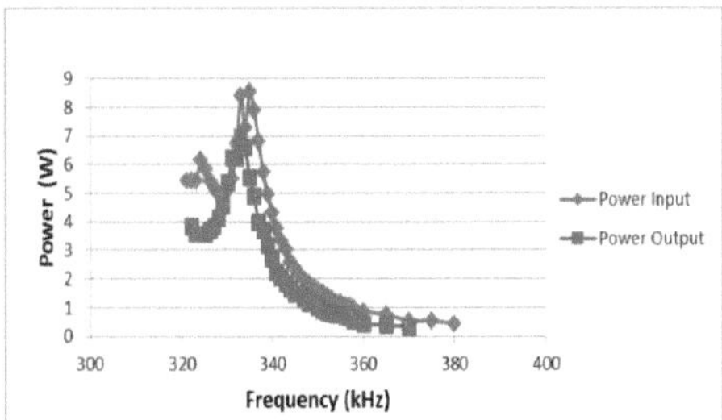

Fig. 7 Power transfer from power source to the transmitter coil in a range of frequency

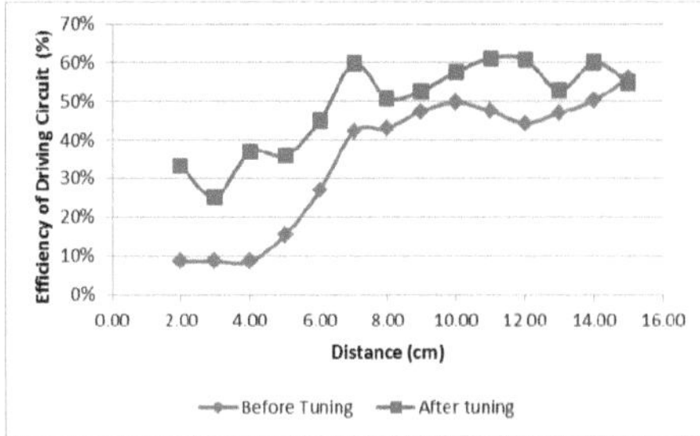

Fig. 8 Efficiency of driving circuit in the transmitter part before and after frequency tuning

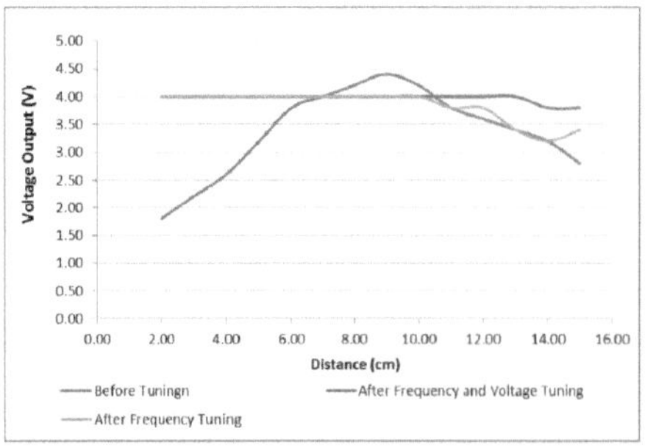

Fig. 9 Voltage Output at receiver for cases of without tuning, with frequency tuning and with both frequency and voltage tuning.

V. CONCLUSIONS

Simple and novel tuning methods have been presented to tune the Class-E power amplifier which is a driving circuit for the primary coil in the WPT system based on the resonant inductive coupling. Operating frequency tuning method helps to tune driving circuit's frequency as close to transmitting coil's resonant frequency as possible, while it ensures the optimal switching condition. This helps to maintain the efficiency of driving circuit, as well as the efficiency of WPT system. Furthermore Voltage tuning method acts as an additional tuning parameter to regulate the voltage output to be constant which is desired condition for powering system. Experimental results showed the validation of the both tuning methods. In the end, the constant voltage output was achieved with both operating frequency tuning and voltage tuning. Both tuning methods needed to work together to have better outcome result. They can be implemented electronically to work as an automated tuning system in the future. It is also suitable to be applied on other implanted medical device.

Future work may include further understanding the performance of WPT system incorporate with tuning parameter such frequency and duty cycle of the driving signal. Besides, further investigation on the capacitance tuning to incorporate with frequency and voltage tuning methods may further improve the outcome result.

ACKNOWLEDGMENT

1. Fundamental Research Grant (FRGS: FP036-2014B)
2. University of Malaya Research Grant (UMRG: RP009D-13AET)

CONFLICT OF INTEREST

The authors declare that they have no conflict of interest.

REFERENCES

1. Basar, M. R., Ahmad, M. Y., Cho, J., & Ibrahim, F. (2014). Application of wireless power transmission systems in wireless capsule endoscopy: an overview. *Sensors, 14*(6), 10929-10951.
2. Guidelines for Limiting Exposure to Time-Varying Electric, Magnetic, and Electromagnetic Fields (up to 300 GHz). International Commission on Non-Ionizing Radiation Protection. Health Phys.1998;74:494–522.
3. ICNIRP Statement on the "Guidelines for Limiting Exposure to Time-Varying Electric, Magnetic, and Electromagnetic Fields (up to 300 GHz)" Health Phys. 2009;97:257–258.
4. Basar, M. R., Ahmad, M. Y., Cho, J., & Ibrahim, F. (2014). Application of wireless power transmission systems in wireless capsule endoscopy: an overview. *Sensors, 14*(6), 10929-10951.
5. Sun, T., Xie, X., & Wang, Z. (2013). Wireless power transfer for medical microsystems. New York: Springer.
6. Basar, M. R., Ahmad, M. Y., Cho, J., & Ibrahim, F. (2014, December). A wireless power transmission system for robotic capsule endoscopy: Design and optimization. In *RF and Wireless Technologies for Biomedical and Healthcare Applications (IMWS-Bio), 2014 IEEE MTT-S International Microwave Workshop Series on* (pp. 1-3). IEEE.
7. Ahn, D., & Hong, S. (2014). Wireless power transmission with self-regulated output voltage for biomedical implant. *Industrial Electronics, IEEE Transactions on, 61*(5), 2225-2235.
8. Li, Y. F. (2012, May). Auto-tuning controller design of class E inverter with resonant components varying. In *Industrial Electronics (ISIE), 2012 IEEE International Symposium on* (pp. 217-221). IEEE.
9. Chen, C. J., Chu, T. H., Lin, C. L., & Jou, Z. C. (2010). A study of loosely coupled coils for wireless power transfer. *Circuits and Systems II: Express Briefs, IEEE Transactions on, 57*(7), 536-540.
10. RamRakhyani, A. K., Mirabbasi, S., & Chiao, M. (2011). Design and optimization of resonance-based efficient wireless power delivery systems for biomedical implants. *Biomedical Circuits and Systems, IEEE Transactions on,5*(1), 48-63.
11. Zhang, F., Hackworth, S., Fu, W., Li, C., Mao, Z., & Sun, M. (2011). Relay effect of wireless power transfer using strongly coupled magnetic resonances.*Magnetics, IEEE Transactions on, 47*(5), 1478-1481.
12. Jegadeesan, R., & Guo, Y. X. (2012). Topology selection and efficiency improvement of inductive power links. *Antennas and Propagation, IEEE Transactions on, 60*(10), 4846-4854.
13. Carta, R., Thoné, J., Gosset, G., Cogels, G., Flandre, D., & Puers, R. (2011). A self-tuning inductive powering system for biomedical implants. *Procedia Engineering, 25*, 1585-1588.
14. Jourand, P., & Puers, R. (2012). A class-e driven inductive power delivery system covering the complete upper body. *Sensors and Actuators A: Physical,183*, 132-139.
15. Aldhaher, S., Luk, P. C. K., & Whidborne, J. F. (2014). Tuning Class E inverters applied in inductive links using saturable reactors. *Power Electronics, IEEE Transactions on, 29*(6), 2969-2978.
16. Sokal, N. O., & Sokal, A. D. (1975). Class EA new class of high-efficiency tuned single-ended switching power amplifiers. *Solid-State Circuits, IEEE Journal of, 10*(3), 168-176.
17. Vasic, M., Garcia, O., Oliver, J. A., Alou, P., Diaz, D., Prieto, R., & Cobos, J. (2012). Envelope amplifier based on switching capacitors for high-efficiency RF amplifiers. *Power Electronics, IEEE Transactions on, 27*(3), 1359-1368.
18. Li, Y. F. (2012, May). Auto-tuning controller design of class E inverter with resonant components varying. In *Industrial Electronics (ISIE), 2012 IEEE International Symposium on* (pp. 217-221). IEEE.

Author: Mohd Yazed Ahmad
Institute: University of Malaya
City: Kuala Lumpur
Country: Malaysia
Email: myaz@um.edu.my

Graphene Hydrogel Novel Nanostructure as a Scaffold

N. Mansouri and S. Bagheri

Nanotechnology & Catalysis Research Centre (NANOCAT), IPS Building, University Malaya, 50603 Kuala Lumpur, Malaysia

Abstract— **To address problems related to current therapeutic and surgical treatments tissue engineering field aims to generate functional tissues in order to improve or restore damaged or diseased tissue function. Graphene hydrogel is capable of to provide groundwork of producing a biomimetic, porous nanocomposite scaffold for effective tissue regeneration. This paper describes graphene's prospective potential as scaffold for tissue engineering. An active cell scaffold based on a graphene hydrogel has been magnificently examined by various characterization methods such as UV–vis, Fourier transform infrared (FT-IR), Raman spectroscopy, x-ray diffraction (XRD), and scanning electron microscopy (SEM). Subsequently, the characterization techniques confirm the capacity of graphene hydrogel as scaffold.**

Keywords— **Tissue engineering, hydrogel, graphene-based scaffold.**

I. INTRODUCTION

Recently, several biomedical applications have been recognized for graphene derivatives. Some of these pioneering applications incorporating graphene biomaterials can be listed as biosensors, regenerative medicine, stem cell engineering, tissue engineered scaffolds and grafts, bioimaging, drug and gene delivery [1], wound healing, cell twitching and signaling [2] and many others. Clinical applications seek for functional scaffold that resembles ECM's microenvironment in order to control cell performance and tissue regeneration rate. Graphene's exceptional properties can contribute to make graphene-based materials as a promising candidate for tissue engineering field. These properties which can drastically lead to superior cell culture are listed below [3]:

(i) Great biocompatibility, chemical inertness and unique surface architecture
(ii) Excellent mechanical properties including high elasticity, flexibility, and adaptability to even or uneven surfaces
(iii) High electrical conductivity which can result in cell behavior including cell adhesion, migration, and orientation
(iv) Possibility for controlled incorporation of biochemical cues onto graphene surface [4]

Hydrogel, a three-dimensional network, comprised of hydrophilic polymers crosslinked through covalent bonds or maintained together by physical intramolecular and intermolecular attractions. There are great potential applications of hydrogels in tissue engineering both directly after fabrications, with or without cell entrapment, and after construction as scaffolds. As mechanical properties of scaffolds made up of hydrogels can be precisely controlled, hydrogel-based scaffold are substantially appreciated due to resembling the in vivo microenvironment of tissues. Hydrogel-based scaffolds' mechanical properties in tissue engineering considerably affect attached and/or encapsulated cells. However, hydrogels have viscoelastic and transport properties provide the resembled in vivo environment of tissues. But, it lacks of mechanical strength which limit their application in tissue engineered products. Accordingly, graphene's extraordinary tailor-made mechanical properties can vastly reinforce hydrogels, biodegradable films, electrospun fibers and further tissue engineering scaffolds [5].

Numerous biocompatible scaffolds based on graphene coatings have been employed to verify potential properties of graphene-based substrates in culturing and proliferation of MSCs [6], iPSC (Induced pluripotent stem cells) [7], neural stem cells (NSC) [8] and circulating tumor cells (CTCs) performance [9]. In this contribution 3D graphene hydrogel was prepared in one step. The prepared 3D graphene hydrogel is suited to serve as tissue engineering scaffolds.

II. EXPERIMENTAL PROCEDURES

A. Production of Graphene Oxide from Graphite

The production of graphene oxide was carried out using the modified Hummer's method. Accordingly, graphite oxide was synthesized through oxidation of graphite flakes with H_2SO_4, H_3PO_4 and $KMnO_4$. Next, $KMnO_4$ was added gradualy to the solution. The solution turned from dark green to dark purplish green. The mixture was left stirring for 72 hours in order to complete the oxidation. During oxidation, the mixture color transformed from dark purplish-green to dark brown. After 72 hours, to halt the oxidation course the mixture was cooled down to below 10°C using ice cubes. This was followed by instant incorporation of Hydrogen peroxide (H_2O_2) solution. The mixture color altered to bright yellow, indicative of a high oxidation level of graphite. After that, the obtained graphite oxide was washed few times with 1 M of aqueous HCl solution and repetitively with DI water to reach pH of 4–5. The washing process was performed via a simple decantation of the

F. Ibrahim et al. (eds.), *International Conference for Innovation in Biomedical Engineering and Life Sciences*,
IFMBE Proceedings 56,
DOI: 10.1007/978-981-10-0266-3_20

supernatant using a centrifugation machine. While washing practice is done with DI water, the graphite oxide goes through exfoliation, which leads to thickening of the graphene oxide solution, producing GO gel.

B. Synthesis of 3D Graphene Hydrogel

A 40 mL of graphene oxide homogeneous aqueous solution with concentration of 1 mg/mL was sonicated for 15 minutes in order to prepare the graphene hydrogel. Next, the final step involves 18 hours of hydrothermal reaction (180 °C) in a 25-mL Teflon-lined autoclave. When the autoclave has been cooled down to room temperature, the obtained reduced graphene oxide hydrogel was moved to a filter paper which assists in eliciting the absorbed surface water. After that, with the purpose of further characterization of the samples, they were shifted to freezer for two days followed by freeze-drying for one day. Overall, freeze-drying step contributed in creating three dimensional structure made by interlinked G-sheets and producing spongy construction. Therefore, all previously-utilized solvents in the aforementioned synthesis procedure of hydrogel have been eliminated and final self-assembled 3D graphene hydrogel were effectively prepared base on the morphology of parent hydrogel.

C. Characterization Techniques

With the aid of Hewlett Packard HP-8453 spectrophotometer, the UV–vis spectroscopy was accomplished. In order to prove the attendance of different functional groups in the GO as well as graphene nanosheets, Fourier transform infrared (FT-IR) spectrometer (Model: Bruker IFS 66/S) has been employed precisely. The Raman spectra were recorded with a RENISHAW (M005-141) Raman system with the laser frequency of 514 nm as excitation source. The laser spot size was 1 lm and power at the sample was below 10 mW, in order to prevent induced heating from laser. Scanning electron microscopy was successfully utilized to monitor the structure and porosity of acquired sample. Furthermore, the microstructure of obtained graphene hydrogel was studied by X-ray diffraction (XRD).

III. Result and Discussion

Fig. 1 shows the FTIR spectra of graphene hydrogel and GO. The FTIR-spectra of graphene oxide represents the characteristic peak at 1639 and 1741 cm^{-1} are corresponding to the C=C stretching vibration and oxygen functionality (C=O), respectively. The peak at 3260 cm^{-1} corresponds to hydroxyl (–OH) groups and the broad range from 2500 to 3500 cm^{-1} confirms the presence of carboxylic (–COOH) group. The FTIR of graphene hydrogel represents the characteristic peaks at 1639 and 3260 cm^{-1} are related to the C=C stretching vibration and hydroxyl group, respectively. Removal of oxygen-containing groups in all graphene hydrogels are clearly indicated by the disappearance of C = O stretching, and carboxylic (–COOH) group.

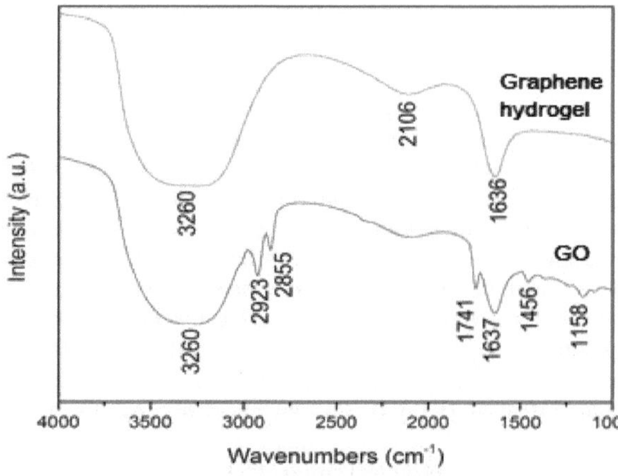

Fig. 1 FTIR spectra of Graphene hydrogel and GO

The UV-Visible spectra of GO and graphene hydrogel are indicated in Fig. 2. The UV-Vis spectrum of the GO solution highlights a peak at 242 nm, which can be due to the π - π* transition of aromatic C–C bonds. Graphene hydrogel's absorption peak corresponds to the π - π* transition of aromatic C–C bond (274 nm).

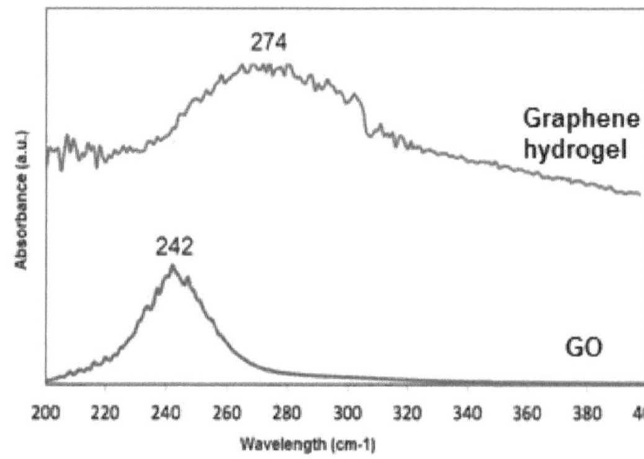

Fig. 2 The UV-Visible spectra of GO and Graphene hydrogel

Fig. 3 demonstrates the XRD patterns of GO and graphene hydrogel. The XRD pattern of GO shows a sharp peak (002) at 2θ=10.7°. After chemical reduction, the sharp (002) peak of GO disappeared while another broad peak of around 2θ = 26.5° shows up in the XRD pattern of graphene hydrogel. In XRD pattern, great level of crystallinity is specified by the slim diffraction peaks defined as sharp peaks.

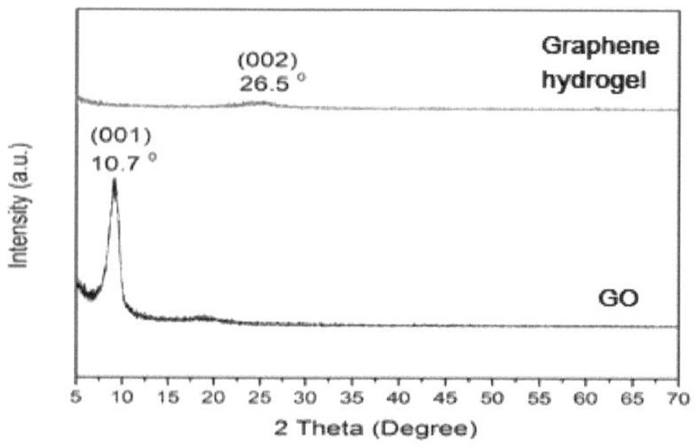

Fig. 3 The XRD patterns of GO and Graphene hydrogel

Fig. 4 indicates Raman spectra corresponds to GO and graphene hydrogel in which GO exhibits a G peak (1590 cm^{-1}) due to the in-plane vibrations of sp^2-bonded carbon atoms. A D band at 1351 cm^{-1} is ascribed to the disordered structures of GO achieved from the wide-ranging oxidation. Besides, the Raman spectrum of graphene hydrogel demonstrates G band as well as D band, while the G peak is red-shifted to 1584 cm^{-1} indicating the recovery of sp^2 domains [10]. This results confirmed the effective utilization of graphene as scaffold material with no observable structural damage.

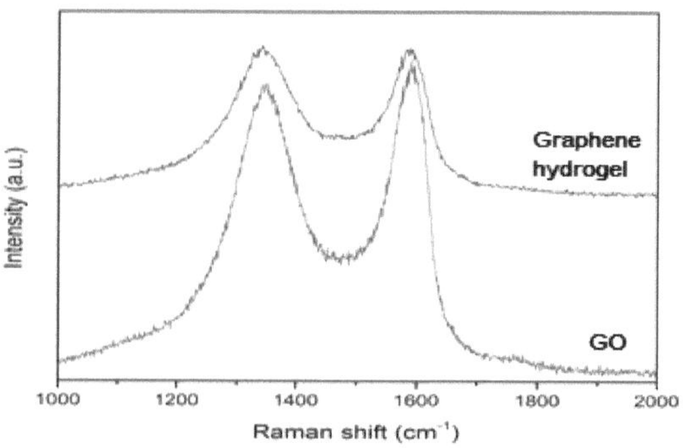

Fig. 4 Raman spectra of GO and Graphene hydrogel

SEM images of the samples have well-organized and interlinked 3D graphene layers which results in a high porous network that look a lot like a loose sponge assembly. Additionally, the porous walls are composed of thin layers of graphene. The supercritical condition resulted in the GO sheets converging, overlapping, and coalescing to produce cross-links, which make the outline of the graphene hydrogel. The GO sheets' mobility in solution is intensely finite, resulting in random orientation of graphene oxide sheets in a hydrogel form [11]. Fig. 5 shows low and high magnifications of graphene hydrogel structure from GO reduction. The three-dimensional structure of the graphene hydrogel can be obviously observed from SEM images possessing interconnected porous structure. The property of graphene oxide and its concentration has a significant impact on property of graphene aerogel including its porosity. High porosity along with graphene hydrogel structure enables the adsorbility of the graphene aerogels towards the metal ions which make it useful for many applications specifically tissue engineering scaffolds. The high degree of interconnected pores are advantageous in order to supply vascularization in vivo and allow for gas exchange [12].

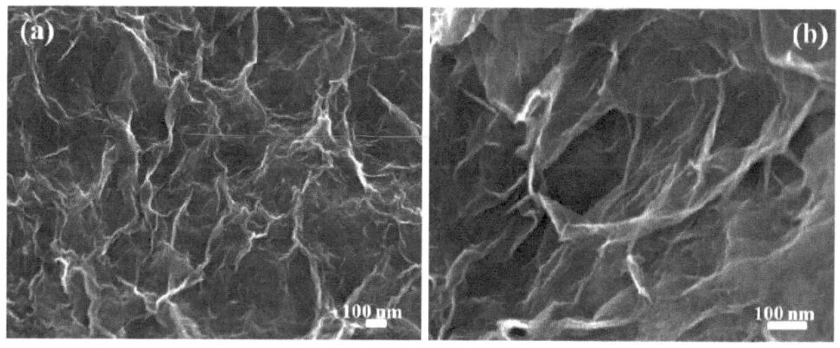

Fig. 5 SEM images of graphene aerogel; (a) low and (b) high magnification, 50,000 and 100,000 respectively.

IV. CONCLUSION

In recent times, graphene bioengineering research area has become comparatively reasonable. Owing to the interesting properties of graphene material, great amount of multidisciplinary research attempts has been extensively increase to facilitate distinct possibilities of graphene as a biomaterial specifically. Cell scaffolds, which regulate cell growth, proliferation, and differentiation, are exclusively one application of graphene promising potential in better tissue regenerative medicine. Noncollapsible 3D graphene hydrogels were productively synthesized with aid of hydrothermal method of graphene oxide. The 3D graphene cylindrical-shaped matrix was recognized as a biocompatible host for hydrogel scaffolds.

CONFLICT OF INTEREST

The authors declare that they have no conflict of interest.

REFERENCES

1. Li, N., et al., Graphene meets biology. Chinese Science Bulletin, 2014. 59(13): p. 1341-1354.
2. Goenka, S., V. Sant, and S. Sant, Graphene-based nanomaterials for drug delivery and tissue engineering. Journal of Controlled Release, 2014. 173: p. 75-88.
3. Ghasemi-Mobarakeh, L., et al., Application of conductive polymers, scaffolds and electrical stimulation for nerve tissue engineering. Journal of tissue engineering and regenerative medicine, 2011. 5(4): p. e17-e35.

4. Georgakilas, V., et al., Functionalization of graphene: covalent and non-covalent approaches, derivatives and applications. Chemical reviews, 2012. 112(11): p. 6156-6214.
5. El-Sherbiny, I.M. and M.H. Yacoub, Hydrogel scaffolds for tissue engineering: Progress and challenges. Global cardiology science & practice, 2013. 2013(3): p. 316.
6. Kalbacova, M., et al., Graphene substrates promote adherence of human osteoblasts and mesenchymal stromal cells. Carbon, 2010. 48(15): p. 4323-4329.
7. Chen, G.-Y., et al., A graphene-based platform for induced pluripotent stem cells culture and differentiation. Biomaterials, 2012. 33(2): p. 418-427.
8. Rossi, F. and E. Cattaneo, Neural stem cell therapy for neurological diseases: dreams and reality. Nature Reviews Neuroscience, 2002. 3(5): p. 401-409.
9. Yoon, H.J., et al., Sensitive capture of circulating tumour cells by functionalized graphene oxide nanosheets. Nature nanotechnology, 2013. 8(10): p. 735-741.
10. Losurdo, M., et al., Graphene as an Electron Shuttle for Silver Deoxidation: Removing a Key Barrier to Plasmonics and Metamaterials for SERS in the Visible. Advanced Functional Materials, 2014. 24(13): p. 1864-1878.
11. Lim, H., et al., Fabrication and characterization of graphene hydrogel via hydrothermal approach as a scaffold for preliminary study of cell growth. International journal of nanomedicine, 2011. 6: p. 1817.
12. Vaquette, C., et al., Effect of culture conditions and calcium phosphate coating on ectopic bone formation. Biomaterials, 2013. 34(22): p. 5538-5551.

Author:	Negar Mansouri
Institute:	Nanocat, Block A, 3RD Floor, IPS building
Street:	University of Malaya
City:	Kuala Lumpur
Country:	Malaysia
Email:	nmansuri@siswa.um.edu.my

Specific Absorption Rate Investigation of Different EBG-M Applicator Structures for Non-invasive Hyperthermia Cancer Treatment Procedure

K. Lias[1], N. Buniyamin[1], and M.Z. Ahmad Narihan[2]

[1] Faculty of Electrical Engineering, Universiti Teknologi MARA, Shah Alam, Selangor, Malaysia
[2] Faculty of Medicine and Health Sciences, Universiti Malaysia Sarawak, Kota Samarahan, Sarawak, Malaysia

Abstract— **The paper presents the results of an investigation of the effects of specific absorption rate (SAR) when applicators (antennas) are used in non-invasive hyperthermia cancer treatment procedure. Several different applicator structures were constructed and the SAR simulation was observed. Rectangular and circular microstrip applicators were initially designed, and then an EBG and water bolus structure was integrated with the applicator. This applicator termed EBG-M applicator was constructed and its SAR simulation was observed using the SEMCAD X solver software. The results indicated that the EBG-M applicators provided better SAR distribution pattern, where the EBG-M with rectangular structures offered better focusing capability, which reduced the unwanted hot spots at the surrounding healthy tissues. Meanwhile, the EBG-M with circular structure obtained better penetration depth. In addition, when the water bolus is added to the structure, it resulted in shaping the SAR contour and provided a cooler environment which then reduced the skin burn problem during the treatment.**

Keywords— **SAR, Non-invasive, Hyperthermia, Applicator, Penetration depth, Focusing**

I. INTRODUCTION

Hyperthermia is an alternative modality for cancer therapy. It provides treatment by using temperature, between 41^0C – 45^0C for a certain period of time. Denaturation of the treated cancerous tissue towards necrotic tissue can be accomplished with applied heat either by hyperthermia itself or as adjuvant with chemotherapy and radiotherapy in cancer treatments, since hyperthermia can increase the sensitivity to chemical drugs and radiation, respectively.

Current technology for non-invasive hyperthermia is basically based on electromagnetic and ultrasound technique. Pros and cons are observed for both techniques. The electromagnetic technique, however, offers safer therapy towards the treated cancerous tissue, whereby ultrasound may result in bone heating injuries due to high heat absorption and penetration.

On the other hand, there are deficiencies of using electromagnetic technique. This must be solved to enhance and increase the efficiency of non-invasive hyperthermia treatment when electromagnetic heat generation technique is used. Despite tremendous research in recent years, major challenges such as lack in penetration depth, especially towards deep-seated cancerous tissue and also the absence of an appropriate applicator that will provide well-controlled temperatures, which will prevent the undesired hot spots from overheating and damaging healthy tissue in the surrounding cancerous region.

One of the parameters that can be used to evaluate and measure the efficiency of a suitable applicator is to observe the heat distribution pattern and heat penetration that it provides through heat absorption into the treated cancerous tissue. This can be measured using the specific absorption rate (SAR) measurement. This paper presents the results of a research that investigates the SAR of different applicator structures. It focuses on a modified microstrip structure that integrates with an EBG structure and water bolus. This integration structure is called as EBG-M applicator.

II. LITERATURE REVIEW

A. *EBG-M Hyperthermia Applicator*

By integrating two main elements, which are microstrip and electromagnetic band gap (EBG) structure, EBG-M hyperthermia applicator is attained. Even though the applicator can be designed in various geometrical shapes and dimensions, there are four basic categories, which are known as microstrip patch, microstrip dipole, printed slot antenna and microstrip travelling-wave antennas [1]. Microstrip antennas can operate within a frequency range of 100MHz to 100GHz.

General advantages of microstrip antenna are low cost, low weight, linear and circular polarization and capability of dual and triple frequency operation. This antenna provides several advantages specifically when used in hyperthermia cancer procedure that includes low development cost, deeper heat penetration and low bandwidth, which is useful for focusing the heat energy into the cancerous tissue area. Examples of research in the area of hyperthermia application using microstrip antennas as an applicator are provided in [2]–[5]. The outcomes obtained were at an acceptable level with penetration depth up to 40mm obtained. Further enhancement is required in order to improve

F. Ibrahim et al. (eds.), *International Conference for Innovation in Biomedical Engineering and Life Sciences*,
IFMBE Proceedings 56,
DOI: 10.1007/978-981-10-0266-3_21

the penetration depth capability and provide heat energy with good focus to treat cancerous tissue.

EBG structures are defined as artificial periodic objects that prevent or assist the propagation of electromagnetic waves in a specified band of frequency for all incident angles and all polarization states [6]. The integration of EBG and microstrip is currently a promising structure to be used to offer solutions in many applications such as in suppressing surface waves substantially, which is described in [7].

Since the integration of EBG with microstrip in various applications has shown promising results, an investigation is carried out to observe the performance of EBG-M applicator when used to treat cancer in hyperthermia procedure. It is expected that this EBG-M applicator will provide heat that will penetrate to a better depth, especially for deep-seated cancerous tissue, which is up to at least 50mm. It is also expected that it will provide better and improve the capability of heat focusing, where unwanted hot spots will reduce significantly. The depth and focusing is illustrated through SAR heat distribution or absorption pattern. The SAR measurement is discussed in the following subsection.

B. Specific Absorption Rate (SAR)

Specific absorption rate (SAR) is a measurement tool to describe the pattern and level of heat absorption into biological tissues. It is the average power per unit mass of tissue dissipated in heat with W/kg as SAR unit and is given by:

$$SAR = \frac{\sigma}{2\rho}|E|^2 (W/kg) \qquad (1)$$

Where σ is the conductivity of tissue (S/m), E is an electric field (V/m), ρ is a density of tissue (kg/m^3). The effectiveness of the hyperthermia treatment could be evaluated through SAR heat distribution and absorption pattern into the cancerous tissue to be treated [8]. Fig 1 shows examples of SAR pattern. Red denote high temperature, while blue is cooler temperature.

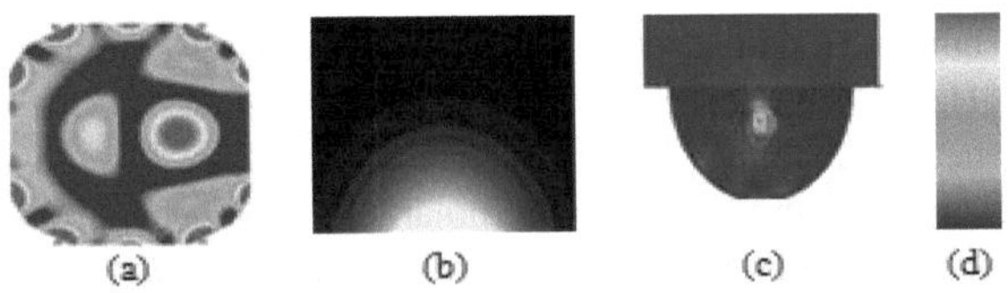

Fig. 1 SAR Distribution (a) [8], (b) [3], (c) [4], (d) SAR level [8]

III. RESEARCH METHODOLOGY

To observe the SAR pattern of different EBG-M applicator structures for non-invasive hyperthermia cancer modality treatment, several steps are required to be implemented. These include designing of EBG-M applicator, depth analysis of mammogram breast cancer images with Digital Imaging and Communication in Medicine (DICOM) software, breast phantom construction and SAR simulation measurement and analysis.

A. Designing of EBG-M Applicator

The EBG-M construction initially starts with microstrip applicator design. Two different microstrip antenna structures are modelled using SEMCAD X simulator, which are known as rectangular and circular toroidal applicator. The rectangular EBG structure is integrated after the microstrip antenna is designed. Details of general microstrip antenna parameters are available in [10]. The substrate to be used is FR4 with electrical properties, relative permittivity, $\varepsilon_r = 4$. The substrate length and width for the structures remain the same, which are 44mm and 52mm, respectively. 2.4GHz is used as the operating frequency, and the power to be investigated are 1W, 10W and 100W.

B. Mammogram Breast Cancer Image Depth Analysis

The data, which are in the form of breast cancer images, are assembled. 17 samples obtained consist of mammogram images of right or left breast that have cancerous tissues. To ensure the position and location of the breast cancer, a through discussion and analysis process with a radiologist and pathologist is conducted.

The Digital Imaging and Communications in Medicine (DICOM) software is used to carry out the breast cancer image depth analysis. DICOM defines the formats for medical images that can be exchanged with the data and quality necessary for clinical use. The mammogram breast cancer image is depicted in Fig. 2.

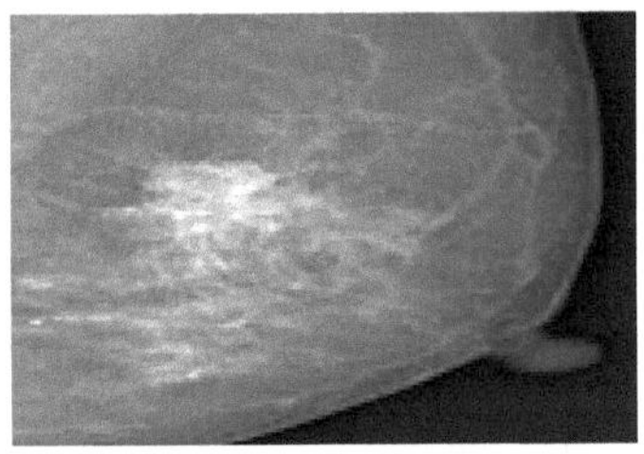

Fig. 2 Mammogram Breast Cancer Image.

Results of the cancer depth analysis are depicted in Fig 3. The highest percentage where cancerous tumors occur is at a depth in the region of 90mm$\leq d \leq$*100mm*. This finding is used to establish one of the specifications of the EBG-M applicator to be constructed. It was decided that heat penetration of up to 50mm is required to provide treatment of deep seated cancerous tissue.

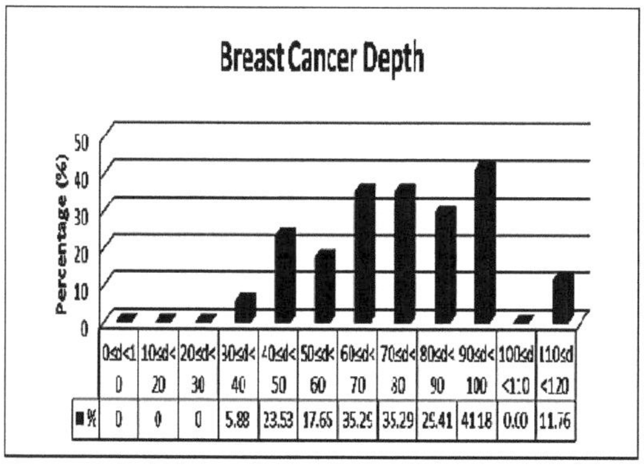

Fig. 3 Breast Cancer Depth

C. Breast Phantom Construction

A Breast phantom with breast cancer at 50mm depth is constructed with SEMCAD X. The electrical and thermal properties of the breast are tabulated in Table 1.

Table 1 Electrical and Thermal Properties for Breast Fat and Breast Cancer

	Relative Permittivity, εr	Electrical Conductivity, σ (S/m)	Density, ρ(kg/m3)	Specific Heat Capacity, C (J/kg/K)	Thermal Conductivity, K (W/m/K)
Breast Fat	5.14	0.125	911	2348.33	0.209
Breast Cancer/ Breast Tumor	5.14	0.125	1911	2352.55	0.789

D. SAR Measurement and Analysis

The final stage in research methodology is to carry out the SAR measurement, analysis and modification. After completing the initial three stages, the SAR pattern is simulated to obtain the SAR distribution pattern as the results of heat delivered and absorbed from the designed EBG-M applicator into the cancerous tissue. SAR for 1g and 10g weight is compared. Also included in research methodology is the water bolus construction for cooling environment during hyperthermia cancer treatment. Water bolus utilized water distilled as a coolant fluid with ε_r, σ and ρ are 76.7, 5e-005 and 1000, respectively.

IV. Result and Discussion

Rectangular and circular toroidal microstrip hyperthermia applicators are designed and constructed with SEMCAD X simulator. The modification of the applicators is subsequently conducted, whereby the EBG and water bolus are integrated with the rectangular and circular microstrip applicator. The SAR distributions of each structure are tabulated in Table 2. As shown, the SAR is different for different applicator structure. Good focusing and penetration depth, of up to 90mm, are obtained for rectangular and circular microstrip applicator, respectively. Furthermore, the SAR pattern has improved positively as the shaped and contour of the SAR pattern shows improved focusing capability, that reduce unwanted hot spots when the water bolus and EBG structures are integrated to the applicator, respectively. On the other hand, different operating powers provide null effect towards the SAR distribution pattern; however, when the operating power is increased, the value of the SAR in mW/g is also increased.

Table 2 SAR for Different Structures of Non-Invasive Hyperthermia Applicator

Hyperthermia Structure	**Specific Absorption Rate (SAR)**	
	1g	**10g**
Rectangular Microstrip Applicator		
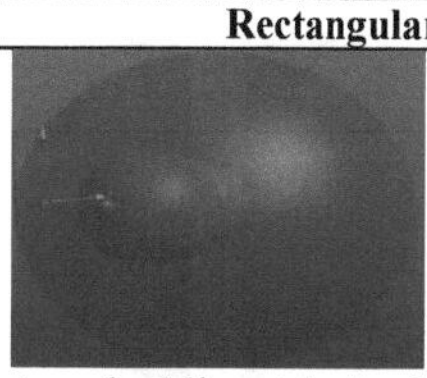 Rectangular Microstrip Applicator		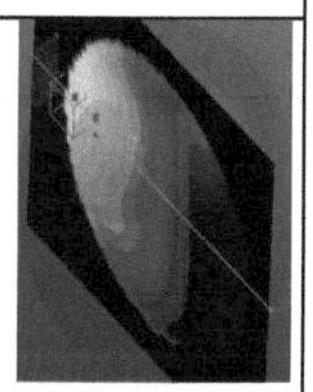
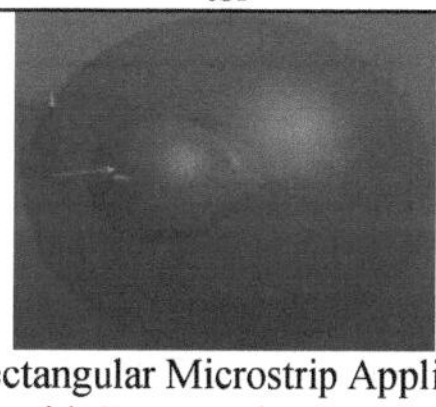 Rectangular Microstrip Applicator with Water Bolus Integration	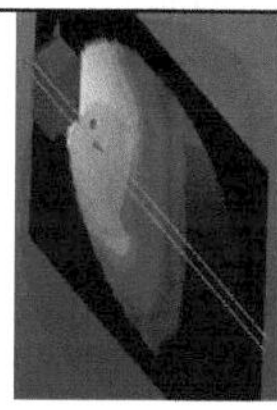	
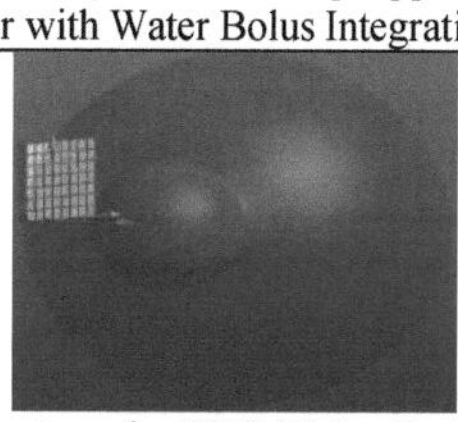 Rectangular EBG-M Applicator	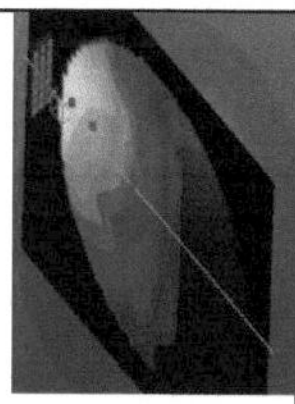	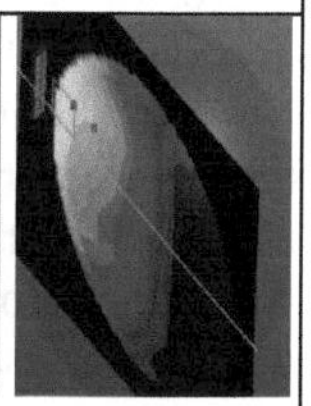
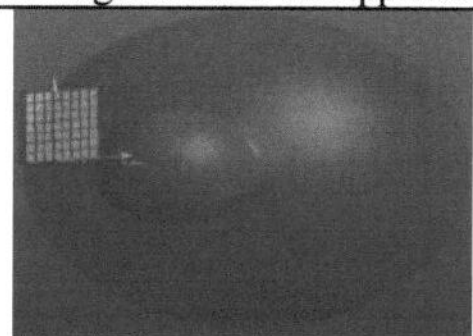 Rectangular EBG-M Applicator with Water Bolus Integration	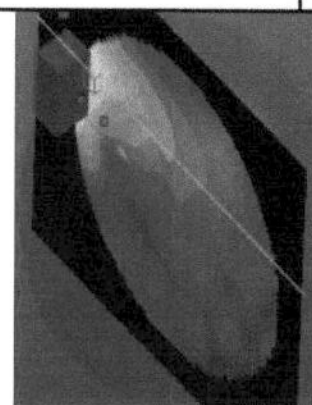	

Table 2 (Continued)

Circular Microstrip Applicator		
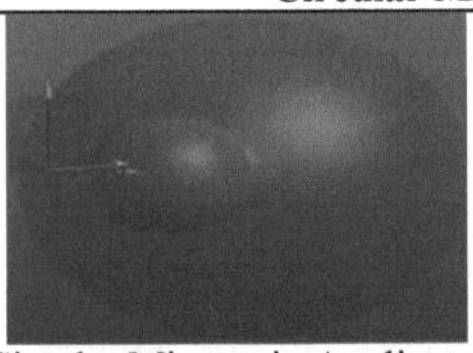 Circular Microstrip Applicator	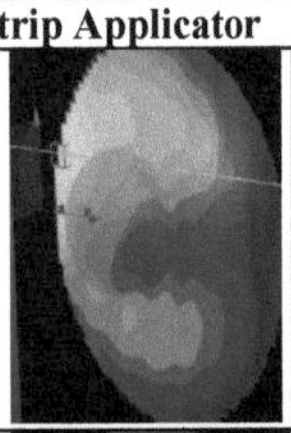	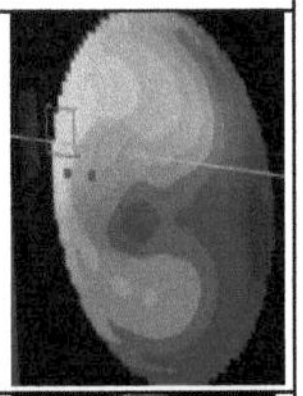
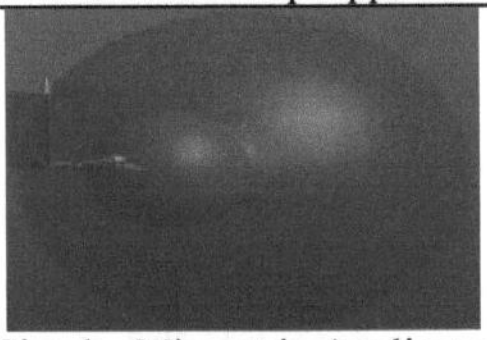 Circular Microstrip Applicator with Water Bolus Integration	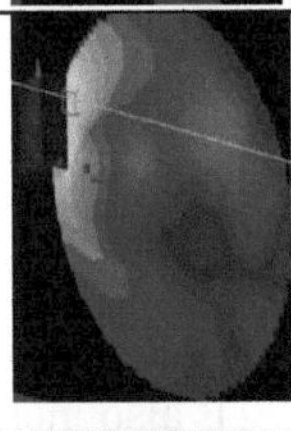	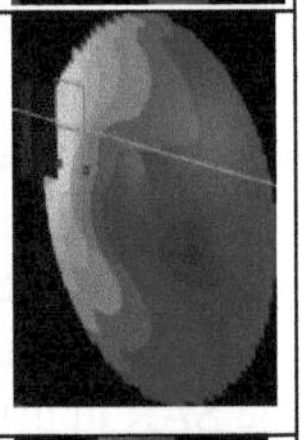
 Circular EBG-M Applicator	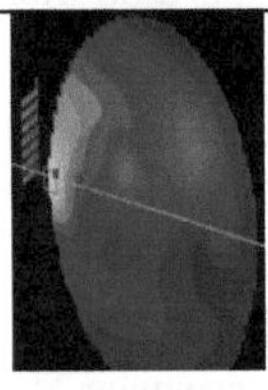	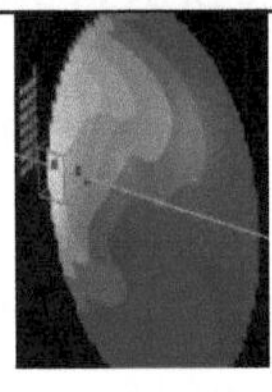
 Circular EBG-M Applicator with Water Bolus Integration	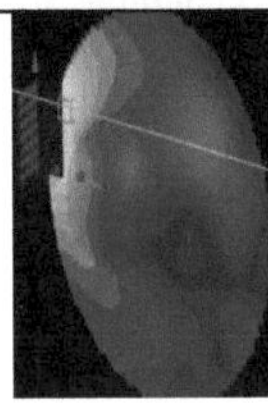	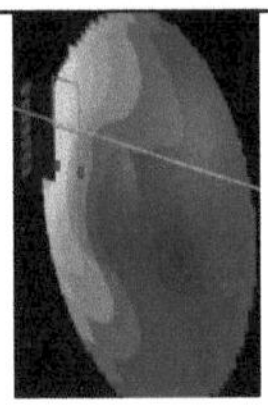

V. CONCLUSIONS

The investigation of SAR pattern with different EBG-M applicator structures for non-invasive hyperthermia cancer procedure is carried out. Rectangular and circular toroidal microstrip applicators provided good focusing and penetration depth, of up to 90mm. However, EBG-M applicator which is the integration of the EBG and water bolus structure with microstrip antenna further modified the SAR distribution where the EBG structure is able to reduce the unwanted hot spots at the surrounding healthy tissue and the water bolus assists in shaping the SAR distribution.

ACKNOWLEDGMENT

The authors gratefully acknowledge and thank University Teknologi MARA (UiTM) for providing the facilities to carry out the research.

CONFLICT OF INTEREST

There is no conflict of interest.

REFERENCES

1. R. Garg, P. Bhartia, I. Bahl, and A. Ittipiboon, Microstrip Antenna Design Handbook.pdf. ARTECH HOUSE, INC, 2001.
2. O. Losito, M. Bozzetti, S. Sterlacci, and V. Dimiccoli, "E-Field Distribution Improvement by New Hyperthermia Applicators," IEEE, 2011.
3. T. Drizdal, P. Togni, and J. Vrba, "Microstrip Applicator for Local Hyperthermia," 2007 Int. Conf. Electromagn. Adv. Appl., pp. 1047–1049, Sep. 2007.
4. D.-X. Yin, M. Li, and J. L.-W. Li, "Non-invasive breast cancer thermotherapy studies using conformal microstrip antennas," Isape2012, pp. 159–162, Oct. 2012.
5. K. Lias and N. Buniyamin, "An Overview of Cancer Thermal Therapy Technology based on Different Types of Antenna Exposure," ICEESE 2013, pp. 90–95, 2013.
6. F. Yang and Y. Rahmat-samii, Electromagnetic Band Gap Structures in Antenna Engineering. 2008.
7. N. Rao and V. Dinesh Kumar, "Investigation of a microstrip patch antenna with EBG structures using FDTD method," 2011 IEEE Recent Adv. Intell. Comput. Syst., pp. 332–337, Sep. 2011.
8. M. J. Ammann, S. Curto, X. L. Bao, and P. McEvoy, "Antenna design considerations for high specific absorption rate in local hyperthermia treatment," 2008 IEEE Antennas Propag. Soc. Int. Symp., no. 1, pp. 1–4, Jul. 2008.
9. J. Stang, M. Haynes, P. Carson, and M. Moghaddam, "A preclinical system prototype for focused microwave thermal therapy of the breast.," IEEE Trans. Biomed. Eng., vol. 59, no. 9, pp. 2431–2438, Sep. 2012.
10. K. B. Lias, M. Z. Ahmad Narihan, and N. Buniyamin, "An Antenna with an Embedded EBG Structure for Non Invasive Hyperthermia Cancer Treatment," 2014 IEEE Conf. Biomed. Eng. Sci. 8 - 10 December 2014, Miri, Sarawak, Malaysia, no. December, pp. 8–10, 2014.

Kasumawati Lias
Faculty of Electrical Engineering,
University Teknologi MARA,
40450, Shah Alam, Selangor, Malaysia
Email:danza252@gmail.com

Norlida Buniyamin
Faculty of Electrical Engineering,
University Teknologi MARA, 40450,Shah Alam, Selangor, Malaysia
Email: nbuniyamin@salam.uitm.edu.my

Mohd Zulkarnaen Ahmad Narihan
Faculty of Medicine and Health Sciences,
Universiti Malaysia Sarawak
94300, Kota Samarahan, Sarawak, Malaysia
Email: nmzulkarnaen@fmhs.unimas.my

MEG Waveform Analysis from Language Cortical Areas in Patients with Brain Tumor – A Preliminary Study

N.H. Sharipudin[1], F. Reza[1], H. Omar[1], and Z. Idris[1,2]

[1] Department of Neurosciences, Universiti Sains Malaysia, Kubang Kerian, Kelantan
[2] Center for Neuroscience Services and Research, Universiti Sains Malaysia, Kubang Kerian, Kelantan

Abstract— **Language plays an important part in human for expressing their thoughts and emotions, where the source of function originated from the brain. This study aims to estimate the early and long latency language-related magnetic fields (LRFs) in patients with brain tumor. The study hypothesized that the brain tumor would alter the patients LRFs as compared to healthy subjects. Three patients with brain tumor which invades the language region in the brain were selected as subjects while two normal healthy subjects were taken as a control group regardless of gender. All the subjects had to undergo a Magnetoencephalography (MEG) scanning while doing a task related to language function. The subjects were shown four characters Malay semantic word and had to do silent reading while asked to lessen their movement (such as blinking) during the whole scanning procedure. The waveform was taken between the latency fields of -150 to 850 ms, and then filters were applied before analyzing the signals. The areas of analysis were done at the Occipital, Temporal, and Frontal region of the brain while the components taken were N100, N200, N400 and N600 with respect to the location. The LRFs were taken from both left and right side of the brain. The results obtained show significant difference of LRFs between the patient and control groups in term of both latency and amplitude. The latency and amplitude readings were inconsistent for all patients while the consistency for the control group was constant. The dominance of brain activation was shown to be on the left side for both control and patient group. As conclusion, there is no difference between patient and control group LRFs due to limitation of study.**

Keywords— **Magnetoencephalography (MEG), language-related magnetic fields (LRFs), brain tumor.**

I. INTRODUCTION

Compared to other functions, language areas in the brain are difficult to analyze of its whereabouts and source location. Generally, there were many previous studies regarding the functions of certain brain regions for language, for example the Wernicke's area, Broca's area and angular gyrus [1]. Furthermore, before the availability of neuroimaging equipment, researchers resort to invasive methods such as the Wada procedure or the "gold standard" techniques [2]. Magnetoencephalography (MEG) is one of the neuroimaging equipment used by recent study to investigate more about language-related function in the brain non-invasively. Moreover, this neuroimaging technique was used for mapping onto patients with neurological disorders or brain damage to further understands the effect of the morbidity to the brain's function [3-6].

This study aim to investigate the functional reorganization of the brain as a result of acquired focal lesions involving language-specific cortex and whether it affects the latency of the language-related magnetic fields (LRFs).The subjects were chosen based on the tumor location situated near the language function area. Previous study had done the mapping of the language area for these types of subjects, however there were still a lot of inquiries of the language function and little were still done using the Malay language. From this preliminary study, more understanding could be achieved regarding the function of the language-related brain area.

II. METHODS

A. Subjects

Consent was first obtained from the subjects who followed the ethical guidelines provided by the Human Research Ethics Committee of Universiti Sains Malaysia (HREC USM), Kelantan. Data was collected from subjects who fulfilled the inclusion criteria, which are 1) patients in whom a tumor invades or is directly adjacent to the expected location of one of the language centers 2) in good health 3) age ranges from 18 to 50 years old regardless of gender and must have 4) Malay as first language. However, subjects whom have 1) high grade tumors, 2) history of significant psychiatric condition, neurological conditions, significant claustrophobic reactions, 3) pregnancy, 4) standard contraindications to MEG examinations and/or 5) been a patient who were treated prior to our MEG recordings were excluded from this study.

B. Measurements

a) LRF components

The LRF components analyzed for this study includes N100 (80-100 ms), N200 (200-250 ms), N400 (375–549 ms) and N600 (550-750 ms) (refer to Figure 3 for example location). The components were taken from the highest peak of each LRF signal with positive value of amplitude and

F. Ibrahim et al. (eds.), *International Conference for Innovation in Biomedical Engineering and Life Sciences*,
IFMBE Proceedings 56,
DOI: 10.1007/978-981-10-0266-3_22

from the electrode of magnetometers at each respective area (refer to Figure 1).

b) LRF sources

Early and Long latency fields between 80ms to 800ms evoked by visual and verbal stimuli were characterized as LRF in the Occipital, Temporal and Frontal cortices.

c) Recording parameters

The MEG equipment and analysis software used were from Elekta Neuromag Ltd, with 306- channels consisting of 204 planar gradiometers and 102 magnetometers. Only 24 magnetometers were analyzed for this study. The locations of these magnetometers are as shown in Figure 1. Throughout data acquisition, the sampling rates were set at a minimum sampling rate of 1000Hz and band pass filters were between 0.01 and 330Hz.

d) Data analysis

Early (30 to 200ms) and Long latency (200 to 800ms) language related magnetic fields were identified. Data was filtered by Low pass (60Hz) and High pass (3Hz) for offline data analysis. The epoch duration was 850ms including a 150ms pre-stimulus interval.

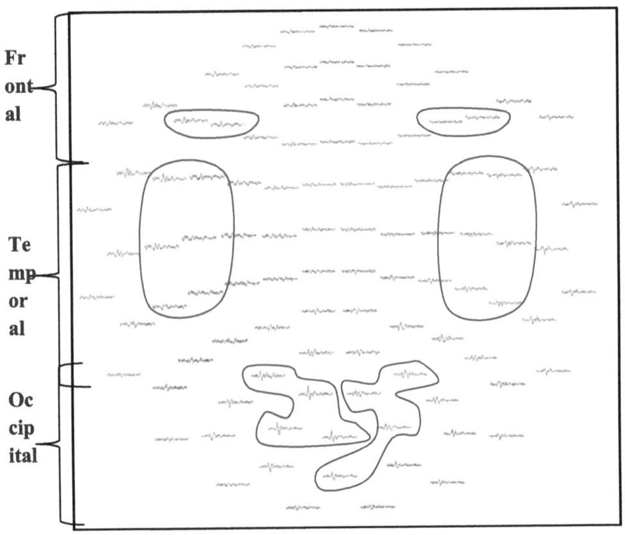

Fig. 1 Channel lay out of 102 magnetometers in MEG. The 24 magnetometers within the selected area were analyzed for data collection located at Frontal, Temporal and Occipital area of the brain.

After filtering, the baseline was applied at 100ms before stimulation. The components taken from Occipital area are N100 and N200, for the Temporal N100, N200 and N400; while for Frontal N100, N200, N400 and N600. Both the latency and amplitude were recorded for each subject. Flat signals or non-active signals are considered as not evoked. The data were recorded to Microsoft Excel and the nonparametric independent t-test was performed. The study used alpha (α) at 0.05, confidence interval of 95 % and P-value < 0.05 was considered to be statistically significant.

C. Procedure

Subjects were first explained of this study objective and consents were given before proceeding with the procedure. Subjects had to properly understand and follow the instructions of the tasks as to reduce errors in data taken. Next, proper measurements were taken with the help of a science officer. To avoid magnetic artifacts, the subjects were first asked to remove all metal objects in their possession. Four Head Position Indicator (HPI) coils were attached to the subjects with skin tape to track the head movements and the subject's head shape was measured as reference for MEG system using Polhemus 3-D digitizer pen.

After digitization, the subjects entered the MEG room (magnetically shielded). Silent reading tasks were performed during MEG, where the subjects sit on a comfortable chair with their heads inserted into the MEG machine. After the presentation of an eye fixation point for 3 seconds, a four characters Malay semantic words were shown for 3 seconds (Figure 2) on an 80-inch rear projection screen located 1.5 m away from the subject inside the same room. Visual stimuli were generated using a visual presentation system, projected by a DLP projector located outside the room.

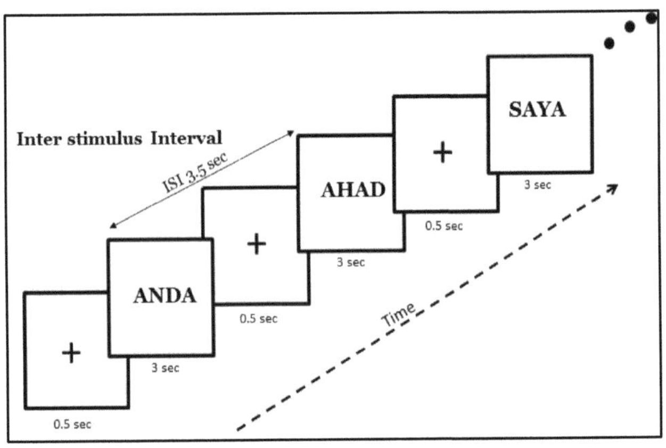

Fig. 2 Words for visual stimulation presented by Presentation software. The words were shown for 3 seconds with inter stimulus interval (ISI) of 0.5 seconds between each word

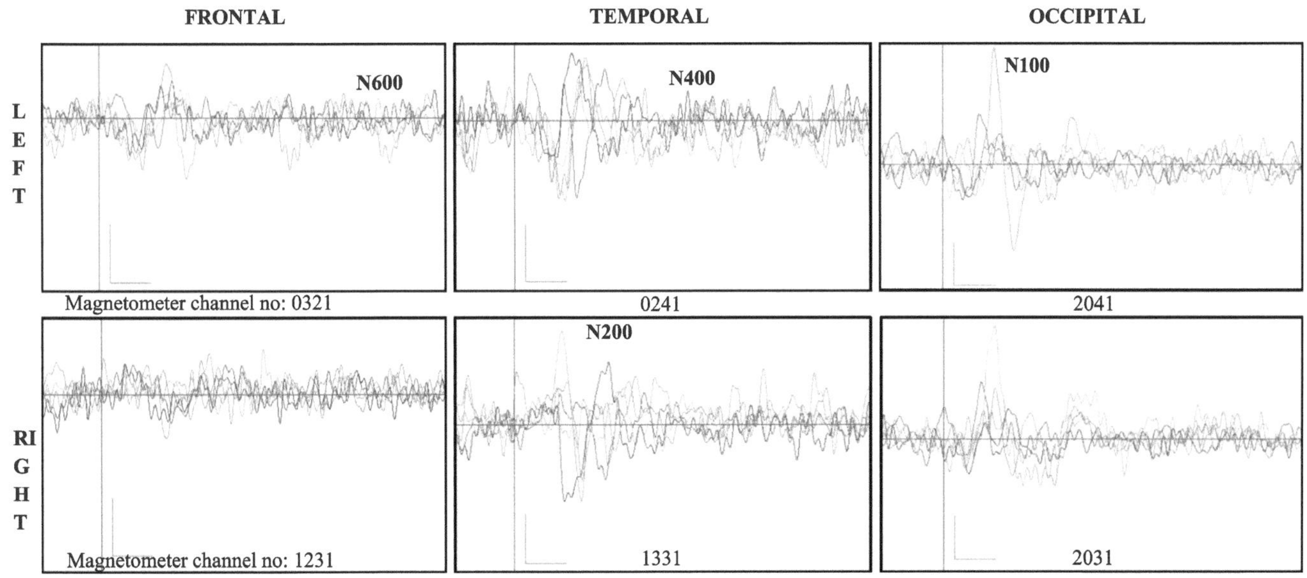

Fig. 3 Example of LRF signals for all groups overlap on top of each other (Control group: purple and green; Patient group: red, orange, blue). Scale bars of signals: x: 100ms, y: 100fT

Table 1 Average value of latency and amplitude between patient and control group ($N_{patient}$ = 3, $N_{control}$ = 2)

Group	Area	Components	Latency (ms)		P-value	Amplitude (fT)		P-value
			Left Region	Right Region		Left Region	Right Region	
Patient	Occipital	N100	133.025 ± 22.593	127.658 ± 23.182	0.414	132.792 ± 72.167	135.800 ± 64.035	0.482
		N200	275.092 ± 74.068	268.258 ± 53.991	0.420	55.400 ± 24.780	71.992 ± 42.999	0.183
	Temporal	N100	163.172 ± 13.416	119.899 ± 22.223	**0.039***	82.461 ± 13.827	62.009 ± 14.354	0.132
		N200	292.556 ± 38.632	256.644 ± 34.983	0.200	30.939 ± 13.119	80.839 ± 16.587	**0.018***
		N400	441.400 ± 16.937	422.294 ± 22.399	0.208	37.111 ± 7.274	47.678 ± 16.418	0.260
	Frontal	N100	169.317 ± 9.254	126.350 ± 72.948	0.053	74.750 ± 31.764	48.700 ± 28.117	**0.011***
		N200	301.050 ± 13.486	275.950 ±159.929	0.159	14.367 ± 11.736	40.375 ± 25.639	0.184
		N400	446.117 ± 21.387	437.850 ±255.336	0.217	37.517 ± 1.097	51.125 ± 35.697	0.440
		N600	644.533 ± 42.067	594.150 ±343.127	0.170	24.717 ± 10.104	34.725 ± 23.175	0.438
Control	Occipital	N100	138.425 ± 3.500	144.000 ± 15.556	0.375	60.138 ± 39.545	74.000 ± 18.738	0.397
		N200	226.425 ± 4.773	316.350 ± 23.865	0.071	46.788 ± 14.089	63.475 ± 0.955	0.162
	Temporal	N100	146.133 ± 37.500	114.900 ± 3.371	0.238	57.742 ± 43.947	118.667 ± 68.047	0.087
		N200	279.058 ± 75.130	288.283 ± 30.358	0.461	61.017 ± 28.049	86.683 ± 8.556	0.251
		N400	462.442 ± 4.349	512.583 ± 13.718	**0.042***	40.500 ± 2.970	48.992 ± 23.181	0.329
	Frontal	N100	142.125 ± 34.047	126.700 ± 18.597	0.375	45.650 ± 17.961	46.750 ± 23.264	0.409
		N200	232.000 ± 0	257.750 ± 0	0.157	38.600 ± 0	30.500 ± 0	0.157
		N400	447.025 ± 15.804	359.300 ± 71.913	0.135	28.650 ± 0.495	47.675 ± 1.591	**0.013***
		N600	562.275 ± 5.339	578.350 ± 22.557	0.206	36.700 ± 1.556	42.875 ± 15.521	0.322

**significant if P-value ≤ 0.05.Frontal N200in control group*

Subjects were tasked to read each word only once and without phonation immediately after presentation of the word. One session consisted of 100 different word presentations. Words were selected from an elementary school dictionary for quick and easy understanding by all subjects. The visual stimuli subtended a horizontal visual angle of 3° and a vertical angle of 1°; thus no eye movements were necessary to scan the presented word. The whole recording took at least 1 hour to finish; however subjects were able to halt the task if they are uncomfortable.

III. RESULTS

This study was able to attain three patients and two healthy volunteers who fulfilled the inclusion and exclusion criteria as subjects. All of the subjects are right-handed. The signal obtained was filtered, analyzed, recorded and then compared between patient and control subjects. The region analyzed were Temporal, Occipital and Frontal region of the brain. Figure 3 showed the signal obtained from each subjects and overlap on top of each other to compare the evoked response and readings based on the brain region.

Meanwhile, Table 1 showed the average value readings for both groups between left and right region of the analyzed brain area. From the results, the synchronizations of signals taken from control group are better compared to patient group. From Figure 3, it can be seen the wave of LRFs components were different for each person in the patient group; while the control group had almost similar wave signals for each component.

In addition, N400 were not evoked at Occipital region and N600 were not evoked at both Occipital and Temporal region. Other than that, the data obtained from latency readings showed control group are more dominant on the left side of the brain. However, for the patient group, the latency readings are earlier on the right side, contrary to the control group. When compared the latency for both groups, the patient groups' latency is latter on the left while control group's latency is latter on the right.

In terms of the amplitude readings, the control group's amplitudes are higher on the right side while the patient group has higher amplitude readings compared to control group for both sides. Lastly, Table 2 showed the evoked LRF signals specifically for patient group as the control group's LRF signals are all evoked. The focus here is to show that there are some areas for certain patient subjects who have non-evoked signals.

Patient 2's right Frontal area was not evoked for the entire LRF components even though the left side was evoked. Meanwhile, for Patient 3, the Frontal right side was not evoked for the N100 component of LRF. For Patient 1, the entire brain region signals had evoked signals. Generally, most of the Frontal right sides were not evoked for the patient group.

Table 2 List of evoked LRFs signals for patient group only

Area	Components	Patient 1		Patient 2		Patient 3	
		Left	Right	Left	Right	Left	Right
Occipital	N100	E	E	E	E	E	E
	N200	E	E	E	E	E	E
Temporal	N100	E	E	E	E	E	E
	N200	E	E	E	E	E	E
	N400	E	E	E	E	E	E
Frontal	N100	E	E	E	**NE**	E	**NE**
	N200	E	E	E	**NE**	E	E
	N400	E	E	E	**NE**	E	E
	N600	E	E	E	**NE**	E	E

Note. E= Evoked, NE = Not Evoked.

IV. DISCUSSION

From Figure 3, significant differences between each subject's amplitude and latency can be seen. Theoretically, signals waveforms from the brain are unique for each person but not at the same level as fingerprints [7]. This is due to the fact that cortex consists of many neurons and the aggregate electrical signals produced are unique depending on the person's mental task; however the main functions for each area are still the same [7, 8]. Therefore, the language area function in the brain for control group showed averagely similar readings between the subjects. Even though this study specified on language functions, other number of areas in the brain are stimulated first such as the Occipital area from the visual stimulation, then the late LRFs, after the primary sensory, were analyzed.

The activation sequence produced by language processing can be seen properly from patient group left brain region LRF's components. The LRF components are gradually becoming latter as it moves from the Occipital to Frontal region, indicating the sequence starting with the subject looking at the visual stimuli first (Occipital), followed by reception of language probably by the Wernicke activity at 400ms (Temporal), then ends at the Frontal region where the possibility of Broca's activity at around 600ms [9].

As stated from the results in Table 1, the signals of the control group are more synchronized compared to the patient group. This mean language-related area of the brain for the control group was almost similar for each subject. However, the patient subjects had inconsistency between themselves for the latency readings. The tumor might have possibilities of disturbing the language function of the brain [10, 11]. Meanwhile for the control group, the results are supported by previous study which stated a healthy subject usually have dominance on the left side of the brain compared to patient subjects [9].

In terms of latency, the control group left hemisphere is more dominant than the right side as its latency is more accurate according to the components' range. This is also supported by the first result stating the synchronization of the control group's signals. Meanwhile, the patient groups had latency earlier on the right hemisphere. This result might be correlated to previous study which mentioned patient with brain tumor had higher activation of intensity on the right side of the hemisphere [12]. However, the previous study focused on the right inferior frontal gyrus. In addition, the amplitude readings in the right hemisphere are also larger for patient group compared to control group, thus further supporting the previous study statement. The higher amplitude might be due to more concentration used by patients compared to control subjects for the tasks. In other words, patient subjects require higher activation of neurons compared to control subjects to complete a language task for this study.

As for the P-value, within the group for either patient or control, between left and right region, there are not many significant differences for the readings. This postulate that left and right region are only slightly contradicting in its level of amplitude and latency. The significant difference of P-value states that there are differences between the left and right value of amplitude or latency, though there are only a few in both groups.

Moreover, as shown in Table 2, some of the readings taken were not evoked for certain components and this might contribute to convoluted data. Most of the non-evoked areas are from patient subject's right Frontal area. The Frontal area is close to the Broca's area which is responsible for speech production [13]. This study required the subjects to read the words but not loudly, so Frontal area should be slightly evoked as presented by the control group. This is proven by the amplitude readings where the Frontal area have the lowest amplitude compared to Occipital and Temporal, making it less evoked compared to other parts of the brain. Besides, the left side is more dominant for speech production as compared to the right side [12]. Thus, accumulating these data, this might be the reason why the right Frontal areas were not evoked for certain patient subjects.

On top of that, this type of study requires the subject to be highly alert, focused and well-performed to produce an evoked data. Therefore, the performance of patient subjects might be one of the study limitations. In addition, the numbers of subjects are also limited for the current preliminary study, thus future improvements are required for better results and understandings.

V. CONCLUSIONS

Overall, the objective of this study which is to estimate the early and long latency LRFs in patients with brain tumor was achieved. Most of the LRFs obtained were parallel with other study results. The LRFs also showed slight differences between control group and patient group in terms of amplitude and latency. However, the activation sequences are still the same for both groups. As for the limitations, more subjects are required; specifically with brain tumors, for future study as it helps strengthen the results and overcome other flaw such as convoluted data, which are produced from subjects' behavior which is common. Localization of LRFs' source are highly recommended for our future research as it can helps develop mapping of language function for patient with brain tumor. Moreover, this study has surgical benefit by locating which brain region still activated for language functions and preserving it. In addition, this method is non-invasive and can be further improve by superimpose of MRI images with the MEG data of the patient.

ACKNOWLEDGEMENT

This study was supported by Short Term Grant (ref: 304/PPSP/61312053) from Universiti Sains Malaysia (USM), and approved by Human Research Ethics Committee (HREC), USM (ref: USMKK/PPP/JEPeM(234.3[8]), Kubang Kerian, Kelantan.

CONFLICT OF INTEREST

The authors declare that they have no conflict of interest.

REFERENCES

1. Papanicolaou, A. C. (2009). Clinical magnetoencephalography and magnetic source imaging. Cambridge University Press. pp 99.
2. Frye, R. E., Rezaie, R., & Papanicolaou, A. C. (2009). Functional neuroimaging of language using magnetoencephalography. Physics of life reviews, 6(1), 1-10.
3. Papanicolaou, A. C., Simos, P. G., Breier, J. I et al. (2001). Brain plasticity for sensory and linguistic functions: a functional imaging study using magnetoencephalography with children and young adults. Journal of Child Neurology, 16(4), 241-252.
4. Simos, P. G., Breier, J. I., Maggio, W. W., Gormley, W. B., Zouridakis, G., et al (1999). A typical Temporal lobe language representation: MEG and intraoperative stimulation mapping correlation. Neuroreport, 10(1), 139-142.
5. Maest ́u F, Saldana C, Amo C, et al (2004). Can small lesions induce language reorganization as large lesions do? Brain Lang; 89(3), 433–438.
6. Pataraia E, Simos PG, Castillo EM, et al (2004). Reorganization of language-specific cortex in patients with lesions or mesial Temporal epilepsy. Neurology; 63(10): 1825–1832.
7. Marcel, Sebastien, and José R. Del Millan (2007). Person authentication using brainwaves (EEG) and maximum a posteriori model adaptation. Pattern Analysis and Machine Intelligence, IEEE Transactions on 29 (4), 743-752.
8. Paranjape, R. B., J. Mahovsky, L. Benedicenti, and Z. Koles (2001). The electroencephalogram as a biometric. In Electrical and Computer Engineering, 2001. Canadian Conference on, vol. 2, pp 1363-1366. IEEE.

9. Grummich, Peter, et al (2006). Combining fMRI and MEG increases the reliability of presurgical language localization: a clinical study on the difference between and congruence of both modalities. Neuroimage 32(4), 1793-1803.
10. Duffau, H., Denvil, D., Capelle, L., (2002). Long term reshaping of language, sensory, and motor maps after glioma resection: a new parameter to integrate in the surgical strategy. J. Neurol. Neurosurg. Psychiatry 72 (4), 511 –516.
11. Grummich, P., Nimsky, C., Fahlbusch, R., Ganslandt, O., (2005). Observation of unaveraged giant MEG activity from language areas during speech tasks in patients harboring brain lesions very close to essential language areas: expression of brain plasticity in language processing networks? Neuroscience Letter. 380 (1), 143–148.
12. Thiel, A., Habedank, B., Winhuisen, L., et al (2005). Essential language function of the right hemisphere in brain tumor patients. Annals of neurology, 57(1), 128-131.
13. Pang, E. W., Wang, F., Malone, M., Kadis, D. S., & Donner, E. J. (2011). Localization of Broca's area using verb generation tasks in the MEG: Validation against fMRI. Neuroscience letters, 490(3), 215-219.

Corresponding author:

Author: Faruque Reza
Institute: Department of Neurosciences, Universiti Sains Malaysia
Street: Jalan Sultanah Zainab 2
City: Kubang Kerian, Kota Bharu, Kelantan
Country: Malaysia
Email: faruque@usm.my

Functional Electrical Stimulation (FES) Based Low-Cost Assistive Device for Foot Drop – A Pilot Study

S.J. Khan[1,2], R. Qureshi[1], S. Jawaid[1], M. Siddiqui[1], U. Sarwar[1], S. Abdullah[3], S.S. Khan[1,2], M.T. Khan[4], and A.Z. Bari[1]

[1] Department of Biomedical Engineering, NED University of Engineering & Technology, Karachi, Pakistan
[2] Department of Biomedical Engineering, University of Malaya, Kuala Lumpur, Malaysia
[3] Healthcare R&D, Tech4Life Enterprises, Karachi, Pakistan
[4] School of Public Health, Dow University of Health Science, Karachi, Pakistan

Abstract— **The objective of this study was to design an efficient and economical FES device. The device was designed for post-stroke foot drop patients who could walk up to 5meters with assistance and could stand up from a sitting position. The device is Arduino-based and can sense the start of the swing phase and able to stimulate the dorsiflexor muscles till the end of the swing phase. The device was tested on 5 male subjects (aged 40-50 years), taking Temporal Spatial Parameters as indicators of efficiency. A significant improvement in the parameters was observed with the usage of the device.**

Keywords— **Gait, Functional Electrical Stimulation (FES), Rehabilitation, Foot drop, Arduino.**

I. INTRODUCTION

'Foot drop' or 'drop foot' is the gait deviation which results in inadequate activation of dorsiflexor muscles of lower limbs as a result of which 'toe drag' occurs depriving the person of the swing phase of the gait cycle. The inability to lift the foot during swing phase makes it problematic to be able to experience curbs, stairs, and uneven surfaces. The findings are the selected phases of excessive plantar flexion with an otherwise normal gait [1]. Stroke is reported to be the most common cause of this disability in the world [2, 3].

Functional electrical stimulation (FES) which is considered to be the most effective treatment of foot drop is the technique of smearing harmless levels of electric current to activate the impaired or disabled neuromuscular system in a synchronized fashion in order to accomplish the lost function. FES recruits a physiological-like stimulation in the intact peripheral nerves, providing restoration of functions of various body structures in the neurologically impaired individuals. FES system may be surface system or percutaneous (implantable). It is clearly evident in many researches already done that FES provides therapeutic effect to patient with time. FES has also been used comprehensively to regenerate the activation pattern of lower extremity muscles to rehabilitate gait and to produce the sequence of lower extremity muscle action desired during a gait cycle.

Previous works on FES (starting from the early 1990's) required more parameters to take into account for accuracy. The methodology adopted by Sabut et al did not consider the time duration of electrical pulses provided to the patient which is major factor for designing FES system and it varies from patient to patient as desire varies [4]. It is too important from the comfort and safety point of view for the patient. Chen et al did not involve the stride frequency and stride length as major parameters so that it could provide the nearest to accurate information about the problem in gait [5]. In another study by Sabut et al FES pulses were applied in the constant mode which might have caused the accommodation phenomena in muscles resulting in the body getting used to the signals and not responding to them accordingly [6].

The aim of this study was to design an FES-based device for foot drop that is safe, cost-efficient, reliable, portable and light weight. The device would detect the start of swing phase and would avoid the accommodation phenomena.

II. DEVICE DESIGN

The process of designing a real time assistive device for foot drop patients is a complex task. The process that is followed for the designing of FES device (Figure 1) starts with the designing of FES stimulator and continues with detection of bending in the knee and end with the controlling of device. The stimulator was designed with the 555 timer based technique.

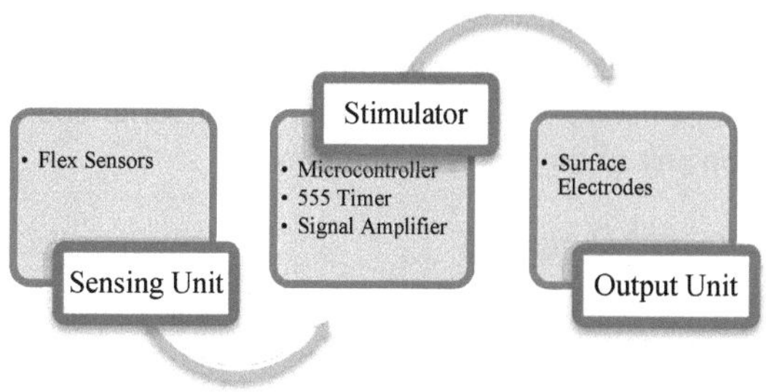

Fig. 1 Flow diagram of FES-based device for Foot drop.

F. Ibrahim et al. (eds.), *International Conference for Innovation in Biomedical Engineering and Life Sciences*,
IFMBE Proceedings 56,
DOI: 10.1007/978-981-10-0266-3_23

The design parameters of the device were carefully selected, keeping in mind the safety threshold and previous researches. The major design parameters as summarized in Table 1 were finalized for the optimal performance of the device.

Table 1 Parameter ranges for device design.

Parameter	Range and Description
Voltage	The voltage requisite lies within the range of 80-110V which will be able to excite the peroneal nerve innervating the dorsiflexors.
Current	The current boundary lies within 90-120mA which produces the depolarizing effect in the muscles.
Frequency	The frequency essential was above 100Hz and it carries tangling effects.
Duty Cycle	The duty cycle was confined to 30% and it depends on the need of subject.

A. Swing Phase Detection

The detection of the start of swing phase is important because this is the point when the stimulation should take place. Flex sensor was selected for detecting the bending movement of knee. It follows the strain gauge principle; the more the knee bends the more the resistance changes at output.

B. Microcontroller

The Arduino-Uno board was used which has microcontroller ATmega328 for the programming and controlling of sensor. Operating voltage is 5V which can be provided through USB or any external source. The board operates on a 6 to 20 volts external supply. However, the external power supply should not be less than 7V, otherwise the 5V would not generate full potential and the board might get unstable.

Conversely, supplying more than 12V would cause the voltage regulator to overheat and damage the board. Therefore a supply range of 7 to 12 volts is recommended.

The Arduino software includes a serial monitor allowing simple textual data to be sent to the Arduino board and vice versa. A Software Serial library allows for serial communication on any of the Uno's digital pins.

C. Stimulator Design

The main component for producing the pulses or waveform is the 555 timer IC, Figure 2. It was used in astable mode to produce pulses. In the first step it produces pulses of low voltage with less current intensity. For voltage amplification 741 op-amp was used to enhance the voltage to certain limit, and current was increased to the required range with the Darlington configuration of power transistors. Further amplification of voltage was still needed for which the present output was provided at the input of power transformers which accomplished the voltage amplification up to 100volts. The waveform it produces is functional monophasic square waveform within the required frequency range and duty cycle completing our requisites. Further sensing and controlling mechanism were implemented to achieve the final task of activating the dorsiflexors of the foot during swing phase. There is a brief description about the components committed in this approach.

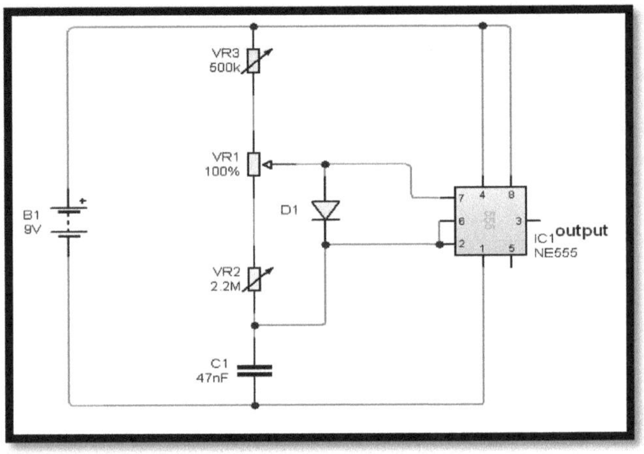

Fig. 2 555 Timer based circuit for generation of pulses

The frequency and duty cycle of 555 timer output can be controlled by the VR3 and VR4 potentiometers respectively. Diode is placed in between pin 7 and 2 to achieve the duty cycle less than 50% which is required.

The 555 timer gives the output consisting of rectangular peaks having amplitude of 9 Volts. This signal is fed to a potentiometer so that we can get the peaks of variable amplitude.

D. Current Amplification

Power transistors were used for the current amplification purpose in this technique as the requirement is of not only high voltage but also high current. 2N3055 is implemented in Darlington pair configuration which provides max output current of 15 ampere which is quite feasible and even much more than the requirement. The Darlington pair provides current gain in value of thousands and works in the combination with resistors and capacitors so that the signal remains smooth after passing through the transistors.

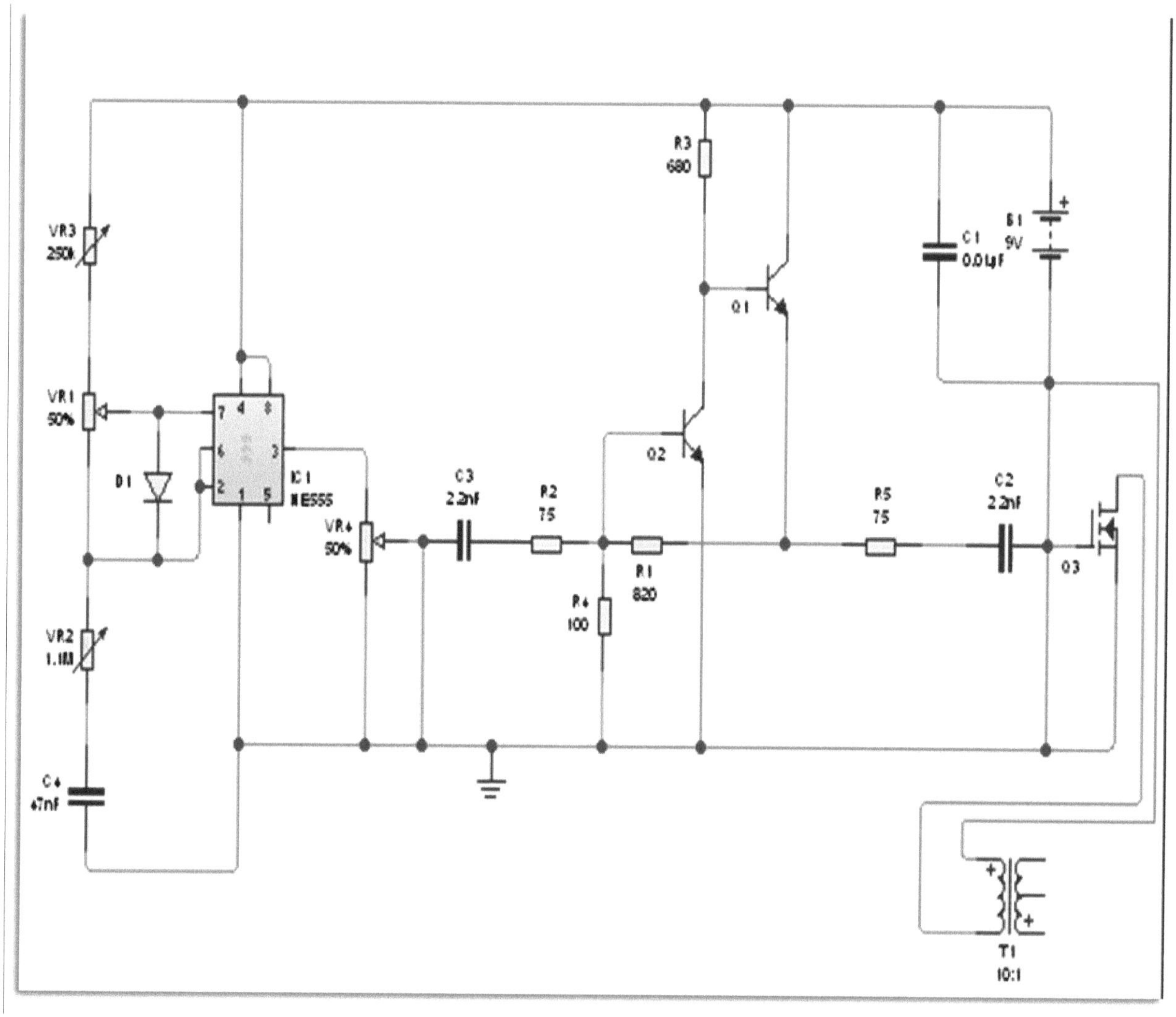

Fig. 3 Complete circuit diagram of the FES-based device

E. Voltage Amplification

Ferrite core transformers were used in the circuit design for stepping up the voltage. They were used due to low power losses at higher frequencies with low noise and circuit works at the higher frequency of 150Hz. These transformers offer high resistivity with low eddy current loss. They are small in size, least expensive and provide minimum loss for high frequency design with current technology.

F. Electrodes

Disposable Adhesive Surface Electrodes are used as a means of transmitting the pulses to peroneal nerve which are light weight and pre-gelled. Special formula gel for a very sticky electrode is used to make it completely adhesive yet easy to remove. Electrodes of size 5x5cm are placed on the motor points of peroneal nerve to provide maximum stimulation with less pain. Silver / silver chloride conductive layer for optimal signal recording. The optimal placement of electrodes is necessary to avoid stimulation and pain in the surrounding area.

G. Mechanism

Initially flex sensor was calibrated by taking different values when the subject performed flexion and extension of the knee. Flex sensor output was connected to the analog pin of Arduino board whereas one LED was placed on digital pin 13 of Arduino board for calibration. When foot underwent swing phase LED turned on, and the LED turns off in standing position or during stance phase of gait cycle, the value recorded at this instant was used as final one which triggers the stimulator circuit.

Flex sensor is placed on the knee. As person starts walking; flex sensor bends due to which voltage measured at analog pin changes and microcontroller compare this measured value with the threshold value present in the program, if it passes the threshold value the logic circuitry based on BJT switches on high as it gets 5 volts and activate the 555 timer circuit which generates signal pulses. The produced signal pulses are transmitted to the peroneal nerve through electrodes placed on the pathway of peroneal nerve. Peroneal nerve gets stimulated and depolarization occurs due to this. This provides activation of the dorsiflexor which enables the patient to lift the foot during swing phase. End of the swing phase will turn off the stimulator and the cycle will repeat for each gait cycle of the patient. Figure 3 shows the entire circuit if the device along with component values.

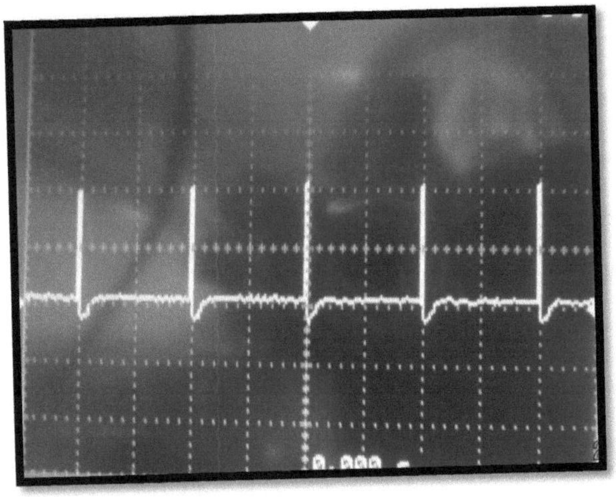

Fig. 4 Resultant Peak-to-Peak Voltages of the device

The generated wave form was triangular peaks having frequency of 100 Hz and voltage of 120V. The peak to peak voltage (Vpp) in Figure 4 is shown as a 10 times multiple of the impedance of the oscilloscope probe that's why the output voltage shown is divided by 10.

III. METHODOLOGY

Five male subjects were considered (aged 40 to 50 years) with a drop foot due to stroke. All the subjects could walk with supports and were able to stand up from sitting position. Only five subjects were considered since the aim was to test the device, which had already provided required results on an oscilloscope. The Temporal and Spatial Parameters (TSP's) were measured initially since they are the identifiers of gait [7]. Written consent was taken from each subject.

A qualitative analysis was done for the subjects by obtaining information about their daily routines, mobility requirements, difficulties in gait due to foot drop etc.

The stimulator was mounted on each subject's knee belt and the two electrodes were attached to the motor ends of the peroneal nerve (on the bellies of Tibialis Anterior, Extensor Digitorum Longus, Extensor Hallucis Longus and Peroneus muscles). The training session continued for a week in which the subjects were briefed about the donning and doffing of the device, precautions for the device, and were encouraged to report any difficulties with the gait if any. After this trial period, the subjects were asked again for an interview and observational gait analysis to test the efficiency of the device.

IV. RESULTS

The subjects reported no difficulties in the accommodation of the device. Ease in mobility was reported from each of the subjects, as well as a restoration of gait towards normal.

The mean values of the TSP's before and after the use of device are summarized in Table 2. An increase in gait efficiency is observed from the values which support the results of the qualitative analysis.

Table 2 Mean values of Temporal & Spatial Parameters pre and post device usage.

	Cadence (steps/ min)	Stride Length (meters)	Walking Speed (meter/ sec)
Standard normal values[7]	110-115	1.4-1.6	1.3-1.6
Without device	53	0.79	0.34
With device	65	0.98	0.53

V. CONCLUSION

The device provides pulses and enables patient to lift the affected foot during swing phase of the gait hence assisting the gait in real time. Assistance in real time provides independence, stability and helps in restoring the mobility. The device is unique in the way that it has three units i.e.; sensing unit, controlling unit and stimulating unit for ease of the patient as it makes the operation of device convenient to the patient.

VI. DEVICE COST

The total cost of the device was USD 200, which means that it is more economical than its competitors and can be reproduced on a budget without compromising on the efficiency.

VII. LIMITATIONS

The biggest limitation of this study is the small sample size. The device needs to be tested on a larger sample of population in order to observe is efficiency. Another drawback is that we have not used cinematographic gait analysis in order to evaluate the kinematics and kinetics with device usage, only observational gait analysis is done. This is because the primary purpose of the study was to present a device design for foot drop patients.

VIII. FUTURE SCOPE

As FES is a developing technique, it covers a vast area of research in rehabilitation engineering and provides new solution to the existent medical problems. The FES device can be used with artificial neural networks (ANN) to make it intelligent enough so that it will be utilized as generalized device for the patients. ANN can be trained to learn the natural coordination pattern that exists between the ankle dorsiflexor and plantar flexor muscles activities of opposite legs during normal gait so that it can triggers the appropriate activity of the impaired dorsiflexor muscles based on the muscle activity of the healthy calf muscle of the opposite leg.

The stimulator and programming section of the device could be advance with the use of Bluetooth technology in order to avoid little inconvenience and it also lightens up the weight of the device system placed at the knee as the programming segment can be placed in the pocket.

This FES based device can be used in combination with measuring the muscle activity of the dorsiflexors of foot. EMG and the sensor used will provide feedback which will give more accurate instant for the onset of stimulator circuit.

ACKNOWLEDGEMENT

The authors would like to acknowledge the supervisor of this project, Dr. Muhammad Abul Hasan (Assistant Professor, NED University of Engineering & Technology, Pakistan) for his support and guidance throughout the project. Also, the authors offer gratitude to Mr. Osama Mazhar (Master Scholar, Universiti Technology Malaysia) for his guidance regarding electronic circuits.

FUNDING

The project was self-funded by the authors.

CONFLICT OF INTEREST

The authors declare no conflict of interest.

REFERENCES

1. Perry J. Gait analysis: Normal & Pathological Function 2009.
2. Rathore S, et al., Characterization of incident stroke signs and sympotoms: findings from the atherosclerosis risk in communities study. Stroke; 2002, 33.
3. Warlow C, et al. Stroke. Lancet; 2003:1211–1224.
4. Sabut SK and Manjunatha M. Neuroprosthesis-Functional Electrical Stimulation: Opportunities in Clinical Application for Correction of Drop-Foot. Proceedings of the First International Conference on Emerging Trends in Engineering and Technology.
5. Chen, M, et al. A Foot Drop Correcting FES Envelope Design Method Using Tibialis Anterior EMG During Healthy Gait With A New Walking Speed Control Strategy. Proceedings of 32nd Annual International Conference of the IEEE EMBS. 2010: Buenos Aires, Argentina.
6. Sabuta SK, *et al.* Functional electrical stimulation of dorsiflexor muscle: Effects on dorsiflexor strength, plantarflexor spasticity, and motor recovery in stroke patients. NeuroRehabilitation, 2011.
7. Kirtley C, *Clinical Gait Analysis*. Vol. 1. Elsevier. 2006.

Author Contact Details:

Saad Jawaid Khan
Department of Biomedical Engineering,
University of Malaya,
Kuala Lumpur, Malaysia, 50603.
Hand phone number: +60129165072
Email: saadjawaidkhan@siswa.um.edu.my
engr.sjkhan@gmail.com

Association Between Urine Uric Acid Levels and Quality of Life in Academicians and Non-academicians in UTAR, Kampar

S.Y. Low[1], A. Loganathan[1,2], S. Sivasangaran[1], and A.L. Samy[2]

[1] Faculty of Science, Universiti Tunku Abdul Rahman, Perak
[2] Faculty of Medicine, University of Malaya, Malaysia

Abstract— **Studies have showed that poor quality of life (QoL) such as work stress and unhealthy diet of high purine rich food elevate uric acid level in urine and blood. High urine uric acid (UUA) level can cause hyperuricemia, which can lead to gout, cardiovascular disease, chronic kidney disease, hypertension, type 2 diabetes and obesity. Our study aimed to investigate the associations between UUA levels and QoL among academicians and non-academicians in Universiti Tunku Abdul Rahman (UTAR), Kampar. A cross-sectional survey using systematic random sampling was conducted among academicians and non-academicians in UTAR. A self-designed socio-demographic form and WHOQOL_BREF questionnaire were distributed together with a urine sample containers (60 ml) among academicians and non-academicians who gave written consent. The collected urine sample was analyzed using Architect c4000 (Abbott), clinical chemistry analyzer. Collected data was analysed using SPSS software (Ver. 19). A total of 100 participants were enrolled in the study with response rate of 60%. The study found positive association between physical domain factors (WHOQOL_BREF) with UUA levels (p-value <0.05). This study findings signal QoL factors to consider in managing UUA level among academicians and non-academicians in higher education institution.**

Keywords— **uric acid, working adult, quality of life (QoL), cross-sectional study**

I. Introduction

Protein-rich foods such as meat type, seafood, and grains contain purines [1]. The purines are metabolized into uric acid in the purine degradation pathway [2]. The uric acid does not oxidize to more soluble compound due to lack of the uricase enzyme in human[3]. Therefore, abnormal secretion of uric acid was significantly associated with high purine-rich foods and various co-morbidities that reflect on poor quality of life [4].

To date, many studies focused on the relationship between uric acid level and co-morbidities [4,5] such as obstructive sleep apnea (OSAS) [6], cardiovascular risk factors such as gout, hyperuricaemia, hypertension, diabetes mellitus, hyperlipidaemia [5], gout [7] and cognitive decline in older people [8]. Other predictors for high uric acid levels are associated quality of life factors among gout patients [7,9]. Research on association of quality of life factors with uric acid level tends to be disease-specific. Furthermore, previous studies used uric acid level in the serum [8] and some studies focused on 24 hours UUA level [10].

However, two studies reported on association between serum uric acid level and quality of life (QoL) in adult at work [11,12]. One of the study focused on adults working in industry and offices [11]. Therefore, this study was aimed to determine the associated QoL factors with UUA levels among academicians and non-academicians in a higher education institution. Hence, an understanding of associated QoL factors with UUA level will help clinicians and researchers find ways to manage UUA level among working people in higher education institution.

II. Methods

A. Study Design

This is a cross sectional study conducted at Universiti Tunku Abdul Rahman (UTAR), Kampar, Malaysia between October 2014 to December 2014. The target population was working adults in UTAR aged 21 years and above. A total of 250 questionnaires and urine collectors were distributed to the working adults using systematic random sampling.

B. Data Collection Tool

The data collecting tools were WHOQOL-BREF and a self-designed questionnaire to assess socio-demographic characteristics such as age, gender, self-reported medical history, exercise practice, eating habit, smoking, alcohol drinking, and working environment. The WHOQOL-BREF instrument is specifically focus on patients' health aspect and well-being followed by reinforcing health care intervention [13]. It was divided into four domains: physical, psychological, social and environment domains. The WHOQOL-BREF was constructed on a likert scale from 1, very poor or very dissatisfied to 5, very good or very satisfied. Questions in WHOQOL-BREF instrument that were categorized under each domain were summed up in a scoring table to yield the final transformed scores.

F. Ibrahim et al. (eds.), *International Conference for Innovation in Biomedical Engineering and Life Sciences*,
IFMBE Proceedings 56,
DOI: 10.1007/978-981-10-0266-3_24

C. Data Collection Method

This study applied systematic random sampling; a sampling technique of choosing individual through a standard sampling frame [14]. The study subjects were consented before enrollment into the study. The academicians' roles were lecturer, assistant professors, associate professors and professors. The non-academician roles were general administration officers, laboratory officers, cleaners and bus drivers. The participants were informed to collect their urine using 60 ml urine container provided. The collected urine samples were analyzed using Automated Clinical Chemistry Analyzer c4000 Architect (Abbott). The UUA assay was selected to analyse the urine sample. The normal UUA level was 1480 *µmol/L* to 4430 *µmol/L.*

D. Data Analysis

The collected data were analysed using SPSS statistics software (Ver.19). Chi-square test or Fisher's exact test used to analyse the associations between socio-demographic background variables and UUA. Normality of distributions was tested using Kolmogorov-Smirnov test. The association between UUA levels, WHOQOL-BREF factors and socio-demographic characteristics were investigated using student's t-test. The scores for WHOQOL-BREF were summed up by using WHOQOL-BREF software.

III. RESULTS

A. A. Socio-Demographic Characteristics

Hundreds academicians and non-academicians were enrolled into this study. The response rate was 60%. Table 1 shows the socio-demographic characteristics.

Table 1 Socio-demographic characteristics

Characteristics	Number (n)
Gender	
Male	57
Female	43
Age Group	
21 – 30	38
31 - 40	45
41 - 60	17
Affiliation	
Academic	86
Non-academic	14

B. Urine Uric Acid Analysis

Table 2 shows that majority of the academicians and non-academicians had a normal range of UUA levels (n=59); the *chi*-square test shown is 31.460 and p=0.0001 (p<0.05). This indicates that there is a significant difference among low, normal and high UUA levels.

Table 2 The Chi-square and p-value of UUA levels

UUA levels	Observed	Expected	Residual	Chi-square test	p-value
Low[a]	26	33.3	-7.3	31.460	0.0001
Normal[b]	59	33.3	25.7		
High[c]	15	33.3	-18.3		
Total	100				

Note:
a = < 1480 µmol/L; b = 1480 µmol/L to 4430µmol/L; c = > 4430 µmol/L

C. Association Between Self-reported Co-morbidities and UUA Levels

Majority of the working adults did not report major co-morbidities such as gout, type 2 diabetes, cardiovascular disease and urinary infection. There was no significant difference in the p-value.

D. WHOQOL-BREF Analysis

Figure 1 shows WHOQOL-BREF analysis. The highest mean score is for physical domain (domain 1) with mean value of 15.73.

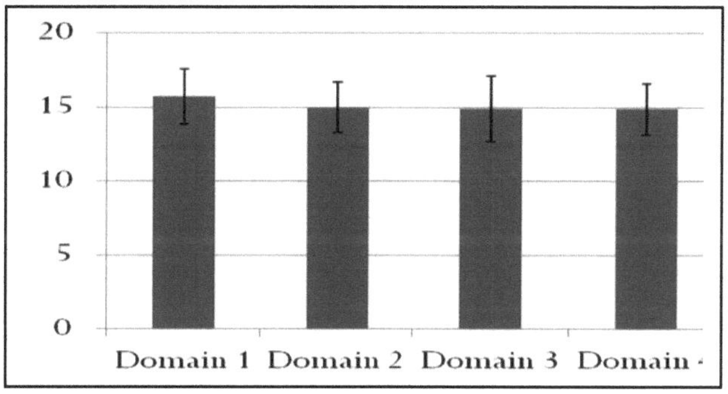

Fig. 1 Mean and standard deviation for *WHOQOL-BREF* domains

E. Association Between QOL-BREF Domains and Affiliation

Table 3 shows that all the four domains of QoL were significantly associated (p-value <0.05) with affiliation. There was significant difference between WHOQOL-BREF four domains and affiliation as the p-value is less than 0.05. The mean and standard deviation value for all the four QoL domains in academicians were higher than non-academicians.

Table 3 Association between QOL-BREF Domains Types with Affiliation

Participants	Academic		Non-academic		*P*-value
	Means	SD	Means	SD	
Domain 1 (physical)	74.75	10.7	64.79	13.35	<0.05
Domain 2 (psychological)	69.71	9.53	62.79	14.56	<0.05
Domain 3 (social	69.37	12.8	60.11	17.04	<0.05
Domain 4 (environment)	69.70	9.00	58.03	14.33	<0.05

*The *p*-value is less than 0.05.

F. Linear Regression Analysis

Table 4 showed the physical domain entered into the regression model at $t < 0.05$. Change in the UUA levels influenced by a change in the physical domain. Correlation between the UUA variable and physical domain is 0.211. The R^2 value 0.044 shows that 4.4% change in the uric acid variable is due to a change caused by the physical domain. The t-test result showed that there is a significant relationship between UUA levels and the physical domain at $t < 0.05$. The psychological, social and environment domains were not included in the regression model as the results were not significant.

Table 4 Linear Regression Analysis

Model	R	R square	t	Significant value
Domain 1	0.211	0.044	2.135	0.035
Domain 2	-0.081	N/A	-0.645	0.520
Domain 3	-0.122	N/A	-1.122	0.265
Domain 4	-0.014	N/A	-0.109	0.913

*The *p*-value of domain 1 is less than 0.05. The association between QOL-BREF domains 1 (physical) with UUA levels is statistically significant.

IV. DISCUSSION

Our study revealed that there were no positive association found between UUA levels and socio-demographic characteristics such as age, gender, and affiliation. There were no association found for self-reported QoL factors such as co-morbidities, diet, alcohol consumption, smoking, exercise and work stress with UUA levels. However, the physical domain, which constitute of physical functioning was significantly associated with UUA levels.

Fifty nine percent working people reported to have normal UUA level. Similarly, a study conducted on normal healthy subjects at Spain had normal range of 24 hours urinary uric acid excretion [2]. This is because the subjects were healthy adults and had no associations with co-morbidities [1,2].

The WHOQOL-BREF analysis showed that the physical domain has the highest mean ± SD. Furthermore, the physical domain was also positively associated with participant's UUA levels (p<0.05). One study reported that the physical domain of quality of life had a lower score in gout patients [15]. This showed that poor physical domain has a significant impact in increasing the UUA level that lead to gout.

In our study, academicians appeared to have a better QoL compared to non-academicians. This may be due to academicians had more flexibility in the working environment compared to non-academicians who are mostly desk-bound. One study reported that academicians have better QoL than non-academicians in a public university by using HRQOL and SF-36 questionnaires [16].

The limitations in our study are most of the participants recruited were academicians as most of the non-academicians were not willing to participate, as they were busy with their job. This resulted in an imbalance number between academic and non-academic participants. Furthermore, the relationships among the four domains of QoL factors was not measured extensively and detailed in this study. Further study is required to examine the relationship among these factors.

The strengths of our study are that the participants were recruited from a wide range of work nature in an academic institution. In addition, the WHOQOL-BREF instrument has been widely used cross-culturally. Furthermore, the WHOQOL-BREF instrument have been proved as a reliable instrument in gathering QoL data in a population to carry out multi-centre quality of life research, which is important for international epidemiological studies.

V. CONCLUSION

Our study revealed that physical domain of WHOQOL-BREF has a significant association with UUA level in academicians and non-academicians. This finding concluded that physical related QoL factors such as activities of daily living, energy, sleep and work capacity were significantly associated with UUA level.

ACKNOWLEDGEMENT

We would like to thank the University Tunku Abdul Rahman (UTAR) for final year project and conference funding. We also would like to thank University of Malaya, Malaysia for research opportunity.

CONFLICT OF INTEREST

The authors do not have conflicts of interest

REFERENCES

1. Choi H (2005) Purine-Rich Foods, Daily and Protein Intake, and the Risk of Gout in Men. New England Journal of Medicine 350: 1093 – 1103.
2. Puig J, Torres R, Miguel E, Sanchez A, Bailen R, et al. (2012) Uric Acid Excretion in Healthy Subjects: A Nomogram to Assess The Mechanisms Underlying Purine Metabolic Disorders. Metabolism 61: 512 – 518.
3. So A, Thorens B (2010) Uric Acid Transport and Disease. Journal of Clinical Investigation 120: 1791-1799.
4. Lee S, Hirsch J, Terkeltaub R, Khanna D, Singh J, et al. (2009) Perceptions of disease and health-related quality of life among patients with gouts. Rheumatology: 1- S5
5. Woo J, Swaminathan R, Cockram C, Lau A, Chan A (1994) Association between serum uric acid and some cardiovascular risk factors in a Chinese population. Postgraduate Medical Journal 70: 486-491.
6. Hirotsu C, Tufik S, Guindalini C, Mazzotti R, Bittencourt L, et al. (2013) Association between Uric Acid level and Obstructive Sleep Apnea Syndrome in a large epidemiological sample. Plos one 8.
7. Becker M, Schumacher H, Benjamin K, Gorevic P, Greenwald M (2009) Quality of Life and Disability in Patients with Treatment Failure Gout. The Journal Of Rheumatology 36: 1041-1048.
8. Schretlen D, Inscore A, Jinnah H, Rao V, Gordon B (2007) Serum uric acid and cognitive function in community dwelling older people. Neuropsychology 21: 136 - 140.
9. Hirsch J, Terkeltaub R, Khanna D, Singh J, Sarkin A, et al. (2010) Gout disease-specific quality of life and the association with gout characteristics Patient Related Outcome Measures 1: 1 - 8.
10. Sila-On A, Pavaro U, Nuchpramool W (1991) Serum and urinary uric acid levels in healthy subjects and in patients with urolithiasis. Journal Medical Association of Thailand 74: 3527-.
11. Oh J, Choi W, Lee M, Han S, Song S, et al. (2014) The Association between Shift Work and Hyperuricemia in Steelmaking Male Workers. Annal of Occupational Environmental Medicine 26: 42 - 47.
12. Bosco J, Greenleaf J, Kaye R, Averkin E (1970) Reduction of Serum Uric Acid in Young Men during Physical Training. The American Journal of Cardiology 25: 46 - 52.
13. World Health Organization (WHO) (2004) The World Health Organization Quality of Life Assessment (WHOQOL): Position Paper from the World Health Organization. Social Science Medicine 41: 1403 – 1409.
14. (2009) Systematic Sampling. [online] Available at: https://explorablecom/systematic-sampling [Accessed 10 March 2015].
15. Roddy E, Zhang W, Doherty M (2007) Is Gout Associated with Reduced Quality of Life? A Case-control study Reumatology (Oxford) 46: 1441 – 1444.
16. Naslina N (2012) A Comparative Study of Health Related Quality of Life Among Academician – A Case Study. Statistics in Science, Business and Engineering (ICSSBE): 1 - 4.

Corresponding author

Annaletchumy Loganathan
Department of Primary Care Medicine, Faculty of Medicine,
University of Malaya, 50603 Kuala Lumpur, Malaysia
Email: annaletchumy@siswa.um.edu.my
Tel: +603-7949 2306
Fax: 603- 7957 7941

Department of Biomedical Science, Faculty of Science,
Universiti Tunku Abdul Rahman, 31900 Kampar, Perak, Malaysia
E-mail: an18na03@yahoo.com
Tel: +605-4688888/4511

Screen Printed Impedance Biosensor for Cytotoxicity Studies of Lung Carcinoma Cells

A.F. Mansor[1], I. Ibrahim[2], I. Voiculescu[3], and A.N. Nordin[1]

[1] Department of Electrical and Computer Engineering, Kulliyyah of Engineering, IIUM Kuala Lumpur, Malaysia
[2] Department of Biotechnology Engineering, Kulliyyah of Engineering, IIUM Kuala Lumpur, Malaysia
[3] Department of Mechanical Engineering, School of Engineering, City College of New York, New York, USA

Abstract— **Electrical Cell-Substrate Impedance Sensing is a powerful tool for monitoring real time cells properties such as adhesion, mobility and cytotoxicity. In this study, a silver/silver chloride screen-printed impedance biosensor was developed to characterize A549 lung carcinoma cells growth in the presence of collagen I, Bovine. Collagen acts as an extracellular matrix (ECM) for A549 and promotes cell attachment. The sensor was incorporated with a culture well which was fabricated from polydimethylsiloxane (PDMS). A549 cells were cultured in the chambers and impedance measurements were taken at 12 hours intervals for 120 hours. Cell Index (CI) were calculated from the impedance data and plotted in comparison with growth profile of the cells in T-flasks for validation of the sensor's functionality. A549 cells were also treated with anti-tumor drug; Paclitaxel and its response were monitored over 5 days. Experimental results show significant change in CI during growth and death after exposure to Taxol, indicating that tumor growth was inhibited in the presence of Taxol.**

Keywords— **Screen Printed Biosensor, A549, Paclitaxel, Silver/Silver Chloride Electrode, ECIS.**

I. INTRODUCTION

Every year, mortality rates due to cancer increase, making it the second leading cause of death around the world. In 2012, there were 8.2 million deaths globally due to cancer [1]. Out of this figure, 19.4% of the deaths were caused by lung cancer making it the highest percentage of deaths amongst other types of cancer. This statistic indicates that lung cancer is the most lethal cancer among top five cancers in man. In Southeast Asia, the mortality rate of lung cancer was approximated to be 95 deaths per 100,000 people [2].

Chemotherapy is a common treatment for cancer that involves infusion of drugs into patients. Paclitaxel or Taxol has been reported to effectively treat rapidly growing cancer cells such as lung and breast cancers [3]. Taxol was sourced from natural products therefore it can be considered as renewable and environmentally safe [4]. It is classified under the mitotic inhibitor type of drugs which prevent protein production of cancer cells needed for mitosis process [5]. Low blood count, hair loss, numbness and muscle pain are common side effects of this drug. Side effect on the patient can be minimized if drug dosage were individualized to each patient rather than using Body Surface Area (BSA) calculation for drug concentration determination [6].

Monitoring response of cancer cells to drug exposure is an important aspect of drug discovery. To avoid subjecting the patient to trial and error therapy, the usage of in vitro cytosensors is suggested to characterize the cells response to drugs. Cytosensors is a biosensor that uses living cells as sensing element and has the ability to detect changes in physiological state of cells based on its electrical characteristic observation [1]. These sensors have been utilized in various research experiments because of its high sensitivities, fast responses and real-time capabilities [7]. Cytosensors have also been successful to monitor primary cardiomyocyte [7], chicken embryonic fibroblasts [8], bacteria [9], and evaluation of breast cancer [10].

In this paper, a screen-printed biosensor was developed to characterize the growth of lung carcinoma cells and its response to Taxol. The growth of the cancer cells can be monitored by observing the impedance measurement. Impedance measurements are directly proportional to cell adhesion and coverage on the electrodes; higher impedance indicates higher coverage of cells on the electrodes. Cell index was calculated from the impedance results and plotted to describe the activities of cells on electrode over the whole duration of the experiment. The aim of the study was to develop a low cost screen printed biosensor with high sensitivity and reliability for monitoring cancer cells and its response to drugs.

II. IMPEDANCE BIOSENSOR

Impedance biosensor is a sensor that converts the biological response into an electrical signal (impedance). Impedance was observed as an electrical parameter that can indirectly monitor cells metabolic activity, cell adhesion on surfaces, responses to drugs and cytotoxicity test [11, 12, 13].

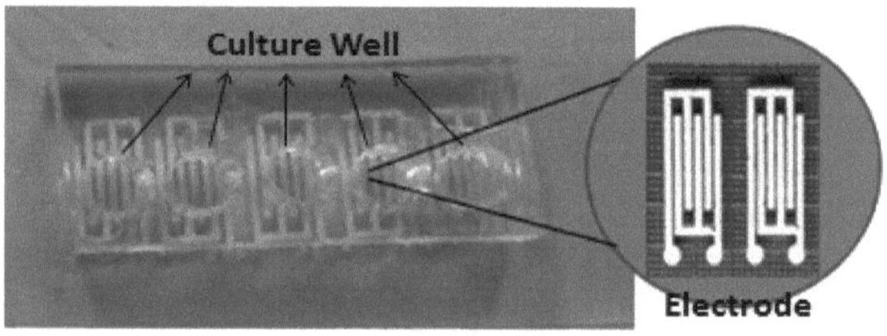

Fig. 1 Screen printed sensor with PDMS culture well.

F. Ibrahim et al. (eds.), *International Conference for Innovation in Biomedical Engineering and Life Sciences*,
IFMBE Proceedings 56,
DOI: 10.1007/978-981-10-0266-3_25

Electrical Cell-Substrate Impedance Sensing (ECIS) is one of the methods introduced by Giaever and Keese to study cell growth based on its attachment with the electrode [14, 15]. With the help of extracellular matrix (ECM) coated on the electrodes, cells will adhere to the electrode surface and over time will proliferate, covering the entire electrode surface. At low frequencies, cell membrane will restrict current flow from the electrodes to its surrounding due to the cell membrane's insulating properties. These cause cells to behave as dielectric particles causing increases in impedance with increased coverage of cells on electrode [16]. Diseased or dead cells will alter cells membrane properties such as losing its adhesion strength and insulating properties making the impedance dropped over time. These make impedance biosensor to be perfect in monitoring cytotoxicity of cancer cells.

In theory, in vitro studies for mammalian cell culture mimic in vivo condition of cells in humans [17]. ECM is frequently used as a coating to the electrode or substrate in order to enhance cells attachment to the surface. For this work, a natural biological ECM was used which is collagen type I, Bovine as the adhesion factor for A549 lung carcinoma.

Silver/silver chloride is the most common type of electrode used in research and industry due to its low cost and simple fabrication. This material is usually made as reference electrode for sensors due to its high stability [18]. Silver/silver chloride was normally used as a reference electrode together with working electrode materials of platinum and gold which has higher conductivity but much more costly compared to silver. In this work, a high conductivity silver/silver chloride paste purposely made for medical diagnostic and environmental sensor application was purchased and solely used as working electrode material without addition of any other material.

Polydimethylsiloxane (PDMS) is silicon based organic polymer that commonly used in microfluidic channel. It has many advantages compared to silicon and glass. Physically, PDMS is a flexible, transparent, easy fabrication and inexpensive material that offer bio-compatibility due to its non-toxic properties [19]. PDMS has been used widely in medical and biological researches including research on cancer cells such as MDA-MB-231 breast cancer cell [20, 21], MCF-7 breast cancer cell [21] and HEP G2 liver cancer cell [22].

III. Materials and Methodologies

A. Cell Culture, Thin Coating Film and Drug Treatment

A549 Lung Carcinoma cell lines, (ATCC® CCL185™) were obtained from the American type Culture Collection (ATCC, Manassas, VA, USA), maintained and cultured in Dulbecco's Modified Eagle Medium, DMEM (Gibco, Paisley, UK) supplemented with 10% (v/v) heat inactivated fetal bovine serum (FBS; Gibco), and 10% concentration of antibiotics. The culture medium was changed every 2-3 days. The A549 cells were incubated under standard cell culture in CO_2 incubator at 37 °C in an atmosphere containing 5% carbon dioxide, CO_2. About $5x10^3$ cells/ml A549 cell lines were added to each sensor well for experimental purposes.

Collagen I, Bovine (Gibco, UK) was diluted to 50 mg/L in 0.02 M acetic acid (Merck, Germany) adjusted to final volume needed. The solution was added on the biosensor at 200 μg per cm^2. The sensor was then incubated at room temperature for 1-2 hours. Excess solution of the collagen was aspirated from the biosensor carefully before adding cells into the well.

The drug used for cytotoxicity testing is the Paclitaxel or known as Taxol (Gibco; Paisley, UK). The final concentration of Taxol with IC50 value 0.857μg/ml was infused immediately to the cell suspension prior to cell seeding and cell culture on the electrodes [1]. The response of cells to the drugs was plotted as cell index number in the next section.

B. Sensor Fabrication

The sensors used in this experiment were screen-printed sensor silver/silver chloride electrodes on a standard glass slide. Screen printing machine, equipment and frame of polyester with 150 threads/inch mesh with 10μm emulsion thickness was obtained from Jujo Technology Resources. The silver/silver chloride paste was obtained from The Gwent Group. An Interdigital Transducer (IDT) design was used as electrode design for the sensor. During the printing process, substrate (glass) will be placed under the design of the mesh. Ink paste was poured on top of the design and pressed using a squeegee. The ink was deposited on the glass surface following shape of the pattern. The ink was cured in a box oven of 80°C for 25 minutes. Each glass slide contained five individual sensors that can be used simultaneously in an experiment.

PDMS was used to create the culture well for the cells. PDMS was weighed and mixed with curing agent in ratio of 10:1. It was left overnight in a close container to make it solid. Once harden, the PDMS was cut to fit the electrodes size, punched to make well on top of the electrode, glued on the glass using another liquid PDMS and left overnight to dried and harden. Once cells have been cultured inside, the wells will be covered by another solid PDMS layer to avoid contamination on the cells.

C. Impedance Measurement

The impedance was measured at frequency of 40 Hz for every 12 hours up to 120 hours by using the Agilent Precision Impedance Analyzer 4294A. The readings obtained were in term of impedance value, Z and phase angle, θ. The data was analyzed and calculated to plot the cell index number.

IV. EXPERIMENTAL RESULTS

The data obtained from impedance analyzer were in the form of impedance and phase. The resistance value can be calculated from these two parameters using the following equations:

$$|Z| = \sqrt{R^2 + X^2} \quad (1)$$

$$\theta = tan^{-1}\left(\frac{X}{R}\right) \quad (2)$$

where Z indicates the impedance, θ indicates phase, R is Resistance and X is Reactance. Cell Index (CI) can be used to express the activities of cells occurred on the surface of the electrode. Determination of resistance value was required to further calculate the CI of the cell growth based on the equation [23]:

$$CI = \max\left[_{i=1,..n} \frac{R_{cell}(f_i)}{R_b(f_i)} - 1\right] \quad (3)$$

where R_b indicates the frequency-dependent resistance of the media, and R_{cell} indicates the frequency-dependent resistance of the cells. Variations of CI will determine the cells attribute of adhesion to electrode surfaces, cells spreading and cell proliferation. CI was plotted as in figure below.

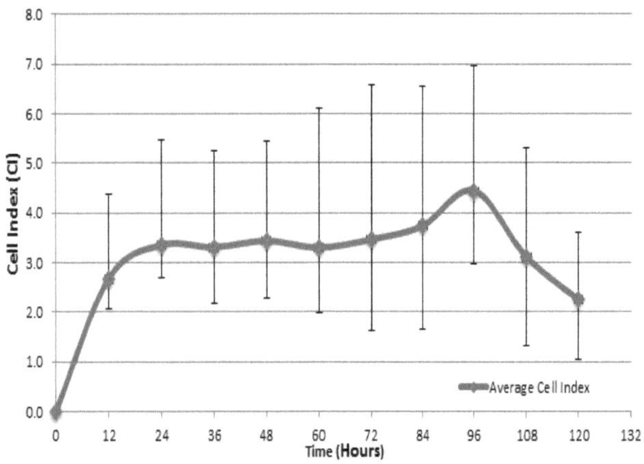

Fig. 2 Cell Index (CI) for A549 Cell Growth. Values are average of 5 sets of experiments at two different times

From impedance measurement and CI observation, the number of cells slowly grew from time 12 hours to 72 hours. Between time 72 hours to 96 hours, cells started to rapidly proliferate, resulting in significant increase of CI. Slightly fluctuation in the growth curve during this time is due to cell motility, which is the migration or movement of cells along the electrode surface or within cells causing current flows differ for each time [15]. The cells achieved full confluence at 96 hours. After 96 hours, impedance started to drop indicating cells undergo death phase, in which the cells alter its membrane properties and start to detach from the substrate. The impedance measurements were also compared with the growth profile obtained using conventional T-flasks measurements as shown in Fig. 3 below. Although cell index values from the impedance measurements were different from the T-flask measurements, the general trend of the graph is similar.

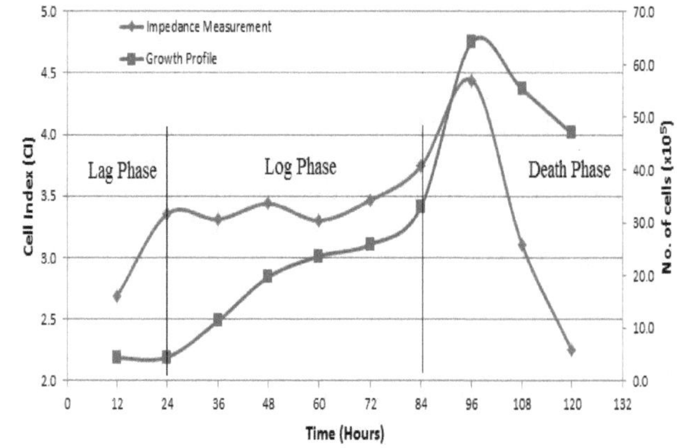

Fig. 3 Comparisons of CI with growth profile of A549

The growth (lag and log phase) is occurs between time 12 hours to 72 hours. During this period, cells attach to the electrode surface and start to proliferate. The highest number of cells counted was at time 96 hours; proportional to the peak CI measurement of the cells. These observations validate the functionality and accuracy of the sensor as a reliable tool for monitoring cell activities. Sensitivity of the sensor may be a bit lower, ranging from 2 to 5 CI measurements compared to currently established cell sensing system such as RT-CES, ranging from 2 to 8 CI measurements for a same set of cells [23]. However, this sensor will greatly reduce the cost as it offers high reproducibility rate and mass fabrication.

The experiment proceeded with drug testing (Taxol) on A549 cancer cells cultured on the sensor. Taxol were infused in suspension A549 cells directly before cultured on the sensor. The CI for treated was plotted as in figure 4 below.

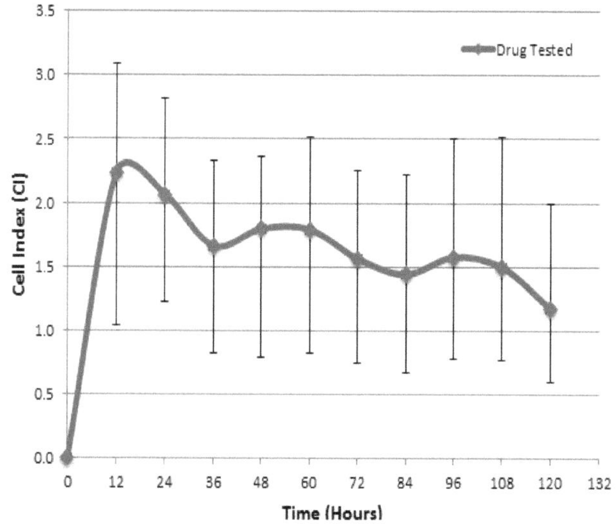

Fig. 4 CI for Drug Treated A549. Values are average of 5 sets of experiments at two different times.

CI for drug testing shows substantial decrease over time. The drug used was an anti-tumor drug that prevents protein production needed by cancer cells to grow allowing the cells to die naturally. Since the growth phase of the cells starts at 24 hours, it is expected that the CI first increases (< 24 hours) and would have enough time to adhere on electrode surfaces.

After 24 hours, graph shows decrement of CI up to 50% from beginning of experiment till the end. This sensor proven to be compatible with drug testing. The large error bars taken over measurements of 5 sets of data indicate that there is still work to be done on to improve the design of the sensor such that such large deviations do not occur between sensors.

V. Conclusions

In conclusion, this study has demonstrated a screen-printed silver/silver chloride impedance biosensor that can monitor A549 lung carcinoma cells activities and cytotoxicity test. Silver/silver chloride electrode has shown compatibility as a working electrode due to its cost effectiveness and good conductivity. However, compatibility of different silver inks need to be tested to ensure that impedance measurements are more reliable with less variation. Screen printed has the potential as personalized sensors as it provides a rapid, real time, continuous, label free, reliable and low cost diagnostic tools for chemotherapy drug screening.

Acknowledgment

This work is supported by Malaysia Ministry of Science and Technology (MOSTI) e-Science Research Grant (SF12-019-0048) and Malaysia Ministry of Education (MOE) FRGS Research Grant (FRGS14-111-0352).

Conflict of Interest

The authors declare that they have no conflict of interest.

References

1. Asphahani, F, & Zhang, M (2011) Cellular impedance biosensors for drug screening and toxin detection
2. Ferlay J, Soerjomataram I, Ervik M, Dikshit R, Eser S, Mathers C, Rebelo M, Parkin DM, Forman D, Bray, F. GLOBOCAN 2012 v1.0 (2013) Cancer Incidence and Mortality Worldwide: IARC Cancer Base No. 11 [Internet]. Lyon, France: International Agency for Research on Cancer
3. National Cancer Institute (2014) A to Z List of Cancer Drugs, Retrieved from http://www.cancer.gov/cancertopics/druginfo/alphalist
4. Newman, David J., and Gordon M. Cragg (2007) Natural Products as Sources of New Drugs over the Last 25 Years. Journal of natural products 70.3 pp 461-477.
5. American Cancer Society (2013) Chemotherapy Principles
6. Clegg A, Scott DA, Sidhu M, Hewitson P,Waugh N (2001) A rapid and systematic review of the clinical effectiveness and cost-effectiveness of paclitaxel, docetaxel, gemcitabine and vinorelbine in non-small-cell lung cancer. Health Technol Assess, vol 5(32),
7. Qiu Y., Liao R. & Zhang X (2008) Real-time monitoring Primary Cardiomyocyte adhesion based on electrochemical impedance spectroscopy and electrical cell-substrate impedance sensing. Analytical Chemistry, Vol. 80, 2008, p. 995.
8. Siddiquei, H. R., Nordin, A. N., Ibrahimy, M. I., Ariffin, M. A., Sulong, N. H., Mel, M., et al (2010) Electrical Cell-substrate Impedance Sensing (ECIS) based biosensor for characterization of DF-1 cells," International Conference on Computer and Communication Engineering (ICCCE). Kuala Lumpur , Malaysia: IEEE
9. Varshney, M. & Li Y (2009) Review Interdigitated array microelectrodes based impedance biosensors for detection of bacterial cells. Biosensors and Bioelectronics, Vol. 24
10. Alexander Jr.F., Price D.T., Bhansali S. Optimization of Interdigitated Electrode (IDE) arrays for impedance based evaluation of Hs 578T
11. Arias, L. R., Perry, C. a, & Yang, L. (2010). Real-time electrical impedance detection of cellular activities of oral cancer cells. Biosensors & Bioelectronics, 25(10), pp 2225–31.
12. Lisdat, F., & Schäfer, D. (2008). The use of electrochemical impedance spectroscopy for biosensing. Analytical and Bioanalytical Chemistry, 391(5), pp 1555–67.
13. Liu, F., & Arifuzzaman, S. (2010). Characterization of endothelial cells using electrochemical impedance spectroscopy. Circuits and Systems, pp 252–255.
14. Wegener, J., C.R., K., & I., G. (2000) Electric cell-substrate impedance sensing (ECIS) as a noninvasive means to monitor the kinetics of cell spreading to artificial surfaces. Experimental Cell Research, pp.158-166.
15. Giaever, I., and C. R. Keese (1991) Micromotion of mammalian cells measured electrically. Proc. Natl. Acad. Sci., Vol. 88, pp. 7896–7900.
16. Bot, C., & Prodan, C (2008) Probing the membrane potential of living cells by Dielectric Spectroscopy
17. Kuncová-Kallio, J., & Kallio, P. J (2006) Lab automation in cultivation of adherent cells IEEE Transactions On Automation Science And Engineering, VOL. 3(2)
18. Waleed Shinwari, M., Zhitomirsky, D., Deen, I. a., Selvaganapathy, P. R., Jamal Deen, M., & Landheer, D. (2010). Microfabricated reference electrodes and their biosensing applications. Sensors, 10(3), pp 1679–1715.
19. Sia, S. K., & Whitesides, G. M. (2003). Microfluidic devices fabricated in poly(dimethylsiloxane) for biological studies. Electrophoresis, 24(21), pp 3563–3576

20. Wang, S. J., Saadi, W., Lin, F., Minh-Canh Nguyen, C., & Li Jeon, N. (2004). Differential effects of EGF gradient profiles on MDA-MB-231 breast cancer cell chemotaxis. Experimental Cell Research, 300(1), pp 180–189.
21. Nguyen, T. A., Yin, T. I., Reyes, D., & Urban, G. a. (2013). Microfluidic chip with integrated electrical cell-impedance sensing for monitoring single cancer cell migration in three-dimensional matrixes. Analytical Chemistry, 85(22), pp 11068–11076.
22. Chiu, Y. T., Lin, C. S., & Chang, C. (2011). In vitro fruiting and seed production in Erycina pusilla (L.) N. H. Wiliams & M. W. Chase. Propagation of Ornamental Plants, 11(3), pp 131–136.
23. Boyd, J. M., Huang, L., Xie, L., Moe, B., Gabos, S., & Li, X. F. (2008). A cell-microelectronic sensing technique for profiling cytotoxicity of chemicals. Analytica Chimica Acta, 615(1), pp 80-87

Author: Anis Nurashikin Nordin
Institute: International Islamic University Malaysia
Street: Jalan Gombak
City: Kuala Lumpur, Selangor
Country: Malaysia
Email: anisnn@iium.edu.my

Prediction of Sepsis Progression in Critical Illness Using Artificial Neural Network

F.M. Suhaimi[1], J.G. Chase[2], G.M. Shaw[3], U.K. Jamaludin[4], and N.N. Razak[5]

[1] Advanced Medical and Dental Institute, Universiti Sains Malaysia, Penang, Malaysia
[2] Department of Mechanical Engineering, University of Canterbury, Christchurch, New Zealand
[3] Department of Intensive Care, Christchurch Hospital, Christchurch, New Zealand
[4] Department of Electronics & Communication Engineering, Universiti Tenaga Nasional, Selangor, Malaysia
[5] Department of Mechanical Engineering, Universiti Malaysia Pahang, Pahang, Malaysia

Abstract— **Early treatment of sepsis can reduce mortality and improve a patient condition. However, the lack of clear information and accurate methods of diagnosing sepsis at an early stage makes it become a significant challenge. The decision to start, continue or stop antimicrobial therapy is normally base on clinical judgment since blood cultures will be negative in the majority of cases of septic shock or sepsis. However, clinical guidelines are still required to provide guidance for the clinician caring for a patient with severe sepsis or septic shock. Guidelines based on patient's unique set of clinical variables will help a clinician in the process of decision making of suitable treatment for the particular patient. Therefore, biomarkers for sepsis diagnosis with a reasonable sensitivity and specificity are a requirement in ICU settings, as a guideline for the treatment. Moreover, the biomarker should also allow availability in real-time and prediction of sepsis progression to avoid delay in treatment and worsen the patient condition.**

Keywords— **sepsis, sepsis score, neural network, ICU.**

I. INTRODUCTION

Sepsis is described as a clinical syndrome with the presence of infection and systemic inflammatory response, SIRS. The infection in sepsis is a pathologic process that caused by the invasion of normally sterile tissue or fluid or body cavity by pathogenic microorganisms [1]. However, the presence of SIRS may or may not indicate the presence of sepsis. This is because SIRS can be triggered by a variety of infectious and noninfectious conditions such as conditions of a patient with burns and pancreatitis.

Organ failure is highly associated with sepsis [2]. It is the condition where the patient may suffer acute respiratory failure, acute renal failure, cardiovascular failure and/or other organ dysfunctions. Additionally, multiple organ failure, MOF is one of the leading causes of ICU mortality [3], regardless of etiology and cause. The number of patients with sepsis increases significantly each year as well as the number of sepsis-related deaths [4, 5].

Early diagnosis is critical because early interventions have been documented to reduce mortality from 46.5% to 30.5% [6] and thus show significant potential. Moreover, early goal-directed resuscitation is recommended for the septic patients, particularly during the first 6 hours after infection recognition and administration of broad-spectrum antimicrobials therapy within 1 hour of the recognition of septic shock [7, 8]. Equally, almost 50% of all sepsis that is clinically defined is culture negative [9]. Hence, the ability to recognize infection and diagnose sepsis as early as possible will consequently improve mortality outcome and patient condition.

However, to initiate antimicrobial therapy and antibiotic administration can cause significant delay due to the process of obtaining appropriate cultures. This process requires incubation and standard procedures that result in a delayed outcome of 2-3 days. Additionally, at least two blood cultures recommended to be obtained prior to antibiotics [7, 8]. Therefore, identifying infection and consequently diagnosing sepsis is a challenging process in critical care. The aim of this study is to test several potential biomarkers and sets of common clinical data used in sepsis diagnosis. An artificial neural network is used to determine their predictive ability. Successful outcomes may provide useful diagnostics, and those sets that do not succeed, indicate data for which no clear pattern or diagnostic efficacy exists.

II. METHODS

A. Clinical Data

Data used in this study consists of sepsis-related clinical data, such as body temperature, respiratory rate, urine output, blood glucose levels, blood culture results, and further laboratory results indicating white blood cell count. These metrics are related to the immune response to infection (temperature, heart rate and respiratory rate) and organ failure (mean arterial blood pressure and urine output), which are the core aspects defining sepsis. Ethics approval to collect, audit and use this data under informed consent was obtained from the Upper South Regional Ethics Committee, New Zealand.

The clinical data were collected during a specific sepsis study from the Intensive Care Unit (ICU) of Christchurch Hospital. There were 30 sepsis patients selected for this study. A total of 7505 data points were gathered from the 30

F. Ibrahim et al. (eds.), *International Conference for Innovation in Biomedical Engineering and Life Sciences*,
IFMBE Proceedings 56,
DOI: 10.1007/978-981-10-0266-3_26

patients in the selected sepsis cohort. These 7505 data points represent a total of 7505 hours of monitoring sepsis patients in the ICU. Clinical data was used to calculate sepsis score and further analysis of the metrics in relation to sepsis score. Table 1 shows the background information and the medical subgroup of the retrospective cohort. 60% of the sepsis cohort is male with a median age of 63 years. The median APACHE II score is 19.

Table 1 Retrospective patient background information

No	Medical Subgroup	APACHE II	Age	Sex	Mortality
1	Septic shock	19	49	F	
2	Septic shock	17	55	F	
3	Pneumonia	18	60	F	
4	Otitis	18	43	F	
5	Pneumonia	15	64	M	Y
6	CAP	17	61	F	
7	CAP	22	74	M	
8	Multiple trauma	19	63	F	
9	CAP	23	52	F	
10	Pneumonia	20	64	M	Y
11	CAP	23	75	M	
12	Pneumonia	21	75	M	
13	Pneumonia	27	70	F	
14	GBS	8	43	M	
15	CAP	24	80	M	
16	Pneumonia	25	71	M	
17	Pneumonia	10	30	F	
18	ARDS	11	63	M	
19	Respiratory failure	17	76	M	Y
20	Type 1 DM	29	46	M	Y
21	Type 2 DM	19	78	F	
22	COPD	17	54	M	
23	CAP	24	88	M	Y
24	Sepsis	19	64	F	
25	Gastrectomy	12	49	M	
26	Pneumonia	18	56	M	
27	Sepsis	16	67	F	
28	COPD	13	55	M	
29	Pneumonia	15	78	M	
30	Sepsis	29	59	M	

a. CAP is Community Acquired Pneumonia
b. GBS is Guillain-Barré Syndrome
c. ARDS is Acute Respiratory Distress Syndrome
d. COPD is Chronic Obstructive Pulmonary Disease

B. *Sepsis Score*

Sepsis score is calculated following criteria classified by American College of Chest Physicians (ACCP)/ Society of Critical Care Medicine (SCCM) definitions of 1992 and 2001 [1]. The criteria calculate a sepsis score including Systemic Inflammatory Response Score (SIRS) and Sepsis-related Organ Failure (SOFA) score [10]. Sepsis score is used to indicate the increased complexity of the disease in patients currently being treated which also represents evaluations and descriptions of the disease. The use of sepsis score provides more accurate monitoring of the patient condition and therapy to be determined by clinicians.

In this analysis, sepsis score is calculated hourly using the available clinical data and blood culture results. The clinical components required for the calculation of sepsis score includes SIRS score factor, infection factor, organ failure factor, fluid resuscitation factor and inotrope usage factor. Although the sepsis score recommended by ACCP/SCCM definition is meant for daily score [10], this analysis aimed to identify potential metrics, which can be used for hourly monitoring and diagnostics. However, the sepsis score is not calculated when patients were admitted due to an unavailability of clinical information. Fig. 1 shows an example of an hourly plot of sepsis score during the patient stay for one of the sepsis patient in this cohort.

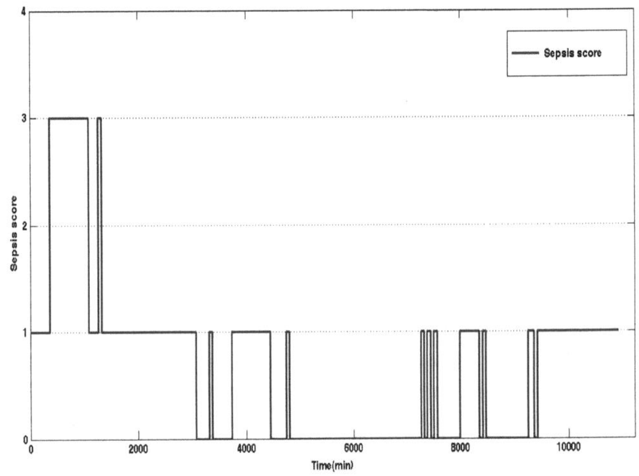

Fig. 1 Hourly plot of sepsis score for Patient 1

C. *Neural Network Analysis*

A sequence of data points, known as a time series of data, is typically measured at successive times spaced at a uniform time intervals. Time series analysis comprises methods for analyzing time series data to extract meaningful characteristics from the data. Forecasting is, therefore, possible by using a model to forecast future events based on known past events and pattern as prediction before the point is measured. Ideally, an hourly prediction of a diagnostic sepsis score can be generated using a model configured with data from one or more previous hours.

In this analysis, a non-linear auto-regressive technique is implemented to get information about the predictive values. In this case, Y is the variable of interest, the sepsis score, and X is the externally determined or measured variable. The information about X is combined using several weights in predicting Y. Fig. 2 illustrates the concept used in getting information about the predictive values. The equation used to obtain Y is shown in (1). Prior values of Y are also used in determining the Y value. L represents the lag of the network system in utilizing prior values of Y. F in Fig. 2 represents the hidden layer functions, a multilayer transfer function that consider weighted input, weighted output feedback, and biases.

In neural networks, a training set is repeatedly used to determine the estimation of weight and biases for a candidate design. Then, validation is required to estimate the performance error of the candidate. The training process stops when the validation error stops decreasing. In contrast, the testing set is used only once on the best artificial neural network configuration to obtain an unbiased estimation of the predicted error of unseen non-training data. Thus, data in the test set is independent of the data in the training set [11]. More specifically, no information on the output is available in the test set, and this knowledge is only used to assess the accuracy of the resulting neural network performance.

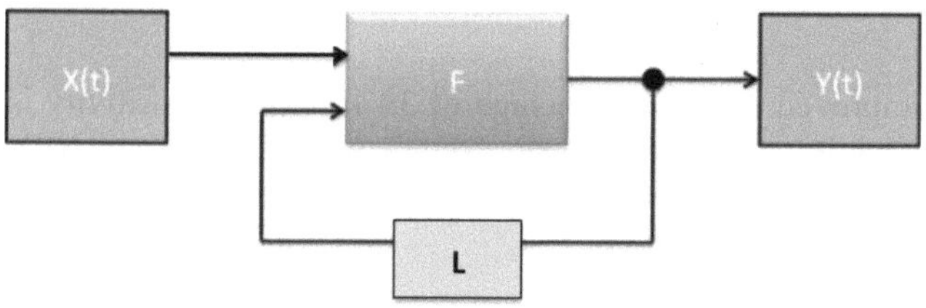

Fig. 2 Block diagram of the auto-regressive technique

$$Y(t) = F(Y_{t-1}, Y_{t-2}, Y_{t-3}, \ldots\ldots\ldots\ldots, X_t, X_{t-1}, X_{t-2}, X_{t-3}) + \varepsilon_t \quad (1)$$

7505 points data were gathered from 30 patients in the sepsis cohort. These data were divided into 5 groups, Group A, B, C, D, and E. These five groups were rearranged to obtain 5 sets of data that were used for neural network training and testing process. For example, in set 1, Groups B, C, D and E were combined to form a training set, and Group A was used for the testing set. For every set of tests, there were 6004 of data points in the training set, and 1501 data points used for the testing, an 80:20 split. The sampling method used is shown in Fig. 3.

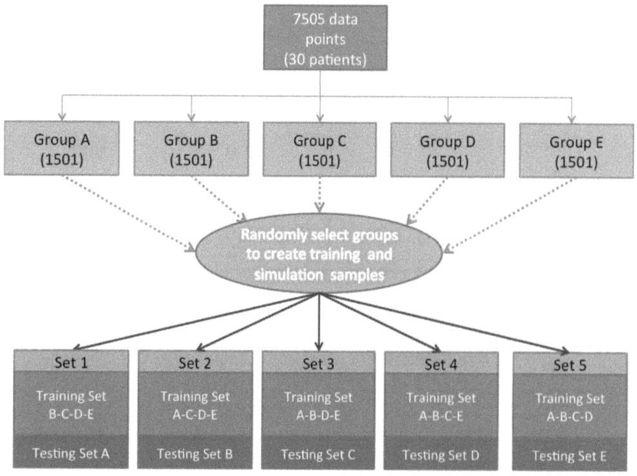

Fig. 3 Group selection for training and testing process

III. RESULTS

Table 2 shows the results of the testing process on the neural network system using several clinical parameters, which are temperature, urine output, heart rate (HR), respiratory rate (RR), and mean arterial blood pressure (MABP) on individual basis. MSE is, mean squared error are shown in Table 2 for each data set. Positive predictive value (PPV) and sensitivity are mostly similar across the data sets and different input variables. In contrast, negative predictive value (NPV) and specificity are much more variable. Very low specificity has been observed for RR clinical input variable with zero specificity for four data sets. Interestingly, Set D has the most promising result compared to other sets, even when using different input variables. In contrast, Set A has the worst result throughout the testing process relatively. These results reflect smaller differences in the specific patients in each set.

Table 3 summarizes the results of individual input tests by comparing the mean average of the testing process across the data sets. MSE, PPV, and sensitivity are almost similar across all the input variables. RR has the smallest specificity followed by urine output, MABP, temperature, and HR. However, HR, temperature, and urine output show an acceptable NPV value for individual testing of 60.3%, 40.3%, and 49.9% respectively. HR has the highest specificity that is 18.6% compared to other clinical variables. Additionally, these might be similar to clinical capability.

Table 4 shows the analysis results of the network system using multiple input variables include temperature, urine output, HR, RR, and MABP as input parameters. Significant improvement can be seen for the whole data set, comprise of Set 1, 2, 3, 4, and 5, particularly in specificity, as compared to results presented in Table III. Specificity is majorly

Table 2 Neural network results using individual clinical input parameters

Input variables	Data set	MSE	PPV	NPV	Sensitivity	Specificity
Temperature	A	0.833	48.2	17.2	93.5	1.3
	B	0.472	77.9	39.1	98.8	2.7
	C	0.792	50.9	39.1	87.6	3.7
	D	0.661	79.4	71.1	97.9	16.8
	E	0.447	81.3	35.1	87.5	25.1
Respiratory rate	A	0.625	49.6	-	100.0	0.0
	B	0.469	77.7	0.0	99.9	0.0
	C	0.682	53.2	0.0	99.8	0.0
	D	0.676	76.9	83.3	99.9	1.4
	E	0.403	78.8	-	100.0	0.0
Heart rate	A	0.990	54.1	90.7	98.1	18.1
	B	0.448	84.5	57.9	91.3	41.8
	C	0.612	57.3	89.2	98.3	16.5
	D	0.772	76.3	12.5	96.3	1.7
	E	0.402	80.8	51.1	96.1	15.0
Urine output	A	0.641	49.7	66.7	99.6	0.8
	B	0.478	77.0	40.0	99.2	1.8
	C	0.672	53.4	100.0	100.0	0.4
	D	0.707	76.8	29.4	97.9	2.8
	E	0.472	78.0	13.4	89.6	6.0
Mean arterial blood pressure	A	0.619	49.6	50.0	99.9	0.1
	B	0.595	77.6	0.0	99.7	0.0
	C	0.730	53.3	-	100.0	0.0
	D	0.998	76.3	22.5	73.4	25.4
	E	0.580	76.1	7.7	80.8	6.0

Table 3 Mean average testing results using individual clinical input

Input variables	MSE	PPV	NPV	Sensitivity	Specificity
Temperature	0.641	67.5	40.3	93.1	9.9
HR	0.645	70.6	60.3	96.0	18.6
RR	0.571	67.2	16.7	99.9	0.3
Urine output	0.594	67.0	49.9	97.3	2.4
MABP	0.704	66.7	16.0	90.8	6.3

Table 4 Neural network results using several clinical parameter

Data set	MSE	PPV	NPV	Sensitivity	Specificity
A	0.647	60.6	75.9	85.3	45.4
B	1.201	81.2	65.2	96.6	22.4
C	0.837	58.4	55.8	70.1	43.1
D	0.867	80.7	45.5	88.9	30.5
E	0.510	85.4	33.1	69.9	55.5
Mean	0.812	73.3	55.1	82.2	39.4

improved to a mean average of 39.4 whereas sensitivity is slightly reduced to a mean average of 82.2. PPV and NPV also show some improvement with a mean average of 73.3 and 55.1, respectively.

IV. DISCUSSIONS

Sepsis and its complications contribute to the high mortality rates in the ICU [12,13]. Predicting sepsis in critical care is a great challenge. To date, there is no specific biomarker developed that can represent or predict the occurrence of sepsis with any better accuracy. Blood culture tests are still considered as the gold standard for confirmation of infection. However, even there, only 51% of sepsis cases are positively identified [4]. Automated culturing systems are also used for bacteria detection in sepsis diagnosis by determining the solution pH or the presence of CO_2. However, this method requires 11-31 hours for detection [9] and is thus neither earlier than clinical practice nor real-time.

In this analysis, several clinical variables are used to identify the relation between the variables and sepsis score. The tests of clinical variables on predicting sepsis score suggest that clinical data, such as hourly urine output, MABP, temperature, RR, and HR add value, but cannot create a significantly improved biomarker. In particular, while PPV values were 66.7-70.6%, NPV values of 16.0-60.3% mean that sensitivity and specificity do not exceed 99.9% and 18.6%. These values are not much better than clinical experience.

It should be noted that the clinical variables recorded might change because of several factors. For example, changes in patient condition or patient reaction towards treatment such as drugs and inotropes, unrelated to sepsis, cleared the issue significantly. Moreover, clinical variables may not change at the same rate for every patient, as their condition evolves. Therefore, and as seen from the results, the recorded clinical variables are, at least sometimes, potentially misaligned to the hourly sepsis score calculated for this analysis.

Additionally, the plot in Fig. 1 suggests that sepsis scores vary and may change rapidly over an hourly interval. Patient 1 had a rapid change from 3 to a score of 1 and vice versa over a short time interval. This may contribute to the instability factor of sepsis score, which might contribute to an insignificant relation of hourly sepsis score to the clinical variables used in the analysis. Moreover, clinical symptoms frequently exist even if the patient is free from infection or sepsis. It is thus a challenge to look at clinical variables, such as temperature, MABP, and RR, as these values may change due to several factors changing in a particular patient and not just due to sepsis. In addition, the change in variables may be different between patients with the same degree of sepsis, creating further difficulty in beating clinical practice.

V. Conclusions

Limited diagnostic methods have prevented a significant improvement in early diagnosis and treatment. Thus, it limits the ability to reduce mortality and cost. Currently, an experienced clinician is the best diagnostic and best able to consider, accurately, all the relevant data. Only a blood culture test is considered a gold standard test for determining sepsis, even though over 50% of diagnosed sepsis is culture negative. Equally, blood culture tests require 1-3 days to obtain the result, which is also far too slow.

Importantly, criteria of a successful biomarker of sepsis must be accountable for diagnosis at a quick rate with a reasonable sensitivity and specificity that exceeds 80% of experienced clinicians [14]. Capturing and analyzing typical ICU data for the diagnostic is a second requirement for a real-time diagnosis, rather than relying on laboratory testing that may take up to 48 hours. Moreover, an additional criterion for a biomarker is to have a reasonable cost. Finally, considering the patient condition and the risk of mortality, a noninvasive diagnostic procedure would also be suitable to be implemented in clinical settings [15].

This study analyzed the potential of several available parameters captured in typical ICU settings as a possible biomarker for sepsis diagnosis. A flexible neural network framework showed that none, either alone or in combination, provided an improvement over current conditions and capability. Hence, it may be at least initially concluded that either a more powerful framework is needed to detect and discriminate sepsis status or that a more optimal biomarker is still needed.

Acknowledgment

The authors wish to thank all the staff nurses from Department of Intensive Care, Christchurch Hospital, Christchurch, New Zealand for their assistance and involvement in this study, Department of Mechanical Engineering, University of Canterbury, Christchurch, New Zealand for the equipment, and Universiti Sains Malaysia for financial support.

Conflict of Interest

The authors declare that they have no conflict of interest.

References

1. Levy MM, Fink MP, Marshall JC, Abraham E, Angus D et al. (2003) 2001 Sccm/Esicm/Accp/Ats/Sis International Sepsis Definitions Conference. Critical Care Medicine 31: 1250-1256
2. Ince C (2005) The microcirculation is the motor of sepsis. Critical Care 9: S13-S19
3. Kumar A, Singh NP (2015) Antimicrobial dosing in critically ill patients with sepsis-induced acute kidney injury. Indian J Crit Care Med 19: 99-108
4. Martin GS, Mannino DM, Eaton S, Moss M (2003) The epidemiology of sepsis in the United States from 1979 through 2000. New England Journal of Medicine 348: 1546-1554
5. Walkey AJ, Wiener RS, Ghobrial JM, Curtis LH, Benjamin EJ (2011) Incident stroke and mortality associated with new-onset atrial fibrillation in patients hospitalized with severe sepsis. JAMA 306: 2248-2254
6. Rivers E, Nguyen B, Havstad S, Ressler J, Muzzin A et al. (2001) Early goal-directed therapy in the treatment of severe sepsis and septic shock. New England Journal of Medicine 345: 1368-1377
7. Dellinger RP, Levy MM, Rhodes A, Annane D, Gerlach H et al. (2013) Surviving Sepsis Campaign: international guidelines for management of severe sepsis and septic shock, 2012. Intensive Care Med 39: 165-228
8. Dellinger RP, Levy MM, Carlet JM, Bion J, Parker MM et al. (2008) Surviving Sepsis Campaign: international guidelines for management of severe sepsis and septic shock: 2008. Intensive Care Med 34: 17-60

9. Carrigan SD, Scott G, Tabrizian M (2004) Toward resolving the challenges of sepsis diagnosis. Clin Chem 50: 1301-1314
10. Bone RC, Balk RA, Cerra FB, Dellinger RP, Fein AM et al. (1992) Definitions for Sepsis and Organ Failure and Guidelines for the Use of Innovative Therapies in Sepsis. Chest 101: 1644-1655
11. Masters T (1995) Neural, novel & hybrid algorithms for time series prediction. New York: John Wiley & Sons
12. Vincent JL, De Backer D (2013) Circulatory shock. N Engl J Med 369: 1726-1734
13. Silva E, Passos Rda H, Ferri MB, de Figueiredo LF (2008) Sepsis: from bench to bedside. Clinics (Sao Paulo) 63: 109-120
14. Holub M, Zavada J (2011) Clinical aspects of sepsis. Contrib Microbiol 17: 12-30
15. Suhaimi FM, Chase JG, Le Compte AJ, Pretty CG, Elliott R, et al. (2012) Assessing microcirculation condition in critical illness using the pulse oximeter's concept. IEEE EMBS Conference on Biomedical Engineering and Science. pp 939-943

Address of the corresponding author:

Author: Fatanah Mohamad Suhaimi
Institute: Advanced Medical and Dental Institute, Universiti Sains Malaysia
Street: Bertam
City: Kepala Batas, Penang
Country: Malaysia
Email: fatanah.suhaimi@usm.my

Iterative Interpolative Pressure Reconstruction for Improved Respiratory Mechanics Estimation During Asynchronous Volume Controlled Ventilation

F. Newberry[1], O. Kannangara[1], S. Howe[1], V. Major[1], D. Redmond[1], A. Szlavecz[2], Y.S. Chiew[1], C. Pretty[1], B. Benyo[2], G.M. Shaw[3], and J.G. Chase[1]

[1] Department of Mechanical Engineering, University of Canterbury, Christchurch, New Zealand
[2] Department of Control Engineering and Information Technology, Budapest University of Technology and Economics, Budapest, Hungary
[3] Department of Intensive Care, Christchurch Hospital, Christchurch, New Zealand

Abstract— **Asynchronous events (AEs) during mechanical ventilation (MV) breathing support can lead to poor respiratory mechanics estimation, as the patient's attempts to breath affects the measured airway pressure and flow. An algorithm that allows improved model-based estimation of respiratory system elastance, E_{rs} during asynchronous volume-controlled MV was developed. This method reconstructs a pseudo airway pressure waveform for each breath, that is similar to a breath that was unaffected by asynchronous efforts. The reconstructed waveforms can be used to estimate true respiratory system mechanics. To test the proposed algorithm, 10 retrospective airway pressure and flow datasets were obtained from 6 MV patients. Each dataset contains 475-500 breaths. Of the 9/10 datasets which contained AEs, 8 experienced a decrease in E_{rs} mean absolute deviation (MAD) and the 5th-95th range (Range90) after pressure reconstruction. The median [maximum (max), minimum (min)] decrease in Range90 divided by median elastance, was 51.3% (67.4%, -16.7%). Additionally, the median elastance for reconstructed breaths in these datasets moved closer to the true, non-asynchronous, elastance value. The median elastance change was 48.7% closer towards the true value, with a maximum shift of 93.4%. The one dataset which did not experience an improvement was found to have a varying pressure amplitude indicative of external factors affecting the MV treatment, rather than a deficiency in the pressure reconstruction. The algorithm demonstrates the ability to consistently enhance elastance estimation in MV patients.**

Keywords— **Respiratory mechanics, asynchrony, mechanical ventilation, parameter identification.**

I. INTRODUCTION

Mechanical ventilation (MV) partially or fully replaces the work of breathing in patients with respiratory failure, to allow the patient to recover from their underlying disease [1]. Patient-ventilator interaction is an important aspect during ventilation where these interactions can influence patient care and outcomes. Poor patient-ventilator interaction results in an increase in the number of asynchronous events (AEs). Hence, the quality of patient-ventilation interaction can be quantified by the number of AEs that occur during MV support [2, 3].

Personalized respiratory support by MV relies on an accurate assessment of respiratory mechanics. However, the accurate identification of respiratory mechanics is inhibited by the presence of spontaneous AEs. AEs occur when a patient's breathing demand is not synchronized with ventilator support. Frequent AEs can lead to sub-optimal MV and other complications including patient discomfort, need for additional sedation, increase the duration of MV and associated with higher mortality [2, 3, 4]. AEs induce random external effects to the airway pressure and flow profile and prevent the correct identification of underlying respiratory mechanics [5, 6]. These effects cannot be isolated without the use of additional invasive measuring tools or specific model-based approach for respiratory mechanics estimation.

When an AE occurs during a volume controlled breathing cycle, the typical airway pressure profile is altered, resulting in an M-wave airway pressure profile, rendering it unsuitable for accurate parameter identification of the respiratory mechanics [7, 8]. In a recent study, Redmond et al. proposed a pressure reconstruction method to reduce the impact of these M-wave asynchronies on respiratory mechanics estimation [9]. This method super-imposes several continuous breathing cycles to extend the segment of a breathing cycle not affected by asynchronies, allowing better model-parameter identification. However, this method only tested on issues identifying respiratory mechanics for breathing cycle with machine induced reverse triggering [10]. Furthermore, it was not able to provide respiratory mechanics estimation if consistent asynchrony occurs during early breathing and does not improve data quality or extend available breath segments. Thus, there is a need of a more robust reconstruction method to identify the underlying respiratory mechanics.

In this study, an iterative interpolative pressure reconstruction (IIPR) method is proposed. This method identifies the presences of AE in the airway pressure during volume controlled ventilation and reconstructs the pressure iteratively through a series of airway pressure filling. This method will yield a pseudo airway pressure unaffected by spontaneous breathing or asynchrony and can be used for respiratory mechanics estimation.

F. Ibrahim et al. (eds.), *International Conference for Innovation in Biomedical Engineering and Life Sciences*,
IFMBE Proceedings 56,
DOI: 10.1007/978-981-10-0266-3_27

II. METHOD

A. Patient Data

Retrospective data from patients who received MV in the Christchurch Hospital intensive care unit (ICU) were used in this study. The patients were ventilated using synchronous intermittent mandatory ventilation (SIMV) volume controlled mode (tidal volume = 6-8 ml/kg) using a Puritan Bennett 840 ventilator [11]. Data from 6 patients were included in this study, providing 10 datasets each having 475-500 breathing cycles. All data were sampled at 50 Hz and processed using MATLAB (R2014a, The Mathworks, Natick, Massachusetts, USA). The use of the data was approved by the New Zealand Southern Region Health and Disability Ethics Committee.

B. Reconstruction Method

This section presents the iterative interpolative pressure reconstruction (IIPR) method. This method extends Damanhuri et al.'s simple airway pressure filling method [12], to improve airway pressure reconstruction. Fig. 1 shows an example of how the airway pressure for an asynchronous breathing cycle is reconstructed iteratively to a 'non-asynchronous' airway pressure that can be used for respiratory mechanics estimation. The sequence of IIPR is shown below:

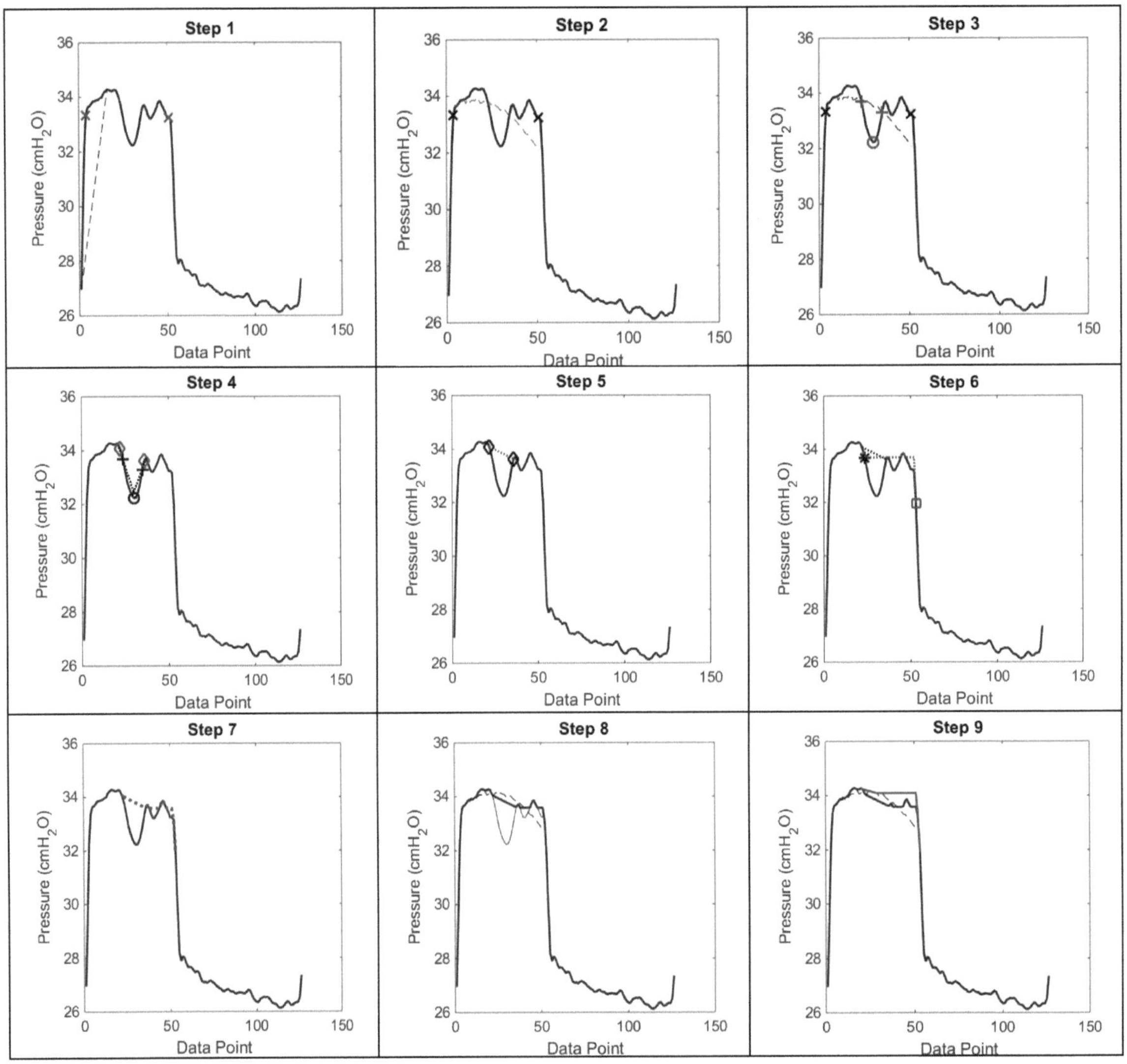

Fig. 1 Airway pressure plots for each step of iterative interpolative pressure reconstruction (IIPR) process

Step 1) Locating the shoulders of the airway pressure:

First, the locations of the left and right shoulders of the airway pressure curve are identified. The right shoulder signifies the end of inspiration and corresponds to the zero-crossing of the flow data. A first approximation to the location of the left shoulder is found by taking the maximum of the shear transform [13] between the first data point and the point of maximum pressure, indicated by the dashed line in Fig. 1, Step 1. The shoulders locations are shown with crosses.

For more reliable identification of the left shoulder, a second shear transform is performed. The left shoulder is defined as the location of the maximum of the shear transform between the first data point and a quarter of the way between the first approximation to the shoulder and the end inspiratory point. For most cases the final identified location of the left shoulder and the first approximation will be the same.

Step 2) First model fitting to single compartment model:

A single compartment linear lung model shown in Equation (1) is then fitted to the data using an integral based linear regression method [14, 15], indicated by the dashed line in Fig. 1, Step 2.

$$P_{aw}(t) = E_{rs}V(t) + R_{rs}Q(t) + P0 \tag{1}$$

Where P_{aw} is the airway pressure, t is time, E_{rs} is the respiratory system elastance, R_{rs} is the respiratory system airway resistance, V is the tidal volume, Q is the air flow and $P0$ is the offset pressure [15] or positive end-expiratory pressure (PEEP) if there is no auto-PEEP. The model is fitted between the two identified shoulders to avoid non-physiological changes in airway resistance and respiratory system elastance induced by the mechanics of the ventilator. The $P0$ or PEEP is defined as the minimum pressure between the first data point and the end inspiratory point.

Step 3) Locating patient-induced effort:

The intersections between the pressure data and the fitted curve are identified. Intersections due to asynchrony are selected based on the gradient of the pressure curve at the crossings. If the gradient is either less than zero or less than the gradient of the dashed line linking the two pressure shoulders, the crossing is detected as asynchronous. If the asynchronous crossing is followed by another crossing, the two crossings are recorded as a pair, corresponding to an asynchronous event. The asynchronous crossings are shown in Fig. 1, Step 3 as '+' signs.

The minimum sheared pressure is then found between the crossings. The sheared minimum is found by first taking the shear transform between the two pressure shoulders to account for the possibility of a non-square pressure wave. This minimum point is shown in Fig. 1, step 3 as a 'o'.

Step 4) Reconstruction by reduction of patient-induced effort:

The lines connecting the sheared minimum point to the identified asynchronous crossings are then extrapolated until intersections with the pressure curve are found. To reduce the gradient of these lines slightly, providing more reliable intersections, these lines begin from the points on either side of the pressure minimum rather than from the minimum point itself. These lines are represented by the dotted lines in Fig. 1, Step 4.

The points on the pressure curve which are a maximum orthogonal distance above these lines are identified as the asynchrony shoulders, seen in Fig. 1, Step 4 as crosses. Hence, these points can be found by taking the shear transform along the dotted lines.

Step 5) Major airway pressure reconstruction:

The identified asynchrony shoulders are then used to construct part of the new airway pressure curve. This step is achieved by joining the asynchrony shoulders with a straight line. This part of the construction is shown by the dotted line in Fig. 1, Step 5.

Step 6) Minor airway pressure reconstruction:

It is evident from Fig. 1 Step 5, that only the largest asynchrony is accounted for with the initial reconstruction. A complementary method of reconstruction is utilized for the right side of the curve. The initial expiratory pressure gradient is projected upwards until it meets the pressure of the last detected asynchronous crossing. This expiratory gradient is evaluated using a first-order forward difference two points after the right pressure shoulder. If this line cuts through the pressure curve, the maximum values of corresponding points of two curves are taken. This early pressure reconstruction is shown in Fig. 1, Step 6.

This method of reconstruction is also carried out for the left hand side of the pressure curve in some situations. If the first identified asynchrony crossing is lower than half way between the left pressure shoulder and the point immediately previous, an early asynchrony is identified. Reconstruction of the left side is carried out only if an early asynchrony is detected, to avoid unnecessary over-construction of pressure.

Step 7) Redefining the non-patient effort induced airway pressure:

The provisional reconstructed pressure is then set as the maximum of the two methods. This is shown in Fig. 1, Step 7 as the dotted line.

Step 8) Single compartment model refitting:

The single compartment lung model can then be fitted to the new reconstructed pressure, displayed in Fig. 1, Step 8. If the area between the newly fitted and the non-constructed pressure data is over a threshold of 4.5% of the area of the reconstructed pressure above PEEP, the breath is classed as asynchronous and an iterative method will ensue. This threshold was chosen by comparing the sensitivity of the algorithm with manual detection. If the breath is not classed as asynchronous, the final pressure curve used to calculate respiratory system elastance, Er_s will be the original raw pressure data. Therefore the correct model fit for a non-asynchronous breath is set as the initial model fit.

Step 9) Iterative airway pressure reconstruction:

In the case of this breath, the area under the new fitted curve is great enough that the breath is classed as asynchronous and therefore more iterations occur. The asynchronous crossings between the newly fitted and the provisional reconstructed pressure are identified. These crossings are used in the same way as before to obtain the next iteration of the reconstructed pressure. This new provisional reconstructed pressure is shown in Fig. 1, Step 9 as the upper solid line.

The algorithm reconstructs pressure by using the model fit to force the pressure reconstruction upwards. Convergence is determined by taking the area between successive iterations. If the area between the new model fit and the raw pressure data, as a percentage of the reconstructed pressure area above PEEP, does not change by more than 4.5% between successive iterations, convergence is deemed to be met. This threshold was chosen to be consistent with the threshold to iniate iterative pressure reconstruction process.

It is important to note that the characteristic of asynchrony, such as timing or magnitude varies intra and inter-patient. Thus, the example shown in this section is only one of many cases that the algorithm will respond to. The final reconstructed pressure is shown in Fig. 2, with the thicker line representing the reconstructed pressure.

C. Data Analysis

Respiratory mechanics: The pseudo airway pressure generated by IIPR and the original airway pressure are used for respiratory mechanics estimation. These estimated respiratory mechanics (mainly E_{rs}) are compared. A two sample Kolmogorov-Smirnov significance test is used to test the difference of E_{rs} distribution, with p-value < 0.05 considered as statistically significant.

Variation and spread analysis: To analyze the spread and variation of the estimated parameters, the mean absolute difference (MAD) and Range90 is calculated. Range90 is the difference between the 95th percentile and the 5th percentile of the parameter distribution. To normalize the results, the Range90 value was divided by the median non-asynchronous E_{rs}. Hence, the spread or the range of the parameter can be quantified and compared within different datasets.

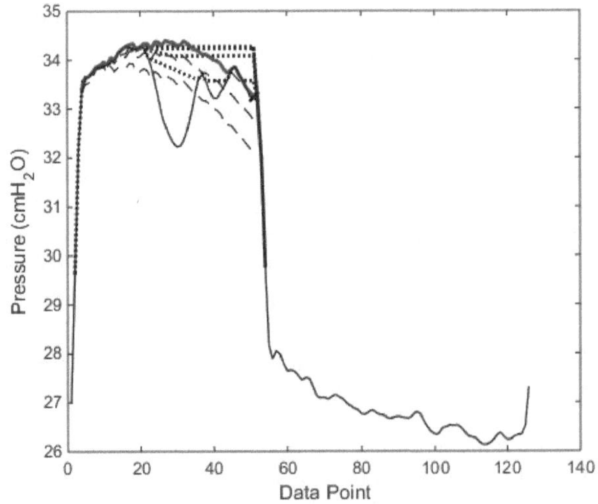

Fig. 2 Final pressure curve reconstructed with IIPR, showing intermediate steps as dotted lines

III. RESULTS

The pressure reconstruction algorithm was tested on ten datasets containing 475 to 500 breaths. Fig. 3 shows different types of airway pressure altered by asynchronous breathing cycles compared to the airway pressure after iterative interpolative pressure reconstruction.

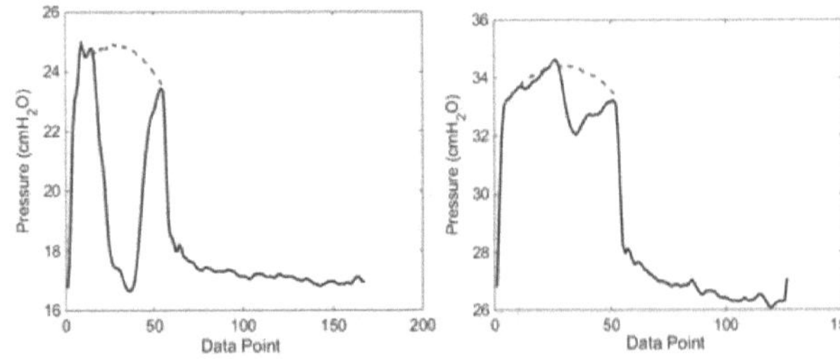

Fig. 3 Four different examples of pressure reconstruction using IIPR from two different patients

Table 1 Pressure reconstruction results

Dataset	Number of AE Breaths (n)	Respiratory System Elastance, E_{rs} (median) Non-AE*	AE (O)	AE (R)	MAD (O)	MAD (R)	Range90/ median (O)	Range90/ median (R)	p-value
1	430	16.40	35.54	22.81	12.03	3.86	2.79	0.91	+
2	437	10.96	33.14	22.33	12.80	5.30	4.05	1.74	+
3	481	20.45	24.00	22.23	12.03	4.48	1.98	0.89	+
4	122	28.88	16.00	17.79	7.72	6.61	1.03	0.68	+
5	358	20.31	30.03	25.50	6.79	5.01	1.29	0.83	+
6	420	23.01	27.42	21.46	8.29	3.99	1.52	0.73	+
7	307	26.13	27.20	24.17	5.95	3.54	1.04	0.55	+
8	0	23.58	NA	NA	0.33	0.33	0.06	0.06	NA
9	118	32.39	43.87	62.52	7.62	11.98	1.38	1.61	+
10	103	16.96	13.45	16.73	1.21	0.63	0.37	0.18	+

* AE: Asynchronous breaths, O: Original airway pressure, R: Reconstructed airway pressure,
+p-value: p-value comparing respiratory system elastance estimated using 2 sample Kolmogorov-Smirnov test

Table 1 displays the associated median respiratory system elastance, E_{rs} of all asynchronous breaths before and after reconstruction alongside the median of the non-asynchronous breaths. Likewise the Range90, and the MAD were calculated for the original and reconstructed set of breaths.

The cumulative distribution function (CDF) plots for E_{rs} of the original and reconstructed breaths of four datasets, Dataset 1, 8, 9 and 10, are displayed in Fig. 4.

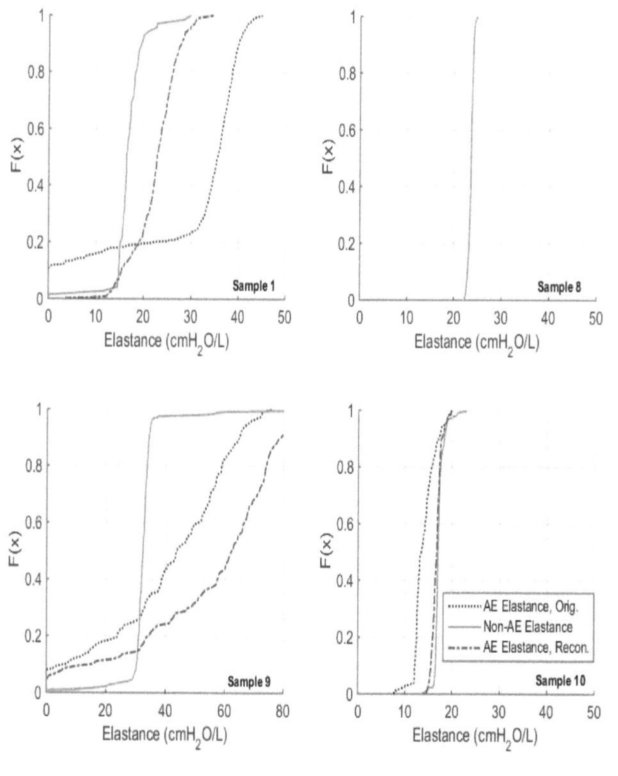

Fig. 4 Cumulative distribution (CDF) plots of the initial and reconstructed E_{rs} for Datasets 1, 8, 9 and 10. Dataset 8 contained no AEs

IV. Discussion

The IIPR method recreates a pseudo airway pressure that was not affected by asynchronous breathing and was used to estimate the breath-specific E_{rs}. Without the application of invasive measuring tools, the true underlying respiratory mechanics (E_{rs}) for an asynchronous breath may not be estimated. However, it is noted that the value of E_{rs} is unlikely to change significantly from breath-to-breath if there is no external disturbance or changes in MV settings. Thus, the validation process is performed by comparing E_{rs} during AE reconstruction and non AE breathing cycles. Additionally, a reduction in variability of elastance after reconstruction indicates an improvement in respiratory mechanics estimation. After iterative interpolative pressure reconstruction (IIPR), the mean absolute deviation (MAD) of E_{rs} was shown to decrease in 8 of 10 datasets containing AEs. A decrease in MAD indicates a reduction in the variability of E_{rs}. This variability is caused by asynchrony and thus, a reduction in variability, or MAD suggests improved patient E_{rs} estimation. The median [maximum (max), minimum(min)] decrease in mean absolute deviation, not including Dataset 8 which contained no AEs, was 39.5% [68.9%, -60%].

For all datasets with a decrease in MAD, it was found that the Range90 for E_{rs} also decreased. The median [max, min] decrease in the 5th to 95th percentile E_{rs} range divided by median E_{rs} not including Dataset 8 which contained no AEs, was 51.3% (67.4%, -16.7%). Dataset 10 contained the least asynchronous events, with 103 breaths requiring reconstruction. Correspondingly, this set has the smallest change in mean absolute deviation and in the 5th to 95th percentile range, indicating that the algorithm has a minimal effect on non-asynchronous breaths.

Dataset 8 contained no asynchronous events and as expected, there was no change in E_{rs} after running this dataset through the IIPR algorithm, as can be seen in Fig. 4 for dataset 8. This result reiterates that breaths which are not identified as asynchronous will not be affected.

Breaths which have a high conformity to the initial model fit are not reconstructed and can be considered non-asynchronous. If no physiological or external changes occur, the median E_{rs} of these breaths gives the true underlying E_{rs} for the data period. The efficacy of the algorithm in removing the effect of patient effort from the breaths can be assessed by comparing the E_{rs} of reconstructed breaths to the true E_{rs}.

For 8 of the 10 datasets, the median E_{rs} after reconstruction moved closer to the true value. The median E_{rs} change was 48.7% closer towards the true value, with a maximum shift of 93.4%. As a result of the reconstructions, the median deviation of determined E_{rs} from the true value decreased from 47.8% to 25.5%.

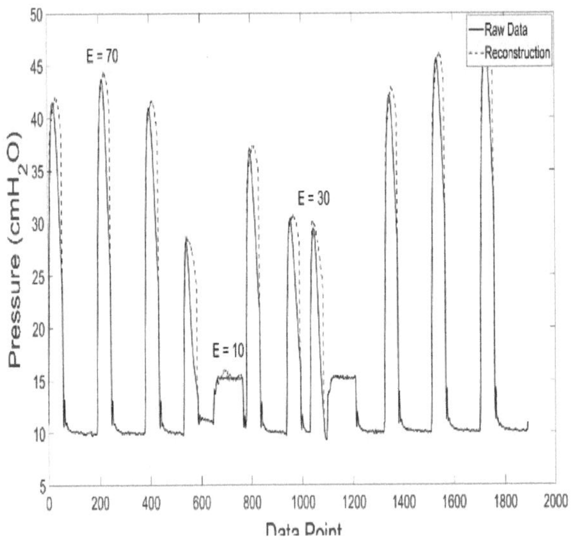

Fig. 5 Breaths with varying pressure, present in Dataset 9

The estimation of E_{rs} for Dataset 9 after running through the IIPR algorithm moved further away from the true E_{rs} value. This poor estimation is because Dataset 9 contained a period of 130 unusual breaths with large breath-to-breath pressure changes and spontaneous breathing. A section of the Dataset 9 with 10 sequential breaths with varying pressure is shown in Fig. 5.

The breaths shown in Fig. 5 do not contain any large asynchronies and are in theory, reconstructed correctly. However the varying airway pressure magnitudes as shown in Fig 5, resulted in oscillating large and small E_{rs} values. Note that spontaneous breaths are included in the identification of patient respiratory mechanics.

The IIPR method reconstructs the airway pressure by extending the asynchronous airway pressure profile to mimic a non-affected airway pressure. This reconstruction pressure iteratively uses a passive single compartment lung model and hence, after reconstruction, it is assumed that the reconstructed pressure is a 'model' airway pressure that was not affected by asynchrony.

V. CONCLUSION

The iterative interpolative pressure reconstruction method is shown to be effective for the wide variety of cases seen in this study. Improvement was shown in the determined E_{rs} values of patient induced asynchronous breaths. Thus, the algorithm exhibits the potential for improvement of real time respiratory mechanics estimation.

ACKNOWLEDGMENT

The authors would like to thank the New Zealand Health Research Council for supporting this research.

CONFLICT OF INTEREST

The authors declared that they have no conflict of interest.

REFERENCES

1. Girard TD, Bernard GR (2007) Mechanical Ventilation in ARDS. Chest 131:921-9
2. Epstein SK (2011) How Often Does Patient-Ventilator Asynchrony Occur and What Are the Consequences? Respiratory Care 56:25-38
3. Schmidt M, Banzett R, Raux M et al. (2014) Unrecognized suffering in the ICU: addressing dyspnea in mechanically ventilated patients. Intensive Care Medicine 40:1-10
4. Poole SF, Chiew YS, Redmond DP et al. Real-Time Breath-to-Breath Asynchrony Event Detection using Time-Varying Respiratory Elastance Model. In: 19th IFAC World Congress; 2014; Cape Town, South Africa; 2014. p. 5629-34.
5. Talmor D, Sarge T, Malhotra A et al. (2008) Mechanical Ventilation Guided by Esophageal Pressure in Acute Lung Injury. New England Journal of Medicine 359:2095-104
6. Brochard L, Martin G, Blanch L et al. (2012) Clinical review: Respiratory monitoring in the ICU - a consensus of 16. Critical Care 16:219
7. Gilstrap D, MacIntyre N (2013) Patient–Ventilator Interactions. Implications for Clinical Management. American Journal of Respiratory and Critical Care Medicine 188:1058-68
8. Akoumianaki E, Lyazidi A, Rey N et al. (2013) Mechanical ventilation-induced reverse-triggered breaths: A frequently unrecognized form of neuromechanical coupling. CHEST 143:927-38
9. Redmond DP, Major V, Corbett S et al. Pressure reconstruction by eliminating the demand effect of spontaneous respiration (PREDATOR) method for assessing respiratory mechanics of reverse-triggered breathing cycles. In: Biomedical Engineering and Sciences (IECBES), 2014 IEEE Conference on; 2014 8-10 Dec. 2014; 2014. p. 332-7.

10. Sundaresan A, Chase JG (2011) Positive end expiratory pressure in patients with acute respiratory distress syndrome - The past, present and future. Biomedical Signal Processing and Control7:93-10311.
11. Szlavecz A, Chiew Y, Redmond D et al. (2014) The Clinical Utilisation of Respiratory Elastance Software (CURE Soft): a bedside software for real-time respiratory mechanics monitoring and mechanical ventilation management. BioMedical Engineering OnLine 13:140
12. Damanhuri N, Chiew Y, Othman N, Docherty P, Shaw G, Chase J (2015) Pressure reconstruction method for spontaneous breathing effort monitoring. Critical Care 19:P259
13. Stevenson D, Revie J, Chase JG et al. (2012) Algorithmic processing of pressure waveforms to facilitate estimation of cardiac elastance. Biomedical engineering online 11:28
14. Chiew YS, Chase JG, Shaw G, Sundaresan A, Desaive T (2011) Model-based PEEP Optimisation in Mechanical Ventilation. BioMedical Engineering OnLine 10:111
15. van Drunen E, Chiew YS, Pretty C et al. (2014) Visualisation of time-varying respiratory system elastance in experimental ARDS animal models. BMC Pulmonary Medicine 14:33

Author: D.O. Kannangara
Institute: University of Canterbury
Street: 20 Kirkwood Avenue, Upper Riccarton
City: Christchurch
Country: New Zealand
Email: dok11@uclive.ac.nz

Effects of Coherent Noise on Ictal Component Selection for EEG Source Imaging

M.A. Habib[1,2,3], F. Ibrahim[1,2], M.S. Mohktar[1,2], S.B. Kamaruzzaman[2,4], and K.S. Lim[2,4]

[1] Department of Biomedical Engineering, Faculty of Engineering, University of Malaya, Kuala Lumpur, Malaysia
[2] Centre for Innovation in Medical Engineering (CIME), Faculty of Engineering, University of Malaya, Kuala Lumpur, Malaysia
[3] Department of Computer Science & Engineering, Chittagong University of Engineering & Technology, Chittagong, Bangladesh
[4] Department of Medicine, Faculty of Medicine, University of Malaya, Kuala Lumpur, Malaysia

Abstract— **This paper examines the effects of coherent noise on the scalp voltage topography, activity power spectra and dipole residual variances of the independent components (ICs) of ictal EEG signals. Eleven different sets of ictal EEG signals are generated by adding various amounts of coherent noises. All of these simulated EEGs are decomposed into their corresponding ICs. Single dipole source that helps to distinguish ictal components from noise components, two-dimensional (2D) topographic map, activity power spectrum, and dipole residual variance were estimated for each of these ICs. Topographic maps show that the number of ictal components decreases with the increase of noise level. Activity power spectrum analysis supports the result of topographic map analysis. The average residual variances not only increase with the increment of noise level but sometimes decrease as well. Simultaneous consideration of these three features is helpful for better selection of ictal components.**

Keywords— **Independent Component Analysis, Ictal EEG, Ictal Component, EEG Source Imaging, Noise.**

I. INTRODUCTION

Electroencephalography (EEG) is used, almost inevitably, for the pre-surgical evaluation of medically intractable focal epilepsy. Possible cortical sources of scalp EEG can be estimated and depicted with the help of a computational technique, namely EEG source imaging (ESI). Either ictal or interictal EEG events are analyzed for ESI based source estimation. Although ictal EEG is comparatively difficult to analyze with ESI because of the low Signal to Noise Ratio (SNR) [1], ictal EEG is believed to be more reliable than interictal EEG in localizing the epileptogenic focus [2].

Ictal EEG measures cortical seizure discharges superposed with various artifacts, external noises and other background brain oscillations. The unwanted parts of ictal EEG can be minimized by utilizing digital filters, various artifact removal/rejection algorithms including independent component analysis (ICA) and sometimes by averaging the selected ictal discharges. The ICA-based ictal ESI studies [3-5] decomposed each set of ictal EEG into a series of spatially fixed and temporally independent components and then analyzed only the selected ictal components for ESI.

The major challenge of ICA-based ictal ESI algorithms is to select the correct ictal ICs that correspond to the actual seizure discharge. There is no concrete rule to identify those ictal ICs. Selection techniques are mostly visual inspection dependent. Jung et al. [3] excluded the ICs of muscle artifacts, eye movements, and 60-Hz noise by visually inspecting the scalp voltage topography and activity power spectra. They localized a single dipole source for each IC using DIPFIT and excluded the ICs of dipole sources located outside the head model or with a residual variance of more than 20%. Yang et al. [4] and Lu et al. [5] used dynamic seizure imaging (DSI) approach for ictal ESI. They also removed the unwanted ICs (e.g. ICs related to eye movement and muscle artifacts) through the visual inspection of scalp voltage topography and activity power spectra. They selected the ictal ICs through statistical tests but with further assistance of visual inspection.

Corresponding ICs for muscle artifacts, eye movement artifacts and line noise have visually noticeable characteristics [6, 7] and comparatively easy to identify, whereas the selection of ictal ICs is not that straightforward. The aim of this study is to ease ictal component selection for ictal ESI. It examines the characteristic changes of ictal ICs with the linear changes of coherent noise level. This study focuses on three different features of ictal ICs, namely 2D scalp voltage topography, activity power spectrum and residual variance of estimated single dipole source.

II. MATERIALS AND METHODS

A. Ictal EEG Generation

EEG data are generated by using a fixed dipole in a four-shell spherical head model [8] with radius of 85 mm. A dipole source of 80 nAm is placed in the mesial temporal lobe as shown in Fig. 1a. A sinusoidal waveform of 5.7 Hz (Fig. 1b) which is typical in patients with mesial temporal lobe epilepsy [9, 10] is used as the source waveform. Thirty three electrodes are used according to 10-10 system. Topographic map of scalp potential distribution, resulting from the dipole source, is illustrated in Fig. 1c. Scalp potentials on each electrode are obtained by multiplying the source

F. Ibrahim et al. (eds.), *International Conference for Innovation in Biomedical Engineering and Life Sciences*,
IFMBE Proceedings 56,
DOI: 10.1007/978-981-10-0266-3_28

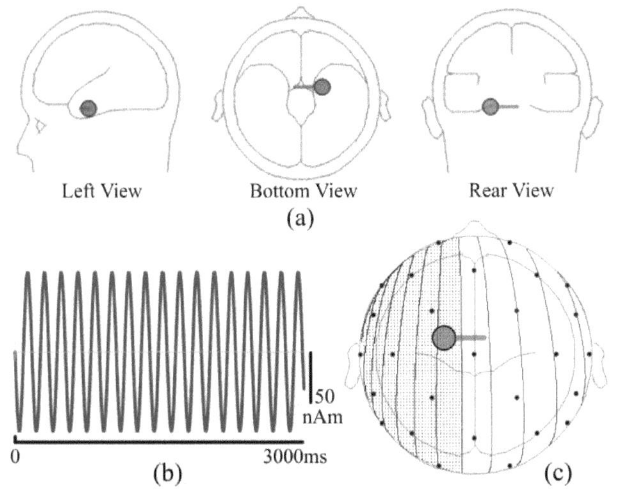

Fig. 1. (a) Three different views of dipole location and orientation. (b) Source waveform. (c) 2D topographic map.

wave by an amplification factor which is computed by solving the forward problem for the dipole source. BESA Simulator and its default values for conductivities and radii of scalp, skull, cerebrospinal fluid (CSF) and brain are used for EEG generation. Simulated EEG data (Fig. 2a) consists of 2 seconds of background EEG followed by 3 seconds of simulated ictal activities. Since no other brain activity or noise is added, initial 2 seconds of EEG are flat. Two seconds of background activities are included because the existing studies [3-5] on ICA based ictal ESI analyzed both pre and post ictal onset data. Generated EEG is considered as noise free and is denoted by N0. The time course of the scalp potentials are stored with sample frequency of 250 Hz.

Ten other sets of EEG data are generated in a similar fashion with same dipole source but with ten different levels of coherent noises. The noise is coherent in the sense that there is quite a high correlation between signal amplitudes from electrodes that are close together. Root mean square (RMS) values of added noises are 0.1 μV, 0.2 μV, 0.3 μV, 0.4 μV, 0.5 μV, 0.6 μV, 0.7 μV, 0.8 μV, 0.9 μV and 1 μV and the generated EEG signals are termed as N1, N2, N3, N4, N5, N6, N7, N8, N9 and N10 respectively. The last EEG data-set (N10), that contains the highest level of noise, is shown in Fig. 2b.

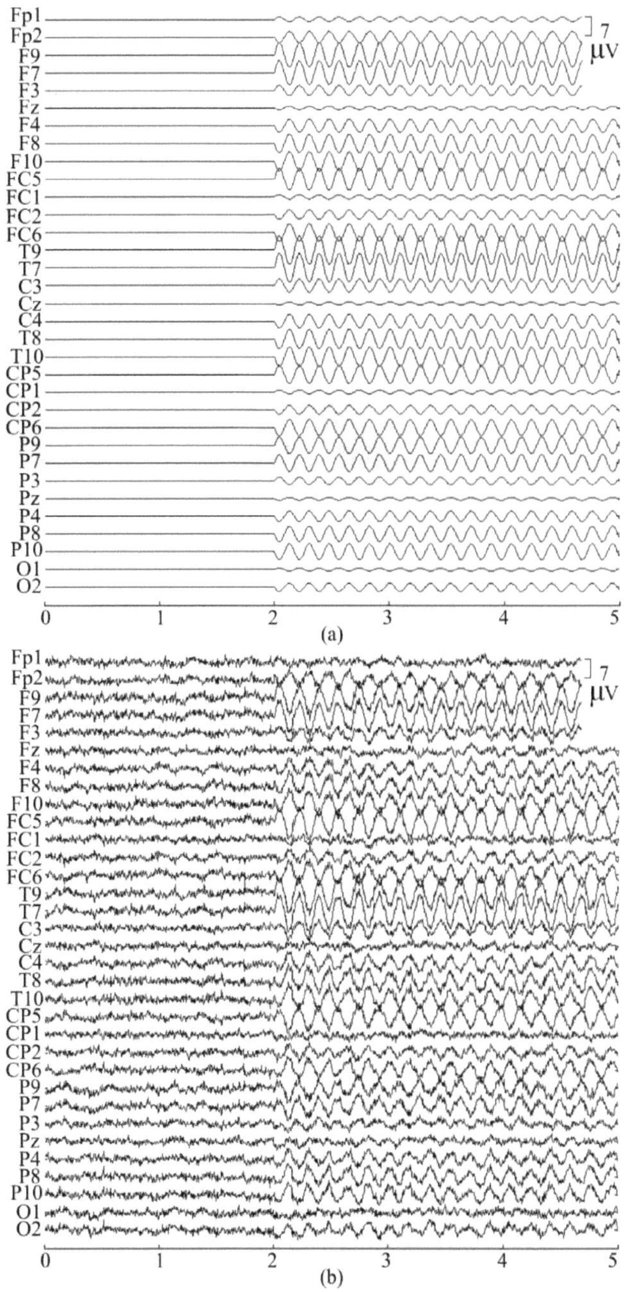

Fig. 2. (a) Noise free simulated EEG (N0). (b) Simulated EEG (N10) with coherent noise.

B. Decomposition of Ictal EEG

Each set of simulated ictal EEG are decomposed into 33 temporally independent, but specially fixed, components. Decomposition is performed by using the open source software tool EEGLAB Version 13.4.3b [11] and its *runica* algorithm, which is implemented based on the logistic infomax algorithm of Bell & Sejnowski [12].

C. Dipole Source Localization of Independent Components

Source locations and corresponding residual variances are estimated for all the ICs by using a single equivalent dipole source model and the DIPFIT function in EEGLAB. A similar head model, as discussed in section II(A), is used for source estimation. EEGLAB plots the dipoles on an average MRI template after accomplishing the transposition of dipole location from the spherical head model to the average MRI template. Estimated dipole locations for the ICs of N0 and N10 are illustrated in Fig. 3. All the dipoles for the ICs of N0 have negligible residual variances and therefore mostly overlapping each other.

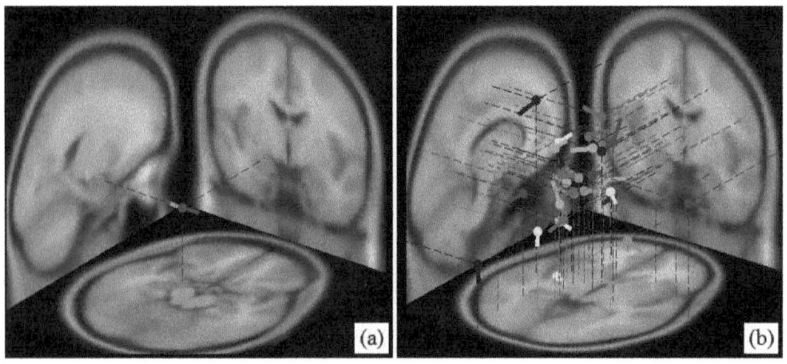

Fig. 3. (a) Equivalent dipole sources for the ICs of noise free EEG set N0. (b) Equivalent dipole sources for the ICs of N10 having 1 μV of noise.

III. RESULTS

A. Topographic Maps

A two dimensional (2D) scalp topographic map is generated for each of the ICs. Corresponding equivalent single dipole sources are also included in the maps. Topographic maps of all the ICs of N0 and N10 are shown in Fig. 4. Potential distributions for all the components of N0 are found similar to that of the original dipole source shown in Fig. 1c, except the orientation of the dipole. On the other hand only few maps for N10 show similarity with the original source. Considering all the 2D maps, generated from the ICs of 11 sets of EEGs, it is evident that the number of dissimilar maps per set increases with the increase of noise level. 2D maps for other nine sets of EEGs cannot be included due to space limitation.

B. Activity Power Spectra

Activity power spectra of ICs exhibit which components contribute most strongly to which frequencies in the data. Power spectra are generated for all the ICs while considering 100% samples and total epoch time range (5 seconds). Component no. 1 of N0 and component no. 2 of N10 has contributed most strongly around the rhythm frequency (5.7 Hz). Power spectra of these two ICs are shown in Fig. 5. Activity power spectra are generated for all the ICs. The number of ICs that contribute highly around rhythm frequency is found decreasing with the increment of noise.

C. Dipole Residual Variance

Residual variances are calculated for all the estimated dipole sources of all the ICs and those are shown in Fig. 6a as a 3D column chart. Higher order ICs are found more likely to have higher residual variances. Average value of the residual variances is also calculated for each set of EEG data and illustrated in Fig. 6b as a line chart. Although in most of the cases the average residual variance have increased with the increment of noise level, opposite relationship is also found in few cases.

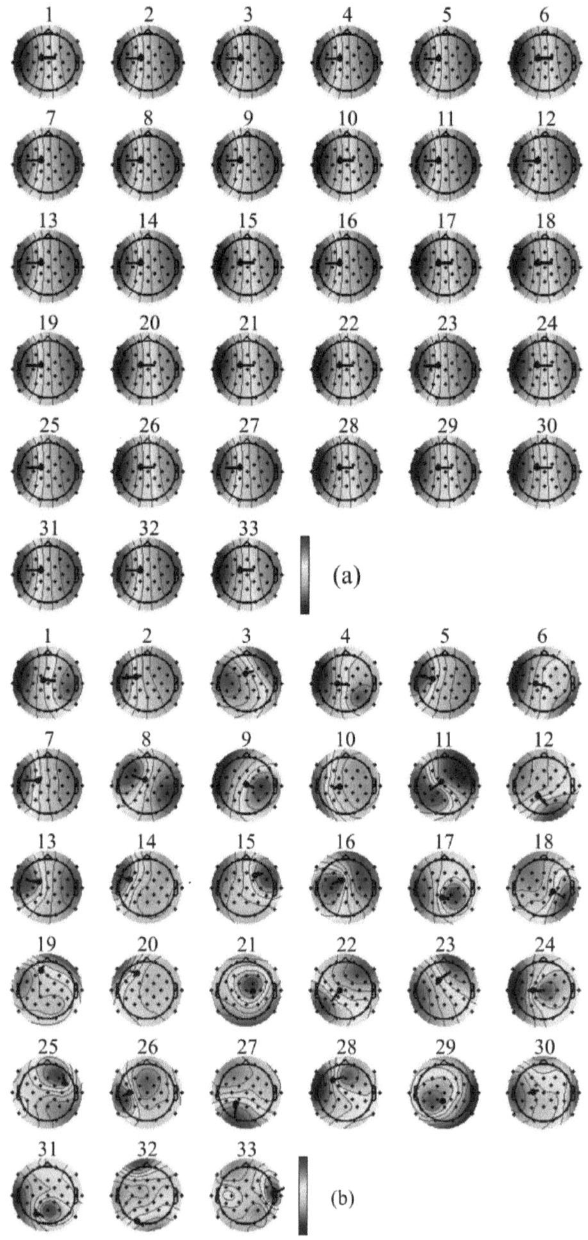

Fig. 4. 2D topographic maps for all the ICs of (a) N0 and (b) N10. Locations and orientations of corresponding equivalent dipole sources are shown. Color bars next to the maps indicate the magnitude and polarity.

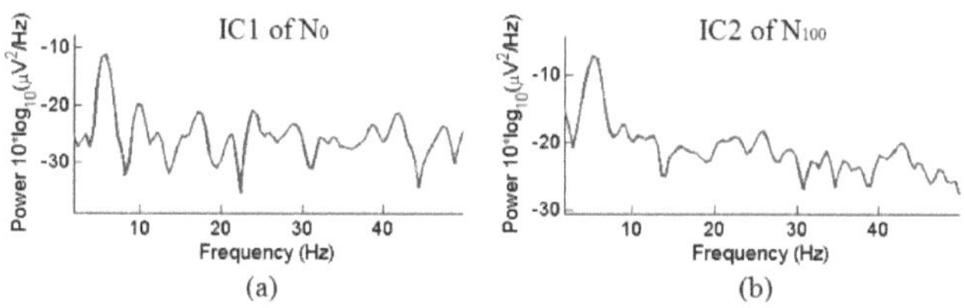

Fig. 5. Activity power spectrum for (a) IC1 of N0 and (b) IC2 of N10.

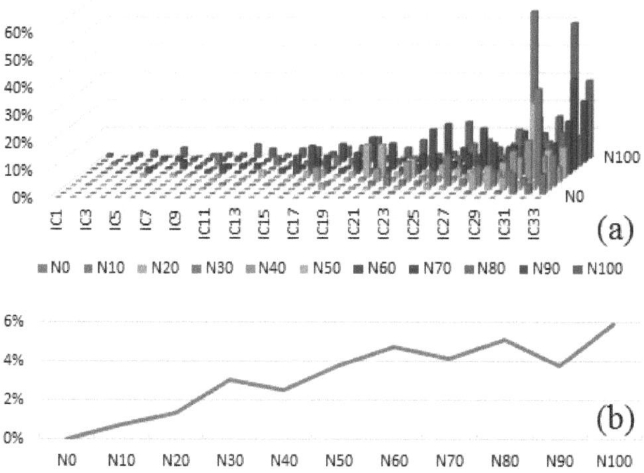

Fig. 6. (a) Dipole residual variances of all the ICs of all the EEG data sets. (b) Changes of average residual variances with noise levels.

IV. DISCUSSION

In parallel with 2D topographic maps, 3D topographic maps can also be used for IC selection. Although 3D maps are more realistic, 2D maps are preferred because of their relevance with the spherical head models that is used for EEG generation. This preliminary study analyzed the most commonly used features of ICs. More rigorous study on the effects of coherent noise on other important features, such as time-frequency representation, is in progress.

V. CONCLUSIONS

Topographic map, activity power spectrum and dipole residual variance are the three major features that are analyzed for IC selection for ictal ESI. The results obtained in this study indicate that the coherent background noise has significant effects on these features. Therefore more attention should be given on coherent noise level and multiple IC features should be considered during the ictal IC selection for the ICA based ictal ESI.

ACKNOWLEDGMENT

This study is fully funded by the University of Malaya / Ministry of Higher Education (UM/MOHE) High Impact Research Grant (UM.C/625/1/HIR/MOHE/ASH/02).

CONFLICT OF INTEREST

The authors declare that they have no conflict of interest.

REFERENCES

1. Kaiboriboon K, Lüders HO, Hamaneh M et al. (2012) EEG source imaging in epilepsy—practicalities and pitfalls. Nat Rev Neurol 8(9):498-507
2. Jayakar P, Duchowny M, Resnick TJ et al. (1991) Localization of seizure foci: pitfalls and caveats. J Clin Neurophysiol 8(4):414-431
3. Jung KY, Kang JK, Kim JH et al. (2009) Spatiotemporospectral characteristics of scalp ictal EEG in mesial temporal lobe epilepsy with hippocampal sclerosis. Brain Res 1287:206-219.
4. Yang L, Christopher W, Benjamin B et al. (2011) Dynamic imaging of ictal oscillations using non-invasive high-resolution EEG. Neuroimage 56(4):1908-1917
5. Lu Y, Yang L, Gregory A et al. (2012) Dynamic imaging of seizure activity in pediatric epilepsy patients. Clin Neurophysiol 123(11):2122-2129
6. Jung TP, Makeig S, Westerfield M et al. (2000) Removal of eye activity artifacts from visual event-related potentials in normal and clinical subjects. Clin. Neurophysiol 111(10):1745-1758.
7. Urrestarazu E, Iriarte J, Alegre M et al. (2004) Independent component analysis removing artifacts in ictal recordings. Epilepsia 45(9):1071-1078
8. Berg P, Scherg M (1994) A fast method for forward computation of multiple-shell spherical head models. Electroenceph clin Neurophysiol 90:58-64
9. De Vos M, Vergult A, De Lathauwer L et al. (2007) Canonical decomposition of ictal scalp EEG reliably detects the seizure onset zone. NeuroImage 37(3):844-854
10. Niedermeyer E, da Silva FL (1987) Electroencephalography: basic principles, clinical applications, and related fields. Urban and Swarzenberg, Baltimore
11. Delorme A, Makeig S (2004) EEGLAB: an open source toolbox for analysis of single-trial EEG dynamics. J Neurosci Methods 134(1): 9-21
12. Bell AJ, Sejnowski TJ (1995) An information-maximization approach to blind separation and blind deconvolution. Neural Comput 7(6):1129–1159

Corresponding Author:	Fatimah Ibrahim
Institute:	University of Malaya
City:	Kuala Lumpur
Country:	Malaysia
Email:	fatimah@um.edu.my

Improved Respiratory Mechanical Estimation During Pressure Controlled Mechanical Ventilation

O. Kannangara[1], F. Newberry[1], S. Howe[1], V. Major[1], D. Redmond[1], A. Szlavecz[2], Y.S. Chiew[1], C. Pretty[1], B. Benyo[2], G.M. Shaw[3], and J.G. Chase[1]

[1] Department of Mechanical Engineering, University of Canterbury, Christchurch, New Zealand
[2] Department of Control Engineering and Information Technology, Budapest University of Technology and Economics, Budapest, Hungary
[3] Department of Intensive Care, Christchurch Hospital, Christchurch, New Zealand

Abstract— **Mechanical ventilation (MV) therapy partially or fully replaces the work of breathing in patients with respiratory failure. Respiratory mechanics during pressure controlled (PC) or pressure support (PS) are often not estimated due to variability induced by patient's spontaneous breathing effort (SB) or asynchronous events (AEs). Proposed is an algorithm which allows for the improvement of respiratory system mechanics estimation during pressure controlled ventilation. For testing, 10 retrospective airway pressure and flow data samples were obtained from 6 MV patients, with each data sample containing 450-500 breaths. All data samples with AE present experienced a decrease in 5th to 95th range (Range90) and mean absolute deviation (MAD) for the estimated respiratory system elastance after reconstruction. These results suggested improved in respiratory mechanics estimation during pressure controlled ventilation. The median [maximum (max), minimum (min)] decrease in MAD was 29.4% (51%, 18.6%), and the median (max, min) decrease in Range90 divided by median respiratory system elastance was 30.7% (48.8%, 6.4%). The algorithm is robust to many different spontaneous breathing efforts, asynchrony shapes and types. The proposed algorithm demonstrates the potential to effectively improve respiratory mechanics.**

Keywords— **Pressure control, Pressure support, Respiratory Mechanics, Model-based, spontaneous breathing.**

I. INTRODUCTION

Respiratory failure patients require mechanical ventilation (MV) for breathing support [1, 2, 3]. To aid recovery from the patient's underlying disease, his/ her work of breathing is partially or fully replaced by the mechanical ventilator [3]. Common MV modes can be mainly divided into volume controlled (VC) or pressure controlled (PC). While VC was able to provide a fixed tidal volume delivery, some clinicians preferred PC mode. This mode can limit the maximum driving pressure delivery during MV. Limiting the maximum pressure can prevent patients from further ventilation pressure induce lung injury, barotrauma [4, 5]. PC ventilation is often extended to pressure support (PS) ventilation, promoting patient to breathe spontaneously that can aid recovery [6].

During PC/ PS ventilation, a predefined driving pressure and a corresponding tidal volume is delivered to the patient. Thus, the delivered tidal volume can be variable, depending on patient's condition. This variability may be beneficial to patients but it affects model-based breath-to-breath respiratory mechanics estimation, causing inconsistency. This inconsistency in respiratory mechanics estimation is further aggravated during PS mode where patient exhibit spontaneous breathing efforts (SB) or asynchrony event (AE) occurs. Thus, during PC/PS ventilation mode, relatively little model-based respiratory mechanics estimation is performed, and it is not used to guide mechanical ventilation.

This study presents an iterative method to improve model-based respiratory mechanics estimation during pressure controlled ventilation. Specifically, an iterative interpolative flow reconstruction method is used. This method operates by identifying whether a pressure controlled or pressure supported breath is distorted by spontaneous breathing effort or an AE. This method then iteratively reconstructs the affected airway flow to a single compartment respiratory model airway flow profile. This method yields a pseudo airway flow profile that is unaffected by spontaneous breathing or asynchrony, allowing estimation of the unaffected, underlying patient-specific respiratory mechanics.

II. METHOD

A. Patient Data

Airway pressure and flow data from MV patients admitted to the Christchurch Hospital intensive care unit (ICU) were used in this study. The patients were ventilated with Puritan Bennett 840 ventilation using synchronous intermittent mandatory ventilation (SIMV) pressure controlled mode to achieve a tidal volume of 6~8 ml/kg [7]. Ten data samples, each having 450-500 breathing cycles (~30 minutes) were extracted from 6 patients included in this study. It is assumed that underlying respiratory mechanics and patient condition do not change significantly over a short period of time ~30 minutes. Therefore, the respiratory system elastance,

F. Ibrahim et al. (eds.), *International Conference for Innovation in Biomedical Engineering and Life Sciences*,
IFMBE Proceedings 56,
DOI: 10.1007/978-981-10-0266-3_29

E identified from reconstructed flow profiles is expected to be for more constant than far those affected by spontaneous breathing or AEs. All data were sampled at 50Hz and processed using MATLAB (R2014a, The Mathworks, Natick, Massachusetts, USA). The use of this data was approved by the New Zealand Southern Regional Ethics Committee.

B. Reconstruction Method

In this section, the sequence of the iterative interpolative flow reconstruction method is presented. Fig 1 illustrates an example of the airway flow reconstruction for an asynchronous breathing cycle in 9 total steps. Each step is specifically defined:

Step 1) Filtering data:

Pressure and flow waveforms are first filtered to remove noise, using a first order zero-phase Butterworth Filter with a normalized cutoff frequency of 0.3 Hz. An example of raw and filtered flow data is shown in Fig 1, Step 1.

Step 2) Locating the shoulders of airway pressure:

The left and right shoulders of the airway pressure curve are identified. The location of these shoulders is important because the fitting of physiological parameters is most dependent on this region. The identified shoulder locations are shown as crosses on the pressure curves in Fig 1, Step 2.

A first approximation to the location of the left shoulder is found by taking the maximum of the shear transform [8]

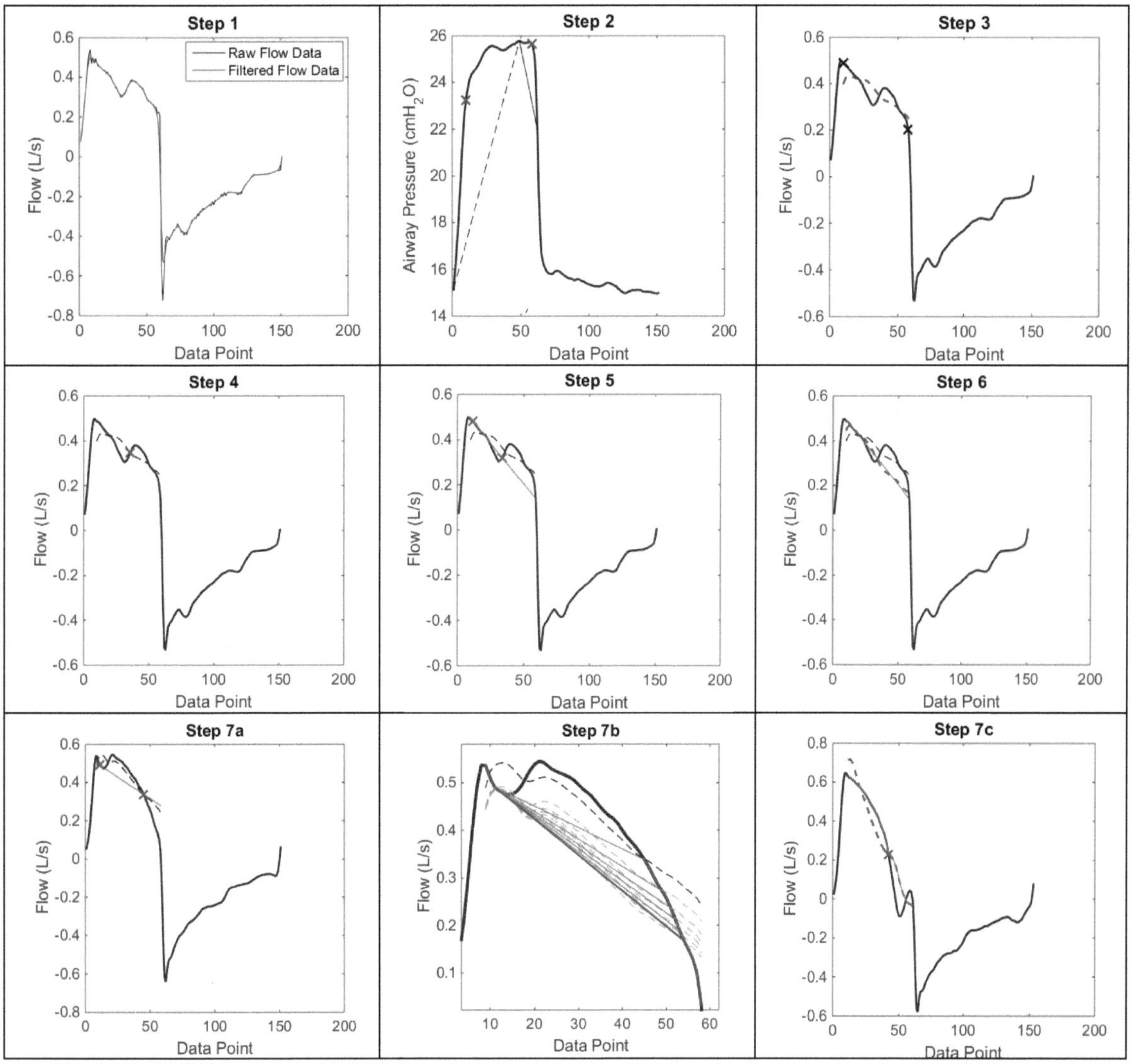

Fig. 1 Airway flow or pressure plots for each step of iterative interpolative flow reconstruction process

between the first data point and the point of maximum pressure. The second shoulder, corresponding to the end of inspiration, is found by taking the maximum of the shear transform between the point of maximum pressure and the point of minimum flow. The shear transform lines for the first and second shoulders are indicated in Fig 1, Step 2 by the dashed and solid lines, respectively.

For a more reliable identification of the left shoulder, a second shear transform is performed. The left shoulder is defined as the location of the maximum of the shear transform between the first data point and a quarter of the way between the first approximation to the shoulder and the end inspiratory point. For most cases, the final identified location of the left shoulder and the first approximation will be the same.

Step 3) First single compartment model fitting:

A single compartment linear lung model is then fit to the data using an integral based linear regression method [9], with the model defined:

$$R\frac{dV(t)}{dt} = (P - P_{PEEP}) - EV(t) \quad (1)$$

Where R is the respiratory system resistance, V is the tidal volume, t is time, P is the airway pressure, P_{PEEP} is the offset pressure or positive end-expiratory pressure (PEEP), E is the respiratory system elastance.

During PC ventilation, pressure is the controlled variable in this case, the model is fit to the volume and flow data and is indicated by the dashed line in Fig 1, Step 3. The model is fitted between the two identified shoulders to avoid non-physiological changes in R and E induced by the mechanics of the ventilator. The PEEP is defined as the minimum pressure between the first data point and the end inspiratory point. If the area between the model fit and actual flow curves is below a user specified threshold, the flow is deemed to be free of patient induced effort or with no AE, the flow reconstruction will not occur.

Step 4) Locating patient-induced effort:

The intersections between the flow data and the model fitted curve are identified. Intersections due to asynchrony are selected based on the gradient of the pressure curve at the crossings. If the gradient is positive, the crossing is detected as asynchronous. The asynchronous crossing is shown in Fig 1, Step 4 as a cross.

Step 5) Reduction of patient-induced effort:

Inspiratory flow unaffected by patient efforts takes a ramp shape. Therefore, the asynchronous crossings are used to estimate the appropriate gradient of the ramp. The flow profile is constructed by using a straight line intersecting two points, referred to from here as 'flow reconstruction points'.

The first flow reconstruction point used is two data points after the first fitted point of the flow data. The first two data points are not used as there is usually some error in the data close to the maximum of the flow due to interaction between the patient and the ventilator.

The second flow reconstruction point used is the point halfway (in time) between the identified asynchronous crossing and the minimum of flow up to that point. This choice results in the estimated true ramp flow being constructed underneath the area of identified asynchrony and, as such, is a more accurate estimate of the unaffected flow waveform. The two flow reconstruction points are shown as crosses in Fig 1, Step 5, and the reconstructed flow is shown as the thin line.

Step 6) Single compartment model refitting

The single compartment lung model is fit a second time to the new reconstructed flow in Fig 1, Step 6. The refitted curve is shown as the dashed line.

Step 7a) First special case - Breaths with early asynchrony, initial reconstruction:

Note that the choosing of the two points in Step 5 for flow reconstruction is only valid if the asynchrony is sufficiently far away from the beginning of inspiration. If the asynchrony is too close to the flow maximum, then there will be no region unaffected by asynchrony between the two points usually used. The estimation of the ramp gradient will thus fail.

Therefore, if the first asynchronous crossing detected is within 8 data points of the first fitted flow point, a modified approach is taken to reconstruct the flow. In this case, the second flow reconstruction point is the point of last crossing in inspiration between the fitted flow and the actual flow data. This region is the most likely to be free of asynchrony if there is an asynchrony near the beginning of the breath. In addition, having the maximum distance between the two flow reconstruction points, it results in the least error. After this point, the reconstructed flow is set as the same as the actual flow. An example of a flow reconstruction with an early asynchrony is shown in Fig 1, Step 7a, with the flow reconstruction points indicated by crosses and thin lines indicating the flow reconstruction.

Step 7b) First special case – Breaths with early asynchrony, iterative reconstruction

The reconstruction method for early asynchronies usually results in the flow curve being reconstructed too high, resulting in an incorrect E estimation. Therefore, an iterative procedure is taken, using the model fit to guide further reconstruction. A new flow curve is constructed using the refitted model curve. The first flow reconstruction point is the same as in Step 6. However, the second flow reconstruction point used is the new intersection at the end of the flow curve between the data and the new model fit. Using this method, the flow is iteratively driven down until convergence or a set number of iterations is reached. This method typically converges within 4 iterations. An example of the iterative procedure is shown in Fig 1, Step 7b, with the final reconstruction indicated by the lowest solid line.

Step 7c) Second special case – Breaths where a linear extrapolation is not suitable:

Another special case is where extrapolating a line would result in negative flow. This issue occurs in situations where either the actual flow becomes negative or where the model fit approaches zero flow early and levels out. In these cases, the flow data is kept the same up to the crossing before the first asynchronous crossing. After this point, the initial model fit is used instead of the raw data. This approach generally acts to raise the final model fit slightly, which does not include a negative flow in inspiration, as seen in the green line in Fig 1, Step 7c.

C. Data Analysis

Respiratory mechanics estimation: The original airway pressure and the reconstructed airway flow generated by the IIFR presented are used for respiratory mechanics estimation. The estimated respiratory mechanics are compared to the respiratory mechanics estimated using the original airway flow prior to reconstruction. A two sample Kolmogorov-Smirnov test is used to test the difference of respiratory mechanics distribution. A p-value of < 0.05 is considered statistically significant.

Respiratory system elastance variation and spread analysis: To analyze the spread and variation of the estimated respiratory system elastance, E, the mean absolute difference (MAD) and Range90 is calculated [10]. Range90 is the difference between the 95th percentile and the 5th percentile of the E distribution. To normalize the results, the Range90 value is divided by the median non-asynchronous E value. Hence, the spread or the range of the parameter can be quantified and compared within different data sets, as it is a form of dimensionless variation.

III. RESULTS

Table 1 displays the associated median E of all asynchronous breaths before and after reconstruction alongside the median of the non-asynchronous breaths. Range90 and mean absolute deviation (MAD) are also calculated for both the original and reconstructed set of breaths. n is the total number of asynchronous breaths in a given sample and p-value is the result of a two-sample Kolmogorov-Smirnov test with $p < 0.05$ implying a significant difference.

The cumulative distribution (CDF) plots for the original and reconstructed E of Data samples, 3, 7, 9 and 10, are displayed in Fig 2.

Table 1 Pressure reconstruction results

Data Sample	Number of AE Breaths (n)	Respiratory System Elastance, E_{rs} (median) Non-AE*	AE (O*)	AE (R*)	MAD (O)	MAD (R)	Range90/ median (O)	Range90/ median (R)	p-value+
1	181	27.42	17.51	23.76	5.71	4.03	0.80	0.57	+
2	141	27.51	18.34	24.37	4.74	3.40	0.69	0.53	+
3	246	24.46	15.58	23.23	5.64	3.32	0.88	0.61	+
4	164	16.23	9.73	15.60	7.74	6.52	2.72	2.55	+
5	194	28.88	23.17	25.96	9.75	7.29	1.65	1.12	+
6	209	25.39	21.20	24.99	10.00	6.05	2.02	1.03	+
7	273	22.35	9.49	16.62	12.63	6.15	2.36	1.22	+
8	238	24.24	23.86	25.84	3.03	2.47	0.63	0.43	+
9	469	36.86	27.28	33.91	8.61	5.21	0.67	0.52	+
10	1	16.16	15.82	16.85	1.03	1.03	0.18	0.18	-

* AE: Asynchronous breaths, O: Original airway pressure, R: Reconstructed airway pressure,
+p<0.05 when comparing estimated E between (R) and (O) using 2 sample Kolmogorov-Smirnov test

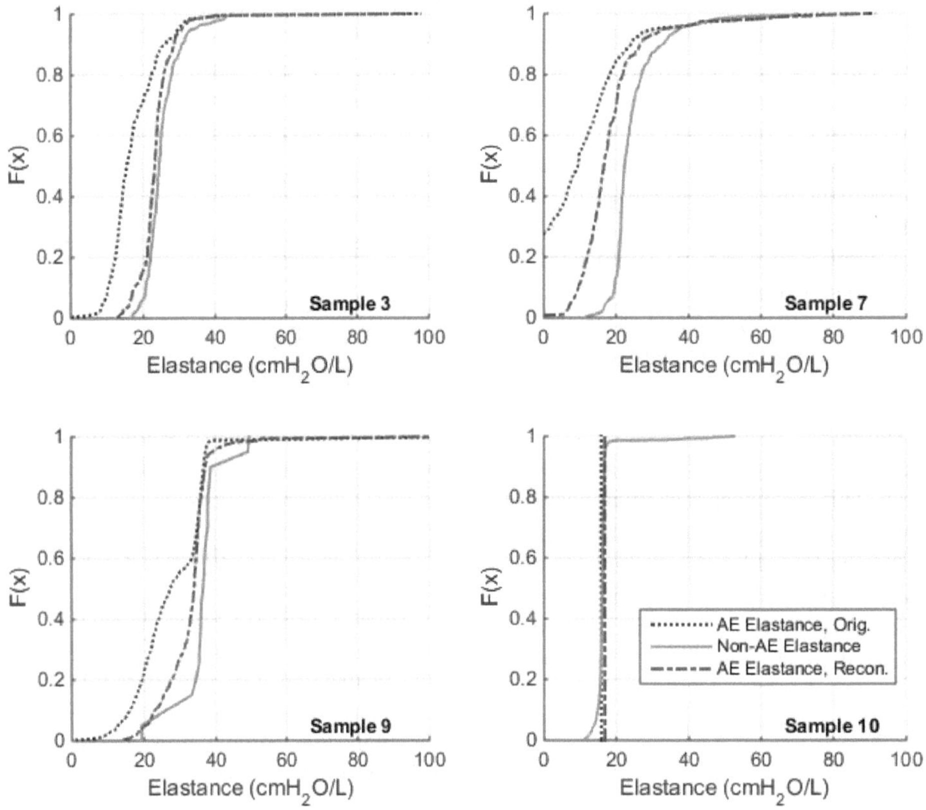

Fig. 2 Cumulative distribution (CDF) plots of the initial and reconstructed respiratory system elastance for Data samples 1, 8, 9 and 10.

IV. DISCUSSION

In this study, all data samples with asynchronous events experienced a decrease in Range90/median E and MAD after flow reconstruction. The extent to which the E changes not only depends on the performance of the algorithm, but also on the number and magnitude of AEs. The median [maximum (max), minimum (min)] decrease in mean absolute deviation across these data samples, not including Data Sample 10, which contained only one AEs, was 29.4% (51%, 18.6%). A decrease in Range90/median or MAD indicates a decrease in variability of E due to asynchrony, and thus, improved the consistency of identification of patient-specific underlying respiratory mechanics. The median (max, min) decrease in the Range90 divided by median E was 30.7% (48.8%, 6.4%).

Data sample 10 contains a minimum amount of AEs and correspondingly has the smallest number of reconstructions. Thus, the Range90/median and MAD of E is minimal, indicating that the algorithm has a minimal effect on non-asynchronous breaths. This result can be found in Fig 2, Sample 10, where all three CDF lines are steep. This result is reiterated in the results of the Kolmogorov-Smirnov test, with all data samples tested in this study having $p < 0.05$ except for the non-asynchronous case of Data sample 10.

Performance of the algorithm in determining the true underlying respiratory system elastance can also be assessed by comparing the E of the reconstructed breaths to the E of non-reconstructed breaths. Breaths that have not been reconstructed are those which do not have upwards fluctuations in flow large enough to result in the model crossing the data at these points, or those which exhibit a high conformation to the model during the initial fit. These breaths are likely to not contain AEs. Therefore, the median E of these breaths gives a measure of the underlying true E, if no physiological or external processes such as MV recruitment manoeuvers or change of ventilator settings occurs over the time period.

For the 9 data samples containing spontaneous breathing or AEs, 8 resulted in the median of the reconstructed E moving closer towards the true E. The median shift was 65.7% closer to the true E, from its original estimate, with a maximum of 89.5%. As a result, E for reconstructed breaths shifted from median 33.3% to 8% deviation from the true E.

After reconstruction, Data sample 8 did not see its E values improved after flow reconstruction, because the median E of the asynchronous breaths was already within 1.7% of the true E value. Thus, the flow reconstruction had a minimal change, as expected in such a case.

While the flow reconstruction has been shown to perform satisfactorily across multiple patients and asynchrony types, the performance of the algorithm has some limitations. Flow reconstruction is hindered by the multitude of possible asynchrony shapes, with the unaffected flow being difficult to determine even by eye. While the algorithm is capable of both lowering and raising apparent E towards a true underlying unaffected E value, it is more effective at raising a low E. This behavior may be due to the majority of 'standard' asynchronies resulting in a lowered apparent E. The more aberrant breaths, which the algorithm is unable to reconstruct effectively, often result in a raised E. Of note is the presence of seemingly normal breaths that have E values significantly higher than the median, appearing in 2.5% of all breathing cycles.

While flow reconstruction is currently not as effective for some types of asynchronies, determination of a true unaffected, underlying E is improved in all cases seen in this study. Therefore, the algorithm exhibits the potential for improvement for guiding MV using model-identified respiratory mechanics. Further improvements may be possible by considering further types of asynchrony and changing the way early asynchronies are handled.

V. CONCLUSION

The proposed iterative interpolative flow reconstruction algorithm is able to improve respiratory system elastance estimation for a wide variety of flow anomaly cases during pressure controlled ventilation. These result have shown the potential use of respiratory mechanics to guide MV therapy in this ventilation mode. The algorithm allows for a more optimized and efficient MV therapy and thereby has a potential to improve patient outcomes in intensive care.

ACKNOWLEDGMENT

The authors would like to thank the New Zealand Health Research Council for supporting this research.

CONFLICT OF INTEREST

The authors declared that they have no conflict of interest.

REFERENCES

1. Ranieri VM, Giunta F, Suter PM, Slutsky AS (2000) Mechanical ventilation as a mediator of multisystem organ failure in acute respiratory distress syndrome. JAMA 284:43 - 4
2. Hasan A (2010) Understanding Mechanical Ventilation: A Practical Handbook. Springer,
3. Girard TD, Bernard GR (2007) Mechanical Ventilation in ARDS. Chest 131:921-9
4. Slutsky AS (1999) Lung Injury Caused by Mechanical Ventilation*. Chest 116:9S-15S
5. Lucangelo U, Pelosi P, Zin W et al. Ventilator-Associated Lung Injury. In: Respiratory System and Artificial Ventilation: Springer Milan; 2008:119-37.
6. Esteban A, Frutos F, Tobin MJ et al. (1995) A Comparison of Four Methods of Weaning Patients from Mechanical Ventilation. New England Journal of Medicine 332:345-50
7. Szlavecz A, Chiew Y, Redmond D et al. (2014) The Clinical Utilisation of Respiratory Elastance Software (CURE Soft): a bedside software for real-time respiratory mechanics monitoring and mechanical ventilation management. BioMedical Engineering OnLine 13:140
8. Stevenson D, Revie J, Chase JG et al. (2012) Algorithmic processing of pressure waveforms to facilitate estimation of cardiac elastance. Biomedical engineering online 11:28
9. Chiew YS, Chase JG, Shaw G, Sundaresan A, Desaive T (2011) Model-based PEEP Optimisation in Mechanical Ventilation. BioMedical Engineering OnLine 10:111
10. Moorhead K, Piquilloud L, Lambermont B et al. (2013) NAVA enhances tidal volume and diaphragmatic electro-myographic activity matching: a Range90 analysis of supply and demand. Journal of Clinical Monitoring and Computing 27:61-70

Author: D.O. Kannangara
Institute: University of Canterbury
Street: 20 Kirkwood Avenue, Upper Riccarton
City: Christchurch
Country: New Zealand
Email: dok11@uclive.ac.nz

Hemodynamic Response to Head-Up Tilt of Patients Implanted with Biventricular Assist Device

K. Nadeem[1], E. Lim[1], B.C. Ng[1], and M. Mubin[2]

[1] Department of Biomedical Engineering, University of Malaya, Kuala Lumpur, Malaysia
[2] Department of Electrical Engineering, University of Malaya, Kuala Lumpur, Malaysia

Abstract— **Head up tilt (HUT) is commonly used to investigate the ability of the heart to regulate blood pressure. During transition through supine to HUT, blood redistributes from the chest to the lower body below the diaphragm, accompanied by rapid fall in the venous pressures and venous return. Patients suffering from biventricular heart failure and assisted with dual rotary pumps (BiVAD) may undergo right ventricular suction as a result of this transition. In this study, we present a validated lumped parameter model of the cardiovascular system for graded HUT integrated with a BIVAD in order to investigate the hemodynamic response during supine and 70° HUT using two different speed settings for left and right ventricular assist devices (VADs). Operating both VADs at the same speed of 2400 rpm during supine resulted in high left atrial pressure (LAP) subsequently causing congestion. This high LAP was relieved when 70° HUT was simulated; however it resulted in reduced central venous pressure, thereby causing right ventricular suction. On the other hand, operating right ventricle assist device at speed lower than left ventricle assist device avoided events of congestion and suction during supine and 70° HUT.**

Keywords— **Heart failure, Biventricular Assist Devices, Head-Up Tilt.**

I. INTRODUCTION

Left ventricular assist devices (LVADs) have tremendously improved the functional capacity and quality of life of the patients suffering from congestive heart failure [13]. However due to increased stress on right ventricle as a result of LVAD implantation, right ventricular failure has been noted in up to 30-50% of these patients, giving rise to higher mortality rate [13,18] . Due to limited availability of right ventricle assist device (RVAD), most of the clinicians implant LVAD as an RVAD to support right ventricle [2,5,18]. As the number of patients implanted with rotary blood pump (IRBP) increases exponentially, there are stronger evidences of long term successful periods of implantation in patients [6]. Thus, the number of patients interacting with unsupervised environment where they are subjected to different physiological activities like exercise, sleep and head up tilt (HUT) has also increased. HUT is a postural condition which causes a decrease in venous return with anecdotal studies indicating some LVAD assisted patients may feel dizziness during the transition from supine to HUT state [8,11]. Despite the high risk of suction, limited studies have been carried out to study the effect of postural change on hemodynamics of patients implanted with IRBP. Muthiah et al [12] conducted a clinical study on patients implanted with IRBP which demonstrated a 10% reduction in pump flow with 60° HUT without causing suction. Lim et al [11] studied the effect of modifying speed, ventricular elasticity, reflex mechanism and total blood volume during 70° HUT using a numerical model integrated with an IRBP. Suction was not observed in supine position however, desensitization of reflex model reduced mean arterial pressure (MAP) significantly upon HUT. Although, these studies have been very useful in providing insights into mechanisms underlying the studied responses of IRBP assisted patients to postural changes, they cannot be directly associated with patients implanted with dual IRBPs. These studies have not taken in account the effect RVAD upon venous return and subsequently cardiac output (CO) at instance of HUT. To date, we are not aware of any studies carried out on patients implanted with biventricular assist devices (BIVADs) who are subjected to HUT or postural change. Therefore, the major motivation for the present study is to investigate the response hemodynamic indices to HUT of patients implanted with BIVADs by using a well validated cardiovascular model for graded HUT which includes arterial and cardiopulmonary reflexes, the auto regulation mechanism of the lower body, diastolic interaction with pericardium as well as local metabolic vasodilation mechanism in active muscles. The present study has been done using two speed settings for LVAD and RVAD. In the first mode, LVAD speed is greater than the RVAD speed while both VADs are operated at same speed in the second mode. The investigations would be focusing upon effect of each mode on MAP, mean pulmonary artery pressure (MPAP), systemic vascular resistance (SVR), CO, VADs flows and atrial pressures.

II. METHODOLOGY

A. Model Description

The model of cardiovascular system (CVS) and rotary blood pumps was adapted from our previous model which

F. Ibrahim et al. (eds.), *International Conference for Innovation in Biomedical Engineering and Life Sciences*,
IFMBE Proceedings 56,
DOI: 10.1007/978-981-10-0266-3_30

was well validated for both healthy and heart failure scenarios during graded head up tilt [8-11]. The model consisted of left and right sides of the heart and pulmonary and systemic circulations with ventricles modeled as time varying elastance functions. Also, a reflex block was included in the model which separately featured the effects of arterial and cardiopulmonary reflexes, the diastolic ventricular interaction with pericardium and auto regulatory mechanism of the lower body.

B. The Head-up Tilt Model

The HUT model consisted of systemic circulation which was distributed into several parallel braches of coronary, splanchnic, upper and lower body, cerebral and renal segments. This was done in order to take in account the different heights of each compartment from hydrostatic point. Moreover, the impact of auto regulatory mechanism on these compartments was also considered. The model has been tested and validated for IRBP-assisted patients undergoing exercise, rest, and 70° HUT. The presented model has produced favorable hemodynamics measurements compared with published literature, in terms of both regional and global hemodynamic indices, at a range of tilt angles [8,11].

C. The BIVAD Model

The LVAD model was adapted from previous model using electrical equation for motor windings, electromagnetic equation for modeling torque transfer function and pump hydraulic equation. The RVAD was the replica of the LVAD integrated into the previously validated numerical model. It consisted of three resistances, two out of which were flow dependent. Suction resistance was also included into IRBP model which is proportional to the difference of pressure between respective ventricle and inlet cannula pressure for each VAD. The VAD component attributed the pressure flow characteristics of VentrAssist (Ventracor Ltd., Chatswood, NSW, Australia), which operated within range from 1600 to 3000 rpm.

D. Simulation Protocols

The model was implemented using Simulink toolbox in MATLAB (The Mathworks, Inc., Natick, MA, USA). The validation of the model was done in following steps:

- Initial simulations were done without an IRBP integrated into the model in order to validate the (NYHA Class IV) biventricular heart failure (BHF) conditions using clinical data reported in the literature. BHF was simulated by tuning the ventricular elastance of both left and right ventricles. This was done in order to reproduce the realistic simulation in terms of MPAP, MAP, left and right atrial pressure (LAP & RAP), CO and SVR for heart failure conditions. The Emaxlv and Emaxrv (mmHg/mL) were tuned to 0.629 and 0.358 respectively in order to agree with the clinically estimated values for ventricular contractility by Konstam et al [3]. The resulting hemodynamics variables for supine position are listed in Table 1.
- After the validation of heart failure scenarios, LVAD was included into the model to simulate supine and 70 ° HUT in order to validate the outcomes with the previous study [11].
- Finally, we integrated the RVAD into the previously validated IRBP-assisted CVS model and simulated supine and 70° HUT to evaluate the response based on two different speed settings which were previously tested for rest conditions only [4,14,18,16,17]. Mode 1 studied the effect of running the LVAD at higher speed of 2400 rpm than the RVAD speed of 1800 rpm. Mode 2 studied the effect of running both VADs at 2400 rpm. Each speed setting was evaluated for supine and 70° HUT on the basis of MAP, MPAP, LAP, RAP, CO, ventricular contractility, VAD flows and SVR.

Table 1 Model simulated and published hemodynamic data for biventriculat heart failure condition during supine

Hemodynamics	Simulation	Literature [8,13,18]
LAP (mmHg)	19.4	14-27
MAP (mmHg)	64.1	60-103
MPAP (mmHg)	27.2	25-38
RAP (mmHg)	10.2	6-19
CO(L/min)	2.7	2-4
SVR (mmHg.s/mL)	1.17	1.09-1.57

III. RESULTS

Comparison between atrial pressure and flow responses during supine and HUT for two different speed modes are presented in Fig. 1 and 2. As expected from the experimental studies focusing on HUT [6,8,11,15], overall LAP, central venous pressure and CO decreased as a result of transition from supine to 70° HUT. It is demonstrated in Fig. 1 that LAP is maintained at approximately 11 mmHg during transition through supine to 70° HUT using Mode 1 while a significant decrement up to 10 mmHg can be observed for Mode 2. Mode 2 showed significantly high LAP of 27 mmHg during supine condition indicating pulmonary congestion which occurs when LAP exceeds 20 mmHg [18]. On the other hand, RAP fell below 2 mmHg during 70° HUT posing a risk of suction. On the contrary, Fig. 3 showed that MAP was maintained above 85 mmHg and increased roughly by 8 mmHg while operating on both modes.

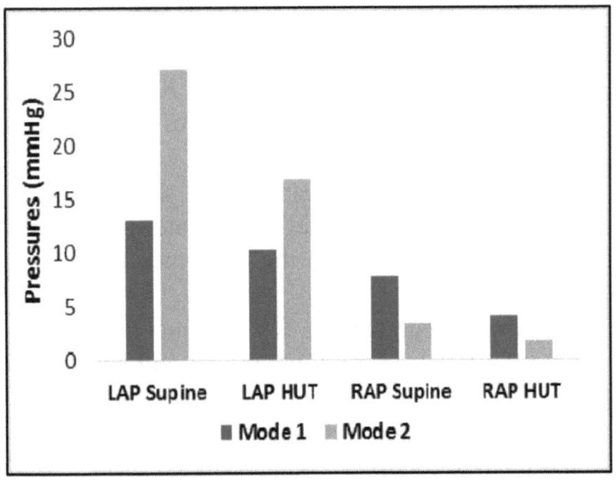

Fig. 1 Simulated hemodynamic systemic and pulmonary pressures for different speed modes during supine and 70° HUT

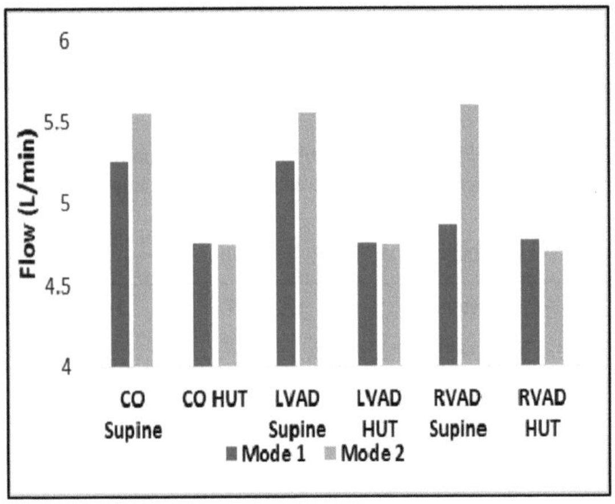

Fig. 2 Simulated cardiac output and VAD flows for different speed modes during supine and 70° HUT

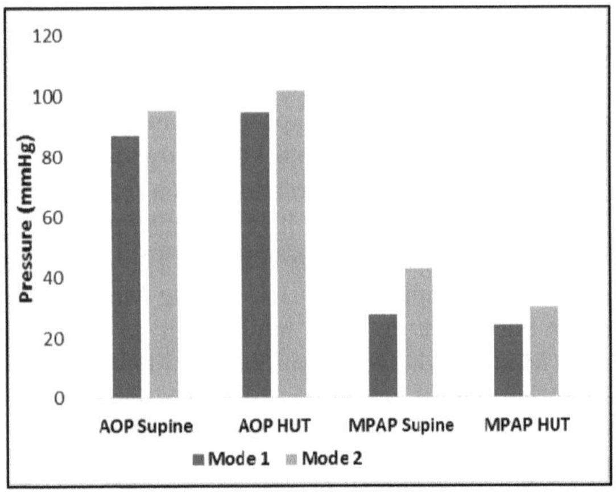

Fig. 3 Simulated hemodynamic atrial pressures for different speed modes during supine and 70° HUT

However, MPAP decreased by ~2 mmHg for Mode 1, while it decreased significantly for Mode 2 by 12 mmHg during 70° HUT. Interestingly, MPAP seemed to have been drastically high (42.65 mmHg) for Mode 2 as oppose to Mode 1 during supine condition.

Fig 4. indicated changes in SVR during supine to 70° HUT for Mode 1 and 2. The systemic resistance changes as a result of auto regulatory response in order to avoid drastic increment in MAP during 70° HUT. There was a considerable correlation between the percentage change of SVR from supine to 70° HUT with 26% increment for Mode 1 and 27% for Mode 2.

IV. Discussion

In normal subjects, sudden upright posture results in peripheral venous pooling leading to a rapid fall in the venous return stroke as well as RAP [6,15]. Contrary to that, venous return is comparatively less sensitive to the postural changes in patients suffering from advanced heart failure due to the reduced venous compliance [1,6,15]. However, the possibility of ventricular suction still resides [11]. The behavior of native CVS, in addition to the IRBP assistance has been studied by Muthiah et al [12] who did not observe any suction events while performing passive HUT study upon nine LVAD implanted patients. Similarly, Lim et al [11] noted a decent decrement in LAP as a result of elevated LV filling pressure but no suction event in her numerical simulation study of the patients implanted with LVAD during 70° HUT. In agreement with that, LAP decreased insignificantly for Mode 1 while it decreased substantially for Mode 2 due to extremely low RV filling pressures in our results for BIVAD implanted patients experiencing 70° HUT (Fig 1). This could be attributed to high RVAD speed and low resistance of pulmonary circulation which lowers differential pressure across right pump as noted by Stevens et al [18] subsequently allowing higher flows than expected during supine (Fig 2). As a result, during 70° HUT lower venous return resulted into RAP below 2 mmHg indicating RV suction and hence the drastic decrease in LAP. Therefore, Mode 2 could result into pulmonary congestion during supine and RV suction during 70° HUT unless there is moderate LV contractility or higher pulmonary afterload as reported from previous studies [18].

Nuerohumoral response is responsible for increases in the SVR in order to maintain MAP during upright posture [7]. Levine et al noted that patients suffering from advanced congestive heart failure did not have a significant increment in norepinephrine and plasma renal activity exhibiting a weakened sympathetic nervous response. Despite of that, SVR rose from 1.2 mmHg.s/mL to 1.425 mmHg.s/mL and

MAP was maintained during orthostatic tilt [7]. Lim et al [8] noted a substantial reduction in increase of heart rate and SVR during 70° HUT. Likewise, SVR increased while operating on both modes with the considerably similar rate of increment (Fig 4). Mode 1 and Mode 2 seemed to have maintained SVR within the physiological range as estimated before [4,11] without significantly obstructing the LVAD flow and maintaining MAP (Fig 2 and 3). However, Mode 2 is seemingly more ideal in case of desensitized reflex model, where paradoxical vasodilation has been noted instead of vasoconstriction during HUT [8] causing a reduction in MAP. This reduced MAP could be maintained by changing the VAD speed because of the availability of wider range for speed change. Also, it has been noted that while operating both VADs at their design speed avoided the risk of thrombus formation and impeller instability [18]. Hence it is essential to take measures in order to make Mode 2 the baseline operating condition.

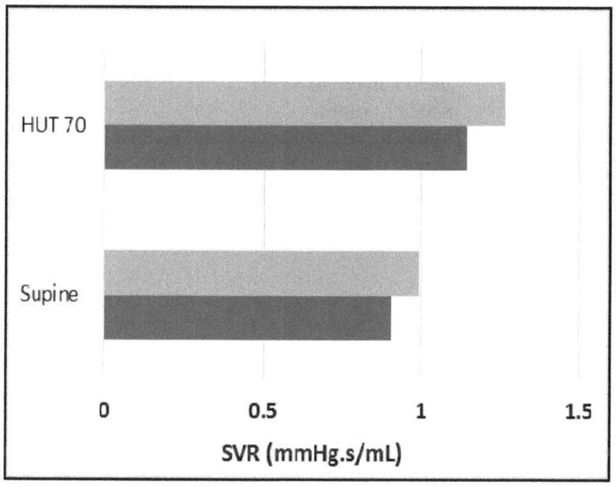

Fig. 4 Simulated systemic vascular resistance change during supine and 70° HUT for different speed modes

Numerous clinical studies have been carried out using Mode 2 for BHF patients under rest conditions [14,16-18]. In one of the studies by Schimitto et al [17], LAP above 20 mmHg was observed when both VADs were operated at same speed for non-fibrillating heart. Similarly, Saeed et al [16] reported high MPAP and LAP in calves implanted with BIVAD which indicated pulmonary congestion for low ventricular contractility state. In consistent with these findings, our results indicated that if both VADs were operated at speed above 2400 rpm, it resulted in pulmonary congestion due to elevated LAP and MPAP during supine. However, if speeds of both VADs were kept below 2400 rpm, it exacerbated the risk of RV suction because of low RAP during 70° HUT. This is attributed to the fact that LVAD is initially designed to be operated against high SVR.

Therefore, when the LVAD is used as an RVAD to work against comparatively lower pulmonary afterload, congestion occurs at high RVAD speed. It has been noted that Mode 2 avoided events of congestion during supine conditions if pulmonary afterload is artificially increased by banding the RVAD outflow cannula [18,13] which could be further tested for HUT.

V. Conclusion

Due to the lack of clinical evidence studying the BIVAD assisted patients undergoing 70° HUT, the present model, in combination with BIVAD integration can provide insights regarding the interaction between CVS and BIVAD during supine and 70° HUT. The predicted results for 70° HUT scenarios could be further validated with clinical and laboratory testing. Furthermore, the model can provide a pioneer platform for development of robust controller for patients implanted with BIVAD who are subjected to different postural changes like rest, exercise, supine and HUT.

Acknowledgment

Authors thank Ministry of Higher Education of Malaysia (UM.C/HIR/MOHE/ENG/50) and University of Malaya Research Grant (RP028-14HTM) for financial support.

Conflict of Interest

The authors declare that they have no conflict of interest.

Refrences

1. Abelmann WH, Fareeduddin K (1969) Increased tolerance of orthostatic stress in patients with heart disease. The American journal of cardiology 23:354-363
2. Hetzer R, Krabatsch T, Stepanenko A, Hennig E, Potapov EV (2010) Long-term biventricular support with the heartware implantable continuous flow pump. J Heart Lung Transpl 29:822-824. doi:DOI 10.1016/j.healun.2010.02.012
3. Konstam MA, Cohen SR, Salem DN, Conlon TP, Isner JM, Das D, Zile MR, Levine HJ, Kahn PC (1985) Comparison of left and right ventricular end-systolic pressure-volume relations in congestive heart failure. J Am Coll Cardiol 5:1326-1334
4. Krabatsch T, Hennig E, Stepanenko A, Schweiger M, Kukucka M, Huebler M, Potapov E, Hetzer R (2011) Evaluation of the HeartWare HVAD Centrifugal Pump for Right Ventricular Assistance in an In Vitro Model. Asaio J 57:183-187. doi:Doi 10.1097/Mat.0b013e3182 11ba2b
5. Krabatsch T, Potapov E, Stepanenko A, Schweiger M, Kukucka M, Huebler M, Hennig E, Hetzer R (2011) Biventricular Circulatory Support With Two Miniaturized Implantable Assist Devices. Circulation 124:S179-S186. doi:Doi 10.1161/Circulationaha.110.011502
6. Kramer B, Massie B, Topic N (1982) Hemodynamic Differences between Supine and Upright Exercise in Patients with Congestive Heart-Failure. Circulation 66:820-825

7. Levine TB, Francis GS, Goldsmith SR, Cohn JN (1983) The Neurohumoral and Hemodynamic-Response to Orthostatic Tilt in Patients with Congestive Heart-Failure. Circulation 67:1070-1075
8. Lim E, Chan GSH, Dokos S, Ng SC, Latif LA, Vandenberghe S, Karunanithi M, Lovell NH (2013) A Cardiovascular Mathematical Model of Graded Head-Up Tilt. Plos One 8. doi:ARTN e77357 DOI 10.1371/journal.pone.0077357
9. Lim E, Dokos S, Cloherty SL, Salamonsen RF, Mason DG, Reizes JA, Lovell NH (2010) Parameter-Optimized Model of Cardiovascular-Rotary Blood Pump Interactions. Ieee T Bio-Med Eng 57:254-266. doi:Doi 10.1109/Tbme.2009.2031629
10. Lim E, Dokos S, Salamonsen RF, Rosenfeldt FL, Ayre PJ, Lovell NH (2012) Numerical Optimization Studies of Cardiovascular-Rotary Blood Pump Interaction. Artificial organs 36:E110-E124. doi:DOI 10.1111/j.1525-1594.2012.01449.x
11. Lim E, Salamonsen RF, Mansouri M, Gaddum N, Mason DG, Timms DL, Stevens MC, Fraser J, Akmeliawati R, Lovell NH (2015) Hemodynamic Response to Exercise and Head-Up Tilt of Patients Implanted With a Rotary Blood Pump: A Computational Modeling Study. Artificial organs 39:E24-E35. doi:Doi 10.1111/Aor.12370
12. Muthiah K, Gupta S, Robson D, Walker R, Macdonald PS, Jansz P, Hayward CS (2013) Effect of Body Position on Continuous Flow Left Ventricular Assist Device Flow Dynamics. J Heart Lung Transpl 32:S233-S233
13. Nadeem K, Ng BC, Lim E, Gregory SD, Salamonsen RF, Stevens MC, Mubin M, Lovell NH (2015) Numerical Simulation of a Biventricular Assist Device with Fixed Right Outflow Cannula Banding During Pulmonary Hypertension. Annals of biomedical engineering. doi:10.1007/s10439-015-1388-2
14. Radovancevic B, Gregoric ID, Tamez D, Vrtovec B, Tuzun E, Chee HK, Moore S, Jarvik RK, Frazier OH (2003) Biventricular support with the Jarvik 2000 axial flow pump: a feasibility study. Asaio J 49:604-607
15. Rudas L, Pflugfelder PW, Kostuk WJ (1990) Comparison of Hemodynamic-Responses during Dynamic Exercise in the Upright and Supine Postures after Orthotopic Cardiac Transplantation. J Am Coll Cardiol 16:1367-1373
16. Saeed D, Ootaki Y, Ootaki C, Akiyama M, Horai T, Catanese J, Fumoto H, Dessoffy R, Massiello AL, Horvath DJ, Zhou Q, Chen JF, Benefit S, Golding LA, Fukamachi K (2008) Acute in vivo evaluation of an implantable continuous flow biventricular assist system. Asaio J 54:20-24. doi:10.1097/MAT.0b013e31815b2d1e
17. Schmitto JD, Burkhoff D, Avsar M, Fey O, Ziehme P, Buechler G, Haverich A, Strueber M (2012) Two axial-flow Synergy Micro-Pumps as a biventricular assist device in an ovine animal model. J Heart Lung Transpl 31:1223-1229. doi:10.1016/j.healun.2012.08.008
18. Stevens MC, Gregory SD, Nestler F, Thomson B, Choudhary J, Garlick B, Pauls JP, Fraser JF, Timms D (2014) In vitro and in vivo characterization of three different modes of pump operation when using a left ventricular assist device as a right ventricular assist device. Artificial organs 38:931-939. doi:10.1111/aor.12289

Author:	K. Nadeem, E. Lim, B.C. Ng, M. Mubin.
Institute:	University of Malaya
Street:	Jalan Universiti
City:	Kuala Lumpur
Country:	Malaysia
Email:	komalnadeemwyne@outlook.com; einly_lim@um.edu.my

Development of a Microfluidic Vibrational Cleaning System for Cleaning Microtissues

K.T. Thong[1], C.F. Soon[2], A.B. Ismail[2], and K.S. Tee[1]

[1] Faculty of Electrical and Electronic Engineering, Universiti Tun Hussein Onn Malaysia, 83000 Batu Pahat, Johor, Malaysia
[2] Biosensor and Bioengineering Laboratory, MiNT-SRC, Universiti Tun Hussein Onn Malaysia, 83000 Batu Pahat, Johor, Malaysia

Abstract— **Microtissues can be cultured on the hydrogel, liquid crystal substrate and scaffolds. Therefore, a cleaner has been developed to clean the microtissue extracted from the culture substrate such as the cholesteryl ester liquid crystal (CELC). Clean microtissue samples were required for the precise experimental output. The cleaner performance was verified by observing the CELC's birefringence properties around microtissue through cross-polarising microscope. In addition, the effects of mechanical vibration that generated by the microtissue cleaner onto the microtissue sample were investigated by live/dead cells assay. Based on the results, 3D microtissue cleaner was effectively cleaned microtissue after three replicated cleaning steps. Two minutes of continuous vibration frequency at 148 Hz and acceleration of 0.89 G_{rms} were suitable to clean the microtissue.**

Keywords— **3D microtissue cleaner, 3D microtissue, cholesteryl ester liquid crystal (CELC), microfluidic chip, mechanical vibration.**

I. INTRODUCTION

Culturing cells in 2 dimensional (2D) or monolayer is a routine technique used in cell biology study before. Culturing cells in monolayer are arguably an unnatural model that is far from the actual biological microenvironment in the tissue [1]. For in-vivo system, cells' morphology and behavior in the tissues are regulated by the mechanical cues, cell-cell communication and biochemistry release in a close encapsulation with other cells. However, in in-vitro system, cells can also be restored to the structures similar to those cells in the in-vivo system by using 3D cell culture techniques.

3D cell culture technique utilised biomaterials as a substrate or scaffold to provide cells with an environment mimicking the in-vivo system. Due to the need for a more realistic model, various methods had been developed to culture cells in 3D construction. Some of the researches involved fabrication of polydimethylsiloxane (PDMS) micromold to contain cells in close association [2], using soft biomaterial such as hydrogel for aggregating the cells [3], hanging drop technique to confined cells in liquid interfacial tension [4], and cholesteryl ester liquid crystal (CELC) substrate for cells self-aggregate into spheroid [5]. For those 3D cells or microtissue cultured on the adhesive biomaterial such as the hydrogel and CELC is suggested to clean the sample properly before conducted the experiment. Previous study had demonstrated that dissolving solution, temperature and mechanical vibration were helpful in remove and clean the adhesive gels from the object's surface [6]. The organic property of both biomaterials is difficult to be dissolved in common aqueous system. It requires interaction with organic medium to be dissolved such as the dissolution of isopropyl alcohol in water [6, 7]. The alternative way is by introducing the vibration on the biomaterial in aqua. The shock wave impact generated by mechanical vibration in the aqua effectively separate the trapped biomaterial from the object surface [6]. The vibration generated by machinery reacts on cell proliferation and differentiation based on their frequency and amplitude [8].This method is easy to produce, control and economical than ultrasonic.

In this work, we are interested to find out the effectiveness of mechanical vibration on separating the CELC from microtissue in aqua and the cell responses on the mechanical vibration.

II. MATERIALS AND METHODS

A. The Design of Microtissue Cleaner

The microtissue cleaning comprised of five main parts that include the water pump, a modified water valve, a pair of vibrator, a microfluidic chip and the electronic control panel as shown in Fig. 1. The function of the water pump system was used to pump in fresh cleaning agent into the microfluidic chip and discard the old media from the microfluidic vibrational cleaning system. The fluid flow of this system was controlled by the modified water valve system to flow in 1.5 ml cleaning agent into the microfluidic reservoir. The microfluidic vibrational cleaning system was designed with a platform for attaching the vibrational motor and the microfluidic chip to clean the sample. The duty cycle of each electric motor's pulse width modulation (PWM) signal can be varied and monitored at the control panel.

F. Ibrahim et al. (eds.), *International Conference for Innovation in Biomedical Engineering and Life Sciences*,
IFMBE Proceedings 56,
DOI: 10.1007/978-981-10-0266-3_31

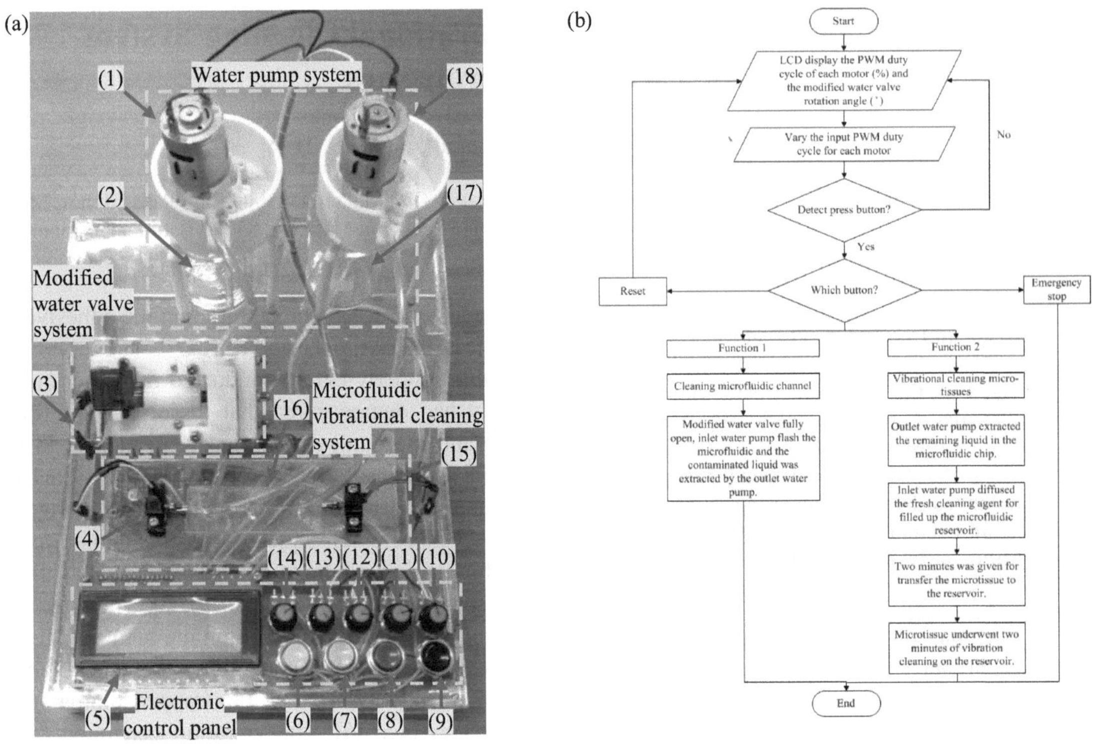

Fig. 1 Image of (a) the 3D microtissue cleaner prototype. The number shown on the 3D microtissue cleaner prototype represent the (1) inlet water pump, (2) water storage tank, (3) modified water valve, (4) left vibrational motor, (5) 20 × 4 LCD, (6) Function 1 button, (7) Function 2 button, (8) emergency stop button, (9) reset button, (10) potential meter for LCD contrast, (11) potential meter for right vibrational motor, (12) potential meter for left vibrational motor, (13) potential meter for outlet water pump, (14) potential meter for inlet water pump, (15) right vibrational motor, (16) microfluidic chip, (17) waste water storage tank, (18) outlet water pump and (b) the 3D microtissue cleaner system flow chart

B. Function of 3D Microtissue Cleaner

An Arduino Mega 2560 R3 microcontroller was programmed to control the water valve, dc water pumps and the vibrational dc motor. Five variable potentiometers were utilized to control the duty cycle of the motors' PWM and the contrast of liquid crystal display (LCD). The parameter setup of each motor must be configured before the cleaning process. In addition, four selective functions were provided in this system. Emergency stop and reset function were ready to take precautions against the system error. Function 1 and Function 2 programs were programmed to perform cleaning of the microfluidic channel and cleaning the microtissue through the vibrational water wave. For Function 1, the flushing took one second to complete. On the other hand, Function 2 produced two minutes of continuous mechanical vibration frequency at 148 Hz and acceleration at 0.89 G_{rms} to exert water vibration force that could remove the CELC attached on the surface of the microtissue. After two minutes of cleaning, the microtissue was removed and the oily media was refilled with fresh media.

C. Live/dead Cells Assay

The live/dead® assay reagent (L-3224, Life Technologies, USA) was used to verify the cell viability of the microtissue after the vibrational cleaning. 20 µl of Ethidium Homodimer-1 (2 mM, Life Technologies, USA) stock solution was added to 10 ml of Hank's Balanced Salt Solution (HBSS, Sigma-Aldrich, UK) and sonicated for 15 minutes. 5 µl of Calcien AM (4 mM, Life Technologies, USA) stock solution was added into the mixture solution and being sonicated for another 15 minutes.

The microtissue in dormant phase was collected from the CELC surface and cleaned with the vibrational cleaner. The microtissue was then transferred to a glass slide and immersed in 0.5 ml of live/dead cell reagent. The microtissue was incubated in the dark condition for 15 minutes at room temperature. After 15 minutes incubation, the sample was rinsed with HBSS three times. The glass slide was covered with a cover slip and sealed with clear fingernail polish for observation. Similar experiment was replicated three times.

III. RESULTS

D. The Clean Microtissue

For pharmacological study, the cleaned microtissues ensure that the treatment of the drug would not be interfered by the residue culture substrate left on the microtissue during harvesting. In this work, the microtissue with residue LC underwent cleaning action within the microfluidic reservoir that was filled up with fresh cleaning agent. It took three repeats of cleaning process to ensure the oil stained was fully removed as shown in Fig. 2. Each cleaning step required two minutes of water wave propagation that created by the motor vibration to wash the microtissue. After two minutes of cleaning, the media with CELC was discarded and the prototype was standby in idle mode. The oil film was gradually removed as observed with the reduction of the birefringence effect of the LCs.

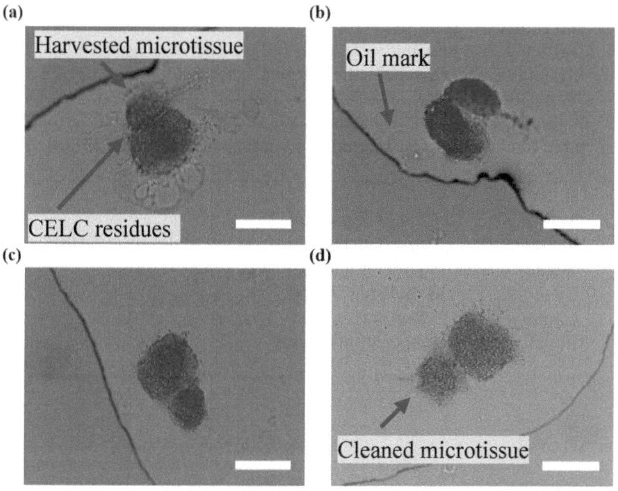

Fig. 2 Cross-polarising micrograph of microtissue (a) harvested microtissue with CELC residues, (b) first cleaning result, (c) second replicated cleaning result and (d) the clean microtissue obtained from third replicated cleaning by the 3D microtissue cleaner. (Scale bar: 100 µm)

E. Cell Viability of Microtissue After Vibrational Cleaning

In order to further explore the cell viability within the microtissue, fluorescence live/dead cells staining was implemented to determine the effects of vibration cleaning induced to the cells within the microtissue. Fluorescence live/dead cells assay presented the live cells in green fluoresce and dead cells in red fluorescence. The staining of microtissue with the live/dead cell assay micrograph is shown in Fig. 3. This result shows that all of the cells were remained alive in the microtissue after three replicated cleaning step. This implied that the intensity of the mechanical vibration was accepted by the microtissue sample.

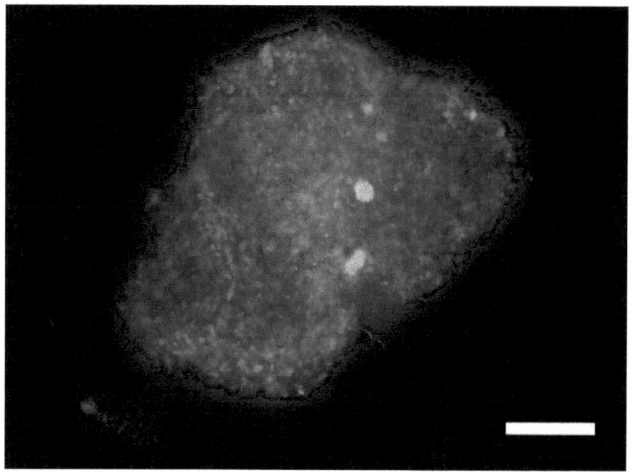

Fig. 3 Micrograph of live and dead cell assay for microtissue after vibrational cleaning. (Scale bar: 100 µm)

IV. DISSUSSION

3D microtissue cleaner is specifically designed to clean off the CELC residues on microtissue which was cultured on the CELC surface based microtissue culture technique. The water vortex caused by the motor vibration is the main factor that used to separate the adhesive CELC gels from the microtissue surface. Although irradiated the ultrasonic in the cleaning agent is faster and more effective to clean the specimen then applied the water vortex but the application of ultrasonic is not encourage on cleaning the microtissue due to the high impact of bubble implosions created by the ultrasonic would break apart the fragility biological tissue into small piece. However, the microtissue spinning in the water vortex help to shed off the CELC gently without harming the cell.

On the study of transmits short term mechanical vibration on microtissue had proved that introduced high frequency and low amplitude vibration did not affect the viability of the cells within the microtissue by fluorescence live/dead cells assay. The dormant phase microtissue that undergone the vibrational cleaning process was strongly supported the verdict by the demonstrated of green fluorescence viable cells were highly distributed around the microtissue especially the critical cell necrosis zone located at the central of microtissue. No cells necrosis was found in the microtissue because cells in dormant phase with sufficient nutrients. This result was different from the previous work [9] but showed higher viability of cells may due to the different cell line and culture technique.

V. CONCLUSIONS

The residues of liquid crystals that encapsulating the harvested microtissue could be removed by repeating the cleaning step with the self develop microtissue cleaner which produced the vibration frequency at 148 Hz and 0.89 G_{rms} vibration acceleration. In addition, high cell viability obtained by the result of live/dead cells assay comes to a conclusion that two minutes of continuous vibration frequency at 148 Hz and 0.89 G_{rms} vibration acceleration were not harmful to the microtissue. Lastly, this cleaner was also suggested applying on routine pre-experiment cleaning for other microtissue culture technique.

ACKNOWLEDGMENT

The authors would like to thank the Malaysia Ministry of Higher Education for providing research funding support under the Research Acculturation Grant Scheme (RAGS Vot No. R027) and Science Fund (Vot. No. S024 or Project No. 02-01-13-SF0104).

CONFLICT OF INTEREST

The authors declare that they have no conflict of interest.

REFERENCES

1. Souza, G. R.,J. R. Molina et al. (2010) Three-dimensional tissue culture based on magnetic cell levitation. Nature Nanotechnology 5(4): 291-296
2. Hsieh, C.-H.,J.-L. Wang et al. (2011) Large-scale cultivation of transplantable dermal papilla cellular aggregates using microfabricated PDMS arrays. Acta Biomaterialia 7(1): 315-324
3. Tsang, V. L.,A. A. Chen et al. (2007) Fabrication of 3D hepatic tissues by additive photopatterning of cellular hydrogels. The FASEB Journal 21(3): 790-801
4. Timmins, N.,F. Harding et al. (2005) Method for the generation and cultivation of functional three-dimensional mammary constructs without exogenous extracellular matrix. Cell and Tissue Research 320(1): 207-210
5. Thong, K. T.,C. F. Soon et al. (2015) The Effects of Enzyme to the Dissociation of Cells in Monolayer and 3D Microtissue on the Liquid Crystal Substrate. 5th International Conference on Biomedical Engineering in Vietnam, Springer
6. Zhuang, X.,W. He et al. (2012) Materials separation from waste liquid crystal displays using combined physical methods. Polish Journal of Environmental Studies 21(6): 1921-1927
7. Okay, O. (2009) General properties of hydrogels. Hydrogel sensors and actuators, Springer: 1-14
8. Gaston, J.,B. Quinchia Rios et al. (2012) The response of vocal fold fibroblasts and mesenchymal stromal cells to vibration. PLoS One 7(2): e30965
9. Folkman, J. and M. Hochberg (1973) Self-regulation of growth in three dimensions. The Journal of Experimental Medicine 138(4): 745-753

Author:	Chin Fhong, Soon
Institute:	Universiti Tun Hussein Onn Malaysia (UTHM)
Street:	86400 Parit Raja
City:	Batu Pahat, Johor.
Country:	Malaysia
Email:	soon@uthm.edu.my

Development of a Joystick Controllable X-Y Translational Stage for an Inverted Microscope

C.F. Soon[1,2,*], S.H. Choo[1], K.T. Thong[1] and K.S. Tee[1]

[1] Faculty of Electrical and Electronic Engineering, Universiti Tun Hussein Onn Malaysia, 83000 Batu Pahat, Johor, Malaysia.
[2] Biosensor and Bioengineering Laboratory, MiNT-SRC, Universiti Tun Hussein Onn Malaysia, 83000 Batu Pahat, Johor, Malaysia.

Abstract— **An inverted microscope is designed to have a stage in which the specimens can be placed for microscopy. Most of the biological microscope is built in with static or manual adjustable stage which cannot provide consistent and precise adjustment to move the specimen during microscopy. As an add-on to the microscope stage, the existing motorized 2-axis linear stage in the market is bulky and heavy. Therefore, a lightweight and joystick controllable bipolar stepper motor X-Y translational stage was proposed to solve the problem. In our design, the joystick functions to provide input signals to the Arduino Mega microcontroller in order to control the position of the X-Y translational stage. For the X-Y translational stage, the stepper motors of a pair of the linear slides were driven by a micro stepping motor driver with full scale drive signal to provide controllable stepping movement. It was found that full scale drive signals could support higher resolution of movement at 99 ± 13 μm. The overall X-Y translational stage weighted approximately 700 g. As a conclusion, the light weight stepper motor based X-Y translational stage allows the user to move the specimen in a small stepping size during microscopy.**

Keywords— **X-Y stage, translational stage, stepper motor, micro-stepping, joystick, inverted microscope**

I. INTRODUCTION

A microscope is a high precision optical instrument that uses a lens or a combination of lenses to produce highly magnified images of small specimens or objects especially when they are too small to be seen by the naked eye. Every microscopes is designed to have a stage where specimens can be placed for microscopy [1]. A 2–axis stage has a clamp that could hold the specimen so that the slide can be placed on it and translate the slide in two directions. The specimen can be contained in glass slides or in a petri dish. The microscope is a durable and long lasting instrument. Modern microscope stage equipped with more functionality and specialisation. Most of the microscope is built up with a static stage plate which limits the view of sample from the eyepieces. In general usage, the user needs to move the specimen by sliding the slide. Hence, the X-Y translational stage is added to enable a static stage to be adjustable. This type of stage is usually made up of coated aluminum and fixed with clamps for glass slides. The stage has a better locking control system for avoiding accidental bumping resulting in defocusing. The manual adjustable stage can be moved in precise incremental movement.

The motorised stage was designed to replace the manual stage of the microscope. The translation movement of the stage is controlled by a stepper motor. As such, the movement of the specimen slide becomes easier and precise. The motor can perform more stable movement with an aid of the controller with buttons control [1]. However, the motorised stage is bulky and expensive in the market [2]. The weight of the X-Y linear stage can go up to 3 kg. Sometimes, it can be oversize to fit into the stage of different brands of inverted microscope.

In this paper, a light weight and joystick controllable X-Y translational stage using two motorized linear stage. The translational stage is controllable by a joystick and the resolution and speed of the translation was evaluated in an inverted microscope. Unlike other optical microscope, the objective lens of the inverted microscope is located down the microscope and the specimen stage is on top of the objectives lens. Hence, the translational stage can be placed on the stage without interference from the rotating objective lens.

II. MATERIALS AND METHODS

A. Circuit design of the system

In this project, a motorised stage to place a petri dish or glass slide was designed and controlled by a two phased bipolar stepper motor (Fig. 1). The stepper motor was connected to an Arduino Mega 2560 Microcontroller. Joystick controller was used to provide movement signal to the microcontroller. Besides that, a variable resistor was added to adjust the step size of the stepper motor.

For the circuit design, each stepper motor was driven by an EasyDriver motor driver independently as shown in Fig. 3. EasyDriver v4.4 from sparkfuns electronics.com was used. The The pin X, pin X, pin Y, and pin $\bar{Y}$ of the stepper motor was assigned to the 4 pins of the port A and B of the EasyDriver, respectively. Input signals from the microcontroller were assigned to the step control pin 1 and 2 (MS1 and MS2) and the directional signals from microcontroller were connected to DIR pin. Step control 1 and 2 control the

F. Ibrahim et al. (eds.), *International Conference for Innovation in Biomedical Engineering and Life Sciences*,
IFMBE Proceedings 56,
DOI: 10.1007/978-981-10-0266-3_32

stepping size while the input signal is the step signal to the EasyDriver.

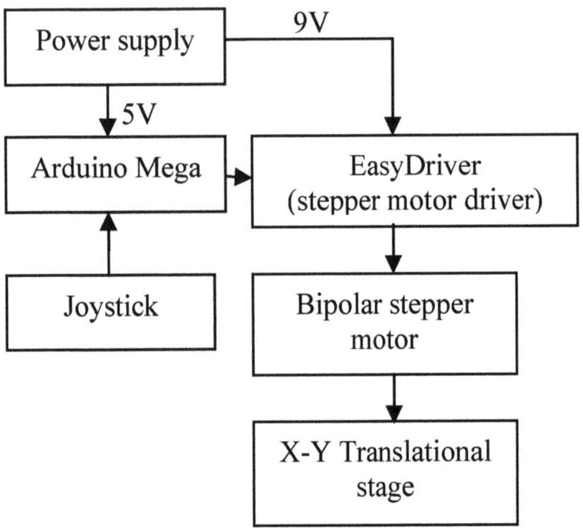

Fig. 1 The block diagram of the x-y translational stage system

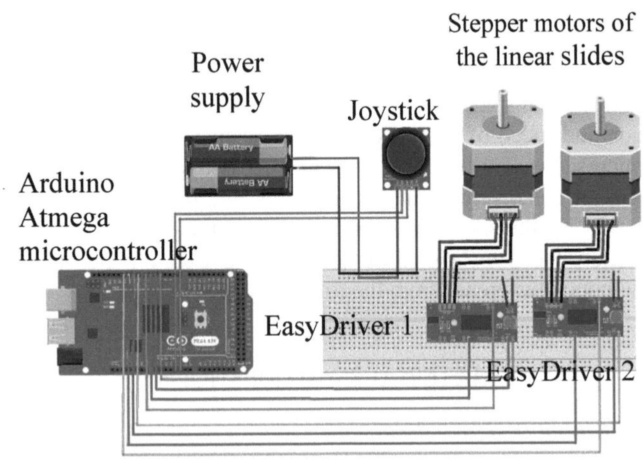

Fig. 2 The schematic connection of the system

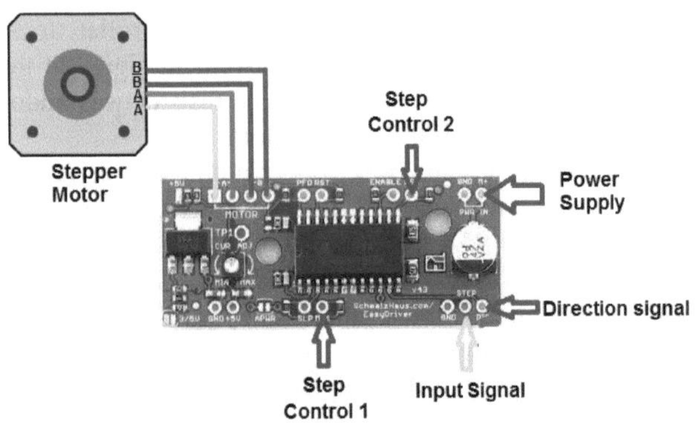

Fig. 3 The connection between EasyDriver motor driver and stepper motor

Fig. 2 shows the overall circuit design of the system. 5V of battery was supplied to the Arduino Mega microcontroller and EasyDriver motor driver. The joystick was assigned 2 wires to the microcontroller for X-direction and Y-direction for signal input. 4 wires were connected from the EasyDriver to the microcontroller. The EasyDriver1 wires were assigned as follows: pin 4 to DIR pin, pin 5 to step pin, pin 6 to MS1 pin and pin7 to MS 2 pin. The EasyDriver2 wires were assigned as follows: pin 8 to DIR pin, pin 9 to step pin, pin 10 to MS1 pin and pin 11 to MS 2 pin. From the power supply board, the microcontroller and EasyDriver were both supplied with 5V and 9V, respectively. The input signal from the joystick to the Arduino controller provides the instruction to the microcontroller. Computer numerical control may imposed higher cost to the system as reported in [2]. Microcontroller processed the signal and sent the information to motor driver (EasyDriver) (Fig. 2). The movement of the stepper motor was then controlled by the EasyDriver signals. Subsequently, the linear lead screw of the linear stage rotates to translate the movement of the stage along the X-axis and Y-axis.

B. Arduino joystick read-out coding

Software development involved the programming of instructions to the Arduino atmega microcontroller. The coding used in Arduino Mega microcontroller was C language. Programming was used to control the input and output signals of the microcontroller. Two output signals from the joystick were sent to the microcontroller. The software library of Arduino software contains standard functions such as analogread() to read the signals from the inputs of th microcontroller. The "analogread()" read the voltage different from the joystick and "map()" was used to convert the values from the analog value range to another digital value range. The signals from the joystick are in analog. In the C code, xAxis=map(analogRead(xpin), 0, 1023, 0, 10) was used to map the joystick variable voltage and convert the equivalent value to a digial value between 0 and 10.

C. Arduino stepper motor control coding

In the motor controlling codes, full step pulse signals were applied to control the input "STEP" signals (Fig. 2). This signal pulses rotate the stepper motor step-by-step. The "DIR" was used to control the direction of the stepper motor rotate. When "DIR" is high, stepper motor rotates clockwise, when "DIR" is low, stepper motor rotates anticlockwise. While the "STEP1_PIN" enabled high was used to produce the signal pulse to rotate the stepper motor step by step. The code digitalWrite(STEP1_pin) was used to generate the signal to the STEP pin as indicated in Fig. 2.

D. The X-Y table platform and X-Y translational stage

An X-Y table was customised according to the size of the NIKON TS-100 inverted microscope stage. In order to provide movement in X and Y directions, the stepper motor was aligned perpendicular to each other. The material chosen for stage was light weight and transparent.

III. RESULTS AND DISCUSSION

A. The hardware of the X-Y translational stage

The hardware consisted of two major parts which were main controller box and X-Y translational stage part (Fig. 5a-b). The main controller box contained all the printed circuit boards (Fig. 5b), EasyDriver motor driver, switches, speed control knob, and joystick breakout controller. The X-Y translational stage can be moved vertically and horizontally depending on the joystick control (Fig. 6a-b). The overview placement of the X-Y translational stage on a Nikon TS-100 inverted microscope is as shown in Fig. 6a. For ergonomic reason the X-Y translational stage was placed on the left side of the petri dish. Fig. 5a shows the X-Y translational stage carrying a petri dish on top of the holder.

B. Resolution analysis

The distance travelled per unit step can be obtained through three repeated measurements using a microscope measurement gauge. A gauge was used to determine the travelling distance or resolution of the translational stage. In addition, the joystick controllable X-Y translational stage can provide a step movement at a resolution of 99 μm ± 13 μm in full scale step. The resolution is subjected to the precision and backlash problem of the lead screw of the linear slide. The overall X-Y translation stage weighted approximately 700 g. The resolution for the full step was approximately 150 μm. It may not provide resolution down to nanometric scale as reported previously [3, 4] but the resolution is within the observation area of 3.142 mm^2 through the eye piece of the microscope using an objective lens of 10x.

C. Speed analysis

The speed control of the stepper motor was assigned to a potentiometer in which the resistance can be varied. The variation of the resistance could be varied between 0 and 2 kΩ in achieving different speed of the stepper motor. Fig. 7 shows the relationship between the speed of the X-Y translational stage and the resistance of the potentiometer under full step control. Based on Fig. 7, the resistance increased linearly with the speed of the stepper motor. The minimum and maximum speed of the stepper motor in full scale step was 260 and 2500 μm/s, respectively. The low speed of the stepper motor did not affect the specimen in the petri dish. At low speed, the stage holder was translated slowly and gradually by the stepper motor. The small vibration of the stage holder due to the stepper motor stepping movement is negligible. While at high speed, the stage holder was translated faster by the stepper motor. The petri dish on the stage holder vibrated gently due to the stepper motor stepping movement. From our test results, the suggested optimum speed is ranged between 140 –330 μm/s in order to minimise the vibration of fluid in the petri dish.

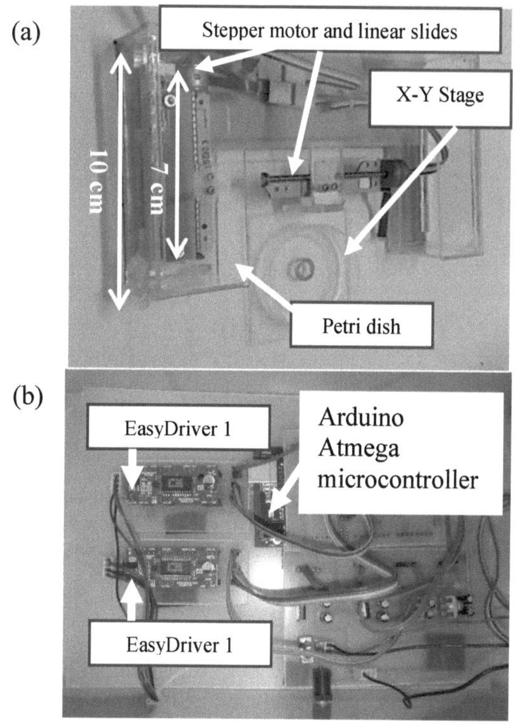

Fig. 4 The hardware system of the (a) the translational X-Y stage and (b) circuit boards connections.

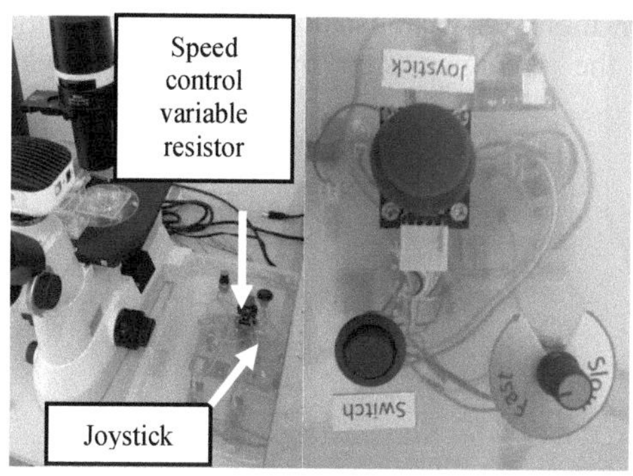

Fig. 5 (a) The translational X-Y linear stage on an inverted microscope stage and (b) switches on the controller box.

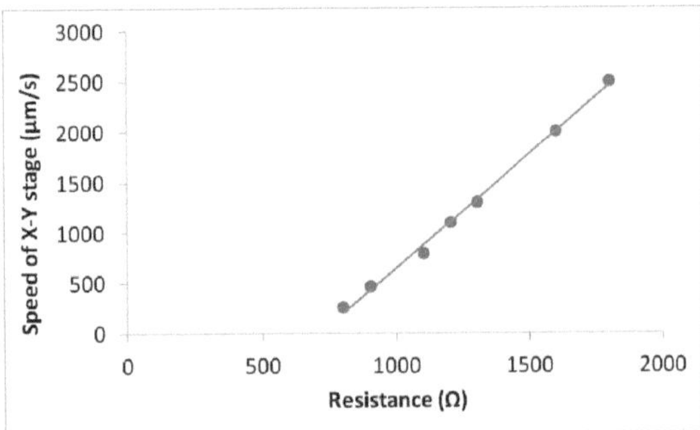

Fig. 6 Graph of speed of X-Y translational stage against speed control resistances

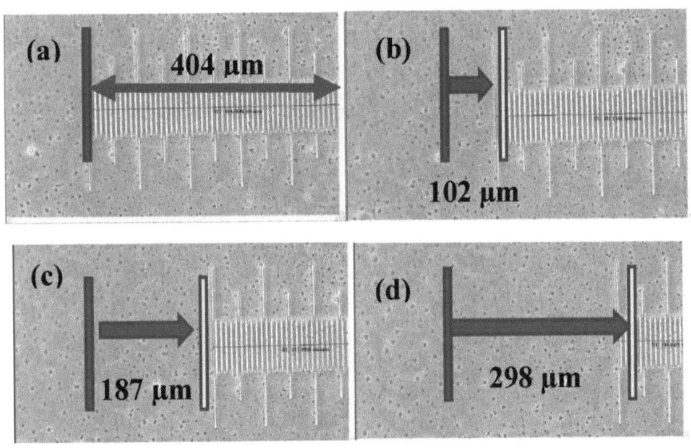

Fig. 7 The inverted phase contrast micrograph of (a) Initial (b) First (c) Second, and (d) Third step movements. The red and yellow vertical lines represent the initial and next positions, respectively.

IV. CONCLUSIONS

The joystick controllable X-Y translational stage presented in this paper has the advantage of being light weight compared to the existing product in the market. The speed of the stage is controllable at a translation resolution of 99 µm $\pm$ 13 µm in full scale step. Nonetheless, the bipolar stepper motor X-Y translational stage allows the user to move the specimen step by step with varying speed. The range of speed to minimise vibration of liquid was determined at 140 – 330 µm/s. Nonetheless, the speed of movement is proportional to the variable resistance of the potentiometer. The varying speed and resolution in micrometer provided by this system is sufficient for routine usage in the cell and tissue biology laboratory.

ACKNOWLEDGMENT

The authors would like to thank the Malaysia Ministry of Higher Education for providing research funding support under the Research Acculturation Collaborative Grant Scheme (RACE Vot No. 1515).

CONFLICT OF INTEREST

The authors declare that they have no conflict of interest.

REFERENCES

1. A. F. Ferreira, J. (2001) Coarse/fine motion control of a teleoperated autonomous piezoelectric nanopositioner operating under a microscope. 2001 IEEE/ASME International Conference on Advanced Intelligent Mechatronics, 1313 - 1318.
2. P. A. Sherring da Rocha Junior, et al. (2013) Stepper motor drive for computer numerical control machines. Brazilian Power Electronics Conference (COBEP), 909 - 914.
3. M. Shigeki, et al. (2009) Nano-Motion Stage for High-Speed and Precision Positioning on an X-Y Plane IEEE Transactions on Magnetics vol. 45: 4972 - 4978.
4. L. Pengbo, et al. (2014) Modeling and control of a novel X-Y parallel piezoelectric-actuator driven nanopositioner American Control Conference (ACC): 1003-1008.
5. K. Z. Zaferullah, et al. (2014) Speed control of stepper motor for collimator jaws positioning based on FPGA implementation. International Conference on Circuits, Systems, Communication and Information Technology Applications (CSCITA), 353 - 357.

Author: Chin Fhong, Soon
Institute: Universiti Tun Hussein Onn Malaysia (UTHM)
Street: 86400 Parit Raja
City: Batu Pahat, Johor.
Country: Malaysia
Email: soon@uthm.edu.my

Prosthetic Hand Controlled by Wireless Flex Sensor on EOD Robot

S. Thanakodi, S.M.H. Azhar, and A. Miskon

Department of Electrical and Electronic Engineering, Faculty of Engineering Universiti Pertahanan Nasional Malaysia, 57000, Kuala Lumpur, Malaysia

Abstract— **EOD or Explosive Ordnance Disposal unit heavily relies upon the EOD robot that has a very simple 2 or 3-fingers gripping tool. This feature itself is a major drawback as the gripper not really suitable for disposing Improvised Explosive Device (IED). This paper proposes a wirelessly controlled prosthetic hand to overcome the problem faced by the current EOD bot and its challenges. The wirelessly controlled prosthetic hand was achieved by utilizing a prosthetic hand, flex sensors, and an Xbee Wireless Module embedded to a hand glove. The proposed design has shown some technical possibility to perform the tasks despite of low angular movements for 2 fingers. In years to come, this hand not only limited for disposing IED, it can be also extended for medical practitioners (specialist) to utilize as a hand for operation conducted from a distance using the fast growing information technology (IT) during any emergencies to save human life.**

Keywords— **EOD, IED, Prosthetic Hand.**

I. INTRODUCTION

EOD team is always acquainted with bomb, juggernaut suit, improvised explosive device (IED) and most of all the EOD robot or in short EOD bot. EOD or Explosive Ordnance Disposal team is a team specialized in disposing away the bombs. In Malaysia, the EOD team is operates under the wing of Royal Army Engineers Regiment. Bomb disposal, is an extremely dangerous job and can claim lives if not being handled properly.

The modus operandi of the bomb disposal team would be normally to stay as far away from their work site, and preferably only interacting via remote controlled robots or EOD bot. The function of these EOD bot cannot be denied especially in sparing the human lives during any disposal mission. Surely in every way it is much better to lose an EOD bot rather than a human's life. These EOD bot can help the technician get an excellent view or an idea of the explosives' nature and its types.

The current robots utilizes various hand like gripper type or hand type in disarming the explosives. The drawback with the current robot hands, is that it's not user friendly and technically not suitable for handling the IED. This is due to the weakness with the gripper that it's not very likely to hold things normally or in other words, nothing much can be done with two or three fingers of the robot. Furthermore, these 2 or 3 fingers cannot perform to the best compared to a human fingers. It is just similar to the industrial fingers where the mechanical robot hands perform the industrial work. Hence, in this paper, prosthetic hand was proposed to replace the EOD bot gripper part.

II. TECHNOLOGIES

In accomplishing the Prosthetic hand controlled by wireless flex sensor, some major materials were used. Prosthetic hand, Flex Sensor, Xbee Wireless Module, and TowerPro MG995 Servomotor are the major devices used in this paper.

A. Prosthetic Hand

Prosthetic hand are more likely similar to human hand as it is a 3D printed out from human hand. This feature enables the handling of IED much easier and better than the gripper type. The prosthetic hand were made using Polylactic Acid (PLA) plastic [3].

B. Flex Sensor

The flex sensor used to control the prosthetic hand directly with the input from human hands. This ensures the synchronization between the user and the prosthetic hand, which enables the prosthetic hand to be controlled easier rather than using a conventional remote control. The flex sensor were knitted on a hand glove and when the user bend the finger, the resistance becomes higher. The higher resistance yields in a higher voltage output. The voltage output were sent to the prosthetic hand wirelessly and the prosthetic hand moves accordingly to the provided voltage signal. There were 5 of 4.5 inches length flex sensor were embedded into the control glove.

C. Xbee Wireless Module

A communication between the glove and the prosthetic hand were made using a wireless module. Xbee were chosen for this purpose as it uses IEEE 802.15.4 Zigbee protocol. The Zigbee supports multiple network topologies such as point-to-point, point-to-multipoint and mesh networks. It also have low duty cycle thus provides a longer battery life. Zigbee protocol also provide low latency making it a better protocol to use in the project.

F. Ibrahim et al. (eds.), *International Conference for Innovation in Biomedical Engineering and Life Sciences*,
IFMBE Proceedings 56,
DOI: 10.1007/978-981-10-0266-3_33

Based on Table 1, the indoor urban range is up to 30m. In this paper, 10m indoor range were tested. Xbee were chosen due to the usage of lower transmitting power. It's a crucial specification as the power supply for the XBee is limited in order to make the glove lighter. Smaller size of XBee is another good feature compared to Xbee Pro, as it optimizes the space usage and made it possible to be embedded at back of the glove. Since the interaction is between the Xbee transmitter and Xbee receiver only, the Zigbee point to point protocol found to be sufficient enough for the project.

Table 1 Xbee Specification

	Xbee	**Xbee PRO**

	PERFORMANCE	
Indoor Urban Range	Up to 30m	Up to 90m
Transmit Power Output	1mW	63mW
Network Topology	Point-to-point, peer-to-peer, point-to-multipoint	Point-to-point, peer-to-peer, point-to-multipoint
RF Data Rate	250 000 bps	250 000 bps
Receiver Sensitivity	-92dBm	-100dBm
	POWER REQUIREMENTS	
Supply Voltage	2.8 - 3.4V	2.8 – 3.4 V
Transmit Current	45mA	55mA
	GENERAL	
Operating Frequency	ISM 2.4GHz	ISM 2.4 GHz
Dimensions	2.438cm x 2.761cm	2.438cm x 3.294cm
Operating Temperature	-40 to 85° C	-40 to 85° C

D. TowerPro MG995 Servomotor

Servomotors were used to move the fingers on the prosthetic hand. The rotational angle of the motors were depended on the output signals from the glove that were transmitted through Xbee. The weight of each TowerPro MG995 is 55gram and with the dimension of 40.6mm x 19.8mm x 42.9mm (length x width x height). Despite of its dimensions, the torque and speed at 4.8V is 10 kg per cm and 0.20 sec/60° respectively. This torque and speed proves to be enough to move the prosthetic hand's fingers.

III. Methodology

A. Overall Process & Components

The system consists of glove embedded with flex sensor, transmitter module, receiver module and prosthetic hand. Both transmitter and receiver utilizes the Xbee module. Table 2 summarizes the component parts used in the design of the wireless prosthetic hand system.

The user wears the glove, and able to move their fingers freely from far while watching the movement of the prosthetic hand through a screen. The Xbee on the glove functions as a transmitter device and transmit the signal to the other Xbee on the prosthetic hand. The receiver Xbee interprets the transmitted data, and produces an appropriate signal to the circuitry on the prosthetic hand and moves the prosthetic hand accordingly. Fig.1 summarizes the overall process.

Table 2 Components Used in Designing the Wireless Prosthetic Hand System

Control Glove	**Prosthetic Hand**
Arduino Uno	Arduino Uno
XBee Coordinator mode	XBee Router mode
4.5" Flex sensor	MG995 Servomotor
Glove	3D Printed Prosthetic Hand
22k Ohm Resistor	Breadboard
9V Battery Holder	Prosthetic hand housing

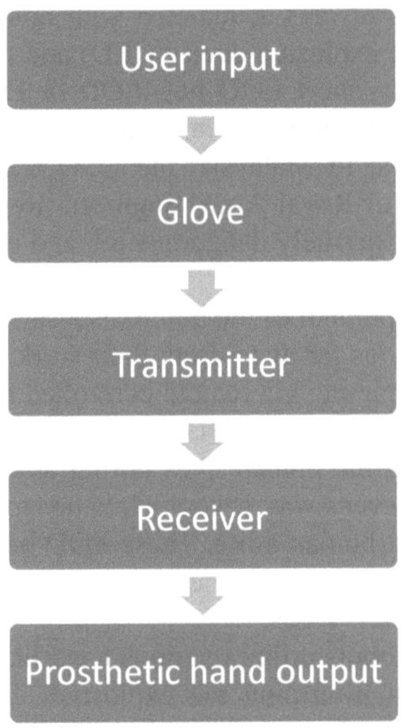

Fig. 1 Overall System Chart

B. Assembling

For the prosthetic hand, it were developed, printed and assembled. The servomotors then, were embedded to the prosthetic hand, and connected to the Arduino. Fig.2 shows the prosthetic hand connections.

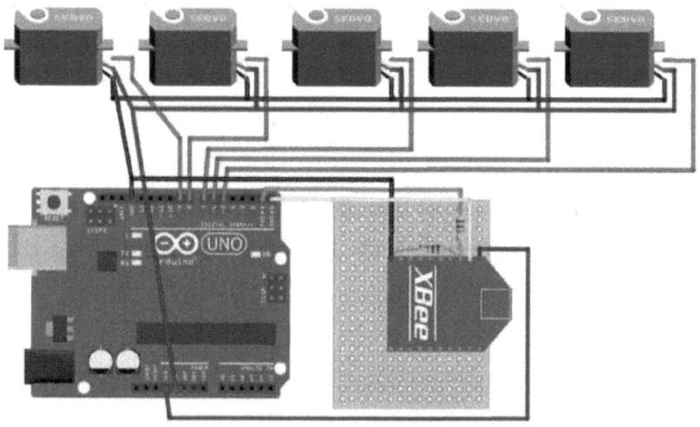

Fig. 2 Prosthetic Hand Schematic Diagram

For the controller glove, the flex sensor were knitted on the back hand side of a glove. The flex sensors then were connected to the Arduino, Xbee and power supply. Fig.3 shows the hand glove's connections.

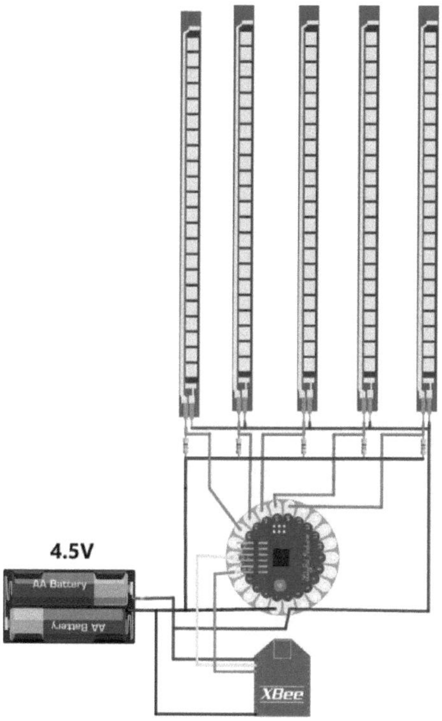

Fig. 3 Glove Schematic Diagram

C. Testing Method

The test run were carried out on the assembled prosthetic hand system by a user sat 10m apart while wearing the hand glove. The communication tests were done to make sure that the prosthetic hand correspond to the control glove. Each prosthetic hand's finger were tested one by one from the 10m distance and, the movement angle of the fingers were tabulated.

IV. RESULTS & DISCUSSIONS

Fig.4 shows the completed glove schematic diagram. The glove consists of five (5) 4.5" flex sensors stitched to it and powered by 9V battery with Arduino uno and Xbee module.

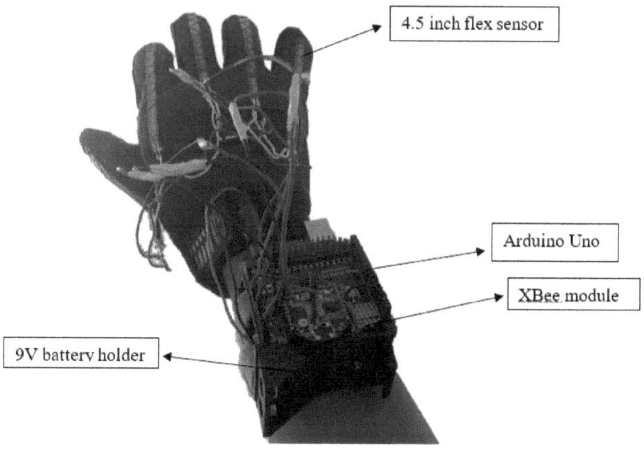

Fig. 4 Glove Schematic Diagram

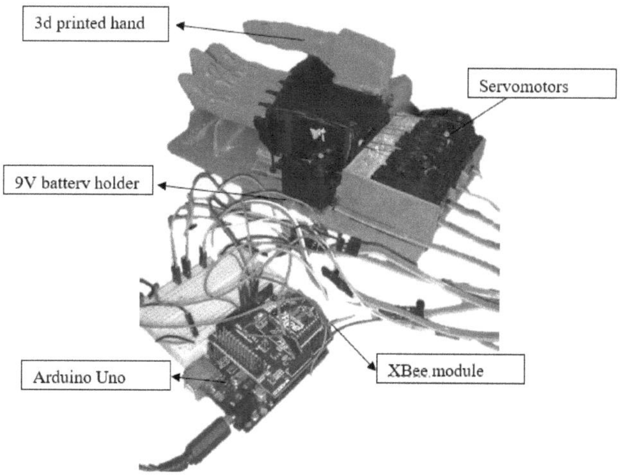

Fig. 5 Prosthetic Hand

Fig.5 shows the assembled prosthetic hand with the servomotors while Fig.6 shows the corresponding of finger and the angular measurement when the glove were used.

Fig. 6 Prosthetic Angular Measurement

Table 3 shows the data that were extracted from the test. The data for the communication part presented firstly and followed by finger movements and the angle of fingers movements. It's found to be satisfactorily and still can be improvised for the middle and index fingers. The rough joints in the prosthetic hand causes the reduction in the both fingers movement angle.

Table 3 Test Results

	YES	NO	COMMENT
	CONNECTION		
Communicate	✓		
Lagging	✓		0.5 sec
	FINGER MOVEMENTS		
1 finger	✓		
2 fingers	✓		
3 fingers	✓		
4 fingers	✓		
5 fingers	✓		
	ANGLE OF MOVEMENT, (°)		
FINGER	**MAX**	**MIN**	**COMMENT**
Pinky	60		
Ring	55		
Middle	5		Rough joints
Index	5		Rough joints
Thumb	34		

For the connection test, based from Table 3, both at distance of 0m and 10m, the glove and prosthetic hand can communicate but the lagging time increased from 0.5sec to 1sec. This is due to the increased range from 0m to 10m.

For the finger movements test, whether at 0m or 10m, the prosthetic hand's finger can directly correspond to the glove movement. Furthermore, the fingers able to move its five fingers simultaneously at 0m and 10m range.

For the maximum angle of movement of the prosthetic's hand fingers, the angle between 0m and 10m distance range did not produce any differences. Based on all the result, it shows that the communication between the control glove and prosthetic hand did have some lagging time but it did not have any effect on the finger test and angle test.

This paper manage to identify the limitations and areas for future improvements. For the wirelessly controlled prosthetic hand to operate, the power consumption need to be reviewed too. Table 4 summarizes the power consumption of each components while Table 5 and Table 6 tabulates the battery life time for the controller gloves and prosthetic hand.

Table 4 Each Components Power Consumption

Component	Voltage, V (V)	Current, I (A)	Power, W (W)	Power Sustained for One Second, P_S (Ws)
Arduino Uno	5	46.5m	0.2325	837
XBee	3.3	0.7m	2.31m	8316
Flex Sensor	5	0.1	0.5	1800
Servomotor	4.8	350m	1.575	5670

Table 5 Control Glove Battery Life Time

Component	Voltage, V (V)	Current, I (A)
Arduino Uno	5	46.5m
XBee	3.3	0.7m
Flex Sensor	5	0.1
Total current, I_T (A)	547.2m	
Minute lasted, L_G (minute)	46.0524	

Table 6 Prosthetic Hand Battery Life Time

Component	Voltage, V (V)	Current, I (A)
Arduino Uno	5	46.5m
XBee	3.3	0.7m
Servomotor	4.8	350m
Total current, I_T (A)	1797.2m	
Minute lasted, L_H	14.022	

From Table 5 and Table 6, the glove's total time it can last, L_G was recorded for 46 minutes while the prosthetic hand's total up time was 14 minutes. This mean that the battery need to be changed every 14 minutes compared to the controller glove that's 46 minutes. This power consumption issue can be improved for a more long lasting usage of the prosthetic hand by replacing for a sustainable energy source or by replacing components with lower power consumptions.

Another limitation with this project is with the weight of the servomotor. The servomotor weighing by 55.2gram each, totalling of 276gram. This weight, when applied on the field, would be a problem for the EOD bot as it has to carry extra weight with it. This can be improvised by using lighter weight external EOD bot material like carbon fibre to compensate the weight of the servomotor.

Apart from the weight issue, accuracy between prosthetic hand and the control glove is another area can be improvised. Even if the hand of the user is fully closed, the prosthetic hand still does not fully close its hand and the angular movement for middle and index finger found to be very deviating. This is due to the rough joints that limits the movements. This is a serious matter for the EOD team as the accuracy play a big role when disarming IEDs. This can be improvised by a smooth design of the prosthetic hand design. Another way of tackling this accuracy issue is by designing the Prosthetic Hand with a closed loop feedback system instead of the current open loop system.

V. Conclusion

The designed wireless prosthetic hand was able to function with some minor glitches on the movement angular. This paper also manage to identify the limitations to the design and suggestions in enhancing it. Since this preliminary design exhibits possibility of the usage prosthetic hand in IED, with the improvised design it can be implemented on EOD bot on consumer level. The same concept can be extended for the medical practitioners' usage, whereby any surgeon could perform surgery using such similar hand via internet protocol and safe life in time without travelling far. Overall the designed wireless flex sensor prosthetic hand have shown a good possibility in for the usage of EOD bot.

Conflict of Interest

The authors declare that they have no conflict of interest.

References

1. Dhillon G.S. and Kenneth W.H., "Direct Neural Sensory Feedback and Control of a Prosthetic Arm," IEEE Transactions on Neural Systems and Rehabilitation Engineering, vol 13 (4), December 2005.
2. Takeda H., Tsujiuchi N., Koizumi T., Kan H., Hirano M., Nakamura Y., "Development of Prosthetic Arm with Pneumatic Prosthetic Hand and Tendon-Driven Wrist," 31st Annual International Conference of the IEEE EMBS Minneapolis, Minnesota, USA, September, 2009.
3. Matsushita K. and Yokoi H., "Robotics Education: Development of Cheap and Creative Emg Prosthetic Applications," IEEE/RSJ International Conference on Intelligent Robots and Systems October, 2009.
4. Gonzalez J., Suzuki H., Natsumi N., Sekine M., Yu W., "Auditory Display as a Prosthetic Hand Sensory Feedback for Reaching and Grasping Tasks," 34th Annual International Conference of the IEEE EMBS San Diego, California USA, September, 2012.
5. Neogi B., Ghosal S., Ghosh S., Bose T.K., Das A., "Dynamic Modeling and Optimizations of Mechanical Prosthetic Arm by Simulation Technique," 1st International Conference on Recent Advances in Information Technology 2012.
6. Blank A., Okamura A.M., and Whitcomb L.L, "Task-Dependent Impedance Improves User Performance with a Virtual Prosthetic Arm," IEEE International Conference on Robotics and Automation Shanghai International Conference Center, May, 2011.
7. Kuchenbecker K.J., Gurari N., and Okamura A.M., "Effects of Visual and Proprioceptive Motion Feedback on Human Control of Targeted Movement," In Proc. ICORR, pp 513–524, June 2007.

Corresponding Author:	Azizi Miskon
Institute:	Department of Electrical and Electronic Engineering, Faculty of Engineering, National Defence University of Malaysia
Street:	Sungai Besi Camp, 57000 Kuala Lumpur
Country:	Malaysia
Email:	azizimiskon@upnm.edu.my

Design and Development of a Lower Limb Exoskeleton for Rehabilitation

Ubaid Ur Rehman, S. Gobee and D. Vickneswari

School of Engineering Asia Pacific University of Technology & Innovation, Kuala Lumpur, Malaysia

Abstract— **This proposed project is a lower limb exoskeleton that is design and develop for rehabilitation while the safety rules and regulation is kept in mind. The developed exoskeleton is made up of purely rehabilitation kits such hip and knee orthoses kits so that users can comfortably wear and do rehabilitation of flexion and extension. The idea is to use inertial measurement unit (IMU) and electromyography (EMG) as a sensor devices where the algorithm is developed to actuate the motors attached at the hip and knee joints from the sensors collected data. The control of the system is done by using Arduino microcontroller where the right lower limb assumed to unsound limb and left lower limb assumed to be sound limb, where from the sensors attached at left sound limb mimic the movement for the right unsound limb.**

Keywords— **Exoskeleton, Lower Limb, Rehabilitation, Inertial Measurement Unit (IMU), Electromyography (EMG).**

I. INTRODUCTION

In everyday routines lower limb plays an important role in day to day physical activities such as flexion and extension, walking motion, sitting and standing position where the joints have significant role. In today's busy and fast life and increasing of senior citizens, people of different age group have numerous amount of paraplegic attack, muscles problem that most of them have less mobility of the leg and rest have certain muscular injuries caused by traffic accidents where it is difficult to move the lower limb [1].

Clinical field of rehab make use of rehabilitation exoskeleton suit which happens to be a special robotic suit that is used to minimize conventional rehabilitation training program and allows physicians to take break from the highly stressful training program exercises. Exoskeleton suits could in fact be activated with the use of patient's movement awareness, enhances the overall self-confidence that patient possesses in relation to rehabilitation by offering them a relaxed and convenient to interact human to machine interface [2]. The exoskeleton suit will be able to assist them in being self-reliant and also fills them with confidence to reintegrate into society.

Pervious study explains it is necessary to understand the function of the exoskeleton, basically exoskeleton is the wearable prototype which has an ergonomic design mechanism that will help the users in movement of limbs [4]. There are different kinds of rehabilitation exoskeleton currently in the market for the rehab requirements [1], [2] and [3].

The development of this project is to aid the rehabilitation process to revive the lower limb by way of flexion and extension by utilizing the Arduino microcontroller to receive feed-back as well as to control electric motors by using inertial measurement unit (IMU) and electromyography (EMG) sensor that is competent at assessing the electrical impulses produced during the muscles activities stated more precisely the flexion and extension of the lower limb. A graphical user interface (GUI) is put together based on the information collected from the EMG sensor to give a real-time biofeedback simulation of a performed flexion and extension exercise.

Furthermore the proposed exoskeleton is to be designs in such a way that it can fit anyone without any assistance of the specialist's physiotherapist which makes it portable and adjustable as well. The flexibility of the leg exoskeletons have number of degree of freedom (DOF), while in this project it makes easy to the rehabilitator by having only 2 DOF that is flexion and extension of the lower limb exoskeleton which is friendly to the patients.

II. METHODOLOGY

The methodology of developing the lower limb exoskeleton is shown in the block diagram in figure 1, where the steps shows that how the lower limb leg can be actuate by DC motor using the signal form the from IMU and EMG muscles sensors. Hence to do the rehabilitation exercise of flexion and extension is shown in the GUI.

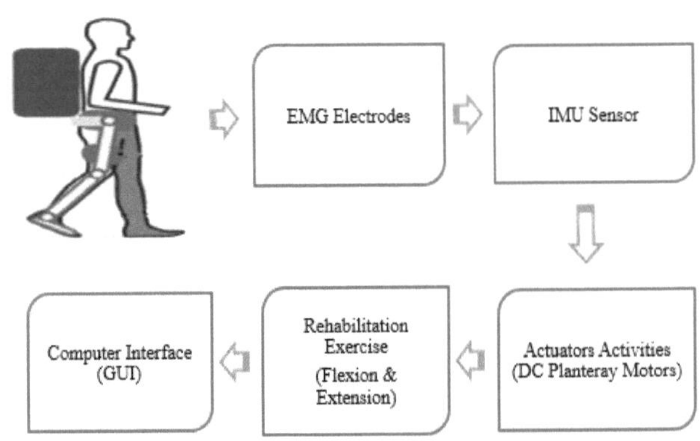

Fig. 1 System Block Diagram

F. Ibrahim et al. (eds.), *International Conference for Innovation in Biomedical Engineering and Life Sciences*,
IFMBE Proceedings 56,
DOI: 10.1007/978-981-10-0266-3_34

A. Design Architecture (Exoskeleton)

First of all the lower limb exoskeleton is only for one leg that is, it attached to the right leg assumed to be unsound limb and the left leg is assumed to be sound limb and. In designing the exoskeleton safety measures also has been taken as per the environment factors, such as the project design architecture is pseudo-anthropomorphic that means the lower limb exoskeleton is similar to the human's, but only includes two degree of freedom that is the walking motion of flexion and extension of the human leg of hip and knee joints.

B. Electromyography (EMG Electrodes)

The electromyography (EMG) sensor which is using the AgCl electrode is placed at the Vastus Medialis areas of the legs that is on the knee thigh area on the sound limb to activate the hip and knee joints actuators respectively on unsound limb. The myo-electric signal is used to control and activate the lower limb exoskeleton.

C. Inertial Measurement Units (IMU)

Inertial measurement units (IMU) sensor consist of three-axial accelerometer, gyroscopes, manometers and thermometers, here the plan is to use 3-axis accelerometer and gyroscope to collect the data to actuate motors such as the system is able to calculate the angles and acceleration ranges of the hip and knee joints so that the maximum and minimum angles will be specified in order to lift the leg for flexion and extension relative to the ground plane. IMU sensors attached at the sound limb area of hip and knee joint to get the accurate angle and hence actuate the motors at the unsound limb.

The accelerometers is working under the principle of output voltage is directly related in proportion to the acceleration, the sensors also capable of finding the exact angle from inclination of the leg to the ground of the exoskeleton.

D. Actuators Activities

The actuators proposed in the project are the DC planetary geared motors with encoders that are placed on the hip and knee joints, so that the hip joint should have more torque which can lift the remaining knee joint motor to activate and hence the walking motion flexion and extension is performed. The motors are controlled through the Arduino microcontroller to receive feedback and to control motors through IMU sensors attached at the sound limb to give angle position as per the normal human being flexion and extension. Whereas the EMG sensor is for the system activation so that when the sound limb do flexion and extension and get the muscles activity thus the system starts. The encoders in the motors will ensure that the motor reaches the same position indicated by the IMU sensors.

E. Rehabilitation Exercise

Lower limb exoskeleton have the potential for rehabilitation of the leg by doing certain exercise, in this project the walking motion exercise is proposed such as the injury, paraplegic, spinal cord injuries and muscles disorders patients can rehabilitate their leg by doing the two degree of freedom exercise of flexion and extension of the limb back to its normal position.

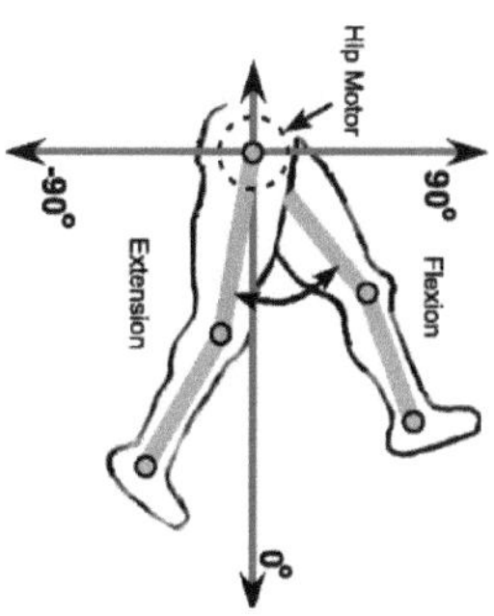

Fig. 2 Angle Orientation at the Lower Limbs [11].

The patients can do the movement exercise in steps, from stretched leg position of unsound limb to the second position of lifting the hip and knee joints simultaneously. Whereas in the project hip joint flexion is limits to 45 degree and coming to the rest position that is the 0 degree to the ground surface as shown in figure 2. Whereas the knee joint flexion is 0 degree and extension is assumed to be -45 degree going backwards while the hip joint is on flexion position. Thus do the normal flexion and extension exercise for lower limb rehabilitation.

F. Computer Interface (GUI)

To track the movement of exoskeleton, a GUI design is done using LabVIEW software, the GUI will simulate the real time lower limb exoskeleton movement as per the users, shows the flexion and extension movement position. The 3D model of exoskeleton is export to 3D picture control in LabVIEW; secondly there will be a graph as well to show the EMG waveform activities.

The GUI will also show the muscles performance and strength including the angles of IMU done by the user exercise, hence the GUI also shows the duration of the exercise and the number of counts done for each walking motion of flexion and extension.

G. Design and System Implmentation

a) Design 3D model of the exoskeleton is shown in figure 3, the exoskeleton is designed in such a way that the normal human body width and height is kept in mind. The lower limb exoskeleton is consists into the two main parts that is using the rehabilitation hip kit and knee thigh kit, hip kit combined placed at both hip and thigh part are connected where the motor is fixed at the hip joint movement for the flexion and extension making overall two degree of freedom. Aluminum strips are used in the design for the connection on the kits to the motors because there will be a need of some strong material as it will help the motor to leg movement.

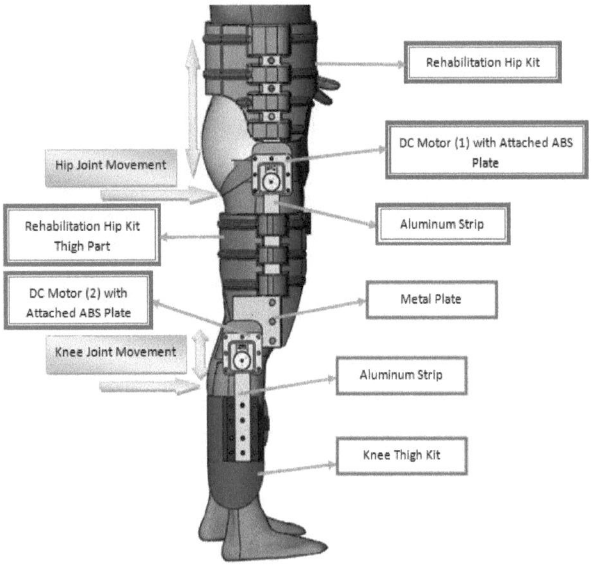

Fig. 3 3D Model of Exoskeleton

b) Prototype The prototype of the lower limb exoskeleton for rehabilitation is built with proper medical rehabilitation kits as shown in figure 4, as the safety of the users comes first as compared to others such as aluminum, carbon fiber, and other materials which can possibly have the skin disease to the users. The DC planetary motors are fixed in a lateral position to the human body to ground plane thus the motors are in distance to the rehab kits. The prototype is built in such a way that it can easily assembled and disassembled from the knee joint therefore it is portable. Moreover the hip joint to the knee joint is adjustable to the limb to enables the exoskeleton to suites more rehabilitators.

c) Walking Algorithm The propose project is the lower limb exoskeleton is where the system is driven by sensors and the motors at both hip and knee joint acts as an actuators. There are total two IMU and one EMG sensors are used whereas all the sensors are placed at left sound limb and the motors are attached to the right unsound limb.

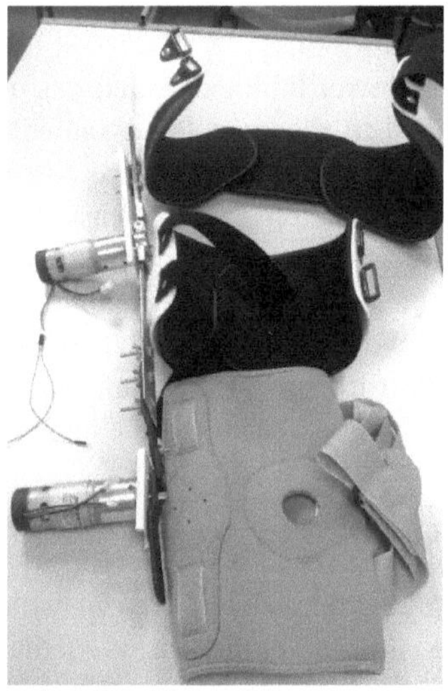

Fig. 4 Lower Limb Exoskeleton Prototype

The methodology, the electrodes are placed on the left sound limb to get the muscles activities through the electrodes placed on the knee thigh of the sound limb so that when the patient lift the left sound limb, hence both the DC motors will activate for right unsound limb where the exoskeleton is attached in the sequence of first hip motor will activate after that the knee motor. It's basically mimicking the movement of the left sound limb to the right unsound limb to rehabilitate the leg.

There are two main position or parts to complete the one cycle of exercise, one is lifting the hip joint part and extension of the knee joint after that the second part reaches to the ground where the hip is of extension position and knee is on flexion position, hence complete the walking motion of the exoskeleton for rehabilitation.

III. Results and Discussion

As the data shows it is hip joint angle to be maximum 45 degree at flexion position while on normal extension position of the lower limb is to be 0 degree. The value from the graph in figure 5 is compared the angles of both left sound and right unsound limb from the motor encoder count, it shows that when the exoskeleton moves that is mimic the IMU angle on left sound limb gives 5 degree difference of actual angle of 20 degree from the IMU as shown in Series 1, thus when do the flexion at hip joint is moves only 15 degree as shown in Series 2. The steps are repeated in the interval of 5 degree and note down the exoskeleton angle real time. At the actual angle of 35 degree the sequence was to properly execute and thus gives 24 degree from unsound limb thus gives the difference of 11 degree.

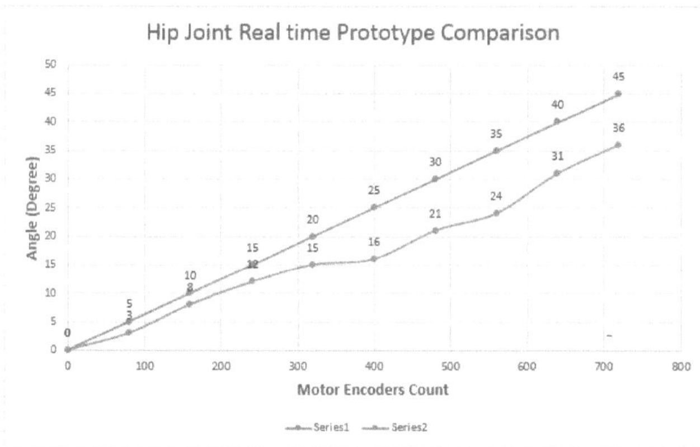

Fig. 5 Graph of IMU Angle with Motor Encoder

IV. CONCLUSIONS

In summary, a simplified and easily assembled and dissembled lower limb exoskeleton for rehabilitation is designed and developed with having adjustable limb that fits to any human body. The exoskeleton for lower limb is successfully developed fully by rehabilitation medical hip and knee orthoses kit for the patients such as muscles disorder problems at lower limb, injuries and for paraplegic attacks, which wants to rehabilitate by doing simple exercise of flexion and extension. Furthermore the IMU's and EMG sensor is attached to the left sound limb and provides data successfully.

The right unsound limb where the medical orthoses kits are attached to the exoskeleton actuate the motors successfully at both hip and knee joint that is mimic the walking motion movements according to the algorithm applied. Relationship of EMG has to be consider to actuate the motors at hip and knee joint in the future to improve the exoskeleton walking motion.

ACKNOWLEDGMENT

I would like to acknowledge APCORE (Asia Pacific Center of Robotic Engineering) for the support in developing the exoskeleton. Also I like to thank Asia Pacific University of Technology and Innovation for funding the research.

CONFLICT OF INTEREST

There is no conflict of interest between the authors or the organization.

REFERENCES

1. Tadej, et al. (2011) Development and Validation of a wearable Intertial mesaurement system for use with lower limb Exoskeleton. In *IEEE-RAS International Conference on Humanoid Robots.* Slovenia , 26th to 28th October 2011. pp. 212-217.
2. Manuel C., Daniel S., Juan C., & Elena G. (2015) An adjustable Compliant Joint for Lower-Limb Exoskeletons. *IEEE Transaction on Mechatronics.* 20(2). p.889-898.
3. Yugan, (2014) Design and Development of Lower Limb Exoskeleton for Rehabilitation. In *International Conference on Biomedical Engineering.* Singapore, 2014. Springer International Publishing Singapore.
4. Renquan Lu., Chun., & Anke. (2014) Development and Learning Control of a Human Limb with a Rehabilitation Exoskeleton. *IEEE Transaction of Industrical Electronics.*61(7). p.776-785.

Author:	Suresh Gobee
Institute:	Asia Pacific University of Technology & Innovation (APU)
Street:	TPM, Bukit Jalil
City:	Kuala Lumpur
Country:	Malaysia
Email:	suresh.gobee@apu.edu.my

Synthetic Biology in Healthcare and Conservation: II. Successful Formulation of a Synthetic Spermatozoa Cryopreservation Medium

J. Ali

Department of Obstetrics and Gynaecology, Faculty of Medicine, c/- KKWK, University of Malaya Medical Center, University of Malaya

Abstract— **Present-day Embryo Culture (ECM), Cell, Gamete and Embryo Cryoprotectant (CM), Stem Cell Culture (SCCM), Cell-based Vaccine Production Media (VPM), etc contain donor serum proteins (DSPs) which carry risk of disease transmission to patients/their babies/healthcare workers. The dependence on DSPs proved difficult to overcome. The European Union recommends avoidance of non-uniform biologicals in healthcare products (EU Tissue Directive No.2004/23/EU) by April 2007. However to-date most manufacturers of healthcare products have not fully complied with this directive. The pioneering research of this author in Australia led to development of synthetic human embryo culture media devoid of DSPs. A clinical trial was performed successfully [Ali, 1997, 2004; Ali et al., 2000] and patented in USA (US Patent 8415094) / PCT protected in Canada/ EU/ Australia/ Russia/ Israel/ worldwide and licensed to Cellcura ASA Norway (www.cellcura.com). The present communication reports development of synthetic spermatozoa cryopreservation medium (SCM). The nature of this intellectual property is proprietary. Patent application is pending. Therefore it shall not be described in detail here but events leading to its development will be presented. In the SCM the mean% +1SD spermatozoa motility pre- and post- freeze-thaw was similar 55.7+17.4 vs 54.9+13.3; n=10 (p>0.05). In previous studies after freeze-thaw using DSP-containing CM the pre and post-thaw spermatozoa motility was significantly different 45+11 Vs 23+12, p=0.047; n=64 that showed a loss of 22% in motility (p=0.047). A prospective study found SCM statistically similar to DSP containing-CM. Proof of principle was demonstrated following human pregnancy after generation of human embryos by intracytoplasmic spermatozoa injection (ICSI) using SCM frozen-thawed testicular spermatozoa. An efficacious SCM for human, possibly mammalian spermatozoa was formulated that may be regulation-compliant, and very safe. The availability of SCM eliminates batch variation and potential for disease transmission in routine spermatozoa cryopreservation in medicine, meat and dairy industries, and species conservation.**

Keywords— **cryoprotectant, medium, protein-free, spermatozoa, synthetic.**

I. INTRODUCTION

Present day culture media for cells, tissues and embryos, and cell-based vaccine production contain donor serum proteins (DSPs) from human or animal donors (human serum albumin (HSA), bovine serum albumin (BSA), foetal calf serum (FCS), maternal serum, cord serum), including the use of xeno feeder cells such as mouse fetal fibroblast cells in the generation of human stem cells, etc. [1-2] carry risk of disease transmission including immunogenic reactions [3]. Stringent purification and sterilization measures cannot exclude with absolute certainty the possibility of transmission of unknown pathogens [4]. It is well recognized prions of BSE /vCJD cannot be destroyed by extreme heat. It is well documented haemophiliacs have been infected with AIDS through blood-derived products. CJD has likewise been transmitted through tissue-derived products and disease transmitted through vaccines [5-7]. The European Union Cell and Tissue Directive No.2004/23/EU [8] came into force in April 2007 that asked member states to move away from healthcare products that contained non-uniform biological BUT DSPs (and mouse feeder cells in stem cell culture), are still used in spite of the risk of disease transmission. It has been reported that almost all existing stem cell lines are contaminated with mouse proteins making them unsuitable for human application [3]. As in cell, embryo and vaccine- production culture media DSPs are also present in cryoprotectant medium (CM) used for the cryopreservation of cells, gametes and embryos in medicine, and species conservation. Diseases can likewise be transmitted to the recipients if the recipients' spermatozoa were cryopreserved in cryoprotectants containing DSPs. The present invention relates to the development of synthetic protein-free spermatozoa CM (SCM) specialized and optimized for human assisted reproduction technology (ART) programs. The usage of protein-free media products will be useful in the prevention of the transmission of protein-bound pathogenic agents to patients undergoing infertility treatment, to newborns conceived from ART and to healthcare workers. The SCM is anticipated to be useful in species conservation and, meat and dairy industries. The objective of the present investigation was to and has successfully developed for the first time a synthetic SCM devoid of DSPs that appears to have overcome the mandatory need for DSPs in CMs that could not be overcome for decades.

F. Ibrahim et al. (eds.), *International Conference for Innovation in Biomedical Engineering and Life Sciences*,
IFMBE Proceedings 56,
DOI: 10.1007/978-981-10-0266-3_35

II. Materials and Methods

A. Spermatozoa Freezing and Thawing

Spermatozoa were equilibrated with SCM according to the protocol developed. The nature and freezing protocol are proprietary, and is currently being considered for patent application. The equilibrated semen was apportioned into Nunc (NUNC, Denmark) 1.8ml cryo vials and frozen according to standard methods in which the vials were held at 14cm above surface of liquid nitrogen in a liquid nitrogen (LN2) tank for 20mins and then at 7cm also for 20mins. At the end of this period the vials were plunged into liquid nitrogen and cryostored at -196°C in LN2 in Dewar specimen tanks for a period of 14 to 60 days (MVE, UK). The cryovials were thawed gently at room temperature. Measured volume of the thawed contents of the cryo vial were transferred into a 15-ml centrifuge tube (Falcon Plastics, USA). The thawed specimen is mixed with 3ml of spermatozoa medium (Origio, Denmark). The suspension was centrifuged at 2000rpm for 5 mins. The supernatant was discarded and the pellet re-suspended in spermatozoa medium and again centrifuged. The supernatant is discarded and the pellet made up to the original volume of the thawed specimen with spermatozoa medium. The pellet is mixed and warmed for about 10mins at 37°C. The motility of the thawed specimen was measured using a Makler chamber as per standard methods.

B. Freezing Human Testicular Spermatozoa

Testicularspermatozoa was extracted as per established protocol following the patient's request. Thirteen oocytes were recovered from the patient's wife. All 13 oocytes were injected with the fresh testicular spermatozoa. Of these, 12 oocytes fertilized (92.3%). The wife became clinically pregnant after transfer of two embryos but later aborted. The remaining testicular spermatozoa were equilibrated in the SCM and apportioned into 4 cryo vials and subsequently frozen-stored as described above. About 1 year and two months later the couple returned for ART treatment. Ten oocytes were recovered from the wife and injected with SCM frozen-thawed testicular spermatozoa.

III. Results

A. Spermatozoa Freezing and Thawing

The mean% +1SD spermatozoa motility pre- and post-freeze-thaw was similar 55.7+17.4 vs 54.9+13.3; n=10 (p>0.05) respectively with a surprisingly very low loss of 0.8%.

C. Freezing Human Testicular Spermatozoa

All ten oocytes (100%) were fertilized after injection with SCM frozen-thawed testicular spermatozoa. Two of the embryos generated were transferred to the patient which resulted in a pregnancy that was on-going at the time this author returned to his home country.

IV. Discussion

For three decades healthcare providers in ART had to use DSP- containing embryo culture media and CMs for spermatozoa, oocyte and embryo cryopreservation in spite of risk of disease transmission [1,2] due to lack of alternatives. Most efforts to develop a synthetic alternative and claims of successful development of chemically defined protein-free medium (PFM) for human application were in fact not truly protein-free, because the sperm meant for fertilization was still prepared in medium that contained DSPs [9-11]. An efficacious and totally synthetic PFM for ART procedures was described by the author in 1997 and communicated in subsequent years [12-15] and patented in USA in 2007 (US 8415094 B2 [16]; PCT protected worldwide including Canada, EU, Israel Australia, Russia, Norway, Singapore, etc, and licensed (www.cellcura.com). The present study is similar to the PFM work directed at developing a synthetic cryopreservation system for spermatozoa. The loss of motility of frozen-thawed spermatozoa in CM using DSP [17,18] was significant. The loss of motility or viability in conventional spermatozoa freezing has been extensively reviewed [17] where the loss of spermatozoa motility following frozen-storage appears to be a common problem in DSP-containing CMs. In a previous study by Menkveldt cited by Brown [18] the drop in spermatozoa motility after freeze-thaw using DSP-containing CM was significant [45+11 Vs 23+12, (p=0.047; n=64] which showed a loss of 22% in motility. On the contrary there is insignificant loss of only 0.8% motility when SCM was used to cryopreserve spermatozoa indicating synthetic CMs devoid of DSP are capable of efficacious cryopreservation of human spermatozoa and that the past challenges faced in overcoming the need for DSPs in CMs is now put to rest. More so after proof of principle was demonstrated following pregnancy from human embryos generated after ICSI using SCM frozen-thawed testicular spermatozoa. A recent study compared the efficacy of SCMs with DSP-containing CMs showed SCM to be statistically similar as DSP-containing CM (Ata'Allah et al., personal communication, 2015).

V. Conclusions

The present investigation has successfully developed an efficacious synthetic spermatozoa cryopreservation medium that overcame past challenges and hurdles associated with the mandatory need to include hazardous DSPs in spermatozoa cryopreservation medium. We now have a complete set of PFM for both embryo generation and spermatozoa cryopreservation.

Conflict of Interest

The author has financial interest in the intellectual property described herein as SCM.

Statement of Informed Consent

Patients have given their informed consent for treatment and procedures performed during this study

References

1. van Os HC, Drogendijk AC, Fetter WP, Heijtink RA, Zeilmaker GH. The influence of contamination of culture medium with hepatitis B virus on the outcome of in vitro fertilization pregnancies. Am J Obstet Gynecol. 1991 Jul;165(1):152-9.
2. Kemmann, E. (1998) Creutzfeldt-Jakob disease (CJD) and assisted reproductive technology (ART). Hum Reprod. 13, 1777
3. Martin MJ, Muotri A, Gage F, Varki A. Human embryonic stem cells express an immunogenic nonhuman sialic acid. Nat Med. 2005;11(2):228-32.
4. Truyen, U., Parrish, C.R., Harder, T.C. et al. (1995) There is nothing permanent except change. The emergence of new viral diseases. Vet. Microbiol. 43, 103-122.
5. Marwick C. FDA calls bovine-based vaccines currently safe. JAMA September 13, 2000. www.jama.ama-assn.org/issues/v284n10/ffull/jmn09133 .html
6. Mercola J. U.K. recalls polio vaccine over 'Mad Cow' fears. October 29, 2000.www.mercola.com/2000/oct/29/polio_vaccine_recall.htm,151].
7. Merten, O.W., *Safety for vaccine(s). 2000.* (19003393) Cytotechnology. 34 (3): p. 181-3.
8. EU Tissue Directive. http://eur-lex.europa.eu/LexUriServ/site/en/oj/2004/l_102/l_10220040407en00480058.pdf 2004/23/EC
9. Caro, C. M. and Trounson, A. (1986) Successful Fertilization, Embryo Development, and Pregnancy in Human in Vitro Fertilization (IVF) Using a Chemically Defined Culture Medium Containing No Protein. J In Vitro Fertil Embryo Transf. 3(4): 215-217
10. Serta, R. S., Sakellariou, M., Kiessling, A. A. et al. (1997) Outcome of human embryos conceived and cleaved in protein-free culture conditions. Proceedings of the 53rd Annual Meeting of the ASRM. October 1997.
11. Parinaud, J., Milhet, P., Vieitez, G. and Richoilley, G. (1999). Use of a medium devoid of any human or animal compound (SMART2I) for embryo culture in intracytoplasmic sperm injection. J. Assist. Reprod. Genet. 16: 13-16.
12. Ali J (1997) Formulation of a protein-free culture system for the culture of human embryos: preliminary findings and pregnancies. In: Abstract book of the 16th Annual Scientific meeting of the Fertility Society of Australia, 2-4 Dec. 1997, Adelaide, Australia, p 34
13. Ali J. (2000) Investigation into the nutrient requirement of the human embryos: Successful formulation and clinical trial of a novel protein-free embryo culture medium. Emirates Med J. 18:195-202
14. Ali, J (2004) Generation of viable human embryos in a protein-free embryo culture (ART-7b) medium enhances clinical pregnancy rate and prevents disease transmission in assisted reproduction. MEFS J. 9:118-127
15. Ali J, Shahata MA, Al-Natsha SD. Formulation of a protein-free medium for human assisted reproduction.Hum Reprod. 2000 Jan;15(1):145-56.
16. USA Patent no. US8415094B2. Inventor: Ali Jaffar bin M Abdullah Protein-free gamete and embryo handling and culture media products.
17. Morris JG, Acton E, Murray BJ, Fonseca F. Freezing injury: the special case of the sperm cell. Cryobiology. 2012;64(2):71-80.
18. Brown S. Men need fertility preservation too. ESHRE Newsletter Pg.15, January 2012.

Author: Jaffar Ali
Institution: Dept of Obstet.Gynaecol, University of Malaya
Street: Jalan Universiti, 59100 Kuala Lumpur
City: Kuala Lumpur
Country: Malaysia
Email: jaffarali@um.edu.my

Numerical Modeling of Blood Flow in Patient-Specific Right Ventricle with Pulmonary Arterial Hypertension Based on MRI

B. Su[1], R.S. Tan[1,2], J.L. Tan[1], K.W.Q. Guo[1], X. Zhao[1], S. Leng[1], J.M. Zhang[1,2], and L. Zhong[1,2]

[1] National Heart Centre Singapore, Singapore
[2] Duke-NUS Medical Graduate School, Singapore

Abstract— **Right ventricle (RV) transposes deoxygenated oxygen to the lungs at much lower pressure but with similar volume as the left ventricle (LV). Unlike the elliptical LV, the RV is funnel-shaped with crescent cross section in the short-axis direction. In part due to its complex anatomy, the studies on the ventricular flow are quite limited, as compared with LV. In addition, RV dilation is expected, when pulmonary arterial hypertension (PAH) is present, and the alternation of ventricular flow has not been studied to the best of our knowledge. In this study, a normal subject and a PAH patient were selected for MRI scans, based on which the time-resolved RV geometries were reconstructed. To model the patient-specific ventricular flow, Computational Fluid Dynamics (CFD) was utilized with dynamic mesh method enabled. The simulation results showed that the ventricular flow in both normal and PAH RVs were smoother than that in LV, especially at the mid and apical regions. The vortex structure indicated that deeper penetration of transtricuspid flow into RV in the normal subject.**

Keywords— **Pulmonary Hypertension, Right Ventricle and Computational Fluid Dynamics.**

I. Introduction

The right ventricle (RV) is funnel-shaped and its inlet and outlet tracks are relatively widely separated by the aorta. It is widely believed that the interaction between the ventricular flow and the myocardium plays a role in the development of certain cardiac dysfunction, though no concrete conclusion has been stated yet. In part due to the complex anatomy of RV, the studies on its internal flow structure is much limited, as compared with the LV.

Direct (i.e. in-vitro) measurement of the ventricular flow is the most desired approach to study the hemodynamics. Using phase contrast magnetic resonance imaging (PC-MRI), Kilner and the coworkers revealed the flow patterns on one slice of the three-dimensional (3D) ventricular flow though both atriums and left ventricle [1]. However, the complex 3D flow structure in the RV could not be depicted solely based on one two-dimensional (2D) planar velocity map, and it was omitted in the article. 4D MRI is the cutting edge tool to visualize the 3D flow evolution over cardiac cycles by combining 3D spatial encoding and 3D velocity-encoded phase contrast method [2]. The videos in [3] demonstrated that the dominant flow feature is a smooth path from the tricuspid to the pulmonary valve with asymmetric vortex ring in the proximity of tricuspid orifice during early and late diastole. In addition, the apical flow has low velocity magnitude. In-vitro measurement is anther manner to have a better understanding of the right ventricular flow, though the RV model is not patient-specific. With the aid of multi-planar image velocemtry, Falahatpisheh and the coworkers summarized that the vortex structure formed during the early filling phase evolved into a single-leg vortex extending toward the pulmonary valve [4]. Computational fluid dynamics (CFD) has been widely adopted for the modeling of LV, and the simulation could be patient-specific with the aid of imaging tools. Mangual and the coworkers utilized the Image ArenaTM software (TomTec Imaging System GmbH, Munich, Germany) to reconstruct the RV geometries from 3D echocardiography [5]. Based on the numerical simulation, they observed a compact vortex structure after the blood passed the tricuspid valve during the rapid filling phase and some irregular vortex structure formed during the atrial contraction phase. Computed tomography (CT) was employed for geometry reconstruction in [6], however, the study was more focused on the LV without elaborating RV flow.

Pulmonary arterial hypertension (PAH) increases the workload of RV and the RV dilatation is observed as a compensation to maintain stroke volume when its contractility is impaired. The alternation of the RV will certainly affect the internal flow field. The primary purpose of this study was to visualize the differences between normal and PAH RVs in terms of flow field using CFD.

F. Ibrahim et al. (eds.), *International Conference for Innovation in Biomedical Engineering and Life Sciences*,
IFMBE Proceedings 56,
DOI: 10.1007/978-981-10-0266-3_36

II. Methods

A. Geometry Reconstruction

The MRI scans were carried out on a Philips 3.0 Tesla system (Philips Healthcare, Netherland). A number of parallel slices of the heart images were acquired in the short-axis direction and the raw images in Digital Imaging and Communications in Medicine (DICOM) format contains the additional information for rending in 3D as demonstrated in Fig. 1a. The contours of endocardium were drawn manually on each slice and the resultant 3D contours (black) were shown in Fig. 1b. The funnel-shaped right ventricle led to contour branching at the basal region and thus a ridge (red) was introduced between the single-contour and double-contour region as indicated in Fig. 1b. Based on the contours and the ridge, the right ventricle was reconstructed as shown in Fig. 1c.

B. Mesh Generation

The reconstructed geometry was meshed using ANSYS ICEMCFD (Version 14.0) and the numerical model had around 1.2 and 1.5 million tetrahedral volumetric cells in the normal and PAH case at the end of diastole, respectively. The temporal resolution of the MRI data was 30 phases per cardiac cycle, and it was insufficient for CFD simulation. Therefore, cubic-spline interpolation was applied to increase the temporal resolution, leading to smooth ventricular deformation. Note that the surface mesh at each time instance had identical number of nodes and connectivity to model the deformation. As shown in Fig. 2, the volume varied significantly from the end of diastole and the end of systole, and the smooth method alone was incapable of tackling large deformation. As a result, remeshing was applied to improve the deteriorated grids.

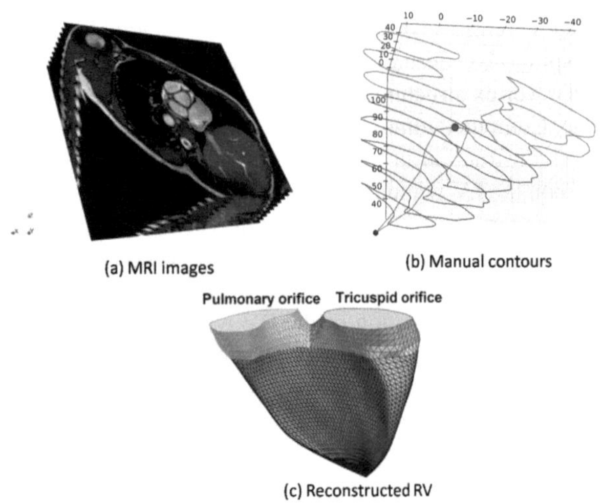

Fig. 1: Geometry reconstruction process.

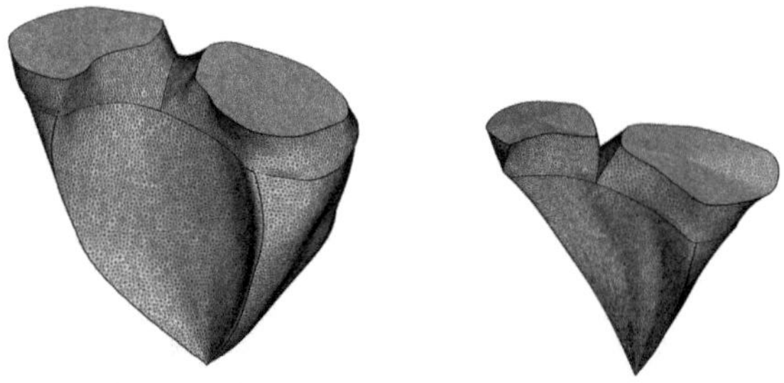
Fig. 2: Geometry reconstruction process.

C. Numerical Modeling

The internal flow of the numerical model was simulated using the commercial CFD solver ANSYS FLUENT (Version 14.0), which utilizes the finite volume methods in arbitrary Lagrangian-Eulerian (ALE) formulation of the Navier-Stokes equations to solve the motion of deforming ventricle. Blood flow was assumed to be laminar and Newtonian with constant dynamic viscosity of 3.5 mPa·s and density of 1,050 kg/m^3. For the ease of numerical modeling, it was assumed that the tricuspid and pulmonary valves were completely closed during the systole and diastole, respectively. Therefore, the flow rate through the corresponding orifices was derived from temporal variation of ventricular volume, according to the principle and mass conservation. The closed valve was represented by wall boundary condition and pressure inlet or outlet boundary was applied to the other opening accordingly.

III. Results and Discussion

Fig. 3 shows the volume profiles of the normal and PAH RV during one cardiac cycle. The unexpected fluctuation of RV volume based on MRI images (black cross line) is observed, due to errors induced during MRI scan processing and the manual segmentation. In order to avoid this issue and resultant unrealistic ventricular flow, only 10 phases were selected in this study (i.e. one in every three consecutive frames) and the simulations were based the geometries reconstructed based on these 10 frames. Therefore, the profiles in both cases are smooth in this study. The normal subject had a typical volume profile, which included a early filling phase, diastasis and atrial contraction after the strong systole.

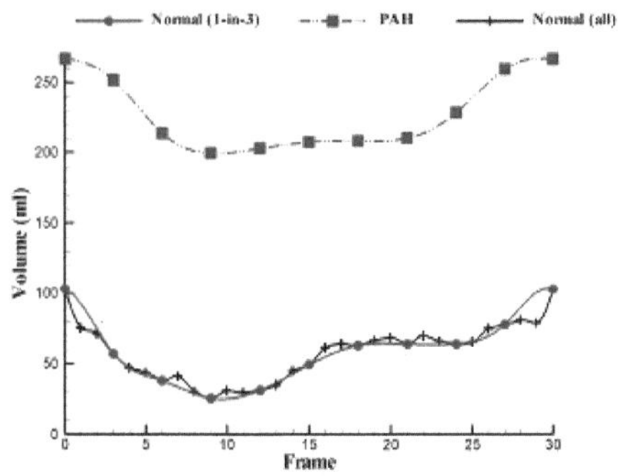

Fig. 3: Volume profiles in the normal and PAH cases in a cycle.

Although similar trend is observed in PAH patient, the RV volume was significantly enlarged by the hypertension. The RV volume varied between 25 ml and 103 ml in the normal RV, leading to the ejection fraction (EF = 1 - VES/VED) of 75%. The corresponding stroke volume was 78 ml. In the PAH patient, the VED and VES were 266 ml and 199 ml, respectively, and the EF was only 25%.

The development of the RV flows in normal and PAH were shown in Fig. 4 and the vector were colored based on its magnitude. During the systole, the tricuspid valve was closed and the blood flowed out of RV through pulmonary valve. Near the end of systole, the flow velocity through pulmonary orifice reduced. The blood was accelerated as it flowed from rear region to the pulmonary orifice, where high velocity flow was observed. During the diastole, the pulmonary valve was closed, thus the blood with high momentum generated during systole was forced to redirect, forming a clockwise swirling flow during the early stage of diastole. The absence of actual tricuspid annulus and the converging duct shaped RV in its proximity retarded the flow separation. Consequently, no conspicuous swirling flow could be observed from the velocity vectors. In this study, the inlet and outlet were assumed as tricuspid and pulmonary valve, respectively. The tricuspid orifice was considerably larger than the pulmonary orifice and resulted in low transtricuspid flow. The vortex structure near the pulmonary valve, which was initiated during systole, finally dissipated during the atrial contraction. Over the entire cardiac cycle, the apical flow was slow and smooth, because the crescent RV suppressed the development of irregular flow at this region.

Fig. 5 shows the evolution of ventricular flow in PAH patient at the same instances as the normal subject for the ease of comparison. Similarly, the dominant flow feature in the PAH was the strong out flow jet through the pulmonary orifice. The dilated outflow track led to slower transpulmonary flow, as compared with the normal subject. For instance, the maximum flow velocity in PAH was 0.23 m/s, while that

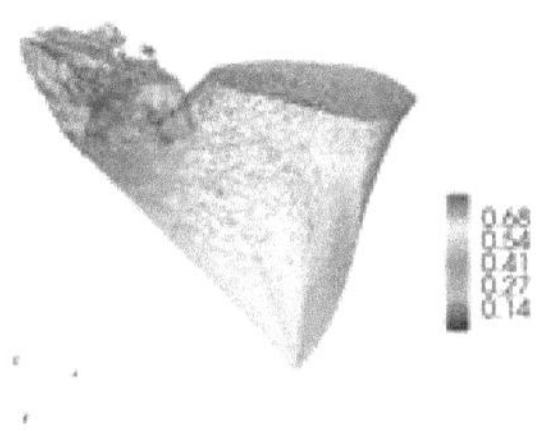

(a) T*=0.12

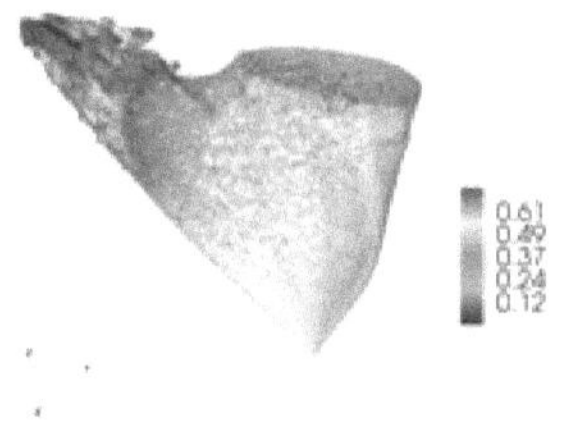

(b) T*=0.20

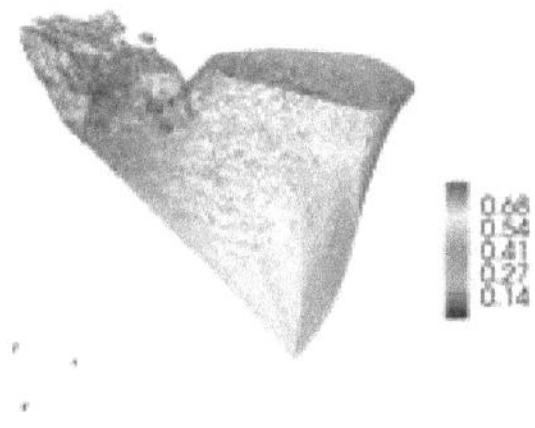

(c) T*=0.28

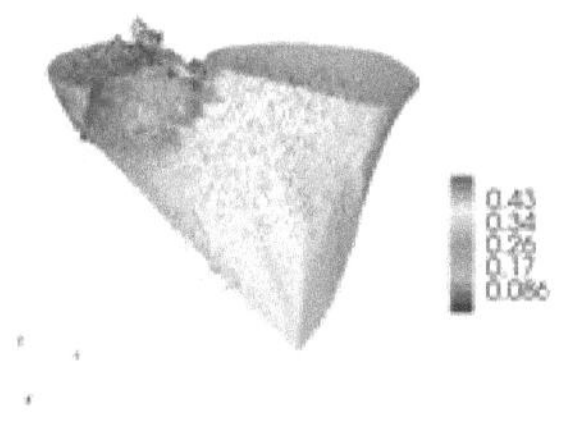

(d) T*=0.32

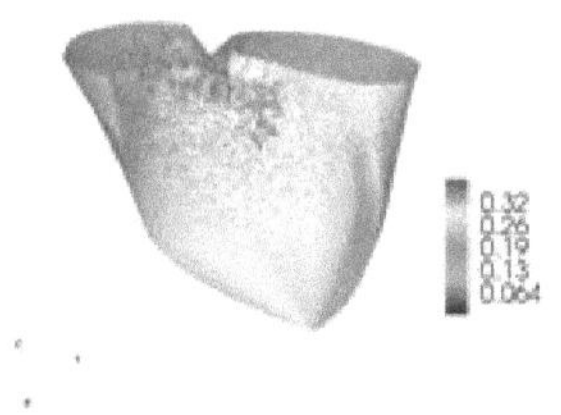

(e) T*=0.68

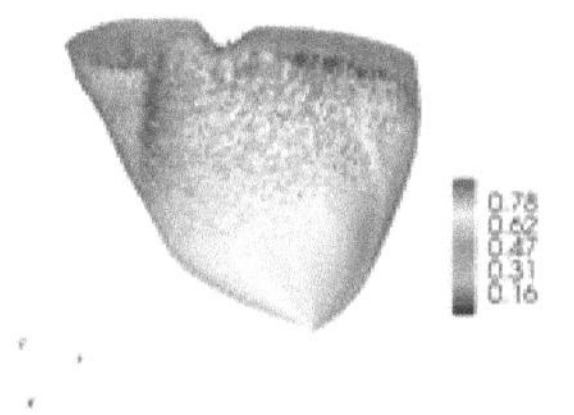

(f) T*=0.96

Fig. 4: Development of intraventricular flow in normal RV during systole (a-c) and diastole (d-f).

in the normal subject was 0.68 m/s. Therefore, the strong swirling flow observed in the normal subject was not present in PAH patient at the early stage of diastole (Figure 5d). With respect to the tricuspid valve, the dilated RV in the proximity region was like diverging duck, and it promoted the flow separation, leading to swirling flow near the orifice.

Since the dominant flow feature during systole was relative simple, the study of the vortex structure was mostly focused on the diastolic phase. As shown in Figs 4 and 5, the flow pattern was basically smooth, thus the magnitude of the vortex iso-surface was small (i.e. 300 $1/s^2$) to illustrate the delicate flow structure near the end of diastole as shown in Fig. 6. In the normal subject, the main vortex structure was arc-shaped with respect to the tricuspid orifice and stretched from the core to the pulmonary orifice along the outflow track. In the PAH patient, vortex structure was observed beneath the tricuspid orifice, which was consistent with the swirling flow in Figure 5f. Compared with the left ventricular in [7–9], the penetration of transvalvular flow into

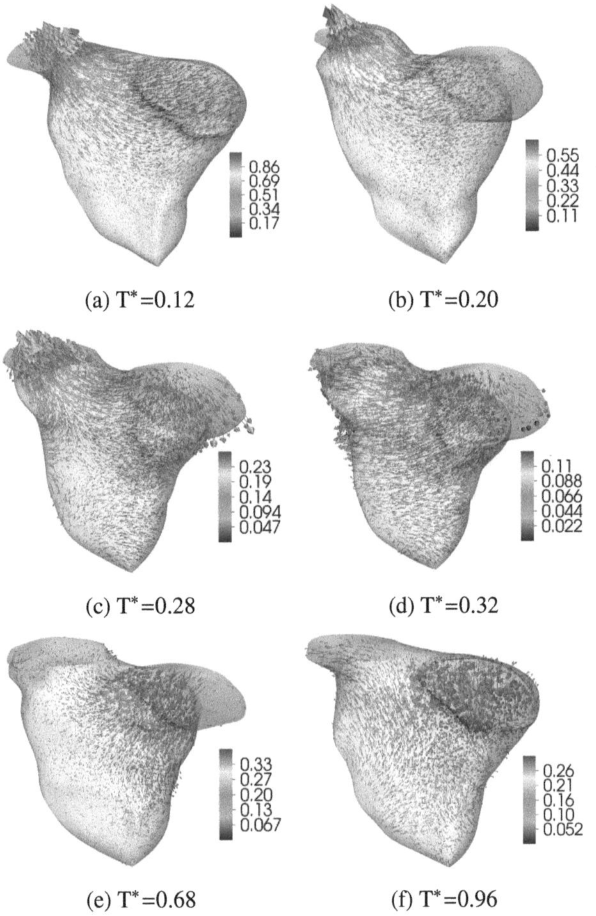

Fig. 5: Development of intraventricular flow in PAH RV during systole (a-c) and diastole (d-f).

ventricle was much shorted, owing to the lower flow velocity, and this phenomenon was also observed in [10].

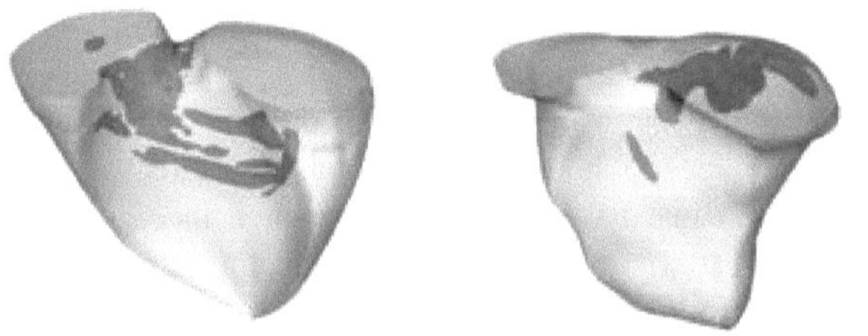

Fig. 6: Vortex structure near ED.

IV. Conclusion

In this study, MRI images were utilized to reconstruct time-resolved RV geometries in normal subjection and PAH patient. To facilitate the mesh generation, the geometries at each frame had the identical topology (i.e. the number of surface nodes and connectivity). In terms of flow field, the blood flows in the mid and apical regions of RV were smooth even when RV was dilated due to PAH. The vortex structure indicated deeper penetration of transvalvular flow into RV in normal subjection.

Conflict of Interest

The authors declare that they have no conflict of interest.

Acknowledgment

The authors would like to acknowledge Goh Cardiovascular Research Award (Duke-NUSGCR/2013/0009) for funding this research.

References

1. Kilner PJ, Yang GZ, Wilkes a J, Mohiaddin RH, Firmin DN, Yacoub MH (2000) Asymmetric redirection of flow through the heart. Nature 404:759-761
2. Markl M, Frydrychowicz A, Kozerke S, Hope M, Wieben O (2012) 4D flow MRI. J Magn Reson Imaging 36(5):1015-1036
3. Fredriksson AG, Zajac J, Eriksson J, et al (2011) 4-D blood flow in the human right ventricle. 2344-2350
4. Falahatpisheh A, Pedrizzetti G, Kheradvar A (2014) Three-dimensional reconstruction of cardiac flows based on multi-planar velocity fields. Exp Fluids. 55:1849
5. Mangual JO, Domenichini F, Pedrizzetti G (2012) De-scribing the highly three dimensional right ventricle flow. Ann Biomed Eng 40:1790-1801
6. Grbic S, Ionasec R, Vitanovski D, et al(2012) Complete valvular heart apparatus model from 4D cardiac CT. Med Image Anal 16:1003-1014
7. Seo JH, Mittal R (2013) Effect of diastolic flow patterns on the function of the left ventricle. Phys Fluids 25:110801
8. Su B, Zhang J-M, Tang HC, Wan M, Lim CCW, Su Y, Zhao X, Tan RS, Zhong L (2014) Patient-specific blood flows and vortex formations in patients with hypertrophic cardiomyopathy using computational fluid dynamics. Biomed Eng Sci (IECBES), 2014 IEEE Conf 276-280
9. Mangual JO, Kraigher-Krainer E, De Luca A, et al (2013) Comparative numerical study on left ventricular fluid dynamics after dilated cardiomyopathy. J Biomech 46:1611-7
10. Mangual JO, Domenichini F, Pedrizzetti G (2012) Three dimensional numerical assessment of the right ventricular flow using 4D echocardiography boundary data. Eur J Mech - B/Fluids 35:25-30

Author: Liang Zhong
Institute: National Heart Centre Singapore
Street: 5 Hospital Drive, 169609
City: Singapore
Country: Singapore
Email: zhong.liang@nhcs.com.sg

Performance Evaluation of Artificial Neural Network Models for the Prediction of the Risk of Heart Disease

Armin Yazdani[1] and Kannan Ramakrishnan[2]

[1] Faculty of Computer Science and Information Technology, University of Malaya, Malaysia
[2] Faculty of Computing &Informatics, Multimedia University, Malaysia

Abstract— **According to world health organization, cardiovascular diseases are the number one cause of death globally and most of them can be prevented by addressing risk factors such as tobacco use, unhealthy diet and obesity, physical inactivity, high blood pressure, diabetes and raised lipids. Hence early and correct diagnosis and administering the appropriate and effective treatment is important. Physicians often make decisions based on current clinical tests and previous experience of diagnosing patients with similar symptoms but it is a difficult task since a lot of factors are contributing to the prediction. In this paper, a clinical decision support system is designed and implemented that can help the doctors in predicting the risk of heart disease. This system is based on the optimal artificial neural network model identified among the different models evaluated using accuracy measures on standard heart disease database. An interface is also developed based on the optimal model to facilitate the doctors in predicting the risk of heart disease.**

Keywords— **Data mining, Prediction, Heart Disease, Artificial Neural Networks.**

I. INTRODUCTION

Even though everyone understands the importance of health due to widespread usage of information available through Internet, mortality rate related to cardiovascular diseases have not decreased as expected. Cardiovascular diseases are a group of disorders of the heart and blood vessels [1]. According to the compilation of data collected from more than 190 countries, around 17.3 million deaths are caused due to heart disease and it is expected to go beyond 23.6 million by 2030 [2]. It is also a fact that in about every 34 seconds, one person has heart attack in the United States alone [3]. Hence, earlier prediction of heart disease and giving appropriate treatment is important in worldwide reduction of mortality rate due to heart disease. This is especially important for the people who are at high cardiovascular risk.

In normal circumstances, doctors predict the risk of heart disease based on their intuition as well as their experience. There is a possibility of variations in diagnosis between the doctors and also within the same doctor at different times. In major hospitals, heart disease database may contain patient details along with records of diagnosis at different times. Decision support system based on data mining algorithms can be designed using this database to aid the doctors in making a decision for prediction of the risk of heart disease. Data Mining is a process to extract hidden knowledge and find patterns among large amounts of data. Artificial Neural Networks (ANN) have been widely used as one of the intelligent techniques for prediction applications.

In this paper, analysis on the performance of ANN models to predict the risk of heart disease is done and optimal model is identified based on prediction accuracy measures using standard database. Section II discusses review on related research. Section III discusses preprocessing of data, followed by research methodology in Section IV. Experimental results are discussed in Section V followed by conclusion in Section VI.

II. LITRUTURE REVIEW

Many researchers have used different intelligent techniques as classifiers for the prediction of heart disease. They have reported good accuracy results and as far as our knowledge, no optimal model has been reported which can achieve 100% accuracy on the chosen dataset.

Sonawane et al (2014) have used multilayer perceptron Neural Network (MLPNN) as a classifier. By using 70% of data for training and 30% of data for testing, they were able to achieve the highest accuracy of 98.58%, with 20 neurons in the hidden layer [4]. Mrudula et al (2010) used support vector machine (SVM) and Artificial Neural Network (ANN) as classifiers and have shown ANN with an accuracy of 97.5% performs better than SVM with an accuracy of 80.41% [5]. Khemphila et al (2010) used logistic regression, decision trees, and neural networks as classifiers and have shown ANN performs best with 80.2% accuracy compared to logistic regression and decision tree approach [6]. Ah Chen (2011) has developed a Heart Disease Prediction System based on ANN and achieved an accuracy of 80% with 13-6-2 architecture [7]. Danger et al (2012) used MLPNN architecture with backpropagation algorithm [8] and they were able to achieve 99.25% accuracy and Jyoti Soni et al (2011) using weight associated classifier were able to achieve an accuracy of 98.58% [9].

From the review, it has been noticed that experiments using ANN as classifier was able to achieve better performances compared to all the other classifiers. But researchers were not able to identify an optimal model which can

F. Ibrahim et al. (eds.), *International Conference for Innovation in Biomedical Engineering and Life Sciences*,
IFMBE Proceedings 56,
DOI: 10.1007/978-981-10-0266-3_37

achieve 100% accuracy on the specified dataset. The proposed approach identifies an optimal ANN model which can achieve 100% accuracy on the standard dataset by extensively searching through the models with different architectural parameters, training algorithms, and different training and testing percentages of dataset.

III. DATASET DESCRIPTION AND ENCODING

Evaluation on the proposed model is done using Cleveland Clinical Foundation Heart Disease dataset, which is part of UCI machine learning repository [10].

Table 1 Description of Input and Output Attributes and their encodings

Attribute	Description	Value
Age	Age in years	< 35: (-2), 35-50: (-1), 51-60:(0), 61-79:(1), >79:(2)
Sex	Gender	Male:1, Female:0
CP	Chest pain type	Typical angina:1 Atypical angina:2 Non-anginal pain:3 Asymptomatic:4
Trestbps	Resting blood Pressure (mm Hg)	< 120: Normal (-1) 120-139: PreHypertension(0) > 139: Hypertension (1)
Chol	Serum cholesterol (mg/dl)	<200 :Desirable (0) 200-239: Bordeline(1) > 240:High (2)
Fbs	Fasting blood sugar > 120 mg/dl	True: 1, False:0
Restecg	Resting ECG Results	Normal:0 Abnormal ST_T wave:1 Left ventricular hypertrophy:2
Thalach	Maximum heart rate achieved	< 100:(-2),100-135:(-1), 136-170:(0),171-200:(1), >200:(2)
Exang	Exercise induced angina	Yes:1, No:0
Oldpeak	ST depression Induced by exercise relative to rest	0-0.99:(0),1-1.99:(1), 2-2.99:(2), 3-3.99:(3), 4-4.99:(4),5-5.99:(5), > 6:(6)
Slope	The slope of the peak exercise ST segment	Up-sloping:1, Flat:2, Down-sloping:3
CA	Number of major vessels colored by fluoroscopy	Values: 0 – 3
Thal	Default type	Normal:3, Fixed defect:6 Reversible defect:7
Num	Diagnosis of heart disease	No risk of heart disease: 0 Possibility of heart disease:1

Table 1 shows the 13 attributes that are given as inputs to ANN model and shows the output (Num) that is to determine the risk of heart disease based on these factors. The attributes were chosen based on their significance in causing the heart disease as reported by other researchers [4-9].

Dataset contains 297 samples after eliminating 6 samples with missing values. Among the 297 samples, 138 belong to 'yes' class (possibility of developing a heart disease) and 159 belong to 'no' class (no risk of developing heart disease). In this dataset non-numeric data such as gender, resting ECG results, exercise induced angina, slope, and thal are converted into numeric data and continuous values of some of the attributes are converted to specific ranges, as reported in Table 1.

Min-Max Normalization technique, which performs a linear transformation on the original data is also applied on this study. Assuming *minA* and *maxA* as the minimum and maximum values of an attribute, *A*, min-max normalization maps a value, v, of A to v' in the range $(new_{maxA};\ new_{minA})$, as given in Equation 1 [11].

$$v' = \frac{v-minA}{maxA-MinA}(new_{maxA} - new_{minA}) + new_{minA} \quad (1)$$

IV. METHODOLOGY

Fig. 1 shows the proposed approach to identify the optimal model of ANN for the prediction of the risk of heart disease. Initially preprocessing is done on the data collected from the standard database as described in section III. Initial architecture of feedforward neural network with a single hidden layer is chosen as 13-7-1. Number of neurons in input layer (ILN) are chosen as 13 corresponding to the selected 13 input attributes. Number of neurons in output layer (OLN) is chosen to be one where zero represents the group with no risk of developing heart disease and one represents the group with possibility of developing heart disease. Initial number of neurons in the hidden layer (HLN) is chosen based on Equation (2).

$$HLN = (ILN + OLN)/2 \quad (2)$$

where ILN, HLN, and OLN represents the number of neurons in input, hidden, and output layers respectively.

Training, Validation and Testing are done using Levenberg-Marquardt (trainlm) backpropagation supervised training algorithm and optimal model is chosen by varying the number of hidden layer neurons and the activation functions of hidden and output layers, until the desired training and testing accuracy is obtained. Experiment is repeated for different training algorithms such as scaled conjugate gradient backpropagation (trainscg), BFGS quasi-Newton backpropagation (trainbfg), and Resilent backpropagation (trainrp). Final optimal model and training algorithm is chosen based on the target accuracy of 100%. Fig. 2 shows the architecture of the feedforward artificial neural network used in this experiment.

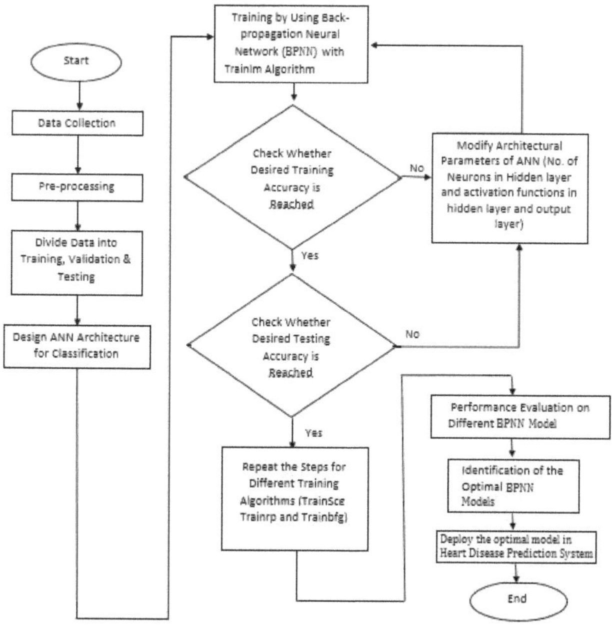

Fig. 1 Proposed approach to identify the optimal ANN model

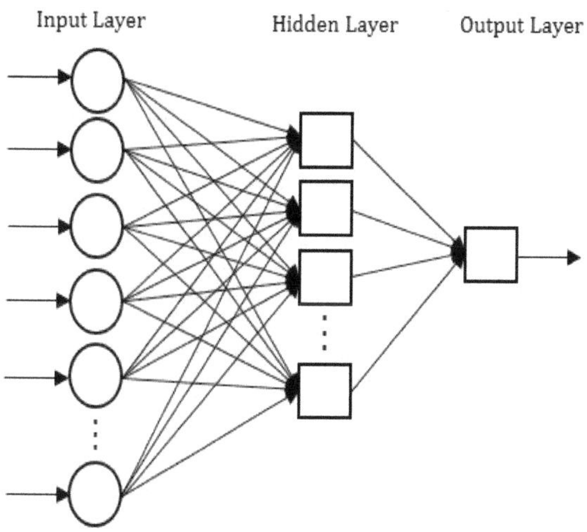

Fig. 2 Architecture of Feedforward ANN

V. EXPRIMENTAL RESULTS

Different performance evaluation measures such as sensitivity, specificity, precision, and accuracy [11] are used to identify the optimal feedforward neural network model and training algorithm. Simulations are done on different training, validation, and testing percentages such as 60-20-20, 70-15-15, 80-10-10 and also on different activation functions for hidden and output layers. Based on empirical results, training-validation-testing percentage of 80-10-10 with hyperbolic tangent sigmoid activation function (tansig) for hidden layer and log-sigmoid activation function (logsig) for output layer are chosen.

Table 2 shows the accuracy results obtained for the different training functions (trainscg, trainrp, trainlm, and trainbfg) by varying the number of hidden layer neurons.

Table 2 Identification of Optimal Training Function

Number of neurons hidden layer	Trainscg	Trainrp	Trainlm	Trainbfg
7	80%	96.7%	96.7%	80%
8	90%	86.7%	93.3%	86.7%
9	93.3%	90%	96.7%	90%
10	90%	96.7%	93.3%	90%
15	80%	80%	93.3%	80%
20	87.7%	93.3%	93.3%	96.7%
30	93.3%	93.3%	**100%**	90%
50	86.7%	80%	96.7%	90%

From Table 2, it can be concluded that the target accuracy of 100% can be obtained using trainlm algorithm with the architecture of 13-30-1. Performance evaluation is also done using different measures with the identified trainlm algorithm and the results are given in Table 3.

Table 3 Performance Evaluation with Different Measures

Number of Neurons in Hidden Layer	Sensitivity	Specificity	Precision	Test Accuracy
7	93.7%	100%	100%	96.7%
8	100%	93.7%	93.3%	96.7
9	94.4%	100%	100%	96.7%
10	90.9%	100%	100%	93.3%
15	88.2%	100%	100%	93.3%
20	87.5%	100%	100%	93.3%
30	**100%**	**100%**	**100%**	**100%**
50	100%	66.6%	94.4%	96.7%

The performance of our optimal model is also compared with the performances of the other models reported in the literature. It is to be noted that the dataset used for evaluating our model is same as that used by other authors and the comparison results are given in Table 4.

Table 4 Performance Comparison with Related Research

	ANN	Decision Tree	SVM
Sonawane et al	98.58%	-	-
Mrudula et al	97.5%	-	80.41%
Khemphila et al	80.2%	79.3%	-
Ah Chen et al	80+-5 %	-	-
Dangare et al	99.25%	-	-
Jyoti Soni et al	98.58%	-	-
Our method	100%	-	-

The identified optimal model is used in the Graphical User Interface (GUI) which has been developed to assist the doctors in predicting the risk of heart disease and is shown in Fig. 3.

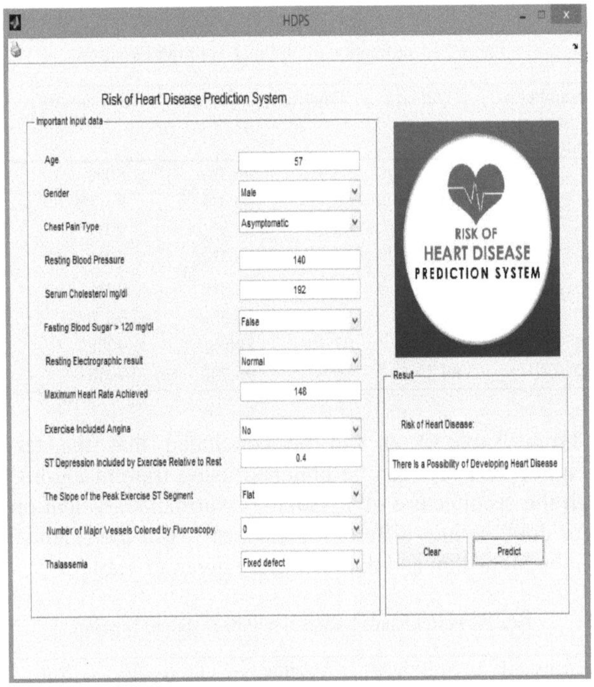

Fig. 3 Risk of heart disease prediction system

VI. CONCLUSION

An optimal model of ANN, which can achieve 100% accuracy on the standard dataset have been identified by varying the architectural parameters, training functions, activation functions and training/validation/testing percentages of dataset. A GUI based on the identified optimal model has been developed to assist the doctors in the prediction of heart disease. However further experiments have to be conducted to test the efficiency of the model on other datasets. This work can be further extended to identify the different levels of the risk of heart disease.

CONFLICT OF INTEREST

There is no conflict of interest.

REFERENCES

1. Cardiovascular diseases (CVDs). Retrieved June 1, 2015. from http://www.who.int/mediacentre/factsheets/fs317/en/
2. American Heart Association statistical report tracks global figures for first time - News on Heart.org. (2014, December 17). Retrieved 2015.
3. Mozaffarian D, Benjamin EJ, Go AS, et al. Heart disease and stroke statistics—2015 update: a report from the American Heart Association. Circulation. 2015;131:e29-322
4. Sonawane, J. S., & Patil, D. R. (2014, February). Prediction of heart disease using multilayer perceptron neural network. In Information Communication and Embedded Systems (ICICES), 2014 International Conference on (pp. 1-6). IEEE.
5. Mrudula Gudadje, Wankhade, K., & Dongre, S. (2010, September). Decision support system for heart disease based on support vector machine and artificial neural network. In Computer and Communication Technology (ICCCT), 2010 International Conference on (pp. 741-745). IEEE
6. Khemphila, A., & Boonjing, V. (2010, October). Comparing performances of logistic regression, decision trees, and neural networks for classifying heart disease patients. In Computer Information Systems and Industrial Management Applications (CISIM), 2010 International Conference on (pp. 193-198). IEEE
7. Ah Chen, Huang, S. Y., Hong, P. S., Cheng, C. H., & Lin, E. J. (2011, September). HDPS: Heart disease prediction system. In Computing in Cardiology, 2011 (pp. 557-560). IEEE
8. Dangare, C. S., & Apte, D. S. S. (2012). A data mining approach for prediction of heart disease using neural networks. International Journal of Computer Engineering & Technology (IJCET), 3(3), 30-40
9. Jyoti Soni, Ansari, U., Sharma, D., & Soni, S. (2011). Intelligent and effective heart disease prediction system using weighted associative classifiers. International Journal on Computer Science and Engineering, 3(6), 2385-2392
10. UCI Machine Learning Repository: Heart Disease Data Set. http://archive.ics.uci.edu/ml/datasets/Heart+Disease
11. Jiawei Han; Micheline Kamber (2006), Data Mining Concepts and Techniques, second edition.

Contact Author: Armin Yazdani
Institute: University of Malaya
Street: Jalan Universiti
City: Kuala Lumpur
Country: Malaysia
Email: armin.ydy@gmail.com

Human Bone Histomorphological Pattern Differences Between Genders: A Review

H. Abdullah[1], F. Mohd Nor[2], and M.M. Abdul Jamil[1]

[1] Biomedical Engineering Modeling and Simulations (BIOMEMS) Research Group, Faculty of Electrical and Electronic Engineering, University Tun Hussein Onn Malaysia, Batu Pahat, Malaysia

[2] Department of Pathology, Universiti Kebangsaan Malaysia, Jalan Yaacob Latif, Bandar Tun Razak, Cheras, Kuala Lumpur, Malaysia

Abstract— **Bone consists of histological parameters or pattern that can be analyzed to determine different characteristics of a person or animal to which it belongs. Human-animal, gender, age, height and weight are some characteristics on which work has been done. Kerley was the first to present microstructural computational method to determine age from human bone sample. This opened doors for research on age, gender and other characteristic determination. In literature main focus of research has been on age estimation. Gender estimation on the contrary has been a byproduct of age estimation or its left open to questions. This paper presents a compilation of researches which has been carried out for gender estimation and compares the microstructural parameters which were found to be different in male and female samples.**

Keywords— **Osteon, Haversian canal, fragmented osteon, bone remodeling and osteoblast.**

I. INTRODUCTION

There are different morphological ways of estimating gender from skeletal remains of human beings [1]. Morphological approaches are efficient but it becomes difficult to estimate gender in case of missing bones or if only fragments of bones are present. Another method of estimating gender is by estimating bone mass of cortical bone. To estimate gender using this method requires information of general cortical bone mass of humans belonging to that race [2]. Bone mass estimation also requires advance systems making this approach arduous.

Histological methods of estimating age and gender from skeletal remains of human came in to consideration when Kerley in 1965 presented age estimation method using microscopic parameters [3]. Kerley's study opened a new aspect of microscopic analysis of bone. Later his study was further extended from only age to composition of gender and age by D.D Thomson in 1981 [4]. He was able to point age graded contrasts in microstructures of males and females. This opened doors for gender estimation from microscopic parameters.

The human skeleton give shape to the body and provides physical support to systems contained within [5]. The structure of bone is optimized so that is relatively strong and light weight. The interior of bone is composed of bone marrow which is surrounded by two major types of bone tissue, cortical bone or the hard outer shell of bone and trabecular bone the spongy looking center. The amount of each type of bone is dependent on the function of that bone. The basic unit of compact or cortical bone is the osteon [5]. It is composed of concentric successive lamellae. This structure contributes to bone structure by resisting bending. Cells called osteocytes are distributed within the concentric lamellae. Osteocytes form a complex network that is thought to important in maintaining the viability and structural integrity of bone. At the center of the osteon there is haversian canal [5]. The canal contain blood vessels and nerves. The blood vessels facilitate the exchange between osteocytes and the blood. Trabecular bone is present in the interior of some bones and resist compression. Osteocytes are also contained within its structure and play an important role in sensing changes [5]. Figure 1 shown the structural components of bone.

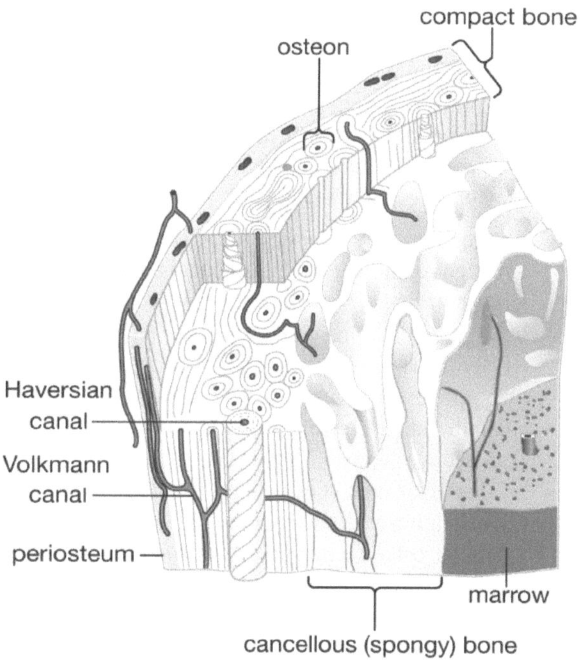

Fig. 1 Structural components human bone [6].

F. Ibrahim et al. (eds.), *International Conference for Innovation in Biomedical Engineering and Life Sciences*, IFMBE Proceedings 56,
DOI: 10.1007/978-981-10-0266-3_38

II. MICROSCOPIC PARAMETERS

A big challenge in bone microscopic analysis for age and gender estimation is differences in selected microscopic parameters and their definitions by authors. Figure 2 shows osteon systems for better understanding of parameters. The result of overall study of bone histological gender and age estimation strongly lies in microscopic parameters selection. Microscopic parameters can be divided into two major categories, observed and derived parameters.

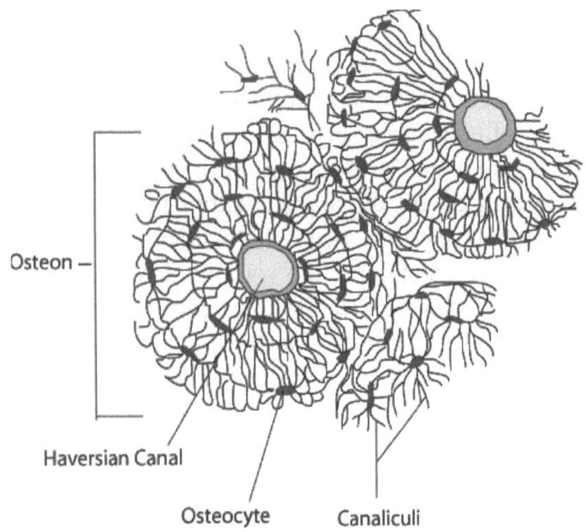

Fig. 2 Osteon system with haversian canal, osteocytes and canaliculi[7].

A. Observed Parameters

Observed parameters are one which do not need to be derived from other parameters. Different authors used different observed parameters with different names, but in this articles generalizes observed parameters. Some of the commonly used observed parameters are:

- Secondary Osteon (SO): Osteon with ninety percent complete haversian canal and bounded by a reversal line.
- Fragmented Secondary Osteon (FSO): Those osteons whose haversian canal is not ninety present complete or have no visible portion of haversian canal rather have only fragments of lamellae present.
- Type II Osteon (OII): Type II osteons are formed when intra-osteonal remodeling occurs and a new smaller osteon emerges within and existing osteon.
- Cortical thickness (CT): The thickness of cortical bone.
- Medullary cavity diameter (MCD): this the diameters of hollow inside of bone containing bone marrow.
- Osteon Count (OC): Total number of osteons.
- Haversian canal diameter (HCD).
- Osteon Diameter (OD).
- Osteon Area (OA).

B. Derived Parameters

Derived parameters are the parameters which are derived from observed parameters. It depends on authors whether they want to use derived parameters. Section III will specify observed and derived microscopic parameters used by researchers in history. Some of the commonly used derived parameters are:

- Secondary Osteon Density (SOD): The sum of complete secondary osteon area within the cross-sectional area.
- Fragmented Osteon Density (FOD): The sum of osteons fragments present in a specific cross-sectional area.
- Secondary Osteon Region (SOR): Ratio of region covered by secondary osteon with region of fragmented osteon plus secondary osteon.

III. BONE HISTOLOGY

Gender estimation using microscopic examination of bone tissue is based upon changes in microstructural features that are products of metabolic process in bone called remodeling. The bone remodeling cycle involves a series of highly regulated steps that depend on interactions between two cell lineages namely, mesenchymal osteoblastic lineage and hematopoietic osteoclastic lineage [8]. Remodeling in osteon, periosteum and endosteum is accomplished by teams of osteoclasts and osteoblasts that work simultaneously in basic multicellular units [9]. Figure 3 shows an osteon system undergoing bone remodeling, with osteoblast and osteoclast. Bone remodeling occurs in multiple locations of the skeleton to remove older bone and replace it with newly-formed bone. Osteon remodeling forms a Haversian system, known as a bone structural unit [10], while remodeling in endosteum and periosteum causes thinning of the periosteum, as formation exceeds resorption on the periosteal surface and resorption exceeds formation on the medullary canal.

These modeling and remodeling stages of bone were analyzed by Kerley and gave a regression equation, based on which human age after death can be calculated. After Kerley presented age estimation from bone histomorphology in 1965, a lot of research started on age estimation from bone samples of human belonging to different races. Some researchers while working on age estimation reported no significant gender change in histology of bones [11]. Based on these report most of the research was focused on age estimation and left a question mark on gender estimation.

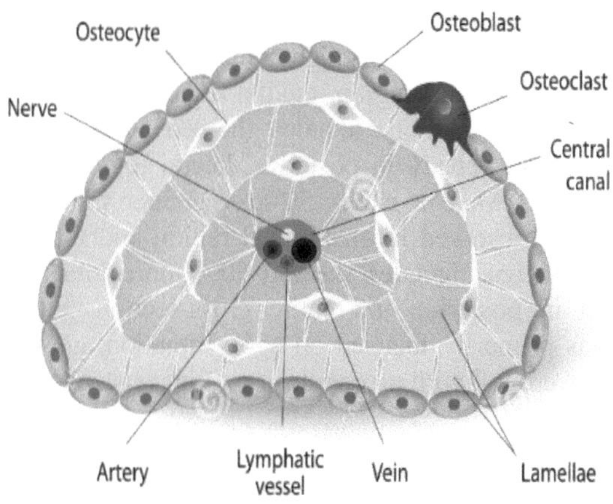

Fig. 3 Osteon system undergoing bone remodeling process with osteoclast and osteoblast [12].

IV. Gender Identification

Singh and Gunberg in 1970 worked on male specified age estimation and tested on some female samples which produced accurate results. But when it was tested on larger number of female samples Samson and Branigan in 1987 produced a poor result of gender estimation [13].

D.D. Thomson in 1979 was the first researcher to make age graded gender identification. He reported with female group to have ($p < 0.05$) values greater average area of haversian canals when compared to males of the same group. However males showed greater haversian canal area (133%) in fifth decade and eight decade while females in these decades showed only increase of (52%). Although males samples showed greater haversian canal area in their fifth and eight decade, females samples overall showed greater summed area of haversian canal ($p < 0.05$) [14]. In 1981 D.D. Thomson further concluded his research on age graded gender identification with more accurate results while working on Yupkil-Inupiaq skeletons. According to his research, Southampton Island eskimo showed haversian canal area averaged $0.084mm^2$ per mm^2 for males and $0.105mm^2$ per mm^2 in females ($p < 0.05$). These values were greater values then other groups studied [15].

David B. Burr in 1990 did research on skeletal remains of an archaic native American population (Pecos Indian) and compared it with modern American population. He observed in Pecos population female had larger osteons then males ($p < 0.2$). Also osteonal mean wall thickness in females was observed to be greater than males. Since osteons in males were smaller in size, they were greater in number and had greater osteon population density compared to females ($p < 0.06$). This insinuates to the historical observation of greater bone remodeling in males compared to females [16].

M.F. Ericksen in 1991 followed D.B. Burr and worked on gender specified age estimation. In his research he divided both genders into age graded groups by decades. With this he observed microstructural parameter, type II osteons (these type of osteons are formed when resorption starts near a haversian canal of a mature osteon. The new osteon is provided with a cement line and gradually fills in with remodeling of previous lamella) increase with age and remain same at a particular age. Similarly non-haversian canals also decreased in both sexes. Further he observed that osteon in males gradually increase with each decade, while in females increase in osteon comes to a hold at ages in sixth decade and remains still till end. He also observed that as females cross their fifth decade their fragments/mm^2 increases than osteons. But in males area of fragments/mm^2 does not increase from of osteon before seventh decade. In terms of area, fragmental bone area increases from osteonal bone in females when she enters in her eighth decade. While in males area of fragmental bone never increases osteonal bone [17].

D.M. Mulhern in 1997 analyzed femoral remodeling patterns in Medieval Nubian Population. The samples were taken from remains of skeletal population dated 1250–1450 AD. Mulhern reported significant changes in microstructural parameters of males and females. According to his findings Males had significantly more intact osteons than females, whereas females had significantly larger osteons than males. Haversian canal dimensions were not statistically significant between the genders. Gender differences in activity patterns in which males were involved in more physically strenuous tasks may have contributed to differences in remodeling variables. The number of intact osteons differs significantly by sex ($p < 0.0001$). The average number of intact osteons for females is significantly lower (6.73/mm^2) than males (9.74/mm^2). The average number of fragmentary osteons also differs significantly between the sexes ($p < 0.05$). Females have (4.68/mm^2), whereas males have (2.59/mm^2). Osteon size also show noticeable between males and females. Females have an average osteon area of $0.040mm^2$ and males average $0.036mm^2$. Mean osteonal cross sectional area and diameter are also significantly different between the sexes ($p < 0.05$). Table 1 shows some of microscopic parameters which showed significant difference in males and females [18].

Table 1 Microscopic parameter differences in males and females [18].

Parameter	Female	Male
Intact osteon	6.73 ± 0.31	9.74 ± 0.39
Fragmented osteon	4.68 ± 0.27	2.59 ± 0.14
Haversian canal area	0.0021 ± 0.0001	0.0022 ± 0.0002
Osteon area	0.040 ± 0.001	0.036 ± 0.002
Average osteon count	12.03 ± 0.47	12.79 ± 0.43

K.L. Bell in 2001 did a research on microradiographic images of femoral mid-shaft cross-section of 66 subjects. He categorized subject according to age and gender in group of two-two decades (21-40, 41-60, 61-80 and >80). In his observation cortical remodeling was greater in female samples, which was noticeable by presence of substantial percentage of new young osteon. Females remodeling was observed to be 34% greater than that in males ($p = 0.034$). When combined both gender showed no notable changes between ages. Also the parameter of osteon density showed no significance with age or gender. Figure 4 shows K.L. Bell statistical analysis of age graded young osteon percentage comparison with gender [19].

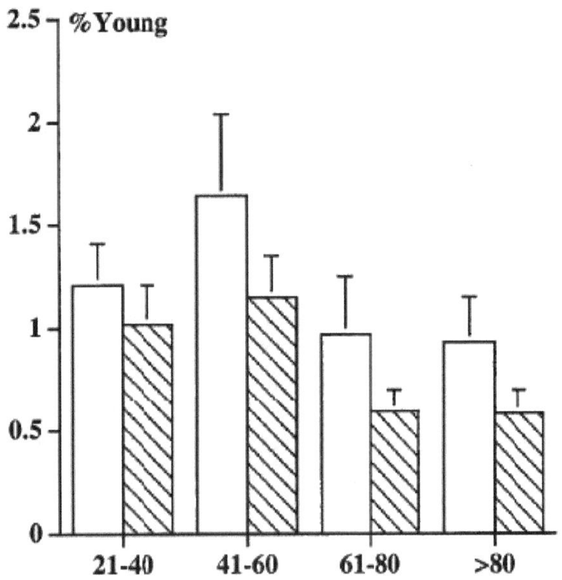

Fig. 4 Age graded young osteon percentage comparison, females (open bars) – males (hatched bars) [19].

Farida Nor in 2009 took 64 human bone samples from Malaysia and analyzed microstructural parameters for age and gender estimation. She concluded that number of osteon showed no significant difference between both genders. This was in acquiescence with research performed by Pfeiffer in 1996 and 1998. However while considering osteon and haversian canal size she observed that females have larger osteon which also makes their number less is observation region. Further haversian canals were larger in females than in males ($p < 0.05$). This was also in agreement with studies conducted by Thompson in 1981 and Mulhern in 1997 [20].

M.T.J.C Hernandez while working on age estimation in 2012 also reported gender differences in femoral cortex remodeling. The research concluded with males having 50% more variation observed in fragmented osteon density by age in regions of anterior region, medial region and total cross-section. On the contrary females showed 50% more fragmented osteon density in posterior, lateral region and total cross-section. Also in males 50% or greater intact osteon density was observed in anatomical region and total cross-section. Whereas females showed 50% more intact osteon variation in posterior and entire cross-section. Since the work was primarily focused on age estimation, gender differences were left with no conclusion of its existence [21].

V. CONCLUSIONS

This paper complies work done by researchers on gender estimation from bone histomorphological patterns or parameters. Main work of bone histological analysis is done on age estimation. While some researchers divided age estimation between two genders. This gave differences in microstructural parameters between males and females. Different microstructural parameters are selected by different researchers. Generally used microstructural parameters (obtained and derived) are discussed in section II. From the literature this paper has given, it can be seen there in no clear view of gender estimation from bone histology. Researchers presented different microscopic parameters which can be focused for gender estimation while examining samples from a specific race and region of humans. Question arises whether their research can be implemented on humans belonging to all races. Bone characteristics change in humans with race and region of earth from which they belong. Up till now very little work has been done on histological analysis of human bones and no differences in microstructural parameters can be considered as standard parameter showing gender differences in humans belonging to all races and region. However studies are being conducted to analyze bone samples of specific regions and races of humans, showing gender microstructural parameter changes for that region and race. A lot needs to be done in this aspect to be precise and have a group of selected parameters which can be analyzed to find gender.

ACKNOWLEDGMENT

The authors would like to thank ORICC and R&D for support and University Tun Hussein Onn Malaysia for providing research opportunity.

CONFLICT OF INTEREST

The authors declare that they have no conflict of interest.

REFERENCES

1. Cho, H., Stout, S.D., and Bishop, T.A.: 'Cortical bone remodeling rates in a sample of African American and European American descent groups from the American Midwest: Comparisons of age and sex in ribs', American Journal of Physical Anthropology, 2006, 130, (2), pp. 214-226
2. Brockstedt, H., Kassem, M., Eriksen, E.F., Mosekilde, L., and Melsen, F.: 'Age- and sex-related changes in iliac cortical bone mass and remodeling', Bone, 1993, 14, (4), pp. 681-691
3. Kerley, E.R.: 'The microscopic determination of age in human bone', American Journal of Physical Anthropology, 1965, 23, (2), pp. 149-163
4. Thompson, D.D., and Gunness-Hey, M.: 'Bone mineral-osteon analysis of Yupik-inupiaq skeletons', American Journal of Physical Anthropology, 1981, 55, (1), pp. 1-7
5. Barnes, E.: 'Introduction': 'Atlas of Developmental Field Anomalies of the Human Skeleton' (John Wiley & Sons, Inc., 2012), pp. 1-6
6. Merriam-Webster Inc., w.M.-W.c.
7. Gray, H.: ' Anatomy of the Human Body', in Editor (Ed.)^(Eds.): 'Book Anatomy of the Human Body' (1918, edn.), pp.
8. Raisz, L.G.: 'Physiology and pathophysiology of bone remodeling', Clinical chemistry, 1999, 45, (8 Pt 2), pp. 1353-1358
9. Jacobs, C.: 'Skeletal Tissue Mechanics R. Bruce Martin, David B. Burr and Neil A. Sharkey (Eds.); Springer, New York, 1998, 392 pages, ISBN: 0-387-98474-7', Journal of Biomechanics, 33, (10), pp. 1339
10. Marieb, E.N.: 'Human anatomy & physiology' (Benjamin Cummings, 2001. 2001)
11. Pfeiffer, S.: 'Variability in osteon size in recent human populations', Am J Phys Anthropol, 1998, 106, (2), pp. 219-227
12. Osteon development and structure. Osteoblast, o., and osteoclast. http://www.dreamstime.com/
13. Singh, I.J., and Gunberg, D.L.: 'Estimation of age at death in human males from quantitative histology of bone fragments', American Journal of Physical Anthropology, 1970, 33, (3), pp. 373-381
14. Thompson, D.D.: 'The core technique in the determination of age at death of skeletons', Journal of forensic sciences, 1979, 24, (4), pp. 902-915
15. Thompson, D.D., and Gunness-Hey, M.: 'Bone mineral-osteon analysis of Yupik-Inupiaq skeletons', Am J Phys Anthropol, 1981, 55, (1), pp. 1-7
16. Burr, D.B., Ruff, C.B., and Thompson, D.D.: 'Patterns of skeletal histologic change through time: comparison of an archaic native American population with modern populations', The Anatomical record, 1990, 226, (3), pp. 307-313
17. Ericksen, M.F.: 'Histologic estimation of age at death using the anterior cortex of the femur', Am J Phys Anthropol, 1991, 84, (2), pp. 171-179
18. Mulhern, D.M., and Van Gerven, D.P.: 'Patterns of femoral bone remodeling dynamics in a Medieval Nubian population', Am J Phys Anthropol, 1997, 104, (1), pp. 133-146
19. Bell, K.L., Loveridge, N., Reeve, J., Thomas, C.D., Feik, S.A., and Clement, J.G.: 'Super-osteons (remodeling clusters) in the cortex of the femoral shaft: influence of age and gender', The Anatomical record, 2001, 264, (4), pp. 378-386
20. Nor, F.M.: 'A comparative microscopic study of human and non-human long bone histology', University of Bradford, 2009
21. Meghan-Tómasita JúRi Cosgriff-Hernández, M.F.S.: 'Histomorphometric Estimation of Age at Death Using the Femoral Cortex: A Modification of Established Methods', The Ohio State University, 2012

Author:	Hadi Abdullah
Institute:	University Tun Hussein Onn Malaysia (UTHM)
City:	Batu Pahat
Country:	Malaysia
Email:	hadi.uthm@yahoo.com

Classification of Healthy Subjects and Insomniac Patients Based on Automated Sleep Onset Detection

C. Dissanayaka[1], H. Abdullah[2], B. Ahmed[3], T. Penzel[4], and D. Cvetkovic[1]

[1] School of Electrical and Computer Engineering, RMIT University, Melbourne, Australia
[2] Razak School of Engineering and Advanced Technology, Universiti Teknologi Malaysia, Kuala Lumpur, Malaysia
[3] Texas A&M University, Education City, Doha, Qatar
[4] Charité Universitätsmedizin, Berlin, Germany

Abstract— **This work aims to investigate new indexes quantitatively differentiate sleep insomnia patients from healthy subjects, in the context of sleep onset fluctuations. Our study included the use of existing PSG dataset, of 20 healthy subjects and 20 insomniac subjects. The differences between normal sleepers and insomniacs was investigated, in terms of dynamics and content of Sleep Onset (SO) process. An automated system was created to achieve this and it consists of six steps: 1) pre-processing of signals 2) feature extraction 3) classification 4) automatic scoring 5) sleep onset detection 6) identification of subject groups. The pre-processing step consisted of the removal of noise and movement artifacts from the signals. The feature extracting step consists of extracting time, frequency and non-linear features of Electroencephalogram (EEG) and Electromyogram (EMG) signals. In the third step, classification was done using ANN (Artificial Neural Networks) classifier. The fourth step consisted of scoring sleep stages (wake, S1, S2, S3 and REM) and produced a hypnogram. In the fifth step, we are detecting sleep onset from our automatic detected hypnogram and identified time of SO reference point and the combination of stages. In the final step we differentiated healthy subjects from insomniac patients based on the parameters calculated in the fifth step.**

Keywords— **Insomnia, Sleep Onset, EEG, EMG, ANN.**

I. INTRODUCTION

Sleep fluctuates between Wake and Rapid Eye Movement (REM) sleep stages cyclically several times during sleep. Identification of sleep stages and these sleep fluctuations can be is essential in classifying sleep disorders. In the past there have been several attempts to evaluate the changes that occur physiologically and behavioral during sleep onset. Automated sleep staging near Sleep Onset (SO) has been attempted in the past studies [1-14]. In our particular study we are using sleep stages to detect and characterise SO. The exact time SO occurs (i.e. reference point) is different for each subject. It is influenced by several factors such as stress and environment effects, sleep deprivation, frequently changing sleep schedules hence it is important to detect it accurately. For the investigation of sleep quality there are already several sleep onset indexes: sleep latency period (time from light out until sleep onset), sleep efficiency (ratio of total sleep time to time in bed), number of awakenings, total sleep time, wake after the sleep onset (WASO) and arousal index (number of arousals per hour).

Normally the recorded Polysomnographic (PSG) signals are examined manually by a sleep technician to produce the hypnogram and score different events on the recording. Manual sleep stage detection is a time consuming task and automatic analysis addresses that problem.

The aim of this study is to automatically detect sleep stages and its combinations characterising fluctuations between W and REM during sleep onset period. Finally, utilise these parameters and features to classify healthy and insomnia subjects.

II. PROCEDURE

A. *Participants and Study Design*

We are using an existing Polysomnographic (PSG) sleep data collected at Charité University (Berlin, Germany). The dataset consists of 20 sleep healthy and 20 sleep primary insomnia patients. Sample characteristics of the insomnia group of patients consisted of 22 to 64 years old (age: 48.8 ± 12.8 years), both male and female. Their mean BMI is 25.7 ± 3.4 kg/m2. The insomnia patients had complaints of falling asleep or non-restorative sleep or maintaining sleep at least three times per week or more for the past 3 months. Their insomnia was not linked to any other sleep disorder or mental disorder (i.e. it was primary insomnia). Patients reported less than 6 hours sleep, had a sleep latency higher than 30 minutes and a wake period of more than 60 minutes after sleep onset. Time in bed was 6.5 to 8.5 hours for at least 5 nights per week during the last 3 months. Regular bedtime during the last 3 months was 21:00 to 24:00 and did not vary by more than 2:00 hours. This was confirmed by a sleep diary completed for the duration of one week prior to the study.

The following sleep stages were manually scored and automatically detected: Wake (W), NREM Stage1 (S1), NREM Stage2 (S2), NREM Stage3 (S3) and REM. Sleep parameters for both groups of subjects are given in Table 1.

F. Ibrahim et al. (eds.), *International Conference for Innovation in Biomedical Engineering and Life Sciences*,
IFMBE Proceedings 56,
DOI: 10.1007/978-981-10-0266-3_39

Table 1 Sleep Parameters

Parameters	Healthy		Insomnia	
	Avg	Std	Avg	Std
Wake	165	133	312	108
Stage1	102	59	111	36
Stage2	365	89	328	74
Stage3	193	51	116	51
REM	167	66	98	44
TST (min)	991.45	52.75	964.45	16.72
TRT (min)	495.50	26.34	481.30	5.36
SL (min)	46.05	63.07	44.65	29.14
SE (%)	83.65	12.71	67.80	11.34

TST= Total Sleep Time, TRT= Total Record Time, SL=Sleep Latency, SE=Sleep Efficiency

B. Signal Pre Processing Analysis

The EEG data was band pass filtered (0.5-70 Hz) and the EMG data was band pass filtered (10-100 Hz). Both of these signals were band stop filtered (47.5-52.5 Hz) to remove power supply artifacts and down sampled to 128Hz. Movement artifacts were removed from all the signals by limiting the signal amplitude to below a threshold of 500μV. All the signals were then normalized in amplitude to the interval [−1, 1].

III. METHODS

Fig.1 shows the steps undertaken by the method section and following are the detailed explanation of those steps.

A. Feature Extraction

The extraction of necessary information from PSG signals can be achieved by many different methods. There are four different groups of features which are related to sleep staging: time, frequency, time-frequency and non-linear features. We are only utilising time, frequency and nonlinear features in this study.

Time Domain Features

Time domain features are usually taken from statistical analysis of the PSG data. For example statistical parameters included the following: mean value, standard deviation, median, skewness and kurtosis can be computed from EEG and EMG signals [15].

Frequency and Time-Frequency Domain Features

In this study we utilised several frequency and time-frequency features to measure the Power Spectral Density (PSD) of EEG signals. These include the absolute power, relative power, Short-Time Fourier Transform (STFT) and Yule-Walker algorithm.

Frequency and time-frequency domain features revealed spectral characteristics of PSG signals. These features were obtained from transformation of time domain function. PSG signals have stochastic characters. Therefore the values of PSD and power spectrum were used to express their spectral structure [16].

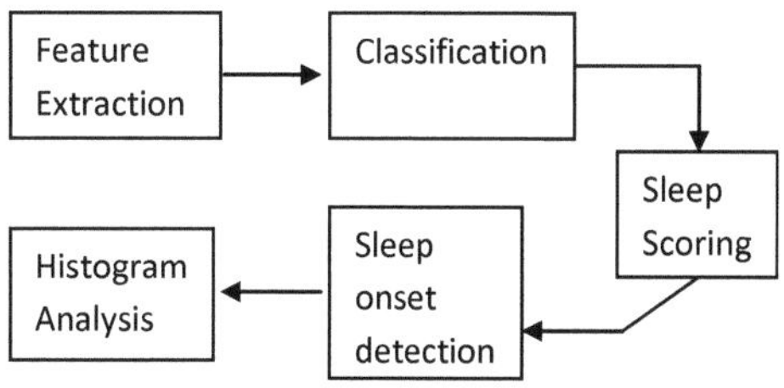

Fig. 1 Block diagram of the steps taken in the complete methodology

There are two main groups for power spectrum estimation, the parametric and non-parametric methods. Parametric methods calculated the PSD directly from PSG signal whereas non-parametric method uses signal origin modeling.

Relative spectral power,

$$S(w) = \frac{F(w) \times F^{*}(w)}{S_{total}} \tag{1}$$

where $F(w)$ is Fourier Transform.

Yule-Walker and Burg methods are examples of parametric methods that were used to solve the task. The periodogram which is based on Welch method [17] is commonly used non-parametric method where easily computed by FFT.

Non-Linear Methods

Non-linear methods have been successfully used on EEG and EMG signals to analyse its non-stationary features. These methods are far more superior to the traditional linear methods, such as power spectral analysis and the Fourier transforms [18]. On the other hand there are non-linear features which are based on chaos theory and non-linear dynamic. According to this concept, biological signals are the outcomes of the chaotic process and can be represented by chaotic parameters.

B. Classification and Automated Hypnogram

Sleep stage classification methods that are commonly used are: Fisher's Linear Discriminant, Quadratic Discriminant, Artificial Neural Networks (ANN) and Support Vector Machines (SVM). In this study we have utilised ANN with one hidden layer. The ANN had 45 input nodes (45 features),

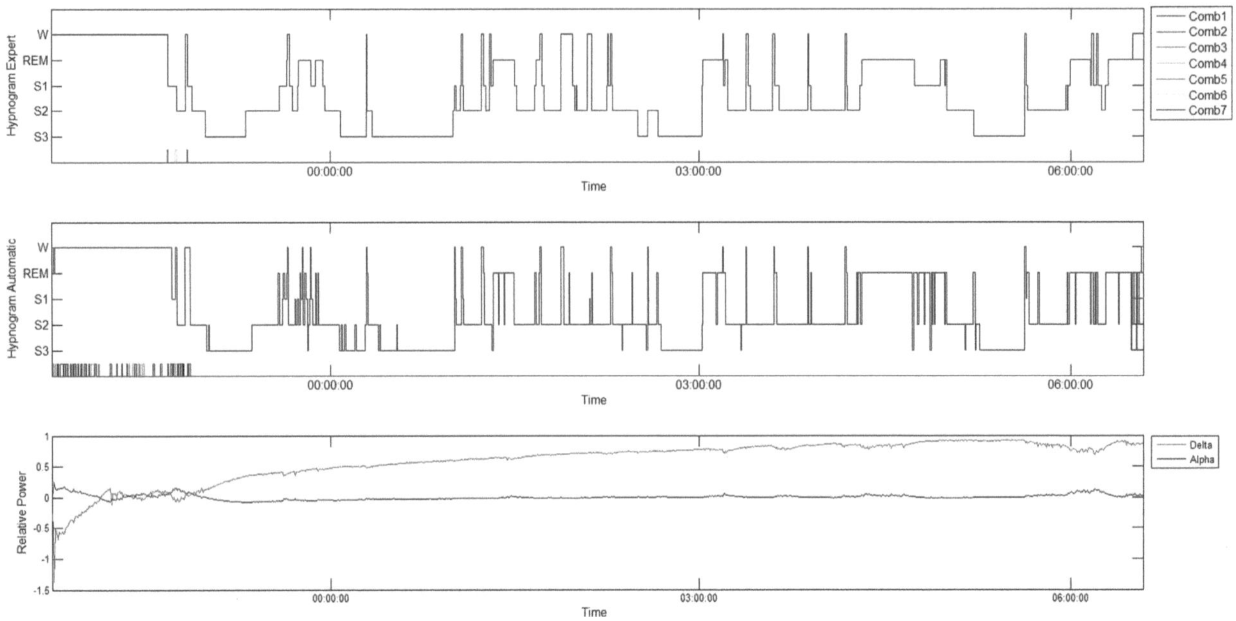

Fig. 2 Hypnograms and combinations in SO period according to relative power of alpha and delta.

5 nodes in the hidden and 5 nodes in the output layer (5 sleep stages). The feature set determine the number of neurons in input layer.

C. Sleep Onset Detection

The SO was detected using automated according to Butkov et al., definition [19] The combinations of sleep stages is given in Table 2. For example, with the combination #1, W>S1>S2>S3, we refer to the order to stages, consecutively. These combinations were identified between the start and the reference point (vertical line) as shown in Fig. 3. These parameters are detected for all healthy and insomniac patients.

Table 2 The combinations of sleep stages characterising fluctuations between W and REM during sleep onset period.

Number (#)	Combination
1	W>S1>S2>S3
2	W>S2
3	W>S3
4	W>REM
5	W>S1>S2
6	W>S1>S3
7	W>S1>REM

D. Identification of Subject Groups

A histogram test was undertaken to investigate the difference between the two subject groups according to the combination distribution detected in previous section. The histograms of the SO combinations were calculated and the results revealed the difference between healthy and insomniac subjects.

IV. RESULTS

All sleep stages are scored using ANN classification. The classification was conducted using Leave One Out (LOO) approach. In order to perform the experiment, the set of 20 Healthy and 20 Insomniac recordings were each split into a Training set (TR) of 19 subjects and a Testing set (TS) of 1 subject. TR was used to determine the best configuration for ANN, which was then tested with the TS set.

A representative of healthy subjects appears in Fig. 3. The subject entered S2 and S3 gradually, starting from S1 to deeper sleep, with no further sleep - wake fluctuations. The change in the alpha and delta power paralleled to this process is given in subplot 3 of Fig. 3. Delta power behaves as a mirror image of alpha, sharply increasing at SO. The reference indicates the crossing of alpha and delta powers which in turn indicates the SO point.

Fig. 4 shows the histograms of the combinations which belong between sleep start point and SO point (x) for the healthy subjects and insomniac subjects. The top two subplots show the expert's SO combinations and bottom two subplots reveal the automated SO combinations. It is clear that the expert's histograms are almost the same and cannot be used to differentiate healthy from insomniac subject.

Table 3 SO points and count of combinations of healthy and insomniac subjects.

	Healthy							Insomniac						
Combination	1	2	3	4	5	6	7	1	2	3	4	5	6	7
Expert Frequency	18	2	0	0	32	0	0	14	3	0	0	34	0	0
Automated Frequency	3	23	18	9	6	1	3	3	51	7	0	5	0	0

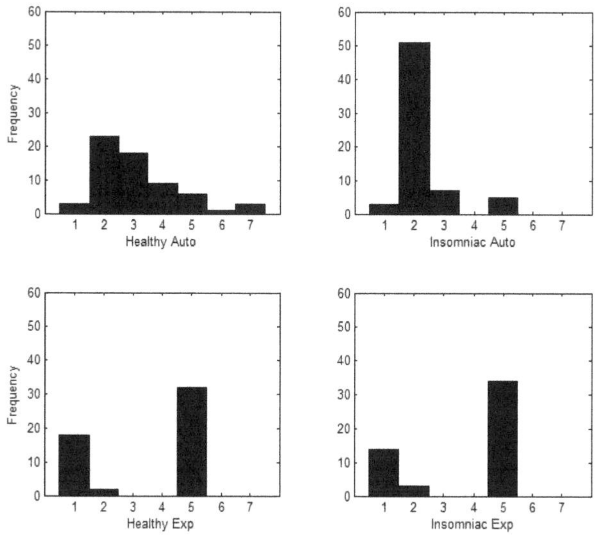

Fig. 3 Histograms of SO combinations of healthy and insomniac subjects for Expert's and Automated hypnograms

On the contrary, the automated histograms are different from each other and can be used to identify healthy subject from insomniac subject. The frequency of SO combination #2 shows a major difference between healthy and insomniac subjects. In healthy subjects all the SO combinations have count value and on the other hand in insomniac subjects SO combinations 4, 6 and 7 have zero count value.

Table 3 shows the number of occurrences (count) that combinations make for expert and automated histograms for healthy and insomniac subjects.

V. Discussion and Conclusion

The present investigation is motivated by the fact that the significance of sleep scoring in SO period for the classification of sleep insomnia and healthy subjects. In this work we have utilized time, frequency and non-linear features for analyzing and processing EEG and EMG signals which are non-linear and non-stationary signals. By using time domain, time-frequency domain and non-linear features, we have classified the W, S1, S2, S3 and REM stages in SO period. In the future we are hoping to employ more classification algorithms such as Fisher's Linear Discriminant, Quadratic Discriminant and Support Vector Machines (SVM) and to statistically analyse the SO combinations and SO reference points. We also plan to utilize other EEG bands to characterise this SO, in a similar manner to alpha and delta. We believe that beta and gamma bands may be significant during SO.

A statistical test can be undertaken to investigate the significance difference between two subject groups according to the combination distribution detected in the Section IV. Statistical analysis can be done using Statistical Package for Social Sciences (SPSS) software. The shape of the histograms of all data sets can be examined to investigate the shape of the distribution. Parametric analysis of variance (ANOVA) can be applied if the data is normally distributed and if most of the data is not normally distributed, the non-parametric Kolmogorov–Smirnov test, with a significance of $p=0.05$, can be applied.

Conflict of Interest

The authors declare that they have no conflict of interest.

References

1. J. R. Smith, M. Negin, and A. H. Nevis, "Automatic Analysis of Sleep Electroencephalograms by Hybrid Computation," *Systems Science and Cybernetics, IEEE Transactions on*, vol. 5, pp. 278-284, 1969.
2. K. Chih-En and L. Sheng-Fu, "Automatic stage scoring of single-channel sleep EEG based on multiscale permutation entropy," in *Biomedical Circuits and Systems Conference (BioCAS), 2011 IEEE*, 2011, pp. 448-451.
3. L. Sheng-Fu, K. Chin-En, H. Yu-Han, P. Yu-Hsiang, and W. Yung-Hung, "Automatic Stage Scoring of Single-Channel Sleep EEG by Using Multiscale Entropy and Autoregressive Models," *Instrumentation and Measurement, IEEE Transactions on*, vol. 61, pp. 1649-1657, 2012.
4. R. Agarwal and J. Gotman, "Computer-assisted sleep staging," *Biomedical Engineering, IEEE Transactions on*, vol. 48, pp. 1412-1423, 2001.
5. J. Fell, J. Röschke, K. Mann, and C. Schäffner, "Discrimination of sleep stages: a comparison between spectral and nonlinear EEG measures," *Electroencephalography and Clinical Neurophysiology*, vol. 98, pp. 401-410, 1996.
6. M. E. Tagluk, N. Sezgin, and M. Akin, "Estimation of Sleep Stages by an Artificial Neural Network Employing EEG, EMG and EOG," *Journal of Medical Systems*, vol. 34, pp. 717-725, 2010/08/01 2010.

7. L. Zoubek, S. Charbonnier, S. Lesecq, A. Buguet, and F. Chapotot, "Feature selection for sleep/wake stages classification using data driven methods," *Biomedical Signal Processing and Control,* vol. 2, pp. 171-179, 2007.
8. P. Haejeong, P. Kwangsuk, and J. Do-Un, "Hybrid neural-network and rule-based expert system for automatic sleep stage scoring," in *Engineering in Medicine and Biology Society, 2000. Proceedings of the 22nd Annual International Conference of the IEEE,* 2000, vol.2, pp. 1316-1319.
9. H. Ren-Jing, C. Shu-Yun, H. Ya-Yun, S. Tung-Sheng, L. Shin-Da, T. Hua, and L. Ching-Hsiang, "Strong Correlation of Sleep Onset between EOG and EEG Sleep Stage 1 and 2," in *Computer, Consumer and Control (IS3C), 2012 International Symposium on,* 2012, pp. 614-617.
10. D. U. K. Shin, H. Sakai, and Y. Uchiyama, "Slow eye movement detection can prevent sleep-related accidents effectively in a simulated driving task," *Journal of Sleep Research,* vol. 20, pp. 416-424, 2011.
11. L. Staner, F. Cornette, D. Maurice, G. Viardot, O. L. Bon, J. Haba, C. Staner, R. Luthringer, A. Muzet, and J.-P. Macher, "Sleep microstructure around sleep onset differentiates major depressive insomnia from primary insomnia," *Journal of Sleep Research,* vol. 12, pp. 319-330, 2003.
12. R. D. Ogilvie and R. T. Wilkinson, "The Detection of Sleep Onset: Behavioral and Physiological Convergence," *Psychophysiology,* vol. 21, pp. 510-520, 1984.
13. R. D. Ogilvie, "The process of falling asleep," *Sleep Medicine Reviews,* vol. 5, pp. 247-270, 2001.
14. E. Huupponen, S.-L. Himanen, J. Hasan, and A. Värri, "Sleep Depth Oscillations: An Aspect to Consider in Automatic Sleep Analysis," *Journal of Medical Systems,* vol. 27, pp. 337-345, 2003/08/01 2003.
15. K. Šušmáková and A. Krakovská, "Discrimination ability of individual measures used in sleep stages classification," *Artificial Intelligence in Medicine,* vol. 44, pp. 261-277, 2008.
16. M. Ronzhina, O. Janoušek, J. Kolářová, M. Nováková, P. Honzík, and I. Provazník, "Sleep scoring using artificial neural networks," *Sleep Medicine Reviews,* vol. 16, pp. 251-263, 2012.
17. H.-J. Park, J.-S. Oh, D.-U. Jeong, and K.-S. Park, "Automated Sleep Stage Scoring Using Hybrid Rule- and Case-Based Reasoning," *Computers and Biomedical Research,* vol. 33, pp. 330-349, 2000.
18. R. Acharya U, O. Faust, N. Kannathal, T. Chua, and S. Laxminarayan, "Non-linear analysis of EEG signals at various sleep stages," *Computer Methods and Programs in Biomedicine,* vol. 80, pp. 37-45, 2005.
19. N. Butkov, T. L. Lee-Chiong, and A. A. S. Technologists, *Fundamentals of Sleep Technology.* Philadelphia, USA: Lippincott Williams & Wilkins, 2007.

Corresponding author:

Author: Dean Cvetkovic
Institute: RMIT University
City: Melbourne
Country: Australia
Email: dean.cvetkovic@rmit.du.au

Automatic Sleep Stage Detection Based on Electrooculography

E. Malaekah[1], H. Abdullah[2], and D. Cvetkovic[1]

[1] School of Electrical and Computer Engineering, RMIT University, Melbourne, Australia
[2] Razak School of Engineering and Advanced Technology, Universiti Teknologi Malaysia, Kuala Lumpur, Malaysia

Abstract— **Scientists concentrate on the assessment of the micro and macro structure of sleep and the associated physiological activities in sleep. Their achievements heavily rely on the use of technology. Utilising the conventional method known as manual sleep stage scoring, is tedious and time-consuming. Thus, there is a significant need to create or develop a new automatic sleep stage detection system to assist the sleep physician in evaluating the sleep stages of healthy or non-healthy subjects. The main aim of this pilot study is to develop an algorithm for automatic sleep stage detection based on electrooculography (EOG) signals. 10 patients with periodic limb movements of sleep (PLMS), 10 patients with sleep apnea hypopnea syndrome (SAHS), and 10 healthy control subjects were utilised in this study. Numerous features were extracted from EOG signals such as cross-correlation, energy entropy, Shannon entropy and maximal amplitude value. K-Nearest Neighbor was used for the classification of sleep stages. An overall agreement between visual and automatic detection of sleep stage was estimated by 80.5% with Cohen's Kappa 0.73. As a result, electrooculography (EOG) signals applied in the automatic sleep stage detection has shown a significant advantage. Knowing that fewer channels can be used to accurately detect sleep stages, it can be applied in ambulatory sleep stage recording and detection.**

Keywords— **Electrooculography (EOG), Sleep stages.**

I. Introduction

Scientific research points out that sleep covers approximately one third of human being`s life. Sleep refers to physical and behavioral state that varies from wakefulness by a loss of reactivity readily and reversibly, in relation to events within one`s environment [1]. Sleep can be divided into two primary and distinct categories: NREM (non-rapid eye movement) sleep and REM (rapid eye movement or R) sleep [2]. NREM can then be classified into light sleep, with stage 1 (N1), stage 2 (N2) (or S1, S2) and deep sleep 3 (N3) (or S3, S4) [3].The abbreviations wakefulness (W), N1, N2, N3, and REM (R) are derived from the new (2007) standard American Academy of Sleep Medicine (AASM) of Iber and colleagues [4]. Polysomnography (PSG) includes a comprehensive sleep study assessing numerous electrophysiological signals of sleep, such as an electroencephalography (EEG), an electrooculography (EOG), an electromyography (EMG) and an electrocardiography (ECG). The manual scoring of sleep stages based on EEG, EOG and EMG is a subjective and time-consuming process; hence there is a need for comprehensive and more accurate automatic techniques that can easily be applied and used in the experimental and clinical sleep research.

Numerous efforts have been made to utilise only a single channel EEG or EOG signal for the detection of sleep stages, or to only one particular sleep stage, such as sleep stage N3 [5, 6]. An EEG automatic detection method has been employed for detecting sleep stages [7]. This technique was comprised of four steps: segmentation, extraction of parameters, analysis of cluster and classification. The feature extraction parameters included the harmonic parameters (center frequency, the bandwidth and the value at the center frequency), Hjorth, and relative band energy [7]. An automatic algorithm used by Liang [8] for detection of SWS, utilised one or two EOG/EEG channels. The result of this study obtained 80% sensitivity and a Cohen's kappa value of 0.755. Another study employed two-channel EOG for automatic sleep stage classification referenced to the left mastoid (M1) [9]. The synchronous EEG activity of sleep stage N2 and N3 were identified by calculating peak to peak and cross-correlation amplitude difference in the 0.5-6 Hz frequency band and between the two EOG channels.

In our pilot study, the main objective was to develop an automatic sleep stage detection method based on two EOG signals, as compared to the standard sleep stage detection criteria described in the AASM, utilising EEG, EOG, and EMG signals.

II. Materials and Methods

A. Participants and Data Collection

An existing PSG data was used in our analysis. This PSG data was acquired in a Belgian sleep hospital using a digital 32-channel polygraph (Brainnet System of MEDATEC, Brussels, Belgium). For our study, this existing data was downloaded from the online database [10]. The PSG signals included three EEG signals (C3-A1, FP1-A1 and O1-A1), two EOG signals (right EOG (REOG) and left EOG (LEOG)), and one submental EMG channel. This PSG data consist of 30 subjects. Table 1 summarises the information of this PSG data. Group 1 consisted of 10 healthy subjects (Control): 7 females aged 20–65 years (average age: 40 years) and 3 males aged 20-27 years (average age: 23.5

F. Ibrahim et al. (eds.), *International Conference for Innovation in Biomedical Engineering and Life Sciences*,
IFMBE Proceedings 56,
DOI: 10.1007/978-981-10-0266-3_40

years). Group 2 consisted of 10 patients with Periodic Limb Movements of Sleep (PLMS): 8 adult males aged 31–71 years (average age: 51 years), and 2 adult females aged 27–69 years (average age: 48 years). Group 3 included 10 patients with Sleep Apnea Hypopnea Syndrome (SAHS): 6 adult males aged 38–73 years (average age: 55 years), and 4 adult females aged 52–74 years (average age: 60 years). The sample frequency was 200 Hz. The visual sleep stage was scored by an expert according to the AASM criteria [4].

Table 1 Summarised the information of the PSG data.

Group of subjects	Group1: Healthy control	Group2: PLMS	Group3: SAHS
Number of subjects (M/F)	10 (3/7)	10 (8/2)	10 (6/4)
Age(years)	42.5 (20-65)	49 (22-46)	56 (38-74)

B. EOG Signal Processing

1. *Pre-processing*

Fig 1 shows the block diagram algorithm of automatic sleep stage detection using EOG signals. The EOG data was segmented into 5-second epochs. The entire EOG vector was processed utilising a zero-phase bandpass filter with a Cascaded Integrator-Comb (CIC) filter of order six, for the following different frequency bands: delta1 (0.5-2 Hz), delta2 (2-4 Hz), theta (4-8 Hz), alpha (8-12 Hz), sigma (12-16 Hz), beta1 (16-20 Hz) and beta2 (20-30 Hz). Since an EOG signal could have been affected by EEG, EMG, and ECG artifacts (and vice versa), applying an algorithm in order to remove the artifacts and noises was necessary. A cascade of three adaptive filters based on a least mean square algorithm was employed, and by these means, ECG, EEG, and EMG artifacts were correspondingly eliminated (more information in [11]).

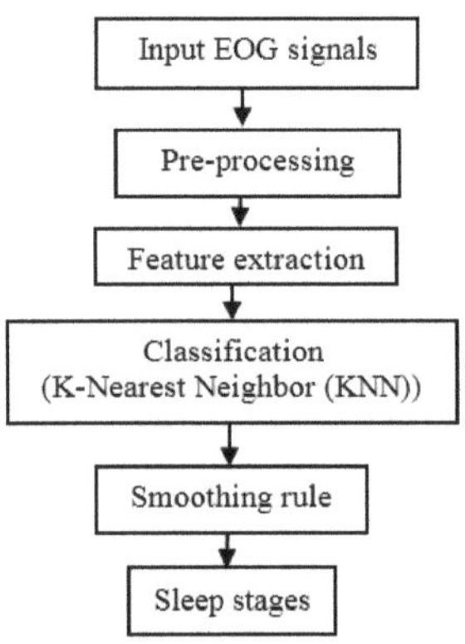

Fig. 1 Block diagram algorithm of automatic sleep stage detection using EOG signals.

2. *Feature Extraction*

Numerous features were extracted from the EOG signal in the time and frequency domain, such as variance, maximal peak amplitude value (MPAV), minimum peak amplitude value (MPAV), total power, energy entropy (i.e. Shannon entropy) and cross-correlation. In order to select the best feature that classified variations in sleep stages and wakefulness, the Sequential Feature Selection (SFS) method was used.

3. *Classification*

In this study, the K-Nearest Neighbor (KNN) was used for classification of sleep and wakefulness stages. The KNN is based on a non-parametric method and can be employing for different pattern classification approach, normally represented as a robust classifier. The KNN classifier is based on a comparison between a new sample (testing data) and baseline (training data). It attempts to find out the KNN within the baseline, and indicates a class which seems to be more normally the nearest neighbor of K. The value of K might need to be diverse in order to detect the corresponding class between the training and testing data. In this paper, the value of K varies from 1 to 5. The Euclidean distance metric was utilised for calculating the distance between the two points. The training and testing data was evaluated based on a 10-fold cross-validation for each group of patients.

4. *Smoothing Rule*

The smoothing rule is a common technique used for increasing the accuracy of detecting the sleep stages. This rule was utilised as in the following example: three consecutive readings of N1, N2, and N2 were replaced as sequence N1, N1, N1 (more information in [12]).

III. RESULTS

In this pilot study, we utilised the EOG signals only for detection of the sleep stages of 30 patients, comprising 10 healthy or Controls (Group 1), 10 PLMS (Group 2) and 10 SAHS (Group 3) patients. Numerous features were extracted from the EOG signal based on different frequency bands as mentioned in the previous section. The overall agreement, sensitivity, and specificity of the detection of sleep stages for Group 1 patients was 83.5%, 85%, and 88% respectively as shown in Table 2. The Cohen's Kappa was 0.79. The general agreement, sensitivity, and specificity for detection of the sleep stages of Group 2 subjects were 80%, 82%, and 86%, respectively. The Cohen's Kappa was 0.71, which was lower than the Cohen's Kappa of the healthy (Control) subjects. The reason for this is that the normal distribution of sleep stages with healthy Controls was much more consistent than that of the PLMS patients. Table 3, shows the confusion matrix, sensitivity, and specificity of the sleep stages of a PLMS patient from Group 2 as

Table 2 The average of agreement, sensitivity, specificity and Cohen's Kappa for each group of patients.

Number of group subject	**Accuracy (%)**	**Sensitivity (%)**	**Specificity (%)**	**Cohen's Kappa**
10 x healthy (group 1)	83.5	85	88	0.79
10 x PLMS (group 2)	80	82	86	0.71
10 x SAHS (group 3)	78	77	80	0.67

Table 3 The confusion matrix of a PLMS patient from Group 2.

		Automatic detection					
		W	***N1***	***N2***	***N3***	***R***	***Sensitivity (%)***
Visual detection	***W***	**524**	6	163	5	28	87
	N1	50	**150**	163	9	36	70
	N2	25	5	**3091**	13	44	80
	N3	10	0	191	**224**	19	85
	R	186	11	277	16	**836**	78
	Specificity (%)	88	88	60	87.6	82.2	

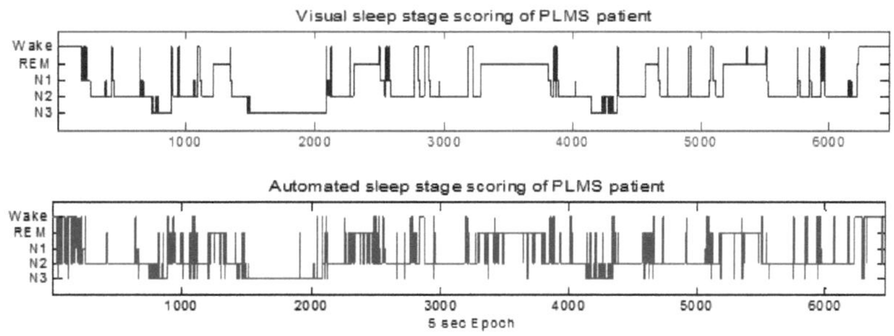

Fig. 2 The hypnogram of visual sleep stage scoring vs. automatic scoring for a PLMS patient.

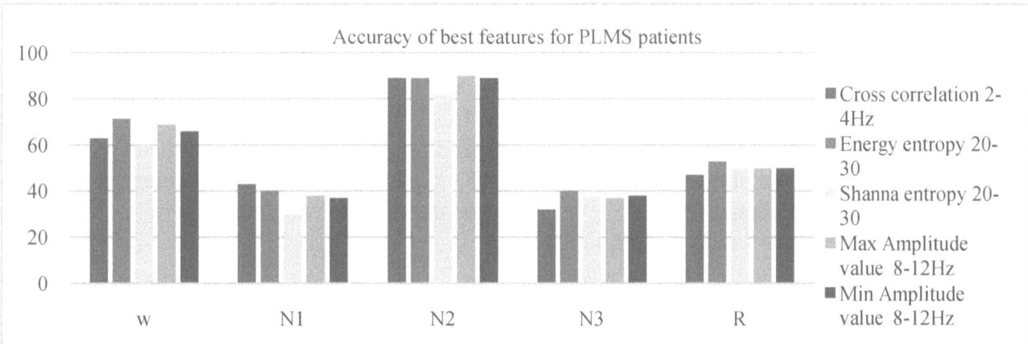

Fig. 3 Show the accuracy of the selected features for detection sleep stages for Group 2.

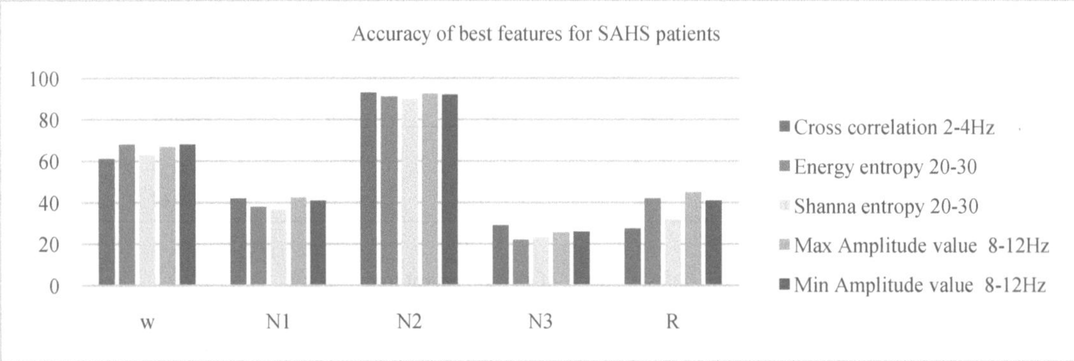

Fig. 4 Show the accuracy of the selected features for detection sleep stages for Group 3.

an example. It is obvious that the total number of sleep stage N2 with PLMS patient was higher than other sleep stages, which decreased the detection of stages N1 and R. On the other hand, the overall agreement, sensitivity, and specificity for detection of sleep stages for Group 3 patients was 78%, 77%, and 80%, respectively. Whilst, the Cohen's Kappa represented was lower than in the other two groups by 0.67. Figure 2, shows the hypnograms of visual sleep stage scoring vs. automatic scoring for a PLMS patient as opposed to with automatic sleep stage detection. Figure 3, 4 shows accuracy of the best features for detection the sleep stages for Group 2 and 3.

IV. DISCUSSION

In this pilot study, our main contribution was in utilising the EOG signals for the automatic sleep stage detection. The overall inter-rate agreement between the visual sleep scoring and automatic sleep stage scoring was 80.5%, with Cohen's Kappa of 0.73. We employed different features which were extracted from the EOG signals and then utilised the KNN classifier for the detection of wakefulness and sleep stages. Some studies used the decision rule based on various thresholds for predicting the sleep stages [9]. However, the results obtained indicated slight an improvement of the specific alpha thresholds for offline applications by accuracy 73%.

Therefore, we used a KNN classifier due to its simplicity and strength in detecting the sleep stages. Several studies employed signals in addition to EOG signals for automatic sleep stage detection, such as EEG and EMG signals [13, 14]. These require more electrodes and more complicated algorithms to increase the accuracy level which has been observed. On the other hand, some studies used only one EEG signal for automatic sleep detection [15].

The number of occurrences of sleep stage N2 in PLMS and SAHS patients were more frequent than in the healthy Control subjects. This was a distinct difference between these three groups. Thus, this led to an overall accuracy of detection sleep stage N2 to be very low, which meant that the KNN classifier was able to predict other sleep stages as sleep stage N2. In Table 3, for example, it was obvious that the total number of occurrences of sleep stage N2 was higher than other sleep stages, which caused a decrease in an overall detection of other sleep stages or wakefulness stage.

Comparable studies have utilised the EOG signal for the detection of sleep stages, or for one particular sleep stage such as N3 [9, 15]. An automatic method was previously developed for detection of SWS based on two EOG channels [12]. This study employed the amplitude criterion for detecting SWS and beta power [18-30 Hz] was utilised to reduce the artefact of EEG signal. The result show inter-rater reliability between the visual and the developed automatic method of 96%, with a Cohen's Kappa value of 0.70. The sensitivity and specificity were 75% and 96%, respectively.

V. CONCLUSIONS

In conclusion, this paper aimed to develop an accurate automatic method for the detection of the sleep stages based on EOG signals. Our results support the idea that automatic sleep stage detection can be implemented based on only two EOG signals and thus be implemented in future ambulatory sleep monitoring and detection.

ACKNOWLEDGMENT

The authors gratefully acknowledge the financial support of the Ministry of Higher Education of the Kingdom of Saudi Arabia. The authors would also like to thank University of MONS - TCTS Laboratory (Stéphanie Devuyst, Thierry Dutoit) and Université Libre de Bruxelles - CHU de Charleroi Sleep Laboratory (Myriam Kerkhofs), which served as the main experimental source for the present research work.

CONFLICT OF INTEREST

All authors have no conflict of interest.

REFERENCES

1. Verhulst SL, Schrauwen N, Haentjens D, et al (2007). Sleep-disordered breathing in overweight and obese children and adolescents: prevalence, characteristics and the role of fat distribution. Archives of disease in childhood. 92(3):205-8.
2. Sadock BJ, Sadock VA. (2010). Kaplan and Sadock's pocket handbook of clinical psychiatry. Lippincott Williams & Wilkins.
3. Miano S, Paolino MC, Castaldo R, Villa MP. (2010) Visual scoring of sleep: A comparison between the Rechtschaffen and Kales criteria and the American Academy of Sleep Medicine criteria in a pediatric population with obstructive sleep apnea syndrome. Clinical neurophysiology. 121(1):39-42.
4. Medicine AAoS, Iber C, (2007). The AASM manual for the scoring of sleep and associated events: rules, terminology and technical specifications, American Academy of Sleep Medicine.
5. Chan M, Estève D, Escriba C, Campo E. (2009) A review of smart homes—Present state and future challenges. Computer methods and programs in biomedicine. 91(1):55-81.
6. Berthomier C, Drouot X, Herman-Stoïca M, et al. (2007) Automatic analysis of single-channel sleep EEG: validation in healthy individuals. Sleep. 30(11):1587.
7. Van Hese P, Philips W, De Koninck J, Van de Walle R, et al. (2001). Automatic detection of sleep stages using the EEG. Engineering in Medicine and Biology Society, 2001 Proceedings of the 23rd Annual International Conference of the IEEE; IEEE.
8. Hang L-W, Su B-L, Yen C-W. (2013) Detecting Slow Wave Sleep via One or Two Channels of EEG/EOG Signals. REM. 17(7.1):17.85-6.6.
9. Virkkala J, Hasan J, Varri A, Himanen S-L, Muller K.(2007). Automatic sleep stage classification using two-channel electro-oculography. Journal of neuroscience methods. 166(1):109-15.
10. The DREAMS Subjects Database. http://www.tcts.fpms.ac.be/~devuyst/Databases/Databa
11. Correa AG, Laciar E, Patiño H, Valentinuzzi M, et al. (2007). Artifact removal from EEG signals using adaptive filters in cascade. Journal of Physics: Conference Series; IOP Publishing.
12. Liang S-F, Kuo C-E, Hu Y-H, Pan Y-H, Wang Y-H. (2012). Automatic stage scoring of single-channel sleep EEG by using multiscale entropy and autoregressive models. Instrumentation and Measurement, IEEE Transactions on. 61(6):1649-57.
13. Garg G, Singh V, Gupta J, Mittal A, Chandra S. (2011).Computer Assisted Automatic Sleep Scoring System Using Relative Wavelet Energy Based Neuro Fuzzy Model. WSEAS Transactions on Biology and Biomedicine. 8(1):12-24.
14. Anderer P, Gruber G, Parapatics S, Woertz M, Miazhynskaia T, Klösch G, et al. (2005). An E-health solution for automatic sleep classification according to Rechtschaffen and Kales: validation study of the Somnolyzer 24× 7 utilizing the Siesta database. Neuropsychobiology. 51(3):115-33.
15. Virkkala J, Hasan J, Värri A, Himanen S-L, Müller K. (2007). Automatic detection of slow wave sleep using two channel electro-oculography. Journal of neuroscience methods. 160(1):171-7.

Author: Emad Malaekah
Institute: RMIT University
Street: Swanston Street
City: Melbourne
Country: Australia
Email: s3321008@student.rmit.edu.au

Influence of Polyvinylalcohol on the Size of Calcium Ferrite Nanoparticles Synthesized Using a Sol-gel Technique

N.H. Sulaiman[1], M.J. Ghazali[1,2], B.Y. Majlis[2], J. Yunas[2], and M. Razali[1,3]

[1] Department of Mechanical and Materials Engineering, Faculty of Engineering and Built Environment, Universiti Kebangsaan Malaysia, 43600 Bangi, Malaysia
[2] Institute of Microengineering and Nanoelectronics, Universiti Kebangsaan Malaysia, 43600 Bangi, Malaysia
[3] Department of Management Office, Faculty of Dentistry, Universiti Kebangsaan Malaysia, 43600 Bangi, Malaysia

Abstract— **In this study calcium ferrite nanoparticles are synthesized using a sol-gel method and the magnetic properties of these nanoparticles are characterized. The influence of process parameters, namely molar ratio of starting materials, calcination temperature and effect of polyvinylalcohol (PVA) on particles size is also investigated. These parameters are also necessary to produce nano-sized calcium ferrite particles with a superparamagnetic behavior for drug delivery systems. In the synthesis a PVA polymer is initially mixed with calcium nitrate and ferric nitrate by using a magnetic stirrer. Citric acid is then used to increase the chelation to produce calcium ferrite nanoparticles. The prepared samples are calcined at 550 °C for 2 h in a vacuum oven, and a furnace. The morphological features of the calcium ferrite nanoparticles examined using scanning electron microscopy. The structure and magnetic properties of these materials are also characterized through X-ray diffraction and vibrating sample magnetometry. SEM images reveal spherical calcium ferrite nanoparticles. These nanoparticles also exhibit an orthorhombic structure with a crystal size of 16.8 nm. The magnetic properties of the sample measured using a vibrating sample magnetometer of the sample measured using vibration sample magnetometer are as follows: superparamagnetic with a coercivity of 43.35 G and a magnetic saturation of 59.3 emu/g. This study will contribute to the development of superparamagnetic nanoparticles for targeted drug delivery systems.**

Keywords— **Magnetic nanoparticle, Superparamagnetism, Polyvinylalcohol.**

I. INTRODUCTION

Magnetic nanoparticles have been considered as important materials in Biomedicine and Biotechnology, Engineering, materials science, and environmental science [1- 5]. As such magnetic nanoparticles are support for magnetic drug targeting for bimolecular sensors, separation and purification, in vivo imaging, hyperthermia treatment. Furthermore magnetic nanoparticles applications have been improved, though the specific control of particle composition, stability and surface functionality [6-18]. The composition, cation distribution, and size of pure nanoferrites, such as Fe_3O_4, $NiFe_2O_4$, $CoFe_2O_4$, $ZnFe_2O_4$, and $MnFe_2O_4$ have play important roles in terms of their properties and applications [19]. Among these ferrites, $CaFe_2O_4$ is the most commonly in various applications, such as oxidation catalysts, high- temperature sensors, and gas absorbers, etc. [20]. $CaFe_2O_4$ also shows relevant physical characteristics, high thermal stability; therefor this ferrite is applicable over a wide temperature range [21, 22]. In the present study calcium ferrite nanoparticles synthesized using a simple sol-gel method with polyvinylalcohol as a surfactant. The structural, morphological and magnetic properties of $CaFe_2O_4$ NPs, the particles are also characterized through X-ray diffraction (XRD), scanning electron microscopy (SEM) and vibrating sample magnetometry (VSM). PVA is the largest synthetic water-soluble polymer produced worldwide [23]. PVA is a hydrophilic and biocompatible synthetic polymer and widely used in various biotechnological and biomedical fields because of its excellent chemical and physical properties, easy processing techniques and low cytotoxicity [24]. In addition, PVA a commonly used polymer surfactant [25], and biodegradable water soluble synthetic semicrystalline polymer, extensively investigated because of its sensitive properties and wide range of applications, such as fiber (vinylon), films in the paper industry, material for textile sizing, modifier of thermosetting resins, pressure sensitive adhesives in plywood manufacturing, and emulsifier. PVA can form hydrogel via different processes such as, cross-linker, and γ- irradiation. Moreover PVA is used in biological applications, including tissue replacement, articular cartilage, artificial skin, wound dressing, drug delivery system, artificial muscle, and actuators [26, 27]. PVA hydrophilic polymer is also selected as a coating of magnetic particles because this polymer its exhibit biocompatibility, biodegradability, functionality. Furthermore, PVA is a good stabilizer of noble metal particles [28- 30].

A. *Superparamagnetism*

Superparamagnetism determines magnetic properties depending on the presence or absence of an external magnetic field [31]. These particles exhibit "superparamagnetism". On an external magnetic field, they are magnetized until their saturation magnetization is achieved; once removed from the magnetic field, they no longer display any residual

F. Ibrahim et al. (eds.), *International Conference for Innovation in Biomedical Engineering and Life Sciences*,
IFMBE Proceedings 56,
DOI: 10.1007/978-981-10-0266-3_41

magnetic interaction. This property is dependent on size and particularly when the size of nanoparticles is as small as 10–20 nm. With this size, these nanoparticles do not exhibit multiple domains, in contrast to those found in large magnet instead, these nanoparticles form a single magnetic domain and exhibits high magnetic susceptibility. In Figure 1 the magnetic field of domain walls in ferrimagnetic materials is removed and is aligned to the direction of the magnetic field. In superparamagnetic materials, the magnetic field is usually defined as a single- domain structure without domain walls, but the magnetic moment is aligned to the direction of the applied magnetic field. Hence, these nanoparticles provide a stronger and more rapid magnetic response than bulk magnets with negligible remanence (residual magnetization) and coercivity necessary field to obtain zero magnetism when a magnetic field is applied (Figure 2) [32, 33]. Unique to nanoparticles, superparamagnetism is essential for drag drug molecules to their target sites. In the body under the influence of an applied magnetic field. Once, the applied magnetic field is removed, magnetic particles do retain no residual magnetism at room temperature and thus unlikely agglomerate; instead, these magnetic particle are easily dispersed, As a consequence, these particles evade uptake by phagocytes and increase their half-life in circulation. Moreover, superparamagnetic particles unlikely cause of thrombosis or blockage of blood capillaries because these particles exhibit a negligible tendency to agglomerate.

II. MATERIAL SYNTHESIS AND CHARACTERIZATION

A. Materials

Calcium nitrate, Ferric nitrate, Citric acid, ethelenglycol, and ethanol were purchased from Accot. Lab. Supplies Sdn. Bhd. Polyviylalcohol was purchased from the Department of Chemical and Process Engineering, Faculty of Engineering and Built Environment, Universiti Kabangsaan Malaysia. Those chemicals were used in sample synthesis. The reagents were of analytical grade with 90% purity.

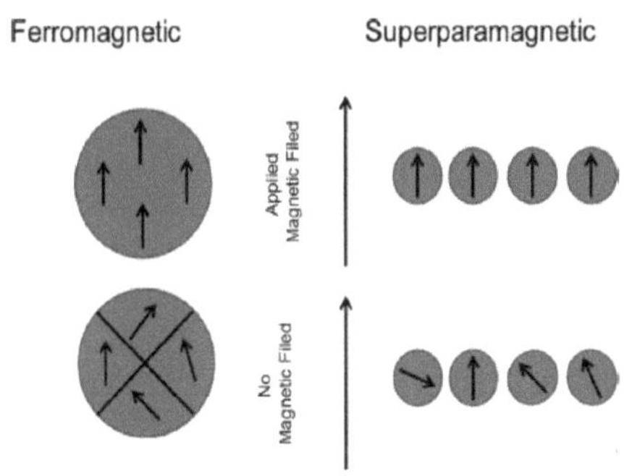

Fig. 1 Magnetic Moment in ferromagnetic and superparamagnetic materials [34]

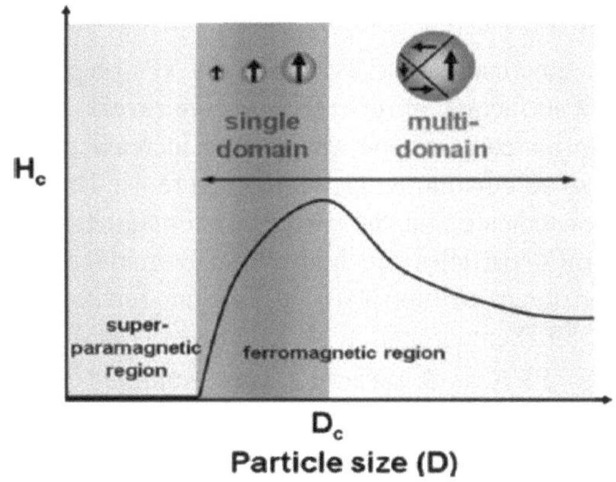

Fig. 2 Schematic of the coercivity size -relation of small particle size [33]

B. Sample Synthesis

Magnetic calcium ferrite NPs were prepared using a sol-gel method. Calcium nitrate $Ca(NO_3)_2$ was initially mixed with $Fe(NO_3)_3$ at a molar ratio of 1:1 M; the mixture was dissolved in 100ml of distilled water with 2M citric acid as a chelating agent. Afterword 3.4gm from PVA dissolved in 30ml of distilled water at room temperature was added to the solution as a surfactant. The solution was continuously stirred for 4 h and heated with a magnetic stirrer at 80 °C. Ethelenglycol (6 mL) was then added to the mixture. The color of the viscous gel changed from orange to brown. The gel was subsequently dried in an oven at 70_80 °C overnight. The resulting particles were collected and calcined in the furnace 550 °C for 2 h a furnace. Calcium ferrite nanoparticle powder was obtained.

III. RESULTS AND DISCUSSION

A. XRD Analysis

Figure 3 shows the XRD pattern of $CaFe_2O_4$ NPs calcined at 550 °C with a molar ration 1:1M of ferric nitrate, calcium nitrate, and citric acid. The magnetic $CaFe_2O_4$ nanoparticles were successfully obtained through the described synthesis. This characteristic is typical for nanocrystalline structures at approximately 2θ =20- 80. All of the peaks indexed to the diffraction peak of the orthorhombic structure are well matched with those found in the standard $CaFe_2O_4$ spectra (JCPDS 78-4321); these peaks are allocated to the typical crystal plane of $CaFe_2O_4$ with good purity [35]. The average crystallite size was calculated using Scherer's equation expressed as follows (1):

$$D = K\lambda/\beta cos\theta \quad (1)$$

Where K is Scherer's constant (K=0.89), λ is the X-ray wavelength, β is the peak width at half maximum, and θ is Bragg's diffraction angle [36]. An average crystal size of 16.8 nm of $CaFe_2O_4$ NPs was obtained.

B. Magnetic Property Analysis

The magnetization curve of $CaFe_2O_4$ NPs prepared with PVA as a surfactant, at room temperature reveals that magnetization increases as the temperature increase to 550 °C because of calcination temperature (Figure 4). The hysteresis curves indicate that the particles are magnetic and that the $CaFe_2O_4$ particles are highly superparamagnetic. The saturation magnetization *(Ms)* of calcium ferrite nanoparticles is 59.38emu/g and their coercivity (Hci) value is 44.373 G. PVA as a surfactant was examined to control crystal size. PVA was also used as a surfactant to determine the optimum mixture composition to achieve a uniform particle size distribution.

C. Characterization of $CaFe_2O_4$

The crystallite structure and size of the synthesized $CaFe_2O_4$ samples were calculated using an X-ray Powder Diffractometer (XRD) model - D8 ADVANCE BRUKER AXS with Cu Ka radiation (1.5406 A) in a 2h scan range of 20–80. The magnetic properties of the powders were examined by the Vibrating sample Magnetometer (VSM) model – LAKESHORE 7404 SERIES and a scanning electron microscope (SEM). The sol-gel process was performed in a clean room for an accuracy purpose.

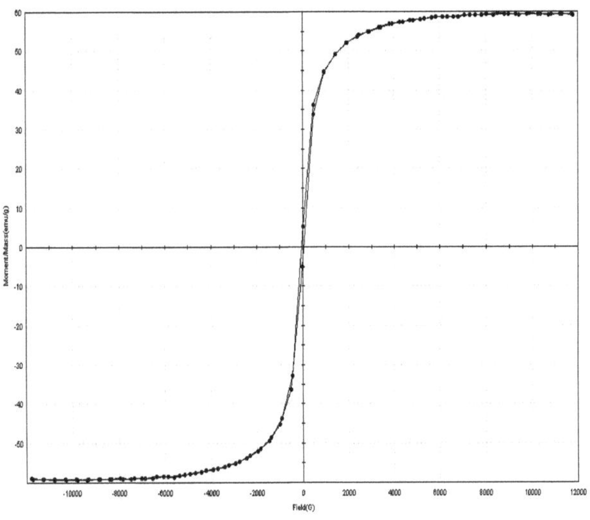

Fig. 3 XRD spectra of calcium ferrite nanoparticles

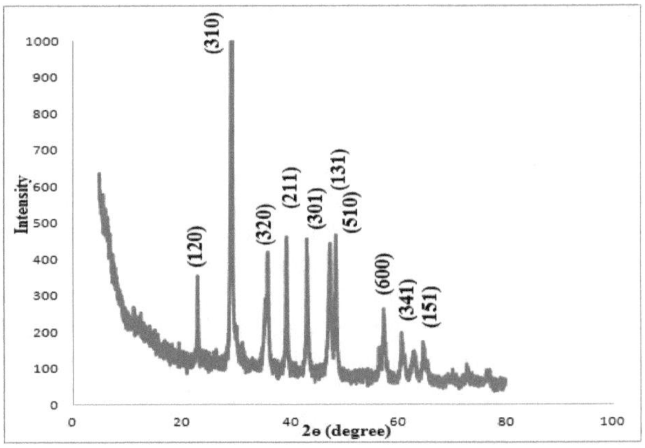

Fig. 4 VSM Characteristic of the synthesized calcium ferrite nanoparticles

D. SEM and EDX Analysis

The SEM & EDX from the images reveal that the morphological characteristics of $CaFe_2O_4$ NPs are slightly affected by the surfactant. Furthermore, the sample is spherical, and the nanoparticles strongly accumulate. After polymerization is completed, the size of the particles increases and the dispersion of the particles is greatly improved [Figure 5(a)]. EDX analysis [Figure 5(b)], confirms that Fe, Ca, and O elements are present in the sample when $CaFe_2O_4$ is prepared.

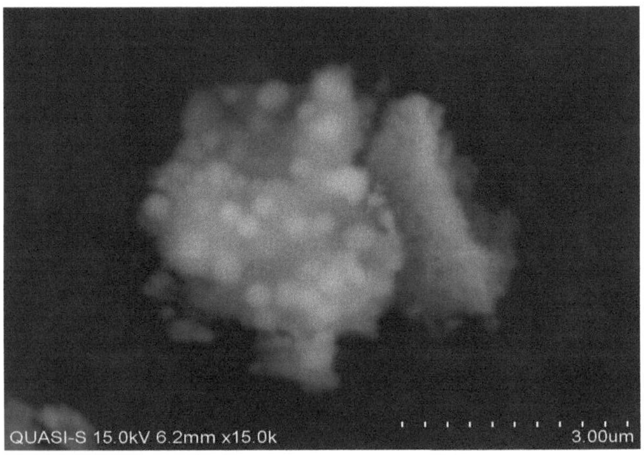

Fig. 5(a) SEM image of the synthesized calcium ferrite nanoparticles

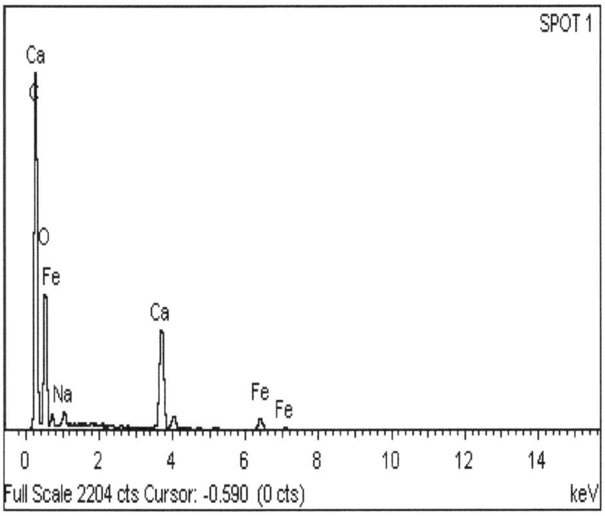

Fig. 5(b) EDX characteristic of calcium ferrite nanoparticles

IV. Conclusions

$CaFe_2O_4$ NPs were successfully prepared via the sol-gel method, and the other materials used in the synthesis are shown in Figure 3. The $CaFe_2O_4$ nanoparticles were calcined at 550 °C. The crystallinity and structure of calcium ferrite nanoparticles were also determine on the basis of the XRD spectrum. The obtained $CaFe_2O_4$ NPs are magnetic and their magnetic saturation is as high as 59.38emu/g. These magnetic properties imply that $CaFe_2O_4$ NPs can be used in biomedical applications such as drug delivery.

Acknowledgment

The author, gratefully thanks to Dr. Muneer M. Ba-Abbad Department Of Chemical And Process Engineering, Faculty Of Engineering And Built Environment, Universiti Kabangsaan Malaysia for his supported.

Conflict of Interest

The author has no conflict of interest.

References

1. Robert C. (2000) Modern Magnetic Materials: Principles and Applications, Wiley, New York.
2. Nicola A. (2003) Magnetic Materials: Fundamentals and Device Appli cations, Cambridge University Press, Cambridge, UK.
3. Rochelle M., Udo S. (2003) The Iron Oxide: Structure, Properties, Reactions, Occurrences and Uses, Wiley-VCH, Weinheim, Germany.
4. Kenneth J. (2001) Nanoscale Materials in Chemistry, Wiley-Interscience, New York.
5. Schmid G. (2004) Nanoparticles, Wiley-VCH, Weinheim.
6. Marcel B. Jr., Mario M., Peter G. et al. (1998) Semiconductor nanocrystals as fluorescent biological labels, Science, 281:2013-2016.
7. WCW C., SM Nie (1998) Quantum dot bioconjugates for ultrasensitive nonisotopic detection. Science, 281:2016-2018.
8. Shaopeng W., Natalia M. et al. (2002) Antigen/antibody immunocomplex from CdTe nanoparticle bioconjugates, Nano Letters, 2:817-822.
9. Cathryn M., Irene Z. et al. (2000) Microsphere- mediated delivery of recombinant AAV vectors in vitro and in vivo, Mol Therapy, 1:S239.
10. D. Panatarotto, CD. Prtidos et al. (2003) Immunization with peptide-functionalized carbon nanotubes enhances virus-specific neutralizing antibody responses, Chemistry & Biology, 10:961-966.
11. Edelstein RL., Tamanaha CR. et al. (2000) The BARC biosensor applied to the detection of biological warfare agents, Biosensors Bioelectron, 14:805-813.
12. Nam JM., Thaxton CC. et al. (2003) Nanoparticles-based bio-bar codes for the ultrasensitive detection of proteins, Science, 301:1884-1886.
13. Rahina M., Jessica P. R. et al. (1995) Protein-sized quantum dot luminescence can distinguish between "straight", "bent", and "kinked" oligonucleotides, Journal American Chemical Society, 117:9099-9100.
14. Mah J., Wong H. et al. (2003) Biomimetic processing of nanocrystallite bioactive apatite coating on titanium, Nanotechnology, 14:619-623.
15. Agustin I., Witold B. et al. (2003) Nanohybrid scratch resistant coating for teeth and bone viscoelasticity manifested in tribology, Materials Research Innovation, 7:110-114.
16. Robert S. M. (1982) Immunospecific ferromagnetic iron dextran reagents for the labeling and magnetic separation of cells. J Immunol Methods, 52:353-367.
17. Weissleder R., Elizondo G. et al. (1990) Ultrasmall superparamagnetic iron oxide: characterization of a new class of contrast agents for MR imaging, Radiology, 175:489-493.
18. Parak WJ., Boudreau R. et al. (200) Cell motility and metastatic potential studies based on quantum dot imaging of phagokinetic tracks, Advance Material, 14:882-885
19. Candeiaa R. A., Bernardib M.I.B et al. (2004) Synthesis and characterization of spinel pigment CaFe2O4 obtained by the polymeric precursor method. Materials Letters 58, 569–572
20. Na-Oki I., Yousuku O. et al. (2005) H2S absorption behavior of calcium ferrite prepared in the presence of coal, Energy Fuels 19, 170–179
21. Daisuke H., Yoichi S. et al. (2006) Mössbauer characterization of calcium ferrite oxides prepared by calcining Fe2O3 and CaO, Hyperfine Interact 167, 809-813
22. Arvind S., S.N. Dolia et al. (2013) Size dependent structural and magnetic behavior of CaFe2O4, Current Applied Physics 13, 830-835.
23. Bin D., Hak-Yok K. et al. (2002) Preparation and characterization of a nanoscale poly(vinyl alcohol) fiber aggregate produced by an electrospinning method. Journal of Polymer Science Part B: Polymer Physics Volume 40, Issue 13, pages 1261–1268.
24. Francesco D., Iiaria A. et al. (2012) Tuning multi/pluri-potent stem cell fate by electrospun poly (L-lactic acid)-calcium-deficient hydroxyapatite nanocomposite, Biomacromolecules, 13(5), 1350–1360
25. Changlu S., Hak-Yonk K. et al. (2003) Fiber mats of poly(vinyl alcohol)/silica composite via electrospinning, Materials Letters 57, 1579 – 1584
26. Jaydevisnh G., Amit B. et al. (2006) Studies on the crosslinking of Poly (Vinyl Alcohol), Journal Polymer Research, 13: 161-169
27. Herman M., Cardina M. et al. (2008) FTIR spectroscopy characterization of poly (vinyl alcohol) hydrogel with different hydrolysis degree and chemically cross-linked with glutaraldehyde, Mater Sci Eng. C, 28:539.

28. Mohapatra S., Pramanik N. et al. (2006) Synthesis and characterization of ultrafine poly vinylalcohol phosphate) coated magnetite nanoparticles, J. Nanosci. Nanotechnol, 6: 823.
29. Petri-Fink A., Chastellain M. et al. (2005) Development of functionalized superparamagnetic iron oxide nanoparticles for interaction with human cancer cells, Biomaterials 26 (2005) 2685.
30. Cavalieri F., Chiessi E. et al. (2008) Novel pva-based hydrogel microparticles for doxorubicin delivery, Biomacromolecules 9: 1967-73.
31. Margarethe H. A., Brigitte et al. (2009) Superparamagnetic nanoparticles for biomedical applications, in:nanostructured materials for biomedical applications. Transworld research network, kearala, india, (2009) pp. 119–149.
32. Dave SR., Gao X. (2009) Monodispersed magnetic nanoparticles for biodetection, imaging, and drug delivery: a versatile and evolving technology. Wiley Interdiscip Rev Nanomed Nanobiotechnol, 1(6):583–609.
33. Kodama RH. (1999) Magnetic nanoparticles, J Magn Magn Mater. 200(1–3):359–372.
34. Young-Wook J., Jin-Sil C., et al. (2007) Heterostructured magnetic nanoparticles: their versatility and high performance capabilities, chemical communications, pp 1203–1214
35. Harold P., Leroy E. (1974) X-ray Diffraction Procedures for Polycrystalline and Amorphous Materials, Wiley, New York.
36. Lavanya K., Verma N. K. (2013) Synthesis, characterization and in vitro cytotoxicity study of calcium ferrite nanoparticles, Materials science in semiconductor processing (16) 1842-1848

Address for correspondence:	M.J. Ghazali, Department of Mechanical and Materials Engineering, Faculty of Engineering and Built Environment, Universiti Kebangsaan Malaysia, 43600 Bangi, Malaysia.
Tel.:	0060192622075;
E-mail:	mariyam.j.ghazali @gmail.com.

A Survey of Human Age Estimation Techniques from Bone Microstructures

I. Khan[1], F. Mohd Nor[2], and M.M. Abdul Jamil[1]

[1] Biomedical Engineering Modeling and Simulations (BIOMEMS) Research Group, Faculty of Electrical and Electronics Engineering, University Tun Hussain Onn Malaysia, Batu pahat, Malaysia
[2] Department of Pathology, Universiti Kebangsaan Malaysia, Jalan Yaacob Latif, Bandar Tun Razak, Cheras, 56000 Kuala Lumpur, Malaysia

Abstract— **The skeleton of human changes with increasing age due to change in bone density. Bone tissues have the capability to store the information of these variations which can be used to predict age from human remains. This study discuss the process and factors that affect human skeleton with increasing age. The techniques to carry out human age estimation are discussed as well, which shows that there is a strong co-relation between increasing age and bone micro structures. Microscopic analysis of osteons, haversian canal and circumferential lamella were analyzed in these techniques to estimate age.**

Keywords— **Human age estimation, bone microstructures, microscopic features, bone remodeling.**

I. INTRODUCTION

Determination of human age after death from their bone cells is an important and frequent requirement in current forensic sciences and in developing demographic profiles.

The biological age of humans at death can be detected by analyzing the characteristic of microscopic features of the bone cross section. Much research has been done on developing methods to estimate human age at death from their remains. Initially, macroscopic, morphological analyses of bone were performed to estimate human age at death. However, due to the fragmentary nature of skeletal remains often encountered in forensic science, age estimation methods based on bone microstructure have proven useful and provide major benefits [1].The researchers have been attempting to develop microscopic methods of estimating human age from humerus, radius, tibia, ulna, fibula and femur bones. The microscopic examination of bone tissue to estimate age is based upon age-associated changes in histomorphological features that occur due to the life-long metabolic process in bone called remodeling. Modeling and remodeling are the two processes in human body that control the shape and structure of bones with increasing age. To understand and discuss the development and application of age estimation methods requires an understanding of basic bone biology.

Section II of this paper will discuss basic bone cells that are responsible for modeling and remodeling. Section III will discuss modeling and remodeling process in bones while section IV will demonstrate different methods of age estimation after human death from bone cells.

II. BONE CELLS

The bones accomplishes its growth, development and maintenance through special cells which include osteoclasts and osteoblasts. Osteoclasts are bone resorbing cells while osteoblasts are bone forming cells. It is the activity of osteoclasts and osteoblasts that is responsible for bone histomorphology and allow for histological estimation of human age after death.

Bone resorption or formation occurs on preexisting bone surfaces or envelopes [2]. These bone envelopes includes periosteal, endosteal, haversian and trabecular envelope. The following section will give an overview of the four major type of cells that are responsible for bone formation and changes in the features of bones that can be used to estimate human age after death.

A. *Osteoclasts, Osteoblasts, Osteocytes and Bone Lining Cells*

Osteoclasts: are multinucleated, highly mobile, primary bone resorbing cells. The size of the osteoclasts is approximately (50-100 μm) in diameter [1]. Osteoclasts cell formation is the result of the fusion of mononuclear progenitors of the monocyte-macrophage family, which originate in the hemopoietic stem cells from bone marrow [2]. Earlier study proposed that the precursor is drawn to bone surface by chemical signals from bone lining cells and change shape to allow osteoclasts access to the exposed extra-cellular matrix [3]. Once differentiated, resorption will occur along the ruffled border of the cell where the osteoclast has attached to the bone matrix. Proton pumps in the ruffled border release protons into the bone-resorbing chamber created where the osteoclast attaches to the bone matrix at the sealing zone. The releasing of proton decreases the pH of the bone resorbing chambers and creates an acidic environment (pH<4) which dissolves the minerals of the bones. Enzymes that are responsible for dissolving the collagenous and non-collagenous matrix of the bone are also released into this chamber which then digest the collagen [4].

F. Ibrahim et al. (eds.), *International Conference for Innovation in Biomedical Engineering and Life Sciences*,
IFMBE Proceedings 56,
DOI: 10.1007/978-981-10-0266-3_42

Osteoblasts: are mononuclear cells that originates from mesenchymal stem cells and are much smaller in size than osteoclasts. Around 100-150 osteoblasts are required to form the amount of bone resorbed by one osteoclast. Osteoblasts are responsible for bone formation, production of osteoid and maintaining skeletal architecture. In addition to maintaining skeletal architecture, osteoblasts also play a role in osteoclasts development and activity.

The osteoblasts nests along the bone surface in areas of active bone formation. They secrete type 1 collagen and non-collagenous proteins. Type 1 collagen is the basic building block of bone while non-collagenous proteins are essential for mineral deposition. After active bone formation, mature osteoblasts ends up in one of three ways: they differentiate further into osteocytes, become quiescent lining cells or undergo apoptosis and die. Approximately 50-70 % mature osteoblasts ends up in apoptosis.

Osteocytes: are considered to be the termination stage of osteoblasts. After the bone formation, some of the osteoblasts are entrapped in small spaces called lacunae. These entrapped osteoblasts in lacunae become osteocytes. They are the most abundant cells in bone tissue (may comprise 90% of all bones tissues). The lacunae of osteocytes are connected by canaliculi, which provides a path for oxygen and nutrients supply from blood capillary to osteocytes. The diameter of canaliculi is considered to be about 0.35μm large. Osteocytes and their numerous cytoplasmic processes forms a three dimensional communication network, which able osteocytes to sense and transmit information about the mechanical stress. Researchers have reported that osteocytes express stretch activated channel and shear-stress responsive element. However, mechanism of signal transduction and genes regulated by mechanical stress are not clarified yet.

Bone lining cells: The bone lining cells are inactive form of osteoblast which have become flattened and remained on the surface after termination of the bone formation process. Bone lining cells are long, slender, and flattened in shape and are about 12μm in length while less than 1μm in thickness. Bone lining cells perform multiple functions similar to those of osteocytes. Bone lining cells are connected to each other via gap junctions. Bone lining cells have the capability of changing and retracting its shape when certain molecular signals are present and play an important role in bone remodeling. Bone lining cells prevents the inappropriate interaction of osteoclast precursors with the bone surface.

III. BONE MODELING AND REMODELING

The skeleton changes across the human life span due to born formation and growth throughout the childhood and gradual loss of bone density in early adulthood. The density of bone is regulated by group of bone cells called osteoclast and osteoblasts. Osteoclasts resorb born while osteoblasts refill the resorbed cavities. Osteoclasts anchors themselves to the bone surface, thereby, creating a micro environment under the cell which is referred to as the sealed zone. Osteoclasts creates an acidic environment within the zone which dissolves the bone minerals content. After the dissolution of bone minerals content, enzymes released from osteoclast remove the remaining collagenous bone matrix to complete the process of resorption. Following resorption, osteoblasts move in to the resorption space and start to produce osteoid. Osteoid, made of collagen, forms a scaffold in which minerals including calcium and phosphate begin to crystalize. Some osteoblasts become trapped in their secreted matrix and thereby become osteocytes, while other go through apoptosis or become bone lining cells which covers the bone surface. This cycle of bone resorption and formation is called bone remodeling. It takes 2 to 5 years for an area on the bone surface to complete one bone remodeling cycle [5]. There is also a process in which bone formation occurs by osteoblasts without prior bone resorption by osteoclast. This refers to bone modeling and it results in increase of bone mass. Bone modeling promotes the growth of bones and is important in maintaining bone strength. While remodeling plays an important role in bone growth by optimizing the bone structure. Distinct phases of bone remodeling are shown in figure 1.

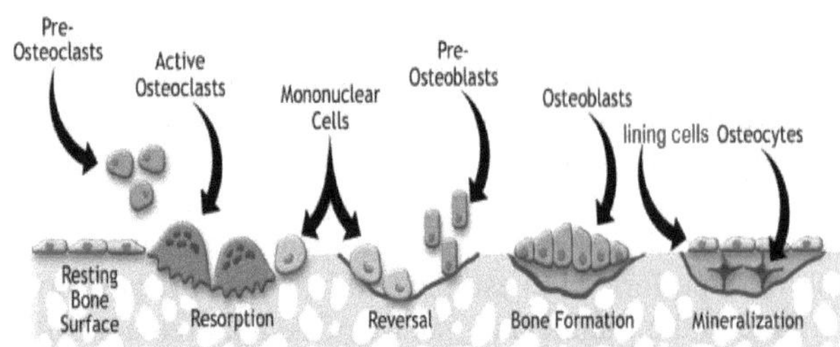

Fig. 1 Schematic presentation of bone remodeling. The osteoclast dissolves bone tissue followed by osteoblast that refill the resorbed cavities with new bone.

IV. REIVEW OF AGE ESTIMATION METHODS FROM BONE CELLS

Balthazard and Lebruneadings: In 1911, Balthazard and Lebrun attempted to estimate age at death using quantitative bone histology and published the first report. The selected and measured haversian canal and attempted to find out the association of increasing age with bone microstructure. They created a hypothesis, that there is an association of bone micro features with increasing age. They could not establish any standard method for age estimation, however, their hypothesis encouraged the development of human age estimation methods using quantitative bone histology [1].

Jowsey: In 1960, Jowsey [6], used microradiographs of femoral cross-sections to study the microstructure of bone and its association with increasing age. Jowsey analyzed and measured the percentage of the bone surface which was occupied by resorption or formation and demonstrated it, as a ratio of the total percentage of bone available. Jowsey reported that bone modeling and remodeling is relatively high in young individuals. This high amount of modeling and remodeling leads to high amount of bone turn over in young age, which results in a high degree of porosity due to the formation of large number of osteons and resorption cavities. Jowsey found a lower rate of turnover in young adults, while a minor increase in the amount of resorption, especially on the endosteal envelope in older adults. Jowsey reported that, up to 25% of the surface of endosteal bone may be occupied by resorption in individuals over 70 years of age. She also reported that there is rare evidence of increase in bone formation or in number of osteons of less than 75% complete mineralization. She reported that individuals over 60 years of age demonstrate high porosity and variation in bone mineral density and there is an increase in the number of osteons that are less than 75% closed.

Kerley: In 1965, Kerley [7] developed a method to estimate human age at death which relied on histomorphometry. Kerley analyzed 126 subjects out of which 115 were European-American and 11 were African-American individuals. 88 subjects out of 126 were males, 29 were females and 9 were individuals of unknown sex. The age range of these individuals were from birth to 95 years and 64 subjects were aged 30 or older. Kerley used a transverse ground cross-section taken from the midshaft of the fibula, tibia, and femur. Kerley analyzed osteon fragments, the percentage of circumferential lamellar bone, absolute counts of intact osteons, , and the number of non-Haversian canals observable in each of four circular fields positioned tangentially to the periosteum. Kerley reported that these four fields were located anteriorly, posteriorly, medially, and laterally and were 1.25 mm in diameter each as shown in Figure 2.

Kerley developed regression equations (linear and curvilinear) for each predicting variable of fibula, tibia and femur. Kerley reported that the number of osteons and osteon fragments in femur increases with increasing age. He also reported that the percentage of circumferential lamellar bone is high in early age but decreases with increasing aget till 55 years of age. After 55 years of age, he reported, it vanishes. Kerley reported that sex and ancestry did not have appreciable effect on the microstructure of bone.

Singh and Gunberg: In 1970 Singh and Gunberg [8] used midshaft of femurs and tibia and considered 1 cm x 1 cm sample of anterior and posterior border. They analyzed 40 subjects out of which 33 were males and 7 were females. They analyzed three variables, the average Haversian canal diameter, the total number of complete osteons, and the average number of lamellae per osteon. They explained that the potential error introduced by counting osteon fragments can be reduced by counting complete osteons only. They reported that average Haversian canal diameter and total number of complete osteons have the strongest correlation with increasing age.

The problem with Singh and Gunberg method is that they choose random location for microscopic fields and they did not give much information about the ancestry of the subjects.

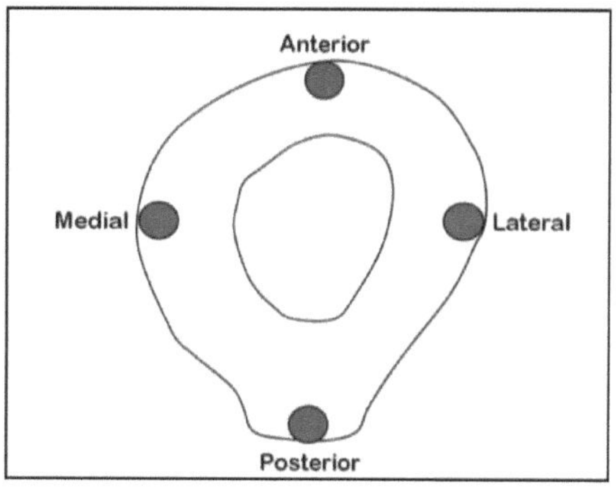

Fig. 2 Cross-section of femoral midshaft demonstrating the location of Kerley's four fields of analysis [7].

Thompson: Thompson [9] in 1979 developed age estimation method using histomorphometry that utilized both the lower and upper extremities. He sampled the anterior portion of femurs and tibia of 116 subjects while the humeri and ulnae of another 31 subjects. He recorded nineteen variables that required both gross and microscopic measurements. He used point-count method [10] with a 10x10 grid eyepiece in four microscopic fields along the anterior periosteal region. Thompson reported that the best perimeter to estimate human age from bone is the percentage of osteonal area. He calculated the percentage of osteonal area by summing the secondary osteon lamellae and the Haversian canal area. He developed both side- and sex-specific equations. In his later study in 1980, he explored sex-specific changes in cortical bone and reported that females undergo an additional 10 years of bone loss compared to males. However, he did not mention the accuracy of sex-specific equations. Later in 1981, he explored the sex-specific method and tested it on 54 subjects which led him to report that femur produces the most accurate age estimation results in white population. However out of these 54, the samples just had 3 Eskimos and 3 African Americans.

Stout: In 1986 Stout [11] developed a histological age estimation method. Stout used cross-sections of the middle-third portion of the sixth rib. Stout's method attempted to reduce complications in sampling error causing from field location and lack of long bones available for sampling. He reported that by sampling the entire cross section, this method can be applied to fragmentary ribs. Stout analyzed several parameters, which included mean osteonal cross-sectional area, cortical area, the number of intact osteons, the ratio of cortical area to total cross-sectional area, intact osteon density, fragmentary osteon density, total visible osteon creations, mean annual osteon creation frequency, effective age of adult compacta and mean annual Haversian bone formation rate. To reduce the complications in differentiating between complete osteons and osteon fragments, he summed the counts of these two variables in each grid area by using a variable called osteon population density. Stout's method indicated that histomorphometric analysis could be used to estimate human age after death using ribs and to interrogate the identity of historically important remains.

Ericksen: In 1991 Ericksen [12] used anterior wedge from the femoral midshaft to estimate human age. The samples included 264 subjects (139 males and 125 females) aging from 14-97 years. Ericksen did not exclude the subjects with pathological conditions because the application of age estimation methods are meant to be used on subjects with unknown medical history. Ericksen evaluated secondary osteons, type II osteons, resorption spaces, non-Haversian canals and osteon fragments, using both microscopically and photographic analysis. The total of the variables from the five fields was divided by the total area evaluated in order to get count per mm^2 (number/mm^2). The variables that were used in photographic images analysis were osteonal bone, unremodeled circumferential bone and fragmentary osteonal bone. To measure the average percent of remodeled, osteonal, and fragmental bone, a 100-space grid was superimposed on the images of the fields. Each space was counted based on the type of bone that covered the majority of that space. However the major problem in photographic image analysis was the different light intensity that did not permit the observer to differentiate structures in the images. Ericksen also developed an equation for pooled sexes.

Cho and Colleagues: In 2002, Cho and Colleagues [13] modified stout's method of histomorphometric aging from ribs. The number of samples that they analyzed were 154 (103 African-Americans, 51 European-Americans). The age range of the subjects were 17-82 years. The mean age of all subjects was 37.8 years. They divided the subjects into two sets (developmental set, validation set). The developmental set contained 69 African-American and 34 European-Americans. The validation set contained 34 African-American and 17 European-American. Cho and colleagues analyzed mean osteonal cross-sectional areas along with the original measurements of Stout's method. They reported that size of osteon is greater in European-Americans samples than in the African-American. The noted differences in osteon population density in different ancestry groups. As like previous methods, this method also came up with the findings that European-Americans have weaker and less dense bones. The validation test of their method produced similar results to the originally developed method.

Maat and Colleagues: In 2006, Maat and colleagues [14] worked on 162 Western European samples out of which 86 were male and 76 were female. The age range was 15 to 96 years old. Maat and colleagues analyzed 3 areas of $1mm^2$ of cortical bone and measured the percentage of non-remodeled circumferential lamellar bone along the periosteal surface. There reported that there is no statistical difference in the percentage of unremodeled bone between males and females.

V. CONCLUSIONS

The basic concepts and techniques of human age estimation from bone cells were highlighted. Different methods proved that bone microstructures can be used and is one of the best predictor of estimating human age after death. The researchers found out, that there is a strong co-relation of bone micro structures such as osteon number, osteon size, non-haversian canal and percentage of circumferential lamella with increasing age. These micro structures can be analyzed and measured using microscopic image processing techniques to estimate human age after death, which can help in many cases, especially in forensic field, reconstruction of population demographics and in individual analysis of human remains. However, the manual techniques often produces subjective results and requires diligent concentration from a highly trained operator. Also manual interpretation of microscopic bone images is error prone because of statistical, structural and temporal variations of objects in a raw bone images. Therefore there is a need of introducing an automatic method to extract micro features from microscopic bone image that do not rely on the expertise and experience of the investigator.

ACKNOWLEDGMENT

The authors wish to express their gratitude to University Tun Hussein Onn Malaysia (UTHM) and ORICC for providing this research opportunity.

REFRENCES

1. Hernández, M.T.J.C.: 'Histomorphometric Estimation of Age at Death Using the Femoral Cortex: A Modification of Established Methods', The Ohio State University, 2012
2. Teitelbaum, S.L.: 'Bone Resorption by Osteoclasts', Science, 2000, 289, (5484), pp. 1504-1508
3. Chambers, T.J., Darby, J.A., and Fuller, K.: 'Mammalian collagenase predisposes bone surfaces to osteoclastic resorption', Cell and tissue research, 1985, 241, (3), pp. 671-675
4. R. Bruce Martin, D.B.B., Lovell, Nancy C.: 'Structure function, and adaptation of compact bone', American Journal of Physical Anthropology, 1990, 82, (1), pp. 116-117
5. Raggatt, L.J., and Partridge, N.C.: 'Cellular and molecular mechanisms of bone remodeling', The Journal of biological chemistry, 2010, 285, (33), pp. 25103-25108
6. J., J.: 'Age changes in human bone.', Clinical Orthopaedics, 1960, 17, pp. 210-218
7. Kerley, E.R.: 'The microscopic determination of age in human bone', American Journal of Physical Anthropology, 1965, 23, (2), pp. 149-163
8. Singh, I.J., and Gunberg, D.L.: 'Estimation of age at death in human males from quantitative histology of bone fragments', Am J Phys Anthropol, 1970, 33, (3), pp. 373-381
9. Thompson, D.D.: 'The core technique in the determination of age at death of skeletons', Journal of forensic sciences, 1979, 24, (4), pp. 902-915
10. Frost, H.M.: 'A model of endocrine control of bone remodelling', Henry Ford Hospital medical bulletin, 1962, 10, pp. 119-170
11. Stout, S.D.: 'The use of bone histomorphometry in skeletal identification: the case of Francisco Pizarro', Journal of forensic sciences, 1986, 31, (1), pp. 296-300
12. Ericksen, M.F.: 'Histologic estimation of age at death using the anterior cortex of the femur', Am J Phys Anthropol, 1991, 84, (2), pp. 171-179
13. Cho, H., Stout, S.D., Madsen, R.W., and Streeter, M.A.: 'Population-specific histological age-estimating method: a model for known African-American and European-American skeletal remains', Journal of forensic sciences, 2002, 47, (1), pp. 12-18
14. Maat, G.J., Maes, A., Aarents, M.J., and Nagelkerke, N.J.: 'Histological age prediction from the femur in a contemporary Dutch sample. The decrease of nonremodeled bone in the anterior cortex', Journal of forensic sciences, 2006, 51, (2), pp. 230-237

Author: Ijaz Khan
Institute: University Tun Hussein Onn Malaysia.
City: Batu Pahat
Country: Malaysia
Email: ijazkhan21sep@gmail.com

The Effects of KGM, Mannose and Co-Supplementation of KGM and Mannose on Mammalian Cells Cultured at Inside and Outside Incubator Conditions

A.B. Marzuke[1], W.S. Wan Kamarul Zaman[1], M. Shahbuddin[2], and S.W. Aung[3]

[1] Department of Biomedical Engineering, Faculty of Engineering, University of Malaya, Kuala Lumpur
[2] Department of Biotechnology Engineering, Kulliyyah of Engineering, International Islamic University Malaysia, Gombak, Selangor
[3] Department of Restorative Dentistry, Faculty of Dentistry, University of Malaya, Kuala Lumpur

Abstract— **Mammalian cells including cancer cells, stem cells and cell lines are important in the application of cells for therapies and research activities. They require the cells to be transported from the laboratory or surgery theater to the bedside, making it a challenge to be exposed at outside, shear stress condition and differences especially in the tropical climate countries. In this research, Konjac Glucomannan (KGM) and D-mannose were examined for their potential use in the delivery of cells when subjected to differences in and outside incubator conditions. The objective of this work was to examine the risk and effects of different type of sugar supplementation in the transportation of cells. In this study, we used mammalian cells; stem cells from human extracted deciduous tooth (SHED), human keratinocyte cell lines (HaCaT) and human breast cancer (MCF-7). We hypothesized that the supplementations of KGM and D-mannose in the culture medium will act as protective agents to the cells due to their unique biological properties interacting via carbohydrate-protein interactions. Experiments were conducted in laboratory conditions to compare the effect of these sugars on the viability of different type of cells when placed at inside and outside incubator condition for 24, 48 and 72 hours. Evaluation of cellular viability and proliferation showed that co-supplementation of KGM and D-mannose inhibited the viability of MCF-7 at both inside and outside incubator conditions while supplementation of these sugars to other cells cultured at outside incubator condition did improve cellular viability and morphology compared to the cells cultured at inside the incubator, elucidating the potential benefit of sugar glycobiology in cellular transportation.**

Keywords— **Mammalian cells, stem cells from human extracted deciduous tooth (SHED), breast cancer cells (MCF-7), keratinocyte cell line (HaCaT), Konjac Glucomannan, D-mannose, cellular viability and proliferation.**

I. INTRODUCTION

The viability of cells such as stem cells, cancer cells and cell lines are important in research activities of tissue engineering and regenerative medicine and also for clinical therapeutics. It is very important to maintain cellular viability while in transportation to be done and thus an effective transportation procedure must be achieved. Factors that were known to affect cellular viability after transportation which are storage, transport temperature and duration was founded in previous study for hematopoietic stem cell [1]. The studies of polysaccharides and plant lectin to assist cellular preservation and transportation have been conducted. For instance alginate, which is a type of plant polysaccharide has the ability to support tissue regeneration as well as cell transportation in the form of cell encapsulation for short term storage of stem cells [2] and mannose rich lectins that were able to preserve human cord blood progenitor cells in suspension cultures up to one month without changing the media [3,4]. Konjac glucomannan (KGM) perennial herbaceous herb and a type of neutral polysaccharide isolated from tubers of *Amorphophallus konjac* K.Koch has the potential use in the transportation of cells due to its ability to increase fibroblast viability in the presence and absence of fetal bovine serum (FBS) [5]. Supplementation of KGM on cell culture also helped to maintain the viability of fibroblasts and adipose derived mesenchymal stem cells in unchanged media up to twenty days and had selective biological effects in the inhibition of keratinocyte viability [5]. KGM' unique biological activities onto different type of cells have the properties of high viscosity, excellent film-forming ability, good biocompatibility and biodegradability, as well as gel performing ability.

Mannose, a simple sugar of unique stereochemistry of C6 glucose plays a vital role in human metabolism, especially in glycosylation of protein [6]. KGM also has the ability to stimulate the viability of mouse fibroblast with the low epichlorohydrine content acting as cross-linker in the hydrogel [7].

The aim of this study was to develop and evaluate KGM and mannose supplementations in medium to facilitate the deliveries of the mammalian cells inter university for research activities as extreme tropical climate in Malaysia with high humidity and rapid weather changes can affect cellular viability and phenotype.

F. Ibrahim et al. (eds.), *International Conference for Innovation in Biomedical Engineering and Life Sciences*,
IFMBE Proceedings 56,
DOI: 10.1007/978-981-10-0266-3_43

II. Materials and Method

A. Subculture of Cells

Cells in flask were washed with 10 ml PBS after the previous media (DMEM) was removed. 5 mL of Trypsin was transferred into the flask and incubated at 37°C. About 5 to 10 minutes was used to allow the cells to detach. After achieving full detachment, 5 mL of DMEM was added into the flask and the solution was transferred into universal tube and centrifuged at 120 rpf for 7 minutes. The pallet formed then resuspended in a known volume of DMEM and transferred into a new flask. The cells were counted before incubation. This experiment used HaCaT cells, TPSC and MCF-7 cells at passage 5-9.

B. Seeding Cells into 12 Well Plates

The media from T-75 flasks containing confluent cells were removed and washed with 10 ml PBS. 5 ml trypsin was added into the flask and incubated at 37°C. The flask was left for 5 to 10 minutes for the cells to detach. After achieving full detachment, 5 mL of DMEM was added into the flask and the solution was transferred into a universal tube and centrifuged at 120 rpf for 7 minutes. The supernatant was discarded and the pallet formed then resuspended in 5 ml DMEM. The cells were counted before seeding into 12 well plates to obtain approximately 20000 cells/ml of media solution in each single well. The media solution was then mixed with cell suspension and transferred into the well plates. After two or three days, the well plates were placed at inside and outside incubator with temperature of 37°C and 25°C respectively.

C. Measurement of Cell Viability Using MTT Assay

The cells in the plates were washed with PBS and 0.2 mL of MTT solution was added into each well. Plates were then incubated for 40 min at 37°C, 5% CO_2 in a humidified atmosphere. During incubation, MTT will be reduced into an insoluble purple-colored formazan product due to mitochondrial dehydrogenase activity. After 40 min, the MTT solution was subsequently removed and 200 µl of acidified isopropanol was added into each well to elute the formazan product from the cells. 100 µl of the isopropanol was then transferred into a 96 well plate and the optical density was read at 540 nm (and reference at 630 nm) using Multiskan Go Thermo Scientific MTT assay machine.

D. Statistical Analysis

Student's t-test was done to determine statistical significance by using means and standard deviation (SD).

III. Results and Discussion

A. The Biological Effects of KGM, Mannose and KGM-mannose on the Viability of SHED

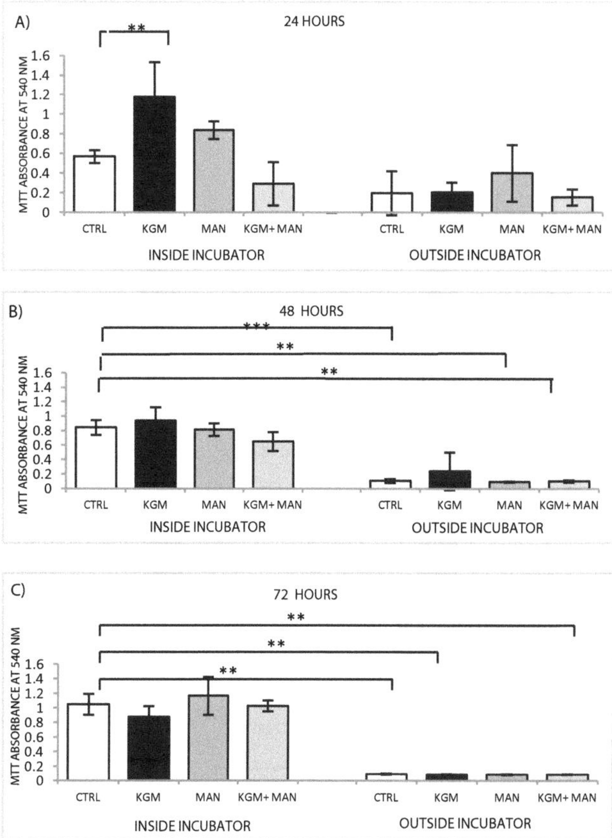

Fig. 1 The effects of supplementation of KGM, mannose, and KGM-mannose on SHED viability inside and outside the incubator. $2x10^5$ teeth pulp stem cells were cultured in 1 mL of 10% DMEM in 12 well plates before being supplemented with medium containing KGM, mannose and KGM-mannose. Cell viability was measured using MTT assay (A) Cell viability after 24 hours (B) Cell viability after 48 hours and (C) Cell viability after 72 hours. Results shown are mean ±SD, (n=2), *P<0.05 significant in comparison to control, **P<0.01 very significant and ***P<0.001 highly significant.

Figure 1 depicts the cellular viability of SHED at inside and outside incubator when supplemented with KGM, mannose and KGM-mannose. The treatment of KGM and mannose increase teeth pulp stem cell viability inside the incubator, however, the viability was inhibited with the co-supplementation of KGM and mannose after 24 hours. Outside incubator, KGM and mannose did not assist in cell proliferation where cellular viability was very low compared to the cells placed inside the incubator.

The supplementation with KGM and mannose did not affect cellular proliferation both inside and outside the incubator after 48 and 72 hours where no significant difference was observed compared to the control. However, cellular viability of the cells place at outside the incubator shows significant decrease compared to control.

B. The Biological Effects of KGM, Mannose and KGM-Mannose on the Viability of HaCaT

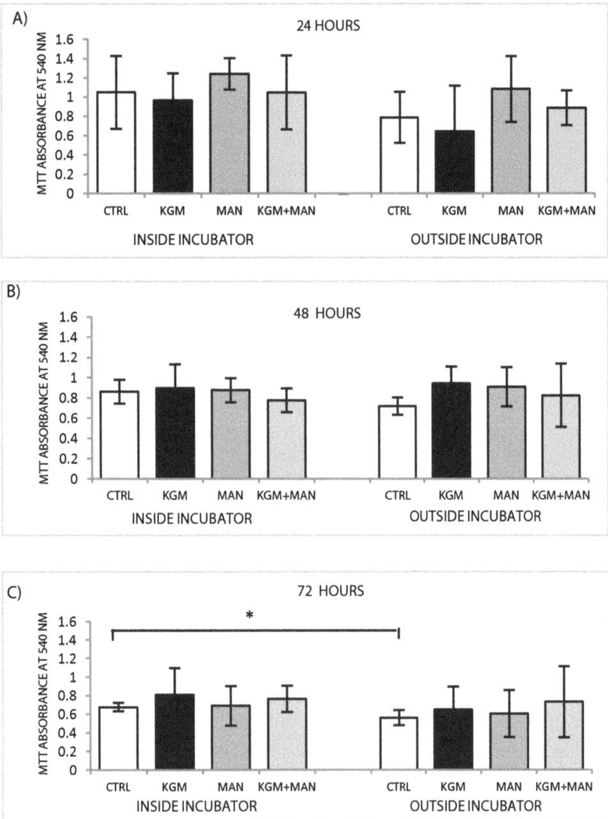

Fig. 2 The effects of supplementation of KGM, mannose, and KGM-mannose on HaCaT viability inside and outside the incubator. $2x10^5$ HaCaT cells were cultured in 1 mL of 10% DMEM in 12 well plates before being supplemented with medium containing KGM, mannose and KGM-mannose. Cell viability was measured using MTT assay. (A) Cell viability after 24 hours (B) Cell viability after 48 hours (C) Cell viability after 72 hours. Results shown are mean ±SD, (n=2), *P<0.05 significant in comparison to control.

The viability of HaCaT cells did not affected by the treatment of KGM, mannose and supplementation of KGM and mannose. There were no significant differences observed both at inside and outside incubator except for slight increases in viability with the treatment of mannose at outside incubator after 48 hours and slight decrease in viability with KGM inside the incubator after 24 hours. Overall, the treatments of sugars did not give any significant difference for both inside and outside incubator indicating the treatment of sugars did not give any effect on HaCaT viability.

C. The Biological Effects of KGM, Mannose and KGM-Mannose on the Viability of MCF-7 Viability

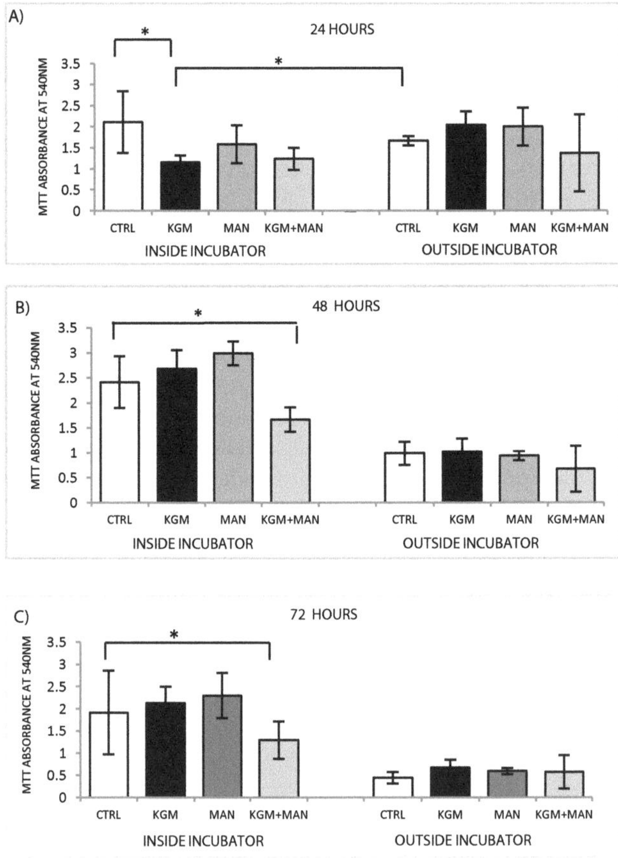

Fig. 3 The effects of supplementation of KGM, mannose, and KGM-mannose on MCF-7 viability inside and outside the incubator. $2x10^5$ MCF-7 cells were cultured in 1 mL of 10% DMEM in 12 well plates before being supplemented with medium containing KGM, mannose and KGM-mannose. Cell viability was measured using MTT assay. (A) Cell viability after 24 hours (B) Cell viability after 48 hours (C) Cell viability after 72 hours. Results shown are mean ±SD, (n=2), *P<0.05 significant in comparison to control.

MCF-7 viability supplemented with KGM, mannose and co-cultured with KGM-mannose placed at outside incubator shows no significant difference when compared to the cells placed inside the incubator after 24 hours with the same supplementations except for the cells supplemented with KGM and KGM-mannose that placed inside the incubator which showed the reduction in viability compared to the same control. The viability of MCF-7 cultured at inside incubator after 48 and 72 hours did not affect by individual treatment with KGM and mannose when compared to control. However, with co-supplementation of KGM and mannose, the viability of MCF-7 cultured inside the incubator significantly inhibited after 24, 48 and 72 hours. All treatments did not give any effect on the simulation or inhibition of MCF-7 viability at outside incubator after 24, 48 and 72 hours.

This effect may be due to the way cells reacted with cell surface receptors that sensitive to both KGM and mannose as cells discriminate sugar isomers differently and this can be due to the anticancer activity caused by the co-supplementation. Cellular viability of the cells placed at outside incubator was not improve with the supplementation of KGM, mannose and both KGM and mannose compared to the control due to almost like hypoxia condition of the surrounding that cause the cells not to be able to simulate or inhibit the cellular viability even with the treatment of the sugars. The outside incubator temperature at 37°C may be not suitable for the cells to survive during transportation. Previous study had found that the optimum temperature for peripheral blood stem cell transportation is 2 to 8°C associated with CD34+ cell viability [1].

IV. Compliance with Ethical Requirements

A. *Conflict of Interest*

The authors declare that they have no conflict of interest.

B. *Statement of Informed Consent and Human Rights*

SHED from deciduous molars were isolated with informed written consent under a protocol approved by University of Malaya Medical Ethics Committee, (DF: CD1411/0087[P]).

V. Conclusion

The effects of KGM, mannose and co-supplementation of KGM and mannose on mammalian cells cultured at inside and outside incubator are not clearly understood. The treatment of these sugars gave different effects on different types of cells. For cellular transportation, we surmise that the supplementation of these sugars onto the cells in some ways increased cellular viability and for cancer cells, shown inhibitory response when cultured at both inside and outside the incubator. The differences of KGM and mannose biological activities on different types of cells can be due to the involvement of specific receptors, glycosylation, the hypoxic effect and also anti-cancer properties of the sugars. Extensive studies on interactions of the sugars to the cells are required to determine their effects on cellular viability and proliferation and thus to gain further understanding on this complex behavior.

Acknowledgement

We thank the University of Malaya for financial support under the University of Malaya Research Grant (Project No: RP040B-15HTM & Project No: RP020B-13AET) and the Ministry of Higher Education (MOHE) for financial support (MOHE High Impact Research Grant: UM.C/625/1/HIR/MOHE/DENT/1).

Conflict of Interest

The authors declare that they have no conflict of interest.

References

1. Watz, E., & et al., (2015). Quality of the Hematopoietic Stem Cell Graft Affects The Clinical Outcome of Allogenic Stem Cell Transplantation. Transfusion, DOI:10.1111/trf.13143.
2. Chen, B., Wright, B., Sahoo, R., & Connon, C.J. (2012). A Novel Alternative to cryo-preserve for the short term storage of stem cells for use in cell therapy using alginate encapsulation. Tissue Engineering Part C : Methods, DOI:10.1089/ten.TEC.2012.0489.
3. Colucci, G., Moore, J.G., Feldman, M., & Chrispeels, M.J.(1999). cDNA Cloning of FRIL, a lectin from *Dolichos lablab*, That Preserves Hematopiotic Progenitors in Suspension Culture, 96, 646-650.
4. Xie, X.Y., & et al., (2004). The Maintenance of Cord Blood CD34$^+$ Progenitor Cells With Plant Lectin FRIL in vitro and The Expression of Related Cell Cycle Modulator HTm4 and HTm4S, 56(3), 306-312.
5. Shahbuddin, M., Macneil, S., & Rimmer, S. (2011). The Potential Use of Konjac Glucomannan for Wound Healing and Cell Transportation, *22*, 43.
6. Kranjčec, B., Dino, P., & Altarac, S. (2014). D-Mannose Powder For Prophylaxis of Recurrent Urinary Tract Infections In Women: A Randomized Clinical Trial. *World Journal of Urology*, 32(1), 79-84, DOI:10.1007/s00345-013-1091-6.
7. Paradossi, G., Finelli, I., Cerroni, B., & Chiessi E., (2009). Adding Chemical Cross-Links to a Physical Hydrogel. Molecules, 14, 3662-3675.

Corresponding Author

Author: Wan Safwani Wan Kamarul Zaman, PhD
Institute: Department of Biomedical Engineering,
Faculty of Engineering, University of Malaya,
Kuala Lumpur, Malaysia
Email: wansafwani@um.edu.my

Efficient Architecture for 3-D Medical Image Compression Using CABAC

A. Muharam[1] and A. Ahmad[2]

[1] VLSI Architecture and Systems Design Research Microelectronic and Nanotechnology Shamsudin Research Centre (MINT-SRC) Universiti Tun Hussein Onn Malaysia (UTHM), Johore, Malaysia
[2] Embedded Computing Systems (EmbCoS) Focus Group,
Department of Computer Engineering, Faculty of Electrical and Electronic Engineering
Universiti Tun Hussein Onn Malaysia (UTHM), Johore, Malaysia

Abstract— **This paper describes the design and implementation of context based adaptive binary arithmetic coding (CABAC) and comparative analysis of grey scale image with discrete wavelet transform (DWT) and without DWT for three dimensional (3-D) medical image compression systems. The proposed system architectures were synthesis in MATLAB. An in-depth performance analysis for simulation is presented. Overall, CT and MRI modalities with DWT outperform in term of compression ratio, PSNR and latency compare with images for CT and MRI without DWT process.**

Keywords— **Discrete wavelet transform (DWT), arithmetic coding, context based adaptive binary arithmetic coding (CABAC).**

I. INTRODUCTION

H.264/AVC is a next generation video compression format to overcome the weaknesses of H.263, and MPEG-2 or MPEG-4 Part-2 in term of bit rate [1], [2]. The newest video compression format was developed by ITU-T Video Coding Experts Group and the ISO/IEC Moving Picture Expert Group. This new format consist of two entropy coding with their own characteristic.

For entropy coding mode equal to '0', represent residual block coded using Context Adaptive Variable Length Coding (CAVLC) and other syntax elements are coded using Exp-Golomb codes. While entropy coding mode equal to '1' represent residual block and other syntax elements are coded using Context Adaptive Binary Arithmetic Coding (CABAC) [3], [4]. Baseline Profile has support for mode '0' only and for main profile and higher profiles support mode '0' and mode '1'.

Moreover than that, CABAC [5] is used in H.264/AVC. Two main parts of CABAC are multiple-context model and a binary arithmetic coder [6], [7]. Binarization of complex symbols in an efficient manner by means of a unary tree is conducted in context-based model [8]. An arithmetic coder encodes each bin according to the selected probability model. There are just have two sub-ranges for each bin corresponding to '0' and '1'.

A-part from that, CABAC offers compression results that are 10-15% better than those obtained with the baseline CAVLC entropy coder [5]. Other researchers claim CABAC outperform universal variable length code (UVLC) by 30-40% [9]. Recently, other work focuses on the practically implementation of CABAC decoding [4]. Where by speed up the decoding process by using pointer chain to retrieve the context model.

An existing implementation focuses on the arithmetic coding (AC) engine found in many compression algorithm due to its high compression efficiency have been reported in [10], [11]. A scheme that can perform CABAC on N-bins (symbols) per cycle is proposed in [10]. Results shows the design occupied 10K bits of memory and 1320 logic cell (1.6% of the total cells) of which 18% were occupied by arithmetic coder and Renormalizer [11].

This research goals towards novel architecture of arithmetic coding for 3-Dimensional medical image compression using CABAC. Software simulation had been using to analyze the architecture with different medical image modalities. An in depth evaluation of transform and CABAC performance in term of compression ratio, peak signal to noise ratio (PSNR) and latency is also addressed.

The composition of this paper as follows. Section 2 discussed the proposed system architecture including transform, quantisation and CABAC blocks. Software simulation results are explains in section 3. Finally, concluding remarks are given in section 4.

II. PROPOSED SYSTEM ARCHITECTURE

Fig. 1 and 2 explain the pre sequence before conducting encoding and decoding process for CABAC. Discrete wavelet transform (DWT) with Haar filter is added in Fig. 2, all decimal values from picture will converted to decimal with new array (72 x 1) before conducting encoding process. Same process repeated at decoding process. In order to evaluate the performance of CABAC, Fig. 1 explains preprocessing without added DWT.

As shown in Fig. 3, illustrated the proposed system for 3-D transposed-based computation used Haar wavelet transform (HWT). Apart from that, by performing the first 1-D HWT along the rows (columns) of the array followed by 1-D HWT along the columns (rows) of the transformed array. The third 1-D HWT is performed on corresponding pixels in each of the N sub-images that constitute the third

F. Ibrahim et al. (eds.), *International Conference for Innovation in Biomedical Engineering and Life Sciences*,
IFMBE Proceedings 56,
DOI: 10.1007/978-981-10-0266-3_44

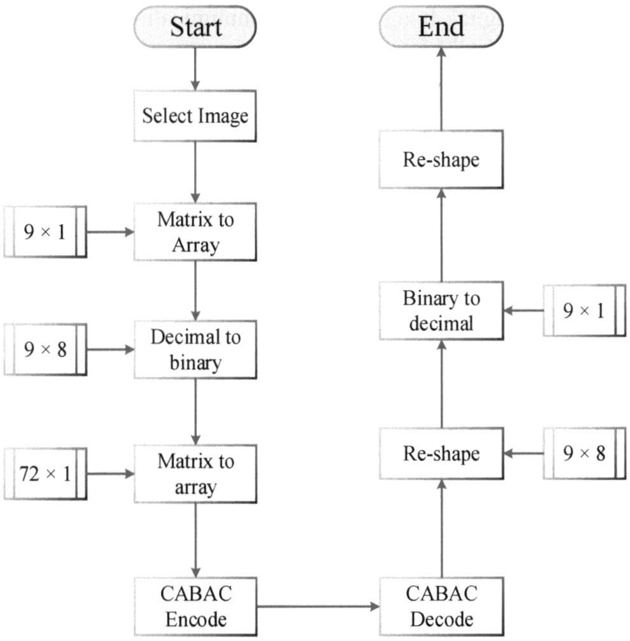

Fig. 1 The flow of encoding and decoding for CABAC without DWT

dimension. All transpositions modules with two memory banks is used to store the transposed coefficients into memory with a fetch unit module that reads back the coefficients for the next 1-D HWT calculation.

Basically, an image is represented as a 2-D array of coefficients, where each coefficient is representing the brightness level of the image. Virtually, the smooth colour variations of an image known as low frequency variations, whilst the sharp variations as high frequency variations. The smooth variations are demanding more importance than the sharp variations because the base of an image is established from the low frequency variations.

On the other hand, the high frequency variations are added to the image to shows the details of the image. Hence, DWT is selected to decompose the variations of the image into sub-images of different size resolution levels.

HWT is one of the simplest types of wavelet that can contribute with the image decomposition. With HWT, it can achieve the optimal compression performance in 3-D medical image compression applications. Moreover, the HWT algorithm computations only take two elements wide at a time, hence the HWT algorithm is exactly reversible without having the edge effects.

The wavelet transform has gained widespread acceptance in signal processing and image compression. Wavelet transform decomposes a signal into a set of basic functions. These basis functions are called wavelets. From a single prototype wavelet called mother wavelet by dilations and shifting, wavelets are obtained. The DWT has been introduced as a highly efficient and flexible method for sub band decomposition of signals.

The 2D-DWT is nowadays established as a key operation in image processing. It is multi-resolution analysis and it decomposes images into wavelet coefficients and scaling function. In DWT, signal energy concentrates to specific wavelet coefficients. This characteristic is useful for compressing images.

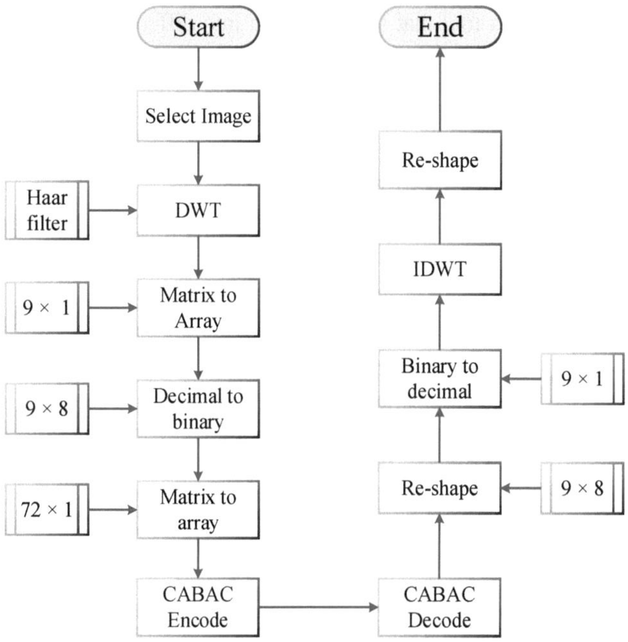

Fig. 2 The flow of encoding and decoding for CABAC with DWT

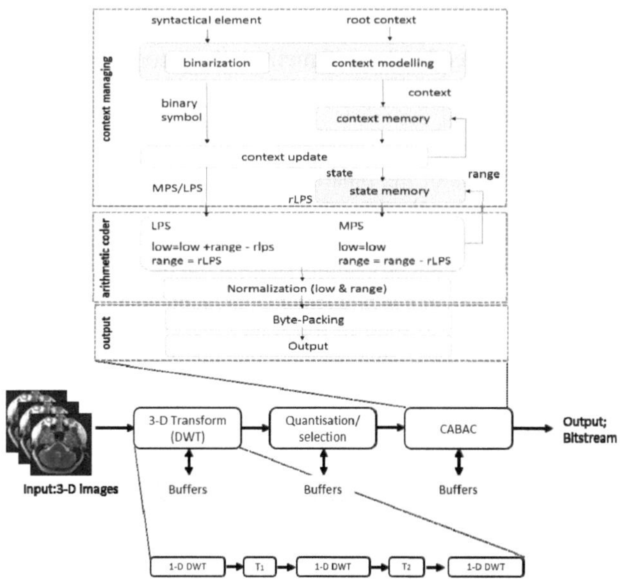

Fig. 3 Proposed system architecture overview

Wavelets convert the image into a series of wavelets that can be stored more efficiently than pixel blocks. Wavelets have rough edges, they are able to render pictures better by eliminating the blockiness. In DWT, a timescale representation of the digital signal is obtained using digital filtering techniques.

A part from that, DWT is a linear transformation that operates on a data vector whose length is an integer power of two, transforming it into a numerically different vector of the same length. It is a tool that separates data into different frequency components, and then studies each component with resolution matched to its scale. DWT is computed with a cascade of filtering followed by a factor 2 subsampling Fig. 4.

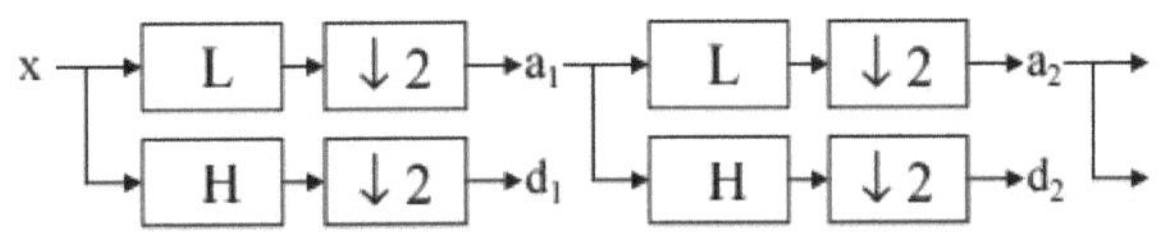

Fig. 4 DWT tree

In order to archived best compression performance for CABAC, three (3) parameters have been considered. Which is by selecting suitable probability models for each syntax element according to the element's context. Next, adapting probability estimates based on local statistic and arithmetic coding. For coding a data symbol involves the following stages.

Binarization process means only accept binary decision either '1' or '0' to proceed for encoding. While the non-binary valued symbols such as transform coefficient or motion vector is converted into binary code prior to arithmetic coding. This process is similar to the process of converting data symbols into a variable length code but the binary code is further encoded by the arithmetic coder prior to transmission. This process will repeated for each bit or bin of the binarized sysmbol.

Next, context model selection is a probability model for one or more bins of the binarized symbols. This model may be chosen from a selection of available model depending on the statistics of recently coded data symbols. The context model only stores the probability of each bin being '1' or '0'.

Arithmetic encoding is a process of arithmetic coder encodes each bin according to the selected probability model. There are only have two sub ranges for each bin corresponding to '0' and '1'. Probability update is updated based on the actual coded value. Example: if the bin value was '1', the frequency count of '1' is increased.

III. RESULT AND ANALYSIS

Simulation was conducted using MATLAB software and two types of modalities MRI and CT with different format, which is JPEG (Joint Photographic Experts Group) and DICOM (Digital Imaging and Communications in Medicine). For DICOM images CT (dental scan), and MRI (standard thoracic), while JPEG images CT (brain), and MRI (weighted human brain)

In term of objective evaluation for N=8, each images are compressed with CABAC entropy coding. Fig. 5 graphically shows the performance of PSNR without DWT and with DWT process. Moreover that, DWT will accelerate the pixel of images depending with proposed transform architecture. Meaning that, higher PSNR generally indicates the reconstruction of higher quality of image. Next, latency process depending on volumes of images have been using during simulation. Different image size will takes differ times to process the image, either same or different modalities. The results for each images with both process as illustrated in Fig. 6 and 7. As we can see at Table 1 and 2, data from CT and MRI modalities with DWT outperform in term of compression ratio, PSNR and latency compare with images for CT and MRI without DWT process. Apart from that, volumes of images reflect the compression ratios during the compression process. More biggest the image size will reduce the effectiveness of compression process due to decrease compression ratio.

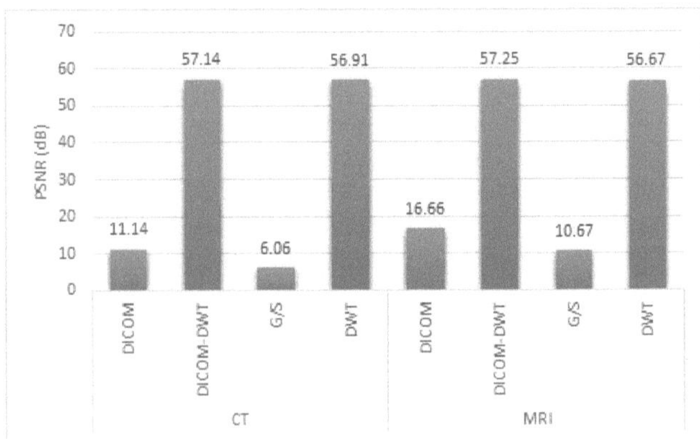

Fig. 5 Comparison of PSNR

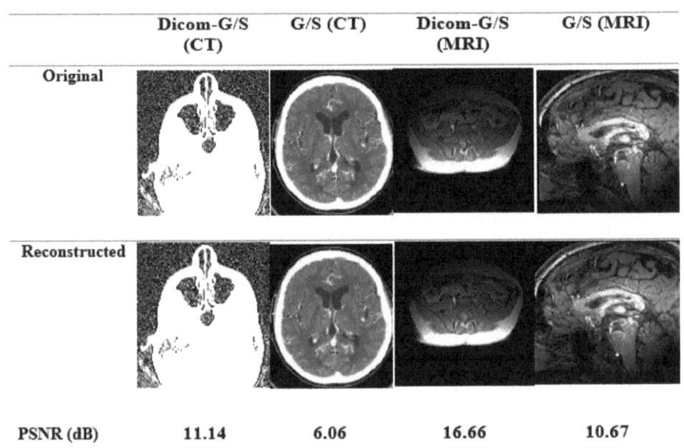

Fig. 6 Comparison of original and reconstructed CT and MRI images for the first slices without DWT

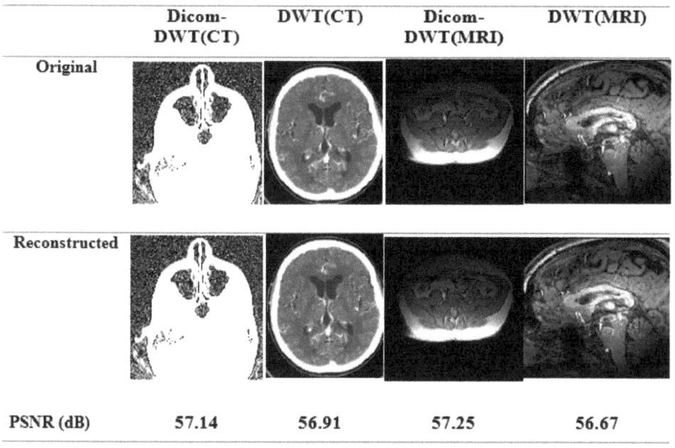

Fig. 7 Comparison of original and reconstructed CT and MRI images for the first slices with DWT

Table 1 Overall performance and comparison with different modalities without DWT

	CT		MRI	
	DICOM	G/S	DICOM	G/S
Compression Ratio (%)	11.3	76.6	78.6	77.1
PSNR (dB)	11.14	6.06	16.66	10.67
Latency (seconds)	172.91	42.12	28.27	327.94

Table 2 Overall performance and comparison with different modalities with DWT

	CT		MRI	
	DICOM-DWT	DWT	DICOM-DWT	DWT
Compression Ratio (%)	56	79.89	83	43.61
PSNR (dB)	57.14	56.91	57.25	56.67
Latency (seconds)	142.24	51.47	24.87	290.44

Some factors such as sharpness and global frequencies that present in the frame of influence the compression ratio and quality of image reconstructed [12]. Since, DWT performs significant results for lossless compression with high quality and significant useful for medical image compression application.

IV. Conclusion

In this paper, we presented a comparison of performance between DWT and without DWT process for 3-D medical image compression with CABAC of CT and MRI modalities. Analysis and performance evaluation of the 3-D images have been conducted. Finding shows images for CT and MRI modalities with DWT outperform in term of compression ratio, PSNR and latency compare with images for CT and MRI without DWT process. Ongoing research focusing on field-programmable gate array (FPGA) and graphics processing unit implementation (GPU).

Conflict of Interest

The authors declare that they have no conflict of interest.

References

1. J. Zhu, D. Zhou, G. He, and S. Goto, "A combined SAO and deblocking filter architecture for HEVC video decoder," 2013 IEEE Int. Conf. Image Process. ICIP 2013 - Proc., pp. 1967–1971, 2013.
2. B. Bae and J.-H. Kong, "A design of pipelined-parallel CABAC decoder adaptive to HEVC syntax elements," IEEE Int. Symp. Consum. Electron., pp. 4–5, 2014.
3. A. P. C. A. High, T. Cabac, V. Sze, and A. P. Chandrakasan, "A high throughput CABAC algorithm using syntax element partitioning Syntax Element Partitioning .," 2014.
4. R. A. Kandalkar and P. M. R. Ingle, "CABAC Entropy Decoding Algorithm Implementation on FPGA For H . 264," vol. 5, pp. 70–75, 2013.
5. R. R. Osorio and J. D. Bruguera, "High-Throughput Architecture for H.264/AVC CABAC Compression System," IEEE Trans. Circuits Syst. Video Technol., vol. 16, no. 11, pp. 1376–1384, Nov. 2006.
6. V. Sze and M. Budagavi, "High throughput CABAC entropy coding in HEVC," IEEE Trans. Circuits Syst. Video Technol., vol. 22, no. 12, pp. 1778–1791, 2012.
7. V. Sze and M. Budagavi, "Parallelization of CABAC transform coefficient coding for HEVC," Pict. Coding Symp., pp. 509–512, 2012.
8. R. R. Osorio and J. D. Bruguera, "Arithmetic Coding Architecture for H.264/AVC CABAC Compression System," Proceeding EUROMICRO Syst. Digit. Syst. Des., 2004.
9. D. Marpe, G. Blattermann, G. Heising, and T. Wiegand, "Video compression using context-based adaptive arithmetic coding," pp. 558–561, 2002.
10. V. Sze, A. P. Chandrakasan, M. Budagavi, and M. Zhou, "Parallel CABAC for low power video coding," Proc. - Int. Conf. Image Process. ICIP, pp. 2096–2099, 2008.
11. H. Shojania and S. Sudharsanan, "A High Performance CABAC Encoder," 2006.
12. A. Ahmad, "Efficient Implementation Of A 3-D Medical Imaging Compression System Using CAVLC," in Proceeding of 2010 IEEE 17th International Conference on Image Processing, 2010, pp. 3773–3776.

Author: Azlan Muharam
Institute: Universiti Tun Hussein Onn (UTHM)
Street:
City: Johore
Country: Malaysia
Email: azlan.muharam.my@ieee.org

Author: Afandi Ahmad
Institute: Universiti Tun Hussein Onn (UTHM)
Street:
City: Johore
Country: Malaysia
Email: afandia@uthm.edu.my

Natural Silk of *Pholcus phalangioides,* a Common Home Spider Species for Wound Healing Applications

M. Shahbuddin, N.A. Puat, M.E.S. Mirghani, and R.A. Raus

Biotechnology-Biochemical Engineering Department, International Islamic University of Malaysia, Malaysia

Abstract— **Spider silk contains peptides and biomolecules that able to stimulate and improve conditions of wound healing. In this study, we report the potential use of natural silk from common home spider, *Pholcus phalangioides,* on human keratinocyte cell line (HaCaT) and teeth pulp stem cell's proliferation and migration. The aim of this study was to examine the range of silk concentrations and their biological effects on the different type of cells. Our study showed that the silk is biocompatible and stimulated the proliferation of HaCaT and teeth pulp stem cells in a dose dependent manner after 24 and 48 hours. Selective effect of cellular migration was observed when the spider silk did not affect the migration of teeth pulp stem cells but only stimulated the migration of HaCaT after 24 hours. The ability of spider silk to stimulate cellular metabolic activity and migration could benefit research and development of biologically active wound dressings.**

Keywords— **HaCaT, teeth pulp stem cells, spider silk, wound healing and cell migration.**

I. INTRODUCTION

Spider silk is a natural material with many favorable characteristics that can be tailored to various forms of architecture and morphology due to its adaptable mechanical properties and stability under a wide range of conditions of humidity and temperature. Spider silks have been used many applications from textiles to wound dressings to regenerative medicine [1, 2]. Spider silk's unique biomolecules compound which comprised of an inner protein core and outer skin covered by protective coating gave its strength and high tensile, which make it suitable material as bioactive wound coverage [3].

Spider silk is biocompatible and has slow biodegradation *in vivo* [4]. Its mechanical properties outperformed most natural and synthetic fibers exhibit an interesting torsional dampening behavior of the dragline thread and able to undergo supercontraction [5, 6]. Native spider silk is biologically active and do not require further modification for wound healing application [7]. In comparison to silkworm silk, spider silk is more favorable due to its less immunogenicity, biocompatibility and stimulation of different type of cellular proliferation kinetics that would be beneficial for tissue engineering and regenerative medicine [8, 9, 10].

In the last twenty years, research on spider silks' biological and mechanical properties explored the potential use of this silk in devices for medical and tissue engineering applications. [11, 12]. Spider silk antimicrobial peptides and bioactive molecules have shown stimulations of specific cells such as monocytes and T-cell which are beneficial for wound healing [13] and on nerve regeneration application [10]. These antimicrobial peptide functions as chemotactic agent to stimulate wound healing process by activation the acquired immune response system [14, 15]. The positive stimulation on fibroblasts and epithelial cells proliferation [16] to induce neo-vascularization [17] and cytokines mobilization [17] is important for skin regeneration. Cell migration is fundamental prerequisite for wound healing [18, 19]. A scratch assay was used to study the cellular migration as it is particularly cheap, easy and well developed for *in vitro* [20]. In this assay, a scratch of 100 μm gap was created on a confluent cell monolayer in a wellplate using a pipette tip. Then, the migration and proliferation of cells into the denuded area was observed over time and quantified quantitatively and qualitatively using a phase contrast microscope and evaluated using ImageJ [20]. Image analysis was used to determine the rate of cell migration by comparing the size of covered size and distance of cellular movement at specified time to the 0hr.

Cell migration is fundamental prerequisite for wound healing [18, 19]. A scratch assay was used to study the cellular migration as it is particularly cheap, easy and well developed for *in vitro* [20]. In this assay, a scratch of 100 μm gap was created on a confluent cell monolayer in a wellplate using a pipette tip. Then, the migration and proliferation of cells into the denuded area was observed over time and quantified quantitatively and qualitatively using a phase contrast microscope and evaluated using ImageJ [20]. The rate of cell migration will be determined by image

F. Ibrahim et al. (eds.), *International Conference for Innovation in Biomedical Engineering and Life Sciences,*
IFMBE Proceedings 56,
DOI: 10.1007/978-981-10-0266-3_45

analysis to compare the covered size and distance of cellular movement at specified time.

In the present study, the migration of the teeth pulp stem cells and HaCaT, human keratinocyte cell lines on a wound scratch assay were studied to quantify the biological effects of *Pholcus phalangioides*, a common home spider species on different type of cells' viability and proliferation.

II. MATERIALS AND METHODS

A. Materials

Ethanol, sodium carbonate (Na_2CO_3), 0.25% of trypsin, penicillin-streptomycin, Accutase, Dulbecco`s Modified Eagle Medium (DMEM), fetal bovine serum (FBS), phosphate buffered saline (PBS), dimethyl sulfoxide (DMSO), isopropanol and MTT (3-(4,5-Dimethylthiazol-2-yl)-2,5-diphenyltetrazolium bromide) assay.

B. Animal Care, Silk Harvesting and Storage

The spiders of the species *Pholcus phalangioides*_were kept in a laboratory conditions in 3 different containers with up to 10 animals per container to avoid cannibalism to ensure the sterilization of the spider web. Webs were sprayed with tap water every day. Additionally, vaporizers were used to moisten the air. Spiders were fed with ants (*Camponotus spp*) three times per week Spider silk was collected by the method recently described in [7]. To keep the room climate ideal, the atmospheric humidity was set to approximately 70%. The bundles of spider web were stored for up to two months before testing.

C. The Culture of HaCaT and Human Teeth Pulp Stem Cells

Human teeth pulp stem cells were isolated from teeth pulp removed during a surgery at University Malaya Medical Center with fully informed patient consent for the use of skin for experimental research.and a gift from Dr. Shamsul Azlin Shamsuddin from Department of Biological Sciences, University of Malaya. Teeth pulp stem cells were isolated from teeth pulp by mincing the biopsy into small pieces, followed by digestion with 0.05% collagenase A in DMEM overnight at 37 C with 5% CO_2. The cell suspension was then centrifuged at 400g and resuspended in medium (DMEM supplemented with 10% fetal calf serum (FCS)). These cells were then cultured in medium in T25 flasks and incubated at 37 C with 5% CO_2. Medium was changed every 2 days and cells were passaged as needed, Teeth pulp stem cells between passage 1 and 3 and HaCaT between passage 7 and 10 were used in the experiments.

D. The Effect of Spider Web on the Proliferation of HaCaT and Human Teeth Stem Cell

1 mL of $4x10^4$ HaCaT cell, teeth pulp and adipose stem cell were cultured in a 12 well plate for overnight. The cells were left to attach for four days and form a confluent monolayer on the well plate. The medium was then removed and washed with PBS. 1 mL of concentrations of spider web (0.1, 0.5 and 1.0 $mg.mL^{-1}$) in culture medium was added into the well. The cell migration at every 0, 12, 24 and 48 hr were then observed 1 mL of fresh medium was added into the culture and then incubated at 37°C with 5% CO_2. Cell viability was measured using MTT assays after 1 and 3 days.

E. The Effect of Spider Silk on the Migration of HaCaT and Teeth Pulp Stem Cells on a Wound Scratch Assay

Cell migration was examined by scratch wound healing assay. 1 mL of $4x10^4$ HaCaT cell, teeth pulp and adipose stem cell were cultured in 12 well plate until reaching its confluency for 72 hours and the scratch wound was introduced using 200 μL pipette tip. The culture was then washed three times with PBS to remove cell debris. 1 mL of fresh medium was added into the culture and then incubated at 37°C with 5% CO_2. 3 independent repeats were captured at 0, 24 and 48 hour time points during cell migration by Olympus microscope camera. The area of cell migration into wound site was quantified using ImageJ and presented as relative migration cells compared with the 0 hr. The magnification of each picture was 4x.

$$(\%\ \text{Area of Relative migration cells}) = \left(\frac{(Area\ of\ 0hr - Area\ of\ 5hr)}{Area\ of\ 0hr}\ x\ 100\right)$$

F. Statistical Analysis

Quantitative data (e.g. MTT optical density readings) were analysed by using Design Expert (Stat-Ease 6.0) and Microsoft Excel (Microsoft Corporation) to obtain means and standard deviation (SD), n = number of independent experiments each with three replicates. Student's t-test was performed to determine statistical significance, indicated in the corresponding figures or tables by: ns (not significant; $p \geq 0.05$), * (significant; $p<0.05$), ** (highly significant; $p<0.01$) and *** (extremely significant; $p<0.001$).

III. RESULT AND DISCUSSION

A. The Effect of Spider Silk on the Proliferation of HaCaT and Teeth Pulp Stem Cells after 1 and 3 Days

Examinations on the biological effects of spider silk on cellular viability and proliferation of HaCaT and teeth pulp stem cells were conducted using MTT assay (Figure 1 and 2). There was no significant differences in the growth rate for both HaCaT and teeth pulp stem cells with 50% increment after 24 and 48 hours of culture. However, an increase in HaCaT cellular viability was observed in the samples treated with spider silk at dose dependent manner (0.5 and 1 mg/ml) with increment of 2 and 3 fold, respectively after 3 days in culture (Figure 1). For teeth pulp stem cells, 0.1 and 0.5 mg/ml spider silk did not give any stimulatory effect to the cellular proliferation and viability after 24 hours but a concentration of 1.0 mg/ml spider silk showed an increment of one fold when the silk was placed in direct contact with cells (Figure 2).

B. The Effect on Spider Web on the Migration of HaCaT and Teeth Pulp Stem Cells in a Wound Scratch Assay after 24 and 48 Hours

Scratch assay method is the simplest method to study cellular migration for wound healing assessment [20]. The migration was observed using a phase contrast microscopy and quantified using ImageJ. The measurement was calculated as the percentage of covered area or total scratch area and the distance of the cells movement towards closing the scratch area. Figure 2 shows the micrographs of increasing concentrations of spider silk on the migration of HaCaT after 0, 20 and 48 hours. From the observations in Figure 3a and 4a, the migration of HaCaT cells after 24 hours was slower compared to teeth pulp stem cells with 15% of area coverage compared to 30%, respectively. Significant differences of nearly 3 fold increase were observed in the migration of HaCaT cells after 24 and 48 hours when treated with 0.5 and 1.0 mg/mL spider web (Figure 3b). There were no significant differences in the rate of migration except for cells treated with 0.5 mg/ml spider silk with 1-2.5 μm/hr rate difference compared to the control.

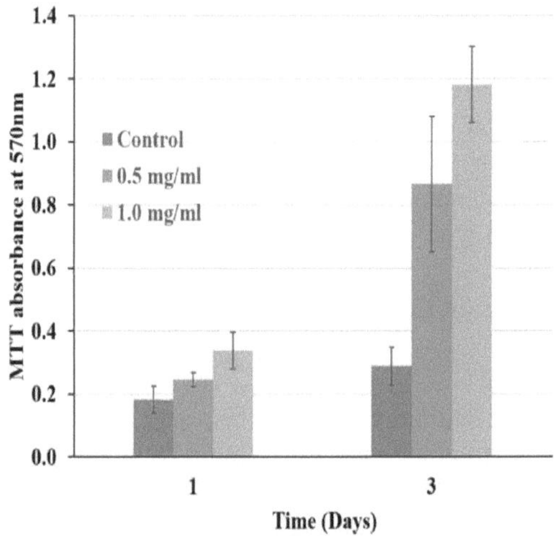

Fig. 1 the effect of varying concentrations of spider web on the proliferation of hacat cell after 1, 3 and 5 days measured using mtt assay. Results shown are mean ±sd, (n=2), ***p<0.001 highly significant, **p<0.01 very significant and *p<0.05 significant in comparison to control

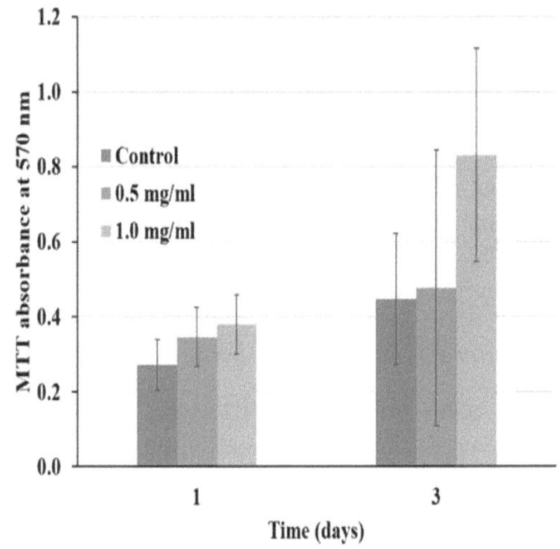

Fig. 2 The effect of varying concentration of the spider web on the proliferation of teeth pulp stem cell after 1 and 3 days measured by MTT assays. Results shown are mean ±SD, (n=2), ***P<0.001 highly significant, **P<0.01 very significant and *P<0.05 significant in comparison to control.

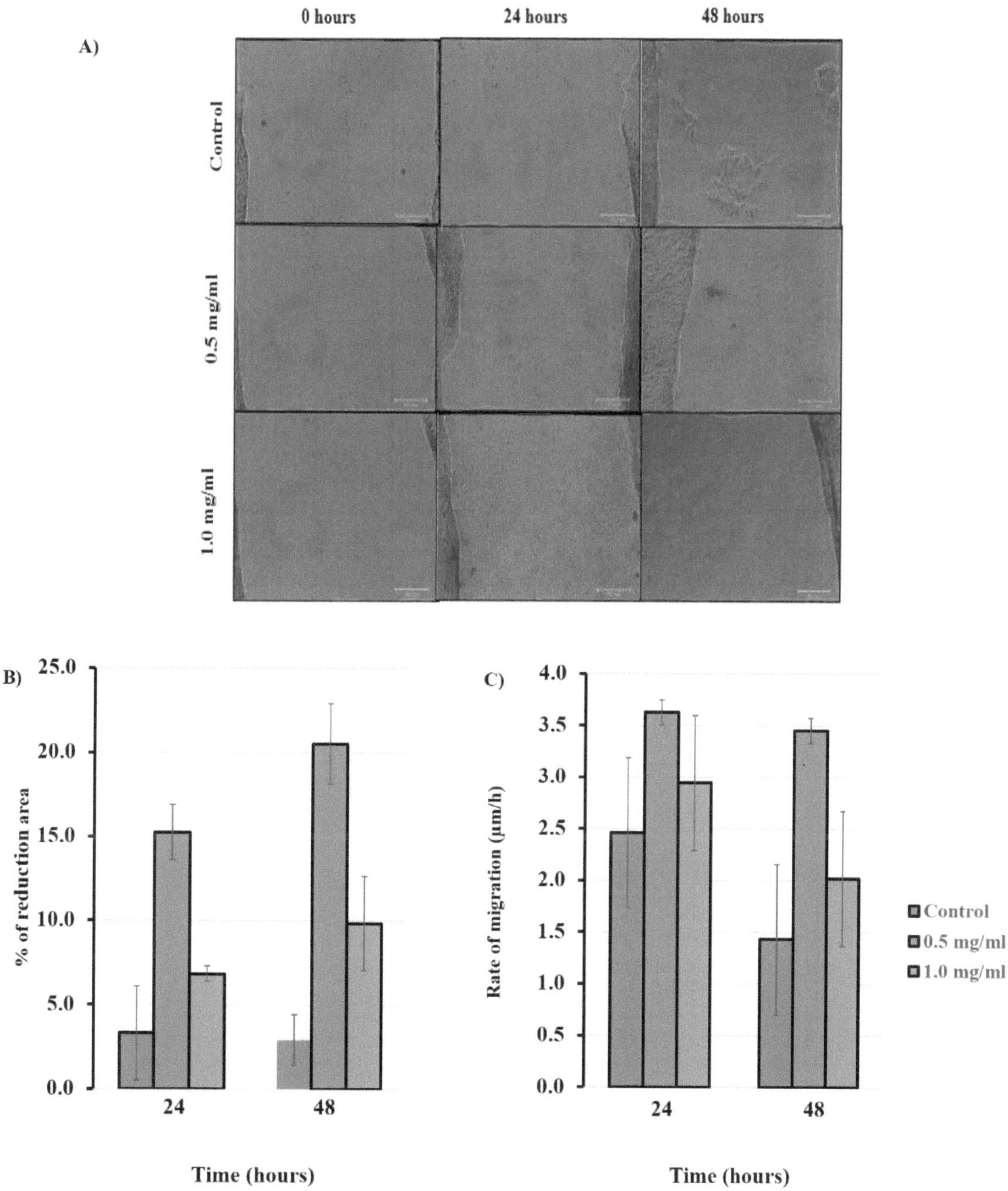

Fig. 3 A) Microphotograph of the biological effect of spider silk on HaCaT cells, and quantitative analysis on B) the reduction of wounded area and C) rate of cellular migration in a wound scratch assay after 24 and 48 hours in a cell culture medium. Results shown are mean ±SD, (n=2), ***P<0.001 highly significant, **P<0.01 very significant and *P<0.05 significant in comparison to control.

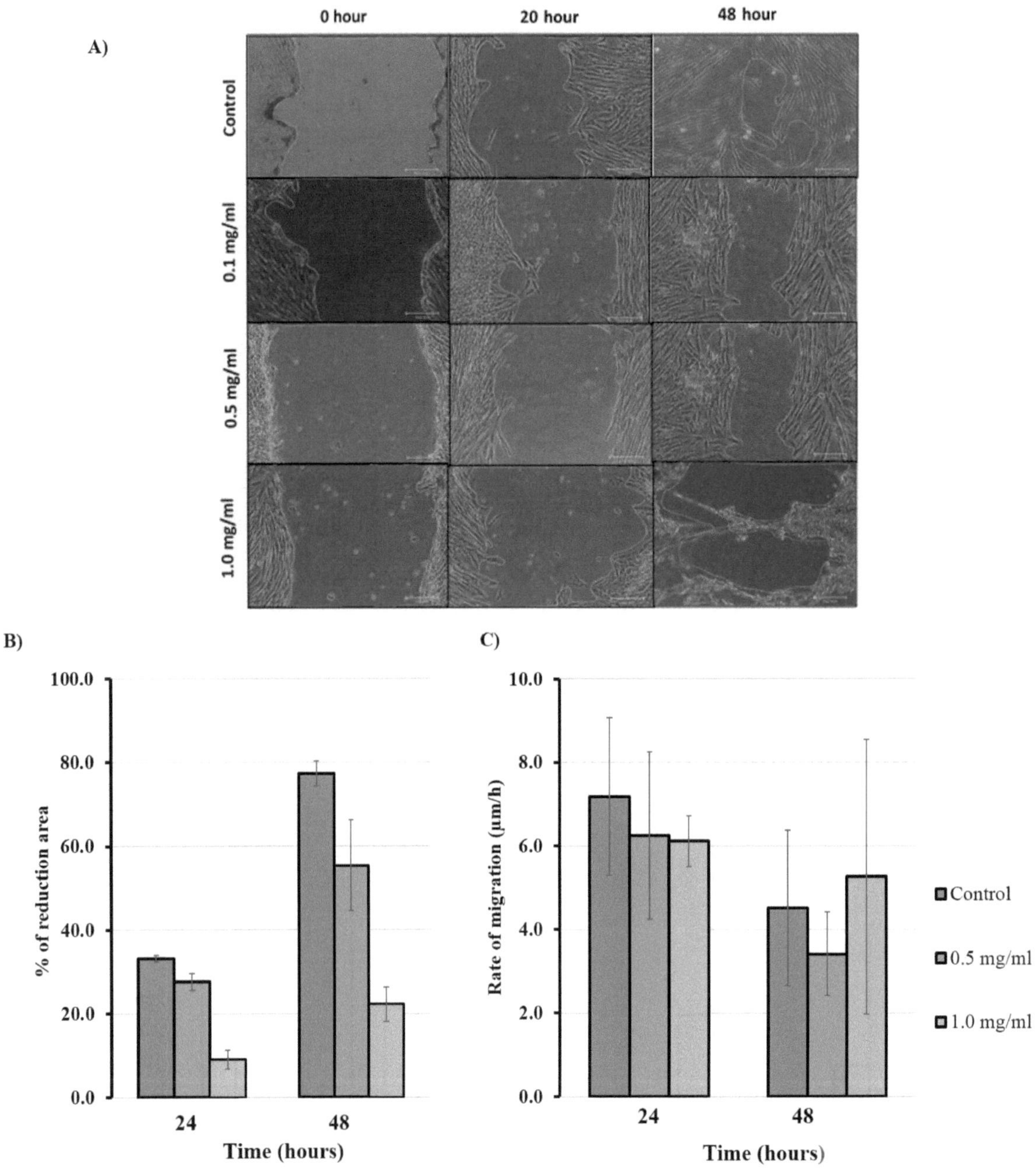

Fig. 4 A) The microphotograph of the biological effect of spider silk on teeth pulp stem cells, and quantitative analysis on B) the reduction of wounded area and C) rate of cellular migration in a wound scratch assay after 24 and 48 hours in a cell culture medium. Results shown are mean ±SD, (n=2), ***P<0.001 highly significant, **P<0.01 very significant and *P<0.05 significant in comparison to control.

IV. Conclusion

This is the first report on the biological effects of home-species natural silk fiber when placed in direct contact with human teeth pulp stem cells and HaCaT. We observed that spider silk kept in sterile and aseptic conditions did not contain any substance that may be cytotoxicity to the cells. The spider silk of ***Pholcus phalangioides*** increased the viability of HaCaT and teeth pulp stem cells in a dose dependent manner. On a wound scratch assay, the spider silk increased the rate of migration and coverage in HaCaT after

24 and 48 hours. Supplementation of spider silk in the culture medium of teeth pulp stem cells did not give any difference to the rate of migration and coverage compare to the control after 24 and 48 hours, citing selective biological effects of the silk towards different type of cells. It was the intention of this research to highlight and explore the potential use of natural spider silk for wound healing application. However, in the context of biomedical research and development there are major limitation and difficulties in the process of cultivation, raring and harvesting silk which are very time consuming and not profitable for a large scale biotechnological production. We hoped that together with this knowledge on the potential use of natural spider silk, this will open doors to further investigations on the silk biological, physical and chemical properties for the development of biomaterials for wound healing.

ACKNOWLEDGMENT

The author wish to thank and acknowledge Dr. Shamsul Azlin Shamsuddin from University of Malaya, Kuala Lumpur, Malaysia for his generosity and donation of teeth pulp stem cells.

CONFLICT OF INTEREST

The author wish to acknowledge full gratitude to the IIUM for funding this work.

REFERENCES

1. Heim, M., Keerl, D. & Scheibel, T. (2009) Spider Silk: From Soluble Protein to Extraordinary Fiber. Angew. Chemie Int. Ed. **48:** 3584–3596
2. Zafar, M. S. & Al-Samadani K. H. (2014) Potential use of natural silk for bio-dental applications. J. Taibah Univ. Med. Sci. **9:**171–177
3. Hakimi, O., Knight, D. P., Vollrath, F. & Vadgama, P. (2007) Spider and mulberry silkworm silks as compatible biomaterials. Compos. Part B Eng. **38:** 324–337
4. Craig, C. L. & Riekel, C. (2002) Comparative architecture of silks, fibrous proteins and their encoding genes in insects and spiders. Comp. Biochem. Physiol. Part B Biochem. Mol. Biol. **133:**493–507
5. Eisoldt, L., Smith, A. & Scheibel, T. (2011) Decoding the secrets of spider silk. Mater. Today **14:**80–86
6. Yang, Q. & Li, G. (2014) Spider-silk-like shape memory polymer fiber for vibration damping. Smart Mater. Struct. **23:**105032
7. Kuhbier, J. W. et al. (2010) Interactions between Spider Silk and Cells–NIH/3T3 Fibroblasts Seeded on Miniature Weaving Frames. PLoS One **5:** e1203
8. Frank Roloff, Sarah Strauß, Peter M. Vogt, Gerd Bicker, and C. R. (2014) Spider Silk as Guiding Biomaterial for Human Model Neurons. Biomed Res. Int.. DOI 10.1155/2014/906819
9. Wang, Y., Kim, H.-J., Vunjak-Novakovic, G. & Kaplan, D. L. (2006) Stem cell-based tissue engineering with silk biomaterials. Biomaterials **27:**6064–6082
10. Allmeling, C. et al. (2008) Spider silk fibres in artificial nerve constructs promote peripheral nerve regeneration. Cell Prolif. **41:** 408–420
11. Huang, J., Wong, C., George, A. & Kaplan, D. L. (2007) The effect of genetically engineered spider silk-dentin matrix protein 1 chimeric protein on hydroxyapatite nucleation. Biomaterials **28:**2358–67
12. Gomes, S. et al. (2012) Biological responses to spider silk-antibiotic fusion protein. J. Tissue Eng. Regen. Med. **6:**356–368
13. Baoyong, L., Jian, Z., Denglong, C. & Min, L. (2010) Evaluation of a new type of wound dressing made from recombinant spider silk protein using rat models. Burns **36:**891–6
14. Grigat, J., Soruri, A., Forssmann, U., Riggert, J. & Zwirner, J. (2007) Chemoattraction of Macrophages, T Lymphocytes, and Mast Cells Is Evolutionarily Conserved within the Human α-Defensin Family. J. Immunol. **179:**3958–3965
15. Aarbiou, J. et al. (2004) Neutrophil Defensins Enhance Lung Epithelial Wound Closure and Mucin Gene Expression In Vitro. Am. J. Respir. Cell Mol. Biol. **30:** 193–201
16. Aarbiou, J. et al. (2002)Human neutrophil defensins induce lung epithelial cell proliferation in vitro. J. Leukoc. Biol. **72:** 167–174
17. Bals, R. (2000) Epithelial antimicrobial peptides in host defense against infection. Respir. Res. **1:**141–50
18. Clark, R. A. F. et al. (1982) Fibronectin and Fibrin Provide a Provisional Matrix for Epidermal Cell Migration During Wound Reepithelialization. J Investig Dermatol **79:**264–269
19. Schneider, L. et al. (2010) Directional Cell Migration and Chemotaxis in Wound Healing Response to PDGF-AA are Coordinated by the Primary Cilium in Fibroblasts. Cell. Physiol. Biochem. **25:**279–292
20. Liang, C.-C., Park, A. Y. & Guan, J.-L. (2007) In vitro scratch assay: a convenient and inexpensive method for analysis of cell migration in vitro. Nat. Protoc. **2:**329–333

Author: Munira Shahbuddin
Institute: International Islamic University Of Malaysia
Street: P.O. Box 10, Jalan Gombak
City: Kuala Lumpur
Country: Malaysia
Email: munirashah@iium.edu.my

Short Review of Electrocardiogram (ECG) Technique Versus Optical Techniques for Monitoring Vascular Health

S. Nur Hidayah Malek, K. Chellappan, and R. Jaafar

Department of Electrical, Electronic & Systems, Faculty of Engineering and Build Environment, Universiti Kebangsaan Malaysia, Bangi, Malaysia

Abstract— **Noninvasive optical techniques have drawn substantial amount of interest among researchers in exploring potential self-monitoring devices. Photoplethysmograpm (PPG) is one of the high potential optical techniques used in vascular health monitoring. It is a low cost and easy to handle device that can measure the volume change in light absorption. Lately there are development in PPG to detect Atrial Fibrillation (AF) that is one of the types of cardiac arrhythmia which initiated by irregular heart rhythms. Electrocardiogram (ECG) is the clinical gold standard measurement for AF. ECG morphology is similar to the second derivatives of PPG signals which has been established in both hypertensive and atherosclerosis studies. This paper reviews on the various AF detection techniques in both ECG and PPG and highlights the potential of PPG as AF detection technique.**

Keywords— **Photoplethysmogram, Atrial Fibrillation, Electrocardiogram, Signal Processing.**

I. INTRODUCTION

Atrial Fibrillation (AF) also known as cardiac arrhythmia. This AF increases the risk factor of cardiovascular and cerebrovascular morbidity and mortality [1]. Atria is located at upper chamber of the heart when irregular, quavering and rapid heartbeats are produced. It happens when the electrical signals of the cardiac muscles contract rapidly and produce unsynchronized rhythms of heartbeats.

The data sets from Physionet, the MIT-BIH Database (AFDB) is a data study for cardiac arrhythmia that can be obtain in an online repository. Fig. 1 shows the RR Interval for AF from the electrocardiogram (ECG) signals [2].

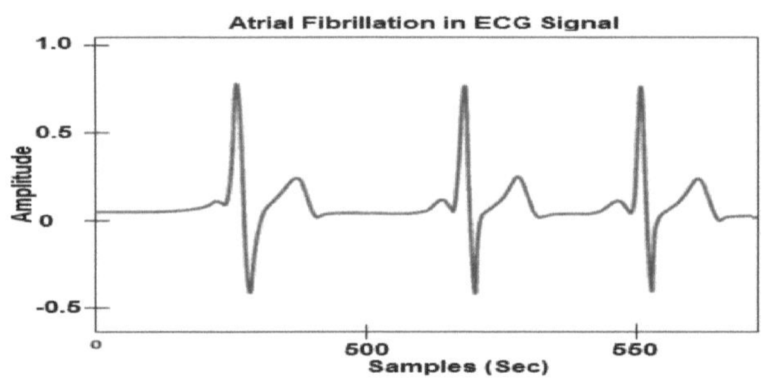

Fig. 1 Illustrations of RR Interval in ECG Signal for AF patients

Photoplethysmogram (PPG) is a non-invasive device that measures the changes of volume in blood flow through human tissue [3]. The most widely used clinical applications of PPG signals are to measure the heart rate, respiration rate and blood oxygen saturation. This device is widely used commercially in a clinical setting because it contains the technologies that is low cost [4]. The wavelength in the blood tissue will be absorbed, scattered and reflected by the light that emits from the photo detector [5]. Fig. 2 refers to the PPG AC component that represents the physiological changes in blood vessel [6].

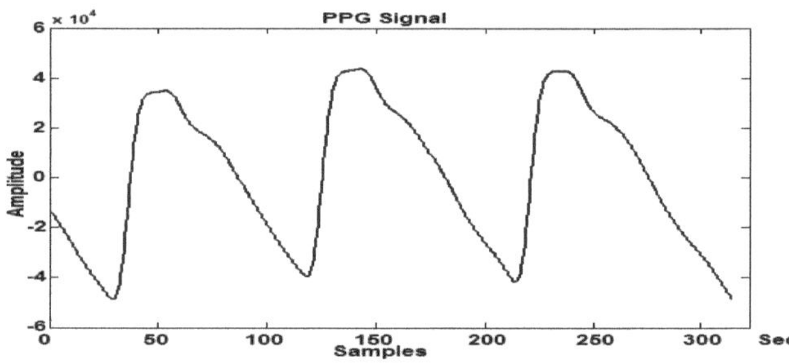

Fig. 2 Illustrations in PPG signal in Cardiac Changes

The single cardiac cycle comes with the morphology of PQRST wave as a baseline of the ECG which have been clinically recognized. Fig. 3 shows the ECG morphology signal that is highlighted on the condition of cardiac activity in the detection of AF [7].

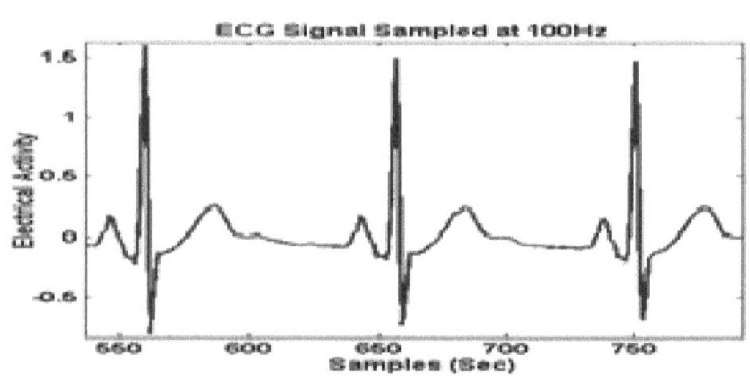

Fig. 3 Illustrations for ECG Morphology Signal of Cardiac Activity

F. Ibrahim et al. (eds.), *International Conference for Innovation in Biomedical Engineering and Life Sciences*,
IFMBE Proceedings 56,
DOI: 10.1007/978-981-10-0266-3_46

This study is to explore the potential role of PPG in the detection of AF versus the ECG monitoring in clinical practice. The most important aspect of the study is to encourage PPG technology to be used in the early detection of AF. It is especially for the sub-clinical or silent AF, which can be miss diagnosed on a standard ECG [8]. The extended recording of ECG techniques is more sensitive, but come with a higher cost and complex setup. A digital recording of a pulse using PPG can be expressed as a waveform that has specific characteristics associated with the different phases of cardiac conduction. These phases are recognized on the ECG waveforms and they are similar to the phases of the PPG signals when the signals are derived into the second derivative signals. The objective of this paper is to review the techniques of AF detection by ECG and to describe the PPG second derivatives in comparison to the ECG morphology of AF.

II. Historical Review of Early Work

In 1887, the first published article on human electrocardiogram (ECG) recorded with a capillary electrometer was initiated by Augustus S. Waller, who have provided a dramatic change in medical science. In this finding, the researcher established the two deflations point downwards, known as V1 and V2 [9]. The newly discovered device was able to capture the electrical activity of the heart and it also can investigate the cause of chest pain. After several years, Willem Einthoven produced the P, Q, R, S and T points referring for one cardiac cycle in Fig. 3. It produces the heartbeat that causes the heart to contract and pump the blood repeatedly with the set of rhythms [10].

Early findings in 1936 by two independent research groups in New Jersey and Stanford have explored the used by non-invasive optical instrument to access blood volume in rabbit ear [11]. After a year, a study of blood volume changes in PPG was continued in human fingers by a research team lead by Alrick Hertzman in St. Louis (USA). This study was a pronounce introduction of the use of PPG in human monitoring [12].

Nowadays, AF is detected by prolonged the ECG monitoring the patient treatment using an event recorder (ER910AF Cardiac Event Monitor, Braemar) that can be recorded in irregularity of RR interval as in Fig. 1. This recorder is a non invasive worn belt with dry-electrodes placed around the chest (Cardia Bio-Systems). It produces a primary stage when the ECG documented data can last within 30 seconds or longer than 90 days in the detection of AF using this device. Another stage is the monitoring condition of AF by using the Holter monitor with minimum monitoring time of 24 hours for AF detection [13].

III. Data Processing Techniques

A. *Signal Pre-processing*

Signal processing for ECG includes principle component analysis (PCA) that is dealing with the compression, beat detection and classification, noise reduction, signal separation, and feature extraction. The noise filtering in ambulatory monitoring is for resting ECG analysis with the case episodes containing excessive noise. The questionnaire will be evaluated based on the feedback that is required in the data recording. The signal processing techniques in PPG as done in other studies include removing motion artifact, through adaptive noise cancelation technique [14], independent component analysis [15], Kalman filtering [16], and time-frequency analysis [17].

B. *Feature Extraction for ECG and PPG*

There are different types of features extracted from both ECG and PPG by researchers in quantifying AF detection or diagnostics in past researches.

a) Atrial Activity (AA) Extraction

AA happens when the P wave is absent with irregular narrowing of the QRS complex. The P wave cannot be seen, it happens when unsynchronized and chaotic at atria produces quivering and fibrillates rather than contracting. As commonly known atria firing rate is extremely high around 300 beats per minute (bpm) for electrical impulses. This fibrillate does not pass through the atrioventricular (AV) node due to refractory properties of the upper part of the heart. AF can be characterized by the absence of P wave that represents the electrical activity of sinoatrial (SA) node and it is obscured by multiple ectopic side [18].

b) Atrial Flutter Extraction

This is one of the types of AF detection in ECG when the rhythm in the atria is more organized and less chaotic than AA. Atrial flutter is really irritable focus with fire rate that is between 250-300 beat per minute (bpm) [19, 20]. It happens when an electrical signal travels around atria in a circular pattern and causes higher contraction while ventricle only can receive a slower rate around 150 bpm. In this stage, AV node has its own mechanism that cannot fire again the electrical signal to ventricle because of refractory period to prevent AV from over firing and ventricle from contracting too quickly. The result comes in multiple P-waves with regular spaces QRS interval with the line that often known as saw tooth pattern that contains a number of waves in every QRS complex for ECG signal [21, 22].

c) RR Interval Extraction

RR Interval (RRI) is the type of detection signal on AF as a one cycle length of variability [23]. RRI invariability happens when different distance between 2 peaks or irregular QRS complex being detected from monitoring ECG collection on PQRST wave [24]. This similarity of RRI in variation from ECG and the interval between peak to peak of PPG can be detected directly and be observed for each beat-to-beat wave signal. This RR interval sample is not uniform in different length and width for one heartbeat of the cardiac cycle and this is the techniques that can indicate the AF condition [25, 26].

d) PPG Features Used in AF Detection

The second derivative of PPG signal is comparable to the ECG waveform as shown in Fig. 4. This has been proven by Takazawa in 1998 and being used by researchers in both hypertensive and atherosclerosis studies. The second derivative wave also has characteristic contours that facilitated the interpretation of PPG in AF detection [27]. PPG has been introduced in AF detection by Jinseok Lee et al through smartphone application. A peak detection algorithm incorporated with filter bank, variable cutoff frequencies, spectral of the heart rate, filters and decision logic was used to estimate the presence of AF statistically [28].

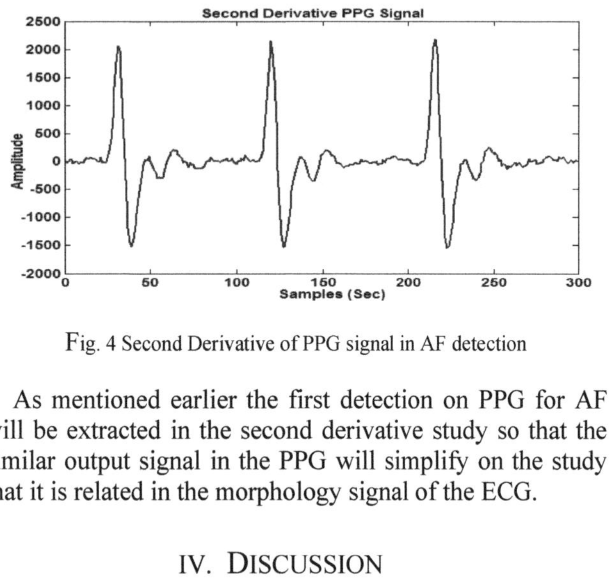

Fig. 4 Second Derivative of PPG signal in AF detection

As mentioned earlier the first detection on PPG for AF will be extracted in the second derivative study so that the similar output signal in the PPG will simplify on the study that it is related in the morphology signal of the ECG.

IV. DISCUSSION

The current AF monitoring technique using ECG found to be unreliable due to the short recording duration [29]. Long term ECG monitoring has been adapted as a solution, but it is found to be complex and costly which is not being a good solution in the low-income countries [30, 31]. A more robust and low cost solution is expected to be a better detection and monitoring system for AF. As such PPG being a choice due to its low cost and non invasive operator independent nature.

V. CONCLUSION

This paper has reviewed on signal processing technique for AF detection using ECG data and the potential of optical technique by the use of PPG signals for monitoring vascular health. AF is one of challenges among the different types of arrhythmia being an interest for many researchers due to its irregular nature and its needs in the long term and frequent monitoring to ensure the managing capacity. PPG has high potential in being a reliable, low cost and robust device in AF monitoring. The operator independent capacity of PPG technology enables further its potential to be a home-based monitoring device.

ACKNOWLEDGMENT

The authors would like to acknowledge the support from MyBrain15 for funding the research conducted at Universiti Kebangsaan Malaysia, UKM. The Study is also partially supported financially by the university grant ERGS/1/2012/TK02/UKM/03/2.

CONFLICT OF INTEREST

The authors declare that they have no conflict of interest.

REFERENCES

1. Elad Anter, et al. (2009) Atrial Fibrillation and Heart Failure Treatment Consideration for a Dual Epidemic. Circulation. 119: 2516-2525
2. Abhay Bajpai, Edward Rowland. (2006) Atrial Fibrillation. Continuing Education in Anaesthesia, Critical Care & Pain. 6(6):219-224
3. Kamal A. A, Hames J. B et al. (1989) Skin Photoplethysmograhy: A Review. Computaional Methods Program Biomedical. 28(4):257-269
4. J. Allen. (2007) Photoplethysmography and its application in clinical physiological measurement. Physiology Measurement
5. Sara Bergstrand, Lars Goran et al. (2009), Blood Flow measurement at different depths using photoplethysmography and laser Doppler techniques. Skin Research and Technology
6. Lindberg LG, Oberg PA. (1991) Photoplethysmography. Part 2 Influence of light source wavelength. Med Biol Eng Comput
7. Sujut Kumar Sahoo et al. (2011) Detection of atrial fibrillation from non-episodic ECG data: A review of methods. Eng. Med. Bio. Society. IEEE. 4992-4995
8. Martin Brond, et al., (2013) Improved detection of Silent Atrial Fibrillation using 72-Hours Holter ECG in patients with Ischemic Stroke, Stroke, 44,:3357-3364
9. Berndt Luderitz. (2003). History Augustus Desire Waller (1856-1922)- The First to Record the Electrical Activity of the Human Heart. J. Int. Cardiac Electrophysiplog y. 9:59-60
10. I. Cruz-gonzalez, et al., (2009) Non-Invasive assessment of myocardial ischemia by using low amplitude osillations of the conventional ECG signal (ECG dispersion mapping) during percutaneous coronary intervention. Acta Cardiol. 64(1):11-15

11. J. Allen, (2007) Photoplethysmography and its application n clinical physiological measurement. Physiol. Meas. 26
12. Edward Sazonoz. Michael R Neuman, et al., Wearable Sensors: Fundamentals, Implementation and Applications. Access Online via Elsevier
13. David J. Gladstone, Melanie Spring et al. (2014) Atrial Fibrillation in Patients with Cryptogenic Stroke. N ENGL J MED. 370(26):2467-2477
14. R. Yousefi, et al., (2014) A Motion-Tolerant Adaptive Algorithm for Wearable Photoplethysmographic Biosensors, IEEE Journal of Biomedical and Health Informatics, 18(2), pp. 670-681
15. B.S. Kim, S.K. Yoo, (2006) Motion Artifact Reduction in Photoplethysmography Using Independent Component Analysis, IEEE Trans. Biomedical Engineering, 53(3), pp. 566-568
16. B. Lee, J. Han, et al., (2010) Improved elimination of motion artifacts from a photoplethysmographic signal using a Kalman smoother with simultaneous accelerometry, Physiological Measurement, 31(12), pp. 1585-1603
17. Y. Yan, C. Poon, Y. Zhang, (2005) Reduction of motion artifact in pulse oximetry by smoothed pseudo Wigner-Ville distribution, Journal of NeuroEngineering and Rehabilitation, 2(3)
18. Steve Goodacre, Richard Irons (2002) Clinical Review on Atrial Arrhythmias. BMJ. 324:594-597
19. Carol Jacobson, Karen Marzlin, Cynthia Webner, (2014) Cardiac Arrhythmias & 12 Lead ECG Intrepretation Module for 23 Continuing Education Credit. Cardiovascular Nursing Education Assoc.
20. Jonathan Rosman, et al., (2009) Atrial Rate and Rhythm Abnormalities in a Patient with Hyperkalemia, 9(3):183-185
21. Melisa Boyer, Bruce A. Koplan (2005) Atrial Flutter. Circulation 334-336
22. Y. Sun, K. L. Chan, and S. M. Krishnan,(2005) Characteristic wave detection in ECG signal using morphological transform, BMC Cardiovasc. Disord. 5:26-28
23. George B. Moody and Roger G. Mark, (1983) A new method for detecting atrial fibrillation using R-R Interval, Computers in Cardiology 227-230
24. N. Srinivas, A. V. Babu, M. et al., (2013) Identification of Cardiac Arrhythmia with respect to ECG Signal by Neural Networks and Genetic Programming 2(2):479–486
25. Alireza Ghodrati, Bill Muray, Stephen Marinello (2008) RR Interval analysis for detection of atrial fibrillation in ECG monitors. IEMBS Vancouvar, British Colombia, Canada. 30: 20-24
26. V. Gokana, et al., (2014) Automatic Detection of Atrial Fibrillation using RR Interval from ECG signals, Book: 15th Int Conf on Biomedical Eng, Vol 43:215-218
27. Kenji Takazawa, et l., (1998) Assessment of Vasoactive Agents and Vascular Aging by the Second Derivative of Photoplethysmogram Waveform. 2nd Internal Medicine 32:365-370
28. Jinseok Lee, et al., (2012) Atrial Fibrillation Detection using Smart Phone, 34th Annual Int. Conf. of the IEEE EMBS 1177-1180
29. Dejan Vukajlovic, et al., (2011) Wireless Monitoring of Reconstructed 12-Lead ECG in Atrial Fibrillation Patients Enables Differential Diagnosis of Recurrent Arrhythmias, 33th Annual Int. Conf. of the IEEE EMBS, 33:4741-4744
30. Denis Jabaudon, et al., (2004) Usefulness of Ambulatory 7-Days ECG Monitoring for the Detection of Atrial Fibrillation and Flutter After Acute Stroke and Transient Ischemic Attack, Stroke, 35:1647-1651
31. Tommaso Sanna, et al., (2014), Cryptogenic Stroke and Underlying Atrial Fibrillation, N EngL J Med, 370:2478-2486

Author1:	Siti Nur Hidayah Binti Ab. Malek
Institute:	Universiti Kebangsaan Malaysia
Street:	43600 Bangi,
City:	Selangor,
Country:	Malaysia
Email:	sitinurhidayahmalek@gmail.com

Author2:	Kalaivani Chellappan
Institute:	Universiti Kebangsaan Malaysia
Street:	43600 Bangi,
City:	Selangor,
Country:	Malaysia
Email:	kckalai@ukm.edu.my

Author3:	Rosmina Jaafar
Institute:	Universiti Kebangsaan Malaysia
Street:	43600 Bangi,
City:	Selangor,
Country:	Malaysia
Email:	rosmina@ukm.edu.my

Non-invasive Assessment of Affective States on Individual with Autism Spectrum Disorder: A Review

N. Rusli, S.N. Sidek, H. Md Yusof, and M.H. Abd Latif

Department of Mechatronics Engineering, International Islamic University Malaysia, Kuala Lumpur, Malaysia

Abstract— **Individuals with Autism Spectrum Disorder (ASD) are identified as a group of people who have social interaction and communication impairment. They have difficulty in producing speech and explaining what they meant. They also suffer from emotional or cognitive states requirement that stance challenges to their interest in communicating and socializing. Hence, it is vital to know their emotion to help them develop better skills in social interaction. Emotion can be derived from affective states and can be detected through physical reaction and physiological signals. There are numerous known modalities available to detect the affective states either through invasive and non-invasive methods. In order to evaluate the affective states of individuals with ASD, amongst the methods used are through electrodermal activity (EDA), electromyographic (EMG) activity, and cardiovascular activity (ECG) and blood flow analyses. Though considered non invasive, these methods require sensor to be patched on to the skin causing discomfort to the subjects and might distract their true emotion. We propose non-invasive methods which is also contactless to address the problem to detect emotion of individual with ASD that is through thermal imaging. Through the impact of cutaneous temperature in blood flow, thermal imprint is radiated and can be detected in this method. To date, no research has been reported of the use of thermal imaging analysis of facial skin temperature on the individuals with ASD. In this paper we will justify the method and also discuss the merits and demerits of other methods.**

Keywords— **Autism, facial skin temperature, physiological signal, affective states, emotion.**

I. INTRODUCTION

Emotions are the reactions or response towards exterior stimuli. The emotions cause changes in the physiological signals, expressive behaviour and states of feeling. These parameters could be useful to identify the affective states. Theoretically, emotions are subset of affective states. It can be detected through physical reaction or physiological signals either in invasive or non-invasive methods. Recognition of human affective states from facial expression or speech for subjects with disabilities is actively going on. However, many challenges still occur when dealing with "uncooperative subjects" like autistic individuals. The autistic individuals have inability to recognize emotions in themselves or in the displays of others. They are also suffered from disability to differentiate different emotions. Physiologically, emotions may change human attention, behaviour and trigger the processing area in brain resulted to elicits disparate biological system and physiological system for optimal affective response. Extracting physiological patterns corresponding to emotional states has its own drawbacks as there are also possibilities of person-stereotypy and also situation stereotypy. This consideration is very significant to autism that has difficulty in facial muscular movement. They might show the same facial expression for different emotions or 'innocent' or too fear facial expression when confronting strangers.

A meta-analysis [1] has done on the individuals with autism spectrum disorder and reported that, most of the tests done on the individual with ASD required the participants to have good speech skills to label the emotion and good emotion recognition of others to match the emotion. The analysis came with a conclusion that autistic individuals did have difficulties in recognizing five basic emotions namely anger, disgust, fear, sad and surprise. The only emotion that can be recognized was happy. Even though, different approaches were used to activate their emotions, the difficulties still lay on the emotion processing and not to the perceptual demands of the different tasks. Hence, based on the theory of having intricacy in facial muscular movement, physiological-based affect discovery could be suitable for autistic individuals. In this paper, the parameter to be analyzed is facial skin temperature (FST). The aptitude to detect the physiological signals may also prove the importance for understanding the physiological mechanisms associated with ASD.

II. LITERATURE REVIEW

This section is organized in the following means. First, it reviews on the definition and characteristic of ASD. Then, it summarizes the findings from neuro-physiological studies of autistic individuals which clearly established association between core impairment in ASD and affective states. Then, explanations on the regulation process of affective states in follows by reviews of non-invasive methods to detect affective states. In Section III, we discuss and justify the proposed method to detect the emotion of the autistic individuals.

F. Ibrahim et al. (eds.), *International Conference for Innovation in Biomedical Engineering and Life Sciences*,
IFMBE Proceedings 56,
DOI: 10.1007/978-981-10-0266-3_47

A. Autism Spectrum Disorder

Autism is defined by the presence of social deficits and communication problems. Meanwhile, Autism Spectrum Disorder (ASD) is called as "spectrum disorder" because autism affects individuals differently and to varying degrees. ASD is a type of neuro-developmental disorder affecting the mental, emotion, learning and memory of an individual. Theoretically defined in APA [2], an individual with ASD has three major deficiencies in himself that related to his relationship with other people. First is the impairment of social skills, which includes the use of several nonverbal behaviors such as eye-to-eye gaze, facial expression, body postures and gestures to regular social interaction. Second is the impairment in communication, which includes delayed development of spoken language and inability to initiate or sustain a conversation with others. Third is the restricted, monotonous and stereotyped pattern of behaviors, interests, and activities, which include the inflexible obedience to specific, nonfunctional routines or formalities.

In general, the social deficits are viewed as the primary and unique characteristic of autism. Meanwhile, emotion is one of the important criteria in having a good social relationship. As stated earlier, the emotion can be identified through physical reaction and physiological signal. However, both processes are most likely originated from brain and it is controls by a limbic system. According to Damasio et al.[3] the basic affective states are strongly dependent on sub-neocortical limbic circuitries. The limbic circuit is composed of amygdala and limbic cortex that plays vital roles to regulate negative and positive emotions respectively. Ominously, in the individuals with ASD, the emotion processing area; amygdala and limbic cortex are found to be different from healthy individual. A post-mortem studies on autism Courchesne et al. [4]; Piven et al. [5]; documented the augmented head size due to excessive size of cortex which was also supported by the MRI findings. In addition, brain studies by Mosconi et al. [6]; Nacewicz et al. [7]; Schumann et al. [8] have found an abnormal developmental trajectory of amygdala volume in autism using structural MRI. Due to deficiency in amygdala, it is believed that it might affect the processing of negative emotion. Fritz U. [9] and Klin et al.[10] confirmed the abnormal brain developments of individuals with ASD resulted to social and communication impairments. Notably, Critchley H.D. et.al [11] stated that the individuals with ASD did not activate a cortical 'face area' when explicitly appraising expressions. The difficulties lead to social impaired between the autistic individuals and others.

B. Emotions Regulation in Individual with ASD

Emotion Regulation (ER) [12] is important not only for understanding the associated features of ASD, but also relevant to the core symptoms of social communication that define ASD as well. It is a complex and dynamic process that broadly refers the modification of biological, subjective, and expressive components of emotional experience [13]. Consciously-deployed changes in emotional responses are mediated via modulating effects in cognitive control on brain regions that involved in arousal and valence in which these two parameters are a subset of affective states. However, another functional magnetic resonance imaging (fMRI) scans of individual with ASD found that they had less change in prefrontal cortex activation compared with their healthy counterparts when undergoing a positive emotion regulation (ER) exercise. However, there are also in print works that did not find any impairment in negative affective states [14] either found deficits in positive affective states [15].

Explanation on amygdala dysfunction in autism remains unclear [16]. Thus, both positive and negative emotion regulation needs an extra research work. Accumulated evidence proposes that the amygdala may be dysregulated in emotional disorders such as anxiety and depression [17]. Likewise, the amygdala deficit might lead the autistic individual to abnormal fear responses such he either shows too little or too much fear or none, compared to non-autistic controls. As example, Vuilleumier et al.[18] showed the MRI findings of fear faces was absent if the amygdala was damaged. It is crucial to determine the impact of specific brain regions such as amygdala to differences in recognition of different emotions.

C. Physiological Signals in Autistic

Many studies have done to detect the emotion of ASD via facial recognition by Dina Tell [19], vocal to label emotion by R. C. M. Philip et.al [20], gesture and posture by Blake et al .[21] ; Hubert et.al.[22] and some use the self-reports [23]. However, self-reports are often not reliable in ASD because of deficits in verbal communication as well as difficulties with emotional awareness and introspection. Their physical difficulties require a new method of assessment to detect transition of emotional states via physiological regulation as stated by Dalton et.al.[24].

The affective states comprise of the signals originating from the autonomic nervous system (ANS). The ANS has two main parts: Sympathetic nervous system (SNS) and parasympathetic nervous system (PNS). The SNS and PNS are responsible to regulate physiological signals; electrodermal (i.e., skin-conductance, skin resistance), thermovascular (i.e., skin blood flow, skin temperature), and respiratory measures. While the PNS promotes restoration and conservation of bodily energy, "rest and digest," the SNS stimulates increased metabolic output to deal with external challenges, so-called "fight and flight." The signals of the ANS are used to identify the emotions of a healthy individual. There has been a large number of published works in the domain of emotion recognition from physiological signals [25]–[27]. Therefore, the same means could be implemented on the individuals with ASD.

Kushki et al. [28] did investigation on ANS towards individual with ASD by applying the common assessment method that used the FDA-approved sensors. The sensors are a breathable tape for measuring thermovascular and respiratory signals, and velcro straps for electrodermal signals (EDA). The experiment has confirmed that the anxiety condition did in fact stimulate an anxiety response in ASD. On the other hand, research done by Rosalind [29] from MIT identified that autistic individuals react and experience differently to emotion. Physiology measurement was done and the result showed that an individual with ASD can appear differently on the outside than what was measured on the inside. Supported research by Goodwin et al. [30] revealed that the individual with ASD is perfectly calm to those who know him, while having an unusually high resting heart-rate, 120 beats per minute or more, instead of the usual 60-80 bpm. Even though the individual with ASD has dysfunctional amygdala, yet from these research works, it shows that the ANS signals can be relied on to detect emotions. Therefore, in this research work we propose the use of physiological signals to detect emotions specifically on the individuals with ASD who has unresponsive facial expression.

D. Thermal Imaging Analysis

Thermal imaging provides solution of non-invasive autonomous monitoring. It permits the recordings of blood flow for cardiac pulse [31], breathe pattern [32], and skin temperature [33][34]. All of these parameters are associated with the affective states and causing impact to the cutaneous temperature. The reliability of these methods have been repeatedly proven with the use of thermal infrared imaging on gold standard methods, such as electrocardiography (ECG), piezoelectric thorax stripe for breathing monitoring, nasal thermistors, skin conductance (EDA), or galvanic skin response (GSR). However, prior to the fact that face is not obscured and direct to social communication and

Table 1 Invasive methods on physiological signals of autistic individuals

Experimental paradigms	Author	Emotion	Outcome
Movie-watching activity and stroop test - wire and probe attached	Kushki (2013)	Anxiety	Changes in EDA and skin temperature were detected
Free activity at home and readings is recorded and tally with diary. wristband attached	Rosalind (2009)	Anxiety	Changes in heart rate and EDA

Table 2 Non-invasive methods on physical reaction of autistic individuals

Experimental paradigms	Author	Emotion	Outcome
Watch the series of pictures. He/she match face or verbally state the answer.	Dina Tell et.al, (2014)	happy, angry, sad, fear, and neutral emotions	Children with autism less accurate in identifying fear and sadness.
Watch the series of image. Then, he is requested to verbally answer to label the emotion.	R. C. M. Philip et. al (2010)	happy, sad, anger, disgust, surprise and fear	Anger was failed to be identified by the ASD group.
Emotion recognition- Match the target stimuli to picture of a body movement without face.	Blake et al (2003); Hubert et al. (2007)	happy, sad, anger, disgust, surprise and fear	ASD cannot identifying happiness and fear through the body movement.
Select a text label from a choice of five to describe the emotion in vocal stimuli.	R. C. M. Philip et. al (2010)	happy, sad, anger, disgust and fear	Found deficits in vocal emotion processing in the ASD group compared to the control group.
Self-reported were collected during exposure and compared to parent reported social ability and stress responsivity	David M Simon et. al (2014)	anxiety, stress	Autisms are able to consistently report their persistent level of anxiety symptoms in stress situation.

interaction, the facial skin temperature is the favorable region to be implemented in this thermal imaging analysis. The disparity in facial skin temperature is caused by periorbital [35] and supraorbital vessels [34] of the face wherein heat increases according to stressors that are believed to facilitate preparedness for rapid eye movement in fight or flight mode [35]. As reported by Asthana et al. [36] change in affective state can cause variation in the facial skin temperature. The changes are due to the phenomena of vasoconstrictive effects in the blood flow from brain to cardiovascular. The variations in blood flow influence the emitted thermal print.. The similar research by Ioannou [37] distinguished the changes in facial skin temperature due to redirected blood to the region of interest (ROI) on face. This variation in temperature is notified for all of the chosen affective states. The blood flow may affected by the affective state process in the brain. Merla et. al [38] claimed that pulsating blood flow produces the strongest variation on the temperature signal.

Therefore, thermal imaging, bind the body's naturally emitted thermal irradiation, allows temperature readings due to the variation in blood flow. This method enables cutaneous temperature recordings to be measured noninvasively, naturally, and contact free. Table 1 summarizes invasive methods that have been used in detecting affective states of autistic individuals. Meanwhile, table 2 reviews of non-invasive methods on physical reactions.

III. FUTURE WORK

Autism is important issue to be addressed these days. As reported in CDC [39], there are more children are being diagnosed with these disorders. From the review, it is shown that the inclusion of facial temperature of autistic individual deserves proper investigation. Temperature changes may be measured and yielded additional information on ANS functioning in the emotion of autistic individual. However, temperature control associated with emotional reactions is far more complex in individual with ASD as it has different brain structure and serves a different purpose. It also has distinct neuro-regulatory systems, and carries its own thermal imprints. Further investigation is needed to find the relation of affective states and the dysfunction of amygdala in autistic brain. Hence, this is important to prove that the ASDs also have emotions towards stimuli regardless of facial expressions. From the recent studies it also shows that the changes of temperature were detected such as in the case of unconscious smile but not in an artificial smile. The same idea could be applied to the individuals with ASD due to difficulty in expressing the emotions physically. Addressing these questions will require more ambitious and large scale studies to provide critical insights into the roots of poor social cognition in autism, and the connection between brain and social information processing.

IV. CONCLUSION

Studies in the field of affective state identification have observed the impact of physiological signals in evaluating the emotions of healthy subjects. However, this is a challenge to individual with ASD due to dysfunction in some components in their limbic system that regulates the affective state. The personal behavior of anti-social due to fearful of eye gaze leads to a non-invasive method to detect the emotion. Thermal imaging is a novel method in the domain of non-invasive method in evaluating the affective state in individuals with ASD. To the best of our knowledge, till date there is no work done to investigate and assess the affective states of individuals with ASD via facial skin temperature.

ACKNOWLEDGEMENTS

The work presented was carried out in the Biomechatronics research laboratory of International Islamic University Malaysia. The authors wish to acknowledge the grant funding from the Ministry of Higher Education Malaysia (RAGS 12-002-0002).

CONFLICT OF INTEREST

The authors declare that they have no conflict of interest.

REFERENCES

1. Mirko Uljarevic • Antonia Hamilton.(2013). Recognition of Emotions in Autism: A Formal Meta-Analysis. J Autism Dev Disord 43:1517–1526
2. American Psychiatric Association (2000). Diagnostic and statistical manual of mental disorders- Text Revision (4th ed-TR). Washington, DC
3. Damasio, A.R., Grabowski, T.j., Bechara, A., Damasio, H., Ponto, L.L., Parvizi, J., et al.(2000). Subcortical and cortical brain activity during the feeling of self-generated emotions. Nature Neuroscience, 3: pp. 1049–1056.
4. Courchesne, E., Karns, C. M., Davis, H. R., Ziccardi, R., Carper, R. A., Tigue, Z. D., et al.(2001). Unusual brain growth patterns in early life in patients with autistic disorder: An MRI study. Neurology, 57. pp. 245–254.
5. Piven, J., Arndt, S., Bailey, J., & Andreasen, N.(1996). Regional brain enlargement in autism: A magnetic resonance imaging study. Journal of the American Academy Child and Adolescent Psychiatry, 35, pp. 530–536.
6. Mosconi MW, Cody-Hazlett H, Poe MD, Gerig G, Gimpel-Smith R, Piven J. (2009) May. Arch Gen Psychiatry Longitudinal study of amygdala volume and joint attention in 2- to 4-year-old children with autism. 66(5). pp. 509-16.
7. Nacewicz BM, Dalton KM, Johnstone T, Long MT, McAuliff EM, Oakes TR, Alexander AL, Davidson RJ. (2006) Dec. Amygdala volume and nonverbal social impairment in adolescent and adult males with autism. Arch Gen Psychiatry. 2006 Dec; 63(12):pp. 1417-28.
8. Schumann CM, Barnes CC, Lord C, Courchesne E. (2009) Nov 5. Amygdala enlargement in toddlers with autism related to severity of social and communication impairments. Biol Psychiatry. pp. 66(10):pp. 942-9
9. Frith U.(2001) Mind blindness and the brain in autism. Neuron 32. pp. 969 – 979.
10. Klin A, Jones W, Schultz R, Volkmar F, Cohen D(2002) Visual fixation patterns during viewing of naturalistic social situations as predictors of social competence in individuals with autism. Arch Gen Psychiatry 59. pp. 809 – 816.
11. Critchley HD, Daly EM, Bullmore ET, Williams SC, Van Amelsvoort T, Robertson DM, Rowe A, Phillips M, McAlonan G, Howlin P, Murphy DG.(2000). The functional neuroanatomy of social behaviour: changes in cerebral blood flow when people with autistic disorder process facial expressions. Brain. 123. pp. 2203-12.
12. Weiss, J. A., Thomson, K., & Chan, L.(2014) A systematic literature review of emotion regulation measurement in individuals with autism spectrum disorder. Autism Research
13. Thompson, R. A. (1994). Emotion regulation: A theme in search of definition. Monographs of the Society for Research in Child Development, 59(2–3), 25–52.
14. Lacroix, A., Guidetti, M., Roge´, B., & Reilly, J (2009) Recognition of emotional and nonemotional facial expressions: A comparison between Williams syndrome and autism. Research in Developmental Disabilities, 30, 976–985.
15. Humphreys, K., Minshew, N., Leonard, G. L., & Behrmann, M.(2007).A fine-grained analysis of facial expression processing in high-functioning adults with autism. Neuropsychologia, 45, 685–695.

16. Mirko Uljarevic • Antonia Hamilton(2012) Recognition of Emotions in Autism: A Formal Meta-Analysis, J Autism Dev Disord 43:pp.1517–1526
17. Davidson RJ, Abercrombie H, Nitschke JB, Putnam K.(1999). Regional brain function, emotion and disorders of emotion. Curr Opin Neurobiol 9:pp. 228-234
18. P. Vuilleumier, M.P. Richardson, J.L. Armony, J. Driver, R.J. Dolan(2004) Distant influences of amygdala lesion on visual cortical activation during emotional face processing Nat. Neurosci. pp. 1271–1278
19. Dina Tell, Denise Davidson, and Linda A. Camras (2014). Recognition of Emotion from Facial Expressions with Direct or Averted Eye Gaze and Varying Expression Intensities in Children with Autism Disorder and Typically Developing Children. Autism Research and Treatment., Article ID 816137, 11 pages
20. R. C. M. Philip.(2010)Deficits in facial, body movement and vocal emotional processing in autism spectrum disorders Psychological Medicine, 40, 1919–1929
21. Blake R, Turner LM, Smoski MJ, Pozdol SL, Stone WL (2003). Visual recognition of biological motion is impaired in children with autism. Psychological Science 14, 151–157.
22. Hubert B, Wicker B, Moore D, Monfardini E, Duverger H, Fonse´ca D, Deruelle C(2007). Brief report : recognition of emotional and non-emotional biological motion in individuals with autistic spectrum disorders. Journal of Autism and Developmental Disorders 37, 1386–92.
23. David M Simon (2014). Examining associations between anxiety and cortisol in high functioning male children with autism .Journal of neurodevelopmental disorders, *5:32 1866-1955-5-32*
24. Dalton,K.M.,Nacewicz,B.M.,Johnstone,T.,Schaefer,H.S., Gemsbacher,M.A., Goldsmith, H.H.,et al.(2005).Gaze-fixation and the neural circuity of face processing in autism. Nat.Neurosci. 8, 519–526.
25. J. A. Healey.(2000).Wearable and automotive systems for affect recognition from physiology. Ph.D. dissertation, MIT.
26. P. Lang, M. Greenwald, M. Bradely, and A. Hamm.(1993). Looking at pictures - affective, facial, visceral, and behavioral reactions. Psychophysiology, vol. 30, no. 3, pp. 261–273.
27. J. Kim and E. Andr´e. (2008). Emotion recognition based on physiological changes in music listening. IEEE Trans. Pattern Anal. Mach. Intell. vol. 30, no. 12, pp. 2067–2083,
28. Azadeh Kushki*, Ellen Drumml, Michele Pla Mobarakl, Nadia Tanell, Annie Dupuis, Tom Chau Evdokia Anagnostou.(2013). Investigating the Autonomic Nervous System Response to Anxiety in Children with Autism Spectrum Disorders. PLoS Journal
29. Rosalind W. Picard .2009. Future Affective Technology for Autism and Emotion Communication. To be appeared in the Philosophical Transactions of the Royal Society B.
30. Goodwin, M.S., et al., Cardiovascular arousal in individuals with autism.(2006). Focus on Autism and Other Developmental Disabilities, 21(2): p.p. 100_123-100_123.
31. Garbey, M., Sun, N., Merla, A., & Pavlidis, I.(2007). Contact-free measurement of cardiac pulse base on the analyses of thermal imagery. IEEE Transactions on Biomedical Engineering, 54, 1418–1426.
32. Ebisch, S. J., Aureli, T., Bafunno, D., Cardone, D., Manini, B., Ioannou, S., Merla, A.(2012). "Mother and child in synchrony: Thermal facial imprints of autonomic contagion". Thermology International,22,121-129
33. Hahn, A. C., Whitehead R. D., Albrecht, M., Lefevre, C. E., & Perret, D. I.(2012). Hot or not? Thermal reactions to social contact. Biology Letters, Physiology, 8, 1–4.
34. Puri, C., Olson, L., Pavlidis, I., Levine, J., & Starren, J.(2005). Stress-cam: Non-contact measurement of users' emotional states through thermal imaging. Proceedings of the 2005 ACM Conference on Human Factors in Computing Systems, 2, 1725–1728.
35. Levine, J. A., Pavlidis, I., & Cooper, M. (2001). The face of fear. *Lancet*, *357*, 1757.
36. Asthana, H.S. and mandal, M.K.(1997). Hemiregional variations in facial expression of emotions. British Journal of Psychology 88, pp. 519-525.
37. Stephanos Ioannou , Sjoerd Ebisch, Tiziana Aureli, Daniela Bafunno, Helene Alexi Ioannides, Daniela Cardone, Barbara Manini, Gian Luca Romani, Vittorio Gallese, Arcangelo Merla (2013). The Autonomic Signature of Guilt in Children: A Thermal Infrared Imaging Study. Journal in PLoS
38. Merla, and I. Pavlidis. Contact-free measurement of cardiac pulse based on the analysis of thermal imagery. IEEE Transactions on Biomedical Engineering.(2007) vol. 54, no. 8, pp. 1418–1426.
39. CDC. Center for Disease Control and Prevention. Prevalence of autism spectrum disorders autism and developmental disabilities monitoring network, 14 sites, United States, 2008. Morbidity and Mortality Weekly Report (MMWR), 61, 2012.

Author:	Nazreen binti Rusli
Institute:	IIUM
Street:	Jalan Sg Pusu
City:	Kuala Lumpur
Country:	Malaysia
Email:	nzrusli@gmail.com

Effective Dose Calculations and Dose Rate Distribution Around Z-Portal Passenger Security Screening

S.I. Farrag

Biomedical Engineering, Faculty of Engineering, Modern University for Technology and Information, Cairo, Egypt

Abstract— **Due to the increased concern over terrorist attacks, new technologies have been developed to improve the efficiency of security screening of passengers. The Z – portal is an inspection system designed to scan vehicles and large trucks, buses, and cargo containers. It has been installed in many sites in Egypt . The passenger should be screened before allowing to pass from a land to another. This technology uses ionizing radiation (X-rays). As the hazards related to ionizing radiation include the well-known carcinogenic risk, as well as other health effects, assessment of public exposure to these new types of security screening of a great importance we are presenting an assessment and distribution mapping of dose rate around two portals, then effective dose calculations and statistical analysis was performed. Methods: Two Z-portal systems are used for screening passenger by vehicles and trucks; Measurement of radiation dose rate was performed using dose rate survey meter, exposed with the two units during ten consecutive exposures each the organ effective doses was calculated. Results: Effective radiation doses ranged between 0.41 µSv and 6.66 µSv for one exposure. Thyroid, bone marrow, colon, lung, stomach, breast absorbed the most radiation for both screening systems with p <0.001. Conclusions: Effective doses of organ screened by the transmission system is significantly higher than the backscatter system however both systems are exceeding the dose limit of the general public per scan, as per ANSI standards of 0.1 µSv per scan for individual, cumulative effect is not considered, both systems should be reevaluated based on the general principles of justification, optimization and dose limitation versus new screening technology like passive millimeter – wave with no radiation emission.**

Keywords— **X-ray security scanners, X-ray transmission, X-ray backscatter, health effects, passenger screening.**

I. Introduction

Perceptions of increased threats from explosives and nonmetallic weapons have prompted the investigation of new passenger screening technologies, including chemical trace-detection techniques and imaging methods that can see through clothing, possible negative heath and public reactions toward many of these new detection technologies will have to be addressed.

A letter of concern was sent from Sedat [20] to the White House's Office of Science and Technology raising concerns about the potential health risks of these type of security screening using X-ray scanners , Their letter to assistant to the US president was "it appears that real independent safety data do not exist." In addition, the authors say: "There has not been sufficient review of the intermediate and long-term effects of radiation exposure associated with airport scanners. There is good reason to believe that these scanners will increase the risk of cancer to children and other vulnerable populations." In brief Sedat said " Society will pay a huge price in cancer because of this,". The European Commission in November 2010 prohibited their use in European airports [19].

The X-ray based security screening technology used in passenger screening relies on two techniques: backscatter or transmission. In the backscatter technique, radiation is reflected from the subject and detected to form an image of the body showing any concealed objects worn on the body. The transmission technique detects X-rays emitted by the equipment that pass through the body of the subject. Any concealed object provides an image by attenuating the radiation. While the backscatter technique can only reveal objects at the surface of the body, the transmission technique also shows objects within the body if their contrast differs sufficiently from the surrounding body fluids or tissue [3].

A review of both biological and biophysical studies on mechanisms of radiation-induced cancer concluded that the cancer risk is likely to occur in direct proportion to dose (with a linear dose-response relationship) even at the lowest doses without a threshold.

The effective dose, which takes into consideration the type of radiation and the sensitivity of the body parts exposed, is the best parameter to assess the health risk from ionizing radiation.

Effective dose, referred as Sieverts, it takes into account where the radiation dose is being absorbed and attempts to estimate the whole-body dose that would be required to produce the same risk as the partial-body dose that was actually delivered in a localized radiologic procedure. It is currently the best measurement available as it allows comparison with other types of radiation exposure, including natural background radiation [8].

Effective dose is calculated for any X-ray technique by measuring the energy absorption in a number of 'key' organs/tissues in the body. Each organ dose is multiplied by a

F. Ibrahim et al. (eds.), *International Conference for Innovation in Biomedical Engineering and Life Sciences*,
IFMBE Proceedings 56,
DOI: 10.1007/978-981-10-0266-3_48

weighting factor that has been determined as a reflection of its radiosensitivity. These are added together, so that the final figure is a representation of 'whole body' detriment.

Table (1), represent relative attributable lifetime risk based upon a relative risk of 1 at age 30 (population average risk). It assumes the multiplicative risk projection model, averaged for the two sexes. In fact, risk for females is always relatively higher than for males, and the lifetime risk attributable to a single small dose of radiation at a given age is greater for children, Beyond 80 years of age, the risk becomes negligible because the latent period between X-ray exposure and the clinical presentation of a tumor will probably exceed the life span of a patient. In contrast, the tissues of younger people are more radiosensitive and their prospective life span is likely to exceed the latent period [10].

Table 1 Risk in relation to age. Data are derived from (ICRP 1990).

Multiplication factor for risk	Age group (years)
x 3	<10
x 2	10-20
x 1.5	20-30
x 0.5	30-50
x 0.3	50-80
Negligible risk	80+

This paper presents actual measurements and mapping of dose rate for Z-portal different screening system as well with the uses of the theory of effective dose calculations organ effective doses for passenger is computed for both security screening systems.

II. MATERIALS AND METHOD

There is two Z- portal gates installed Fig (1) for inspection of all vehicles, trucks, cars before passing from land to another, all vehicles, cars, trucks are screened and inspected while passenger are sited inside. All passenger inside the cars are screened, men's, women's, pregnant women's and all children's, kids, newborn. Z-portal Backscatter was used for screening cars, vans, small trucks, and their cargo. Threats and contraband are stopped in their tracks. The Z -Backscatter Inspects vehicles for threats, imaging cars from three sides simultaneously with high energy X-ray, imaging provides photo-like images from the top and sides and quickly highlights organic threats and contraband, such as stowaways, explosives, currency, drugs, and alcohol.

Z-portal transmission was designated to Buses and trucks drive through. It is engineered to deliver unprecedented penetration into dense objects like cargo containers, tankers, and large vehicles. Its high-energy transmission X-ray imaging represents a way to detect contraband and threatening materials such as weapons, drugs, and explosives.

Fig. 1 Shows the two Z-portal systems extracted from Google Earth

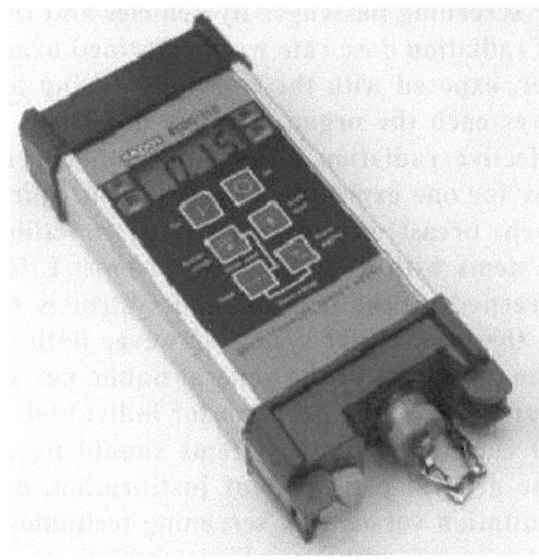

Fig. 2 RDS-110 Multi-purpose Survey Meter for dose rate measurements .

Measurements of dose rate was using Multi-purpose Survey Meter RDS-110 Fig (2) with an accuracy of ± 5% which enables measurement of low background dose rate registering even the smallest meaningful changes in dose rate values.

Measurements were done by operating each screening system , and mapping dose rate results in the spaces, in cars, vehicles, trucks and around the Z-portal systems Fig (3), ten sites locations were measured for each system in each scan, Exposure was performed ten times to gain enough doses to ensure proper measurement for each system. Average dose rate was then calculated and organ effective dose was then calculated for screened passengers exposed to the primary beam. Radiation weighted doses to individual organs were summed using (E2007) ICRP 103 tissue weighting factors, to calculate the effective dose [9].

The effective dose E expressed in μSv was calculated for the Thyroid, bone marrow, oesophagus, skin, bone surface, brain, colon, gonads, lung, stomach, bladder, breast, liver, by multiplying the tissue weighting factor W<u>T</u> by the mean absorbed dose at the irradiated area H_T. And summed over all of the tissue /organ exposed as per equation (1);

$$E = \Sigma W_T \times H_T . \quad (1)$$

The tissue weighting factors as defined by the ICRP in 2007 are shown in table (2), it represents the relative radiosensitivities of the organ and therefore the contribution of that organ to overall risk. The effective dose E allows the risk to the whole body to be expressed, thereby giving a broad indication of the level of detriment to health from the exposure.

Table 2 Tissue weighting factors as defined by ICRP 2007

Organ	W_T	ΣW_T
Thyroid	0.04	
Bone marrow	0.12	
Oesophagus	0.04	
skin	0.01	
bone surface	0.01	
brain	0.01	
Colon	0.12	
Gonads	0.08	
Lung	0.12	
Stomach	0.12	
Bladder	0.04	
Breast	0.12	
Liver	0.04	
Remaining tissues	0.12	
Total body		1

Whereas :
W_T= tissue weighting factor.
ΣW_T= sum of tissue weighting factors.

III. RESULTS

The measured dose rate using RDS-110 for the backscattered and the transmission system is summarized in table (3, 4). The effective doses were persistently higher in case of the Z portal transmission system compared to the Z- portal backscatter system the magnitude was around 16 times.

Results of Effective dose calculated to the specific organ is summarized in table (6), The ratio was constant and was

Table 3 The measured dose rate for the Z-portal vehicles backscatter system

Location	Dose rate μSv/hr
In the car	5
Pedestal	3
Outside the hanger	0.08
Control unit	0.08
Background radiation	0.08

Table 4 The measured dose rate for the Z-portal trucks transmission system

Location	Dose rate μSv/hr
In the truck	50 - 35
Pedestal	9 - 2
Outside the hanger	1.2 -0.15
Control unit	0.23
Background radiation	0.08

persistently around 6.01 %. The body total exposure amounted to 6.66 and 0.41 μSv using the Z-portal transmission system and backscatter system respectively.

The major contributor was the bone marrow (50.0%) with the lung, colon, stomach and breast It is interesting to note that the skin, bone surface brain were contributing equal portions (4%) each.

The statistical analysis performed with SPSS shows that: The overall mean for the exposure in case of the Z- portal backscatter system is (0.026 μSv) amounted to 5.9% of the corresponding value of the Z- portal transmission system (0.44 μSv). The difference was as expected was highly significant the P is shown in the table <.0.001. A descriptive statistical analysis is presented in table (5).

A relationship between both Z-portal systems is deducted from table (7) as per equation (2). It could be concluded that the exposure to the radiation was much higher in case of transmitted type compared to the transmission type by 16 times whereas transmission type will have better penetration and more detective power.

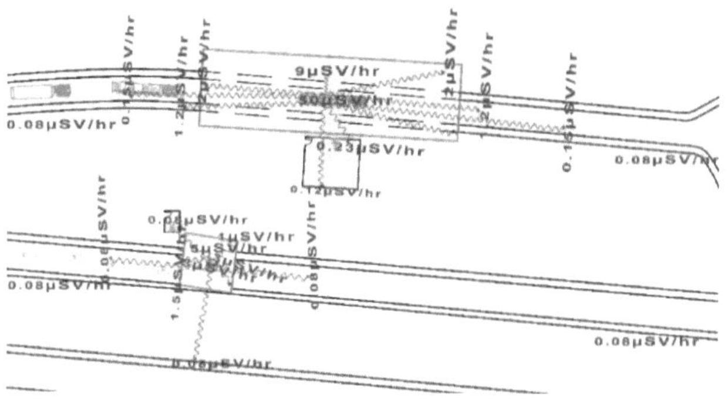

Fig. 3 Shows measured dose rate mapping for both transmission and backscatter systems

The statistical analysis performed with SPSS shows that: The overall mean for the exposure in case of the Z- portal backscatter system is (0.026 μSv) amounted to 5.9% of the corresponding value of the Z- portal transmission system (0.44 μSv). The difference was as expected was highly significant the P is shown in the table <.0.001. A descriptive statistical analysis is presented in table (5).

A relationship between both Z-portal systems is deducted from table (7) as per equation (2) . It could be concluded that the exposure to the radiation was much higher in case of transmitted type compared to the transmission type by 16 times whereas transmission type will have better penetration and more detective power.

Table 5 Descriptive statistics

	Minimum	Maximum	Mean	Std. Deviation
Z-portal backscatter	.004	.050	.0274	.020
Z-portal transmission	.066	.800	.4455	.316
contribution	.010	.122	.0668	.048
Valid N (listwise)				

Table 6 Organ effective doses from both security screening Z-portal systems

Organ	Z-portal backscatter μSv	Z-portal transmission μSv	Ratio %
Bone marrow	0.05	0.8	16.00
Colon	0.05	0.8	16.00
Lung	0.05	0.8	16.00
Stomach	0.05	0.8	16.00
Breast	0.05	0.8	16.00
Gonads	0.03	0.53	17.67
Thyroid	0.016	0.266	16.63
Oesophagus	0.016	0.266	16.63
0Nb Bladder	0.016	0.266	16.63
Liver	0.016	0.266	16.63
skin	0.004	0.066	16.50
bone surface	0.004	0.066	16.50
brain	0.004	0.066	16.50
ΣWT for the whole body	0.41	6,66	

Table 7 Coefficient

	Unstandardized Coefficients		Standardized Coefficients		
	B	Std. Error	Beta	t	Sig.
(Constant)	.010	.007		1.483	.166
Z-portal backscatte	15.902	.204	.999	78.036	.001

Dependent Variable: Z-portal transmission

$$\mathbf{Z_t = 0.01 + 15.902\, Z_b} \qquad (2)$$

Whereas :

Z_t is the Z-portal transmission

Z_b is the Z-portal backscatter

IV. Discussion

While X-ray based security scanners are currently used in the USA and as a trial in one UK airport, several Member States (e.g. Italy, France, Germany and Austria) prohibit the use of ionizing radiation for non-medical purposes, whereas due to many terrorist attacks Egypt has installed many security screening scanners for vehicles, trucks in many sites. Many of the passengers are children they are going to had a weekend on a touristic places and their risk is higher. The risk is approximately twice that of an adult patient [13,14].

Calculations presented here are greater than the proposed ANSI standard of 0.1 μSv per screening [2]. As well than NCRP [13] safety standard limits to a dose per screening of 0.25 μSv (25 μrem) reference effective dose for general-use full-body security screening systems. The annual dose limit is 250 μSv (25,000 μrem) over a 12-month period. Individual members of the general public, results breaches this guidelines

As per our study some travelers are passing twice and more which exceeds thousands times annually.

At 0.4 μSv per screening as per doses results of Z- portal backscatter system a total of 500 scans would be necessary to reach the administrative dose limit of 0.25 mSv per year for a member of the general public.

The higher effective doses calculated in this study were similar to other study of Rez [18] which fall between 0.4 μSv and 9 μSv. Boetticher et al [4] evaluated circumstances under which the application of the new model yields relevant dose differences compared to the prevailing model. As an example the Effective doses were calculated and compared from the measured organ doses according to several studies [3, 12,17] total body effective dose is 0.160 μSv for a transmission system and 0.025 μSv whereas in this study was 6.66 and 0.41 μSv respectively.

The effective radiation dose resulting from the Z- portal backscatter was found to be equal to 0.41 μSv which breaches all guidelines of NCRP of 0.25 μSv and ANSI of 0.1 μSv, Z-portal transmission was found to be equal to 6.66 μSv which highly exceeds the guidelines. Both doses ΣW_T for the whole body exceeds the guidelines of NCRP.

The mean radiation effective dose to organs resulting from the Z-portal backscatter (0.0274 ± 0.0198 μSv) differed from the mean radiation effective dose to organs resulting from the Z-portal transmission (0.445 ± 0.315 μSv) the difference proved to be statistically significant ($P<0.001$), the same calculation was done using non parametric statistics Wilcoxon signed rank test and proved to be statistically significant as well ($P<0.001$).

A recent paper [22] estimated doses to the skin as high as 2.5 μGy for 50 kVp X-rays and 0.68 μGy for 50 kVp X-rays (effective doses of 0.9 and 0.8 μSv, respectively). Whereas in this study effective dose to the skin was 0.004 and 0.066 μSv respectively The results of mentioned study were higher than this study as it is based on a different approach from the other studies.

One should state here, that it is difficult to give reliable and meaningful estimates of effective doses for children up to the age of at least 14 years, since the variations in stature and size are even greater than for adults. In addition, there is still no reference set of real anatomy based voxel phantoms representative of the various sizes of children.

Dose limitation means setting standards for limiting radiation exposure to each individual, to avoid or minimize any health risks. Optimization entails reducing radiation exposure within practical constraints, as low as reasonably achievable. This means that just complying with the dose limits is not sufficient.

V. CONCLUSION

The international standards of radiation protection are based on the general principles of justification, optimization and dose limitation. Justification requires consideration of the benefits obtained from use of radiation relative to the potential risks, to preclude any unnecessary radiation exposure. The benefits and risks can include a variety of gains (for example, security scanners mainly improved safety) and adverse effects (health risks, economic costs, etc.), which are not directly commensurate and therefore the weighing of the trade-off is not straightforward. Whether a technology or its application for a certain purpose is deemed acceptable is ultimately not a scientific issue, but a political decision influenced by social factors.

Public screening systems using X-ray technology should be reevaluated versus the appearance of new screening technologies that does not uses radiation e.g. passive millimeter wave systems.

The cumulative effective dose from a screening scanners transmission or backscatter scanners to a person who uses this travel daily should be taken into consideration , further studies should include dose –effect curve in order to build a the potential dangers, to build a sufficient data of intermediate and long term effects of radiation exposure associated with security screening scanners.

CONFLICT OF INTEREST

The author declares that she do not have any conflict of interest.

REFERENCES

1. American College of Radiology (ACR) Statement on Airport Full-body Scanners and Radiation (2010)
2. ANSI/HPS N43.17- (2009) Radiation Safety for Personnel Security Screening Systems Using X-Ray or Gamma Radiation
3. Assessment of the Rapiscan Secure 1000® Body Scanner for Conformance with Radiological Safety Standards. July 21, (2006), produced for TSA, measurements made at FDA, report completed at NIST
4. Boetticher H, Lachmund J, et al; (2007) recommendations of the ICRP change basis for estimation of the effective dose: what is the impact on radiation dose assessment of patient and personnel?
5. CDRH response to Center for Study of Responsive Law inquiry on people screening, November 5, (2010)
6. Committee to Assess Health Risks from Exposure to Low Levels of Ionizing Radiation NRC. Health risks from exposure to low levels of ionizing radiation: BEIR VII Phase 1. Washington (DC): National Academies Press; (2005)
7. Committee to Assess Health Risks from Exposure to Low Levels of Ionizing Radiation NRC . Health risks from exposure to low levels of ionizing radiation: BEIR VII Phase 2. Washington (DC): National Academies Press; (2006.)
8. HPS Position Statement Use of Ionizing Radiation for Security Screening Individuals (2009)
9. ICRP ; ICRP publication 103 ; The (2007) Recommendations of the International Commission on Radiological Protection
10. ICRP ; ICRP publication 119 ; The (1990) Recommendations of the International Commission on Radiological Protection
11. ISCORS TECHNICAL REPORT (2008) -1 July United States , Guidance for Security Screening of Humans Utilizing Ionizing Radiation.
12. National Research Council Commission on Engineering and Technical Systems report, Airline Passenger Security Screening New Technologies and Implementation Issues Publication NMAB—482—1 (1996)
13. NCRP Statement 10, Recent Applications of the NCRP Public Dose Limit Recommendation for Ionizing Radiation (2004)NCRP Commentary No. 16 - Screening of Humans for Security Purposes Using Ionizing Radiation Scanning Systems (2003)

14. NCRP released a press release May (2010) regarding Commentary No. 16
15. NCRP report no. 116 Limitation of Exposure to Ionizing Radiation (1993) recommends annual limits on radiation dose for the general public. Report 116 also introduces the concept of a negligible individual dose.
16. NCRP report no. 160, Ionizing Radiation Exposure of the Population of the United States (2009)
17. Radiation Safety Engineering Assessment Report for the Rapiscan Secure 1000 in Single Pose Configuration, Johns Hopkins University Applied Physics Laboratory, Assessment for TSA, revised August (2010)
18. Rez P, Kenneth L. ; (2011), The Dose from Compton Backscatter Screening; Radiat Prot Dosimetry. Apr;145(1):75-81
19. Scientific Committee on Emerging and Newly Identified Health Risks , European Community ,(2012) Health effects of security scanners for passenger screening (based on X-ray technology)
20. Sedat J. ; (2010) , letter of concern to the white house about the potiential health risks for whole body backscatter scanner

Author:	Sherine Farrag
Institute:	Modern University for technology and information
City:	Cairo
Country:	Egypt
Email:	sherine.farrag@gmail.com

The Effect of a Different Motion Speed on the Induced Current in the Body of Female Nurses Working in the Vicinity of MRI Scanners

S.I. Farrag

Biomedical Engineering, Faculty of Engineering, Modern university for technology and information Cairo, Egypt

Abstract— **Medical staff working near magnetic resonance imaging (MRI) scanners are exposed both to the static magnetic field (SMF) itself and also to electric currents that are induced in the body by motion within the SMF. Currently available data on induced electric currents for realistic movements are limited. This study computationally investigates the movement-induced electric fields for realistic movements in the magnetic field of a 1.5 T MRI scanner.**

Computation of induced fields by motion at different velocities in SMF at the vicinity of MRI scanners is based on magnetic field measurements for both the motion of the worker and the fringe field of the MRI scanner. The induced fields and the related specific absorption rate are computationally determined using the Matlab script method.

Results shows that SAR values and induced electric currents has a significant correlations with the motion speed with $p<0.001$.

Keywords— **Occupational exposure, MRI safety, induced current density, Specific Absorption Rate, Static magnetic fields.**

I. INTRODUCTION

Humans moving in SMF of magnetic resonance imaging (MRI) systems may experience mild sensory effects, including vertigo, nausea, headache, metallic taste in mouth, and magnetophosphenes [4,8,20]. For patients undergoing computational dosimetry of induced electric fields during realistic movements in the vicinity of a 3 T MRI scanner during MRI examination, the most frequently reported sensation was vertigo [7,14].

The effects of static magnetic fields are caused by two related factors. The first factor is the exposure to a static magnetic field itself. As shown by Roberts et al [19], a strong static magnetic field can directly affect the vestibular function, which could explain sensations of vertigo attributed this effect to the Lorentz force that acts on ionic currents that constantly flow in the endolymphic fluid of the labyrinth of the inner ear.

The second factor is the exposure to the electric fields/currents that are induced in the human body moving in a static magnetic field. If the induced currents are strong enough, they may stimulate excitable nerve, muscle, or receptor cells. Preventing the stimulation of the central nervous system or peripheral nerves has been the biological basis for deriving the basic restrictions in the ICNIRP (2010) [11] guidelines for the exposure to low-frequency electromagnetic fields.

Weak direct currents can also affect the vestibular function as is evident from the studies of galvanic vestibular stimulation (GVS) [3,7]. Consequently, if the magnitude of movement-induced currents were comparable to the currents used in GVS, they could electrically stimulate the vestibular system resulting in sensations of vertigo. As discussed by Glover et al [6], the magnetic-field related vertigo might actually be attributable to the combination of both static magnetic field acting on the vestibular system and GVS-like stimulation by movement-induced currents.

The exposure of workers to the static and gradient fields of MRI scanners have been successfully investigated using measurements [2,9,14]. However, for assessing the exposure to movement-induced electric fields, computational techniques are needed.

Several studies has clearly indicated that the magnitude of movement induced electric fields can be within the same order of magnitude as the basic restrictions set by international guidelines or standards [11,12]. Various different models for the human anatomy have been used: homogeneous model of the whole body was employed in [1], 4 or 6 mm resolution whole-body model in Liu et al [3,17] , a three-domain model of the head in Cobos et al [2], and four different heterogeneous anatomical models of the head with 2 mm or 1 mm resolution in Laakso [15, 16]. Some of the studies have used realistic models for the magnetic field of the MRI magnet [1,3,17] ,while others have employed simple analytical expressions for the magnetic field. However, all previous studies have only considered idealized translational or rotational movements.

In this study, movement induced fields are computationally investigated for realistic motion in the vicinity of an MRI scanner. The path of movement is based on a realistic model , where nurses go through during patient positioning in 1.5 T MRI scanner .

The computed induced electric fields were then compared with the exposure limits of ICNIRP , and the signific-

F. Ibrahim et al. (eds.), *International Conference for Innovation in Biomedical Engineering and Life Sciences*,
IFMBE Proceedings 56,
DOI: 10.1007/978-981-10-0266-3_49

ance of the results on occupational exposure assessment was discussed. A derived formulae correlating the worker motion speed and the induced currents as well the Computed specific absorption rate are then presented.

II. MATERIALS AND METHODS

A. Theory

The electrical behavior of the biological matter can be described by the conductivity (σ) which interacts with the electric field (E) applied to the tissue. The current (J) is obtained by (1):

$$J=\sigma E \tag{1}$$

Where J is the induced current in (A/m^2)

When an electric conductor, such as the human body, moves in spatial heterogeneous static magnetic fields B, it induces an electrical field E expressed by the equation below [8].

$$Rw = IDRx frac WU{}^{!}_{/}DRo \tag{2}$$

Where E is the electric field in (v/m) k is a geometry factor for a given subject, a radius of 0.64 m is assumed for a typical current loop in the body, while for the head a radius of 0.07 m [8]. Worker walking speed components (v_x or v_y) are vectors. We chose one path for all nurses shown in Fig (2, 3). A study of seven nurses working in MRI patient positioning are considered with age ranges from 22-50 years old. The walking speed is then measured and calculated for each nurse taking into consideration that the path is divided into three parts the first from the control room to the MRI room door where the speed increase linearly, the second is till the MRI gantry where speed remains constant and equal to the maximum and then decrease linearly up to stop point for patient positioning and then the third part when returning back to the control room where speed increase linearly and then remains constant and equal to the maximum value to permit the worker to go out the scanner room. Three maximum value of the walking speed were calculated for each nurse and they are in table (1): walking speed depends on the age , size of the nurses, only movement in the ground plane that is along x, y directions is considered. A computational model for Specific Absorption Rate (SAR) and induced current for both head and trunks then created through Matlab, based on organs and tissue electrical conductivity.

$$SAR = \sigma E^2 / \rho \tag{3}$$

Where SAR is the specific absorption rate in (W/kg), σ is the electrical conductivity of human tissues, and ρ is the mass density of human body assumed 1000 Kg/m^3 [22]

B. Methods

C. Computational Modeling and plotting of MRI field distribution around 1.5 T.

The iso-gauss line map (fringe field) of the MRI scanner brand Sigma GE 1.5 Tesla was actually measured using a Gaussmeter brand OXFORD Instruments using the HALL effect with resolution of 0.01mT (0.1 Gauss).

The knowledge of ellipse geometric theorem as a and b as the minor and major radii respectively of an ellipse is used to estimate the magnetic fringe field of the MRI Applying the Pythagorean identity (4) Ellipse circumference is defined by (6).

At any point located in the contour of each ellipse has same static magnetic field, we defined objects in the X and Y directions for any geometric points in the SMF field of the scanner, a polynomial interpolation function between all ellipses contour points and static magnetic field (B) was then generated using polyfit method.:

$$\frac{x^2}{a^2} + \frac{y^2}{b^2} = 1 \tag{4}$$

$$\frac{a_i}{b_i} = constant \tag{5}$$

$$\text{Circumference} \approx 2\pi \sqrt{\frac{1}{2}(a^2 + b^2)} \tag{6}$$

A Matlab script was created to estimate the static magnetic field value (B) at each point of the MR room in the x, y Axis.

The induced current density associated with worker's movements during conduction of any MR procedure is dependent on the walking speed and on the spatial gradient fields associated with a specific path.

The detailed pathway starts from the control room 'point A' from which the worker walks to the control room door' point B' then to the bore of the magnet ' point F ' passing by 'D' then ' E' points where he/she stops to start patient positioning. Then takes the gradient coil from 'point G' stopping at 'point F' and finish patient positioning to start the procedure. Finally, he/she exits the room and return back to the control room through the pathway. The total pathway length as described by Fig. 5. To go forward from the control room to patient positioning is 13.65 meters including gradient coil placement. Then for going backward into the control room. The total pathway is 11.3 meters. In her round trip, the walking speed of the operator is calculated. Three values of walking speed for each subject are calculated, the range of walking speed was based on the age and size for each nurse the maximum was for a young, thin nurse was at 1.6, 1.4 and 1.2 m/s. and the minimum was for an old nurse suffering from obesity is 0.6, 0.7, 0.8 m/s The induced current density *J* due to operator movements along the chosen walking path for each walking speed was then calculated and a computational modeling of the tissues and the organs SAR values was then constructed.

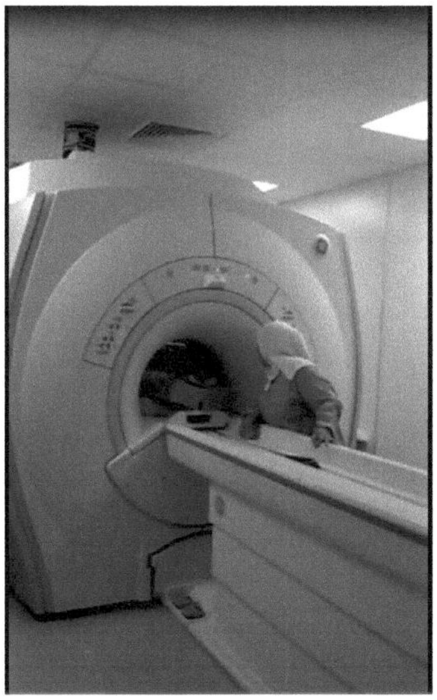

Fig. 1 Shows operators during patient positioning

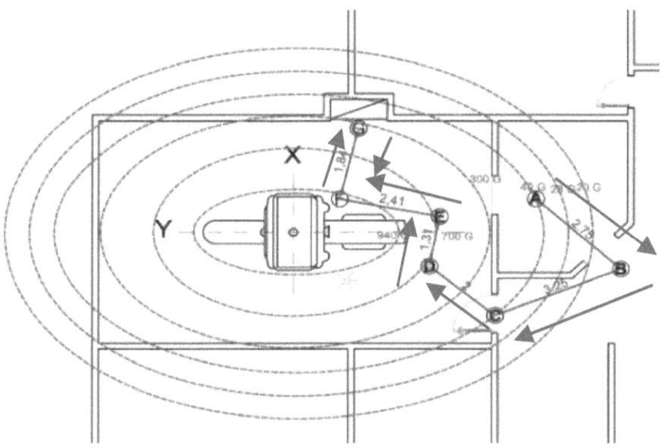

Fig. 2 Shows operator path during patient positioning along x, y directions.

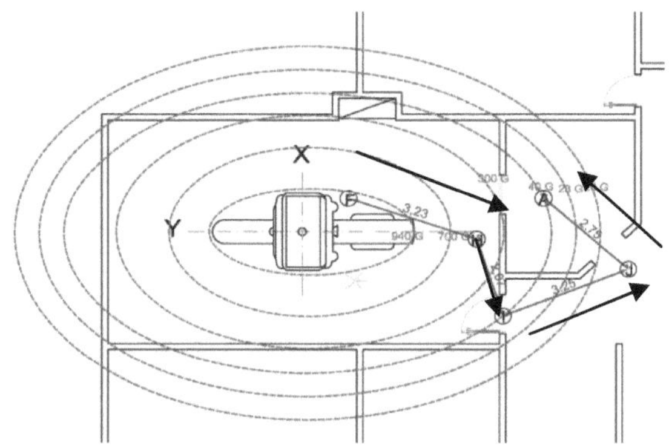

Fig. 3 Shows operator path, returning back after patient positioning along x, y directions.

Table 1 Motion speed vectors for each subject

Case number	Speed vectors m/s
1	1.2 , 1.4 , 1.6
2	0.9, 1.1, 1.2
3	0.6, 0.7, 0.8
4	0.7, 0.8, 0.9
5	0.65, 0.7, 0.82
6	0.73, 0.85, 0.98
7	0.66, 0.75, 0.88

III. RESULTS

Table (2) shows computed induced current (J) and Specific absorption rate (SAR) for each worker's motion speed.

Maximum induced current reported was 1.6 A/m^2, and maximum SAR value was 1.26 mW/kg.

The regression statistics of the velocity of induced current are demonstrated by Fig (3)

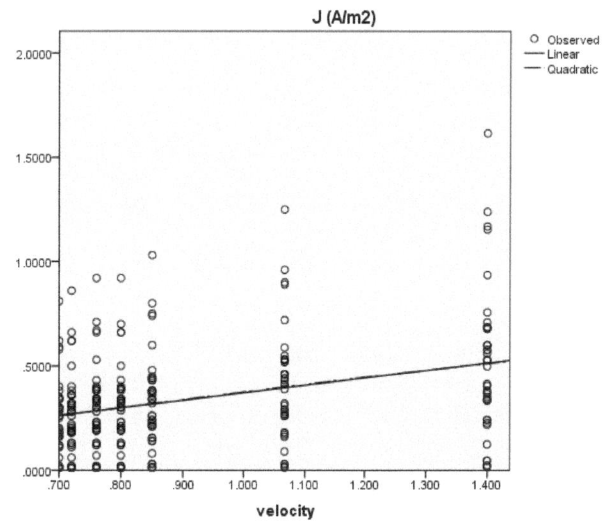

Fig. 4 Shows Regression statistics of velocity on induced current (J)

The statistical analysis performed with SPSS shows that there is a linear positive correlation between motion speed and induced current as well between motion speed and specific absorption rate table (3) shows correlations coefficient between motion speed (velocity) with J and SAR which show high significance p< 0.001.

Table 2 Shows computed induced current (J) and Specific absorption rate (SAR) for the seven cases motion speed

Tissue Name	J (A/m²)	SAR (mW/kg)	J (A/m²)	SAR (mW/kg)	J (A/m²)	SAR (mW/kg)	J (A/m²)	SAR (mW/kg)	J (A/m²)	SAR (mW/kg)	J (A/m²)	SAR (mW/kg)	J (A/m²)	SAR (mW/kg)
Aorta	0.335	0.261	0.26	0.15	0.16	0.06	0.19	0.08	0.18	0.07	0.21	0.1	0.19	0.08
Bladder	0.218	0.17	0.16	0.1	0.11	0.04	0.12	0.05	0.11	0.049	0.14	0.07	0.12	0.05
Blood	0.936	0.73	0.72	0.44	0.4	0.18	0.5	0.23	0.5	0.211	0.6	0.3	0.53	0.24
Blood vessel	0.335	0.261	0.26	0.15	0.16	0.06	0.19	0.08	0.18	0.07	0.21	0.1	0.19	0.08
Body fluid	1.17	0.913	0.9	0.55	0.59	0.23	0.66	0.29	0.62	0.26	0.75	0.37	0.67	0.3
Bone cancellous	0.124	0.097	0.09	0.05	0.06	0.24	0.07	0.03	0.06	0.028	0.08	0.04	0.07	0.03
Bone Cortical	0.046	0.036	0.03	0.02	0.02	0.009	0.02	0.01	0.02	0.01	0.03	0.01	0.02	0.012
Bone marrow	0.015	0.0122	0.012	0.007	0..007	0.003	0.008	0.003	0.008	0.003	0.01	0.005	0.008	0.004
Breast fat	0.023	0.018	0.018	0.011	0..011	0.004	0.01	0.005	0.012	0.005	0.015	0.07	0.013	0.006
Cartilage	0.351	0.274	0.27	0.165	0.177	0.06	0.2	0.089	0.188	0.07	0.22	0.11	0.2	0.09
CSF	1.615	1.26	1.25	0.76	0.81	0.32	0.92	0.411	0.86	0.36	1.03	0.51	0.92	0.41
Cervix	0.561	0.438	0.43	0.26	0.28	0.111	0.32	0.14	0.3	0.12	0.36	0.18	0.322	0.14
Colon	0.499	0.389	0.388	0.23	0.25	0.09	0.285	0.12	0.26	0.11	0.32	0.16	0.28	0.12
Duodenum	0.686	0.536	0.53	0.32	0.34	0.13	0.39	0.17	0.36	0.155	0.44	0.22	0.39	0.17
Fat	0.023	0.018	0.018	0.011	0.011	0.004	0.013	0.005	0.01	0.005	0.015	0.007	0.013	0.006
Gall bladder	0.757	0.59	0.588	0.35	0.38	0.15	0.43	0.19	0.4	0.17	0.48	0.24	0.43	0.19
Gall bladder bile	1.155	0.901	0.89	0.544	0.58	0.22	0.66	0.29	0.62	0.26	0.74	0.37	0.66	0.29
Gland	0.6	0.469	0.46	0.28	0.3	0.11	0.34	0.15	0.32	0.13	0.38	0.19	0.34	0.15
Heart	0.53	0.414	0.41	0.25	0.26	0.1	0.3	0.13	0.28	0.11	0.34	0.17	0.3	0.13
Kidney	0.577	0.45	0.44	0.27	0.29	0.11	0.33	0.14	0.31	0.13	0.37	0.18	0.33	0.14
Liver	0.343	0.268	0.266	0.16	0.17	0.06	0.19	0.08	0.18	0.07	0.22	0.11	0.19	0.08
Lung deflated	0.413	0.322	0.32	0.19	0.2	0.08	0.23	0.1	0.22	0.09	0.26	0.13	0.23	0.1
Lung inflated	0.226	0.176	0.17	0.1	0.11	0.044	0.12	0.05	0.12	0.05	0.14	0.07	0.12	0.05
Lymph	0.6	0.469	0.46	0.28	0.3	0.11	0.34	0.15	0.32	0.13	0.38	0.19	0.344	0.15
Mucous membrane	0.382	0.298	0.29	0.18	0.19	0.07	0.21	0.09	0.2	0.08	0.24	0.123	0.219	0.098
Muscle	0.53	0.414	0.41	0.25	0.26	0.1	0.3	0.13	0.28	0.11	0.34	0.17	0.3	0.13
Nail	0.046	0.036	0.03	0.022	0.02	0.009	0.02	0.01	0.02	0.01	0.03	0.015	0.02	0.012
Nerve	0.241	0.188	0.18	0.11	0.12	0.04	0.13	0.06	0.13	0.05	0.155	0.077	0.13	0.06
Oesophagus	0.679	0.529	0.52	0.31	0.34	0.13	0.388	0.17	0.36	0.15	0.43	0.218	0.38	0.17
Ovary	0.53	0.414	0.41	0.25	0.26	0.1	0.3	0.13	0.28	0.11	0.34	0.17	0.3	0.13
Pancreas	0.6	0.469	0.46	0.28	0.3	0.11	0.34	0.15	0.32	0.13	0.38	0.19	0.344	0.15
Prostate	0.686	0.536	0.53	0.32	0.34	0.13	0.39	0.17	0.36	0.155	0.44	0.22	0.39	0.17
Skin dry	0.335	0.261	0.26	0.15	0.16	0.06	0.19	0.08	0.18	0.07	0.21	0.107	0.19	0.08
Skin wet	0.374	0.292	0.29	0.17	0.18	0.07	0.21	0.09	0.2	0.08	0.24	0.12	0.214	0.09
Small intestine	1.24	0.968	0.96	0.58	0.62	0.2	0.7	0.31	0.66	0.28	0.8	0.39	0.71	0.31
Spinal chord	0.241	0.188	0.18	0.11	0.12	0.04	0.13	0.06	0.13	0.05	0.155	0.077	0.13	0.06
Spleen	0.577	0.45	0.44	0.27	0.29	0.11	0.33	0.14	0.31	0.13	0.37	0.18	0.33	0.14
Stomach	0.679	0.529	0.52	0.31	0.34	0.13	0.38	0.17	0.36	0.15	0.43	0.218	0.38	0.17
Tendon	0.366	0.286	0.28	0.17	0.18	0.07	0.32	0.09	0.19	0.08	0.23	0.117	0.2	0.09
Testis	0.686	0.536	0.533	0.32	0.34	0.13	0.39	0.17	0.36	0.155	0.44	0.22	0.39	0.17
Thymus	0.6	0.469	0.46	0.28	0.3	0.11	0.34	0.15	0.32	0.13	0.38	0.19	0.34	0.15
Trachea	0.405	0.316	0.31	0.19	0.2	0.08	0.23	0.1	0.21	0.09	0.26	0.13	0.23	0.1
Uterus	0.71	0.554	0.55	0.33	0.35	0.14	0.4	0.18	0.38	0.16	0.45	0.22	0.4	0.18

Table (4), (5) is the derivation of the formulae correlating the motion speed with the induced current and the specific absorption rate .

In some tissue type listed in Table (2), for example CSF, blood, body fluids,..etc the maximum current densities are found to exceed the exposure limit of 10 mA/m^2 suggested byICNIRP guidelines. [11,12].

It is well justified that such organs and tissues (e.g. CSF, Blood, blood fluids,.etc) as shown in table (2) has the maximum current and SAR values as they contain free water at the molecular level whereas liver for example shows a less induced current and SAR values due to their high proportion of bound water.

A formulae correlating the velocity and induced current (7), the velocity and SAR (8) are Then predicted using SPSS program

Table 3 Correlations

		J (A/m2)	SAR (mW/kg)
Velo	Pearson Correlation	.345**	.583**
	Sig. (2-tailed)	.001	.001
	N	301	301

**. Correlation is significant at the 0.01 level (2-tailed).

Table 4 Velocity and Induced current correlation

Model		Unstandardized Coefficients B	Unstandardized Coefficients Std. Error	Standardized Coefficients Beta	t	Sig.
1	(Constant)	.011	.053		.207	.836
	velo	.360	.057	.345	6.360	.001

Dependent Variable: J (A/m2)

$$\mathbf{J = 0.011 + 0.360 \quad V} \tag{7}$$

Where J: is the induced current
V: is motion speed

Table 5 Velocity and SAR correlation

Model	Unstandardized Coefficients B	Unstandardized Coefficients Std. Error	Standardized Coefficients Beta	t	Sig.
1 (Constant)	-.198	.032		-6.271	.001
velo	.422	.034	.583	12.422	.001

Dependent Variable: SAR (mW/kg)

$$\mathbf{SAR = -0.198 + 0.422\ V} \tag{8}$$

Where SAR : is the specific absorption rate
V: is the motion speed

IV. DISCUSSION

Movement induced fields were investigated in seven working nurses in the vicinity of MRI. The workflow for determining the induced electric field consisted of magnetic field measurements, modelling the magnetic field of the MRI scanner, modelling the trajectory of movement based on the measured and modelled magnetic fields, and finally numerically calculating the induced current and SAR values using the Matlab script method. The induced electric current in the head and trunk were approximately linearly dependent on the average motion speed of the candidate , the estimates for the maximum induced current is 1.6 A/m2 with a SAR value of 1.26 mW/kg For the CSF tissues.

For the path of movement of this study, the induced electric current is exceeding the basic restrictions. It was also found that the induced currents are within the same order of magnitude as the motion velocity This supports the assumption that the movement induced electric fields have a significant role in the generation of magnetic field induced. For the comparison of the assessed exposure to limit values in international guidelines, we used the recent ICNIRP guidelines for time varying (ICNIRP 2010) electric and magnetic fields, and the electric field induction by a movement in a static magnetic field (ICNIRP 2012). The basic restrictions for the induced electric fields in the ICNIRP (2010) guidelines are given in terms of root-mean-square values; they can be converted into peak values by multiplying them by √2.It is important to mention that guideline did not consider the dB/dt gradient which takes into account the motion speed of torso or head of workers

Calculations for SAR values and induced electric currents clearly shows that the slower the motion speed, the lesser the SAR and induced currents. This notion may help to reduce the annoying sensory impulses for MRI staff whose work necessitates repeated helping of patients and supporting them in the vicinity of scanners. Recommendations for a slower movements of torso and head should be well-known to staff.

The derived formulae (7) and (8) describe well the relationship between the induced current , SAR and the velocity where they are linearly correlated. The statistical analysis, shows high significance $p<0.001$.

V. CONCLUSION

The model used for estimating induced current density and SAR is the same adopted by ICNIRP guidelines, and can be used as a first approximation of the workers exposure in any MRI facility. A very important area of the use of presented method is while providing workers with occupational safety and health education. It can be used for training MRI workers to avoid high exposure, and change as much as possible, their motion behavior in the scanner room. A pathway drawn on floor of scanner room with time limits for each subsequent movement may prove to be helpful for staff to adjust to slower motion speed during entry and retreat from the scanner room. The least speed motion was 0.6-0.7 and 0.8 m/s that decrease significantly the induced current was for old obese nurse this speed cannot be achieved usually for normal young nurses so the median speed 0.7 , 0.8 and 0.9 m/s is the best speed model which decrease the induced current and can be achieved by normal motion.

Due to limited room space only one trajectory is used for patient positioning.

Decreasing exposure time is one of the parameters that can help reducing exposure whereas it was found unpractical to consider in this study, as per nurses discussion, there is sometimes some problem with the patient that needs longer time for positioning, specially for children and elderly people positioning.

A further detailed dose –effect study will be of high importance. Whereas motion speed had a clear effect on SAR values and induced currents in bodies of MRI staff in the vicinity of MRI scanners.

CONFLICT OF INTEREST

The author declares that she do not have any conflict of interest

REFERENCES

1. Chiampi M and Zilberti L 2011 "Induction of electric field in human bodies moving near MRI: An efficient BEM computational procedure" IEEE Trans. Biomed. Eng. 58(10) 2787–93
2. Cobos S´anchez C, Glover P, Power H and Bowtell R 2012 "Calculation of the electric field resulting from human body rotation in a magnetic field" Phys. Med. Biol. 57(15) 4739–53
3. Crozier S and Liu F 2005 "Numerical evaluation of the fields induced by body motion in or near high-field MRI scanners" Prog. Biophys. Mol. Biol. 87 267–78
4. de Vocht F, Muller F, Engels H and Kromhout H 2009 "Personal exposure to static and time-varying magnetic fields during MRI system test procedures" J. Magn. Reson. Imaging 30(5) 1223–8
5. de Vocht F, van Drooge H, Engels H and Kromhout H 2006 " Exposure, health complaints and cognitive performance among employees of an MRI scanners manufacturing department" J.Magn. Reson. Imaging 23(2) 197–204
6. Glover P, Cavin I, Qian W, Bowtell R and Gowland P 2007 " Magnetic-field-induced vertigo: A theoretical and experimental investigation" Bioelectromagnetics 28(5) 349–61
7. Glover P M and Bowtell R 2008 " Measurement of electric fields induced in a human subject due to natural movements in static magnetic fields or exposure to alternating magnetic field gradients" Phys. Med. Biol. 53(2) 361–73
8. Hartwig V, Giovannetti G, Vanello N., et al,: 2009" Biological effects and safety in magnetic resonance Imaging: a review," . Int J Environ Res Public Health , , 6:1778–1798.
9. Hartwig V, Giovannetti G, et al. 2011,"A novel tool for estimation of magnetic resonance occupational exposure to spatially varying magnetic fields," Magn Reson Matter Phy, 24:323-330.
10. Hartwig V, Giovannetti G, et al, 2010 "Numerical calculation of peak-to-average specific absorption rate on different human thorax models for magnetic resonance safety considerations". Appl Magn Reson, , 38:337–348
11. ICNIRP 2010 "Guidelines for limiting exposure to time-varying electric and magnetic fields (1 Hz to 100 kHz) " Health Phys. 99(6) 818–36
12. ICNIRP 2012 'Draft guidelines for limiting exposure to electric fields induced by movement of the human body in a static magnetic field and by time-varying magnetic fields below 1 Hz." ' URL: http://www.icnirp.org/OpenMovement/ICNIRPMovementConsultationDraft2012.pdf
13. IEEE 2002 "IEEE Standard for Safety Levels with Respect to Human Exposure to Electromagnetic Fields, 0-3 kHz" , C95.6-2002 (New York: Institute of Electrical and Electronics Engineers)
14. Kannala S, ToivoT, et al 2009,"Occupational exposure measurements of static and pulsed gradient magnetic fields in the vicinity of MRI scanners." Phys Med Biol, 54:2243–2257.
15. Laakso I and Hirata A 2012 "Computational analysis of thresholds for magnetophosphenes " Phys.Med. Biol. 57(19) 6147–65
16. Laakso I, Kännälä S, Jokela K. 2013 " Computational dosimetry of induced electric fields during realistic movements in the vicinity of a 3 T MRI scanner" , Phys. Med. Biol Apr 21;58(8):2625-40
17. Liu F, Zhao H and Crozier S 2003 "Calculation of electric fields induced by body and head motion in high-field MRI" J. Magn. Reson. 161 99–107
18. Proposal for a Directive of the European parliament and of the council on the minimum health and safety requirements regarding the exposure of workers to the risks arising form physical agents (electromagnetic fields)
19. (XXth individual Directive within the meaning of Article 16(1) of Directive 89/39/EEC). European Commission, 14 June 2011.
20. Roberts D, Marcelli V, Gillen J, Carey J, et al , 2011 "MRI magnetic field stimulates rotational sensors of the brain Current" Biology 21(19) 1635–40
21. Shellock F G and Crues J V 2004 " MR procedures: Biologic effects, safety, and patient care " Radiology 232 635–52
22. Farrag S. , 2014 , "Numerical computation of specific absorption rate and induced current for workers exposed to static magnetic fields of MRI scanners ." The IEEE conference on biomedical engineering and sciences , IEEE – EMBS Malaysia Chapter – Malaysia .

Author:	Sherine Farrag
Institute	Modern University for technology and information
Country:	Egypt
Email:	sherine.farrag@gmail.com

Identifying Dynamic Effective Connectivity States in fMRI Based on Time-varying Vector Autoregressive Models

S.B. Samdin[1,2], C.M. Ting[1,2], S.H. Salleh[1,2], M. Hamedi[1,2], and A.M. Noor[1]

[1] Center for Biomedical Engineering (CBE), Institute of Human Centered Engineering, Universiti Teknologi Malaysia, Johor, Malaysia
[2] Faculty of Biosciences & Medical Engineering, Universiti Teknologi Malaysia, Johor, Malaysia

Abstract— **We propose a framework to estimate the transition of effective connectivity states in functional magnetic resonance imaging (fMRI), with the changing experimental conditions. The fMRI effective connectivity is traditionally assumed to be stationary across the entire scanning time-course. However, recent evidence shows that it exhibits dynamic changes over time. In this study, we employ a non-stationary model based on time-varying autoregression (TV-VAR) to capture the dynamic effective connectivity, and K-means clustering to identify the change-points of the connectivity states. The TV-VAR parameters are estimated sequentially in time using the Kalman filtering and the expectation-maximization (EM) algorithm. The extracted directed connectivities between brain regions are then used as features to the K-means algorithm to be partitioned into a finite number of states and to produce the state change-points, assuming the task condition boundaries are unknown. Experimental results on motor-task fMRI data show the ability of the proposed method in estimating the state-related changes in the motor regions during the resting-state and active conditions, with low squared estimation errors. The estimated brain-state connectivity also reveals different patterns between the healthy subjects and the stroke patients.**

Keywords— **Dynamic Brain Connectivity, fMRI, Kalman Filters, State-space Models, Vector Autoregressive Model.**

I. Introduction

The effective connectivity in brain networks analysis refers to the influence that one neuronal region exerting over another [1]. This connectivity is directed causal influence that manifests as the information flow between regions. Estimation of effective connectivity in functional magnetic resonance imaging (fMRI) time series with vector autoregressive (VAR) model has been extensively studied in [2]–[5]. This approach can represent the integration within the system in time and frequency domains [6]. However, standard VAR model assumes the system to be stationary, whereas the nature of fMRI time series is weakly non-stationary [7].

Recent study has reported that functional connectivity during resting-state was highly dynamic with time [8]. Thus, studies have been done for estimating the dynamic functional connectivity using sliding window based methods [9], [10], multivariate volatility models [11] specifically in resting-state condition. These methods however have limitations where window-based solution is ineffective with the existence of abrupt changes while the volatility models are not appropriate for slow varying state changes.

Similar to the functional connectivity, the effective connectivity is also shown to be dynamic and time-dependent earlier in [12]. Dynamic VAR model has been proposed in [13] for estimating the directed functional connectivity during motor task experiment using wavelet expansion functions which are locally stationary processes. Later, time-varying VAR model with Kalman filtering were used for fMRI and EEG effective connectivity estimation in [14]–[16]. These studies showed that the time-varying VAR model gives better estimates compared to standard time-invariant VAR. It is useful for evaluating brain state dynamics especially during resting state. Variety mental states were experienced throughout resting condition which cannot be pre-specified. Thus, the stable connectivity states and change-points need to be learned from the data itself [10]. Allen *et al.* has proposed a framework for estimating dynamics pattern of functional connectivity by using independent component analysis, sliding window and clustering method for resting dataset. The study addressed the cross-dependence between distinct brain region without assessing causal influence (directed) connectivity. Furthermore, sliding window technique suffer from poor temporal resolution where too long window size choosen, fast changes cannot be modeled [17].

In this article, we propose a framework to assess the state-related dynamic effective connectivity of state-related fMRI data using non-stationary VAR model. The framework consists of time-varying VAR, clustering method and stationary VAR.

II. Methods

A. Stationary VAR Model

Vector autoregressive model (VAR) is a time series analysis method which is a multivariate generalization of univariate AR models. The VAR model is commonly used to capture the temporal inter-dependence between multivariate signals. VAR model frequently used to

F. Ibrahim et al. (eds.), *International Conference for Innovation in Biomedical Engineering and Life Sciences*,
IFMBE Proceedings 56,
DOI: 10.1007/978-981-10-0266-3_50

characterize dynamic systems. The fMRI time series can be modeled into a VAR model by denote the series as, $\boldsymbol{y}_t = (y_1, \ldots, y_N)'_t$ be a $N \times 1$ vector of the fMRI time series measured from N brain regions of interest (ROIs) at time t, for t=1,…,T a VAR (p) model that characterize the current observation as a linear dependence on its p past observation vectors, $\boldsymbol{y}_t$, as

$$\mathbf{y}_t = \sum_{k=1}^{p} \mathbf{A}_k \mathbf{y}_{t-k} + \mathbf{v}_t \tag{1}$$

where $\{\mathbf{A}_k\}_{k=1}^{p}$is a $N \times N$ matrix of stationary AR coefficients at time lag k and $\{\mathbf{v}_t\}$ are $N \times 1$ i.i.d. Gaussian noise with zero mean and $N \times N$ covariance matrix $\mathbf{R}$.

Model (1) can be formulate in the form of multivariate linear model,

$$\mathbf{Y} = \mathbf{X}\boldsymbol{\beta} + \mathbf{E} \tag{2}$$

where $\mathbf{Y} = [\mathbf{y}'_1, \ldots, \mathbf{y}'_T]'$ fMRI time series $T \times N$ dimension, $\mathbf{X}$ is deterministic vectors of previous observation data with $T \times Np$, $\boldsymbol{\beta} = [\mathbf{A}'_1, \ldots, \mathbf{A}'_p]'$ is $Np \times N$ containing the AR coefficients matrix with $\mathbf{E} = [\mathbf{v}'_1, \ldots, \mathbf{v}'_T]'$ is the covariance noise matrices. The model parameters of multivariate linear model were estimated by least square estimator (LSE) process. VAR model only provides information on the temporal properties of a stationary linear system. The conditional least squares (LS) estimator of $\boldsymbol{\beta}$ and $\mathbf{R}$ with time points $t = p + 1, \ldots, T$ are defined respectively as $\widehat{\boldsymbol{\beta}} = (\mathbf{U}'\mathbf{U})^{-1}\mathbf{U}'\mathbf{Y}$ and $\widehat{\mathbf{R}} = (1/(T - p))^{-1}(\mathbf{Y} - \mathbf{U}\widehat{\boldsymbol{\beta}})'(\mathbf{Y} - \mathbf{U}\widehat{\boldsymbol{\beta}})$ with L lag of VAR coefficients.

B. *Time-varying VAR Model*

An extension of the stationary model is the time-varying VAR model that can describe instantaneous correlation in time series , by allowing the coefficient matrix in Eq. (1) to change with time

$$\mathbf{y}_t = \sum_{k=1}^{p} \mathbf{A}^c_{kt} \mathbf{y}_{t-k} + \mathbf{v}_t \tag{3}$$

where $\{\mathbf{A}^c_{kt}\}_{k=1}^{p}$ are $N \times N$ matrices of time-varying AR coefficients at time t and $\mathbf{v}_t$ is a $N \times 1$ i.i.d. Gaussian observational noise with mean zero and covariance matrix $\boldsymbol{R}^c$, $\mathbf{v}_t \sim N(\mathbf{0}, \mathbf{R}^c)$. The TV-VAR model can be formulated into state-space form as in [18]

$$\mathbf{y}_t = \boldsymbol{C}_t \mathbf{a}_t + \mathbf{v}_t \tag{4}$$

$$\mathbf{a}_t = \mathbf{a}_{t-1} + \mathbf{w}_t \tag{5}$$

with, $\boldsymbol{C}_t = \mathbf{I}_N \otimes \mathbf{X}'_t$ where $\mathbf{I}_N$ is a $N \times N$ identity matrix, $\mathbf{X}_t = [\mathbf{y}'_{t-1}, \ldots, \mathbf{y}'_{t-p}]'$ and $\otimes$ denotes the kronecker product. Defining $\mathbf{a}_t = vec([\mathbf{A}^c_{1t}, \ldots, \mathbf{A}^c_{pt}]')$ a $pN^2 \times 1$ state vector of TV-VAR coefficients at time point t, re-arranged from the matrices $\{\mathbf{A}^c_{kt}\}_{k=1}^{p}$, process (2) is re-written in a compact form (4) with a linear mapping $\boldsymbol{C}_t$ of the past observations, as the observation equation. In the state equation, the hidden state $\boldsymbol{a}_t$ is assumed to follow a first-order Gauss-Markov process as in (5) and $\mathbf{w}_t$ is a Gaussian state noise with mean zero and $pN^2 \times pN^2$ covariance matrix $\mathbf{Q}^c$. We denote the superscript c indicating the time-varying parameters and $\boldsymbol{\theta} = (\boldsymbol{R}^c, \boldsymbol{Q}^c)$ the model parameters of TV-VAR state-space model.

The standard form of VAR model assumes the underlying series to be stationary and their effective connectivity is constant over time regardless of task conditions. Thus, this motivates use of TV-VAR model for connectivity analysis to capture the varying connectivity at every time points and effectively estimate the abrupt changes of a signal. In task based fMRI, the connectivity parameters vary smoothly over time where the connectivity remains constant within task conditions without variation over multiple instances [19]. Our aim is to develop a method that can identify both the smooth and abrupt changes, which we shall describe in the following section.

C. *Estimation Algorithm*

The feature extraction step involves fitting the TV-VAR model to obtain preliminary estimates of the VAR coefficients. These VAR coefficients characterize the stochastic properties of the fMRI time series. The second step is the identification of the states and their transition points by K-means clustering the TV-VAR coefficients. Parameters estimation procedure are as follows:

1) *Feature Extraction*

In feature extraction we use the TV-VAR model as in (3) with formulation into state-space model (SSM) (4)-(5). We estimate the model parameters, $\{\boldsymbol{A}_{kn}\}_{k=1}^{p}$ and $\boldsymbol{\theta}$ of TV-VAR from formulated state-space form by using the closed form solution by Kalman filter with the Rauch–Tung–Striebel (RTS) smoother also known as Kalman smoother and the EM algorithm, where details can be found in [15], [20], [21]. The TV-VAR coefficients, $\mathbf{a}_t$ can be recursively estimated by the Kalman filter algorithm since the model is in a linear-Gaussian form.We also iteratively estimate the model parameter $\boldsymbol{\theta} = (\mathbf{R}^c, \mathbf{Q}^c)$ by using expectation maximization (EM) algorithms where the details algorithm can be found in [15]. The obtained time-varying connectivity features are defined as $\{\hat{\mathbf{a}}_{1:T}\}_{p \times p}$.

2) *K-means Clustering*

The extracted features of fMRI series are partitioned using k-means clustering technique to identify the initial states and their transition points. Let $\boldsymbol{X} = \{\boldsymbol{a}_t\}, t = 1 \ldots T$ be the set of pN^2 dimensional time varying AR coefficients to be clustered into a set of K clusters, $\boldsymbol{C} = \{\boldsymbol{c}_k\}, k = 1 \ldots K$. In this study, we used Manhattan distance as suggested in [10] which is more effective for high dimensional data compared to the Euclidean distance. Let μ_k be the median of cluster $\boldsymbol{c}_k$. The sum absolute differences between μ_k and the points in cluster $\boldsymbol{c}_k$ over all K clusters is defined as

$$J(C) = \sum_{k=1}^{K} \sum_{\boldsymbol{a}_t \in \boldsymbol{c}_k} |\boldsymbol{a}_t - \mu_k|. \quad (6)$$

Basically, the K-means algorithms includes; 1) selecting an initial partition with K clusters; 2) generate a new partition by assigning each pattern to its closest cluster center; 3) compute new cluster centers; while steps 2-3 were repeated until cluster membership stabilizes, as cited in [22].

III. RESULTS AND DISCUSSION

We assess the effectiveness of the proposed method for identifying the dynamic changes in effective connectivity of fMRI dataset. The motor-task fMRI data were provided by University of California, Irvine [23] which consist of two group of subjects they are healthy subjects and stroke patients. Both data set are based on a block-design experimental paradigm with alternating resting and task conditions, which is expected to induce piecewise stationary regimes of directed dependence between the fMRI signals. The subjects performed two conditions during fMRI which hand grasp-release movement (active-rest condition) where the tasks always start with rest then active and alternating by 12 scans toward the end with total of 48 fMRI scans.

Functional images were acquired on a Philips Achieva 3.0T with high-resolution T1-weighted images were acquired using a 3D MPRAGE sequence (TR/TE = 8.5/3.9ms, FA= 8, FOV =256×204×150, slices = 150, voxel size = 1×1×1mm3) and the blood oxygenation level dependent (BOLD) images were acquired using a T2-weighted gradient-echo echo planar imaging sequence (TR = 2000ms, TE = 30ms, flip angle = 70, FOV =240×240×154, slices =31, voxel size =2×2×2mm3).

In this study, we analyzed five sets of healthy and stroke data with five regions of interest (ROIs) which are primary motor cortex (M1), dorsal premotor cortex (PMd) and a midline supplementary motor area (SMA), known as the main region involved in task conditions experiments. The connectivity was analyzed by fitting the models to the subject-specific residuals. Details of regions selection and preprocessing can be obtained in [23]. The optimum order for time series was determined by Schwarz's Bayesian Criterion that has been implemented in [24]. The optimum order obtained, $p_{opt} = 1$ with the smallest mean- squared prediction error. Thus, we apply a 5-dimensional TV-VAR(1) model to extract the time-varying directed coefficients of 5 motor ROIs for both dataset.

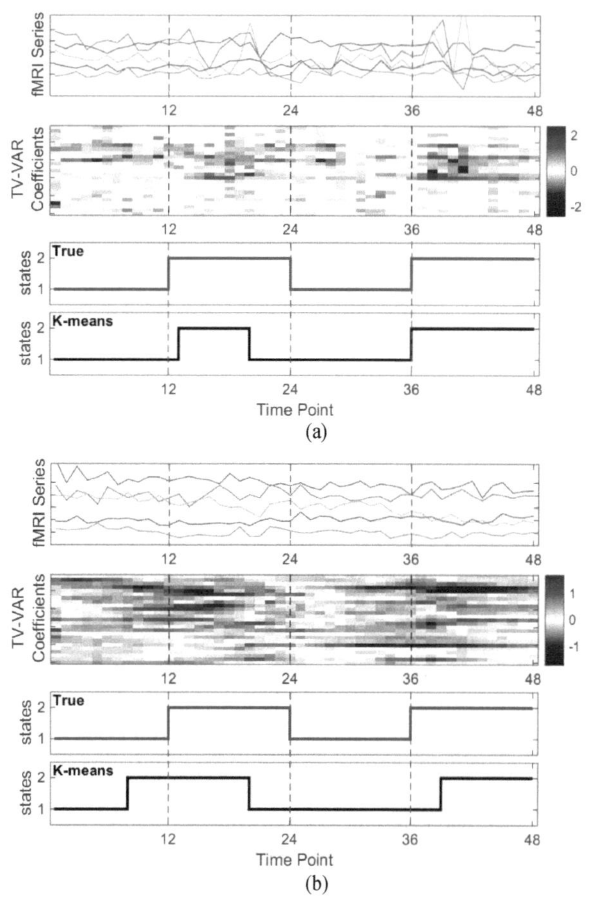

Fig. 1 Temporal dynamic effective connectivity of fMRI where (a) a healthy subject, and (b) a stroke patient. For (a) and (b), 1st subplot is BOLD fMRI mean time-series of five motor-related ROIs from a healthy subject. 2nd subplot is sequence of TV-VAR(1) coefficient matrices between the ROIs. 3rd Subplot is state sequences assumed known from the experimental designs (red) and estimates by the K-means clustering of the TV-VAR coefficients (black).

Results for Healthy Subjects: The TV-VAR cients $\hat{\mathbf{a}}_{1:T} = vec([\mathbf{A}_{1t}, \ldots, \mathbf{A}_{pt}]')$ from one of healthy subjects estimated by Kalman algorithms are shown in Fig. 1(a). It clearly seen that the directed connections between the ROIs are vary across pairs of motor-task and also change throughout time course. The dynamic patterns shows that these changes tend to congregate into distinct piece-wise stable regimes according to the states, i.e., time points t = 1-12 and t = 25-36 for resting state; t = 13-24 and t = 37-48 for active state. Compared to that block design, connectivity increase during active condition while during rest the connectivity low or weaker in strength. The classification results of those time points feature into two states regime shown in stairs plot of Fig.1 (a). Red-line is the true states and black-line is the estimated states by k-means

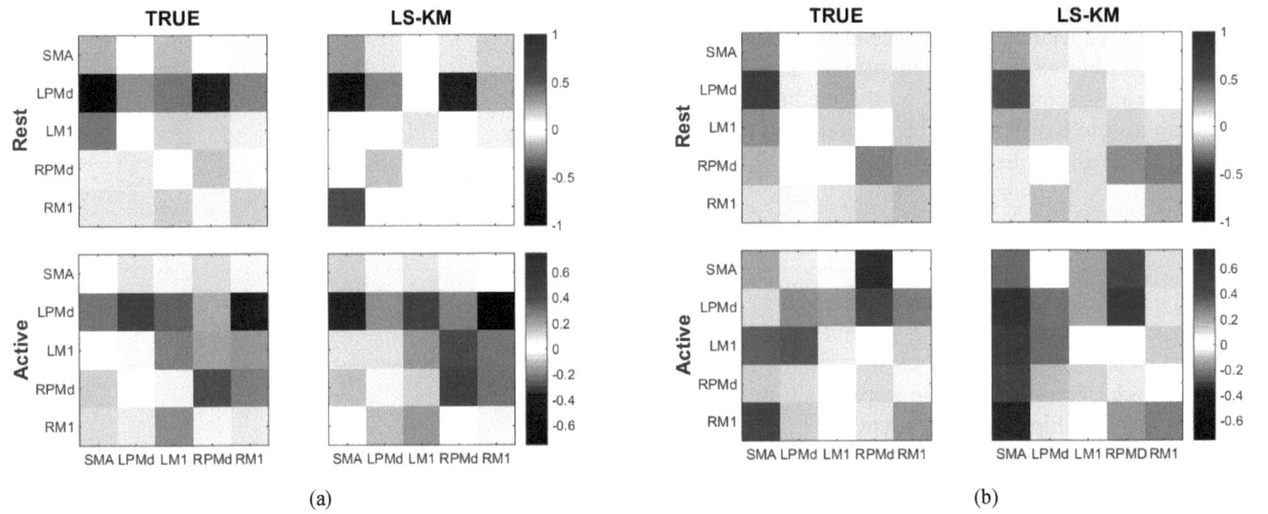

Fig. 2 Estimated VAR connectivity matrices between five motor brain regions during the resting and the active state with the averaged ground truth, (TRUE) and the averaged least square with k-mean clustering (LS-KM). (a) Healthy subjects, (b) stroke patients.

clustering where 90% of the actual states correctly classified. It shows that the state-dependent change points in connectivity regimes are efficiently detected.

Results for Stroke Patients: The strokes patients involved in this study are right side body affected (left ipsilesional). The estimated states and TV-VAR(1) coefficients for stroke patients shown in Fig. 1(b). The directed networks are varying with time even though they seem to be collective to each regime. 77 % of the estimated were correctly classified by k-mean clustering. These clustered states, were then aligned with fMRI series and we applied the stationary VAR model to estimate directed effective connectivity of each regime using least square estimator.

Fig. 2 shows the estimated VAR coefficient matrices between the motor ROIs for the resting (top) and the task-activated state (bottom). We computed the least-squares estimates based on the regime boundaries provided by the experimental design as ground-truth (TRUE) for comparison. It is shown that the estimates for both estimated healthy subjects Fig 2(a) and stroke patients Fig 2(b) reveals different effective connectivity pattern for both brain states. The LS-KM estimates give a close resemblance to the ground-truth for each brain state. However, based on the observation inspection the TRUE and LS-KM estimation VAR are much similar to each brains states. As low squared estimated errors shown in Table 1, it reflected well to the effective connectivity results in Fig. 2. The obtained results show that the proposed methods are capable of detecting the state-related changes of dynamic brain connectivity. However, further investigation need to be carried out to overcome the limitation of K-means algorithm which provides a 'hard' assignment of time points into states and also does not account for the temporal structure.

Table 1 Mean squared errors for directed connectivity matrices $\mathbf{A}_{[R]}$ and $\mathbf{A}_{[A]}$ during the resting and task-activated brain states. The results are average over 5subjects of each group.

Datasets	MSE	
	$\mathbf{A}_{[R]}$	$\mathbf{A}_{[A]}$
Healthy Control	2.69(0.72)	2.55(2.10)
Stroke	1.52(0.53)	2.79 (1.98)

Five subjects from both datasets healthy and stroke were evaluated. The connectivity estimation performance is assessed with square errors between the estimated VAR connectivity matrices and the ground-truth connectivity matrix $\widetilde{\mathbf{A}}$ defined as $\|\widehat{\mathbf{A}} - \widetilde{\mathbf{A}}\|_F^2$, where $\|\mathbf{M}\|_F = [tr(\mathbf{M}^T\mathbf{M})]^{1/2}$ denotes the Frobenius norm. The mean square errors of each datasets were presented in Table 1. The estimated squared errors were sufficiently low along with its standard deviations for each group.

IV. CONCLUSION

We proposed a framework to estimate dynamic effective connectivity and brain-state condition. The framework combines stationary VAR and time-varying VAR models and the k-means clustering. The state boundary estimated by classifying the AR coefficients gives relatively good estimation of the switching brain states in healthy and stroke fMRI data, with low squared estimation errors. The proposed method able to identify changes in connectivity structure of brain networks as characterized by the transition between distinct underlying brain states. This motivates us to apply the framework for the resting state or default mode brain to investigate the cognitive states in the future work.

ACKNOWLEDGEMENT

The authors would like to thank Prof. Dr. Hernando Ombao and Dr. Steven C. Cramer from University of Carlifornia, Irvine for the datasets. This work was supported by the Universiti Teknologi Malaysia and the Ministry of Education Malaysia under Fundamental Research Grant Scheme R.J130000.7809.4F668.

CONFLICT OF INTEREST

The authors declare that they have no conflict of interest.

REFERENCES

1. K. J. Friston, "Functional and effective connectivity: a review.," *Brain Connect.*, vol. 1, no. 1, pp. 13–36, Jan. 2011.
2. L. M. Harrison, W. D. Penny, and K. Friston, "Multivariate autoregressive modeling of fMRI time series," *Neuroimage*, vol. 19, no. 4, pp. 1477–1491, Aug. 2003.
3. R. Goebel, A. Roebroeck, D.-S. Kim, and E. Formisano, "Investigating directed cortical interactions in time-resolved fMRI data using vector autoregressive modeling and Granger causality mapping," *Magn. Reson. Imaging*, vol. 21, no. 10, pp. 1251–1261, Dec. 2003.
4. P. a Valdés-Sosa, J. M. Sánchez-Bornot, et al, "Estimating brain functional connectivity with sparse multivariate autoregression.," *Philos. Trans. R. Soc. Lond. B. Biol. Sci.*, vol. 360, no. 1457, pp. 969–81, May 2005.
5. G. Deshpande, K. Sathian, and X. Hu, "Assessing and compensating for zero-lag correlation effects in time-lagged Granger causality analysis of FMRI.," *IEEE Trans. Biomed. Eng.*, vol. 57, no. 6, pp. 1446–56, Jun. 2010.
6. B. P. Rogers, S. B. Katwal, et al, "Functional MRI and multivariate autoregressive models," *Magn. Reson. Imaging*, vol. 28, no. 8, pp. 1058–1065, 2010.
7. B. Gaschler-Markefski, F. Baumgart, et al, "Statistical methods in functional magnetic resonance imaging with respect to nonstationary time-series: Auditory cortex activity," *Magn. Reson. Med.*, vol. 38, no. 5, pp. 811–820, Nov. 1997.
8. C. Chang and G. H. Glover, "Time-frequency dynamics of resting-state brain connectivity measured with fMRI.," *Neuroimage*, vol. 50, no. 1, pp. 81–98, Mar. 2010.
9. N. Leonardi, J. Richiardi, M. Gschwind, S. Simioni, J.-M. Annoni, M. Schluep, P. Vuilleumier, and D. Van De Ville, "Principal components of functional connectivity: a new approach to study dynamic brain connectivity during rest.," *Neuroimage*, vol. 83, pp. 937–50, Dec. 2013.
10. E. a Allen, E. Damaraju, S. M. Plis, E. B. Erhardt, T. Eichele, and V. D. Calhoun, "Tracking whole-brain connectivity dynamics in the resting state.," *Cereb. Cortex*, vol. 24, no. 3, pp. 663–76, Mar. 2014.
11. M. a Lindquist, Y. Xu, M. B. Nebel, and B. S. Caffo, "Evaluating dynamic bivariate correlations in resting-state fMRI: a comparison study and a new approach.," *Neuroimage*, vol. 101, pp. 531–46, Nov. 2014.
12. A. Aertsen and H. Preissl, "Dynamics of activity and connectivity in physiological neuronal networks," in *Nonlinear dynamics and neuronal networks*, vol. 13, Ḥ. G. Schuster, Ed. VHC Verlag, 1991, pp. 281–302.
13. J. R. Sato, E. A. Junior, D. Y. Takahashi, M. de Maria Felix, M. J. Brammer, and P. A. Morettin, "A method to produce evolving functional connectivity maps during the course of an fMRI experiment using wavelet-based time-varying Granger causality.," *Neuroimage*, vol. 31, no. 1, pp. 187–96, May 2006.
14. M. Havlicek, J. Jan, M. Brazdil, and V. D. Calhoun, "Dynamic Granger causality based on Kalman filter for evaluation of functional network connectivity in fMRI data.," *Neuroimage*, vol. 53, no. 1, pp. 65–77, Oct. 2010.
15. S. B. Samdin, C. Ting, S. Salleh, M. Hamedi, and A. B. M. Noor, "Estimating Dynamic Cortical Connectivity from Motor Imagery EEG using Kalman Smoother & EM Algorithm," in *2014 IEEE Workshop on Statistical Signal Processing (SSP) ESTIMATING*, 2014, pp. 181–184.
16. M. Hamedi, S. Salleh, C. Ting, S. B. Samdin, and A. M. Noor, "Sensor Space Time-Varying Information Flow Analysis of Multiclass Motor Imagery through Kalman Smoother and EM Algorithm," in *2015 International Conference on BioSignal Analysis, Processing and Systems (ICBAPS)*, 2015, pp. 118–122.
17. Ü. Sakoğlu, G. D. Pearlson, K. a. Kiehl, Y. M. Wang, A. M. Michael, and V. D. Calhoun, "A method for evaluating dynamic functional network connectivity and task-modulation: Application to schizophrenia," *Magn. Reson. Mater. Physics, Biol. Med.*, vol. 23, no. 5–6, pp. 351–366, 2010.
18. M. Arnold, W. H. Miltner, H. Witte, R. Bauer, and C. Braun, "Adaptive AR modeling of nonstationary time series by means of Kalman filtering.," *IEEE Trans. Biomed. Eng.*, vol. 45, no. 5, pp. 553–62, May 1998.
19. J. F. Smith, A. Pillai, K. Chen, and B. Horwitz, "Effective Connectivity Modeling for fMRI: Six Issues and Possible Solutions Using Linear Dynamic Systems.," *Front. Syst. Neurosci.*, vol. 5, no. January, p. 104, Jan. 2011.
20. R. H. Shumway and D. S. Stoffer, "An approach to time series smoothing and forecasting using the EM algorithm," *J. Time Ser. Anal.*, vol. 3, no. 4, pp. 253–264, 1982.
21. Z. Ghahramani and G. E. Hinton, "Parameter Estimation for Linear Dynamical Systems," *Univ. Toronto Tech. Rep. CRGTR962*, vol. 6, pp. 1–6, 1996.
22. A. K. Jain, "Data clustering: 50 years beyond K-means," *Pattern Recognit. Lett.*, vol. 31, pp. 651–666, 2010.
23. C. Gorrostieta, M. Fiecas, H. Ombao, E. Burke, and S. Cramer, "Hierarchical vector auto-regressive models and their applications to multi-subject effective connectivity.," *Front. Comput. Neurosci.*, vol. 7, no. November, p. 159, 2013.
24. T. Schneider and A. Neumaier, "Algorithm 808: ARfit---a matlab package for the estimation of parameters and eigenmodes of multivariate autoregressive models," *ACM Trans. Math. Softw.*, vol. 27, no. 1, pp. 58–65, Mar. 2001.

Author: S. Balqis Samdin
Institute: Center for Biomedical Engineering, Universiti Teknologi Malaysia
Street:
City: 81310 Johor Bahru
Country: Johor
Email: sbalqis4@live.utm.my

Noninvasive Monitoring of Temporal Variation in Transcutaneous Oxygen Saturation for Clinical Assessment of Skin Microcirculatory Activity

A.K.C. Huong, S.P. Philimon, and X.T.I. Ngu

Faculty of Electrical and Electronic Engineering, Universiti Tun Hussein Onn Malaysia, Johor, Malaysia

Abstract— **This work demonstrated the use of Extended Modified Lambert Beer model (EMLB) for continuous monitoring of one's transcutaneous oxygen saturation based on reflectance spectroscopic data collected from the developed noncontact optical system. This quantification technique relies on spectral signature of hemoglobin components in the wavelength range of 520 − 600nm to give the best guess of transcutaneous oxygen saturation value. We conducted spectroscopic measurement on right index finger of fourteen recruited Asian volunteers in resting condition and after an application of pressure to their upper right arm for demonstration work. The obtained results revealed time- and subject-averaged transcutaneous oxygen saturation, S_tO_2, of 91.2 ± 5.4% and 12.3 ± 8.9%, respectively, for volunteers at rest and blood flow occlusion experiments. The range and variation in the S_tO_2 value observed in this work agreed reasonably well with that presented in most of the literature. This work concluded that the detected fluctuation in S_tO_2 value is likely due to respiratory and vasomotion activity, and the proposed technique could potentially be used to clinically assess oxygen demand in local tissues and regional microcirculatory system.**

Keywords— **Skin microcirculation, Extended Modified Lambert Beer model, transcutaneous oxygen saturation, spectroscopy, skin oximetry.**

I. INTRODUCTION

Noninvasive technologies for continuous measurement of blood flow and tissues oxygen consumption have gained increasing focus in understanding vascular condition of a patient. A rich microcirculation system is commonly associated with high survival rate of mammalian cells in part due to the fast gas exchange rate and sufficient transport of nutrients to tissues [1]. This process is, however, interrupted in patients with sepsis and systemic sclerosis owing to the high responsivity of microcirculatory function to inflammatory response [2]. In addition the microcirculation process can also vary with one's systemic blood pressure, physical and mental activities. Many works have been carried out to assess regional blood circulation and their metabolism process using blood flow and blood oxygen saturation as parameters to monitor microcirculation changes, vascular and hemodynamic response [3]. These findings are important for understanding the development of human tumor [4], sclerosis and cerebral activities [3], and the progression of diabetic foot disease [5]. The relationships between vascular activity and blood flow, and oxygen consumption are previously investigated by means of laser speckle contrast technique, Doppler Flowmetry and magnetic resonance imaging technique [6]. Meanwhile transcutaneous oxygen saturation, S_tO_2, which reflects the ratio of oxygenated hemoglobin to total hemoglobin in the microcirculation system of a volume of tissues [7], is found by means of interpretation of absorption spectroscopy data [8].

Various techniques have been proposed to find a person's blood oxygen saturation using absorption spectroscopy measurement on different skin regions of volunteers. These include the use of either an analytic model such as Modified Lambert Beer law (MLBL) [9], Extended Modified Lambert Beer model (EMLB) [10], cubic law model [11] and power law model [12], or nonlinear fitting of the measured reflectance signals using a library of data given from Monte Carlo model or Diffusion approximation [13]. While most of these models estimate the value of the required parameters using a fitting process [10, 11, 14], some are via simultaneous solution of the model [9, 15].

This paper aims to demonstrate the use of EMLB for continuous measurement of transcutaneous oxygen saturation and to identify factors affecting its temporal variability. We assume oxyhemoglobin and deoxyhemoglobin are the only absorbing components in blood and a pressure of 140mmHg applied to upper arm of a healthy human subject with systolic blood pressure value of 120mmHg for 60s is able to stop arterial blood flow into the arm below the cuff. This paper is organized as followed: in section II(A), the employed point spectroscopy system and experimental procedure are described. This is followed by a description of EMLB and quantification strategy in section II(B). Section III presents the time varying transcutaneous oxygen saturation value obtained in this work. Also included in this section is the comparison between values presented in this study and that in the literature. This is followed by a discussion of these results in section IV and conclusion in section V.

II. MATERIALS AND METHODS

A. Point Spectroscopy System and Experimental Procedure

Figure 1 illustrates the non-contact spectroscopy system used in this study. We employed a 9W white light emitting

F. Ibrahim et al. (eds.), *International Conference for Innovation in Biomedical Engineering and Life Sciences*,
IFMBE Proceedings 56,
DOI: 10.1007/978-981-10-0266-3_51

diode (LED) (model no. Lumileds, Philips) to illuminate right index finger of the recruited volunteers. The light source is placed at a distance of 80mm from the sample and an angle of 20° from normal axis. The detection system consists of an optical fiber connected directly to a commercially available spectrometer (Ocean Optics USB4000). Light reflected from skin and collected into the optical fiber is diffracted by the diffraction component to produce an intensity spectrum with spectral resolution of 0.2nm across wavelength range of 200 − 850nm before being detected. The tip of the optical fiber is positioned at a distance of 8mm from the sample and 15° from the normal. The data are then transferred to a computer via universal serial bus (USB) port for further processing and analysis.

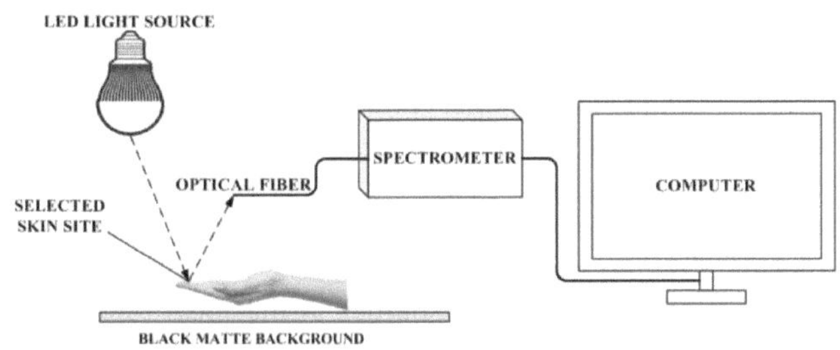

Fig. 1 Non-contact reflectance spectroscopy system

This study recruited fourteen Asian volunteers (aged 24.3 ± 2 years) for demonstration work. The blood pressure of these volunteers is measured at systolic and diastolic values (mean ± standard deviation) of 118 ± 5.5mmHg and 78 ± 4.3mmHg, respectively, using a blood pressure monitor (Model: Maestros IRIS-50). The subjects declared no serious underlying illness such as pulmonary disease, heart disease and anemic, and gave their written informed consent prior to measurement. These volunteers are instructed to relax, remained still in chair in a dark, quiet room before the spectroscopy measurement is performed on their right index finger. Meanwhile for the arterial blood flow occlusion experiment, a pressure of 140mmHg is applied by inflating blood pressure cuff on upper right arm of the subjects for 60s before spectroscopy data are collected from the same digit. The integration time of each sampling point is 100ms and data are collected for a total duration of 100s. These settings are chosen to verify the feasibility of the employed analytic technique to detect changes in a person's S_tO_2 with external intervention, and to compare temporal variability of this value for the performed experiments.

The wavelength dependent light attenuation, $A_{cal}(\lambda)$, used in the estimation of S_tO_2 is calculated using intensity data from selected skin sites, spectralon and with optical fiber tip blocked [10]. We assume light probed at local tissues, arterioles, venules and capillaries in skin before backscattered and reached the detection system, therefore the predicted transcutaneous oxygen saturation value is the mean of oxygen saturation across these vessels and tissues.

B. Extended Modified Lambert Beer Model and Iterative Fitting Procedure

The Extended Modified Lambert Beer model (EMLB) in Eq. (1) is previously developed by Huong and Ngu [10] based on the knowledge of light absorption and scattering across different skin layers. The performance of this model was evaluated using Monte Carlo method, and the authors observed a low mean absolute error of less than 1% in the estimated blood oxygen saturation value.

$$A(\lambda) = G_0 + \mu_a(\lambda) d_0 + G_1\lambda + \lambda \exp(-\mu_a(\lambda) d_1) \quad (1)$$

The first and second terms in Eq. (1) are from MLBL. While the term $G_1\lambda$ is to represent light attenuation contributed by absorption and scattering processes in epidermal layer, the intertwined effects of these processes in dermal layer on the total light attenuation is represented by the exponential term in Eq. (1). The light absorption, μ_a, is given from extinction coefficients of oxyhemoglobin (ε_{OxyHb}) and deoxyhemoglobin (ε_{Hb}) from the report of Zijlstra [16] as followed:

$$\mu_a(\lambda) = \left(\left(\varepsilon_{OxyHb}(\lambda) - \varepsilon_{Hb}(\lambda)\right) S_tO_2 + \varepsilon_{Hb}(\lambda)\right) T \quad (2)$$

where T represents total hemoglobin concentration and S_tO_2 is the transcutaneous oxygen saturation.

The measured light attenuation is fitted using the EMLB in Eq. (1) to give the best guess of S_tO_2. The fitting process started by initialized all the linear and nonlinear fitting parameters in Eq. (1) with value of '1'. This algorithm required priori knowledge of extinction coefficients of hemoglobin components in the wavelength range of 520 − 600nm for calculation of $\mu_a(\lambda)$ in Eq. (2), and hence the wavelength dependent attenuation values in Eq. (1). The optimum value of these fitting parameters (including S_tO_2) are searched using *fminsearch* function in MATLAB based on the absolute mean difference between the measured light attenuation and the values given from EMLB, ΔE. The fitting process is terminated when either ΔE is less than 1×10^{-20} or when number of iteration has reached 1000.

III. RESULTS

The mean and standard deviation of time dependent percent transcutaneous oxygen saturation predicted for the fourteen human subjects participated in the present study for at rest and arterial blood flow occlusion experiments is shown in Figure 2. These transcutaneous oxygen saturations averaged over a period of 100s in Figure 2 for the performed experiments are calculated and tabulated in Table 1. There is no gold standard with which the value estimated in this work can be compared, therefore transcutaneous oxygen saturation value reported in previous works in Table 1 is used here to validate the obtained value.

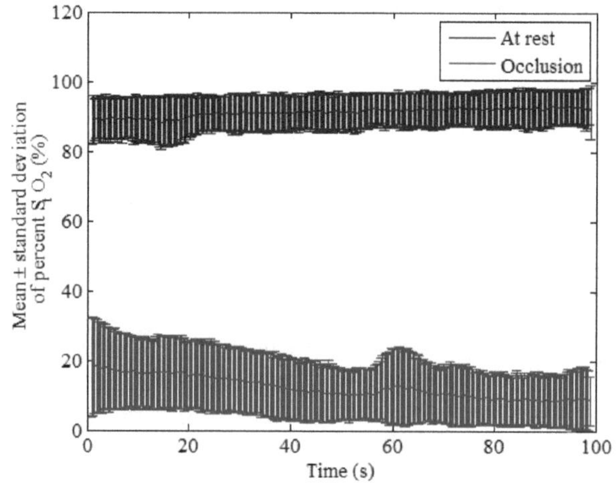

Fig. 2 Mean and standard deviation of temporal variation in percent transcutaneous oxygen saturation, S_tO_2, estimated for the recruited fourteen volunteers in resting and arterial blood flow occlusion conditions for a total duration of 100s

Table 1 Comparison of time-averaged mean and standard deviation of percent transcutaneous oxygen saturation, S_tO_2, estimated for volunteers at rest and during arterial blood flow occlusion condition obtained in this study and in the literature

	Investigator [ref]	**S_tO_2**
At rest experiment	Caspary *et al.* [17]	92 ± 2.6%
	Zhang *et al.* [18]	93 ± 1%
	Kobayashi *et al.* [11]	68 ± 6%
	Thorn *et al.* [8]	63 ± 11%
	This work	91.2 ± 5.4%
Blood flow occlusion experiment	Kobayashi *et al.* [11]	48%
	Vogel *et al.* [12]	35%
	Ferrari *et al.* [19]	20%
	This work	12.3 ± 8.9%

IV. DISCUSSION

The S_tO_2 value shown in Figure 2 estimated for the recruited volunteers revealed a lower mean percent S_tO_2 for arterial blood flow occlusion experiment compared to that for at rest condition. The mean percent S_tO_2 value is shown to fluctuate at a value close to 90% when volunteers are in resting condition owing to the presence of arteriovenous anastomoses in fingertips that provide direct shunting of arterial blood into venous system. These values are similar to that observed in Caspary *et al.* [17] and Zhang *et al.* [18] shown in Table 1 for measurement on acral skin (i.e. palm and fingertip). The value estimated by Kobayashi *et al.* [11] and Thorn *et al.* [8] is, however, lower than that predicted in the present study. This could be resulted from the underestimation of this value given from MLBL and the inappropriate assumptions made on skin thickness and scattering related parameters during analysis as discussed in Huong and Ngu [10]. Figure 2 revealed high frequency temporal changes in S_tO_2 when the volunteers are in resting condition, which is shown by the time-averaged standard deviation value of 5.4% in Table 1. This is probable to be related to spontaneous noise from respiratory system, changes in regional carbon dioxide and cardiac cycle.

A low time-averaged mean percent S_tO_2 of 12.3% is shown in Figure 2 with an application of pressure of 140mmHg to upper right arm of the recruited volunteers. These S_tO_2 values are comparable to that from the report of Ferrari *et al.* [19], but lower than that in the study by Kobayashi *et al.* [11] and Vogel *et al.* [12]. This is likely due to differences in the applied pressure in the corresponding articles. Interestingly, a plateau of mean S_tO_2 value of approximately 12% can be seen in Figure 2 starting from time point of 40s into the arterial blood flow occlusion experiment indicating the oxygen sources in the local blood medium have not been deprived within the examined time span. In addition, the variability in the percent S_tO_2 value increased in blood flow occlusion measurement compared to that of the at rest condition as indicated by the observed rise in the time-averaged standard deviation value from 5.4% in resting condition to 8.9% during blood flow occlusion experiment in Table 1. This increased in oscillation is most probably a result of skin microcirculatory vasomotion, which activity increases with the applied pressure to arterial wall.

V. CONCLUSION

We have demonstrated the feasibility of using Extended Modified Lambert Beer model and the developed iterative fitting to measure temporal changes in transcutaneous oxygen saturation of the recruited volunteers with different experimental conditions, and hence its potential application to monitor a person's skin microcirculatory and hemodynamic activities. The S_tO_2 value is shown to drop from time-averaged mean value of 91.2% to a consistent value of 12.3% after an application of pressure of 140mmHg to upper right arm of the recruited volunteers for more than 1 minute. This work concluded that temporal variation in S_tO_2 in resting condition is likely due to respiratory and cardiac cycle, whereas the oscillation in this value during blood flow occlusion condition is contributed by vasomotion activity.

ACKNOWLEDGMENT

The authors would like to thank Universiti Tun Hussein Onn Malaysia (GIPS U165) and Ministry of Education Malaysia (RAGS R015 and R016) for financially supporting this work.

CONFLICT OF INTEREST

The authors declare that they have no conflict of interest.

REFERENCES

1. Thorn, C.E., Kyte, H., Slaff, D.W., and Shore, A.C.: 'An association between vasomotion and oxygen extraction', American Journal of Physiology-Heart and Circulatory Physiology, 2011, 301, (2), pp. H442-H449
2. Le Pavec, J., Girgis, R.E., Lechtzin, N., Mathai, S.C., Launay, D., Hummers, L.K., Zaiman, A., Sitbon, O., Simonneau, G., and Humbert, M.: 'Systemic sclerosis–related pulmonary hypertension associated with interstitial lung disease: impact of pulmonary arterial hypertension therapies', Arthritis & Rheumatism, 2011, 63, (8), pp. 2456-2464
3. Boas, D., Strangman, G., Culver, J., Hoge, R., Jasdzewski, G., Poldrack, R., Rosen, B., and Mandeville, J.: 'Can the cerebral metabolic rate of oxygen be estimated with near-infrared spectroscopy?', Physics in medicine and biology, 2003, 48, (15), pp. 2405
4. Vaupel, P., Kallinowski, F., and Okunieff, P.: 'Blood flow, oxygen and nutrient supply, and metabolic microenvironment of human tumors: a review', Cancer research, 1989, 49, (23), pp. 6449-6465
5. Greenman, R.L., Panasyuk, S., Wang, X., Lyons, T.E., Dinh, T., Longoria, L., Giurini, J.M., Freeman, J., Khaodhiar, L., and Veves, A.: 'Early changes in the skin microcirculation and muscle metabolism of the diabetic foot', The Lancet, 2005, 366, (9498), pp. 1711-1717
6. Kamshilin, A.A., Teplov, V., Nippolainen, E., Miridonov, S., and Giniatullin, R.: 'Variability of microcirculation detected by blood pulsation imaging', PloS one, 2013, 8, (2), pp. e57117
7. Siegelaar, S.E., Barwari, T., Hermanides, J., van der Voort, P.H., Hoekstra, J.B., and DeVries, J.H.: 'Microcirculation and its relation to continuous subcutaneous glucose sensor accuracy in cardiac surgery patients in the intensive care unit', The Journal of thoracic and cardiovascular surgery, 2013, 146, (5), pp. 1283-1289
8. Thorn, C.E., Matcher, S.J., Meglinski, I.V., and Shore, A.C.: 'Is mean blood saturation a useful marker of tissue oxygenation?', American Journal of Physiology-Heart and Circulatory Physiology, 2009, 296, (5), pp. H1289-H1295
9. Pittman, R.N., and Duling, B.R.: 'A new method for the measurement of percent oxyhemoglobin', Journal of applied physiology, 1975, 38, (2), pp. 315-320
10. Huong, A., and Ngu, X.: 'The application of extended modified Lambert Beer model for measurement of blood carboxyhemoglobin and oxyhemoglobin saturation', Journal of Innovative Optical Health Sciences, 2014, 7, (03)
11. Kobayashi, M., Ito, Y., Sakauchi, N., Oda, I., Konishi, I., and Tsunazawa, Y.: 'Analysis of nonlinear relation for skin hemoglobin imaging', Optics Express, 2001, 9, (13), pp. 802-812
12. Vogel, A., Chernomordik, V.V., Demos, S.G., Pursley, R., Little, R.F., Tao, Y., Gandjbakhche, A.H., Yarchoan, R., Riley, J.D., and Hassan, M.: 'Using noninvasive multispectral imaging to quantitatively assess tissue vasculature', Journal of Biomedical Optics, 2007, 12, (5), pp. 051604-051604-051613
13. Pifferi, A., Taroni, P., Valentini, G., and Andersson-Engels, S.: 'Real-time method for fitting time-resolved reflectance and transmittance measurements with a Monte Carlo model', Applied Optics, 1998, 37, (13), pp. 2774-2780
14. Huong, A.K., and Ngu, X.T.: 'In situ monitoring of mean blood oxygen saturation using Extended Modified Lambert Beer model', Biomedical Engineering: Applications, Basis and Communications, 2015, 27, (01), pp. 1550004
15. Huong, A.K., Stockford, I.M., Crowe, J.A., and Morgan, S.P.: 'Investigation of optimum wavelengths for oximetry', in Editor (Ed.)^(Eds.): 'Book Investigation of optimum wavelengths for oximetry' (International Society for Optics and Photonics, 2009, edn.), pp. 736811-736811-736816
16. Zijlstra, W.G., Buursma, A., and van Assendelft, O.W.: 'Visible and near infrared absorption spectra of human and animal haemoglobin: determination and application' (VSP, 2000. 2000)
17. Caspary, L., Thum, J., Creutzig, A., Lubbers, D., and Alexander, K.: 'Quantitative reflection spectrophotometry: spatial and temporal variation of Hb oxygenation in human skin', International Journal of Microcirculation, 1995, 15, (3), pp. 131-136
18. Zhang, R., Verkruysse, W., Choi, B., Viator, J.A., Jung, B., Svaasand, L.O., Aguilar, G., and Nelson, J.S.: 'Determination of human skin optical properties from spectrophotometric measurements based on optimization by genetic algorithms', Journal of Biomedical Optics, 2005, 10, (2), pp. 024030
19. Ferrari, M., Binzoni, T., and Quaresima, V.: 'Oxidative metabolism in muscle', Philosophical Transactions of the Royal Society B: Biological Sciences, 1997, 352, (1354), pp. 677-683

Author:	Audrey K. C. Huong
Institute:	Universiti Tun Hussein Onn Malaysia
Street:	86400 Parit Raja
City:	Batu Pahat, Johor
Country:	Malaysia
Email:	audrey@uthm.edu.my

Passive Hand Rehabilitation: Soft-Actuated Finger Mobilizer

R. Sulaiman and N.A. Hamzaid

Department of Biomedical Engineering, Faculty of Engineering, University of Malaya, Kuala Lumpur, Malaysia

Abstract— **Stroke is one of the top five diseases in Malaysia contributing to major morbidity and mortality not only in Malaysia but in the whole world. Severe stroke can lead to death while the stroke survivors will face weakness (flaccid), spasticity and decrease in proprioceptive sensation depending on the affected part of the brain. The cost of rehabilitation session as well as the restriction in mobility demotivates the patient to undergo rehabilitation therapy in occupational therapy. A portable, lightweight and low cost device was proposed for finger mobilizer specifies for these patients. In order to develop the device, we need to consider muscle conditions of the patient which is muscle spasticity and muscle flaccid. After considering the condition of the patient and the efficiency of the device used, the most suitable design which incorporated the concept of soft actuated finger mobilize is chosen. A glove-like device which uses the application of air pressure as the mobility mechanisms was identified as the most suitable concept for this project. The final design was fabricated and tested on five normal subjects with no pathological condition of the hand. Quantitative analysis was conducted to acquire public opinion on the prototype produced. Nearly all of them agreed on this concept although a lot of improvement needs to be done regarding the design of the prototype.**

Keywords— **stroke, soft-actuated, muscle spasticity.**

I. INTRODUCTION

Among the chronic disease that occurs throughout the world, stroke can be grouped into as one with among the greatest disease burden due to the recovery needed to attain normal limb function. According to the literature, stroke is one of the disease contributes to major morbidity and mortality in the world. In Malaysia, it is one of top five leading causes of death after ischemic heart disease, septicemia, malignant neoplasms, and pneumonia. Since 2005, the percentage of stroke has been increasing from 6.6% to 8.4%. This increment shows the severity in stroke among Malaysians which the highest risk factor turns to be hypertension [1].

After recovery, stroke survivors often face further complications requiring great effort and support. In order to address the flaccidity and weakness of their hands, often the patients were intensively practicing with familiar objects and movement to accommodate with disability. Hand rehabilitation for example, encourages the patient to move their muscle actively in a normal ROM with the help of therapist; ensuring the proper use of muscle and ROM to avoid any abnormal function of hand.

Nowadays, robotic devices were commonly incorporated to stroke patient rehabilitation. Robotic rehabilitation often used in repetitive tasks for stroke patient, replacing labor work done by therapists. There is lots of robotic rehabilitation that can be found in market or still under development, designed to aid rehabilitation either for elbow, wrist or finger. There are three classes of robotic devices for finger rehabilitation which are endpoint control, actuated object and exoskeleton [2].

For the exoskeleton device, aside from using linkages or rigid frame, a concept of soft robotics which incorporate the use of pneumatic soft actuators, memory shaped alloy and wire mechanisms as an actuation mechanism. The concept of pneumatic soft actuators is always aligned with the concept of tendon driven mechanisms where the actuator used resembles the function of human tendon mechanisms [3]. This concept satisfies the requirements for hand exoskeleton where the safety and user friendly criterion are implemented [4].

The basic operation of soft robotics depends on the regulation of inner pressure control, manipulating the expansion and contraction of the elastic material [4]. When the flexible or under actuated is applied to the exoskeleton, it might looks like there is a lack in the control of finger joints. However, according to Heo and colleagues, the lack in control of coincidence of the centre of rotation of the device is compensated with the skeletal structure of the wearer's hand, providing a skeletal structure guiding the motion of the exoskeleton device [5].

In this paper, a soft-actuated finger mobilizer incorporating a glove-like structure is proposed that allows finger extension in a passive mode. The device was tested on subjects to determine the degree of functionality of the soft-actuated concept used is working or not.

II. DESIGN CONSIDERATIONS

A. Exoskeleton Design

Three exoskeleton designs which use linkages, gear and motor as actuation mechanism were designed and tested using Solidwork software to determine the accuracy in

F. Ibrahim et al. (eds.), *International Conference for Innovation in Biomedical Engineering and Life Sciences*,
IFMBE Proceedings 56,
DOI: 10.1007/978-981-10-0266-3_52

power transmission and dimensions of the design. After receiving professionals review regarding the designs, the final design was chosen incorporating a glove-like structure exoskeleton device with pneumatic soft actuator position at the back of the glove. The final design satisfies crucial design requirement for robotic hand rehabilitation which is light weight, low cost and non-invasive.

B. *Air Flow Mechanisms*

Flow of air between the pump used and the rubber material used as the actuator needs to be efficient with a minimal pressure drop while at the same time direct the air into individual finger without delay. Three designs were tested to determine the most suitable design to be used for the glove. For the final design, rubber glove will be used as the actuator to extend the finger since it can direct the flow of air into individual finger effectively.

C. *Glove Design*

The glove also needs to be designed such that it will fit the user hand considering the pathological condition of the user's hand. Feature such as zip or velcro strap were considered. The best feature that can withstand the expension of the rubber glove inside the glove will be chosen.

III. METHODOLOGY

A. *Prototype Testing*

Upon completion of the prototype development, two different gloves with different feature were tested out to determine the most suitable design that can deliver the most effective finger extension function. The prototype was tested on a normal subject to test the workability of the device. Glove design 1 and design 2 shown in Fig. 1 was tested on the subjects. This method was also conducted to analyze the drawbacks of the device based on the subject opinion and the questionnaires given.

B. *Patient Testing*

Firstly, the subject was given a brief explanation about the device until the subject understands the overall procedure of the experiment. After that, the subject was asked to fill in the first part of the questionnaire regarding the first impression on the device. While the subject fills in the questionnaire, the prototype was set up.

Design 1 glove was worn by the subject and air will be pumped into the rubber glove until the finger is fully extended. The pump then will be stopped and the procedure will be repeated with a design 2 glove. Opinions from the subject regarding the workability of the device will be noted and the questionnaire will be analyzed.

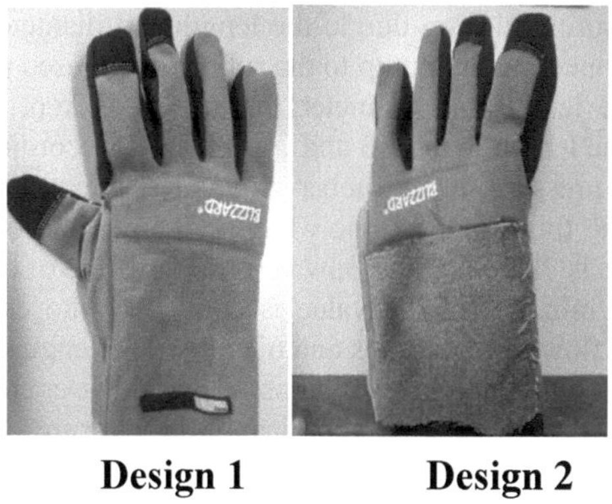

Fig. 1 Two different glove designs that was tested on subjects.

C. *Kinovea Analysis*

The efficiency of finger expansion can be analyzed using Kinovea software where video was taken before and after experiment was conducted. The difference between the initial and final angle as well as the time taken to achieve the final angle will be analyzed. To use the Kinovea software, firstly, the video was exported into the software. An angle option is chosen and placed on the PIP joint of the subject's hand. The other two points was placed vertically on the DIP joint and MCP joint of patient hand. The angle was then automatically calculated by the software. Stop watch was placed at the side of the glove to indicate the expansion time.

IV. RESULTS AND DISCUSSION

A. *Theoretical Result*

Overall success of the design depends greatly on the air flow mechanisms from the compressor pump to the glove. Investigation on pressure drop occurrence in the air flow mechanism will determine a suitable tube length to be used in the air flow design in order to ensure the effectiveness of air flow mechanisms. Fig. 2 shows the overall mechanisms of the design.

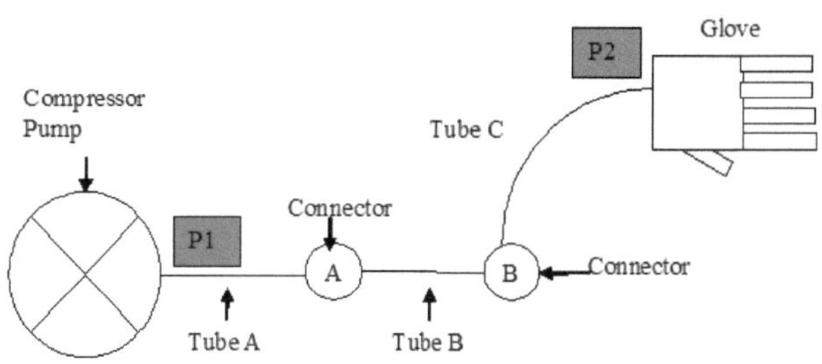

Fig. 2 Overall designs of air flow mechanisms to the rubber actuator.

Pressure will drop due to the length and diameter of the tube connecting the pump to the rubber glove used as actuator. The length and diameter for tube A is kept constant while the length in tube B and C is changed according to the desired pressure in the rubber glove. The internal diameter of all the tubes is the same which is 0.5 cm and the initial pressure is 7kg/cm^2. The flow volume flow rate of the pump is 40 L/min. When the value is converted into cm^3/s, the volume flow rate becomes 666.67cm^3/s. The length adjusted must not hinder the flow of air as well as the user comfort in wearing the glove. The significance in pressure drop was investigated using formula below.

$$Volume\ flowrate = \mathcal{F} = \frac{P_1 - P_2}{\mathcal{R}} \quad (1)$$

Where,

$$\mathcal{R} = \frac{8\eta L}{\pi r^4} \quad (2)$$

From the calculation using the above equation, the final air pressure taken at P2 is 6.9994 kg/cm^2. This value indicates that the length of the tube used only have small significance to the pressure drop in the overall design.

B. Patient Testing

Glove design 1 and design 2 was tested out on 5 normal subjects to determine the workability and effectiveness of the device in providing finger extension. From the testing video analyzed using Kinovea software, the results is tabulated in the Table 1 and Table 2.

Table 1 Extension of finger in design 1

Subject	Initial angle (°)	Final angle (°)	Angle difference (°)	Duration of expansion (s)
1	142	179	37	3.54
2	139	173	34	3.06
3	145	176	31	2.39
4	142	180	38	2.33
5	141	175	34	2.33
	Average		34.80	2.73

Table 2 Expansion of finger in design 2

Subject	Initial angle (°)	Final angle (°)	Angle difference (°)	Duration of extension (s)
1	141	176	35	2.79
2	145	179	34	1.63
3	138	173	35	2.43
4	143	177	34	1.88
5	140	178	38	2.60
	Average		35.20	2.27

From the data collected, the finger extension achieved using glove design 1 and design 2 is within 30° to 40° in the duration between 2 to 4 seconds. The difference in the extension angle between the 2 designs is only 0.4°; hence we can conclude that within this extension range, the lower time achieved to expand the finger has more effective air flow and was more efficient to be used in the final glove design. In design 1, the average duration for the maximum finger extension is 2.73s while in design 2 is 2.27s. From the result obtained, we can conclude that glove design 2 is more effective in allowing finger extension.

Design 2 as shown in Fig 4 was sewed with an inelastic fabric at the back of the glove to inhibit further expansion of the glove when the air flown in. Fig 4 shows how design 2 glove can achieve finger extension faster than design 1 glove. In design 1 glove, since the rubber glove is inserted at the back of biker's glove, due to the elastic fabric of the glove, the expansion of rubber glove will be directed in upward direction instead of flowing to each individual finger. In design 2 glove, since inelastic fabric is sewed at the back of the glove, the air will be directed to each individual finger.

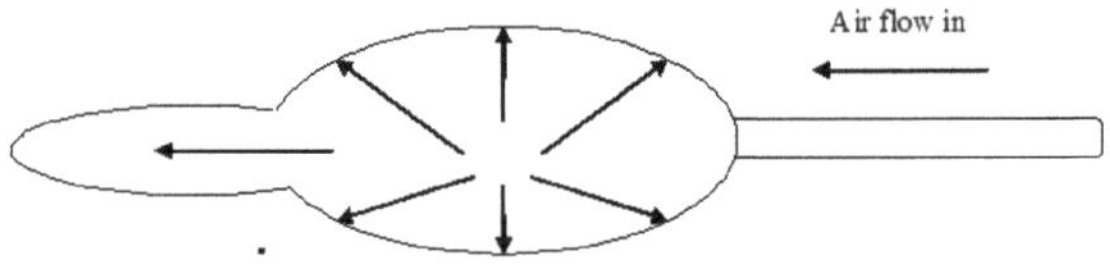

Fig. 3 Cross-sectional view of the rubber glove when the air is flown in

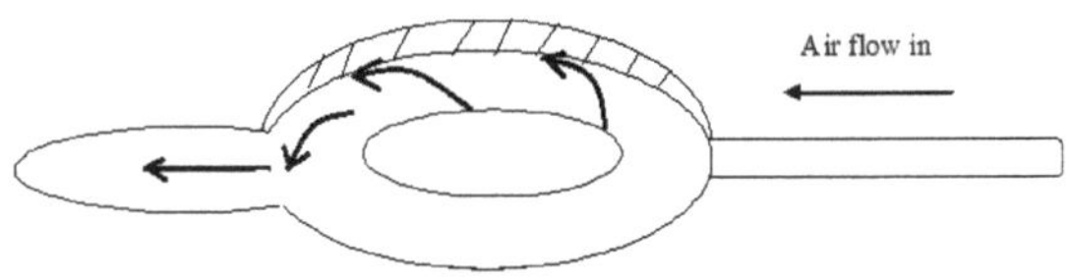

Fig. 4 Cross-sectional view of the rubber glove inside when inelastic fabric is used to prevent the overexpansion of glove

V. COMPLIANCE WITH ETHICAL REQUIREMENTS

A. Conflict of Interest

Since the glove is manually controlled, testing on patient can be quite difficult with the possibility of finger over expansion. Considering pathological condition of patients especially stroke patient where muscle contracture exist, the design needs to be alter to ensure safety of the user.

VI. CONCLUSION

After conducting a few analyses, the desired concept of finger mobilizer for the robotic rehabilitations was achieved.

Based on tests performed on the prototype developed, several limitations have been detected and further modifications needs to be done to increase performance of the prototype developed.

VII. LIMITATION AND FUTURE IMPROVEMENTS

One of the limitation is the prototype could not be controlled automatically. Although manual control over the equipment gives the advantages to the therapists in controlling frequency and intensity of rehabilitation, an automated system can be designed to fulfill the rehabilitation requirement such as intensity and frequency of training while at the same time gives exact power and movement required by the patient. In addition to that, when the system is being controlled manually, there is no visual device to monitor every parameter govern by the system such as velocity of air applied and force delivered.

Other than that, instead of controlling the finger extension manually using compressor pump, electrical pump is more futuristic for commercialization while at the same time increases the function as well as the efficiency in aiding stroke patient rehabilitation. Conventional compressive pump sold in current market has specifics specification and is difficult to modify and manipulated to suit the design criteria for soft actuated finger mobilizer. The electrical pump can be developed using pneumatic shaft connected to the microcontroller. The air delivery from the pump needs to be further studied to ensure the optimal air delivery to the system.

Other than that, the material and fabric used for the glove also needs to be further studied and modified. The glove used in the prototype developed could not fully flex the finger due to the existing feature in biker's glove where the palm part of the glove is sewed with sponge to prevent friction and gives better grip for bikers. A more flexible glove, with a suitable material and fabric needs to be studied to provide maximal function of finger extension and flexion.

ACKNOWLEDGMENT

This study was supported in part by a grant from High Impact Research (HIR) University of Malaya.

CONFLICT OF INTEREST

The authors have no commercial or other financial conflict of interest associated with the study.

REFERENCES

1. In, H. K., Cho, K. J., Kim, K. R., & Lee, B. S. (2011). Jointless structure and under-actuation mechanism for compact hand exoskeleton. *IEEE International Conference on Rehabilitation Robotics*, (pp. 1-6).
2. Heo, P., Gwang, M. G., Lee, S.-j., Rhee, K., & Kim, J. (2012). Current Hand exoskeleton technologies for rehabilitation and assistive engineering. *International Journal of Precision Engineering and Manufacturi, 13*(5), 807-824.
3. Loo, K. W., & Gan, S. H. (2012). Burden of Stroke in Malaysia. *International Journal of Stroke, 7*, 165-167.
4. Noritsugu, T. (2005). Pneumatic Soft Actuator for Human Assist Technology. *Proceedings of the 6th JFPS International Symposium on Fluid Power.* Japan: Okayama University.
5. Schabowsky, C. N., Godfrey, S. B., Holley, R. J., & Lum, P. S. (2010). Development of pilot testing of HEXORR: Hand EXOskeleton Rehabilitation Robot. *Journal of Neuroengineering and Rehabilitation, 7*(36), 1-16.

Author:	Ramizah bt Sulaiman
Institute:	University of Malaya
Street:	Jalan Universiti
City:	Wilayah Persekutuan Kuala Lumpur
Country:	Malaysia
Email:	ramizah303@gmail.com

Development of Flexible 2D Ultrasound Arrays for Scoliosis Assessment

Queenie T.K. Shea, Patrick Y.M. Yip, and Yong Ping Zheng

Interdisciplinary Division of Biomedical Engineering, The Hong Kong Polytechnic University, Hong Kong, China

Abstract— **Conventional way of assessing scoliosis requires taking X-ray radiograph on the coronal plane of the spine. Rapid radiograph examinations on scoliosis patients could produce radiation hazards and increase the risk of cancer. Recently, it has been shown that ultrasound imaging could produce reliable Cobb's angle measurement for scoliosis assessments. However, this method requires moving the ultrasound probe manually by physician's hand and could produce error if subjects moved during the imaging process. More importantly, it is very difficult to take images when subjects are wearing back braces. This study has aimed to overcome the above issues by developing a flexible ultrasound transducer arrays which could stay on subjects' back during the examination. 4mm diameter piezoelectric transducer elements were soldered onto flexible copper Printed Computer Board (PCB). The transducer surface was then emerged into soft silicone gel to eliminate possible air gaps and provide comfortable cushioning between transducer and the subject's back. Accelerometers and electromagnetic spatial sensors were explored to measure the elements' location and orientation. The Preliminary results showed that the structure of the spinous process could be identified with the flexible transducer array by comparing A-mode signals with B-mode images taken with conventional ultrasound probe. It was also suggested that the larger angular coverage of the flexible transducer array could be helpful for studying the orientation of reflecting surface. This could be done by finding the angle of reflection of the echoes by capturing signals with nearby elements during single element stimulation.**

Keywords— **scoliosis, flexible 2D ultrasound arrays, spine imaging, coronal plane.**

I. INTRODUCTION

Scoliosis is a spine deformity disease which results in a lateral curvature of the spine on the coronal plane. It is often coupled with vertebral rotation of the spine as they are considered in most cases idiopathic [1]. In Hong Kong, there are around 3.1% of children suffering from scoliosis [2]. Scoliosis can be treated by surgery, physical therapy or by wearing TLOS (Thoraco-Lumbo-Sacral-Orthosis) brace and etc. [1]. To provide diagnosis for this disease, the gold standard is to measure the Cobb's angle of patient's spine on the coronal plane [3]. This is usually done by X-ray examinations on the thoracic plus abdominal region of the body. Once scoliosis is detected, regular and rapid X-ray assessments on patients might be required to monitor curvature progression, treatment outcomes and especially to ensure correct fitting of the TLOS brace. Rapid assessments using X-ray could lead to radiation hazards and increase in risk of cancer [4]. Although a recent imaging device (EOS imaging) has been shown to reduce X-ray doses to patient when performing spine imaging examinations [5], it is still more desirable to minimize or eliminate the radiation hazard in scoliosis assessments.

It has recently been demonstrated that 3D ultrasound imaging is feasible imaging method for the assessment of scoliosis [6] [7]. This new imaging system has eliminated the radiation hazards on spine imaging as ultrasound imaging is non-ionizing. This imaging system requires moving the ultrasound probe vertically down the subject's back by physician's hand while the ultrasound probe is installed with an electromagnetic spatial and orientation sensing device [8]. The location of the spinous process and the transverse process were then identified from the ultrasound images for calculation of the Cobb's angle. However, the scanning process normally takes a few minutes and the major limitation of this is the possible error caused by patients' movements during the scanning processes. It is also very difficult to perform the imaging on subjects' back when subjects are wearing a TLOS brace. As the brace will have to be opened at the back for ultrasound scanning, this could reduce the brace pressure on the body and hence affect the brace performance.

In order to overcome these issues, we proposed to develop a flexible 2D ultrasound transducer array which can fit and stick onto the back of the subject during ultrasound scanning. This flexible array will potentially provide an ultrasound imaging method on the spine where no manual or mechanical movements of the probe will be needed. More importantly, it can be used when brace are worn normally by subjects, which helps providing accurate examination on the brace performance.

Another advantage of this flexible ultrasound transducer array is the increase in angular coverage on the body surface [9]. This can increase the collection of acoustic information by collecting reflected acoustic signals with the adjacent elements. This can potentially increase the signal contrast and can be used for determining the orientation of the reflecting surface i.e. the spine surface at specific region.

F. Ibrahim et al. (eds.), *International Conference for Innovation in Biomedical Engineering and Life Sciences*,
IFMBE Proceedings 56,
DOI: 10.1007/978-981-10-0266-3_53

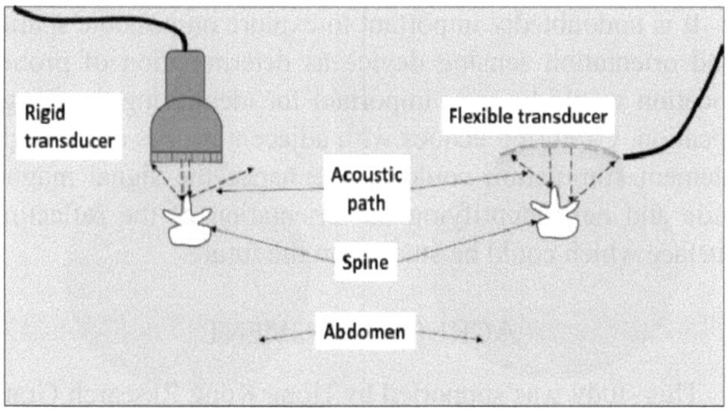

Fig. 1 Illustration of increase in angular coverage using flexible ultrasound transducer array on a curved body surface compare to conformal rigid ultrasound transducer array [9].

Different spatial sensors and/or bending sensors will be explored to find an optimized solution for this imaging method.

II. TRANSDUCER ARRAY DESIGN

A. Design Requirements

The two most important requirements of the transducer arrays design are to be flexible and sticky enough to stay on the subject's back during the examinations. It should also preferably cover the whole back from the first thoracic vertebrate (T1) to the fifth and also last lumbar vertebrate (L5), although smaller prototypes will be made at this stage.

Another important requirement would be to minimize the thickness of the transducer arrays in order to fit in the narrow gap between the TLOS brace and the back of the subject.

B. Arrays Components

a) Piezoelectric Transducer Probes

Piezoelectric (PZT) material were used in this transducer design although other type of ultrasound generating method such as laser-generated ultrasound [10] could be a possible alternative.

The transducer probes used in this design have 4mm diameter of transmission surface and probe thickness of 5mm. Centre of transmission frequency is 5.18MHz.

As we aimed to design a transducer which can cover the full length of the back of the spine, using small conventional ultrasound elements would greatly increase the total number of elements and the complexity of the circuitry. Hence relatively larger transducer elements were used in this design to reduce the amount of elements needed. However, this would also reduce the image resolution due to wider spacing between probes.

b) Flexible Printed Circuit Board (PCB)

The PCB is made of thin copper sheet coated with a plastic layer. Total thickness of the flexible PCB is 0.2mm. The distance between centers of piezoelectric probes when soldered is 6mm. 6x6 probes were soldered onto each flexible PCB sheet.

c) Spatial and Orientation Sensors

In order to determine the position or more importantly the relative position and orientation of each PZT probes during the imaging process, different spatial sensors were explored. Accelerometers were proposed to be used due to their few advantages; small and no limitation to the maximum amount of sensor. Although they can only provide orientation information, combining them with electromagnetic spatial sensing device could provide enough information for the imaging process.

d) Solid Gel Pads

During scoliosis assessments, subjects can be standing up, leaning forward or in other postures. Hence it is important that the ultrasound transducers can attach on subjects' back stably using solid gel pads. The gel pads must also have low acoustic attenuation and similar acoustic impedance as body tissue to maximize the transmission of acoustic energy from the transducers to the body.

As any small air gap between transducer and the skin will produce strong reflection of ultrasound wave due to impedance mismatch, the gel pads should ideally be soft enough to fit perfectly onto the irregular surface of the back. Soft gel pads can also provide cushioning between the rigid transducer and subjects' skin especially when brace were worn on top of the transducers.

Silicone gel were used due to its flexibility and stability over long period of time.

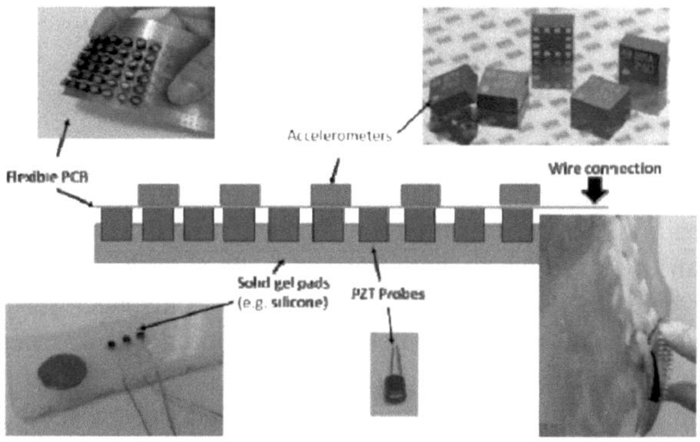

Fig. 2 Figure illustrated the design of flexible ultrasound array. The top left and bottom right corner of this figure has shown the flexibility of the first prototype

III. METHOD AND RESULTS

The first goal is to provide enough acoustic information with the flexible ultrasound array to accurately determine the Cobb's angle. Bony features such as the spinous process and the transverse process were attempted to be identified using transducers probes in single line array. A full length spine phantom from the C1 vertebrate to the coccyx embedded into silicone gel was used in this study. A B-mode ultrasound image was taken using 128 elements transducer array on the T2 vertebrate where the structure of the spinous process could be observed clearly on the image, see Fig. 3.

Flexible transducer array were placed on the same location where elements were stimulated separately on a single line at the T2 vertebrate. By comparing with the B-mode image obtained before, the reflecting bony structure corresponding to each echo were identified manually. Using this method, the location of the bony structure could be identified along the whole spine for Cobb's angle examination.

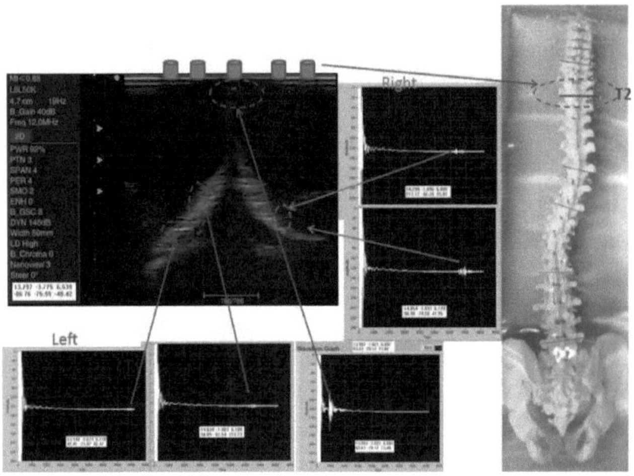

Fig. 3 B-mode image of T2 vertebrae, and the corresponding A-mode signals measured on the same location with flexible ultrasound transducer.

IV. Discussion and Conclusion

This preliminary study demonstrated that the design of this ultrasound transducer array could provide flexible bending on the transmission surface. The surface of the transducer could fit onto surface with complex geometry given with its flexibility. Emerging the transducer surface into silicone gel with sticky surface could ensure secure connection between the arrays and the back of the subject and eliminating possible air gaps created during subject's movements.

Although the separations between probes of this flexible ultrasound transducer array are relatively large compare to conventional ultrasound array, it has been shown that the echoes obtained by stimulating individual element separately could be used to identify the spinous structure by comparing the A-mode signal with B-mode images taken with conventional ultrasound imaging transducer.

It is undoubtedly important to explore on possible spatial and orientation sensing device as determination of probes location would be very important for identifying the image location. Capturing echoes with adjacent probes after single element stimulation could also enhance the signal magnitude and help identifying the orientation of the reflecting surface which could be studied on the future.

Acknowledgment

This study was supported by Hong Kong Research Grant Council (152220/14E) and the Hong Kong Polytechnic University.

Conflict of Interest

The authors declare that they have no conflict of interest.

References

1. A. Bessette and C. M. Rousseau, (2012) Human Anatomy and Physiology : Scoliosis: Causes, Symptoms and Treatment. Nova Science Publishers, Inc., New York
2. Tang S.P. et al. (2003) Adolescent Idiopathic Scoliosis (AIS): An Overview of the Eliology and Basic Management Principles. HK J Paediatr, pp. 8:299-306
3. Dickman, D et al., (2001) Assessment of scoliosis with Ortelius 800 preliminary results. Clin Appl Notes, Vols. April:1-7
4. Shu X.O et al. , (1994) Diagnostic X-ray and ultrasound exposure and risk of childhood cancer. British Journal of Cancer, vol. 70, pp. 531-536
5. Luo T. D. et al., (2015) Cumulative Radiation Exposure With EOS Imaging Compared With Standard Spine Radiographs. *Spine Deformity,* vol. 3, no. 2, pp. 144-150
6. C. Cheung, G. Zhou, Law SY, T. Mak, K. Lai and Y. Zheng, (2015) Ultrasound volume projection imaging for assessment of scoliosis. IEEE Transaction on Medical Imaging , vol. 8, pp. 1760-1768
7. M. Li, J. Cheng, M. Ying, B. Ng, Y. Zheng, Lam TP, W. Wong and M. Wong, (2010) Application of 3-D ultrasound in assisting the fitting procedure of spinal orthosis to patients with adolescent idiopathic scoliosis. Studies in Health Technology and Informatics, vol. 158, pp. 34-37
8. C.-W. J. Cheung and Y. Zheng, (2013) Development of 3-D Ultrasound System for Assessment of Adolescent Idiopathic Scoliosis (AIS). Engineering in Medicine and Biology Society (EMBC), 2013 35th Annual International Conference of the IEEE.
9. R. S. e. a. Singh, (2011) Conformal Ultrasound Imaging System. Acoustical Imaging, Springer Netherlands, pp. 211-222.
10. R. J. Colchester, (2014) Laser-generated ultrasound with optical fibres using functionalised carbon nanotube composite coatings. Appl. Phys. Lett., vol. 173502, p. 104

Author: Queenie TK Shea, Patrick YP Yip, Yong Ping Zheng
Institute: The Hong Kong Polytechnic University
Street: Yuk Choi Road
City: Hong Kong
Country: China
Email: queenie.shea@polyu.edu.hk

Electronic Communication Device for Elderly Patients with Speech Impaired Geriatric Population

M.F. Azmi, Y. Ahmad, K.M. Lee, and N.A. Hamzaid

Department of Biomedical Engineering, University of Malaya, Kuala Lumpur, Malaysia

Abstract— **The importance of communication in human daily lives is undeniable as it is a mean of expressing ideas, thoughts and most importantly the basic needs. Unfortunately, there are elderly patients with speech disabilities, commonly caused by aphasia. Aphasic patients can lose their ability to obtain basic needs and healthcare services as they are unable to communicate. There is no electronic communication device or Voice Output Communication Aid (VOCA) that is made locally to help local stroke patients to communicate. Imported VOCA is extremely expensive and do not work in multi languages and dialects which usually are the native languages for local patients. Hence, there is a demand for such device to be designed. The design criteria and features of the device are decided based on literature review and surveys done. The device consists of 6 basic needs buttons, output voice messages such as eat, and drink, toilet, thank you, yes and no. It costs less than RM400, can be operated with minimum technical and reading skills and is portable. Additional features for this device included the option for any 3 languages, optional recordings of user's family voice as the voice messages and a built in alarm for urgent use. The circuit of the device was tested to prevent over current, burn and reduce power consumption. The overall design of the prototype is good according to the healthcare professionals and patients' evaluations. Minor improvements can be done in the future to bring the device into the market. Finally, approaches to commercialise the device and the business model are briefly explained.**

Keywords— **Stroke, Aphasia, VOCA, Geriatric, Speech, Low Cost.**

I. INTRODUCTION

Aphasia is a condition common to elderly with stroke. They are in general unable to communicate thus will require a suitable communication device. The current devices are expensive and are mostly complicated to use.

There are a lot of diseases and reasons for one to lose their communicating ability, and some are able to regain their ability by going through therapy sessions. Nevertheless, there are a group of people who require more attention and help before they can get back to normal again. They are the elderly patients with speech disabilities associated with stroke (Figure 1). According to The Star Online, 6 new cases of stroke occur every hour in Malaysia and there are about 52,000 of Malaysians suffering from strokes annually. Stroke is one of the top five leading causes of death and one of the top ten causes for hospitalization in Malaysia [1].

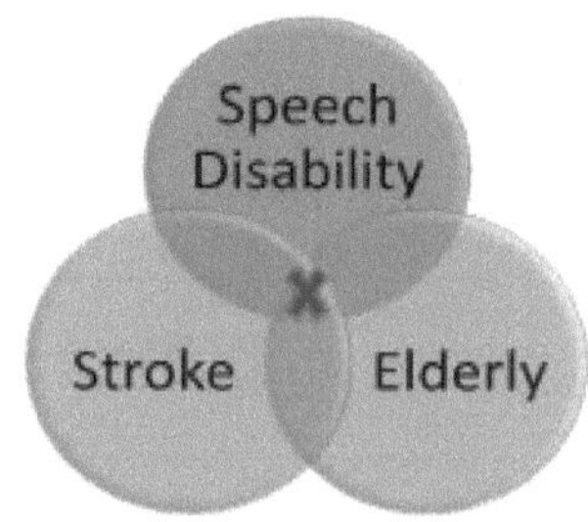

Fig. 1 Targeted user population is intersection of three groups marked in red

US statistics source by Centers for Disease Control and Prevention [2] said that nearly three-quarters of all stroke cases occur in people over the age of 65. The risk of having a stroke will be doubled by each decade after the age of 55.

The brain controls everything including interpretation and understanding of speech. A stroke can lead to speech disorder if it damages the part of the brain that is responsible for language. For most people, the language function is controlled by the left hemisphere of the brain. A stroke can causes speech disabilities in many different ways. The common post stroke condition associated with speech disabilities is aphasia [4].

Aphasia is a condition which causes an individual to have difficulties in processing language despite that he or she has a normal intelligence [5]. Individuals with aphasia can frequently have difficulties in retrieving words, understand the verbalization and combined words into sentences and phrases. They also have to face a lot of challenges to be able to read or write [6].

They have to rely on augmentative alternatives communication (AAC) devices but those devices are getting more tech-savvy, targeting children and younger adults as their users. Along with the intercept of advance technology, these devices available in the market are also extremely expensive, adding a burden for the patients in Malaysia to own one. On top of that, due to the fact that the device will be

F. Ibrahim et al. (eds.), *International Conference for Innovation in Biomedical Engineering and Life Sciences*,
IFMBE Proceedings 56,
DOI: 10.1007/978-981-10-0266-3_54

marketed to a multilingual society such as Malaysia, the multilingual option serves as a strong point.

Hence, an ideal AAC would include the above points and designed suitably especially for the elderly in order to assist them in their communication.

II. METHODOLOGY

Two surveys were conducted to identify the design consideration, criteria and unique features for the prototype. The findings concluded the requirements to build the prototype.

Table 1 Summary of requirements from literature

Requirements	Description
Simple Interface	Prevent unwanted slowness and complexity in communication
Easy to use	Required minimum skills either technical or reading skills to operate the device
Portable	Can be moved around
Easy to maintain and repair	Prevent unwanted downtime
Comfortable	Looks pleasant and not odd-looking to others; Comfortable to hold everyday

Firstly, 30 survey forms had been distributed to patients with speech problems, doctors, nurses, and allied health professionals, caretakers and member of public in PPUM. Secondly, 30 survey forms were distributed in National Stroke Association of Malaysia (NASAM) in Petaling Jaya. Hence, there were a total of 60 survey forms distributed.

Forms were given to both parties, the patients and caretakers as most patients cannot talk or express themselves in a proper or comprehensible manner. This has brought a wider perspective on the matter.

A. Design Considerations, Criteria and Unique Features

The design and the features of the electronic communication device are decided based on the literature review, surveys done and logic consideration.

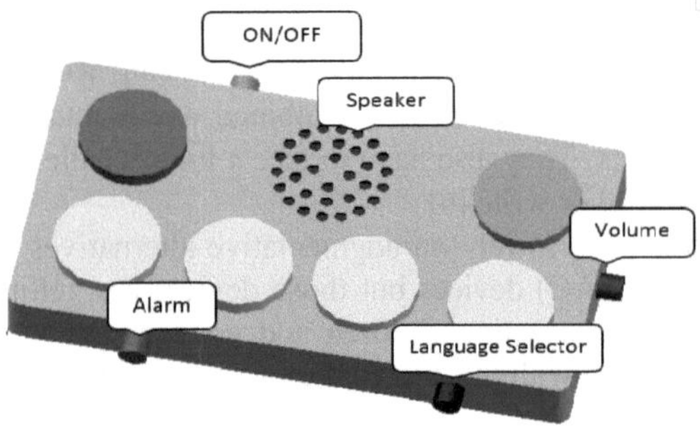

Fig. 2 Illustration of prototype

The targeted users are patients at old age and with speech disabilities, that is possibly aphasia. At the same time, they might have acute stroke condition where their arms have little range of movements. In this situation, a VOCA is the best choice in helping them to communicate with people compared to unaided AAC such as gestures, hand movements and face expressions. Their faces, arms and hand muscles might not be able to move a lot. VOCA is suitable as it can produce a synthetic voice to take the place of the original voice.

A flow chart (Figure 4) has been constructed as below to present this intuitively.

B. System Development

The block diagram of the device is as shown in Figure 3. The microcontroller and audio decoder integrated circuit (IC) are its main components which are the Arduino Uno and Sparkfun's MP3 Shield. Microcontroller has a role in assigning correct data files to different inputs. Meanwhile, audio IC decodes data to audio for its output. In addition, the different buttons displayed comes from individual switches which will complete the circuit when pressed upon which will produce voice. On top of that, different languages can be chosen using the language selector switch. Voice data can be stored within a micro SD card and inputs conveyed to the micro controller. The power supply can be cut off using the on off button when the device is not in use. An amplifier is used to raise the audio amplitude before its output at the speaker.

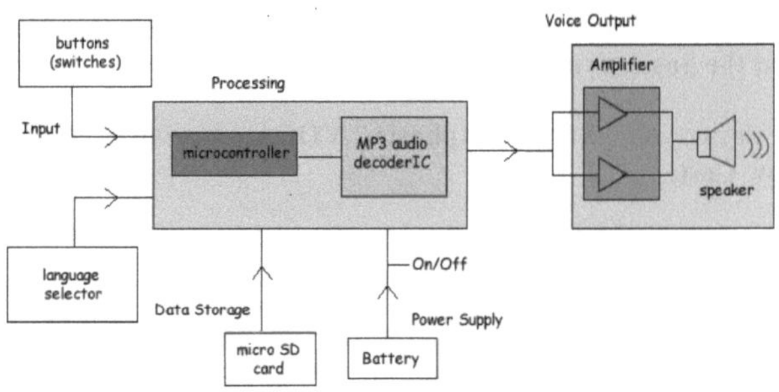

Fig. 3 Block diagram of electronic communication device

C. Prototype Functionality Evaluation

Following completion of design prototype, users' evaluation was obtained by interviewing 4 healthcare professionals and 2 elderly speech impaired patients in University of Malaya Medical Centre (UMMC). Objective of the interview was to identify the functionality of the prototype from the perspective of healthcare professionals and patients' perspective.

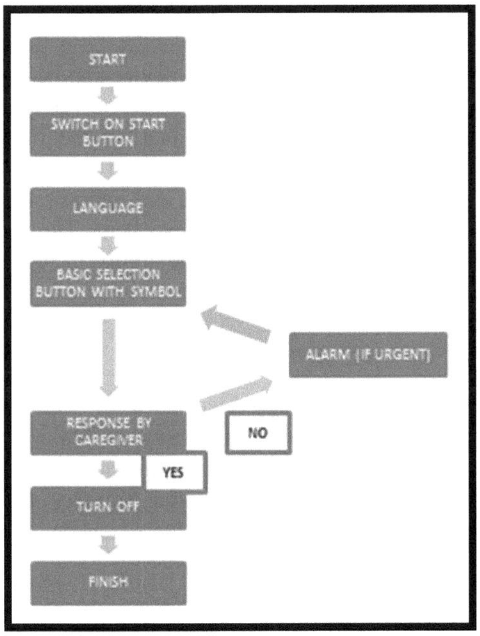

Fig. 4 Operation Flow of the Prototype

III. RESULTS AND DISCUSSION

A. Outcome of the Prototype

Figure 5 shows the finished prototype. On completion, it was able to produce audio which were audible and adjustable to volume when functional buttons were pressed. It is functional in 3 main languages which are Bahasa Malaysia, English and Mandarin languages conveying 6 different messages. An oscilloscope was used to observe the output signal from the circuits before the amplifier output to speaker was connected. This process proved that the device was fully functional. The tact switch buttons were used instead of the large push buttons for function testing. The following Table 2 are the results of the test.

Fig. 5 The external appearance of the prototype

B. Results of Prototype Functionality

Part 1 Interview with healthcare professionals

In their opinion, the device is attractive in terms of colour, contrast of the functional buttons and the housing. This helps patients to locate the buttons easily as the buttons are systematically arranged. However, one the therapist prefers the Yes and No button at the bottom of the device and the functional buttons to be on top. She believe that the patients will find the Yes and No button very convenient at the bottom as they use it quite often.

Table 2 Outcome of prototype built

Criteria and Specifications	**Description**	**Completed**
1.Audible and adjustable sound	• Within a 2 meters range and low background noise	Yes
2. Three different languages	• Bahasa Malaysia • English • Mandarin	Yes
3.Four functional buttons	• I want to eat • I want to drink • I want to go to toilet • Thank you	Yes
4. Alarm button		Yes
5. 45mm diameter round buttons with symbols on top of them	• Easy to press •Minimize reading and technical knowledge	Yes
6. Simple components	• Easy to repair and maintain • Reduce downtime	Yes
7. Yes and No buttons	• For questions answering	Yes
8. Program own sound		Yes
9.Low cost	• Less than RM400	Yes
10.Lightweight	• Less than 1kg	Yes
11. Portable	• Can operate using battery	Yes

Part 2 Interview with patients with communication problems

According to Table 3, the overall functionality of the prototype is proven to work at its best. The patients were able to respond to the questions in a short time period. For example, the toggle switch for the former and push button switch for the latter. The 'thank you' symbol can be modified by adding some words on it. The prototype is light weight and portable. More patient testing should be carried out to plot a reasonable inference.

Table 3 Functionality evaluation of patients on the prototype

Test Patients	**Malay Female, 80** Unclear speech, thumbs and index finger on the left cannot move. Could only understand Malay.	**Malay Male, 85** Slightly unclear speech, fingers have low range of movements. Could only understand Malay.
	Questions	
Is it easy to turn on	Yes (11.65s)	Unable to turn on
Can you hear the message	Yes (5.39s)	Yes He repeated all the messages (4.02s)
Is it easy to press the button	Yes (3.86s)	Yes (3.21s)
Is it easy to adjust the volume	Yes (3.90s)	Unable to adjust
Can you understand the symbol	Only understand eat, drink and toilet	Yes (3.54s)
Do you like the Yes and No buttons	Yes (3.61s)	Yes (3.66s)
Do you like the housing of the device	Yes She is able to lift up the device (4.12s)	Yes He explained that his grandchildren will like it (3.94s)

*(s) indicates the time (measured by using stopwatch) of subject took to perform or respond to the tasks or questions.

IV. Conclusion

The prototype was developed according to the design criteria, consideration and extracted from literature review and surveys done. It has simple interface that does not require the user to navigate through a menu of what they wanted. Suitable features for the elderly such as the big buttons, symbol representation and reprogrammable sound messages are additional credits. It is reprogrammable but initially it has six different messages. Patient can rely on this device to improve their speech ability. With the assistance of this device, patients can slowly learn six different words to the extent of delivering the words by themselves with speech. So those words can then be re-placed with new words. Learning bit by bit is a strategic method to help them recover. The second objective is met as the device is built below at a cost of RM400. In addition, the device can be used with battery type AA, 9V and Li-Polymer to power up the device. Li-Polymer is the best choice as it has the longest battery lifespan. Interview with healthcare professionals, speech and occupational therapists, and patients were conducted. The overall feedback was excellent. The prototype is well suited for locals as it has the ability to record messages in multi-languages and dialects.

Acknowledgment

The authors acknowledge Assoc Prof Dr. Tan Maw Pin of Geriatric Medicine from University of Malaya for her significant recommendations and continuous support. This research was fully supported by the Ministry of Higher Education, Malaysia and University of Malaya through UMCIC Grant No. **RU019C-2014C.**

Conflict of Interest

The authors declare no conflict of interest.

References

1. Stroke Association. (n.d.). Common problems after stroke. About stroke Retrieved 16 Nov, 2014, from http://www.stroke.org.uk/about/common-problems
2. Brust, J., Shafer, S., Richter, R., & Bruun, B. (1976). Aphasia in acute stroke. Stroke, 7(2), 167-174.
3. Hodge, S. (2007). Why is the potential of augmentative and alternative communication not being realized? Exploring the experiences of people who use communication aids. Disability & Society, 22(5), 457-471.
4. Jacobson, N. (2007). Dignity and health: a review. Social Science & Medicine, 64(2), 292- 302
5. Woolhead, G., Tadd, W., Boix-Ferrer, J. A., Krajcik, S., Schmid-Pfahler, B., Spjuth, B., . . . Dieppe, P. (2006). "Tu" or "Vous?": A European qualitative study of dignity and communication with older people in health
6. Dreher, B. B. (2001). Communication skills for working with elders: Springer Publishing Company.

Author: Mohd Faiz bin Azmi
Institute: Department of Biomedical Engineering,
Faculty of Engineering, University of Malaya
Street: Jalan Universiti
City: Kuala Lumpur
Country: Malaysia
Email: faizspencer@gmail.com

Camera Based Arm Motion Tracking for Stroke Rehabilitation Patients

A. Akhavizadegan and M.Y.I. Idris

University of Malaya, Kuala Lumpur, Malaysia

Abstract— **Human motion tracking for rehabilitation has been a progressive research area since the 1980s. It has been driven by the increased number of patients who have suffered a stroke, or some other motor function disability. Rehabilitation is a dynamic process in which it enables patients to regain their normal functional capabilities. To achieve this goal, a patients' activities need to be continuously monitored, and subsequently corrected. However, due to economic pressures, stroke patients are receiving less therapy and are sent home sooner, so the potential benefit of the therapy is not completely realized. Thus, it is essential to develop a rehabilitation technology that allows individuals who had suffered a stroke to practice intensive movement training without the need of an always-present therapist. This paper introduces a real-time human arm motion tracking system specifically intent to be used for home-based rehabilitation. The system consists of an inexpensive webcam and a laptop. Then our algorithm is able to extract the patient's arm, track the motion of any point on the arm over time, evaluate its velocity and measure the angular motions, namely "elbow flexion", "elbow extension", "wrist flexion" and "wrist extension" movements, furthermore, the trajectory of any point of interest on the arm can be evaluated and visualized for the patient or doctor for further analysis.**

Keywords— **Stroke, rehabilitation, arm motion tracking, skin segmentation, arm skeleton extraction.**

I. Introduction

Human motion tracking for rehabilitation has been a progressive research area since the 1980s. It has been driven by the increased number of patients who have suffered from motor function disability. Rehabilitation is a dynamic process which allows patients to regain their functional capability once more [25]. Approximately 75% of patients with stroke have difficulty in performing basic daily activities and over 50% would have difficulty in walking. Therefore, travelling to physiotherapy centers can be one of the major setbacks faced by patient with stroke and their caregiver. The travelling also leads to several other problems such as the travelling cost, the travelling time, the limited parking space and etc. To overcome such problems, a simple camera based arm motion tracking system alongside image processing techniques has been implemented in order to give the system a standalone and portability capabilities which makes it stand out from any existing commercial technologies available to this day. In this project, we propose a vision based arm motion tracking system using a single camera for patients with stroke, the camera will detect and track the motion of the arm, then measures parameters such as the velocity, angular motion and the trajectory of the specific point(s) on the arm, the information extracted from the motion of the arm such as the velocity, angular motion alongside the trajectory coordinate points we are then able to decide whether the motion performed by the patient was accordingly to the criteria of that specific treatment or not, furthermore, a visual feedback will provide the patient with his/her performance progress based on these information. The objective of this study is to introduce an inexpensive easy to use monitoring tool as an alternative to the traditional methods that can be both motivational and intuitive, overall, the system will allow patients in a more advanced phase of recovery, to carry out their physiotherapy at home or anywhere desired.

II. Literature Review

Visual tracking and analysis of human motion is currently one of the most popular research topics in computer vision [4, 8, 11, 19]. This popularity is due to its wide range of applications such as athletic/medical performance analysis, surveillance tracking, and perceptual interface. Most current visual tracking systems are classified into two categories: marker-based visual tracking systems and marker-free visual tracking systems. Using markers to show the region of interest moderately simplifies the human motion tracking problem, there are a number of commercial marker-based systems available on the market, while marker-based visual tracking systems are sufficient enough to be employed successfully in many areas such as athletic performance analysis and the motion capture for animation, however, they have a downside that they only run in supervised environment and might not be suitable for daily usage, therefore, it is usually desirable to develop a marker-free visual tracking system instead of the intrusive marker-based. Marker free based visual tracking systems furthermore get divided into two groups, namely multiple camera configuration and single camera configuration. Multiple camera configuration systems are considered expensive and it is also difficult to construct. L. Enrique et al [10]

F. Ibrahim et al. (eds.), *International Conference for Innovation in Biomedical Engineering and Life Sciences*,
IFMBE Proceedings 56,
DOI: 10.1007/978-981-10-0266-3_55

developed a prototype called "Gesture therapy" which basically used two webcams where one is placed in front of the patient and another webcam is placed above the patient's head, then, patients were asked to interact with a virtual environment by moving their impaired arm accordingly. The system located and tracked the hand using color and motion information in a 3D space, the coordinates of the hand were sent to a simulator that simulates the daily life activities. On the other hand, single camera configuration system is considered cheap and desirable, however it can be affected by many constraints and requires prior knowledge about the appearance of the subject, the geometry of the subject, the kinematics and dynamics of the subject. Regarding this issue in single camera configurations, Polana and Nelson [16] proposed an approach that can spatially and temporally normalize, segment and recognize a repetitive motion such as walking or any repetitive motion activity without having any prior knowledge of specific parts, or classification of the actor using a basis based on a bottom up processing. In 2001 Sminchisescu and Triggs [3] tackled the difficulty of 3D human body tracking by carefully designing a robust matching-cost metric that combines robust optical flow, edge energy and motion boundaries together to reduce the correspondence ambiguities. For reconstructing the 3D pose and motion from a single camera view Bowden et al. [15] used the 2D silhouette of a human in motion and the corresponding 3D skeletal structure, that were encapsulated within a non-linear point distribution model. Later in 2003 a simple, efficient and robust method for recovering 3D human motion from a sequence of images from an uncalibrated camera was presented by Barron and Kakadiaris [2], another interesting simple, inexpensive and portable image processing system for kinematic analysis of human gait in real-time was also proposed by Yeasin and Chaudhuri [13]. A recent study done by Ming Du et al [6] in 2013, showed how they track the human motion using a monocular camera, by taking advantage of the DE-MC (Markov chain) algorithms, to approximate complicated distributions, additionally, they have also applied their algorithm to solve the 3D articulated model-based human motion tracking problem. In 2014 Ding, et al [5] proposed a method that estimates the relative 3D coordinates of skeleton joints from a monocular video sequences, but their system had two general limitations 1) their model is applicable to the situation which the target depth value is much smaller than the distance between the object and the camera, 2) the human movement should be in a way that is facing the camera. Liu et al [13] presented a full-body human motion tracking system using exemplar-based conditional particle filter for monocular camera configuration. A method which can be applied to a stationary or moving camera platform and being able to perform in real-time in cluttered environments has been proposed by Huang et al [8], they performed several experiments and their method showed quite promising results. Spruyt et al [18], presented a method that uses an unsupervised method to automatically learn the context in which a hand is placed. More recently Adams et al. [1] used a Kinect sensor and a high-fidelity virtual world interface to acquire the depth image of the scene and used an unscented Kalman filter-based tracking algorithm to estimate upper extremity joint kinematics in real-time during performance of virtual activities of daily living, moreover they were able to generate metrics related to speed and smoothness of motion of the subject. Tian et al [21] addresses the limitations of Kinect and inertial measurement unit (IMU) sensors used in trajectory tracking systems and proposed a method that fuses IMU and Kinect data to provide a robust hand position information. They achieve the hand position by three fusion strategies: double integration of IMU internal sensors, IMU internal sensor fusion with geometrical constraints and unscented Kalman filter (UKF) based fusion of IMU and Kinect. Tanaka et al [20] proposed a portable six degree of freedom motion tracking system which comprises of high-accuracy augmented reality markers that would be attached to the subject arms and a single camera configuration. Their system was able to estimate the pose of the arms with an error of 5mm in space and about 2^0 (degree) in orientation. Ligorio and Sabatini [12] developed a novel Kalman filter for human motion tracking by fusing the data from a triaxle gyroscope and a triaxle accelerometer, they used Stereo photogrammetric data as a reference their method achieved, on an average, a root mean square attitude error of 3.6° in manual activities and 1.8° in locomotion tasks.

Other types of systems are also available for tracking the human motion. They usually integrate visual sensors with other types of sensors. Many hybrid approaches have been proposed to accomplish tracking performance such as visual–audio [14], visual–radar [17] and visual–inertial approaches. In [23], Y. Tao integrated a visual sensor with an inertial sensor to track human arm motion in telerehabilitation program, a similar technique also has been used in [24] to track the motion of the arm for a rehabilitation program.

III. SYSTEM CONFIGURATION

The aim of this study is to track the motion of any desired point of a subject's arm, and measure the velocity, angular motion and trajectory of that specific part(s) using only a simple webcam a computer and image processing algorithms. Figure 1, shows the conceptual diagram of our arm motion tracking system

Fig. 1 Conceptual diagram of webcam based motion tracker

The system which consists of only a laptop and a webcam, uses the skin color from the skin regions of the detected face as a cue to automatically isolate the arm of the subject (assuming the subject is not wearing a long sleeve clothing) from the background, the system then extracts the contour of the hand and uses the coordinate points of the contour as to track any point(s) on the arm, to evaluate how fast or how slow the motion has been performed, also to measure the angle of motion performed, specifically "elbow flexion", "elbow extension", "wrist flexion" and "wrist extension" , and last but not least it will allow for the subject or physiotherapist to visualize the trajectory of the performed motion for any further analysis.

IV. SKIN SEGMENTATION

A. Color Space

The RGB color space is commonly the default color space for most image formats, and thanks to colorimetriy, computer graphics and signal transmission standards a lot of color-spaces with different properties have been developed, however high correlation between the Red, Green and Blue channels creates an overlap between skin and non-skin points thus producing high number of false positive, hence making it not a very favorable choice for color analysis and color based recognition algorithms to decrease the overlap between the skin and non-skin pixels we therefore transform the RGB color space into a more reliable with less overlapping of its channels namely YCrCb.

B. Updating the Threshold Values Using Feedback Loop

In order to estimate the threshold values for our skin segmentation technique we make use of the popular and rapid face detection method introduced by Viola and Jones [22], its ability to perform very fast while achieving high detection accuracy thanks to its cascade of Haar-based feature detector implementation is a key feature of its attractiveness among other face detection algorithms. We take advantage of this real-time face detection technique to find the face of the person in the scene. By doing so we are able to eliminate some of the challenges present in skin color detection techniques, namely, ethnicity, individual characteristics. Next, we divide the entire image into two separate images, one image that contains only the detected face and represents the skin pixels, and secondly the entire image excluding the detected face that represents the non-skin pixels with in the scene (Figure 2). These two images are then thresholded to create a binary representation of skin and non-skin pixels of the scene and would be updated and altered automatically.

Fig. 2 The scene divided into two separate images, entire image excluding the face (left image), the detected face (right image

C. Skin Region Growing

To segment the skin regions, we use the binary image created from the previous stage, the binary image shows us the regions that are certain to be skin pixels, thus

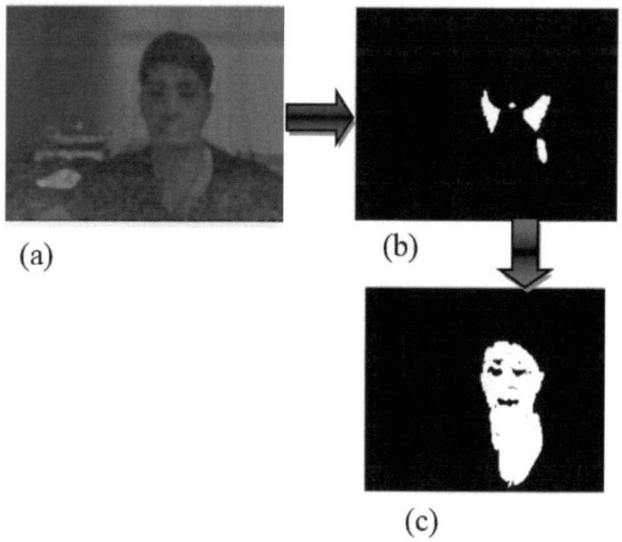

Fig. 3 (a) HSV color space input image, (b) certain skin regions, (c) region growing from the location of stage (b)

based on this fact we extract the coordinate points of each skin pixel from the binary image and use it as a reference point for the region growing stage of our algorithm. Seed fill also called the flood fill, is an extremely useful function that is commonly used in isolating connected regions to a given point in a multi-dimensional array for further processing or analysis. Moreover, it is considered an fundamental algorithm in computer graphics and used in a widespread of applications such as Computer Aided Design, Realistic graphics, Geographic Information System, Image processing, and so on [7]. A traditional seed filling algorithm has been implement in our algorithm, in which we color the neighboring pixel of the seed points if it is within a specified range of the original seed point value. In this case the algorithm is able fill the skin regions with in the scene using multiple seed points.

V. Arm Region Extraction

A. Contour Extraction

After segmenting the skin regions from the background we apply contour extraction algorithm to extract the boundaries of the segmented regions and use the contour size information to eliminate unwanted small regions (such as unwanted background noise) and unwanted big regions (such as the head portion).

B. Arm Skeleton Extraction

Close approximation of the joints on the arm, namely the elbow, shoulder and the fist is estimated by taking into account the proportions of the subjects arm and the size of the extracted contour.

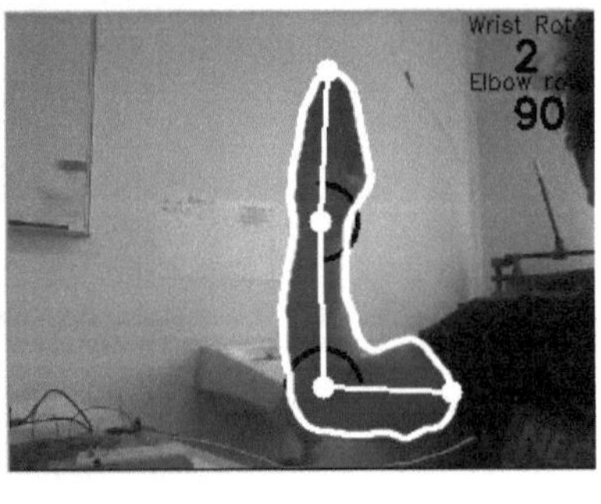

Fig. 4 The white line and filled circle shows the estimation of the joints and skeleton of the arm

VI. Motion Tracking

A. Coordinate Based Tracking

In order to track any point (s) on the arm we use the coordinates points of the arm contour, the uniqueness of this tracking method is that each point on the contour will represent a region of interest on the arm, so for example if we tend to track the thumb on the subjects arm we just assign the corresponding point to be tracked over time. Furthermore, this technique for tracking a point in space shows more promising results than popular tracking algorithms such as optical flow and Kalman tracking.

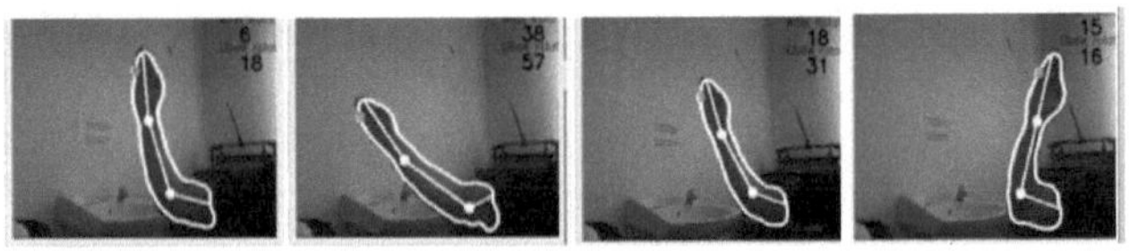

Fig. 5 The point on the arm being tracked robustly regardless how fast the arm is moved

B. Velocity

By determining the point that is desired to be tracked over time, we are also able to measure the velocity of that specific point on the arm. Evaluating the speed of each desired point, is accomplished by taking into account the displacement and duration that it takes for the point on the arm to move from point A to point B. since the velocity that we obtain has a unit of pixels per second a calibration procedure has been deployed to know the exact displacement in space.

$$distance = \sqrt{(x_2 - x_1)^2 + (y_2 - y_1)^2} \tag{1}$$

$$Velocity = \frac{distance}{duration} \tag{2}$$

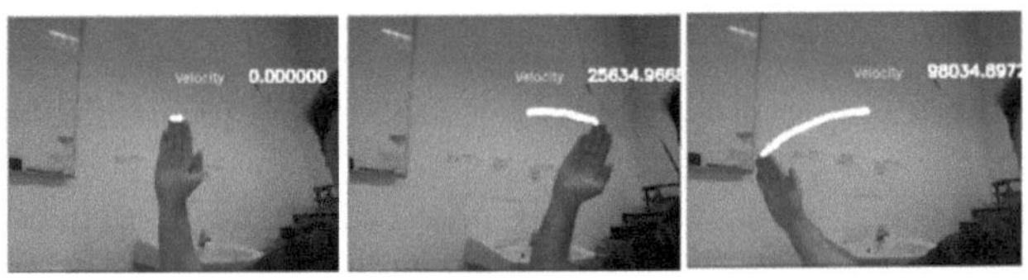

Fig. 6 Two different classes of velocity (a) no movement; (b) Slow movement; (c) Fast movement

Fig. 6 shows some results of the velocity evaluation procedure, and can tell the subject if he/she is performing the specific exercise slowly or fast based on the exercise criteria. The velocity is represented as a float number, so basically the greater the number the faster the motion was performed and vice versa.

C. Angular Motion of the Arm

The angular motion of the joints are calculated using tangent half-angle formula as described below. Figure 7 shows the angular motions of the joints.

$$Angular\ Motion = 2arctan\frac{y}{\sqrt{x^2+y^2}+x} \quad (3)$$

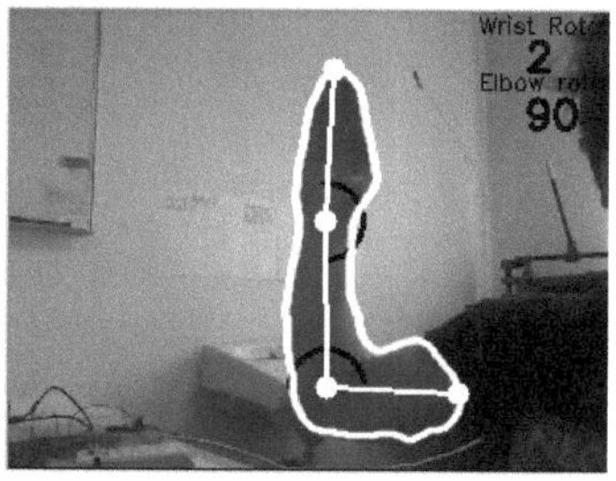

Fig. 7 Up right corner shows the values that corresponds to the angular displacement of the joints

D. Trajectory

A precise trajectory visualization of the motion performed by the targeted point on the arm is also evaluated and visualized as seen in figure 8.

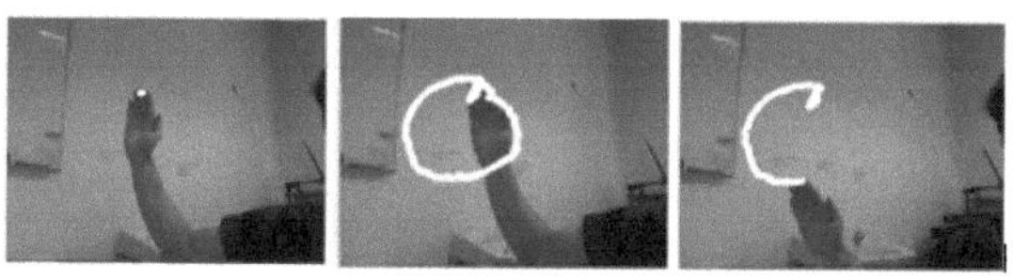

Fig. 8 Trajectory visualization result

VII. EXPERIMENTAL RESULTS

In order to evaluate the algorithm's performance, we collaborated with the University Malaya Medical Centre, and carried out a set of experiments to see how accurate and reliable our algorithm would perform in real-life scenarios. The experiments were conducted on the data gathered from subjects performing "elbow flexion", "elbow extension", "wrist flexion" and "wrist extension" tasks. One set of results are given in figure 9.

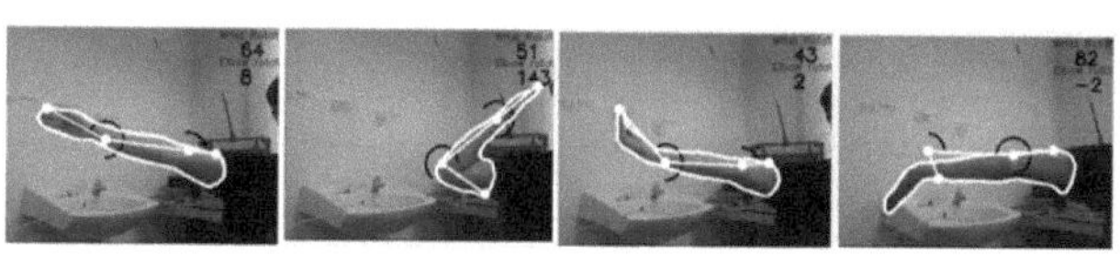

(a) (b) (c) (d)

Fig. 9 (a) Elbow extension; (b) elbow flexion; (c) wrist flexion; (d) wrist extension

In our experiments, we compared each results obtained from a goniometer device (figure 10) with our computer vision measuring algorithm to evaluate the angles of the elbow extension, elbow flexion, wrist flexion and wrist extension. Figure 10 shows the measurements we made with the goniometer for our reference.

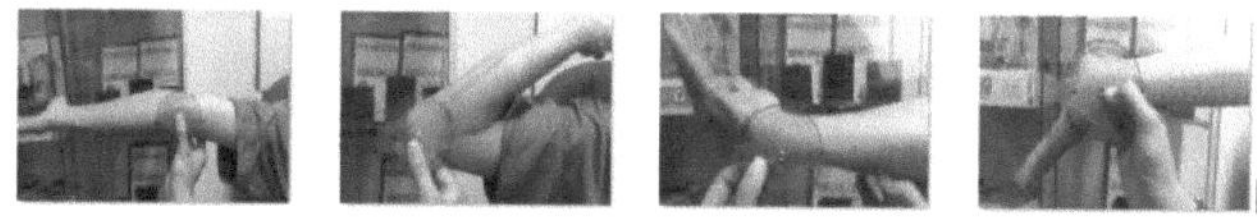

(a) (b) (c) (d)

Fig. 10 Goniometer measurements (a) elbow extension; (b) elbow flexion; (c) wrist flexion; (d) wrist extension

VIII. CONCLUSION

In this paper, we proposed a simple monocular camera based arm motion tracking method for stroke rehabilitation patients that is able to segment and track the patients arm in real-time video sequences captured in semi-controlled environment, and moreover we are able to measure the velocity, angular motion and the trajectory of any point on the arm as the subject is performing his/her task. In the future work we plan to further improve the accuracy of the measured parameters and make it applicable and robust to any environmental situation.

CONFLICT OF INTEREST

The authors declare that they have no conflict of interest.

REFERENCES

1. Adams, Richard J., et al. "Assessing upper extremity motor function in practice of virtual activities of daily living." *Neural Systems and Rehabilitation Engineering, IEEE Transactions on* 23.2 (2015): 287-296.
2. C. Barron and I. Kakadiaris. A convex penalty method for optical human motion tracking. In IWVS'03, volume Nov, Berkeley, pages 1–10, 2003.
3. C. Sminchisescu and B. Triggs. Covariance scaled sampling for monocular 3d body tracking. In Proceedings of the Conference on Computer Vision and Pattern Recognition, pages 447–454, Kauai, Hawaii 2001.
4. D. M. Gavrila, The Visual Analysis of Human Movement: A Survey – . Journal of Computer Vision and Image Understanding, Vol.73, No1, pages 82-98, 1999.
5. Ding, Jianhao, Zhijie Lin, and Hongbiao Xie. "A method of 3D recovery of human skeleton from monocular." Management Innovation and Information Technology61 (2014): 379.
6. Du, Ming, Xiaoming Nan, and Ling Guan. "Monocular Human Motion Tracking by Using DE-MC Particle Filter." Image Processing, IEEE Transactions on22.10 (2013): 3852-3865.
7. Duo-le, Feng, and Zhao Ming. "A New Fast Region Filling Algorithm Based on Cross Searching Method." Advances in Computer Science and Education Applications. Springer Berlin Heidelberg, 2011. 380-387.
8. Huang, Cheng-Ming, Yi-Ru Chen, and Li-Chen Fu. "Visual Tracking of Human Head and Arms Using Adaptive Multiple Importance Sampling on a Single Camera in Cluttered Environments." (2014): 1-1.

9. J. K. Aggarwal, and Q. Cai, Human Motion Analysis: A Review– Journal of Computer Vision and Image Understanding. 1999.
10. L. Enrique Sucarl, Gildardo Azcárate, Ron S. Leder, David Reinkensmeyer, Jorge Hernández, Israel Sanchez, and Pedro Saucedo.: Gesture Therapy: A Vision-Based System for Arm Rehabilitation after Stroke.
11. L. Wang, W. Hu and T. Tan, Recen Developments in Human Motion Analysis– PR(36), No. 3, March 2003, pp. 585-601
12. Ligorio, Gabriele, and Angelo Sabatini. "A Novel Kalman Filter for Human Motion Tracking with an Inertial-based Dynamic Inclinometer." (2015).
13. Liu, Jigang, Dongquan Liu, and Justin Dauwels. "3D Human motion tracking by exemplar-based conditional particle filter." Signal Processing (2014).
14. P. Prez, J. Vermaak, and A. Blake, "Data fusion for visual tracking with particles," Proc. IEEE, vol. 92, no. 3, pp. 495–513, Mar. 2004.
15. R. Bowden, T. Mitchell, and M. Sarhadi. Reconstructing 3d pose and motion from a single camera view. In BMVC, pages 904–913, Southampton 1998.
16. R. Polana and R. Nelson. Low level recognition of human motion. In Proc. of Workshop on Non-rigid Motion, pages 77–82, 1994.
17. S. C. Thomopoulos and L. Nillson, "3-D motion tracking using stereo camera and range radar," in Proc. SPIE—Sensing Reconstruction Three-Dimensional Objects Scenes, B. Girod, Ed., Jan. 1990, vol. 1260, pp. 21–35.
18. Spruyt, Vincent, Alessandro Ledda, and Wilfried Philips. "Robust Arm and Hand Tracking by Unsupervised Context Learning." Sensors 14.7 (2014): 12023-12058.
19. T. Moeslund, and E. Granum, "A Survey of Computer Vision-Based Human Motion Capture", Computer Vision and Image Understanding (81), No 3, pages 231-268, 2001.
20. Tanaka, Hideyuki, Yasushi Sumi, and Yoshio Matsumoto. "A portable 6-DOF motion tracker using high-accuracy AR markers—First report on the feasibility."*Machine Vision Applications (MVA), 2015 14th IAPR International Conference on*. IEEE, 2015.
21. Tian, Yushuang, et al. "Upper limb motion tracking with the integration of IMU and Kinect." *Neurocomputing* 159 (2015): 207-218.
22. Viola, Paul, and Michael J. Jones. "Robust real-time face detection."International journal of computer vision 57.2 (2004): 137-154.
23. Yaqin Tao and Huosheng Hu.: "A Novel Sensing and Data Fusion System for 3-D Arm Motion Tracking in Telerehabilitation" IEEE Trans. On Instrumentation and Measurement, vol. 57, no. 5, May 2008
24. Yaqin Tao, Huosheng Hu, Huiyu Zhou.: Integration of Vision and Inertial Sensors for Home-based Rehabilitation. InerVis 2005, the 2nd Workshop on Integration of Vision and Inertial Sensors, ICRA'05, IEEE International Conference on Robotics and Automation, Barcelona, Spain 18 April 2005.
25. Zhou, Huiyu, and Huosheng Hu. "Human motion tracking for rehabilitation—A survey." Biomedical Signal Processing and Control 3.1 (2008): 1-18.

Author: Alireza Akhavizadegan
Institute: University of Malaya
Street: Jalan Universiti
City: Kuala Lumpur
Country: Malaysia
Email: alireza_a110@hotmail.com

A New Approach for Reagent Storage-Releasing on Centrifugal Microfluidic Platforms Using Bubblewrap and Latex Membrane

M.M. Aeinehvand[1,2], F. Ibrahim[1,2], and M.J. Madou[1,2,3,4]

[1] Department of Biomedical Engineering, Faculty of Engineering, University of Malaya, Kuala Lumpur, Malaysia
[2] Centre for Innovation in Medical Engineering (CIME), Faculty of Engineering, University of Malaya, Kuala Lumpur, Malaysia
[3] Department of Biomedical Engineering, University of California, Irvine, USA
[4] Department of Mechanical and Aerospace Engineering, University of California, Irvine, USA

Abstract— **The implementation of latex membranes on centrifugal microfluidic platform has led to the development of advanced liquid handling features, with the exception of a proper reagent storage-releasing technique. This paper presents a new approach for reagent storage-releasing technique using bubbles of a bubblewrap with latex membrane tabs. The bubble was first filled with a liquid sample, sealed with a biocompatible paraffin wax, and then was placed in a chamber of the microfluidic disc. The chamber was then covered with a latex membrane tab similar to single microballoon pump. To release the stored liquid the membrane was manually deflected into the chamber to burst the bubble. The sealed bubbles are leak-free up to 106 kPa, and the average reagent recovery rate is 92%.**

Keywords— **Centrifugal microfluidic platform, On-board reagent storage, Bubble wrap, Latex membrane, Microfabrication.**

I. INTRODUCTION

In Lab-on-a-Chip (LOC) microfluidic platforms micropumps (e.g., syringe pumps) and microvalves are used to automate many types of laboratory functions (e.g., mixing, metering and separation) required for conducting various analytical assays [1]. Centrifugal microfluidic platforms are a type of LOC devices in which the liquid propulsion force is generated by spinning the fluidic platform [2, 3].

Pre-storing and releasing the required reagents prior to conducting an assay instead of pipetting them into a microfluidic platform is an essential feature for further automation but remains a significant challenge [4]. Glass ampules [5], miniature stick packs made of aluminum composite foil [6], and tubes sealed with Ferrowax [7] are just some examples of the reagent storage containers developed for use on integrated centrifugal microfluidic platforms. Each of those types of containers requires a unique actuation mechanism to release the stored reagents. For example, the liquid in the glass ampule is released as it is manually broken, the stick pack is opened by centrifugal pressure and the Ferrowax is melted with the aid of an external laser heater. In a more recently developed liquid storage approach for LOC microfluidics, David Bwambok *et al* employed a bubble of a bubblewrap sheet as a reagent storage reservoir [8]. In general, bubble wrap sheet is fairly water impermeable, widely available, and is inexpensive. Furthermore due to the manufacturing method employed, bubblewrap sheets are sterile and free of contamination [8].

Latex membranes, because of their elasticity, have helped enhance the flexibility of the microfluidic disc platform in handling liquids at low spin rates [9-11]. In this study, we have integrated a latex membrane tab and a bubble on a microfluidic disc to develop a reagent storage-releasing technique. A bubble wrap was charged with a liquid sample and sealed with paraffin wax. The bubble wrap was embedded in a chamber of a microfluidic disc, and then the chamber was sealed with a latex membrane. The actuation mechanism of the device is based on the deflection action of the latex membrane by mechanical force that leads to bursting the bubble. This paper describes and demonstrates the details of the fabrication procedure and the operating mechanism of the newly introduced liquid storage-releasing device. The reagent recovery rate of the liquid storage bubbles as well as their robustness are evaluated. In general, the elasticity of the latex membrane facilitates actuation of the reagent storage containers that require mechanical force to release the reagent.

II. METHODOLOGY

In this new reagent storage-releasing approach, bubbles of a bubblewrap sheet are used as reagent storage reservoirs. A corner of cylinder-shaped bubbles were cut with a sharp blade. The bubbles were filled with liquid samples and sealed by a low melting temperature paraffin wax (Fig. 1a). The bubbles were then placed inside the source chambers of a microfluidic disc in such a way that the sealed orifices were aligned directly with microchannels in the disc. The chambers with the liquid-filled bubbles were then sealed by latex membrane tabs. The multilayer microfluidic disc, which was used to evaluate the reagent recovery rate of the bubbles and their robustness, consisted of three Polymethyl-methacrylate (PMMA) and two pressure sensitive adhesive (PSA) discs (Fig. 1b). The disc contained five microfluidic units, each comprised of a source chamber,

F. Ibrahim et al. (eds.), *International Conference for Innovation in Biomedical Engineering and Life Sciences*,
IFMBE Proceedings 56,
DOI: 10.1007/978-981-10-0266-3_56

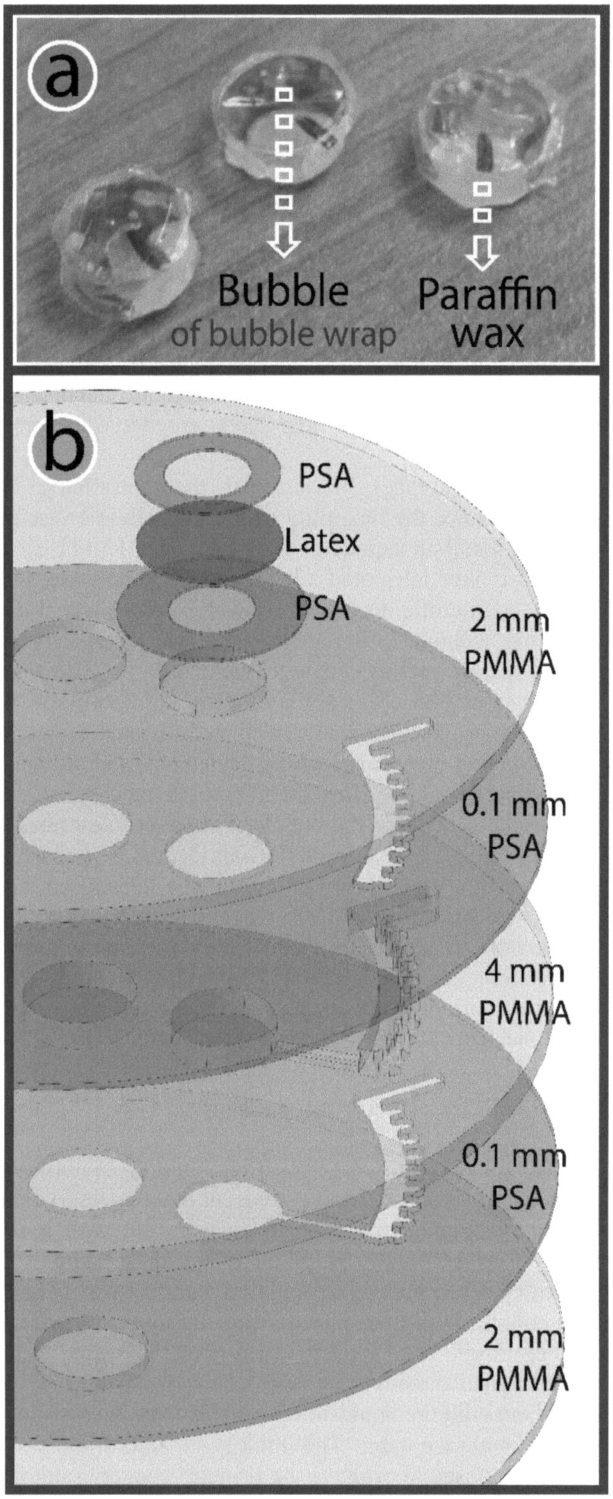

Fig. 1 (a) Bubbles of a bubble wrap were filled with a liquid sample and sealed with paraffin wax. (b) Breakdown of the five layer microfluidic disc and the three layer latex membrane tab.

a microchannel and a metering chamber. Each of the latex membrane tabs was made from two PSA ring and one small latex disc as shown in the figure 2.b. The details of the fabrication procedure of the microfluidic disc with latex-membrane tabs (i.e., latex microballoons) have been previously reported [9]. Briefly, the process involves using a computerized numerical control (CNC)-cutting/engraving machine and a cutter plotter for machining and cutting out the microfeatures in the PMMA and PSA discs, respectively. After fabrication of the five-layer disc, the liquid-filled bubbles were placed in the chambers and the chambers were sealed with the three-layer membrane tabs.

Two set of experiments were conducted to evaluate the robustness and the reagent recovery rate of the bubbles. To test the robustness of the sealed bubbles, the disc spin rate was continuously increased until the liquid sample leaked out of the bubble. In a second set of experiments, 210 μl of colored deionized water (DI) water (same as the volume of liquid stored in each of the bubbles) was charged into the source chamber of a control unit (Fig 2.a). The bubble then was manually burst to release the stored liquid sample. Because the orifice of the bubble is facing the microchannel in the disc, a portion of liquid bursts into the metering chamber (Fig. 2b). The disc was spun at up to 1500 RPM, and the induced centrifugal force propelled almost the entire liquids samples into the metering chambers (Fig. 2c).

III. RESULTS AND DISCUSSION

The latex membrane in the disc enabled actuating the liquid-filled bubbles with no need for any external actuator or spinning the disc. Although the external mechanical force required for bursting the bubble was manually applied, a mechanical system can be developed to automate the actuation procedure. In general, the latex membrane can used for actuating various reagents container, which could burst under mechanical force e.g., glass capsule and stick packs.

Many rigorous tests are required to evaluate on-board reagent storage devices including leak tests, reagent recovery tests, reagent loss through evaporation over time tests and various types of reagents tests. The preliminary two tests were carried out to demonstrate the possibility of using the bubbles as widely available reagent storage containers in microfluidic disc, and using latex membranes to easily actuate the bubbles prior to a progress. The bubbles were leak-free at least up to 106 kPa (equivalent to 6500 RPM in this disc).

Most of the fluidic procedures (e.g., mixing, separation and sedimentation) on disc require spin rate below 6500 RPM. The amount of liquid transferred into the metering chamber of the control unit was considered a 100% recovery rate. The experimental analysis show that with the bubbles we reach an average liquid recovery rate of 92±4%, which means that 16.8±8.2 μl of the liquid was trapped in the bubble. Further experiments are required to evaluate the ability of the bubbles to remain leak-free over specific time periods.

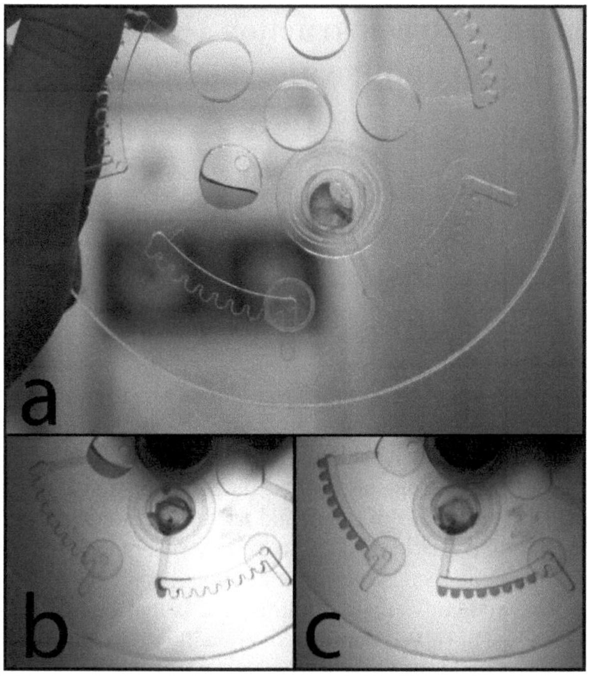

Fig. 2 (a) The liquid sample was pipetted into the source chamber of the left microfluidic unit, while the right microfluidic unit was equipped with a reagent storage device i.e., a liquid-filled bubble with a latex membrane. (b) The bubble was burst with the help of an external mechanical force. (c) The disc was spun and the liquids were transferred into the metering chambers.

IV. CONCLUSION

Previous studies have demonstrated that latex membranes enable precise valving [11], pumping [9] and mixing [12] of liquids on centrifugal microfluidic platforms. In this study, we have utilized the latex membrane to develop a new approach for the actuation of the reagent storage containers. Bubbles of a bubble wrap were used as low cost and widely available reagent containers. The elasticity of the latex membrane allowed for actuating the bubbles to release the pre-stored liquid samples with no need for using any external actuator or inducing a high centrifugal pressure. The experimental analysis demonstrated that the bubbles are leak-free at least up to 106 kPa, and have average reagent recovery rate of 92±4%. To verify the practicality of the bubbles for real life applications, further experiments are required to evaluate their compatibility with various reagents and their robustness and impermeability over a long period of time. However, the elasticity and the durability of the latex membrane has helped paved a new way of actuating reagent pre-storage containers in centrifugal microfluidic platforms. This allows for easy actuation of various reagent containers that can be actuated by mechanical forces.

ACKNOWLEDGMENT

This research was supported by University of Malaya (UM) High Impact Research Grant UM-MOHE UM.C/625/1/HIR/MOHE/05 from the Ministry of Higher Education Malaysia (MOHE), and University of Malaya Research Grant (UMRG RP009A-13AET). Fatimah Ibrahim would like to acknowledge the Yayasan Sultan Iskandar Johor for funding a Special Equipment Grant. Marc Madou acknowledges support of the National Institute of Health (NIH) Grant 1 R01 AI089541-01.

CONFLICT OF INTEREST

The authors declare that they have no conflict of interest.

REFERENCES

1. Yafouz B, Kadri N.A, and Ibrahim F. (2014) Dielectrophoretic Manipulation and Separation of Microparticles Using Microarray Dot Electrodes. Sensors 14(4):6356-6369
2. Strohmeier O, Keller M, Schwemmer F et al. (2015) Centrifugal microfluidic platforms: advanced unit operations and applications. Chem Soc Rev 44:6187-6229
3. Madou M, Zoval J, Jia G et al. (2006) LAB CHIP. Annu Rev Biomed Eng 8:601-628
4. Kong L.X, Parate K, Abi-Samra K and Madou M. (2015) Multifunctional wax valves for liquid handling and incubation on a microfluidic CD. Microfluid Nanofluid, 18(5-6):1031-1037
5. Lutz S. Weber P, Focke M et al. (2010) Microfluidic lab-on-a-foil for nucleic acid analysis based on isothermal recombinase polymerase amplification (RPA). Lab Chip 10(7):887-893
6. van Oordt T, Barb Y, Smetana J et al. (2013) Miniature stick-packaging–an industrial technology for pre-storage and release of reagents in lab-on-a-chip systems. Lab Chip 13(15):2888-2892
7. Hwang H, Kim Y, Cho J et al. (2013) Lab-on-a-disc for simultaneous determination of nutrients in water. Anal. Chem 85(5):2954-2960
8. Bwambok, D.K., et al., Adaptive use of bubble wrap for storing liquid samples and performing analytical assays. Analytical chemistry, 2014. 86(15): p. 7478-7485
9. Aeinehvand M.M, Ibrahim F, Harun S.W et al. (2014) Latex micro-balloon pumping in centrifugal microfluidic platforms. Lab Chip 14(5):988-997
10. Aeinehvand M.M, Ibrahim F, Harun S.W et al. (2015) Biosensing Enhancement of dengue virus using microballoon Mixers on centrifugal microfluidic Platforms. Biosens Bioelectron 67:424-430
11. Aeinehvand M.M, Ibrahim F, Harun S.W et al. (2015) Reversible thermo-pneumatic valves on centrifugal microfluidic platforms. Lab Chip 15(16):3358-3369

Author: Fatimah Ibrahim
Institute: University of Malaya
City: Kuala Lumpur
Country: Malaysia
Email: fatimah@um.edu.my

Author: Mohammad Mahdi Aeinehvand
Institute: University of Malaya
City: Kuala Lumpur
Country: Malaysia
Email: m.aeinehvand@yahoo.com

Quaternion Based Freehand 3D Baby Phantom Reconstruction Using 2D Ultrasound Probe and Game Controller Motion and Positioning Sensors

F. Mohamed[1,2], W.S. Mong[1], and Y.A. Yusoff[1,2]

[1] Faculty of Computing, Universiti Teknologi Malaysia, Johor, Malaysia
[2] Media and Game Innovation Centre of Excellence (MaGIC-X), UTM-IRDA Digital Media Centre, Institute of Human Centred Engineering, Universiti Teknologi Malaysia, Johor, Malaysia

Abstract— **This project aims to investigate and design a low-cost 3D ultrasound system from a standard 2D ultrasound setting and an existing game controller position tracker (i.e., PlayStation Move). Ultrasound imaging plays an important role in diagnosis, especially in medical fields, which has a large impact on clinical analysis. However, there is limited number of 3D ultrasound machines in the market and most of the medical practices are unable to use this kind of facilities to conduct their analysis. Thus, the shelf game controller that equipped with the position tracking device can be added into the existing 2D ultrasound machine to increase the capability of the 2D machine to visualize ultrasound images in 3D. Position tracking with the PlayStation Move does not only provide the low-cost alternative, but it is a portable device as well. From the collected data, Quaternion technique is selected to approximate the movement in 3D. Quaternion provides a convenient mathematical notation for representing orientations and rotations of objects in three dimensions. Quaternion method is numerically more stable and efficient because it can avoid the problem of gimbal lock. By using PlayStation Move, the medical operative can scan the subject by using the existing 2D ultrasound probe in 3D.The information such as the positions and orientations of the image at a certain location is integrated with 2D ultrasound imaging to generate a 3D ultrasound imaging.**

Keywords— **Ultrasound, Medical Imaging, 3D reconstruction, Medical visualization, 2D ultrasound probe, 3D imaging.**

I. INTRODUCTION

Ultrasound imaging plays an important role in diagnosis, especially in medical fields. Ultrasound as a diagnostic device is a widespread, non-invasive technique used to investigate the interior of the human body. Unlike magnetic resonance imaging (MRI) and computerized tomography (CT), ultrasound offers interactive visualization of the underlying anatomy in real time. It uses high-frequency sound waves to visualize images of the human body such as bone, muscle and other internal organs to capture their size and structure. Additionally, ultrasound does not utilize ionizing radiation or require specialized facilities, which are safe and easy to use.

Ultrasound can provide useful information on medical diagnosis. Computer graphic techniques are used to assist the diagnosis process by visualizing the ultrasound data. Computer graphics techniques aided in extracting meaningful information from ultrasound data and displaying it by using different types of filters and processing methods.

Ultrasound has been used in variety of clinical settings, including bone reconstruction, cardiology, and cancer detection. The main advantage of ultrasound is that certain structures can be observed non-invasively without compromising the subject safety such as exposure to radiation. Ultrasound also can be done at faster rate than X-rays or other imaging techniques. The designed system is tested by reconstructing a baby phantom as a case study.

Quaternion provides a convenient mathematical notation for representing orientations and rotations of an object in three dimensions. A quaternion expression is a typical form of three-dimensional rotations that consists of a scalar and 3D vector component.

Standalone machines become increasingly complex, while a new segment within the engineering community is trying to make these machines smaller, more power efficient, and less costly than previous machines [2]. The 3D ultrasound system combined the 2D ultrasound images together with the positions and orientations data by using the quaternion method to visualize a 3D baby phantom structure. The size of the PlayStation Move that is small and easy to move together with ultrasound probe provides a better experience for the user. In order to give a more realistic experience, the output of the system used the position mapping of the scan surface are presented in a 3D view.

II. LITERATURE REVIEW

Ultrasound imaging is a real-time and free from harmful radiation. In other word, conduct an ultrasound scans will not harm the human body. As a result, it becomes popular in medical use for all patients. Ultrasound is a non-invasive procedure, very safe, does not involve ionizing radiation, and provides real-time imaging [11]. Apart from that, most ultrasound scans are quick and painless, which suitable for

F. Ibrahim et al. (eds.), *International Conference for Innovation in Biomedical Engineering and Life Sciences*,
IFMBE Proceedings 56,
DOI: 10.1007/978-981-10-0266-3_57

all people. It also can provide a real time image of body movement and blood vessel.

Ultrasound scan is a widely used and safety technique in the medical field to help doctor diagnoses illness from the patient. Although ultrasound scan has many advantages, it has some limitations such as the scanning cost is too expensive. Besides, ultrasound systems are costly [1]. Some people who come from poor family or poor people from wealthy country cannot conduct such scan because they can't afford it. This makes the poor people who suffered from illness cannot undergo diagnosis and analysis by doctor. It is true that ultrasound scan will benefit a patient if the scanning cost can be reduced and available at all area.

An ultrasound scan can be used in various ways, such as diagnosing illness of a patient, monitoring an unborn baby, guiding a surgeon during certain procedure/ and providing educational media for the clinician. Ultrasound scans are a routine procedure for pregnant women. Pregnant women should perform ultrasound scans at least 2 times during a pregnancy period to know more about fetal growth [4]. Pregnant women are now monitored by ultrasound scan to examine the health and position of the unborn baby. An ultrasound scan can be done at any time during pregnancy and it does not harm the unborn baby. Ultrasound scan can be used to diagnose organ such as liver and bone structure. Another common application for ultrasound phantom is in educational sectors [6].Many undergraduate students can use ultrasound scan to learn the structure of the human bone, liver, and other organs.

Ultrasound imaging becomes more important in diagnosis. However, ultrasound system is only available in certain country.People who live at the rural area and poor country do not have ultrasound facility. Apart from that, ultrasound requires a highly experienced and skilled operator to detect a malignant lump, which make this scan only available at certain large hospital and modern country. The results of ultrasound scan dependence on the experience and knowledge of the diagnostician to manipulate the ultrasound transducer [3]. It is true that ultrasound scan requires experience clinician to get an accurate result. As a result, only the certain place can provide ultrasound facility to the patient. Training becomes an important component as it can create more experience and knowledgeable trainee to meet the market requirement. Highly experienced and knowledgeable diagnostician can provide a better analysis on ultrasound image, which is important to the patient.

3D ultrasound scan has become more popular in the medical field. Major steps in 3D reconstruction for ultrasound imaging modalities are images acquisition, images segmentation, volumes reconstruction, and display [7]. Each step in 3D ultrasound reconstruction is important and takes their role to produce the 3D ultrasound images. Image acquisition is taking 2D ultrasound image from the ultrasound machine. A series of 2D ultrasound machine will be taken and save on a laptop for 3D reconstruction purpose. The purpose of the image segmentation is to change the typical way an image represent, which is very meaningful [10]. Volume reconstruction is a combine series of 2D ultrasound image together with the position and orientation for 3D reconstruction purpose.

The image created by the 3D ultrasound scan has a better visualization than 2D ultrasound scan. Image acquisition is one of the key issues in 3D ultrasound imaging. Image acquisition in ultrasound imaging modalities more or less affect the qualities of the images [12]. The better the image acquires, the better the qualities of image. The scanning experience and knowledge of people to take an ultrasound image are important as it will affect the qualities of the image taking from an ultrasound machine. Different body parts require different probe scanning pattern method in order to acquire best image [6]. In order to acquire the image of ultrasound, different scan pattern at different position and orientation will be applied. The pattern such as linear scan, rotational scan, volume scan, freehand scan, and others will be used at different conditions.

III. METHOD

Quaternion method [8] is the most significant technique used in this research. This method will be used to coordinate the rotations and orientations during 3D ultrasound reconstruction. The orientations in quaternion value achieved from the PlayStation Move will be saved on a laptop in order to make a 3D ultrasound reconstruction. A quaternion is a mathematical concept that related to number theory and algebra. The method is numerically more stable and efficient because it avoids the problem of gimbal lock. Gimbal lock is the loss of one degree of freedom in a three-dimensional because both yaw and roll rotate about the vertical axis [5].

Quaternion method provides a way to represent orientations and rotations of the objects in three dimensions besides having better performance than other. Quaternion is used as a descriptive of the axis direction of rotation and total angle of rotation around axis [9]. Direction of an axis of rotation and total angle of rotation are represented using the equation below:

$$Angle = cos\ (qw/2)$$
$$Axisx = qx\ /\ sqrt(1\text{-}qw*qw)$$
$$Axisy = qy\ /\ sqrt(1\text{-}qw*qw)$$
$$Axisz = qz\ /\ sqrt(1\text{-}qw*qw)$$

From the equation, Angle is the total angle of rotation. *Axisx*, *Axisy*, and *Axisz* are the direction of an axis of rotation while values of *qx, qy,* and *qz* are the representative of quaternion value and *sqrt* is the representative of square root. Fig.1 shows the direction of axis rotation and angle of rotation around axis.

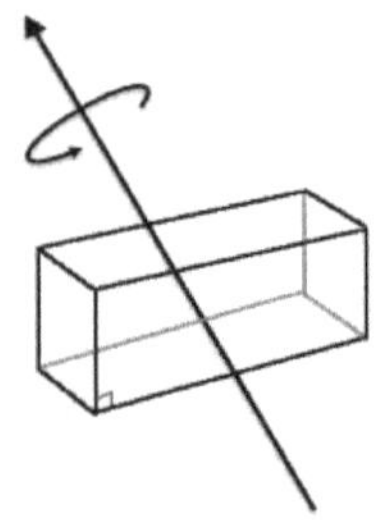

Fig. 1 Direction of axis of rotation and angle of rotation around axis

IV. SYSTEM DESIGN

A. Hardware Setting

A system developed in this research consists of an ultrasound machine that is connected to a laptop using a frame grabber. The ultrasound machine will be used to scan the structure of the baby phantom. Fig.2 shows the overall of the system design.

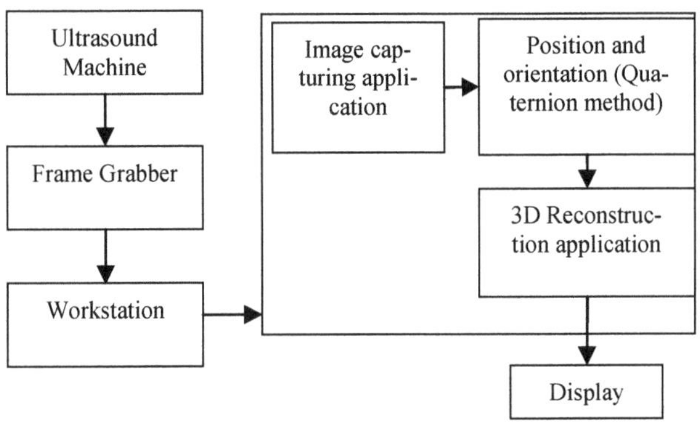

Fig. 2 System design

The PlayStation Move is connected to a computer by using Bluetooth communication. The PlayStation Move internal sensors are used to acquire the positions and orientations of ultrasound image from a certain angle. At the same time, the PlayStation Eye is connected to a laptop using USB cable. It is important that the PlayStation Eye is integrated with PlayStation Move to acquire accurate position of the PlayStation Move. In order for the PlayStation Eye to operate, an installation of a driver (CL eye test) in a laptop is essential. Calibration need to be carried out before using the Play Station Move and PlayStation Eye. The calibration can be done by rotating PlayStation Move at different directions in front of PlayStation Eye.

B. Software Design

There are two components that are used in this study. First, the acquisition component that enables to acquire and save 2D ultrasound image, position and orientation to a laptop. The ultrasound image that is transferred from the ultrasound machine to a laptop screen will be saved in this stage for 3D reconstruction purposes. This component uses PlayStation Move existing button to synchronize and save the ultrasound image as bitmap format. Fig. 3 shows the flowchart of image capturing.

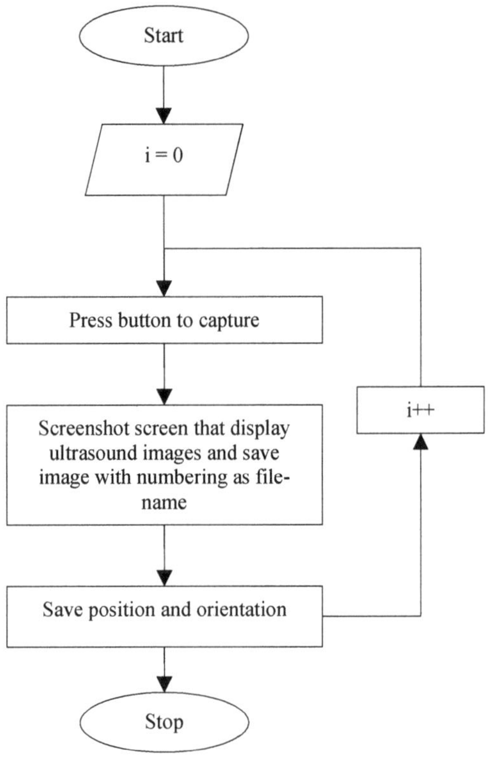

Fig. 3 Flowchart of image capturing process

The system starts to capture the screen that displays ultrasound images when user presses button on PlayStation Move and save image with numbering i as a file name. At the same time, positions and orientations in quaternion will be saves on a text file. The system is developed by using C # programming language and Microsoft Visual Studio 2010.

Second, the main focus of the research is to develop a 3D reconstruction software based on 2D ultrasound image. Software used to perform 3D reconstruction is the core value of the entire research. The developed software will read image files that have been saved in a laptop previously, together with the text file that consists of the positions and orientations of the image. It will start to read the positions and orientations, later save them into an array. Next, the software creates a plane based on the number of 2D ultrasound image. Finally, it will get the texture of every image and put it on the plane for 3D reconstruction. It is important that the position and orientation of every data slice must be accurate to get a precise 3D object. The quaternion method will be used to coordinate the rotations and orientations during the 3D ultrasound reconstruction. Fig.4 shows the flowchart of the 3D reconstruction software.

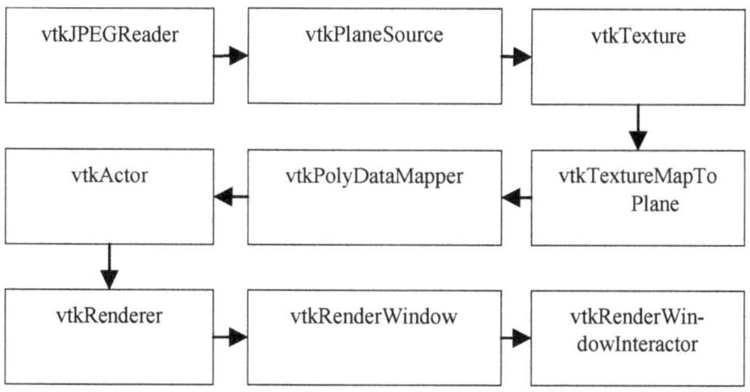

Fig. 4 Flowchart of 3D reconstruction process

V. Research Design

A baby phantom was used to test the capabilities of the 3D reconstruction system. The baby phantom is then scanned and tagged using the developed system. The output dataset comprises frame of the 2D ultrasound image. Each frame is named in sequence such as img0, img1, imgn. Each frame is accompanied with the respective positions and orientations in text file. The system is developed by using Microsoft Visual Studio 2010 with Visualization Toolkit (VTK). The Visualization Toolkit (VTK) is an open-source, freely available software system for 3D computer graphics, image processing and visualization. The VTK libraries are used to visualize the reconstructed 3D images of baby phantom. Fig.5 shows the schematic diagram of the research design.

In order to compare the reconstructed result with a real object, images of phantom acquired from different positions and orientations are needed. Fig.6 shows the structure of the baby phantom used in this research.

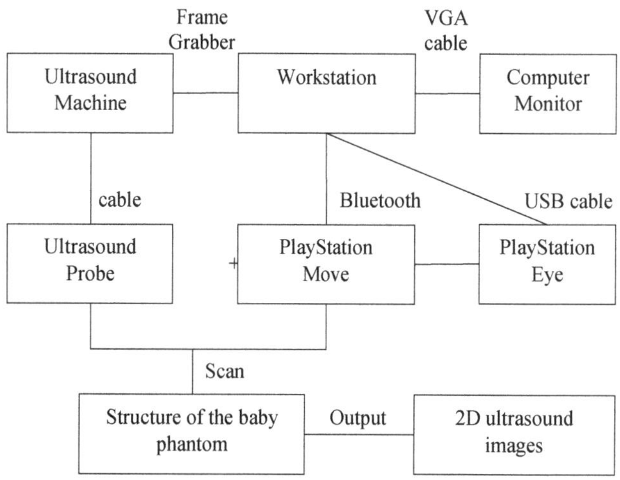

Fig. 5 Schematic diagram of the research design

Fig. 6 Structure of baby phantom

VI. Result and Discussions

The 2D ultrasound image of baby phantom is acquired from the ultrasound machine and stored into a laptop for 3D reconstruction. The 2D ultrasound images are made up of only thin slices that provide flatter and low-resolution images. The 2D ultrasound scan gives you the outlines and flat-looking images, which are difficult to understand. People who are not familiarize with the 2D ultrasound imaging might not able to view and analyze the results. The 2D image of baby phantom is taken at different positions and orientations to produce a good condition of the 3D image reconstruction.

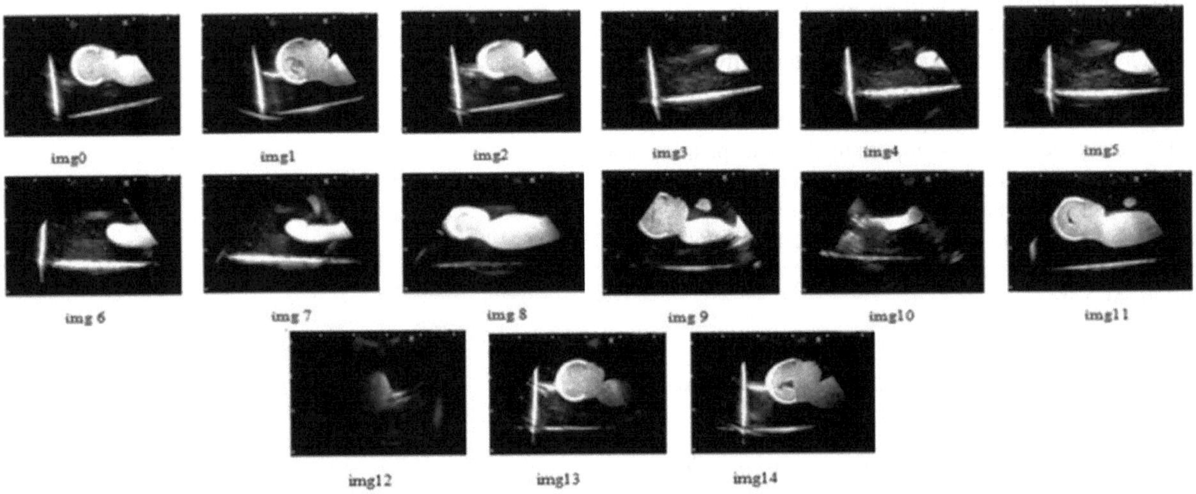

Fig. 7 2D ultrasound images taken by using ultrasound machine

Conducting a 2D ultrasound scan is easier and simpler as user can sweep the probe manually over the baby phantom while a focused image is displayed on the ultrasound machine and the image is transferred to a laptop by frame grabber. User that has basic experience and skills is able to conduct 2D ultrasound scan. 2D ultrasound image contains a certain part of the baby phantom at a certain position and angle. As a result, 2D ultrasound image does not have the general shape of the baby phantom. Fig.7 shows the output of the 2D ultrasound image taken by using the ultrasound machine.

This research is able to justify and provide the solid information whether the utilization of a low cost device is convincing enough to produce 3D image. The 3D ultrasound has the same principle as 2D ultrasound but the only difference is it integrates with the position and orientation to produce 3D image. The 3D ultrasound reconstruction will collect series of 2D ultrasound images and put into 3D reconstruction software to obtain 3D images. The 3D ultrasound scan is easier to understand as it produces a real shape of the object. The accuracy of the position and orientation is important during the 3D reconstruction. The 3D ultrasound image reconstruction can show the structure of the baby phantom clearly. With the 3D ultrasound, it required more advanced and complex computer software. As a result, the cost of 3D ultrasound is more expensive than 2D ultrasound. The 3D image reconstruction in this research will combine all the 2D ultrasound images acquired from ultrasound machine and build into a baby phantom shape. Fig.8 shows the 3D image reconstructed view from 15 multi-orientation ultrasound images.

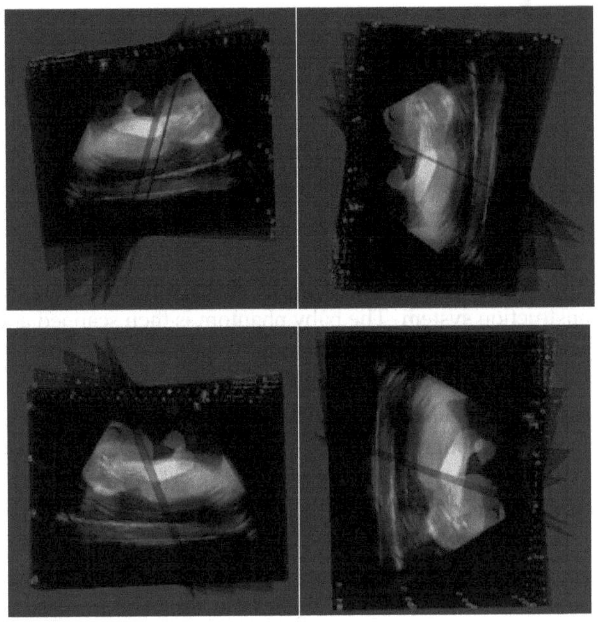

Fig. 8 3D image reconstructed view from 15 multi-orientation ultrasound images

Comparison between the baby phantom and the 3D reconstructed image were made by observing visible features of the baby phantom (see Figure 9).

The white color part indicates the structure of baby phantom while the black color part represents the background .The 3D image has successfully reconstruct the head, hand, and body of the baby phantom which is similar with the targeted object. Ultrasound only shows the general shape of the baby phantom as it is unable to penetrate and move through the structure of baby phantom. A reconstructed image more and less has the general shape of the baby phantom.

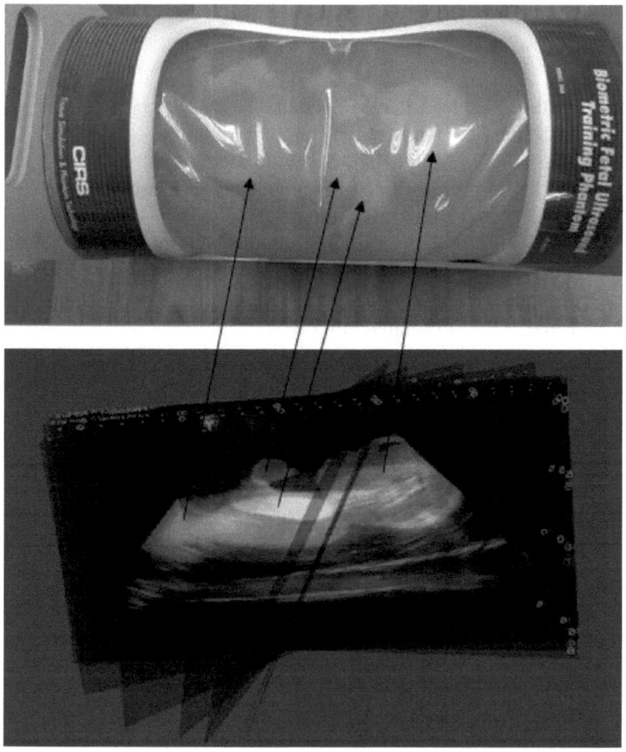

Fig. 9 Visual comparison between the baby phantom and the 3D reconstructed images

There might have some errors between the baby phantom and 3D reconstructed image due to the minor inaccuracy of position and orientation value during data acquisition of 2D ultrasound images. The accuracy of the position and orientation is important to provide a better result and output. Besides, the method of data acquisition also affects the quality of the 3D reconstruction object. The scanning method used in this research is freehand scan and might cause some inaccuracy problem as some parts of the baby phantom are not scan by ultrasound probe. Even though the 3D reconstructed image is shorter than the leg of the baby phantom, which is not similar to the baby phantom, the overall comparison show that the 3D reconstruction software is capable to illustrate a baby phantom shape. This indicates the success of this research and proves that the 3D reconstruction could be performed by using the low-cost device such as PlayStation Move and PlayStation Eye.

At the same time, the number of slices taken also affects the 3D image reconstructed. The number of 2D ultrasound frames taken should be as many as possible.

VII. CONCLUSION AND FUTURE WORK

Based on the output of the study, it can be concluded that the position tracking system for 3D reconstruction can be developed by using the low-cost game controller. The design of the system keeps the cost of the components and hardware to an absolute minimum. The system is designed in a systematic way which enable the implementation in any 2D ultrasound machine.

These are some future works that can be applied on this project in order to improve the overall system. Marching cubes algorithm can be considered to design the 3D reconstruction as it is a surface rendering method for 3D reconstruction. Marching cubes algorithm will take eight neighbor locations at a time and determine the polygon needed to represent the isosurface. It is a surface rendering method which can creates the isosurface based on the intensity value from each 2D ultrasound image.

Another aspect that needs to consider is the involvement part of human body for the testing purpose. This study has shown the capability of the system with the use of baby phantom. It is recommended that further studies need to be carried out for bone scans and moving structures.

ACKNOWLEDGMENT

This work is supported by the Fundamental Research Grant Scheme of Ministry of Education of Malaysia under the grant FRGS/1/2014/ICT07/UTM/03/1.

CONFLICT OF INTEREST

The authors declare that they have no conflict of interest.

REFERENCES

[1] Abd Ellah, M. M., et al. (2009). Software development for low cost, high quality, real-time 4D ultrasound imaging system on laptops (PCs). Radio Science Conference, 2009. NRSC 2009. National.

[2] Baran, J. M. and J. G. Webster (2009). Design of low-cost portable ultrasound systems: Review. Engineering in Medicine and Biology Society, 2009. EMBC 2009. Annual International Conference of the IEEE.

[3] Fenster, A. and D. B. Downey (1996). "3-D ultrasound imaging: a review." Engineering in Medicine and Biology Magazine, IEEE 15(6): 41-51.

[4] Megalingam, R. K., et al. (2013). Assistive technology for pregnant women health care: Rural area, mobile ultrasound scan system (using ASTM E1384–07 standard). Global Humanitarian Technology Conference: South Asia Satellite (GHTC-SAS), 2013 IEEE.

[5] Min-Xiu, K., et al. (2013). Application of orientation interpolation of robot using unit quaternion. Information and Automation (ICIA), 2013 IEEE International Conference on.
[6] Rizqie, M. Q., et al. (2013). Handmade bone phantom as educational media for image processing studies. Teaching, Assessment and Learning for Engineering (TALE), 2013 IEEE International Conference on.
[7] Rizqie, M. Q., et al. (2013). Prototype of 3D reconstruction system for ultrasound imaging modalities: Case study handmade bone phantom. Instrumentation, Communications, Information Technology, and Biomedical Engineering (ICICI-BME), 2013 3rd International Conference on.
[8] Shoemake, K. (1985). "Animating Rotation with Quaternion curves." Computer Graphics 19.
[9] Skehan, D. P. (2011). Virtual Training System For Diagnostic Ultrasound. Faculty of the WORCESTER POLYTECHNIC INSTITUTE. Degree of Master of Science: 121.
[10] Ting, Y. and S. Huazhong (2011). Ultrasound image segmentation by spectral clustering algorithm based on the curvelet and GLCM features. Electrical and Control Engineering (ICECE), 2011 International Conference on.
[11] Torres, P. M. B., et al. (2012). 3D femur reconstruction using a robotized ultrasound probe. Biomedical Robotics and Biomechatronics (BioRob), 2012 4th IEEE RAS & EMBS International Conference on.
[12] Nadhirun N. N., Tissue Phantom And Bin-Filling For 3d Ultrasound Image Reconstruction (2014), UTM FBME BEng Project Report.

Author: Farhan Mohamed
Institute: Universiti Teknologi Malaysia
Street: UTM-IRDA Digital Media Centre,
Institute of Human Centred Engineering,
UIRL Building, T03,
Universiti Teknologi Malaysia
City: Skudai Johor
Country: Malaysia
Email: farhan@utm.my

Characterization of Impedance Spectroscopy for Single-Walled Carbon Nanotubes with Ionic Liquid of Dielectrophoretic Assembly Method

A.S. Mohamad[1] and M.P. Hughes[2]

[1] Section of Medical Engineering Technology, Universiti Kuala Lumpur British Malaysian Institute, Selangor, Malaysia
[2] Centre for Biomedical Engineering, Department of Mechanical Engineering Sciences, University of Surrey, Guildford, United Kingdom

Abstract— **An ionic liquids have been demonstrated as solvents in the development of green chemistry to reduce environmental impact. The use of ionic liquids has been reported in organic synthesis, chemical separation and electrochemical applications of 1-Butyl-3-Methylimidazolium Trifluoro methanesulfonate (BMIM-OTf) have attracted attention as highly useful and versatile solvents because of their many unique properties. They have high ionic conductivities, wide liquids range and solubilities characteristics. More than that, they are tolerant of strong acids, purified and reused. The usage of ionic liquids is considered environmentally-friendly for dielectrophoresis (DEP) experiments.**

Keywords— **Dielectrophoresis, ionic liquid, SWNT, green chemistry.**

I. INTRODUCTION

Ionic liquids based on the BMIM-OTf are salt with a liquid phase at room temperature. BMIM-OTf cation is the most commonly used, offering a wide range of properties. It is a versatile and stable ionic liquid. The discovery BMIM-OTf to investigate the surface and the interface of various media shown OTF anion strongly interacts with [bmim]+ cations [1].

Chakraborti et al. demonstrated, BMIM cations is influenced by the structure of the imidazodium moiety with examined the effect of various neutral, acidic and basic ionic liquids. They found that BMIM cations is useful to prove catalytic efficiency and their behaviour [2].

Zyto et al. reported the crystal structure of metal-organic compound have an interaction with two OTf anions. Thus, the outer coordination sphere and bridge the coordination cations through nitrogen, hydrogen and oxigen (N-H...O) interactions to form two symmetry-independent hydrogen-bounded chain [3].

II. PREPARATION OF MATERIAL

The DEP assembly method of SWNT nanoparticle was carried out on micro-tip electrodes with glass substrates. The biodevices of substrates composed with gold (Au) and chromium (Cr). The gap is 8 μm between the micro-tip electrodes as shown in Figure 1. The DEP microelectrodes were fabricated using standard photolithography and etching processes. All the processes of fabrication were conducted at clean room and vacuum laboratory facility in the Advance Technology Institute, Faculty of Engineering and Physical Science, University of Surrey. The SWNT propeties are 50-70% carbon basis, 10 nm diameter and have length of 1 μm. SWNTs material was obtained from Sigma-Aldrich Company Limited in the United Kingdom. The BMIM-OTf has been used as a medium of conductivity for SWNTs experimental. This ionic solvent was purchased from Sigma Aldrich respectively. The nanoparticles were suspended in 25 ml of BMIM-OTf containing SWNTs were weighed at 0.0106g. Samples then sonicated for 30 minutes in room temperature and stored in the fridge at 8°Celcius.

In conducting BMIM-OTf experiment, the non-organic nanoparticles were weighted at 0.20 mg/ml of SWNT, 2.34 mg/ml was suspended with BMIM-OTf (Sigma Aldrich UK) in each sample. A 10 ml suspension was then stirred with rotamixer (Tucker UK Limited) for 30 seconds and sonicated for at least 30 minutes.

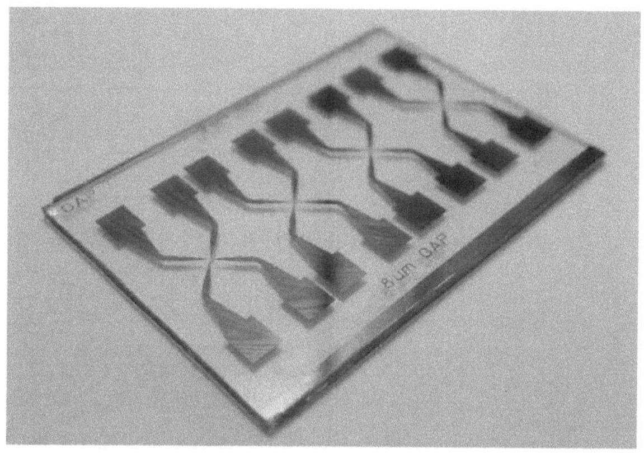

Fig. 1 Chromium and gold micro-tip electrode after the photolithography fabrication process

III. EXPERIMENTAL SETUP

Impedance spectroscopy of SWNT impedance work used NumetriQ Impedance Analyser PSM1735. Micro-tip electrode was connected to analyser and interfaced with Impedance analysis N4L. The cable type of RS232 connected to

F. Ibrahim et al. (eds.), *International Conference for Innovation in Biomedical Engineering and Life Sciences*,
IFMBE Proceedings 56,
DOI: 10.1007/978-981-10-0266-3_58

personal computer operated with Window 7 platform. The CommVIEW version 1.15 (Newtons4th Limited UK) has been used to analyse the data as shown in Figure 2. The data of resistance of the micro-tip electrode gap at 1 second intervals. The analyser measurement was setup by setting the parameters of 10 Vp-p, at five per decade between 100 Hz and 6.3 MHz frequencies. The data analysis process began prior to the SWNT samples being placed on the micro-tip electrodes. The work performed for typically 180 seconds continuously. The time constant of the change of resistance as a function of time was then acquired and stored in the computer. The experiment was repeated three times to verify the consistency of the data.

Moreover, the experiments of BMIM-OTf were using micro-tip 8 μm electrode to employ positive DEP forces. The work concentrates on nano-particles into focused electrode gap (bridge). The NumetriQ phase multimeter and impedance analyser were calibrated and setup. Low frequency dielectrophoresis was employed and the frequency was set to 100 Hz with 5 Vp-p. A drop of the nanoparticles suspension (5 μl) was applied to the middle of paired microelectrode after 30 seconds. The time constant of the change of resistance as a function of time was obtained after 300 seconds. A typical deposition time of 5 minutes was then followed by rinsing sample with methanol and gently blowing the drop off the surface. The experiment was repeated on needle-shaped electrodes energized with a 5 Vp-p with frequencies of five per decade between 100 Hz and 630 kHz.

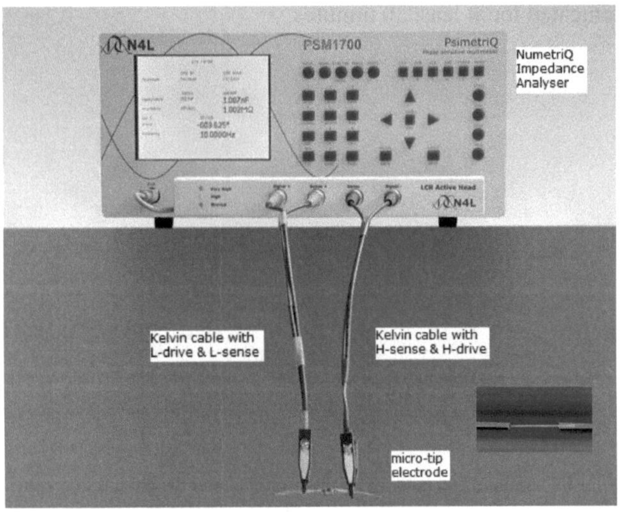

Fig. 2 NumetriQ impedance analyser experiment setup

IV. EXPERIMENTAL RESULTS

The challenge of achieving the bridging of SWNT across electrodes provide further understanding when using a technique based on dielectrophoresis deposition as developed by Krupke et al. [4]. As such a setup in two-point resistance geometry has been presented, the electrical measurement of impedance was observed. The SWNT is placed on top of Cr/Au layer where ends lie on it. This Cr/Au configuration normally displays much higher contact resistance for metallic and semiconducting SWNTs are measured for the designed micro-tip electrodes. Metallic SWNTs displays a resistance from the measurement of impedance using NumetriQ analyser.

Table 1 shows the properties of BMIM-OTf which are very useful to interpret data of impedance spectroscopy inorganic particles.

The plot of the reciprocal collection time constant shows the behaviour of a heterogeneous mixed. Figure 3 shows the solid line that represents the best fit for two populations of SWNTs on semiconducting and metallic. A spherical model indicates conductivity of 208 mS/m and 9.1 mS/m for rod model of SWNTs. The dielectric dispersion centered approximately at 1 kHz and 16 kHz. Meanwhile, a relative permittivity is 66.7ε_0 for population 1 and 48.3ε_0 for population 2. Their further relative population is 73.3 ±0.5% and 26.7 ±0.1% respectively.

Table 1 Physical properties of BMIM-OTf ionic liquids [10]

Molecular weight	Molecular formula	Conductivity (mS/m)	Dielectric Permittivity
288.3	$C_9H_{15}F_3N_2O_3S$	30.49	12.1

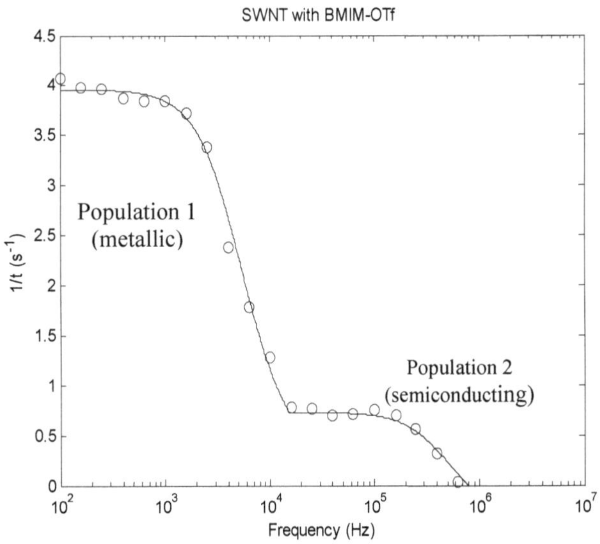

Fig. 3 The impedance measurement of purified SWNTs with BMIM-OTf

Table 2 The populations of SWNTs for the dielectrophoresis exerted on a 8 μm micro-tip electrode shape using BMIM-OTf solvent with relative standard deviation

SWNT	Permittivity	Conductivity	Relative Permittivity
Population 1	71.7 ±4.0%	208.0 ±1.4%	73.3 ±0.5%
Population 2	56.7 ±5.1%	9.1 ±7.1%	26.7 ±0.1%

The time constant is inversely proportional to the force of DEP assembly population of impedance. This observation strongly suggests a sequential assembly mechanism of SWNTs at the gap between the electrodes under AC electric fields. Experimental data was analysed through MATLAB programming of impedance measurement.

V. DISCUSSION

The assembly of nanowires in between electrodes is essential for making useful devices. The growth of nanobridges of SWNT, using DEP is a potential solution. A positive DEP on the micro-tip electrode was demonstrated. The ionic liquids of BMIM-OTf have been characterized two populations from the measurement of SWNT nanowires. A plot of reciprocal collection time constant shows two dielectrics dispersion centered approximately at 5 kHz and 15 kHz for SWNTs. Thus, two characteristics of the results have been known for nanomaterials on BMIM-OTf.

SnO_2 nanopowder can be formed as a ceramic insulator, or a crystalline, n-type semiconductor [5]. The observation of ionic liquids BMIM-OTf to the organic solvent DMF, has found extraction efficiency was to be higher than DMF. Their permittivity, electrical conductivity and viscosity have been investigated as BMIM-OTf highly sensitive to temperature and conductive material [6,7].

Modelling indicates that particles suspended in BMIM-OTf have a lower effective permittivity than those suspended in DMF. It can be seen that impedance of SWNT increase with permittivity for the both population of $5\varepsilon_0$ for conducting and $3.4\varepsilon_0$ for semiconducting. This behaviour indicates that the role of nanowire is more conductive for DEP of inorganic particles using ionic liquids. The catalyst of bimetallic SWNT in population 2 may decrease the effect of SWNT behaviour. The permittivity is considerably higher than the value of SWNT using aqueous of $40\varepsilon_0$ as reported by Mureau et al. [8]. This clearly shows that the enhanced performance and conducting of the BIMIM-OTf compared to DMF in non-aqueous solvents. This also suggests the addition of specific amount nanoparticles is expected to provide significant characteristics of SWNT impedance in DEP system.

Property of DEP force can be defined by the DEP spectrum and crossover frequency in which the geometry of the particle can be calculated. For dielectrophoretic force on a spherical-shaped particle can be determined using Equation (1) and (2). For a prolate ellipsoid with major axis r_1 and minor axis r_2 are given by the following equations [8]:

$$F_{sphere} = 2\,\pi\, r^3\, \varepsilon_m\, Re\,(\frac{\varepsilon_p^* - \varepsilon_m^*}{\varepsilon_p^* + 2\varepsilon_m^*})\, \nabla E^2 \quad (1)$$

$$F_{rod} = \frac{2\,\pi\, r_1\, r_2^2 \varepsilon_m}{3}\, Re\,(\frac{\varepsilon_p^* - \varepsilon_m^*}{\varepsilon_m^*})\, \nabla E^2 \quad (2)$$

where ε_m is the permittivity of the suspending medium, E is the electric field, ∇ is the differential vector operator, ε^*p and ε^*_m are the complex permittivities of the particle and medium respectively.

Moreover, the results on conductivity of DMF and BMIM-OTf represent the solvent as the medium of conductivity. The minimum value conductivity of this work indicates SWNT using DMF is 1.96 mS/m. Therefore, high of ionic liquids has an impact on conductivity because high salts usually give high conductivity [9] Particles in non-aqueous media have a higher apperent conductivity and permittivity than those suspended in aqueous media.

The anion ionic liquids contribute to the polarizability of nanowire assemblies. This feature can be used to characterize the particles. The anions at the medium conductivity interface were in homogeneous environment [1]. The nanowires have been assembled between electrodes gap from their non-aqueous solution via DEP. The mechanism causing field enhancement was assumed when electrode at the tips of growing leads to an increase the electric field applied.

It can be hypothesized that the time taken for the impedance to change is due to the collection of nanowires is inversely proportional to the force behaving on those tubes. Therefore, the reciprocal of the time constant indicate the magnitude of the force. Over a population of nanowires, the reciprocal of the time constant also indicates the relative population of subpopulation nanowires.

VI. CONCLUSION

In conclusion, the successful assembly of SWNT nanomaterial using DEP technique on the micro-tip electrode has been presented. Based on the experiment, the assembly of nanomaterial for performing dielectric spectroscopy mainly depends on the voltage, frequency and solution density. The dielectric characteristic of nanomaterial has been determined. In addition, it is possible to use DEP to manipulate nanowires across electrode gaps in non aqueous solutions of BMIM-OTf. Impedance measurements of the nanowires have been characterized for SWNT. The ability of nanowires interacts with DEP manipulation with real time measurement provides a technique for automated collection of nanowires. The time response and frequency applied are important because changes in ionic strength may lead to slow changes in the background signals. To accomplish this experiment in ionic liquids solvent is significant due to salt concentration in this range are typically needed for biomolecular recognition. Hence, the ability to perform dielectric properties in BMIM-OTf solvent is expected to enable the development of new types of biomedical devices applications.

ACKNOWLEDGMENT

The work is supported by the Majlis Amanah Rakyat (MARA) of UniKL BMI Attachment Scheme. The authors also thanks to Dr. Kai Hoettges and Dr. Fatima Labeed for their kind help with the experiments.

CONFLICT OF INTEREST

The authors declare that they have no conflict of interest in this research work.

REFERENCES

1. Iwahashi, Takashi, et al. "Anion configuration at the air/liquid interface of ionic liquid [bmim] OTf studied by sum-frequency generation spectroscopy." The Journal of Physical Chemistry B 112.38 (2008): 11936-11941
2. Chakraborti, Asit K., et al. "Catalytic application of room temperature ionic liquids:[bmim][MeSO 4] as a recyclable catalyst for synthesis of bis (indolyl) methanes. Ion-fishing by MALDI-TOF-TOF MS and MS/MS studies to probe the proposed mechanistic model of catalysis." Green Chemistry 10.10 (2008): 1111-1118.
3. Żyto, Magdalena, Barbara Wicher, and Maria Gdaniec. "Bis (di-2-pyridylamine-κ2N2,N2) silver (I) trifluoromethanesulfonate: polar arrangement of trifluoromethanesulfonate anions in a pseudo-centrosymmetric framework of coordination cations." Acta Crystallographica Section C: Crystal Structure Communications 66.11 (2010): 355-357.
4. Krupke, Ralph, et al. "Separation of metallic from semiconducting single-walled carbon nanotubes." Science 301.5631 (2003): 344-347.
5. Schierbaum, K., Weimar, U., & Göpel, W. (1992). Comparison of ceramic, thick-film and thin-film chemical sensors based upon SnO2. Sensors and Actuators B: Chemical, 7(1-3), 709-716.
6. Mohamad, Ahmad Sabry, Kai F. Hoettges, and Michael Pycraft Hughes. "Impedance spectroscopy of Zinc Oxide nanoparticles in dielectrophoresis biotechnology." Applied Electromagnetics (APACE), 2014 IEEE Asia-Pacific Conference on. IEEE, 2014. 10.1109/APACE.2014.7043759
7. Hapiot, Philippe, and Corinne Lagrost. "Electrochemical reactivity in room-temperature ionic liquids." Chemical Reviews 108.7 (2008): 2238-2264.
8. Mureau, Natacha, et al. "In situ and real time determination of metallic and semiconducting single-walled carbon nanotubes in suspension via dielectrophoresis." Applied physics letters 88.24 (2006): 243109.
9. Clarke, Richard W., et al. "Surface conductivity of biological macromolecules measured by nanopipette dielectrophoresis." Physical review letters 98.19 (2007): 198102.
10. Min, Y., Akbulut, M., Sangoro, J. R., Kremer, F., Prud'homme, R. K., & Israelachvili, J. (2009). Measurement of Forces across Room Temperature Ionic Liquids between Mica Surfaces. The Journal of Physical Chemistry C, 113(37), 16445-16449.

Address of the corresponding author:

Author:	Dr. Ahmad Sabry Mohamad
Institute:	Universiti Kuala Lumpur British Malaysian Institute
Street:	Bt.8 Jln Sg. Pusu, 53100 Gombak
City:	Selangor
Country:	Malaysia
Email:	sabry@unikl.edu.my

Principal Component Analysis of Honey Spectrum Obtained from Surface Enhanced Raman Spectroscopy

M.F. Raduan, W. Mansor, Khuan Y. Lee, and A.R. Mohd Radzol

Faculty of Electrical Engineering, Universiti Teknologi MARA, Shah Alam, Selangor, Malaysia

Abstract— **Honey spectrum produced from Surface Enhanced Raman Spectroscopy has to be passed through feature extraction process before pure honey can be recognised. This paper describes feature extraction of honey spectrum obtained from Surface Enhanced Raman Spectrometer using Principal Component Analysis. Honey was diluted in distilled water before the samples were passed through Surface Enhanced Raman Spectrometer to obtain honey spectrum. Baseline filtering, smoothing and normalization were then performed to remove interferences and enhance the Raman spectrum. Principal Component Analysis was carried out using three criteria; eigenvalue-one criterion, scree test, and cumulative variance to extract significant components in honey spectrum. Results obtained from the analysis show that Principal Component Analysis is able to reveal significant coefficients that reside in honey samples.**

Keywords— **Pure honey, PCA, Raman Spectroscopy.**

I. INTRODUCTION

Honey offers several health benefits such as provides energy, increases immunity, fights bacteria, heals cuts and burn and others. Because of these, most people prefer pure honey than adulterated honey. Adulteration in honey has become commonplace in the market today, therefore discrimination between pure and impure honey is very important [1]. Pure honey can be identified using Surface Enhanced Raman Spectrometer (SERS), a technique that relies on inelastic scattering of monochromatic light to observe low-frequency modes [2].

SERS is well known for its use in medical researches such as cancer detection, bacteria detection and even virus detection. SERS provides amplification of Raman signal by a factor of 10^4 to 10^{10} when molecules bounded to the surface of noble metal (Au, Ag and Cu). This makes it capable of detecting biological molecules such as DNA, RNA and even pathogen [3].

Classification of pure honey requires feature extraction to be performed to produce inputs for the classifier. One of the popular feature extraction tehniques used is Principal component analysis (PCA). It is a simple method that provides reduction of a complex data set to a lower dimension to reveal the sometimes hidden, simplified structures [4].

This paper describes feature extraction of honey spectrum using PCA based on three criteria; eigenvalue-one criterion, scree test, and cumulative variance. The honey spectrum was obtained using SERS and enhanced through filtering and normalization.

II. METHODOLOGY

Overall process of extracting honey features is shown in Figure 1. The process includes sample preparation, SERS analysis of honey, Raman spectrum of honey processing and PCA analysis.

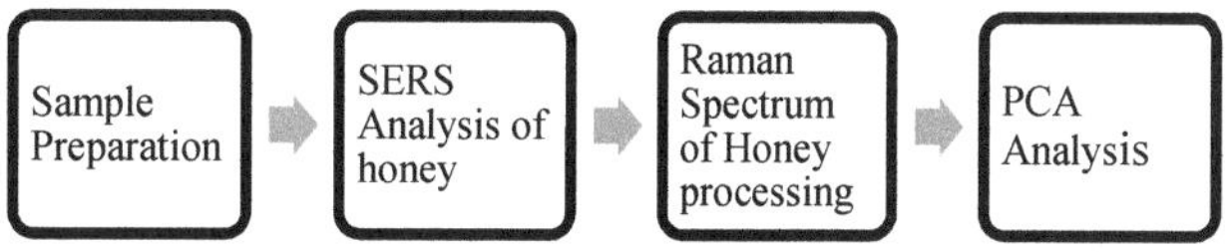

Fig. 1 Overall process of extracting honey features

A. SERS Analysis of Honey

Four samples of honey, labeled Honey A, Honey B, Honey C and Honey D, were used in this work. Before SERS analysis was carried out, a few samples were prepared. To obtain solute concentration, weight per weight (w/w) method shown in Equation (1) was used.

$$W_t = \frac{m}{s} x100\% \tag{1}$$

Where W_t is percentage of weight, m is mass of solute and sis mass of solution.

Distilled water (5% and 25% (w/w)) was used to dilute each sample of honey and to ensure fully diluted, the samples were left for two days. A blank gold slide was used to hold 10ml of each of the samples for analysis.

SERS analysis was then performed to obtain the spectrum of the raw undiluted honey, diluted honey and glucose. The purposes are also to figure out its basic component and to determine the spectrum regions that contain the required information.The average gold slide spectrum was then computed and used in baseline filtering.

F. Ibrahim et al. (eds.), *International Conference for Innovation in Biomedical Engineering and Life Sciences*,
IFMBE Proceedings 56,
DOI: 10.1007/978-981-10-0266-3_59

B. PCA Analysis of Honey Spectrum

Before the honey spectrum can be analysed, high or low frequency interferences have to be removed. In this work, the spectrum of the blank glass slide and background was removed from the honey spectrum using baseline filtering. Smoothing was carried out using low pass filter to eliminate high frequency interferences. This was followed by normalization to reduce the data size.

After removing noise, PCA analysis was then performed on the honey spectrum using three criteria; eigenvalue-one criterion, scree test, and cumulative variance to obtain significant features. Using eigenvalue-one criterion, any components which have eigenvalue greater than 1 will be retained and interpreted. Values which are less than 1 will be discarded. Here, Equation (2) was used to calculate the eigenvalue. In the equation, λ is commonly used to describe the multiplier in eigenvalue [5].

$$Eigenvalue = \begin{bmatrix} v1-\lambda & cov(x,y) \\ cov(x,y) & v2-\lambda \end{bmatrix} \quad (2)$$

where

$$cov(x,y) = \sum \frac{(x-\tilde{x})(y-\tilde{y})}{n-1}$$

$$x = \sum \frac{(x-\tilde{x})^2}{n-1} \qquad y = \sum \frac{(y-\tilde{y})^2}{n-1}$$

$$\tilde{x} = \sum \frac{x}{n} \tilde{y} = \sum \frac{y}{n}$$

The SCREE test is a method of displaying the eigenvalues on a graph for the purpose of finding a "break" between components that have large or small eigenvalue [6]. It retains the components that appear before the break and does not retain the one that appear after the break. The third criterion which is proportion variance, is used to keep components that contain 5% to 10% of the total variance. The proportion variance, $P_v$was calculated using Equation (4).

$$P_v = \frac{Eigenvalue\ for\ the\ component\ of\ interest}{Total\ eigenvalues\ of\ the\ correlation\ matrix} \quad (4)$$

where the total eigenvalues of the correlation matrix refers to the total number of variable analyzed [6].

III. RESULTS AND DISCUSSION

A. Eigenvalue-One-Criterion

Table 1 to Table 4 shows the eigenvalue for each honey sample. For all samples, coefficients 1 to 9 produce eigenvalue which is greater than 1, which indicates they are significant data.

Table 1 Eigenvalue of Honey A sample

Coefficient	Eigenvalue	Difference	Proportion	Cumulative
1	15.212 3	7.9625	0.3381	33.8096
2	7.2505	2.9940	0.1611	49.9233
3	4.2565	0.4918	0.0946	59.3831
4	3.7647	0.4126	0.0837	67.7499
5	3.3521	0.3673	0.0745	75.1998
6	2.9848	0.1025	0.0663	81.8333
7	2.8824	0.1446	0.0641	88.2392
8	2.7378	0.1836	0.0608	94.3236
9	2.5541	0	0.0568	100
10	0	0	0	100
Total	44.995 9	12.6588	1	750.4618

Table 2 Eigenvalue of Honey B sample

Coefficient	Eigenvalue	Difference	Proportion	Cumulative
1	6.7829	4.1416	0.3446	34.4607
2	2.6413	0.7326	0.1342	47.8801
3	1.9087	0.1714	0.0969	57.5774
4	1.7373	0.0941	0.0883	66.4037
5	1.6432	0.1345	0.0835	74.7519
6	1.5087	0.2199	0.0767	82.4170
7	1.2889	0.1299	0.0655	88.9652
8	1.1590	0.1460	0.0589	94.8534
9	1.0130	0	0.0515	100
10	0	0	0	100
Total	19.6830	5.7699	1	747.3094

Table 3 Eigenvalue of Honey C sample

Coeffcient	Eigenvalue	Difference	Proportion	Cumulative
1	92.8266	83.1552	0.7415	74.1537
2	9.6713	5.3474	0.0773	81.8796
3	4.3240	0.2168	0.0345	85.3337
4	4.1072	0.6031	0.0328	88.6147
5	3.5041	0.2896	0.0280	91.4140
6	3.2145	0.3025	0.0257	93.9819
7	2.9120	0.4190	0.0232	96.3081
8	2.4930	0.3646	0.0199	98.2997
9	2.1285	0	0.0170	100
10	0	0	0	100
Total	125.1813	90.6981	1	909.9855

Table 4 Eigenvalue of Honey D sample

Coeficient	Eigenvalue	Difference	Proportion	Cumulative
1	7.43451	2.6434	0.2816	28.1597
2	4.7911	1.8970	0.1815	46.3070
3	2.8941	0.1533	0.1097	57.2689
4	2.7408	0.5815	0.1038	67.6501
5	2.1593	0.1012	0.0818	75.8288
6	2.0581	0.4534	0.0780	83.6241
7	1.6047	0.1604	0.0608	89.7021
8	1.4442	0.1697	0.0547	95.1725
9	1.2745	0	0.0483	100
10	0	0	0	100
Total	26.4012	6.1600	1	743.7130

B. SREE Test Criterion

Figure 2 shows the scree test plot of the four honey samples Honey A, Honey B, Honey C and Honey D. Based on the theory stated previously, significant components are those which have high eigenvalue and large gap between each other. It is found that for Honey A, Honey B and Honey D honey samples, coefficients 1 to 6 are significant whereas for Honey C, only coefficients 1 and 2 are significant.

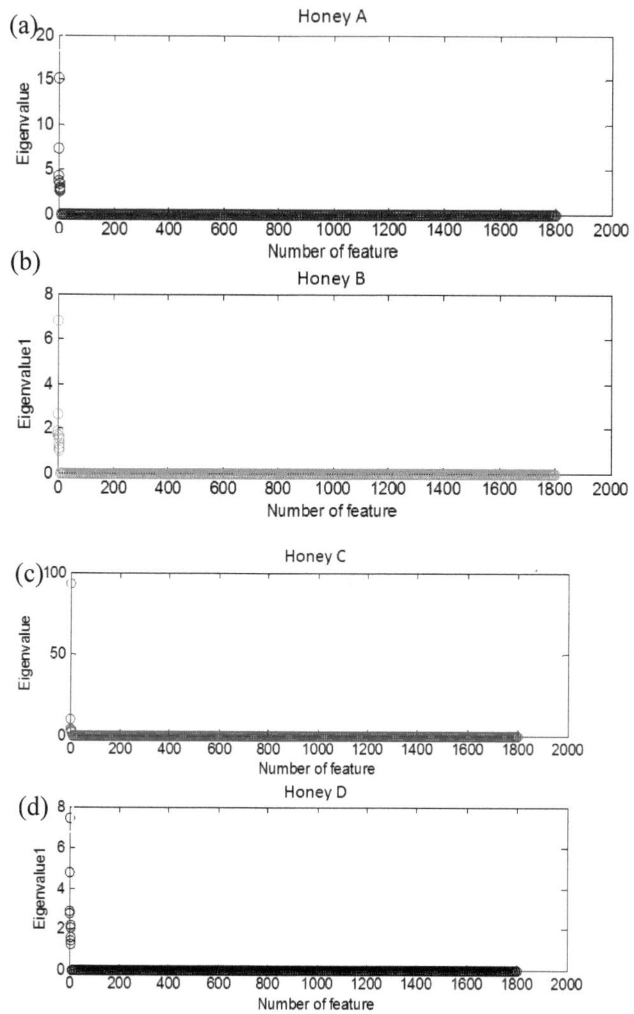

Fig. 2 Scree test plot of (a) Honey A (b) Honey B (c) Honey C (d) Honey D

C. Proportion Variance

As described previously, proportion variance retains significant components that contain 5% to 10% of the total variance. As shown in Table 1, significant components of Honey A are coefficient 1 to 9 since they have 100% of cumulative variance which is the same as Honey B samples. Honey C with coefficient 1 and 2 produces 82% of cumulative variance and Honey D sample produces 95% of cumulative variance from coefficient 1 and 2. Components other than the one that have variance between 5% and 10% are discarded.

D. Scatter Plot

Figure 3 shows cumulative variance obtained from coefficients 1 and 2 for all samples plotted on the same graph. These coefficients were chosen due to greater proportion in all honey samples compared to other coefficients.

From Figure 3, the spot clearly show that Honey A, Honey B and Honey D belong to the same group except Honey C. The feature extraction result obtained from this work will be used to classify pure honey and to find the relationship between these four samples.

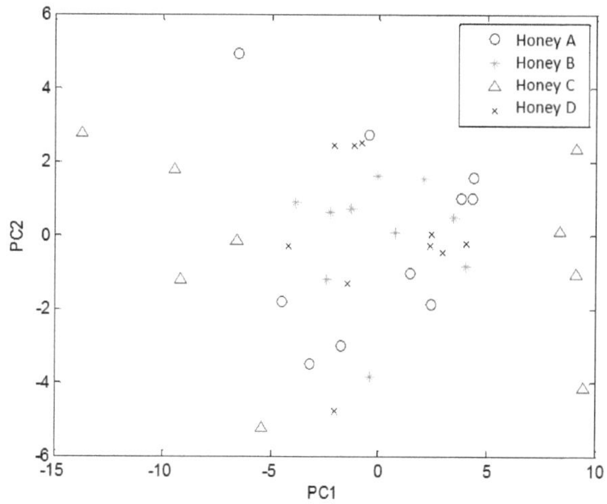

Fig. 3 Scatter plot of coefficients 1 and 2 cumulative variance of all samples

IV. CONCLUSIONS

The feature extraction of four honey samples using PCA has been described in this paper. Using PCA, significant components in honey sample could be extracted. For all honey samples, different PCA criteria reveal different sets of coefficient as significant components. Overall, Honey C and Honey D samples have less significant components compared to Honey A and Honey B samples. Scatter plot shows that Honey C honey is not in the same group as the other three honey samples. The features of honey extracted from PCA can be used as input to a classifier to identify pure honey.

ACKNOWLEDGMENT

The author would like to thank the Ministry of Education (MOE), Malaysia, for providing the research funding 600-RMI/ERGS 5/3 (20/2013); the Research Management Institute, Universiti Teknologi MARA, Malaysia, for the Research Intensive Fund 600-RMI/DANA 5/3/RIF (40/2012), 600-RMI/DANA 5/3/RIF (79/2012) and 600-RMI/DANA 5/3/RIF (373/2012); the Faculty of Electrical Engineering, Universiti Teknologi MARA, Malaysia, for the support and assistance given to the authors in carrying out this research.

CONFLICT OF INTEREST

The authors declare that they have no conflict of interest.

REFERENCES

1. S. Sivakesava and J. Irudayaraj, "Detection of inverted beet sugar adulteration of honey by FTIR spectroscopy," *Journal of the Science of Food and Agriculture,* vol. 81, pp. 683-690, 2001.
2. K. Kneipp, H. Kneipp, I. Itzkan, R. Dasari, and M. Feld, "Surface-enhanced Raman scattering: A new tool for biomedical spectroscopy," *Curr. Sci,* vol. 77, pp. 915-924, 1999.
3. J. D. Driskell, K. M. Kwarta, R. J. Lipert, M. D. Porter, J. D. Neill, and J. F. Ridpath, "Low-level detection of viral pathogens by a surface-enhanced Raman scattering based immunoassay," *Analytical chemistry,* vol. 77, pp. 6147-6154, 2005.
4. J. Shlens, "A Tutorial on Principal Component Analysis," Center for Neural Science, New York University New York CityApril 22, 2009.
5. A. R. M. Radzol, Y. K. Lee, W. Mansor, and F. M. T. Tawi, "Automated Extraction Of Principal Components Of Non-Structural Protein 1 From Sers Spectrum."
6. K. Labib and V. R. Vemuri, "An application of principal component analysis to the detection and visualization of computer network attacks," in *Annales des télécommunications*, 2006, pp. 218-234.
7. M. Paradkar and J. Irudayaraj, "Discrimination and classification of beet and cane inverts in honey by FT-Raman spectroscopy," *Food Chemistry,* vol. 76, pp. 231-239, 2002.
8. M. J. Goetz Jr, G. L. Cote, and M. Motamedi, "Detection of glucose using Raman spectroscopy," in *Engineering in Medicine and Biology Society, 1994. Engineering Advances: New Opportunities for Biomedical Engineers, Proceedings of the 16th Annual International Conference of the IEEE*, 1994, pp. 816-817.
9. "Carbohydrates and the Sweetness of Honey," National Honey Board2001.

Author:	Wahidah Mansor
Institute:	Universiti Teknologi MARA
City:	Shah Alam 40450
Country:	Malaysia
Email:	wahidah231@salam.uitm.edu.my

Attenuation of 3-Dimensional Epicardial Strain from Cardiac Magnetic Resonance Associated with Obstructive Hypertrophic Cardiomyopathy

X.D. Zhao[1], R.S. Tan[1,2], H.C. Tang[1,2], S. Leng[1], J.-M. Zhang[1,2], B.Y. Su[1], and L. Zhong[1,2]

[1] National Heart Centre Singapore, Singapore
[2] Duke-NUS Medical Graduate School, Singapore

Abstract— **Hypertrophic cardiomyopathy (HCM) is a genetic disease that leads to left ventricle (LV) hypertrophy with or without the presence of LV outflow tract obstruction. The aim of this study was to assess endocardial and epicardial strains based on cardiac magnetic resonance (CMR) images for control subjects and patients with and without obstruction. CMR scans were performed for 19 control subjects and 19 HCM patients. Endocardial and epicardial contours were delineated in cine two-, three-, and four-chamber MR images by CMRtools. Endocardial (resp. epicardial) length at end-diastole $L_{ED,endo}$ (resp. $L_{ED,epi}$) and end-systole $L_{ES,endo}$ (resp. $L_{ES,epi}$) were calculated and endocardial and epicardial strains were defined as $S_{endo} = |\ln(L_{ES,endo}/L_{ED,endo})| \times 100\%$ and $S_{epi} = |\ln(L_{ES,epi}/L_{ED,epi})| \times 100\%$, respectively. S_{endo} and S_{endo} were significantly different between control and HCM patients (both $P < 0.05$) except three-chamber S_{endo}, and they were significantly smaller for obstructive HCM patients compared with control. Moreover, ROC analysis found that 3-dimensional average epicardial strain > 17.2% had the best sensitivity 94.7% and 94.7% accuracy to differentiate HCM from normal subjects (AUC = 0.972), and 3-dimensional average endocardial strain > 18.8% had the best sensitivity 76.9% and 100% accuracy to differentiate HCM with and without obstruction (AUC = 0.859). Therefore, 3-dimensional endocardial and epicardial strains may be useful in triage and management of HCM patients with and without obstruction.**

Keywords— **Ventricular strain, hypertrophic cardiomyopathy, magnetic resonance imaging**

I. Introduction

Hypertrophic cardiomyopathy (HCM) is the most genetic cardiovascular disease [1, 2], and the estimated prevalence in the general adult population with phenotypic evidence of HCM is 1:500 [3,4]. Sudden death is the most common cause of death in HCM and occurs more often in young asymptomatic or only mildly symptomatic patients. HCM is a genetic disease that leads to hypertrophy of the left ventricle (LV) with or without the presence of LV outflow tract (LVOT) obstruction. The clinical presentation as well as degree of obstruction may vary depending on the location and extent of hypertrophy within the heart as well as the presence of other abnormalities such as mitral regurgitation, myocardial ischemia, and diastolic dysfunction.

The presence of systolic anterior motion (SAM) of the mitral valve may lead to dynamic LVOT obstruction due to impingement of the mitral valve leaflets upon the hypertrophied basal septum. In all patients with hypertrophic cardiomyopathy, although altered diastolic filling is evident, it is the high contraction load imposed by the obstruction that significantly worsens ventricular filling and relaxation [5]. Moreover, obstruction is associated with distortion of the mitral valve apparatus, and this results in secondary mitral regurgitation, and further elevates left atrial pressure, and substantially contributes to severe symptoms of dyspnea [6].

Doppler echocardiographic findings of a late-peaking, high-velocity signal from the LVOT have been considered to be an accurate method to measure the LVOT severity. However, it requires skilled expertise to measure the outflow tract obstruction. Cardiac magnetic resonance (CMR) imaging is the best imaging choice for assessing the LV morphology and function, such as regional curvedness, wall stress and area strain [7–10]. It has proven to be an important tool for the evaluation of patients suspected of having HCM, because it can readily diagnose those with phenotypic expression of the disorder and can potentially identify the subset of patients at risk of sudden cardiac death [1]. This study aims to develop an easy and useful marker associated with HCM patients with obstruction based on CMR images in two-, three-, and four-chamber views.

II. Methods

A. Subjects

In this study, 19 control subjects and 19 HCM patients (13 patients without obstruction and 6 patients with obstruction) were prospectively enrolled at the Department of Cardiology at the National Heart Centre Singapore. The study protocol was approved by the Institutional Review Board. All subjects

F. Ibrahim et al. (eds.), *International Conference for Innovation in Biomedical Engineering and Life Sciences*,
IFMBE Proceedings 56,
DOI: 10.1007/978-981-10-0266-3_60

were recruited without consideration of gender or ethnicity, and gave informed consent. None of control subjects had any of the following: 1) significant valvular or congenital cardiac disease, 2) history of myocardial infarction, 3) coronary artery lesions, or 4) abnormal LV pressure, end-diastolic volume or ejection fraction. Demographics data were collected at the time of scan.

B. MRI scans and data processing

All subjects underwent MRI scanning, which was performed by using steady-state free precession (SSFP) cine gradient echo sequences. The subjects were imaged on a 1.5T Siemens scanner (Avanto, Siemens Medical Solutions, Erlangen, Germany). Some preliminary short-axis acquisitions were used to locate the plane passing through the mitral and aortic valves. The following planes (ventricular two-, three- and four- chamber and short-axis planes with 12-14 equidistant slices covering both ventricles) were acquired. Each slice was acquired in a single breath hold, with 25 temporal phases per heart cycle. Fig. 1 depicted the sampled segmented trueFISP two-dimensional cine MR images of two- (first column), three- (second column) and four-chamber (last column) at end-diastole (ED) phase in a 50 years old healthy male subject (top row), 65 years old female HCM patient without obstruction (middle row) and a 48 years old male HCM patient with obstruction.

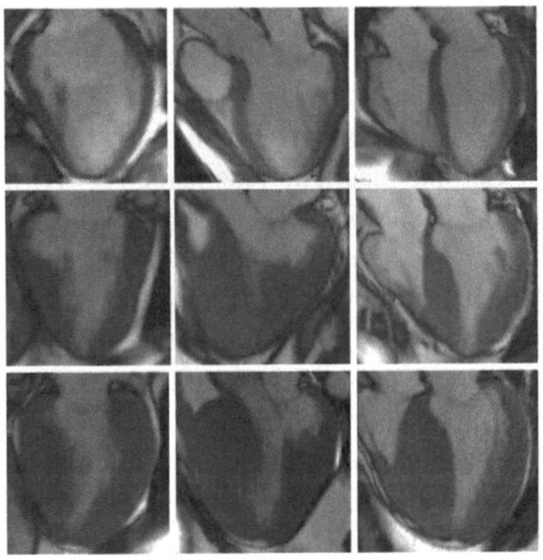

Figure 1: Sampled segmented trueFISP two-dimensional cine MR images cine MR images of two- (first column), three- (second column) and four-chamber (last column) at end-diastole (ED) phase in a 50 years old healthy male subject (top row), 65 years old female HCM patient without obstruction (middle row) and a 48 years old male HCM patient with obstruction.

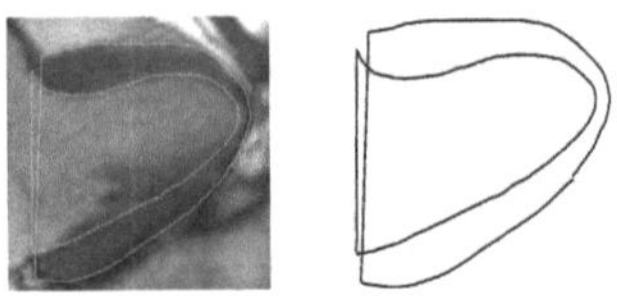

Figure 2: Left: segmented endocardium and epicardium of two-dimensional cine MR image at end-diastolic phase; Right: discrete contours of endocardium and epicardium output from MATLAB.

The MRI data were processed by using CMRtools (CVIS, London, UK). We observed that CMRtools would automatically use a straight line to connect the starting and ending points, if the delineated curve was not closed. Therefore, in order to properly compare the endocardium and epicardium surfaces, we delineated their contours separately, as shown on the left of Fig. 2 for two-dimensional cine three-chamber view, and the corresponding contour output in MATLAB was given on the right. The endocardial length was the contour path from one valve to the other valve, similar for the epicardial length.

C. Endocardial and epicardial strains

We defined the endocardial and epicardial strains as follows:

$$S_{\text{endo}} = \left|\ln\left(\frac{L_{\text{ES,endo}}}{L_{\text{ED,endo}}}\right)\right| \times 100\%; \; S_{\text{epi}} = \left|\ln\left(\frac{L_{\text{ES,epi}}}{L_{\text{ED,epi}}}\right)\right| \times 100\%,$$

where $L_{\text{ED,endo}}$ (resp. $L_{\text{ED,epi}}$) was the length of endocardium (resp. epicardium) contour from one valve to the other valve at ED phase; and similarly for $L_{\text{ES,endo}}$ and $L_{\text{ES,epi}}$. This formula was applied to cine two-, three- and four-chamber, respectively.

D. Statistical analysis

All statistical analysis was performed using SPSS Software. All continuous variables were presented as mean ± standard deviation (SD), while categorical data were presented as frequencies in terms of percentage. Statistical significance of differences between three or more groups was investigated by one-way analysis of variance (ANOVA) with a post hoc Bonferroni correction. A value of $P < 0.05$ was considered statistically significant.

III. Results and discussion

The demographics of control and HCM patient with and without obstruction groups was tabulated in Table 1. Both HCM groups had normal diastolic and systolic blood pressure, but LV end-diastolic volume (LVEDV) index became higher in patients with obstruction, and LV end-systolic volume (LVESV) index was smaller in patients without obstruction, although not significantly. Furthermore, HCM patients with obstruction had significantly higher LV mass index than control group and patients without obstruction ($P < 0.05$).

A. Comparison of length and strain

Endocardial and epicardial lengths and strains of two-, three-, and four-chamber at ED and ES phases for 19 normal subjects and two HCM patients groups were calculated, and tabulated in Table 2, together with 3-dimensional lengths and strains (average of two-, three-, and four-chamber). We can see that HCM patients with obstruction had longer endocardium and epicardium at ES than normal subjects in three-, and four-chamber views ($P < 0.05$), and four-chamber ES epicardium was significantly longer than HCM patients without obstruction. For endocardial and epicardial strains, control group was significant higher than both HCM patient groups ($P < 0.05$), and global endocardial strain (average endocardial strain in two-, three- and four-chamber views) for patients with obstruction was significantly lower than control and patients without obstruction.

Table 1: Demographics of control subjects and HCM patients with and without obstruction.

Variables	Control (n =19)	HCM without obstruction (n = 13)	HCM with obstruction (n = 6)	P value
Age, yr	53 ± 11	52 ± 13	47 ± 15	0.582
Gender, Male/Female	12/7	3/10	4/2	0.056
Weight, kg	67 ± 14	62 ± 14#	84 ± 24#	0.026
Height, cm	163 ± 11	160 ± 11	167 ± 13	0.404
Body surface area, m^2	1.72 ± 0.24	1.66 ± 0.22	1.97 ± 0.35	0.060
Diastolic blood pressure, mmHg	74 ± 8	71 ± 14	76 ± 10	0.572
Systolic blood pressure, mmHg	129 ± 17	133 ± 20	130 ± 28	0.862
LVEDV index, ml/m^2	73 ± 12	73 ± 12	86 ± 18	0.104
LVESV index, ml/m^2	24 ± 8	19 ± 7	29 ± 13	0.068
LV ejection fraction, %	67 ± 6	74 ± 8	67 ± 10	0.046
LV mass index, g/m^2	54 ± 10	88 ± 23*#	129 ± 64*#	<0.001

*Note: Data are expressed as mean ± SD or as number (percentage). HCM = hypertrophic cardiomyopathy; LVEDV = left ventricle end diastolic volume; LVESV = left ventricle end systolic volume; LV = left ventricle. * there is a significant difference between control group and HCM patients with and without obstruction; # there is a significant difference between HCM patients with and without obstruction.*

B. Differentiating HCM vs Normal

LV ejection fraction was one of the indices to differentiate patients from normal. In Fig. 3(a), we plotted receiver

Table 2: Length and strains for controls and HCM patients with and without obstruction.

Variables	Control (n =19)	HCM without obstruction (n = 13)	HCM with obstruction (n = 6)	P value
Two-chamber				
ED endocardium length, mm	132.1 ± 37.3	119.2 ± 12.0	130.5 ± 23.9	0.459
ED epicardium length, mm	139.6 ± 38.0	126.1 ± 12.9	143.3 ± 26.8	0.373
ES endocardium length, mm	101.4 ± 29.4	97.1 ± 12.6	110.5 ± 22.5	0.529
ES epicardium length, mm	113.3 ± 32.1	110.6 ± 11.5	129.3 ± 26.4	0.332
S_{endo}_2ch, %	26.6 ± 5.3	20.8 ± 5.3*	16.9 ± 3.7*	<0.001
S_{epi}_2ch, %	21.2 ± 4.0	13.2 ± 3.4*	10.5 ± 4.2*	<0.001
Three-chamber				
ED endocardium length, mm	121.4 ± 16.2	121.7 ± 11.7	133.2 ± 17.9	0.234
ED epicardium length, mm	127.1 ± 17.6	126.4 ± 13.6	140.6 ± 25.8	0.233
ES endocardium length, mm	97.8 ± 14.0	101.2 ± 10.1	105.5 ± 17.7*	0.028
ES epicardium length, mm	104.1 ± 13.5	110.8 ± 13.2	126.7 ± 28.3*	0.020
S_{endo}_3ch, %	21.7 ± 4.4	18.5 ± 4.9	14.5 ± 4.2*	0.005
S_{epi}_3ch, %	19.9 ± 5.4	13.2 ± 3.3*	11.1 ± 4.5*	<0.001
Four-chamber				
ED endocardium length, mm	123.8 ± 15.3	119.5 ± 11.1	134.1 ± 16.3	0.127
ED epicardium length, mm	129.9 ± 16.9	127.5 ± 11.2	146.0 ± 24.1	0. 077
ES endocardium length, mm	95.5 ± 13.3	99.0 ± 9.1	116.1 ± 19.4*	0. 007
ES epicardium length, mm	103.5 ± 14.0	111.6 ± 12.9#	130.9 ± 27.0*#	0.004
S_{endo}_4ch, %	26.2± 4.4	19.8 ± 4.7*	15.0 ± 7.6*	<0.001
S_{epi}_4ch, %	22.7 ± 3.0	13.6 ± 5.8*	11.6 ± 5.4*	<0.001
Mean				
ED endocardium length, mm	125.8 ± 16.4	120.2 ± 10.7	132.61 ± 18.8	0.246
ED epicardium length, mm	132.2 ± 17.5	126.7 ± 11.6	143.27 ± 24.8	0. 158
ES endocardium length, mm	98.2 ± 13.5	98.8 ± 9.3	114.0 ± 19.0*	0. 040
ES epicardium length, mm	107.0 ± 14.7	111.0 ± 11.8	129.0 ± 26.6*	0.022
Global S_{endo}, %	24.8 ± 3.0	19.7 ± 2.7*#	15.5 ± 4.4*#	<0.001
Global S_{epi}, %	21.3 ± 3.0	13.3 ± 3.7*	11.1 ± 4.1*	<0.001

Data are expressed as mean ± SD. HCM = hypertrophic cardiomyopathy; ED = end-diastole; ES = end-systole; S_{endo} *= strain of endocardium and* S_{epi} *= strain of epicardium. * there is a significant difference between control group and HCM patients with and without obstruction; # there is a significant difference between HCM patients with and without obstruction.*

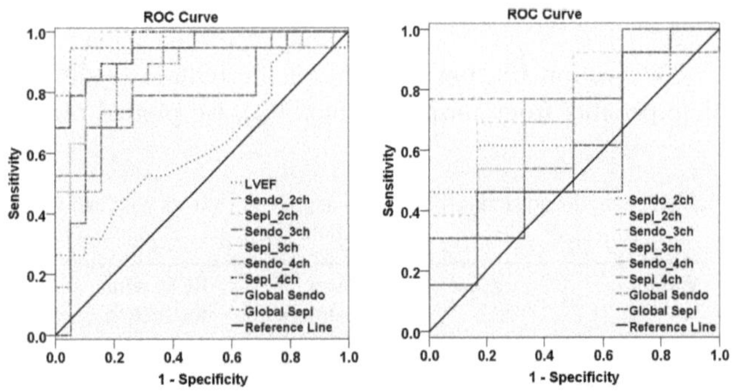

(a) Control and HCM groups (b) HCM and HOCM

Figure 3: (a): Receiver operating characteristic (ROC) curves for LVEF, Sendo_2ch, Sepi_2ch, Sendo_3ch, Sepi_3ch, Sendo_4ch, Sepi_4ch, mean of endocardium strain (Global Sendo) and mean of epicardium strain (Global Sepi). The areas under ROC curve are 0.623, 0.834, 0.975, 0.773, 0.898, 0.859, 0.953, 0.939 and 0.972 for each of these parameters, respectively. (b): ROC curves for Sendo_2ch, Sepi_2ch, Sendo_3ch, Sepi_3ch, Sendo_4ch, Sepi_4ch, Global Sendo and Global Sepi. The areas under ROC curve are 0.718, 0.679, 0.769, 0.590, 0.718, 0.628, 0.859 and 0.603 for each of these parameters, respectively. HCM = hypertrophy cardiomyopathy; HOCM = hypertrophic cardiomyopathy with obstruction.

operating characteristic (ROC) curves for LVEF, Sendo_2ch, Sepi_2ch, Sendo_3ch, Sepi_3ch, Sendo_4ch, Sepi_4ch, mean of endocardial strain (Global Sendo) and mean of epicardial strain (Global Sepi). The areas under ROC curve were 0.623, 0.834, 0.975, 0.773, 0.898, 0.859, 0.953, 0.939 and 0.972 for each of these parameters, respectively. Global Sepi > 17.15% had the best sensitivity 94.7% and 94.7% accuracy to differentiate HCM from normal subjects (AUC = 0.972).

C. Differentiating HCM with and without obstruction

Receiver operating characteristic (ROC) curves for Sendo_2ch, Sepi_2ch, Sendo_3ch, Sepi_3ch, Sendo_4ch, Sepi_4ch, Global Sendo and Global Sepi were plotted in Fig. 3(b). The areas under ROC curve were 0.718, 0.679, 0.769, 0.590, 0.718, 0.628, 0.859 and 0.603 for each of these parameters, respectively. Global Sendo > 18.80% had the best sensitivity 76.9% and 100% accuracy to differentiate HCM patients with and without obstruction (AUC = 0.859).

IV. Conclusion

In this study, we found that HCM patients with obstruction had significantly lower endocardial and epicardial strains compared with control subjects. 3-dimensional epicardial strain had the best sensitivity 94.7% and 94.7% accuracy to differentiate HCM from normal subjects, and 3-dimensional endocardial strain had the best sensitivity 76.9% and 100% accuracy to differentiate HCM with and without obstruction.

Conflict of Interest

The authors declare that they have no conflict of interest.

Acknowledgment

This research is supported by the SingHealth Foundation (SHF/FG453P/2011), the Goh Cardiovascular Research Grant (Duke-NUS-GCR/2013/0009), and the National Medical Research Council, Singapore, under its Cooperative Basic Research Grant (NMRC/EDG/1037/2011), Singapore-China Joint Research Programme (14/1/32/24/002).

References

1. Hansen MW, Merchant N. MRI of hypertrophic cardiomyopathy: part I, MRI appearances. American Journal of Roentgenology 2007; 189: 1335-1343.
2. Hansen MW, Merchant N. MRI of hypertrophic cardiomyopathy: part II, differential diagnosis, risk stratification, and posttreatment MRI appearances. American Journal of Roentgenology 2007; 189: 1344-1352.
3. Kovacic JC, Muller D. Hypertrophic cardiomyopathy: state-of-the-art review, with focus on the management of outflow obstruction. Internal Medicine Journal 2003; 33: 521-529.
4. Elliott P, McKenna WJ. Hypertrophic cardiomyopathy. Lancet 2004; 363: 1881-1891.
5. Nishimura RA, Ommen SR. Hypertrophic cardiomyopathy: the search for obstruction. Circulation 2006; 114: 2200-2.
6. Maron MS, Olivotto I, Zenovich AG, etc. Hypertrophic cardiomyopathy is predominantly a disease of left ventricular outflow tract obstruction. Circulation 2006; 114: 2232-2239.
7. Zhong L, Su Y, Yeo SY, Tan RS, Ghista DN, Kassab G. Left ventricular regional wall curvature and wall stress in patients with ischemic dilated cardiomyopathy. Am J Physiol Heart Circ Physiol 3: H573-H584, 2009.
8. Zhong L, Su Y, Gobeawan L, Sola S, Tan RS, Navia JL, Ghista DN, Chua T, Guccione J, Kassab GS. Impact of surgical ventricular restoration on ventricular shape, wall stress and function in heart failure patients. Am J Physiol Heart Circ Physiol 5: H1653-H1660, 2011.
9. Zhong L, Gobeawan L, Su Y, Tan JL, Ghista D, Chua T, Tan RS, Kassab G. Right ventricular regional wall curvedness and area strain in patients with repaired tetralogy of Fallot. Am J Physiol Heart Circ Physiol 302: H1306-16, 2012.
10. Teo SK, Vos FJA, Tan RS, Zhong L, Su Y. Regional ejection fraction and regional area strain for left ventricular function assessment in male patients after first-time myocardial infarction. J R Soc Interface 12: 20150006, 2015.

Author: Liang Zhong
Institute: National Heart Centre Singapore
Street: 5 Hospital Drive, 169609
City: Singapore
Country: Singapore
Email: zhong.liang@nhcs.com.sg

Mg-1.0Zn-xCa-Based Biodegradable Cardiovascular Stent Alloy: A Microstuctural and Textural Studies

K.D. Permana[1,2], B. Ariwahjoedi[2,3], and A.S. Shuib[4]

[1] Chemical Engineering Department, Faculty of Engineering, Universiti Teknologi PETRONAS, Tronoh, Malaysia
[2] Centre for Imaging and Signal Research, Universiti Teknologi PETRONAS, Tronoh, Malaysia
[3] Fundamental and Applied Science Department, Faculty of Science and Technology, Universiti Teknologi PETRONAS, Tronoh, Malaysia
[4] Research and Consultation Department, Iqurah Sdn. Bhd., Ipoh, Malaysia

Abstract— **Magnesium (Mg) alloys have been used for stent application since decade ago due to its biocompatibility. However, in chloride-abundant environment like human body, the degradation is considered too rapid. Alloying is one of options to reduce its rapid corrosion rate. Zinc (Zn) and Calcium (Ca) were chosen as alloying elements regarding their biocompatibility to human body, reduction in the corrosion rate and improvement on the grain size. The microstructures of three alloys Mg-1.0Zn-xCa (x=0.0, 1.0, 2.0wt%) were studied by field-emission scanning electron microscopy (FESEM). Surface morphological analyses exhibited that the presence of Ca led to the uniformity and the formation second phase, reduction of grain size and the boundary more visible.**

Keywords— **microstructure, morphology, stent, magnesium stent, cardiovascular disease.**

I. INTRODUCTION

Cardiovascular disease (CVD) has been marked as first-rank killer in the world. The cause of CVD is blockage in the lumen on blood vessels by fat deposition. Depending on the stage of blockage, there are many kind of treatments for CVD. At early stage, changing lifestyle would be sufficient, but at final stage, the vessels have a risk to rupture. Stent implantation is one of the treatments [1–3].

Stent is a tubular shape device inserted by surgeon using balloon catheter to open up and hold up the vessel. An implanted stent is required to last for 6-12 months to enable vessel healing [3]. After that time, the presence of a stent in the blood vessel does not provide any more benefit. Another surgery is required to remove permanent metallic stents after healing, resulting in additional risk to the patient and incurring additional cost. The implanted stent could cause long term vessel wall dysfunction, blood clotting and an inability to adapt to vessel growth, necessitating further surgery [4].

These considerations led to the interest in the development of biodegradable metallic stents. Biodegradable stents can maintain the integrity of the artery wall during the healing process and then dissolve in the body fluid without any surgery is required after the artery wall heals [5]. These stents provide temporary vessel opening until the vessel remodeling is complete, the stent in turn will dissolve and disappear. Biodegradable stents have mechanical strengths similar to those of permanent implants, making them able to maintain the integrity of the artery wall during the healing process [4–8].

Magnesium (Mg) alloys have good biocompatibility and biodegradability in physiological environment. Therefore, Mg alloys have been used widely as implant, especially in bone and stent application. Furthermore, the alloys also exhibit good tensile strength and Young's modulus such that it could be used to open up weakened and narrowed arteries. Based on their mechanical strengths, Mg are suitable for stent application. However, rapid degradation of most Mg alloys limits its capability and function [1], [9–10] (Shaw, 2003).

Zinc (Zn) has been used in Mg-xZn system. This alloying element would improve the corrosion resistance of Mg. The maximum improvement is the alloy with 1.0 wt% Zn content. Further addition of Zn will decrease the corrosion resistance [11–13]

On the other hand, calcium (Ca) has no effect directly on corrosion resistance. But the presence of Ca would improve grain size refinement. The Mg alloy becomes more compact and it would reduce the access of water in contact with Mg [14].

Three essential elements could be used together as an alloy; Mg used as based metal and Ca and Zn as alloying elements. Mg-Ca-Zn is assumed to having great potential applicability as cardiovascular stent material. It has been tested as bone implant materials in certain Mg-Zn-Ca compositions, but application for stent has not been pursued yet due to vast possible compositional ranges [15].

II. MATERIALS AND METHODS

A. Materials and Processing

The studied alloys were Mg–1.0Zn–0.0Ca, Mg–1.0Zn–1.0Ca, and Mg–1.0Zn–2.0Ca (in wt%) alloys as prepared by using high purity 99.99 wt% Mg, high purity 99.99 wt% Zn,

F. Ibrahim et al. (eds.), *International Conference for Innovation in Biomedical Engineering and Life Sciences*,
IFMBE Proceedings 56,
DOI: 10.1007/978-981-10-0266-3_61

high purity 99.99 wt% Ca. The respective alloys were heated at 750°C under argon flow with dwelling time 2 hours inside a horizontal tube furnace.

B. Microstructural Characterization and Chemical Analyses

Three samples of the alloy, Mg-1.0Zn, Mg-1.0Zn-1.0Ca and Mg-1.0Zn-2.0Cawere ground with SiC emery paper up to 1200 grits and finely polished with 0.5 µm diamond paste for microstructure observation. Then they were etched by using acetic-picral as etching solution. The samples was analyzed by using Field-Emission Scanning Electron Microscope (FESEM) with energy dispersive x-ray spectroscopy (EDS) capability. The images were captured at 500x magnification and gathered to analyze in terms of their morphologies. Three spots area in Mg-1.0Zn, Mg-1.0Zn-1.0Ca, and Mg-1.0Zn-1.0Ca, respectively, was analyzed by EDS. These spots was represent three different are on the captured images.

III. RESULTS AND DISCUSSION

The microstructures of the Mg-1.0Zn consist of α-Mg phase. The image of Mg-1.0Zn shows no distinct grain boundaries in the alloy, as shown in Figure 1(a). Zhang [16], stated that maximum Zn solubility limit in Mg is 2%. Below 2%, Zn was solid solution in Mg matrix. It was confirmed by EDS results that highest content of Zn was in Mg matrix (grey area) as shown in Figure 2(a) and Table 1. Chemical composition in the dark area was oxide 13.4 wt% and Zn 0.4 wt%. Three spots in Figure 2 (a), (b), and (c) prove that Zn is distributed almost evenly on the matrix of Mg.

On the other hand, Mg-1.0Zn-1.0Ca image has distinctive grain boundary phase, as shown in Figure 1(b). Ca addition on the alloy makes the boundary phase more visible. It was confirm by EDS that on the boundary it has higher Ca content than in the inner grain. In Mg-1.0Zn-2.0Ca, the grain size are relatively smaller and the inter-dendrite phase more visible, as shown in Figure 1(c). In Mg-1.0Zn the second phase was in the form polygonal particles and then in the Mg-1.0Zn-1.0Ca it changed to smaller round particles. Finally in Mg-1.0Zn-2.0Ca, the second phase was appeared as lamellar structures.

IV. CONCLUSIONS

The presence of Ca on Mg-1.0Zn-xCa alloy led to the presence of the second phase. On the other hand, the Ca also induced the formation of more grain boundaries and refined the grain size. Grain refinement could leads to improvement in mechanical properties of the alloy. The effect of grain refinement to corrosion rate was not investigated. However, presumably refined microstructures could lower sub-lattice strain and might lead to corrosion rate decrement.

ACKNOWLEDGMENT

The work was supported by Short-Term Internal Research Fund (STIRF) – cost center 0153AA-C95, Department of Chemical Engineering, and Department of Fundamental and Applied Science, Universiti Teknologi PETRONAS.

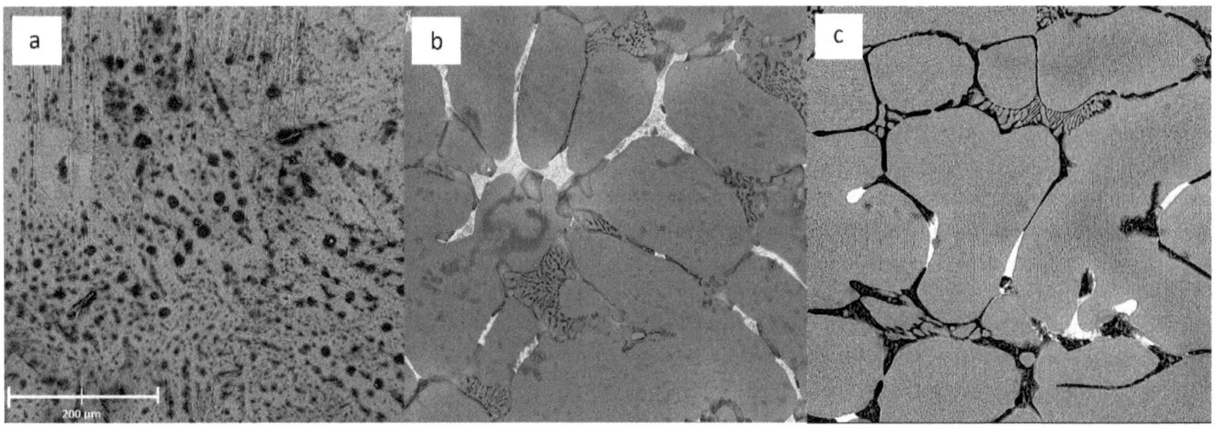

Fig. 1 FESEM image of (a) Mg-1.0Zn, (b) Mg-1.0Zn-1.0Ca, (c) Mg-1.0Zn-2.0Ca

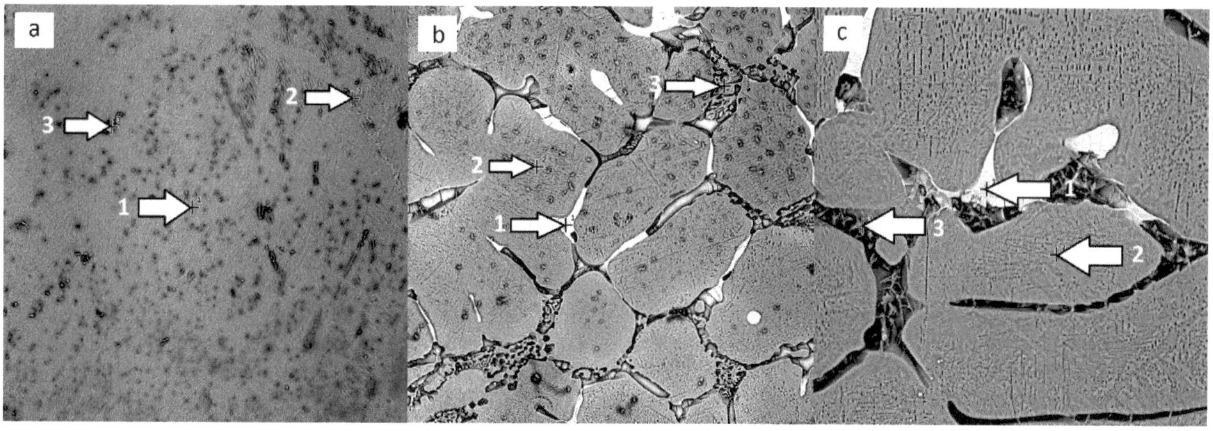

Fig. 2 Element Analyses of (a) Mg-1.0Zn, (b) Mg-1.0Zn-1.0Ca, (c) Mg-1.0Zn-2.0Ca

Table 1 Element Analyses of Mg-1.0Zn, Mg-1.0Zn-1.0Ca, and Mg-1.0Zn-2.0Ca

Elements	Symbol	Percentage wt%								
		[a]Mg-1.0Zn			[b]Mg-1.0Zn-1.0Ca			[c]Mg-1.0Zn-2.0Ca		
		1	2	3	1	2	3	1	2	3
Magnesium	Mg	79.5	74.8	81.3	65.0	84.3	96.6	64.9	93.6	78.2
Zinc	Zn	0.6	0.5	0.4	18.9	0.6	0.9	22.8	1.2	2.5
Calcium	Ca	0	0	0	4.5	0.2	1.3	4.5	0.4	8.1
Oxygen	O	18.6	21.9	13.4	11.5	13.8	0	7.7	3.9	11.3

[a, b, c] Each composition associated with the numbered spot on Figure 2.

CONFLICT OF INTEREST

The authors declare that they have no conflict of interest.

REFERENCES

[1] B. a Shaw, "Corrosion Resistance of Magnesium Alloys," *ASM Handbook, Vol. 13A Corros. Fundam. Test. Prot.*, vol. 13, pp. 692–696, 2003.

[2] H. Hermawan, "Biodegradable Metals: State of the art," *Biodegrad. Met.*, pp. 13–23, 2012.

[3] H. Hermawan, D. Dubé, and D. Mantovani, "Developments in metallic biodegradable stents," *Acta Biomater.*, vol. 6, no. 5, pp. 1693–1697, 2010.

[4] M. Haude, R. Erbel, P. Erne, S. Verheye, H. Degen, D. Böse, P. Vermeersch, I. Wijnbergen, N. Weissman, F. Prati, R. Waksman, and J. Koolen, "Safety and performance of the drug-eluting absorbable metal scaffold (DREAMS) in patients with de-novo coronary lesions: 12 month results of the prospective, multicentre, first-in-man BIOSOLVE-I trial," *Lancet*, vol. 381, no. 9869, pp. 836–844, 2013.

[5] T. S. N. Sankara Narayanan, I. S. Park, and M. H. Lee, "Strategies to improve the corrosion resistance of microarc oxidation (MAO) coated magnesium alloys for degradable implants: Prospects and challenges," *Prog. Mater. Sci.*, vol. 60, no. August 2013, pp. 1–71, 2014.

[6] F. Witte, N. Hort, C. Vogt, S. Cohen, K. U. Kainer, R. Willumeit, and F. Feyerabend, "Degradable biomaterials based on magnesium corrosion," *Curr. Opin. Solid State Mater. Sci.*, vol. 12, no. 5–6, pp. 63–72, 2008.

[7] F. Witte, J. Fischer, J. Nellesen, H. A. Crostack, V. Kaese, A. Pisch, F. Beckmann, and H. Windhagen, "In vitro and in vivo corrosion measurements of magnesium alloys," *Biomaterials*, vol. 27, pp. 1013–1018, 2006.

[8] J. E. Moore, J. S. Soares, and K. R. Rajagopal, "Biodegradable Stents: Biomechanical Modeling Challenges and Opportunities," *Cardiovasc. Eng. Technol.*, vol. 1, no. 1, pp. 52–65, 2010.

[9] N. Hort, Y. Huang, D. Fechner, M. Störmer, C. Blawert, F. Witte, C. Vogt, H. Drücker, R. Willumeit, K. U. Kainer, and F. Feyerabend, "Magnesium alloys as implant materials-Principles of property design for Mg-RE alloys," *Acta Biomater.*, vol. 6, no. 5, pp. 1714–1725, 2010.

[10] J. Kuhlmann, I. Bartsch, E. Willbold, S. Schuchardt, O. Holz, N. Hort, D. Höche, W. R. Heineman, and F. Witte, "Fast escape of hydrogen from gas cavities around corroding magnesium implants," *Acta Biomater.*, vol. 9, pp. 8714–8721, 2013.

[11] X. N. Gu and Y. F. Zheng, "A review on magnesium alloys as biodegradable materials," *Front. Mater. Sci. China*, vol. 4, no. 2, pp. 111–115, 2010.

[12] G. Wang, S. Ge, Y. Shen, H. Wang, Q. Dong, Q. Zhang, J. Gao, and Y. Wang, "Study on the biodegradability and biocompatibility of WE magnesium alloys," *Mater. Sci. Eng. C*, vol. 32, pp. 2190–2198, 2012.

[13] Y. Sun, B. Zhang, Y. Wang, L. Geng, and X. Jiao, "Preparation and characterization of a new biomedical Mg-Zn-Ca alloy," *Mater. Des.*, vol. 34, pp. 58–64, 2012.

[14] H. R. B. Rad, M. H. Idris, M. R. A. Kadir, and S. Farahany, "Microstructure analysis and corrosion behavior of biodegradable Mg-Ca implant alloys," *Mater. Des.*, vol. 33, pp. 88–97, 2012.

[15] K. D. Permana, A. S. Shuib, and B. Ariwahjoedi, "A Review of Magnesium Alloys for Use in Biodegradable Cardiovascular Stents," *World Appl. Sci. J.*, vol. 30, pp. 375–381, 2014.

[16] B. P. Zhang, Y. Wang, and L. Geng, "Research on Mg-Zn-Ca alloy as degradable biomaterial.," *Biomater. - Phys. Chem.*, pp. 184–204, 2011.

Author:	Kusniar Deny Permana
Institute:	Universiti Teknologi PETRONAS
Street:	Bandar Seri Iskandar
City:	Tronoh
Country:	Malaysia
Email:	kusniardenypermana@gmail.com, kusniar_g02061@utp.edu.my

Virtual Trial Protocol Analysis of Nursing Workload Intensity within ICU

N.N. Razak[1], A.A. Razak[1], C.G. Pretty[2], N.H Ahamad[1], F.M. Suhaimi[3], and U. Jamaluddin[4]

[1] College of Engineering, Universiti Tenaga Nasional, Kajang, Malaysia
[2] Centre of Bioengineering, University of Canterbury, Christchurch, New Zealand
[3] Advanced Medical and Dental Institute, Universiti Sains Malaysia, Penang, Malaysia
[4] Faculty of Mechanical Engineering, Universiti Malaysia Pahang, Pahang, Malaysia

Abstract— **Currently, effective glycaemic control protocols consume significant nursing time, which may be unsustainable as the number of patients requiring control increases with increasing rates of diabetes. This paper investigates the safety and efficacy of basal insulin therapy as a means to reduce nurse workload associated with glycaemic control in intensive care patients with stress hyperglycaemia. Validated virtual trial simulations (N = 40 patients) of a successful glycaemic control protocol (SPRINT) using 1-2 hourly interventions and a modified version using 4 hour interventions augmented with basal insulin therapy using Glargine. An additional model was used to capture the kinetics of Glargine. Workload was assessed by counting the total number of interventions (BG measurements, changes to insulin and nutrition rates) per day. Glycaemic performance was assessed by time in the target band (4.4-7.0 mmol/L) and number of severe hypoglycaemic episodes (BG<2.2 mmol/L). Workload reduction is around 30% (p<0.001) due to basal insulin therapy. Glycaemic control performance was slightly reduced from 86% to 80% (p=0.006) time in the target band using basal insulin therapy and 4 hourly interventions. However, safety was maintained with 0 incidence hypoglycaemia. Basal insulin therapy enables glycaemic control protocols with reduced intervention frequency while maintaining performance and safety. Reduced intervention frequency directly translates into reduced nurse workload associated with glycaemic control.**

Keywords— **Nursing effort, Glycaemic control, Model-based Protocol, Glargine.**

I. INTRODUCTION

Stress-induced hyperglycaemia is relatively common in the critically ill and may occur in patients without any prior history of diabetes [1-3]. Studies have shown that controlling glycaemia to normal levels can reduce mortality and morbidity in the intensive care unit (ICU) [4-6]. With the prevalence of diabetes increasing rapidly [7], the clinical burden of implementing glycaemic control in the ICU could become unsustainable with current methods.

Successful tight glycaemic control (TGC) protocols for critical care have typically relied on average blood glucose (BG) measurement intervals of 1-2 hours, to avoid the risk of hypoglycaemia [6,8]. However, measurements and interventions at this frequency can consume significant nurse workload [9-11]. Gartemann et al. reported that TGC activities consumed 7.1% (42 mins) of nurse work time during a 12-hour shift [10]. With an increasing number of patients entering the ICU with impaired glucose tolerance, strategies to reduce the nurse workload associated with TGC are essential.

One potential method for reducing nurse workload associated with glycaemic control is to treat ICU patients with hyperglycaemia more like ambulatory diabetics and use basal insulin therapy. Our hypothesis is that with long-acting basal insulin, patient glycaemia will be more stable and require fewer changes to infused insulin rates and enteral/parenteral nutrition rate. The validity of this hypothesis can be effectively tested and potential protocols refined in simulation prior to a clinical pilot trial.

This paper presents an *in-silico* proof of concept study investigating the efficacy and safety of basal insulin therapy with Glargine for reducing nurse workload associated with glycaemic control. Simulations are conducted using a validated virtual trial method [12,13] coupled with a 4-compartment model of Glargine [14] to accurately capture the insulin kinetics. The paper-based SPRINT glycaemic control protocol [4] is tested, where 4-hourly interventions are used in conjunction with basal insulin therapy, rather than the standard 1-2 hours.

II. METHODOLOGY

The virtual trial method used in this study relies on a physiological model of the glucose-insulin system and real patient data. The model used for this study is that of Lin et al. [15]. For simulation of the behaviour of subcutaneous insulin, an additional model is required and the model of Wong et al. [14] is used.

Retrospective data from 40 patients in ICU from Intensive Care Unit, Christchurch Hospital totaling 8100 hours were used for this study. These patients all had more than 8 hours of 2U of insulin per hour and insulin sensitivity profiles with a low variability, to mimic the type of patient who might benefit from basal insulin therapy. Table 1 shows the median and interquartile range (IQR) of age, APACHE II score and length of stay (LOS). Patient cohorts were divided into three categories of LOS which were less than 5 days, 5 to 10 days and more than 10 days.

F. Ibrahim et al. (eds.), *International Conference for Innovation in Biomedical Engineering and Life Sciences*,
IFMBE Proceedings 56,
DOI: 10.1007/978-981-10-0266-3_62

Table 1 Patient Demographic

Demographic	Median [IQR]
Patient (n)	40
Age (years old)	59 [44 71]
Gender (female:male)	19:21
APACHE II score	19 [17 27]
Length of stay (LOS) in days	6 [4 11]

The patient's time-varying insulin sensitivity metric (*SI*) was fitted to the actual clinical data using an integral fitting method [16]. The resulting time varying *SI* profiles represent time-varying metabolic status for individual patients. Testing new interventions with this profile, in simulation, provides new outputs. Thus, the profile of *SI* can be used to create "virtual patients" for testing insulin protocols.

The modified protocol called 4-Hour protocol was simulated and compared with actual clinical data of patients receiving intensive insulin therapy under SPRINT protocol. The frequency of BG measurements, changes in feed rates and intravenous (IV) insulin boluses are governed by the SPRINT protocol. SPRINT requires current and previous blood glucose measurements, the amount of previous hour IV insulin bolus and nutrition given in the previous hour, all to determine nutrition and insulin bolus for the next interval. In 4-Hour protocol, instead of 2-hourly BG measurement when patient is stable, BG measurement frequency is reduced to 4-hourly. Patient is categorized as stable with 3 consecutive measurements within 4.0-6.1mmol/L. The recommended insulin bolus from SPRINT controller is also reduced by 1 unit. As SPRINT operates on the basis of estimating patient's apparent insulin sensitivity, the protocol is still applicable with a background infusion. Virtual trials are performed using updated SPRINT with daily dose of glargine.

In this newly simulated protocol, the dosing frequency of Glargine is once per 24 hours. The first dose is given at 12 hours after ICU admission. The size of initial Glargine bolus is the sum of SPRINT boluses administered during the previous 12 hours. The following Glargine is calculated as being half of the total daily insulin (IV boluses+Glargine) from the previous day. Each Glargine bolus is capped at 40 U/daily for patient safety.

Safety and performance of the protocols are evaluated by number of hypoglycaemic events (BG < 4.0 mmol/L), median and IQR of BG measurements level, percentage time spent in desired band (4.4-7.0 mmol/L), amount of insulin prescribed (IV boluses+Glargine), amount of nutrition given and nursing workload intensity based on number of interventions. Specifically, interventions that involved measuring BG levels, adjusting feed rates, administering SPRINT IV and Glargine bolus.

III. RESULTS

Figure 1 shows a cumulative distribution frequency (CDF) comparison between two protocols. The 4-Hour protocol is indicated with solid line while SPRINT clinical with dashed line. Analysis by cohort depicts an almost similar performance in terms of BG control between the two protocols.

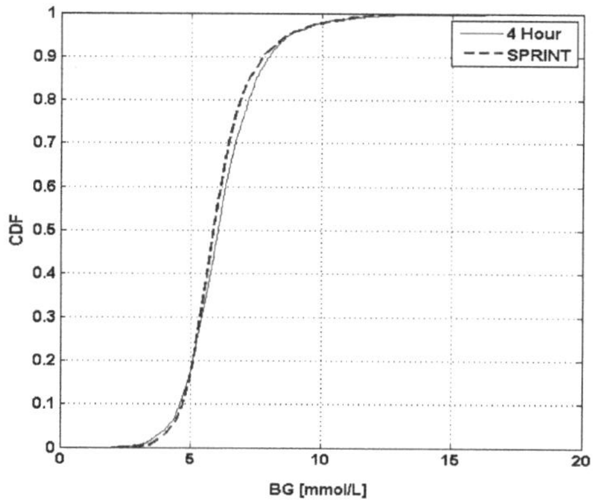

Fig. 1 Cumulative distribution function of BG measurement levels by cohort analysis for SPRINT clinical and 4-Hour protocol.

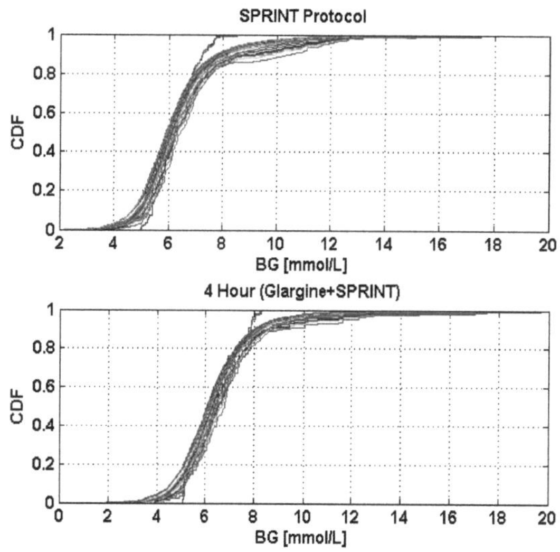

Fig. 2 Per Patient BG measurement levels of SPRINT Clinical and simulated 4-Hour protocol.

For a closer look at the effectiveness of protocols individually, Figure 2 depicts the per-patient BG measurements simulated for 4-Hour protocol and actual records from SPRINT clinical. Differences in tightness of control can be seen between SPRINT clinical and 4-Hour protocol. SPRINT shows a tighter control with more than 80% of patients having BG measurement levels under 7.0mmol/L. This is closely followed by the 4-Hour protocol at around 70% patients with BG levels under 7.0mmol/L. More importantly, per-patient variability is not an issue with just a minimal number of outliers.

Table 2 shows median and [IQR] of BG [mmol/L], insulin sensitivity (SI) [mU.min/L], time band within 4.4-7.0 (mmol/L) [%], amount of insulin bolus (IV) [U/day], number of intervention [N/day] for nursing workload and feed (nutrition intake) daily [mmol/min] for SPRINT clinical and 4-Hour protocol. The p-value for each assessment is included in the table.

Table 2 Comparison of Simulated 4-Hour protocol and SPRINT Clinical.

Median & IQR	SPRINT Clinical	4-Hour Protocol	P-value (significant when <0.05)
BG [mmol/L]	5.33 [4.96 5.67]	5.39 [4.99 5.89]	p=0.035
IV [U/day]	46.47 [39.52 51.12]	32.45 [28.77 40.87]	p<0.001
Intervention [N]	39.15 [37.14 40.74]	27.25 [26.32 28.50]	p<0.001
TimeBand [%]	86.80 [77.99 91.57]	80.77 [72.86 85.49]	p=0.006
Feed [mmol/min]	0.87 [0.87 1.09]	0.87 [0.79 0.87]	Not significant

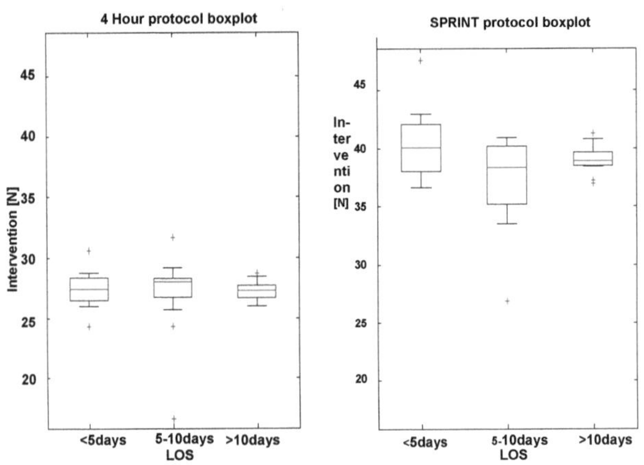

(a) (b)

Fig. 3 Cohorts intervention boxplot (a) 4-Hour protocol (b) SPRINT Clinical

Figure 3 is the box plot and whisker comparison for the 4-Hour and SPRINT clinical. Number of interventions was compared to LOS less than 5 days, 5 to 10 days and more than 10 days. This is done to assess if there is a relationship between nursing effort and LOS.

IV. Discussion

The virtual trial results of a combination and modified protocol referred to as 4-Hour protocol demonstrated a safe and promising protocol. An almost similar control both by cohort and per-patient analysis is achieved by the 4-Hour protocol as compared to SPRINT clinical. Significant clinical effort reduction is obtained with good performance in control quality and patient safety. Median BG measurement level, percentage within desired time-band (4.4-7.0mmol/L) and avoidance of hypoglycaemia achieved control as good as in SPRINT clinical. Feed rate in both protocols are similar with difference in upper and lower quartile. Interventions daily in SPRINT clinical are at 39.15 [37.14, 40.74] compared to 27.25[26.32, 28.50] in 4-Hour protocol. This reduction of around 12 daily interventions is meaningful once translated to minutes or hours saved. A study showed that for every hour nurses need to locate a glucose metre, perform a finger stick, record and adjust readings and perform appropriate rate adjustments which take around 5 minutes per patient [17]. Thus, a reduction of 12 units is roughly 60 minutes saved.

Results from Figure 3(a) of intervention box plot show that nursing effort once grouped per LOS is only slightly higher for LOS between 5-10 days. This might be attributed to the characteristics of Glargine build up that usually takes 3 days or longer before insulin is observed in plasma (Lehmann et.al., 2009). However, study is needed as to why less effort is required during the first 5 days. A different trend is recorded in SPRINT clinical where nursing effort is highest for LOS <5 days. Understandably, patients are much more dynamic at the start of ICU treatment. As reported by Carayon et al., patients with shorter LOS have slightly higher number of intervention [18].

Limitations to this study included number and criteria of patients simulated. Larger cohort would generate more statistics and enable thorough analysis. This virtual analysis only included 40 patients; therefore its results are a positive proof of concept and not conclusive. The insulin requirement of patients in this study is generally stable and consistent hourly which might contribute to the positive outcome.

In overall, this virtual trial results give a closer look at the potential benefits from Glargine as basal insulin therapy. More importantly, the outcome of the simulation opens a possible clinical proof of concept to demonstrate that reduction of nursing effort does not compromise patient safety and glycaemic control quality.

V. CONCLUSION

This study successfully demonstrates a safe and effective approach in reducing nursing effort within an ICU setting while maintaining the benefits of tight glycaemic control.

ACKNOWLEDGMENT

We are thankful to Ministry of Education Malaysia for the grant to enable this research. Thank you to Centre of Bioengineering, University of Canterbury and Christchurch ICU for the collaboration. Ethics were granted by Ministry of Health, New Zealand.

CONFLICT OF INTEREST

The authors declare that they have no conflict of interest.

REFERENCES

1. Capes, S.E., Hunt, D., Malmberg, K., & Gerstein, H.C. (2000) Stress hyperglycaemia and increased risk of death after myocardial infarction in patients with and without diabetes: a systematic overview. Lancet 355(9206):773-8.
2. Chase, J.G., Le Compte, A.J., Suhaimi, F., et al. (2011) Tight glycemic control in critical care--the leading role of insulin sensitivity and patient variability: a review and model-based analysis. Computer Methods and Programs in Biomedicine 102(2):156-71
3. Dellinger, R. P., Levy, M. M., Carlet, J. M., et al (2008). Surviving Sepsis Campaign: international guidelines for management of severe sepsis and septic shock: 2008. Critical care medicine Vol. 36.
4. Chase, J.G., Shaw, G., Le Compte, A., et al (2008) Implementation and evaluation of the SPRINT protocol for tight glycaemic control in critically ill patients: a clinical practice change. Critical Care 12(2): R49.
5. Krinsley, J. S. (2004) Effect of an intensive glucose management protocol on the mortality of critically ill adult patients. Mayo Clinic Proceedings 79(8):992-1000
6. Van den Berghe, G., Wouters, P., Weekers, F., et al (2001) Intensive Insulin Therapy in Critically Ill Patients. The New England Journal of Medicine 345(19):1359-1367.
7. Mastura, I., Chew, B. H., Lee, P. Y., et al (2011) Control and Treatment Profiles of 70,889 Adult Type 2 Diabetes Mellitus Patients in Malaysia - A Cross Sectional Survey in 2009. International Journal of Collaborative Research on Internal Medicine & Public Health 3(1):98-113.
8. Evans, A., Shaw, G. M., Le Compte, A., et al (2011). Pilot proof of concept clinical trials of Stochastic Targeted (STAR) glycemic control. Annals of Intensive Care 1(1):38.
9. Aragon, D. (2006) Evaluation of nursing work effort and perceptions about blood glucose testing in tight glycemic control. American Journal of Critical Care 15(4):370-7.
10. Gartemann, J., Caffrey, E., Hadker, N., et al (2012). Nurse workload in implementing a tight glycaemic control protocol in a UK hospital: A pilot time-in-motion study. Nursing in Critical Care, 17: 279–284.
11. Malesker, Mark A., Pamela A. Foral, Ann C. McPhillips, et al (2007) An Efficiency Evaluation of Protocols for Tight Glycaemic Control in Intensive Care Units. American Journal of Critical Care 16(6):589-598.
12. Chase, J. G., Suhaimi, F., Penning, S., et al (2010) Validation of a model-based virtual trials method for tight glycemic control in intensive care. Biomedical Engineering Online 9:84
13. Stewart, K. W., Pretty, C. G., Tomlinson, H., et al. (2015) Stochastic Model Predictive (STOMP) glycaemic control for the intensive care unit: Development and virtual trial validation. Biomedical Signal Processing and Control 16(2015):61-67
14. Wong, J., Chase, J. G., Hann, C. E. et al (2008) A subcutaneous insulin pharmacokinetic model for computer simulation in a diabetes decision support role: validation and simulation. Journal of Diabetes Science and Technology 2(4): 672-80
15. Lin, J., Razak, N. N., Pretty, C. G., et al (2011) A physiological Intensive Control Insulin-Nutrition-Glucose (ICING) model validated in critically ill patients. Computer Methods and Programs in Biomedicine, 102(2):192–205.
16. Hann, C. E., Chase, J. G., Lin, J., et al (2005) Integral-based parameter identification for long-term dynamic verification of a glucose-insulin system model. Computer Methods and Programs in Biomedicine 77(3):259-70
17. Goldberg, P. A., Sakharova, O. V., Barrett, P. W., et al (2004) Improving glycemic control in the cardiothoracic intensive care unit: clinical experience in two hospital settings. Journal of cardiothoracic and vascular anesthesia 18(6):690-697.
18. Carayon, P.,& Gurses, A.P. (2008) Nursing workload and patient safety - A human factors engineering perspective. In Patient safety and quality: An evidence-based handbook for nurses 2:203-216

Author: Normy Norfiza Binti Abdul Razak
Institute: Universiti Tenaga Nasional
Street: Jalan Ikram-Uniten, 43000 Kajang
City: Selangor
Country: Malaysia
Email : normy@uniten.edu.my

Agreement Between Eyes in Wide-field Fluorescence Lifetime Imaging Ophthalmoscopy Measurements at the Human Retina in Healthy Volunteers

M. Klemm[1], E. Nagel[1,2], A. Dietzel[1], L.K.W. Lai[3], E. Supriyanto[4], and D. Schweitzer[5]

[1] Institute of Biomedical Engineering and Informatics, Ilmenau University of Technology, Ilmenau, Germany
[2] Ophthalmic Practice Ankermedicum, Rudolstadt, Germany
[3] Department of Biomedical Engineering, University of Malaya, Kuala Lumpur, Malaysia
[4] Department of Clinical Science and Engineering, Universiti Teknologi Malaysia, Johor, Malaysia
[5] Experimental Ophthalmology, University of Jena, Jena, Germany

Abstract— **Fluorescence lifetime imaging ophthalmoscopy (FLIO) is a new imaging technique for measuring the time-resolved in vivo autofluorescence generated by endogenous fluorophores in the ocular fundus. The aim is to assess the metabolism of the retina. Based on the fluorescence lifetime, different fluorescent compounds in the eye can be distinguished. Pathologic changes may be observed by detecting changes in the fluorescence lifetime. Until now, FLIO is performed using 30° fundus images. For certain diseases section of the fundus is of interest. In this work, observational clinical study in young healthy volunteers was performed to examine the usability of 55° wide-field fluorescence lifetime images. The time-resolved retina autofluorescence was measured (scanning laser ophthalmoscope: 55° of fundus, 62x62 µm² per pixel; excitation: diode laser with pico-second pulses, 473nm, 80MHz repetition rate; detection: spectral channels 498-560nm (ch1) and 560-720nm (ch2), time-correlated single photon counting method) in both eyes of 11 healthy volunteers (28.7±3.6 years). Three repetitive measurements on different days within one week at a similar time have been performed. All subjects had a crystalline lens and an undilated pupil. A modified 3-exponential approach was applied to determine the mean fluorescence lifetimes τ_m. τ_m were computed on the ETDRS grid and in a 15x15 pixel region 25° superior to the fovea. The Wilcoxon rank-sum test was used to test for statistical significant differences between left and right eye. The coefficient of variation in the superior region is 5.6% (right) and 4.7% (left) in ch1 and 2.5% (right) and 2.1% (left) in ch2. No statistical significant differences have been found between right and left eye. FLIO measurements in young healthy volunteers using a 55° lens are repeatable in both eyes and show no significant differences. Spectral channel 2 should be preferred because of its lower variability. The data analysis was done with FLIMX (available at http://www.flimx.de).**

Keywords— **FLIM, FLIO, FLIMX, autofluorescence, retina.**

I. INTRODUCTION

Pathological changes in the ocular fundus are preceded by changes in metabolism. Shifted balances in the metabolism can be treated before permanent damage occurs. The fluorescence of some of the substances involved in the metabolism can be excited, such as lipofuscin / N-retinyliden-N-retinylethanolamin (A2E), advanced glycation end products, collagens, flavines and possibly nicotinamide adenine dinucleotide in its reduced form (NADH). In total, at least 10 different substances in the ocular fundus are known to be fluorescent. The excitation and emission spectra of these substances strongly overlap, so that a separation is not possible solely on the basis of spectral information. With the help of another property of fluorescence, the fluorescence lifetime, these substances can be analyzed. Thus, using the method of time-resolved autofluorescence measurements, it is possible to investigate substances in the human eye which are involved in the metabolism [1]. Fluorescence lifetime imaging ophthalmoscopy (FLIO) is a new imaging technique and was developed by Schweitzer et al. [2]. It has been shown to produce repeatable measurements [3, 4]. Consequently, early detection of eye disease based on the fluorescence lifetime is conceivable. Lately, fluorescence lifetimes have been determined for subretinal deposits of metabolic byproducts, called drusen, retinal pigment epithelium (RPE) cells, and Bruch's membrane in histological sections of a human donor eye [5]. Changes in fluorescence lifetime parameters in vivo in humans have been found in patients with diabetes [6], glaucoma [7], macular holes [8] and patients with Alzheimer's disease [9]. All in vivo measurements have been performed using a 30° field of view of the fundus. Recently, there is a trend in ophthalmology to image larger section of the fundus. Thus the feasibility of a 55° lens for FLIO measurements is investigated. In this work, the agreement between both eyes and the repeatability of the FLIO measurements using a 55° lens are evaluated.

II. METHODS AND MATERIALS

A. Instrumentation

To measure the time-resolved autofluorescence at the fundus a confocal scanning laser ophthalmoscope is used. The schema of the instrumentation is shown in Figure 1A. A 473 nm pulse laser with a repetition rate of 80 MHz and a pulse width of about 70 ps (full width at half maximum) is utilized for the excitation of the fluorescence. The radiation

F. Ibrahim et al. (eds.), *International Conference for Innovation in Biomedical Engineering and Life Sciences*,
IFMBE Proceedings 56,
DOI: 10.1007/978-981-10-0266-3_63

power in corneal plane is about 150 µW. The fluorescence photons are distributed in two spectral channels (498–560 nm and 560–720 nm) by a beam splitter with a cutoff wavelength of 560 nm. For each spectral channel, the incoming photons are detected by a hybrid detector (HMP-100-40, Becker & Hickl GmbH, Berlin, Germany). The signals of the detectors are processed by the principle of time-correlated single photon counting (TCSPC). The time-resolved fluorescence measurement is spatially resolved in a 55° x 55° area of the fundus with a lateral resolution of circa 62 x 62 square microns. The instrumental response functions (IRF) for both spectral channels are shown in Figure 1B. The full width at half maximum of the IRFs are 172 ps and 153 ps, which is sort enough to determine fluorescence lifetimes well below 100 ps.

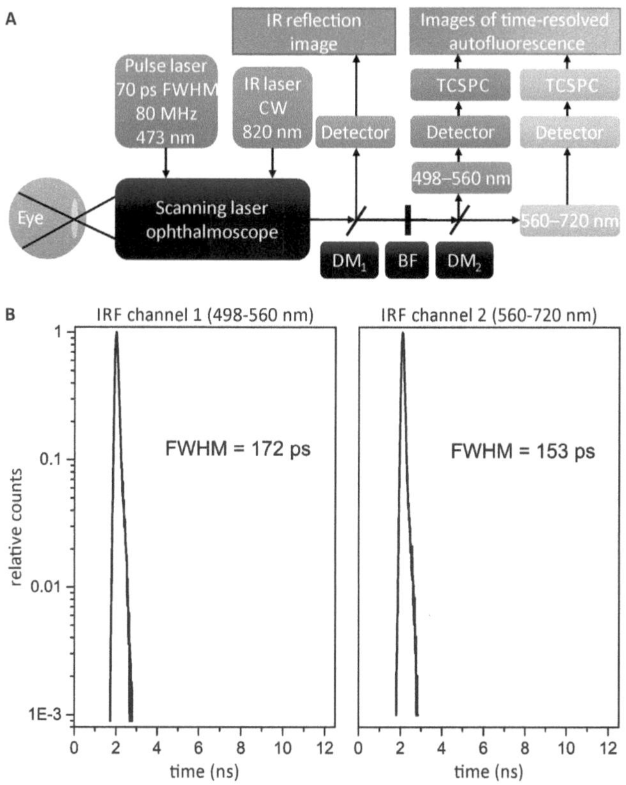

Fig. 1 Scheme of the FLIO instrumentation (A) and plot of the instrumental response functions (B) for both spectral channels [10]

To obtain a sufficiently large number of photons per pixel, acquisition times in the range of minutes are required. This is made possible by an infrared image recorded in parallel to the fluorescence, which is used for registration and correction of eye movements. The TCSPC technique [11-13] generates time-, space- and spectrum-resolved fluorescence decay datasets. More details of the FLIO instrumentation have been described elsewhere [3].

B. *Study Design*

Three repetitive FLIO measurements have been taken from both eyes of 11 healthy volunteers aged 28.7±3.6 years (9 male, 2 female) on different days within one week at a similar day time. Both eyes were measured in random order. All subjects had a crystalline lens and an undilated pupil. All research procedures were performed according to the Declaration of Helsinki. Approval for the study was obtained from the ethics committee of the Jena University Hospital. Written informed consent was obtained from each volunteer prior to participation in the study.

C. *Data Analysis*

A currently often used approach to model the fluorescence lifetime of TCSPC data is the multi-exponential model [11, 14]:

$$\frac{I_1(\mathrm{t})}{I_0} = IRF * \sum_i \alpha_i \cdot e^{-\left(\frac{t}{\tau_i}\right)} + b \tag{1}$$

where I is the fluorescence intensity, IRF is the instrumental response function, α is the amplitude, t is the time, τ is the fluorescence lifetime, b is the offset and the asterisk denotes a convolution integral. Schweitzer et al. [15] enhanced the multi-exponential model to take the layered structure of the eye into account, which causes fluorescence photons from structures that are farther away from the laser scanner ophthalmoscope to arrive later because of the additional travel time required for the excitation light and the fluorescence photons:

$$\frac{I_1(\mathrm{t})}{I_0} = IRF * \sum_i \alpha_i \cdot e^{-\left(\frac{t-tc_i}{\tau_i}\right)} + b \tag{2}$$

where tc is a time shift parameter.

Three exponential functions and one time shift parameter tc_3 are used to approximate the fluorescence lifetime. The determination of the model parameters is performed by a combination of linear and non-linear (differential evolution [16], Nelder-Mead [17]) minimization algorithms.

To assess the goodness of the fit, the reduced χ^2-error is used:

$$\mathrm{X}^2 = \frac{1}{m-p} \cdot \sum_{j=1}^{m} \frac{\left[I_M(t_j) - I_C(t_j)\right]^2}{I_M(t_j)} \tag{3}$$

where m is the number of time channels of the photon histogram, $I_M(t_j)$ is the number of measured photons in time channel j, $I_c(t_j)$ is the number of calculated photons using the multi-exponential model discussed above and p is the number of free parameters in the model.

For some applications, the mean fluorescence lifetime τ_m is a good overview parameter which can be derived from the fluorescence amplitudes and lifetimes:

$$\tau_m = \frac{\sum_i \alpha_i \cdot \tau_i}{\sum_i \alpha_i} \quad (4)$$

The ETDRS grid [18] and a 15x15 pixel square region 25° superior to the fovea were applied to each measurement by an expert. The outer ring and the superior region were used for a statistical comparison based on the Wilcoxon rank-sum test.

The data analysis was performed with FLIMX [10] which is documented and freely available for download online under the open source BSD–license (http://www.flimx.de).

III. RESULTS

An image of the fluorescence lifetime of spectral channel 1 from the left eye of a healthy volunteer in shown in Figure 2. The shortest fluorescence lifetimes occur in the center of the macula while the longest fluorescence lifetimes occur in the optic disc.

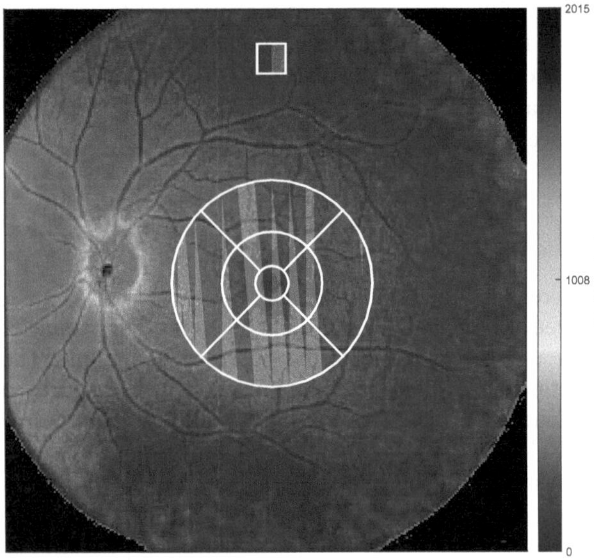

Fig. 2 Mean fluorescence lifetime of spectral channel 1from the left eye of a healthy volunteer. The EDTRS grid and the superior region are colored in white. For better orientation, an overlay of the fluorescence intensity was added.

The descriptive statistics (mean and standard deviation) over all volunteers are given in Tables 1 for the right eye and in Table 2 for the left eye.

Table 1 Mean fluorescence lifetime of the **right** eye for each repetitive measurement and both spectral channels. Values are mean value ± standard deviation.

Measurement	Spectral channel	outer ring ETDRS grid	Superior region
1	1	350±25ps	348±20ps
2	1	357±27ps	357±20ps
3	1	350±26ps	327±17ps
1	2	255±11ps	260±7ps
2	2	257±12ps	261±7ps
3	2	257±12ps	254±5ps

Table 2 Mean fluorescence lifetime of the **left** eye for each repetitive measurement and both spectral channels. Values are mean value ± standard deviation.

Measurement	Spectral channel	outer ring ETDRS grid	Superior region
1	1	350±25ps	332±17ps
2	1	344±24ps	325±16ps
3	1	348±25ps	327±15ps
1	2	252±10ps	260±6ps
2	2	250±10ps	261±5ps
3	2	251±10ps	254±5ps

The mean fluorescence lifetime of repetitive measurements at different days are within the standard deviation. Similarly, the mean fluorescence lifetime of the ETDRS grid's outer ring and the superior region are within the standard deviation of each other.

The mean values of each section (outer ring ETDRS grid, superior region) over all measurements from each eye are compared against each other. The p values of the Wilcoxon rank-sum test are given in Table 3. No p value is below the significance threshold of 0.05.

Table 3 P values of the Wilcoxon rank-sum test comparing right and left eye based on the mean fluorescence lifetime averaged over the segmented regions per measurement.

Spectral channel	outer ring ETDRS grid	Superior region
1	0.57	0.87
2	0.27	0.66

IV. DISCUSSION

Repeatability of FLIO measurements has been shown for the 30° lens before [3, 4] and could be confirmed for the 55° lens, even at the outer regions. Agreement between eyes has also been analyzed by Dysli et al. [4] before. In contrast to Dysli et al., three exponential functions are used in this work to approximate the fluorescence lifetime while Dysli et al. used only two exponential functions.

V. Conclusion

FLIO measurements in young healthy volunteers using a 55° lens are repeatable in both eyes and show no significant differences between eyes. Spectral channel 2 should be preferred because of its lower variability.

Acknowledgments

Research in part supported by the German Federal Ministry of Education and Research (Grant No. 03IPT605A), the German Research Council (DFG HA 2899/19-1) and the German Academic Exchange Service (DAAD, Grant No. 57040277).

Conflict of Interest

The authors declare that they have no conflict of interest.

References

1. Schweitzer D, Schenke S, Hammer M, Schweitzer F, Jentsch S, Birckner E, et al. Towards metabolic mapping of the human retina. Microsc Res Techniq. 2007;70(5):410-9. doi: 10.1002/Jemt.20427. PubMed PMID: ISI:000246408100004.
2. Schweitzer D, Hammer M, Schweitzer F, Anders R, Doebbecke T, Schenke S, et al. In vivo measurement of time-resolved autofluorescence at the human fundus. J Biomed Opt. 2004;9(6):1214-22. doi: 10.1117/1.1806833. PubMed PMID: ISI:000225697400014.
3. Klemm M, Dietzel A, Haueisen J, Nagel E, Hammer M, Schweitzer D. Repeatability of Autofluorescence Lifetime Imaging at the Human Fundus in Healthy Volunteers. Curr Eye Res. 2013;38(7):793-801. doi: 10.3109/02713683.2013.779723. PubMed PMID: WOS:0003197 54700011.
4. Dysli C, Quellec G, Abegg M, Menke MN, Wolf-Schnurrbusch U, Kowal J, et al. Quantitative analysis of fluorescence lifetime measurements of the macula using the fluorescence lifetime imaging ophthalmoscope in healthy subjects. Invest Ophth Vis Sci. 2014;55(4):2106-13. PubMed PMID: Medline:24569585.
5. Schweitzer D, Gaillard ER, Dillon J, Mullins RF, Russell S, Hoffmann B, et al. Time-Resolved Autofluorescence Imaging of Human Donor Retina Tissue from Donors with Significant Extramacular Drusen. Invest Ophth Vis Sci. 2012;53(7):3376-86. doi: 10.1167/Iovs.11-8970. PubMed PMID: ISI:000306181200014.
6. Schweitzer D, Deutsch L, Klemm M, Jentsch S, Hammer M, Peters S, et al. Fluorescence lifetime imaging ophthalmoscopy in type 2 diabetic patients who have no signs of diabetic retinopathy. J Biomed Opt. 2015;20(6):061106. doi: 10.1117/1.JBO.20.6.061106.
7. Ramm L, Jentsch S, Augsten R, Hammer M. Fluorescence lifetime imaging ophthalmoscopy in glaucoma. Graef Arch Clin Exp. 2014;252(12):2025-6. PubMed PMID: WOS:000345589300022.
8. Sauer L, Schweitzer D, Ramm L, Augsten R, Hammer M, Peters S. Impact of Macular Pigment on Fundus Autofluorescence Lifetimes. Invest Ophthalmol Vis Sci. 2015;56(8):4668-79. doi: 10.1167/iovs.14-15335. PubMed PMID: 26207302.
9. Jentsch S, Schweitzer D, Schmidtke KU, Peters S, Dawczynski J, Bar KJ, et al. Retinal fluorescence lifetime imaging ophthalmoscopy measures depend on the severity of Alzheimer's disease. Acta ophthalmologica. 2014. Epub 2014/12/09. doi: 10.1111/aos.12609. PubMed PMID: 25482990.
10. Klemm M, Schweitzer D, Peters S, Sauer L, Hammer M, Haueisen J. FLIMX: A Software Package to Determine and Analyze the Fluorescence Lifetime in Time-Resolved Fluorescence Data from the Human Eye. Plos One. 2015;10(7):e0131640. doi: 10.1371/journal.pone.0131 640. PubMed PMID: 26192624; PubMed Central PMCID: PMC 4507995.
11. Lakowicz JR. Principles of Fluorescence Spectroscopy. 3rd ed. New York: Springer; 2006. 954 p.
12. Becker W. Advanced time-correlated single photon counting techniques. Berlin: Springer; 2005. 401 p.
13. Becker W. The bh TCPSC Handbook. 5th ed. Berlin: Becker & Hickl GmbH; 2012. 690 p.
14. O'Connor DV, Ware WR, Andre JC. Deconvolution of Fluorescence Decay Curves - Critical Comparison of Techniques. J Phys Chem-Us. 1979;83(10):1333-43. PubMed PMID: ISI:A1979GV58600019.
15. Schweitzer D, Klemm M, Hammer M, Jentsch S, Schweitzer F. Method for simultaneous detection of functionality and tomography. Clinical and Biomedical Spectroscopy. 2009;7368:736804. doi: 10.1364/ECBO.2009.7368_04.
16. Storn R, Price K. Differential evolution - A simple and efficient heuristic for global optimization over continuous spaces. J Global Optim. 1997;11(4):341-59. doi: Doi 10.1023/A:1008202821328. PubMed PMID: WOS:A1997YF22800001.
17. Nelder JA, Mead R. A Simplex-Method for Function Minimization. Comput J. 1965;7(4):308-13. PubMed PMID: ISI:A1965CLA3200015.
18. Grading Diabetic-Retinopathy from Stereoscopic Color Fundus Photographs - An Extension of the Modified Airlie House Classification - ETDRS Report Number 10. Ophthalmology. 1991;98(5):786-806. PubMed PMID: WOS:A1991FL41900005.

Author: Matthias Klemm
Institute: Technische Universität Ilmenau
Street: Gustav Kirchhoff-Str. 2
City: Ilmenau
Country: Germany
Email: matthias.klemm@tu-ilmenau.de

Graphene as a Smart Material for the Recognition of DNA Biomolecule

M.M. Rahman[1], P.K. Lee[1], and S.B. Abd Hamid[1]

[1]Nanotechnology and Catalysis Research Centre, University of Malaya, Kuala Lumpur, Malaysia

Abstract— **DNA molecule is the building block for most living organism. Recently, the detection of specific DNA target is a potential field not only for organism species authentication but also in disease diagnostics and gene therapy. DNA molecule has unique attribute for its molecular recognition. The two-dimensional, graphene oxide material provides excellent physio-chemical properties for utilization in the detection of specific DNA molecule. Hence, here we bring forward the structure of the DNA biomolecule. We have described the graphene and its synthesis methods, molecular binding pattern with DNA and the different detection techniques along with the potential application and use of nanomaterial. Finally, we outlined the conclusion and future perspectives in this area.**

Keywords— **Graphene, DNA detection, Electrochemical, Gold nanoparticles.**

I. INTRODUCTION

After six decades of DNA discover with its structural and functional properties, now it is not limited to its canonical role in biology but also in the development of molecular detection techniques. The recent applications of DNA molecule recognition is now extended for detection of organism species, bioimaging, diagnostics and sensing for drug delivery [1, 2, 3]. Current DNA hybridization detection technique based on polymerase chain reaction assay and molecular fluorophores labeling is suffering from the accuracy due to wide absorption and emission bands and various rates of fluorophore photo bleaching [4]. Graphene based materials such as Graphene oxide (GO) and reduced Graphene Oxide rGO, provides excellent physio-chemical properties for the detection of specific DNA molecules. Conjugations of DNA with the graphene sheet facilitate the DNA hybridization detection with enhanced signal amplification. Furthermore, the advent of Au NPs as a sensory element provided a broad spectrum for the detection of specific DNA target. Thus here we described the DNA molecule, its recognition technique using the graphene as an enhanced sensing material.

II. DNA

DNA-Deoxyribonucleic acid is the bio-molecular structure that architects the genetic materials for the development and functioning of living organisms such human, animal, plants, bacteria and some viruses [5]. The well established chemical structure of DNA is composed of a long chain, made up of alternate sugar and phosphate groups, joined together in regular 3' 5' phosphate di-ester linkages (Fig. 1). Each sugar is attached with any of the four nitrogenous base adenine (A), guanine (G), thymine (T) or cytosine (C) [6]. Thus, the building blocks of DNA are made of one deoxyribose sugar, one phosphate, and one of these four bases called as nucleotide (Fig. 1). To form a native complete DNA, single nucleotides join to make chains that come together with complementary pairs using hydrogen bonding and form long double strands helical structure (Fig. 1). The physical properties of the double stranded DNA helix (dsDNA) like natural or synthetic polymer [7]. The characteristics of molecular recognition model of complementary strands of DNA allow construction of nano-platform such as DNA tags for organizing the assembly of colloidal particles [8]. Thus the recent technological advancement made molecular recognition of specific DNA target as an emerging field for species authentication, drug discovery, forensic investigation and biodiagnostics [3].

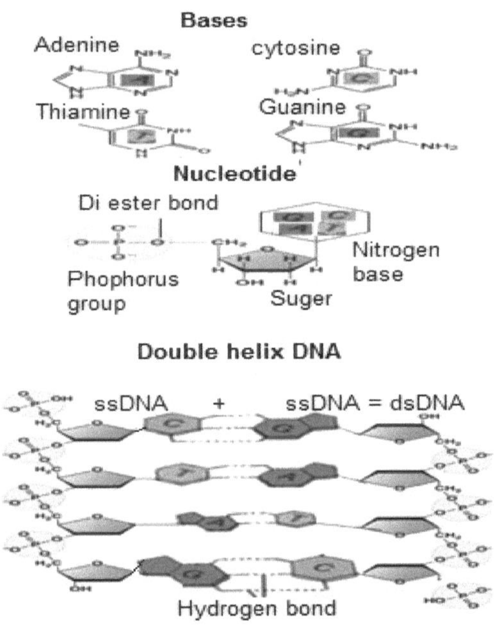

Fig. 1 Structure of DNA

F. Ibrahim et al. (eds.), *International Conference for Innovation in Biomedical Engineering and Life Sciences*,
IFMBE Proceedings 56,
DOI: 10.1007/978-981-10-0266-3_64

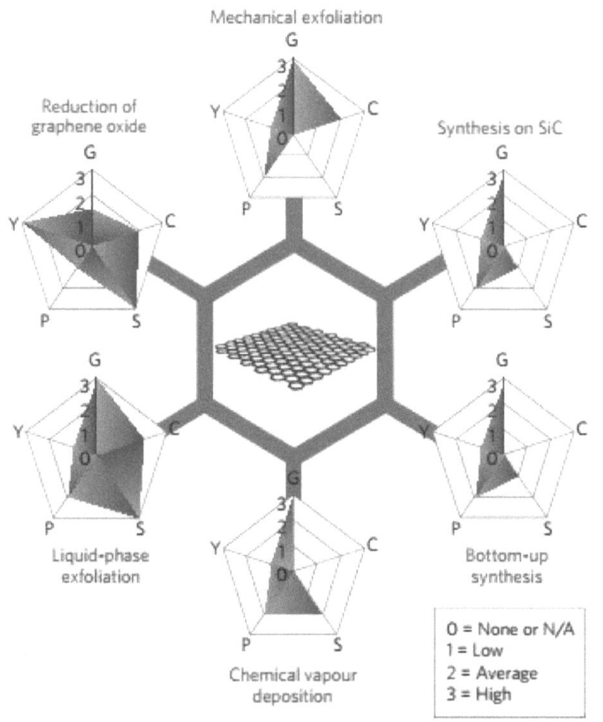

Fig. 2 Schematic of the common synthesis protocol for graphene. Each method has been evaluated in terms of graphene quality (G), cost aspect (C; a low value corresponds to high cost of production), scalability (S), purity (P) and yield (Y) of the overall production process [9].

III. GRAPHENE

Graphene is the basic structural bits and pieces of graphitic materials and is a flat monolayer of tightly packed carbon atoms into a two-dimensional honeycomb pattern. The first ten decades after the discovery of graphene it was used to describe various carbon based materials. However, in last two decades the graphene is utilized as an excellent condensed-matter analogue of (2+1)-dimensional quantum electrodynamics and as an integral part of 3D materials. Graphene oxide is a composed carbon, oxygen, and hydrogen in different ratio. It is a unique material with a single monomolecular layer of graphite with different oxygen containing functional group (epoxide, carbonyl, carboxyl and hydroxyl). Structurally, the reduced graphene oxide (rGO) resembles graphene but contains structural defects, residual oxygen and other heteroatoms. The recent application of graphene oxide (GO) or reduced graphene oxide (rGO) has been fascinated for the fabrication of DNA-based sensors, due to their excellent dissolving capability either in water or various other solvent for sensing. Thus, graphene based materials can be used in the biosensors not only for the detection of specific DNA target but also to detect metal ions and proteins [1].

IV. SYNTHESIS OF GRAPHENE

Several synthesis protocols have been reported to obtain graphene. The most common methods for graphene synthesis and their evaluation is incorporated in Figure 2 [9]. In the scientific research laboratory and for the proof-of concept devices mostly scotch-tape method of graphene synthesis is used. This method was first described by Geim and coworkers to obtain pristine perfect structured graphene layer(s) using mechanical exfoliation (repeated peeling) of highly oriented pyrolytic graphite. The drawback of this method, it cannot be utilized for mass production. The mild exfoliation of graphite may produce defect-free/defect-less graphene but until now it is suffering for the production of lower yield. For mass production and for application in electronic devices, preparation of Graphene by thermal decomposition of SiC wafer under ultrahigh vacuum (UHV) conditions or by CVD growth on metal substrates (ruthenium , Ni , and Cu) or by substrate free CVD have been proposed [10]. For the preparation of graphene to use in electrochemistry are produced by chemical or thermal reduction of graphite oxide GO. Graphene from GO reduction, which is also called functionalized graphene sheets or chemically reduced graphene oxide usually has abundant structural defects and functional groups which are advantageous for DNA sensor application. It is also considered to be the most economical way to produce graphene [10].

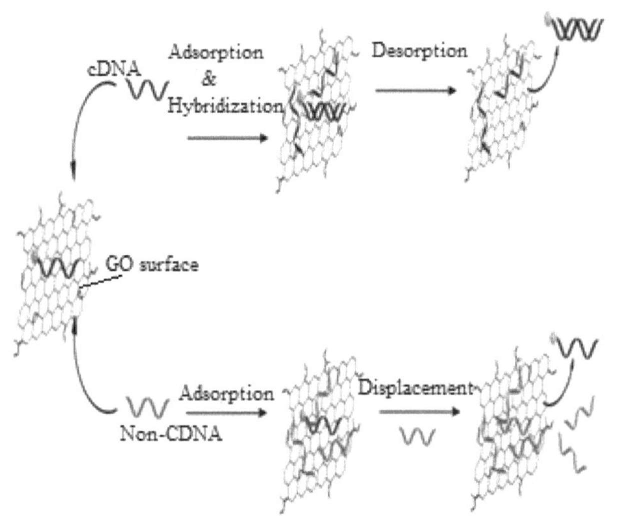

Fig. 3 Desorption mechanism of the DNA Adsorption on GO surface [12].

V. DNA CONJUGATION WITH GRAPHENE

As GO composed of different functional group, it provides strong attraction to water and various solvent [11]. Thus the GO is hydrophilic in nature and very stable in aqueous dispersion and can aid for sensing DNA. The single-stranded DNA (ssDNA) molecule interacts with GO through π–π stacking interaction. Desorption of the GO-adsorbed DNA molecule may occur either via facilitated desorption following hybridization with cDNA on the GO surface or via nonspecific simple displacement by non-cDNA according to the law of mass action (Fig. 3) [12].When the dye-labeled ssDNA binds to GO the fluorescence of the dye quenched. However, restorations of fluorescence of dye signal occur while the specific DNA target binds to complimentary sequence by forming a dsDNA. GO provide excellent physical and electrochemical properties for the detection DNA. It allows the oxidative detection of DNA which is the simplest form of electrochemical DNA sensor. Furthermore, simple conjugation and sustainability with DNA molecule with certain nanomaterials such as AuNPs act as ideal transducers for nano scale sensing.

VI. GRAPHENE BASED DNA DETECTION

A. Optical Detection

Certain disease condition or DNA mutation can be determined by using the disruption of π–π stacking interactions of graphene with DNA hybridization feature and fluorescence dye quenching (Fig. 4.a). Thus Luo, et al. [13] have reported significant chemiluminescence (CL) emission with formation of dsDNA with complimentary target and increased peroxidatic activity of a horseradish peroxidase mimicking DNAzyme. The detection limit of the assay was 34 pM. Zhao, et al. [14] described a higher-sensitive detection technique using Exonuclease III aided signal amplification with the fluorescence quenching efficiency of GO. The detection limit of the assay was 20 pM. Molecular beacon probe can also be used for the detection of specific DNA target or single nucleotide polymorphisms (SNPs) using GO as a nanoquencher for fluorescence. The of long-range resonance energy transfer (LrRET) based label free DNA sensor with the graphene was introduced by Jiang, et al. [15] using the Ethidium bromide (EB). Detection of the specific DNA target was determined by the fluorescence intensity of gradually recovered EB form the formation of complimentary dsDNA. SYBR Green (SG) can be used as an intercalator molecule for an effective DNA hybridization detection of HIV-1 DNA with a detection limit of 0.31 nM [1].

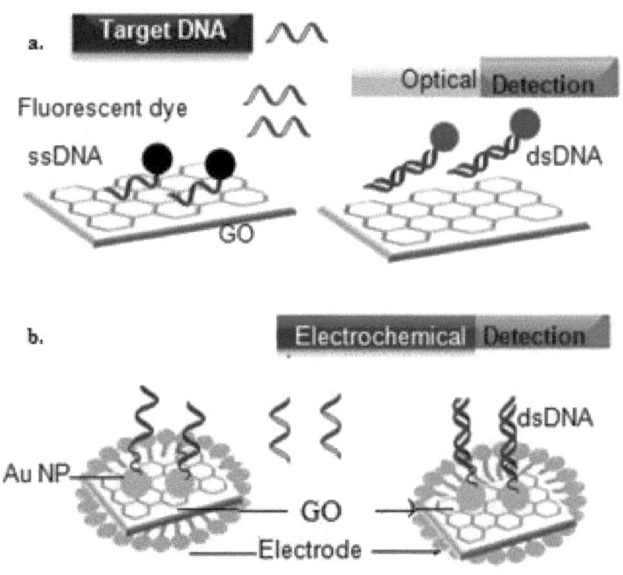

Fig. 4 Graphene based DNA detection

B. Electrochemical Detection

Zhou, et al. [16] reported the oxidation potentiality of DNA nucleobases (A, T, C and G) to distinguish different species. The assay describes the chemical reduction of GO (CR-GO) electrodes for a label-free electrochemical DNA sensing. It can be used for the SNP detection. The specific G base could be determined by a reduced graphene oxide (rGO)– chitosan electrode doped with Fe3O4 nanoparticles using the electron transfer resistance property Yin, et al. [17]. Huang [18] reported the electrochemical detection G and A by rGO with detection limits of 2.5x108 M and 5.0x108 M respectively. Li, et al. [19] initiated the application of the electrochemical DNA sensor for anti cancer drug development by ssDNA probe immobilization on electrode surface for hybridization with the target DNA and subsequent methylation and cleaving by HpaII endonuclease. The electrochemical response of the reporter thiamine was used for the detection of DNA methylation at the site of CpG and single-base mismatched sequence. The detection limit of the assay was 0.05 ± 0.02 U mL-1. at a signal/noise of 3. For the detection of methicillin resistant S. aureus (MRSA) DNA, Wang, et al. [20] described the use of GO with glassy carbon electrode modified with (3-aminopropyl) triethoxysilane (APTES). The detection limit of the assay was 10-13 M.

Gold nanoparticles (AuNPs) are very attractive nanomaterials with unique optical properties along with highly resonant particle plasmons [4]. Thus, Zhang and Jiang [21] decorated graphene sheets with AuNPs for the detection of specific DNA target using the electrode with modified graphene (Fig. 4.b). The immobilization of the probe DNA with AuNPs was done by Au S bond (Fig. 4.b).

The intensity of the peak currents for effective DNA hybridization with the probe was measured by using the DNA intercalating dye adriamycin. Thus the detection limit of the AuNP aided assay was 3.5x10 -14 M under an estimated signal/noise ratio of 3. Furthermore, modification of electrode with ERGO-d (GT) and coupling of Au NPs with horseradish peroxidase (HRP) functionalized carbon spheres (CNS) as the tracer can be used for an enhanced signal amplification of DNA detection [22]. The detection limit of the assay for sensing DNA was 5 aM.

VII. CONCLUSIONS AND FUTURE PROSPECTIVE

Graphene has shown the potentiality with its specific physio-chemical properties for the detection of DNA biomolecule. Thus DNA sensing based on graphene is noteworthy. However, although different synthesis protocols of graphene have been described but need to work for economical production approach with high yield. The interactions between graphene with DNA and electrochemical absorption of different analytes need to explore more to obtain highly sensitive device and sensors.

ACKNOWLEDGEMENT

The authors acknowledge the University of Malaya grant TR002B-2014B of Prof Sharifah Bee Abd Hamid.

CONFLICT OF INTEREST

The authors do not have any conflict of interest to publish this manuscript.

REFERENCES

1. Gao L, Lian C, Zhou Y, Yan L, Li Q, et al. (2014) Graphene oxide–DNA based sensors. Biosens Bioelectron 60: 22-29.
2. Tjong V, Tang L, Zauscher S and Chilkoti A (2014) "Smart" DNA interfaces. Che. Soci Rev 43: 1612-1626.
3. Rahman M, Ali M and Abd Hamid SB (2015) Gold Nanoparticles-An Enhanced DNA Sensing Tools Using Surface Enhance Raman Scattering. Advanced Materials Research. Trans Tech Publ. pp. 439-443.
4. Ali M, Rahman M, Hamid SBA and Hashim U (2014) Nanoscale DNA Sensing-Potential and Prospects. Advanced Materials Research. Trans Tech Publ. pp. 486-489.
5. Mckay A (2009) Genetically modified foods: facts, worries, policies and public confidence.
6. Watson JD and Crick FH (1953) The structure of DNA. Cold Spring Harbor Symposia on Quantitative Biology. Cold Spring Harbor Laboratory Press. 1953: 123-131.
7. Bustamante C, Bryant Z and Smith SB (2003) Ten years of tension: single-molecule DNA mechanics. Nature 421: 423-427.
8. Yurke B, Turberfield AJ, Mills AP, Simmel FC and Neumann JL (2000) A DNA-fuelled molecular machine made of DNA. Nature 406: 605-608.
9. Raccichini R, Varzi A, Passerini S and Scrosati B (2015) The role of graphene for electrochemical energy storage. Nat Mater 14: 271-279.
10. Wu H, Liu J, Aksay I and Lin Y (2010) Graphene based electro chemical sensors and biosensors. Electro Analy 22: 1027-1036.
11. Du D, Wang L, Shao Y, Wang J, Engelhard MH, et al. (2011) Functionalized graphene oxide as a nanocarrier in a multienzyme labeling amplification strategy for ultrasensitive electrochemical immunoassay of phosphorylated p53 (S392). Anal Chem 83: 746-752.
12. Park JS, Goo N-I and Kim D-E (2014) Mechanism of DNA Adsorption and Desorption on Graphene Oxide. Langmuir 30: 12587-12595.
13. Luo M, Chen X, Zhou G, Xiang X, Chen L, et al. (2012) Chemiluminescence biosensors for DNA detection using graphene oxide and a horseradish peroxidase-mimicking DNAzyme. Chem Commun 48: 1126-1128.
14. Zhao X-H, Ma Q-J, Wu X-X and Zhu X (2012) Graphene oxide-based biosensor for sensitive fluorescence detection of DNA based on exonuclease III-aided signal amplification. Analytica Chimica Acta 727: 67-70.
15. Jiang Y, Tian J, Chen S, Zhao Y, Wang Y, et al. (2013) A Graphene Oxide–Based Sensing Platform for The Label-free Assay of DNA Sequence and Exonuclease Activity via Long Range Resonance Energy Transfer. J fluor 23: 697-703.
16. Zhou M, Zhai Y and Dong S (2009) Electrochemical sensing and biosensing platform based on chemically reduced graphene oxide. Anal Chem 81: 5603-5613.
17. Yin H, Zhou Y, Ma Q, Ai S, Chen Q, et al. (2010) Electrocatalytic oxidation behavior of guanosine at graphene, chitosan and Fe 3 O 4 nanoparticles modified glassy carbon electrode and its determination. Talanta 82: 1193-1199.
18. JimmyáHuang P-J (2011) Synergistic pH effect for reversible shuttling aptamer-based biosensors between graphene oxide and target molecules. J Mat Chem 21: 8991-8993.
19. Li W, Wu P, Zhang H and Cai C (2012) Signal amplification of graphene oxide combining with restriction endonuclease for site-specific determination of DNA methylation and assay of methyltransferase activity. Anall Chem 84: 7583-7590.
20. Wang Z, Zhang J, Chen P, Zhou X, Yang Y, et al. (2011) Label-free, electrochemical detection of methicillin-resistant staphylococcus aureus DNA with reduced graphene oxide-modified electrodes. Biosens Bioelectron 26: 3881-3886.
21. Zhang Y and Jiang W (2012) Decorating graphene sheets with gold nanoparticles for the detection of sequence-specific DNA. Electrochimica Acta 71: 239-245.
22. Dong H, Zhu Z, Ju H and Yan F (2012) Triplex signal amplification for electrochemical DNA biosensing by coupling probe-gold nanoparticles–graphene modified electrode with enzyme functionalized carbon sphere as tracer. Biosens Bioelectron 33: 228-232.

Author: Md. Mahfujur Rahman
Institute: Universiti Malaya
Street: Jalan Universiti
City: Kuala Lumpur
Country: Malaysia
Email: dr_mahfuj@yahoo.co.uk

Author: Sharifah Bee Abd Hamid
Institute: Universiti Malaya
Street: Jalan Universiti
City: Kuala Lumpur
Country: Malaysia
Email: sharifahbee@um.edu.my

Quantification of Coronary Artery Cross Section Lumen Area and Area Stenosis with 3D Centerline-Centric Straightening

H.F. Cui[1], D.S. Wang[1], M. Wan[2], J.M. Zhang[3], X.D. Zhao[3], S.Y. Tan[3,4], A.S.L. Wong[3,4], R.S. Tan[3,4], W.M. Huang[5], W. Xiong[5], Y.P. Duan[5], J.Y. Zhou[5], Y.L. Chi[5], and L. Zhong[3,4]

[1] School of Physical and Mathematical Sciences, Nanyang Technological University, Singapore
[2] School of Information Engineering, Nanchang University, Nanchang, Jiangxi, P.R.China
[3] National Heart Centre Singapore, Singapore
[4] Duke-NUS Medical Graduate School, Singapore
[5] Institute for Infocomm Research (I2R), Singapore

Abstract— **Area stenosis (AS) is emerging as new index to assess the severity of coronary artery disease (CAD). Coronary artery lumen area and its derived area stenosis can be obtained noninvasively from computed tomography angiographic (CTA) images. In this study, we developed a novel approach with 3D centerline-centric straightening method, which consisted of original segmentation, smoothing, fast marching (FM) method and central cutting, to quantify coronary lumen area. We tested the differences of area stenosis in a human patient with suspected CAD by varying the combination of methods (Group 2: original segmentation + no smoothing + FM +central cutting; Group 3: Original segmentation + smoothing + thinning + central cutting; Group 4: denser segmentation + smoothing +FM + central cutting; Group 5: original segmentation + smoothing + FM + one-sided cutting). The data with our method demonstrated patient had area stenosis of 37.5% with low intra-observer variability (1.1%). Group 2 to Group 4 underestimated the area stenosis (AS = 32.3%, 36.8% and 32.0%, respectively), however, Group 5 overestimated the area stenosis (AS = 45.6%). All groups 2 to 5 had much higher intra-observer variability than our method. The computational time per data set is approximately 25 minutes, which demonstrates our methods clinical potential as a real-time cardiac assessment tool.**

Keywords— **Area stenosis, centerline extraction, centerline-centric straightening.**

I. Introduction

One of the aims of 3D cardiac image processing is to automatically and accurately reconstruct coronary arteries, for the detection and quantification of potential coronary artery stenosis. Currently, the severity of a coronary artery stenosis can be evaluated by two main different approaches: 1) by measuring the reduction in longitudinal diameter (angiography imaging) and 2) by evaluating lumen reduction (anatomo-pathological section or CT imaging) [1]. In clinical practice, the percentage stenosis is most commonly evaluated according to the degree of vessel diameter reduction resulting from the stenosis compared to the proximal normal diameter [2]. But unfortunately, accurate measurement of the focal minimum is a challenging task since it is dependent on the projection plane and reference site [3]. Lumen area measurement has distinct advantages and priorities over lumen diameter estimation, i.e., less dependency on viewing angle and less reliance on reference site selections. Additionally, lumen area is directly related to the hemodynamic properties of vessels, i.e., coronary blood flow or myocardial perfusion assessment, and is therefore expected to be more consistent and more significant [4].

Current approaches for cross sectional lumen area (CSLA) measurement are composed of several processing steps: 3D coronary artery segmentation, centerline extraction and cross sectional lumen area computation. In this study, we will focus on the quantitative assessment and investigation of each step affecting the accuracy of patient coronary artery CSLA, which is of utmost importance in clinical diagnosis and treatment planning of CAD. Furthermore, we introduce the concept of centerline-centric straightening and propose a new method for generating straightened segment of coronary artery along its centerline. The method can be used to facilitate quantitative assessment of the severity of coronary stenosis in an accurate and straightforward manner.

The remaining part of this paper is organized as follows. Section 2 presents our method for coronary artery CSLA measurement, which is described in three parts: 3D segmentation, centerline extraction and lumen area measurement. Our new approach for generating straightened segment of coronary artery is also explained in detail. In Section 3, we show the result of our approach over a coronary CT angiography (CTA) dataset from one patient. Conclusions are presented in Section 4.

F. Ibrahim et al. (eds.), *International Conference for Innovation in Biomedical Engineering and Life Sciences*,
IFMBE Proceedings 56,
DOI: 10.1007/978-981-10-0266-3_65

II. Methodologies

A. Coronary artery segmentation

Frangi's vesselness filter is applied to detect coronary arteries in 3D CT images. After that, morphological operations are conducted to extract all the remaining contrast filled structures within the heart. Then a threshold based on the intensities of coronary arteries is used to segment the complete coronary artery tree. A denser mesh is also generated in order to demonstrate the effect of prior segmentation result on the accuracy of CSLA.

B. Centerline extraction

There exists a large number of techniques for extracting coronary artery centerline, both from geometric and volumetric data. First, we apply a topological thinning method together with a pruning operation at voxel level. The skeleton consists of a set of segments, which need to be connected by using the minimum spanning tree algorithm, in order to create the final centerline. More detailed information about centerline extraction can be found in [5]. Moreover, we employ a number of fast marching method propagations to extract the centerline at subvoxel precision [6].

C. 3D cross sectional lumen area measurement

To analyze the 3D centerline of coronary artery, one skeleton point is defined as $\mathbf{P_0} = (x_0, y_0, z_0)$. The tangent vector to the centerline can be obtained by central or one-sided finite difference approximation and denoted as $\mathbf{N} = (a, b, c)$. Assume that $\mathbf{P} = (x, y, z)$ is any point in the normal plane perpendicular to the centerline. This plane has the equation

$$a(x - x_0) + b(y - y_0) + c(z - z_0) = 0 \quad (1)$$

Based on the extracted centerline, cross sectional plane can be cut at each skeleton point over the 3D surface mesh, eventually yielding a circular disk-like polygon. When cutting a set of vertices, V_i $(i = 1, 2, ..., n)$, at one skeleton point, a set of n triangles are formed by connecting each vertex V_i to the skeleton point, which decompose the polygon. For each single 2D triangle T_i, its area can be calculated based on the original CT images resolution. Therefore, the 3D cross sectional lumen area of this normal plane can be approximated by $Area(S) = \sum_{i=1}^{n} Area(T_i)$.

D. 3D centerline-centric straightening

Assume there are n skeleton points $p_i^{\mathcal{P}}(0,0)$ given in physical space $\mathcal{P}$ [7]. For each point $p_i^{\mathcal{P}}(0,0)$, the **uvn** basis $\mathbf{B_i}$ is defined by

$$\mathbf{u} = \mathbf{b_i},\ \mathbf{v} = \mathbf{n_i},\ \mathbf{n} = \mathbf{t_i};\ 1 \le i \le n \quad (2)$$

where $\mathbf{t_i}$, $\mathbf{n_i}$ and $\mathbf{b_i}$ are traditional normalized tangent, unit normal and unit binormal respectively. In [8] Daae Lampe et al. introduced a constant, user-specified up vector $\mathbf{u}$

$$\mathbf{b_m}(s) = \frac{\mathbf{u} \times \mathbf{t}(s)}{\|\mathbf{u} \times \mathbf{t}(s)\|} \quad \text{and} \quad \mathbf{n_m}(s) = \mathbf{t}(s) \times \mathbf{b_m}(s) \quad (3)$$

This technique is not applicable at those points where the tangent is parallel to $\mathbf{u}$. We extend this method by using this $\mathbf{u}$ only at those points where the tangent is not parallel to $\mathbf{u}$. But for those points where $\mathbf{t}(s) \parallel \mathbf{u}$, the binormal is computed by using the normal in the previous position as the "suggested" up vector.

Then vessel space $\mathcal{V}$ can be traced by the moving frame, which generates a curvilinear grid. Based on $\mathcal{V}$, a mapping from $\mathcal{P}$ to the straightened space $\mathcal{S}$ of a point $[x, y, z]_i^{\mathcal{P}}$ lying on the normal plane $\mathbf{P_i}$ is defined by

$$[x, y, i]^{\mathcal{S}} = \mathbf{B}_i^{-1}([x, y, z]_i^{\mathcal{P}} - p_i^{\mathcal{P}}(0,0)) + [0, 0, i] \quad (4)$$

The basis matrix $\mathbf{B_i}$ can be thought of as consisting of the columns $(\mathbf{b_i}|\mathbf{n_i}|\mathbf{t_i})$. As we are dealing with discrete skeleton points, the basis matrix must be calculated with finite differences.

Moreover, (4) defines a mapping from the straightened space $\mathcal{S}$ to physical space $\mathcal{P}$, $f: \mathcal{S} \to \mathcal{P}$ as

$$f(x, y, i) = p_i^{\mathcal{P}}(0,0) + x\mathbf{b_i} + y\mathbf{n_i} \quad (5)$$

An important property of this transformation is that it preserves distances orthogonally out from the center. For example, given two points $\mathbf{p_1}$, $\mathbf{p_2}$ in the i-th orthogonal plane in $\mathcal{S}$, which can be decomposed to $[x_1, y_1, i]$ and $[x_2, y_2, i]$ respectively. Then it can be shown that

$$\|f(\mathbf{p_2}) - f(\mathbf{p_1})\| = \|\mathbf{p_2} - \mathbf{p_1}\| = \|(x_2 + y_2) - (x_1 + y_1)\| \quad (6)$$

We utilize this planar coherence to create a fast implementation of curve-centric straightening by slice-by-slice-based integration along the centerline.

III. Results and Discussion

In this study, a connected components based coronary artery segmentation approach was applied into a cardiac CTA dataset acquired by a MDCT scanner with ECG gating in DICOM format. The data size was 512×512×560, and the resolution was 0.427mm×0.427mm×0.25mm. Moreover, the centerline and CSLA were explored to quantify the anatomic significance of coronary artery stenosis [3]. All the experiments were performed on a HP Z800 Windows Server. Fig. 1(a) shows the obtained segment of Left Anterial Descending (LAD) in space $\boldsymbol{S}$. Fig. 1(b) depicts the centerline obtained by using the fast marching method, which is formed by connecting 191 skeleton points. Fig. 1(c) shows an example of curvilinear grid generation in 3D. The lesion site is located at the 103-th skeleton point and the 127-th skeleton point is selected as the proximal reference site. The area stenosis is define as $(Area_{ref} - Area_{les})/Area_{ref} * 100\%$.

Table 1 summarizes the different method combinations in this study for quantitative assessment of area stenosis. Group 1 consisted of connected components based segmentation, centerline extraction by fast marching method, and cross sectional plane cutting by using central finite difference, which currently produces the best result regarding a good visual result of centerline and lumen contour. The remaining four groups of CSLA were sampled, each of which changed only one factor and kept the other factors fixed compared to Group 1. Fig. 2 shows the cross sectional lumen profile trend along the branch for five different groups of data. It can be seen that Group 2 produces larger CSLA at almost all the skeleton points. This is because the centerline and the cross sectional plane are explored over the coronary artery surface mesh without a smoothing operation. Moreover, Group 5 gives the CSLA with larger oscillation along the segment of LAD, since one-sided finite difference approximation of tangent vector along the centerline is less accurate and reliable.

In Table 2 we compare the obtained minimal area, proximal area and area stenosis between five groups. As shown in Fig. 2, Group 2 overestimates the cross sectional lumen areas at both lesion site and reference site, but with larger proportion at lesion site. It therefore underestimates the area stenosis. Group 5 gives larger area stenosis since the gradient vector approximated by one-sided finite difference produces larger CSLA at reference site. Table 3 was obtained by repeating exactly the same procedure as in Table 2. It can be seen that area stenosis in Group 1 has the least intra-observer

Table 1: Different method combinations for CSLA measurement.

	Segmentation		Smoothing		Centerline		Cutting	
	Original	Denser	Yes	No	Thinning	FM	Central	One-Sided
Group 1	✓		✓			✓	✓	
Group 2	✓			✓		✓	✓	
Group 3	✓		✓		✓		✓	
Group 4		✓	✓			✓	✓	
Group 5	✓		✓			✓		✓

Note: FM–Fast marching method, CSLA–Cross sectional lumen area.

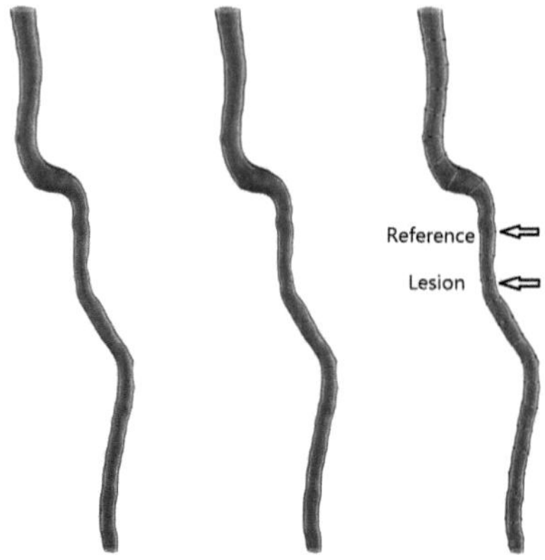

(a) (b) (c)

Fig. 1: (a) Segment of LAD. (b) The centerline obtained by fast marching method. (c) A curvilinear grid is constructed along and around the centerline, and after the reformation, this grid becomes a new Cartesian grid.

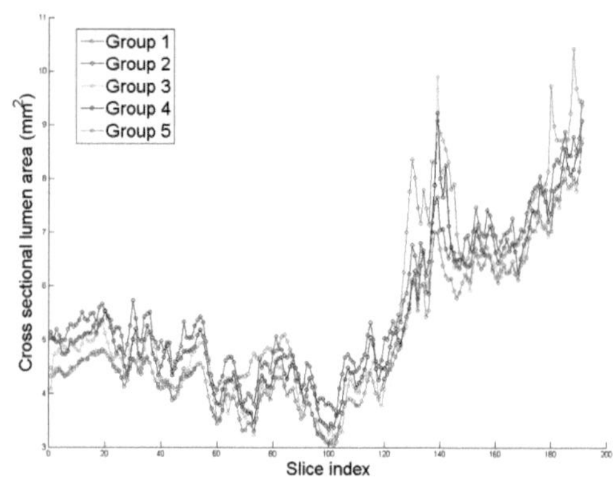

Fig. 2: A comparison of cross sectional profile along the segment of LAD.

variation. Group 4 also gives small intra-observer variation but it underestimates the area stenosis. Fig. 3 depicts the obtained cross sectional lumen contours at the same lesion site for different groups. It is worth noting that Group 5 totally changes the geometric feature of the lumen contour at lesion site. Moreover, the lumen contour of Group 3 is shifted due to the fact that extraction of centerline at voxel level may change the tangent vector along the centerline. In Fig. 4 we plot the straightened segment near the lesion site for different groups. It can be seen that Group 2 and 4 give less reliable straightened artery and cause misleading severity of coronary artery stenosis.

IV. Conclusion

In this study, we developed a novel method to quantify coronary stenosis using area stenosis index. This method

Table 2: Minimal area, proximal area, and area stenosis for each group.

	Cross Sectional Lumen Area - 1st time		
	$Area_{min}$ (mm^2)	$Area_{pro}$ (mm^2)	Area stenosis
Group 1	3.0396	4.8646	37.5%
Group 2	3.6099	5.3313	32.3%
Group 3	3.1357	4.9645	36.8%
Group 4	3.3401	4.9137	32.0%
Group 5	3.0753	5.6540	45.6%

Note: $Area_{min}$–Minimal Area, $Area_{pro}$–Proximal area.

Table 3: Minimal area, proximal area, and area stenosis for each group.

	Cross Sectional Lumen Area - 2nd time		
	$Area_{min}$ (mm^2)	$Area_{pro}$ (mm^2)	Area stenosis
Group 1	3.0727	4.9483	37.9%
Group 2	3.7170	5.6254	33.9%
Group 3	3.5436	5.2441	32.4%
Group 4	3.6061	5.3800	32.9%
Group 5	3.0704	6.2047	50.5%

Note: $Area_{min}$–Minimal Area, $Area_{pro}$–Proximal area.

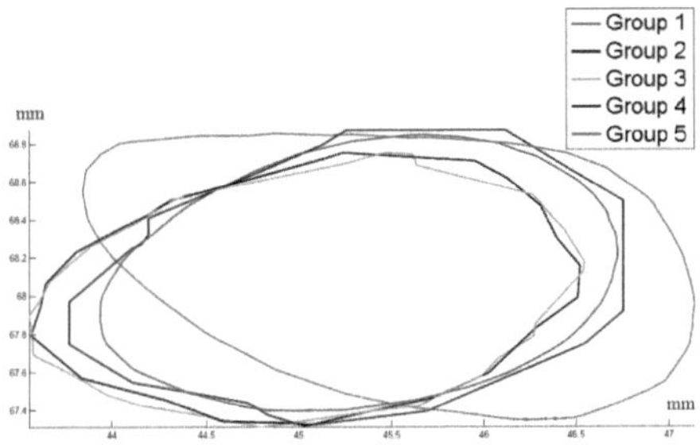

Fig. 3: A comparison of lumen contour at one specific lesion site for five different groups .

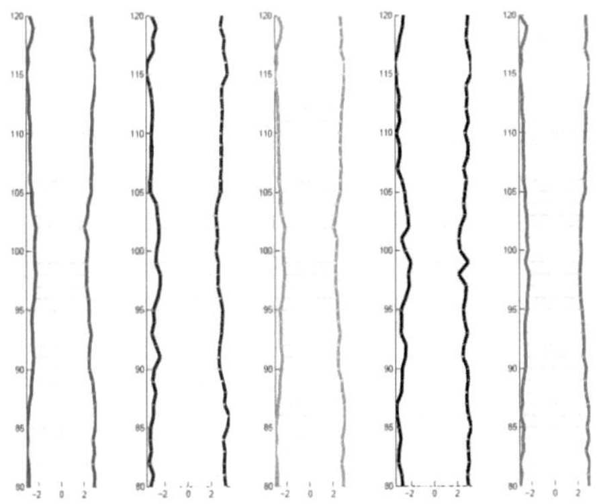

Fig. 4: A comparison of straightened segment at lesion site for different groups.

consists of original segmentation, smoothing, fast marching (FM) method and central cutting. The data demonstrates that this method carries very minimal intra-observer variation in time-efficient manner. However, investigation on larger population is necessary before making a final conclusion. This method holds potential for downstream computational fluid dynamics analysis of coronary flow simulation in clinical bedside.

Conflict of Interest

The authors declare that they have no conflict of interest.

Acknowledgment

The study is partially supported by Biomedical Research Council Research Grant (14/1/32/24/002) and research grant from the Agency for Science, Technology and Research (A*STAR), SERC Biomedical Engineering Programme (1321480008). We are very grateful to the National Heart Centre Singapore for the DICOM data set.

References

1. Nieman Koen, Cademartiri Filippo, Lemos Pedro A, Raaijmakers Rolf, Pattynama Peter MT, Feyter Pim J. Reliable noninvasive coronary angiography with fast submillimeter multislice spiral computed tomography *Circulation.* 2002;106:2051–2054.
2. Samuels Owen B, Joseph Gregg J, Lynn Michael J, Smith Harriet A, Chimowitz Marc I. A standardized method for measuring intracranial arterial stenosis *American journal of neuroradiology.* 2000;21:643–646.
3. Zhang Jun-Mei, Zhong Liang, Su Boyang, et al. Perspective on CFD studies of coronary artery disease lesions and hemodynamics: A review *International journal for numerical methods in biomedical engineering.* 2014;30:659–680.
4. Topol Eric J, Nissen Steven E. Our preoccupation with coronary luminology the dissociation between clinical and angiographic findings in ischemic heart disease *Circulation.* 1995;92:2333–2342.
5. Cui Hengfei, Wang Desheng, Wan Min, et al. Coronary artery segmentation via Hessian filter and curve-skeleton extraction in *Biomedical Engineering and Sciences (IECBES), 2014 IEEE Conference on*:93–98IEEE 2014.
6. Van Uitert Robert, Bitter Ingmar. Subvoxel precise skeletons of volumetric data based on fast marching methods *Medical physics.* 2007;34:627–638.
7. Angelelli Paolo, Hauser Helwig. Straightening tubular flow for side-by-side visualization *Visualization and Computer Graphics, IEEE Transactions on.* 2011;17:2063–2070.
8. Lampe Ove Daae, Correa Carlos, Ma Kwan-Liu, Hauser Helwig. Curve-centric volume reformation for comparative visualization *Visualization and Computer Graphics, IEEE Transactions on.* 2009;15:1235–1242.

Author: Liang Zhong
Institute: National Heart Centre Singapore
Street: 5 Hospital Drive, 169609
City: Singapore
Country: Singapore
Email: zhong.liang@nhcs.com.sg

Curvedness-Based Imaging of the Heart: From Bench to Clinical Cardiology

L. Zhong[1,2], Y. Su[3], and R.S. Tan[1,2]

[1] National Heart Centre Singapore, Singapore
[2] Duke-NUS Graduate Medical School, Singapore
[3] Institute of High Performance Computing, A*STAR, Singapore

Abstract— **Heart disease is a major health care burden. Early diagnosis of heart disease and its underlying etiology is paramount. Doctors use diverse imaging tests– echocardiography, nuclear scintigraphy and cardiac magnetic resonance (CMR) imaging –for diagnosis and prognostication. CMR is arguably the most accurate and reproducible. We have pioneered the concept of surface curvedness as a descriptor of local left ventricular (LV) function that does not require a frame of reference in contrast to other measures, and further developed curvedness-based imaging (CBI) by combining MRI and advanced computational methodologies. CBI has been used in a wide range of cardiac-related applications, such as i) assessment of left ventricular remodeling in patients with ischemic cardiomyopathy and in patients after myocardial infarction; ii) assessment of right ventricular remodelling in congenital heart disease (e.g., repaired Tetralogy of Fallot); iii) quantification of ventricular dyssynchrony, and iv) evaluation of surgical efficacy in heart failure (e.g., ventricular restoration surgery). Our developed software for CBI analysis greatly reduces the time a radiologist or cardiologist spends in analyzing cardiac images. It produces comprehensive curvedness-based measures of cardiac remodeling. Incorporating such a technology into the clinical management pathways may improve treatment choice and prognostication, potentially reducing overall healthcare costs.**

Keywords— **Heart Function, Regional remodeling, Cardiac magnetic resonance, Cardiology**

I. INTRODUCTION

Heart failure (HF) is a major health care burden on society and suffering for the individual [1]. In the United States, HF affects 5 million patients annually. It is the leading cause of hospitalization for people 65 years and the rate of hospital readmission within 6 months ranges from 25% to 50% [2]. In Singapore, HF accounted for 4.5% of all hospital admissions and 2.5% of overall mortality in the geriatric age group [3]. It confers an annual mortality of 10% [4]. HF can occur with either preserved or reduced left ventricular (LV) ejection fraction (EF), depending on different degrees of cardiac remodeling [5-10]. Cardiac remodeling can be defined as a change in shape, size and function of the heart due to physiological and pathological conditions (e.g., after a heart attack). Quantification of cardiac remodeling is important both for HF diagnosis and monitoring of progress in clinical cardiology.

Diverse heart imaging modalities can be used to assess global and regional myocardial function to aid in HF diagnosis: echocardiography, nuclear emission computed tomography and cardiac magnetic resonance (CMR). All these modalities can determine LVEF and volumes. Echocardiography is largely operator-dependent and, due to acoustic window limitations, suffers from poor inter-study reproducibility [11]. Nuclear imaging employs ionizing radiation and possesses inferior spatial and temporal resolution [12]. CMR provides superior resolution and reproducible results, but current techniques fail to exploit the full potential of the rich image dataset. Frequently, there is lack of agreement between modalities. The disparate threshold values and normal ranges can render the assessment of patients with borderline LVEF difficult. Furthermore, inter-study reproducibility can range from about 5% to 7% for CMR to about 15% to 20% for echocardiography, making serial assessment of HF problematic, especially if different modalities had been used. Therefore, reproducibility, automated, comprehensive assessment of systolic and diastolic ventricular performance behavior, and physiological meaningful are needed. In our previous work, we have demonstrated ventricular curvedness and curvedness-based imaging as a novel and promising clinical approach for assessment of global and regional myocardial LV function in diverse heart diseases.

II. METHOD

The key concept behind our methodology is using curvedness and its derivations as quantitative indices of cardiac remodeling. As such, it is important to introduce the notion of curvedness with respect to a three-dimensional (3D) surface.

A. *What is Surface Curvedness?*

Consider the general case of a set of connected points in 3D space such that they form a 2-manifold discrete surface. To estimate the analytical form of the underlying surface geometry, a local surface fitting approach can be used. In this approach, for a point on the surface under consideration, 20 neighborhood points in its vicinity are used to estimate the surface geometry by an osculating paraboloid (Fig. 1). The coefficient, z, of the paraboloid at each neighborhood

F. Ibrahim et al. (eds.), *International Conference for Innovation in Biomedical Engineering and Life Sciences*,
IFMBE Proceedings 56,
DOI: 10.1007/978-981-10-0266-3_66

point, with respect to the centre point, is obtained. The estimated surface is then compared against the original surface, such that the divergence of the estimated paraboloid and the data points are minimized. This is similar to minimizing the difference between the calculated value and measured value of z.

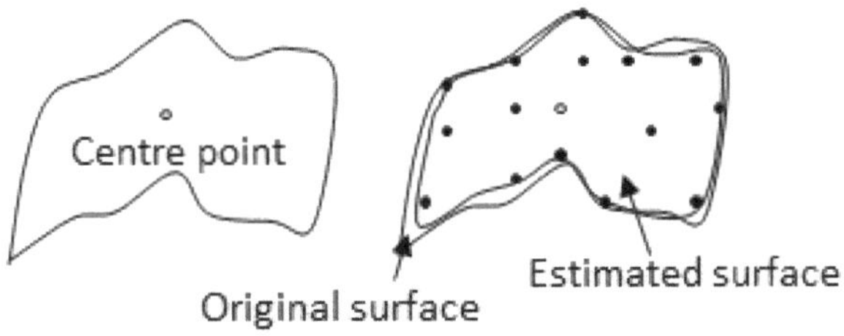

Fig. 1 Local surface estimation using a local patch fitting approach

The maximum and minimum principal curvatures at a point on a surface, expressed as k_1 and k_2, respectively, are obtained from the roots of:

$$\det = \begin{vmatrix} L-kE & M-kF \\ M-kF & N-kG \end{vmatrix} = 0$$

where E, F, G, L, M and N are components of the first and second fundamental forms of the surface at that point.

The curvedness, C, is defined as:

$$C = \sqrt{\frac{k_1^2 + k_2^2}{2}}$$

The change in curvedness from end-diastole to end-systole, ΔC, is defined as:

$$\Delta C = \frac{C_{end-systole} - C_{end-diastole}}{C_{end-systole}} \times 100$$

B. Pipeline for Curvedness Determination

The details of the procedure for determining the LV endocardial surface curvedness from cardiac MR images has been described in our previous publications [13-15], and can be summarized as follows:

1. Interactively segment the LV contours from the CMR images.
2. Reconstruct the LV surface by triangulation of the LV contour points derived from the segmentation process.
3. For each vertex on the LV surface, select a set of n-ring neighboring vertices.
4. At each selected vertex, perform quadric surface patch fitting and compute the surface curvedness and curvedness-based parameters (i.e., wall thickness, area strain, wall stress).

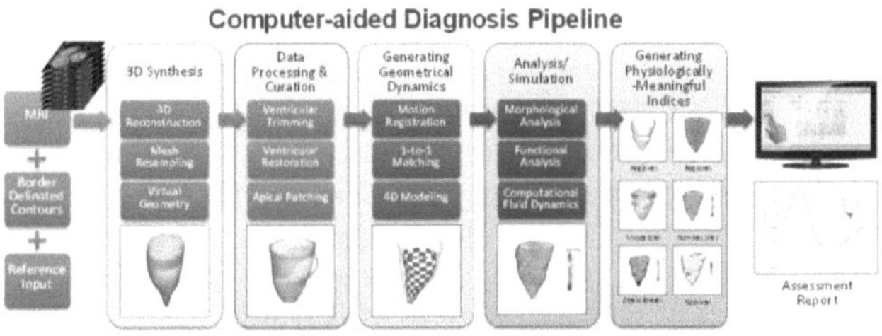

Fig. 2 Computer-aided diagnosis pipeline for curvedness determination

III. APPLICATIONS

Through extensive research, scientists at the National Heart Research Institute of Singapore (NHRIS) and the Institute of High Performance Computing (IHPC), A*STAR have developed the Cardiowerkz software to automatically segment, reconstruct three-dimensional cardiac structure and derive quantitative measures to assess cardiac remodeling (i.e., curvedness, wall stress and area strain). The Cardiowerkz software is a user-friendly platform with accurate and robust algorithms. The complete computational pipeline (as shown in Fig. 2) involves a series of complex algorithmic procedures: (1) 3D synthesis, (2) data processing and curation, (3) generation of geometrical dynamics, (4) analysis processes, and (5) generation of cardiac indices. The pipeline takes about 2-5 mins to analyze a full cardiac cycle on a commodity computer system.

The Cardiowerkz software has been used in a wide range of cardiac-related applications, such as

- assessment of left ventricular remodeling in patients with ischemic cardiomyopathy and in patients after myocardial infarction [16-18],
- assessment of right ventricular remodeling in congenital heart disease (e.g., repaired Tetralogy of Fallot) [19],
- quantification of ventricular dyssynchrony, and
- evaluation of surgical efficacy in heart failure (e.g., ventricular restoration surgery) [20].

IV. CONCLUSIONS

The Cardiowerkz software greatly reduces the time a radiologist or cardiologist spends in analyzing cardiac images. It produces comprehensive curvedness-based measures of cardiac remodeling. Incorporating such a technology into the clinical management pathways may improve treatment choice and prognostication, potentially reducing overall healthcare costs.

ACKNOWLEDGMENT

The study is partially supported by the Biomedical Research Council Research Grant (14/1/32/24/002) and research grant from the Agency for Science, Technology and Research (A*STAR), SERC Biomedical Engineering Programme (1321480012), and the Goh Cardiovascular Research Grant (Duke-NUSGCR/2013/0009).

CONFLICT OF INTEREST

The authors declare that there is no conflict of interest.

REFERENCES

1. Hunt AS (2005). ACC/AHA 2005 guideline update for diagnosis and management of chronic heart failure in the adult: a report of the American College of Cardi0ology/American Heart Association Task Force on Practice Guidelines (Writing Committee to Update the 2001 Guidelines for the Evaluation and Management of Heart Failure). J Am Coll Cardiol 46:e1-82. Erratum in: J Am Coll Cardiol 47:503-505.
2. Heart Disease and Stroke Statistics-2004 Update (2004). Dallas, Tex: American Heart Association.
3. Ng TP, Niti M (2003). Trends and ethnic differences in hospital admissions and mortality for congestive heart failure in the elderly in Singapore, 1991-1998. Heart 89:865-870.
4. Krumholz HM, Chen YT, Wang Y, Caccarino V, Radford MJ, Horwita RI (2000). Predictors of readmission among elderly survivors of admission with heart failure. Am Heart J 139:72-77
5. Stefan Neubauer (2007). The failing heart – an engine out of fuel. N Engl J Med 356:1140-1151
6. Hogg K, Swedberg K, mcMurray J (2004). Heart failure with preserved left ventricular systolic function: epidemiology, clinical characteristics, and prognosis. J Am Coll Cardiol 43:317-327.
7. Ahmed A, Rich MW, Fleg JL, Zile MR, Young JB, Kitzman DW (2006). Effects of digosin on morbidity and mortality in diastolic heart failure: the ancillary digitalis investigation group trial. Circulation 114:397-403.
8. Yusuf S, Pfeffer MA, Swedberg K, Granger CB, Held P, McMurray JJ (2003). Effects of candesartan in patients with chronic heart failure and preserved left ventricular ejection fraction: the CHARM-Preserved Trial. Lancet 362:777-781.
9. Cleland JG, Tendera M, Adamus J, Freemantle N, Polonski L, Taylor J (2006). The perindopril in elderly people with chronic heart failure (PEP-CHF) study. Eur Heart J 27:2338-2345.
10. Massie BM, Carson PE, McMurray JJ, Komajda M, McKelvie R (2008). Irbesartan in patients with heart failure and preserved ejection fraction. N Eng J Med 359:2456-2467.
11. Bellenger NG, Burgess MI, Ray SG (2000). Comparison of left ventricular ejection fraction and volumes in heart failure by echocardiography, radionuclide ventriculography and cardiovascular magnetic resonance; are they interchangeable? Eur Heart J 21:1387-1396.
12. Brenner DJ, Hall EJ (2007). Computed tomography: an increasing source of radiation exposure. N Engl J Med 357:2277-2284.
13. Zhong L, Yeo SY, Su Y, Le TT, Tan RS, Ghista D (2007). Regional assessment of left ventricular surface shape from magnetic resonance imaging. Conf Proc IEEE Eng Med Biol Soc 2007:884-887.
14. Yeo SY, Zhong L, Su Y, Tan RS, Ghista DN (2007). Analysis of left ventricular surface deformation during isovolumic contraction. Conf Proc IEEE Eng Med Biol Soc 2007:787-790.
15. Yeo SY, Zhong L, Su Y, Tan RS, Ghista DN (2009). A curvature-based approach for left ventricular shape analysis from cardiac magnetic resonance imaging. Med Biol Eng Comput 2009;47(3):313-322.
16. Zhong L, Su Y, Yeo SY, Tan RS, Ghista DN, Kassab G (2009). Left ventricular regional wall curvedness and wall stress assessment in patients with ischemic dilated cardiomyopathy, Am. J. Physiol. Heart Circ. Physiol.; 296(3): H573-H584.
17. Su Y, Zhong L, Lim CW, Ghista DN, Chua T, Tan RS (2012). A geometrical approach for evaluating left ventricular remodeling in myocardial infarct patients, Comput. Methods Programs Biomed.; 108(2):500-510.
18. Teo SK, Vos FJ, Tan RS, Zhong L, Su Y (2015) Regional ejection fraction and regional area strain for left ventricular function assessment in male patients after first-time myocardial infarction. J R Soc Interface 12(105).
19. Zhong L, Gobeawan L, Su Y, Tan YL, Ghista DN, Chua T, Tan RS, Kassab GS (2012). Right ventricular regional wall curvedness and area strain in patients with repaired Tetralogy of Fallot, Am. J. Physiol. Heart Circ. Physiol.; 302(6):H1306-H1316.
20. Zhong L, Su Y, Gobeawan L, Sola S, Tan RS, Navia JL, Ghista DN, Chua T, Guccione J, Kassab GS (2011) Impact of surgical ventricular restoration on ventricular shape, wall stress and function in heart failure patients, Am. J. Physiol. Heart Circ. Physiol.; 300(5):H1653-60.

Corresponding author:

Author:	L. Zhong
Institute:	ational Heart Centre Singapore
Street:	5 Hospital Drive 169609
City:	Singapore
Country:	Singapore
Email:	Zhong.liang@nhcs.com.sg

Quantitative Evaluation of Spinal Coronal Curvature for Scoliosis Using a Fast 3-D Ultrasound Projection Imaging Method

W.W Jiang, G.Q. Zhou, K.L. Lai, and Y.P. Zheng

Interdisciplinary Division of Biomedical Engineering, The Hong Kong Polytechnic University, Kowloon, Hong Kong, China

Abstract— **In this study, a novel fast 3-D ultrasound projection imaging (FUPI) method to provide quantitative evaluation of spinal coronal curvature for scoliosis patients was proposed. Unlike conventional 3-D volume rendering approaches, this method directly projected the raw images to form the coronal images. The non-planar rendering method, following the natural curve of the spine, was utilized to contain the complete spine information into the projection images. Based on 30 patients with scoliosis (ages of 16.3±2.9 years), the comparison study between the new method and the conventional 3-D rendering method was performed. The processing time and the projection images were both compared. The average processing times for the two imaging methods were 15.07±0.03s and 149.50±33.44s, respectively. There were high correlations between the measurement results using the images obtained by the two 3-D imaging methods (y = 0.9733x, r=0.970 for thoracic region, y = 1.0224x, r=0.968 for lumbar region). The above results demonstrated the new method could greatly reduce the processing time while preserving the comparative image quality. It can be expected that the developed FUPI method may help to provide fast 3-D ultrasound diagnosis of scoliosis in clinics.**

Keywords— **scoliosis, spine, coronal image, 3-D ultrasound, projection imaging.**

I. INTRODUCTION

Scoliosis is a complex three-dimensional spine deformity associated with vertebral axial rotation and the lateral deviation. Adolescent idiopathic scoliosis (AIS) is the most common form of scoliosis, with the prevalence of 2% to 4% in the United States [1] and 3.08% in Hong Kong [2]. It was reported that AIS could cause a series of health problems including the thoracic insufficiency syndrome, back pain, spine degeneration, and psychosocial issues such as increased depression and lower self-image [3-5]. For the teenagers with AIS, the situation is even worse because the risk of curve progression is very high for the immature skeleton. However, the prognostics of AIS is still far from satisfactory, though many factors have been studied such as Risser sign, Cobb angle, standing heights, sitting heights, and curve pattern. Regular observation is important to monitor the curve progression, and it is a necessary step for the follow-up diagnosis, surgery planning, and the treatment outcome assessment.

In clinics, standing X-ray radiography is the most common diagnosis method for scoliosis curvature assessment [6]. However, it was reported that the frequent exposure to X-ray could raise the risk of cancer, especially breast cancer, because the breast was exposed to X-ray during scoliosis diagnostic radiography and the radiation dose cumulated to the breast with the frequent examinations [7]. In addition to the breast cancer, it was reported that the patients of AIS with radiation exposure had a 7.5% increased risk of developing lung cancer [8]. And it was reported that the risk was obviously increased with the cumulative radiation dose. Another drawback of X-ray radiography examination is that the intra-rater and inter-rater variation of curve deformation measurement using Cobb angle can be up to 3-5° and 6-9° respectively [9, 10]. In clinics, an increase of 5° or more is used as the indicating curve progression. So the measurement variation can cause problems to the clinical diagnosis based on the angle measurement results, especially for the long-term monitoring. Except for the standing X-ray radiograph, the surface topographic methods are also used to estimate spine curvature using stereo cameras or finger palpation. Measurement systems of surface topography are radiation free and cost effective. However, these methods are fraught with limitations of inability to provide accurate measurements and to visualize the bony structure.

Compared with the above assessment methods, ultrasound has advantages of being radiation-free, low-cost, real-time, and it allows visualizing the accurate bony architecture. Therefore, the approach of using ultrasound for scoliosis has attracted more and more attentions over the past decade. The feasibility of ultrasound measurement in scoliosis was investigated in as early as 1989 [11]. In recent years, freehand 3-D ultrasound, combining conventional 2D ultrasound with position sensor, has been advanced to overcome the limitations of 2D viewing and measuring of 3-D anatomy, and a number of such systems have been reported for scoliosis assessment. One approach for scoliosis measurement is to measure the spinal curvature on the 2D ultrasound images with the help of the positional tracking. The images containing the transverse processes were either picked from a pile of recorded 2D raw B-mode images [12] or captured in real-time while locating the target from observations [13]. However, this approach is time-consuming

F. Ibrahim et al. (eds.), *International Conference for Innovation in Biomedical Engineering and Life Sciences*,
IFMBE Proceedings 56,
DOI: 10.1007/978-981-10-0266-3_67

because of the identification of the sonographic landmarks of the transverse process from dozens of acquired images. In addition, this method does not generate 3-D representation of the spine anatomy, lacking the ability of viewing the whole spine anatomy. Another approach to utilize the positional information of images is to form the whole spine image. One study was to use the maximum intensity projection method to get the coronal images and use landmarks of laminae to measure the spinal curvature [14]. In this method, the sagittal images were firstly generated and used as the guidance. This method was relative tedious and time-consuming for the manual marking. Another study to generate the coronal spine image was conducted by Cheung et al [15]. In this study, a 3-D volume was reconstructed according to the acquired data set. And then the coronal images were generated from the volume data. Compared with the above method, this volume projection imaging method can provide coronal image without the manual intervene. However, the 3-D volume reconstruction is still time-consuming because of the large raw image data (2000 scans, 640*480 pixels, about 600 MB).

In this study, a fast 3-D ultrasound projection imaging method was proposed to provide the spinal coronal images. To demonstrate the system performance, its processing time and the generated coronal images were compared with that of an earlier reported volume rendering method.

II. METHODS

A. System Overview

A freehand 3-D ultrasound imaging system, named Scolioscan system (Model SCN801, Telefield Medical Imaging Ltd, Hong Kong), was developed with industrial and ergonomic designs of the hardware for the assessment of scoliosis. The system is shown in Fig. 1. A rigid frame, consisting of the chest board, hip board and four supporters, was used to help the subject to maintain a stable posture during the ultrasound scanning. Before scanning, the subject stood on the platform and the two boards moved up and down along the side panel to be repositioned according to the height of the subject. Two supporters on the chest board were relocated to align with clavicle anterior concavities, whereas two supporters on the hip board were relocated to align with bilateral anterior superior iliac spines, the length of supporter's shafts on both boards were adjusted until they came in contact with the patient. A liner 2D ultrasound probe (frequency: 4-10 MHz; width: 10 cm) was used to scan the subject. An electromagnetic position sensor was attached to the probe to obtain the spatial information of ultrasound images. The electromagnetic transmitter was put in the transmitter box. Two LCD monitors were utilized in this system. The operator monitor was used for setting scanning parameters, saving and retrieving data, performing reconstruction, displaying images, conducting measurement, and generating reports. The patient monitor was to provide information for patients, including a green eye spot with location set according to the height of patient to facilitate him/her to keep a stable posture for head and neck during scanning.

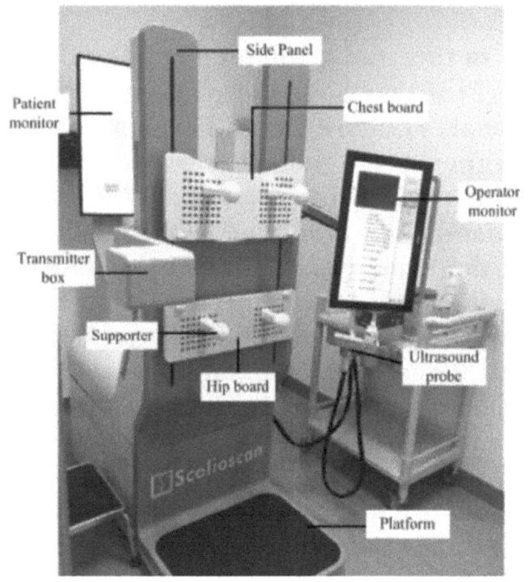

Fig. 1 The Scolioscan system and it components

B. *Data Acquisition and the Fast 3-D Ultrasound Projection Imaging Method (FUPI)*

Before scanning, the subject was asked to remove all metallic items. He/she wore a gown with back opened to the operator and stood in front of the boards. The operator adjusted the chest and hip boards, and the four supporters as required to keep the subject in a natural stable posture. Then the ultrasound gel was applied to the scanning area to ensure the good coupling between the probe and the skin. The operator put the probe on the subject and adjusted the ultrasound parameters according to the tissue condition. In this study, the scanning covered the spine area from the fifth lumbar vertebra (L5) to the first thoracic vertebra (T1). The operator put the probe on L5 and T1 to record the lower and upper boundaries of the scanning, respectively. Finally, the operator moved the probe to L5 to initiate the scanning. The probe was steered to cover the spine vertically. The ultrasound images together with their corresponding position data were recorded. When the probe passed through the upper boundary, the data collection would be stopped automatically. It took about 30 seconds to complete the whole scanning and about 2000 2D ultrasound images were recorded for the further processing.

In conventional 3-D ultrasound imaging method, two steps are usually utilized to visualize the spine anatomy. The first step is to reconstruct the tissue volume according to the 2D ultrasound raw data and positional information.

Then the volume is visualized using the planar or non-planar reslicing method. An earlier reported method for scoliosis, named volume projection imaging (VPI) method, just adopted the above two-step imaging method to get the spinal coronal images [15]. However, in 3-D ultrasound imaging system for scoliosis, the purpose is to provide the coronal images to further assess the spine deformity. So the first step of volume reconstruction can be skipped to improve the speed of visualization. In this study, a fast 3-D ultrasound projection imaging method (FUPI) which bypassed the procedure of volume reconstruction was proposed to generate the volume projection images. This method was based on a narrow-band volume rendering method introduced by Gee et al [16]. In the proposed FUPI method, the rendered coronal image was obtained by directly projecting a layer with a certain thickness of the raw image data to the coronal plane. To follow the shape of the spine, a non-planar layer which was defined using the distance to the skin, was applied to be projected to the coronal plane, as shown in Fig. 2a. A coronal image coordinate system with a regular pixel array was defined according to the data set. Based on the spatial information and the calibration matrix, each pixel on the defined layer was transformed to the new coordinate system. The final value of the pixel on the new coronal plane was determined on the basis of all pixels falling into its region. The average blending method was used to determine the final intensity of the pixel by using all pixels in the region. After images were processed, the interpolation was implemented to fill the gaps in the projection image. In this study, considering the computation time and the interpolation performance, the common bilinear interpolation method was adopted. The average of the non-empty pixels in the nearest 2-by-2 neighborhood was set as the value of the empty pixel. Finally, the image was further enhanced with Histogram equalization that is one of the most useful image enhancement techniques.

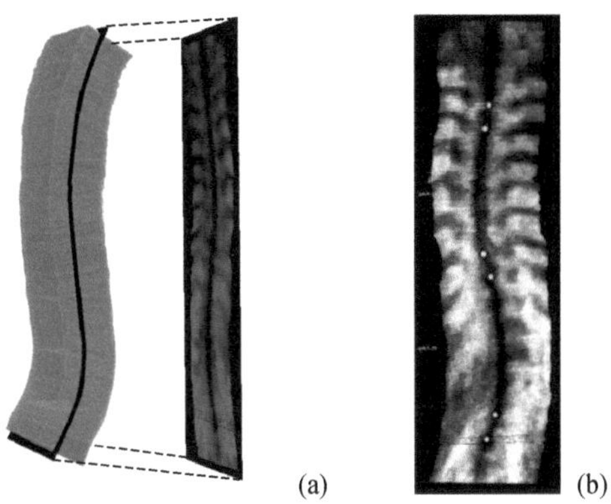

Fig. 2 Illustration of the non-planar rendering approach and curvature angle measurement method. (a) The non-planar rendering method; (b) The angle measurement method.

To evaluate the spine deformity, the angle measurement method using the shadow generated by the spinous was developed by Cheung et al [15]. The principle of this measurement method was to identify the inflection point of the spinous shadow line and the vertebrae containing the inflection point was treated as the most tilted one. As shown in Figure 2b, three short lines were manually drawn. They located in the middle of the shadow line and covered the corresponding vertebra. Then the angle between two lines were calculated and used as the spine curvature angle. There are two curvature angles in Figure 2b, the first angle between the upper two lines is the angle in thoracic regions and the second angle between the lower two lines is the angle in lumbar regions.

C. Comparison Study between the Fast 3-D Ultrasound Projection Imaging (FUPI) Method and the Conventional Volume Rendering Method

To investigate the performance of the FUPI method, a comparison study between the new method and the earlier VPI method was performed. The two imaging systems ran on the same computer with configuration of Intel Core i5 and 32.0GB of memories. 30 patients with scoliosis (7 male and 23 female; mean age, 16.3 ± 2.9 years; BMI, 18.5 ± 1.6 kg/m^2) were recruited in the Department of Orthopaedics and Traumatology of The Chinese University of Hong Kong. This study was approved by the human subject ethics committee of the institution and all patients (or their parents for those patients under 18 years old) gave the informed consent. The recruited patients were scanned by the sonography operator using the Scolioscan system and the raw data including the images and their positional information were saved. The raw data were processed by the two imaging methods respectively. The comparison study included the processing time comparison the curvature angle measurement results comparison. The processing times of the two imaging methods were recorded respectively. For each patient, two images were obtained using the two methods. And there were totally 60 images (30 patients * 2 processing methods) to be measured in this test. The same operator measured the angles on all projection images using the previously described method, as shown in Figure 2b.

D. Data Analysis

The processing times using the FUPI method and the VPI method on 30 patients was analyzed. The correlation relationship of the processing time and the data size was also presented. The angles on the measurement images were manually drawn and values were compared to observe the difference between the two imaging methods. The Pearson correlation coefficient r was used to assess the correlation of the two methods. All statistical evaluations were performed using the statistical software (SPSS for Windows, version 17.0; SPSS, Chicago, IL, USA).

III. Results

A. Processing Time Comparison

The results showed that the processing times on 30 subjects were 149.50±33.44s (range: 102.07-248.59s) and 15.07±0.03s (range: 15.02-15.15s) for the VPI method and the FUPI method, respectively. The VPI method takes, on average, 10 times of processing time compared with the developed method. In addition, the standard deviation (SD) values is much larger (33.44s) for the time of the VPI method compared with the time of the FUPI method (0.03s), which means the processing time varies largely for different subjects. This is further demonstrated by Figure 3, which shows the relationship between the processing time and the acquired data size. As shown in Figure 3a, there is a clear linear relationship between the time and the processing images numbers for the VPI method. This tread indicates that the processing time increases along with the size of acquired raw data. While for the developed method, there is no relationship between the time and the acquired image number. In this study, for a data set with about 4000 2D images, the processing time of the VPI method was 248.59s, which was more than 16 times of the time (15.15s) used by the FUPI method. The large difference on the processing time indicates the great advantage of the proposed method on time saving, especially for the large image data.

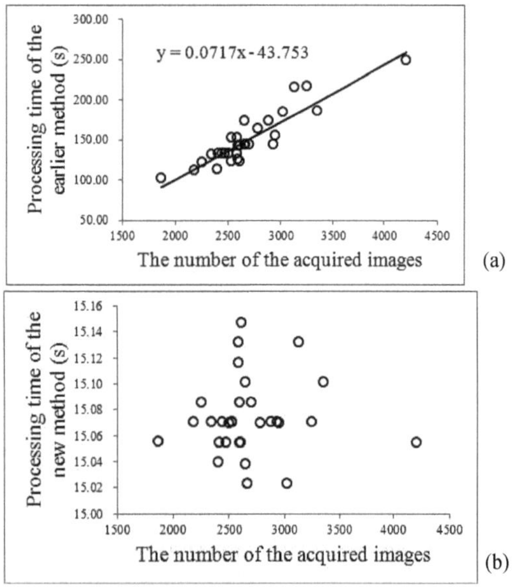

Fig. 3 The relationship between the processing time and the number of the acquired image. (a) The relationship for the VPI method. (b) The relationship for the FUPI method.

B. Angle Measurement Results Comparison

Fig. 4 shows three typical measurement images with the curvature angles from small to large. For each subject, two measurement images were generated using the VPI method and the FUPI method, respectively. The correlation plots between two measurements for 30 patients are shown in Fig. 5. Very good linear correlations were obtained for the measurements both in thoracic region (y = 0.9733x, r = 0.970) and in lumbar region (y = 1.0224x, r = 0.968). The results demonstrate that the two methods have the similar performance in the spine deformity illustration.

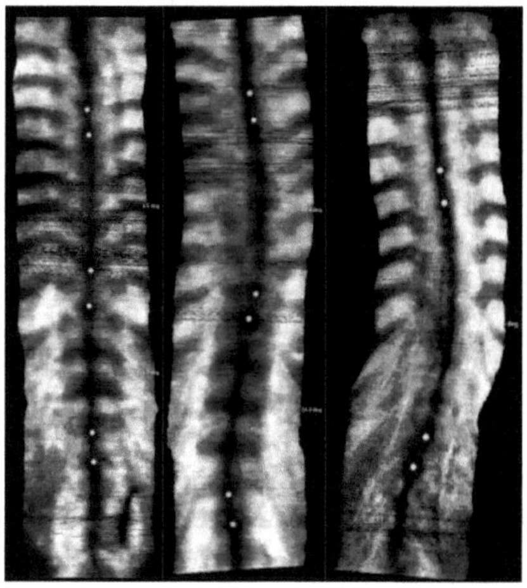

Fig. 4 The typical angle measurement results (Left to right: small angle to large angle)

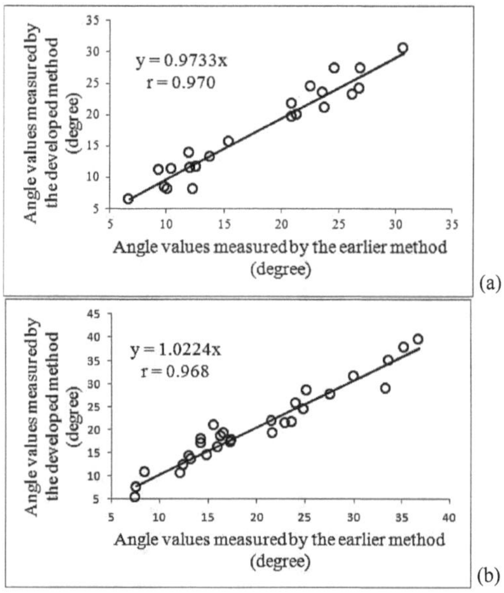

Fig. 5 The scatter plots of angle measurement results using the VPI method and the FUPI method. (a) The relationship for the thoracic regions. (b)The relationship for the lumbar regions.

IV. Discussion and Conclusion

In this study, a fast 3-D ultrasound projection imaging method for scoliosis assessment has been developed. Instead of the volume generation in conventional 3-D rendering method, this system directly projected the raw images to the coronal plane to get the projection images. One of the most important advantages of the new projection method is that it can greatly save the processing time. Compared with the conventional 3-D rendering method, this method could save as much as 10 times of processing time on average. The time saving is very useful for the clinical application of the Scolioscan system. With the processing time of about 15s, this system allows the patients to get the spine measurement images at once when the scanning is finished. The short processing time also allows the operator to review the projected image right after the scanning. If there are problems in the projected image, for example the missed region during the scanning, the operator can perform the scanning again immediately.

In conclusion, this study presented a fast 3-D ultrasound projection imaging method for scoliosis. The in-vivo experiment revealed that this method could greatly reduce the processing time while preserving the comparative image quality. It can be expected that this new method may help to provide fast 3-D ultrasound diagnosis of scoliosis in clinics.

Acknowledgements

This study was supported by the Research Grant Council of Hong Kong (PolyU5332/07E, PolyU152220/14E) and the Hong Kong Innovation and Technology Fund (UIM213). The authors would like to thank the generous supports from staff of Telefield Medical Imaging Limited for providing Scolioscan system and technical supports. Thanks are also given to Jack Chun-Yiu Cheng and Tsz-Ping Lam of The Chinese University of Hong Kong for their help in recruiting subjects and patients who participated in this study.

Conflit of Interest

The author Zheng YP owned a number of patents related to the Scolioscan system, which have been licensed to Telefield Medical Imaging Limited for commercialization. Zheng YP was currently a consultant for this company for the improvement of Scolioscan system.

References

1. Horne JP, Flannery R, Usman S (2014) Adolescent Idiopathic Scoliosis: Diagnosis and Management. Am Fam Physician 89: 193-198.
2. Tang SP CJ, Ng BKW, Lam TP (2003) Adolescent Idiopathic Scoliosis (AIS): An Overview of the Etiology and Basic Management Principles. HK J Paediatr 8:299-306.
3. Campbell RM, Smith MD, Mayes TC, et al. (2003) The characteristics of thoracic insufficiency syndrome associated with fused ribs and congenital scoliosis. J Bone Joint Surg Am 85A: 399-408.
4. Weinstein SL, Dolan LA, Spratt KF, et al. (2003) Health and function of patients with untreated idiopathic scoliosis - A 50-year natural history study. Jama-J Am Med Assoc 289: 559-567.
5. Rushton PRP, Grevitt MP. Comparison of Untreated Adolescent Idiopathic Scoliosis With Normal Controls A Review and Statistical Analysis of the Literature. Spine 38: 778-785.
6. Cobb J (1948) Outline for the study of scoliosis. American Academy of Orthopaedic Surgeons Instructional Course Lectures5:15.
7. Doody MM, Lonstein JE, Stovall M, et al. (2000) Breast cancer mortality after diagnostic radiography - Findings from the US Scoliosis Cohort Study. Spine 25:2052-2063.
8. Nash CL, Gregg EC, Brown RH, et al. (1979) Risks of exposure to X-ray in patients undergoing long-term treatment for scoliosis. J Bone Joint Surg Am 61:371-374.
9. Carman DL, Browne RH, Birch JG (1990) Measurement of scoliosis and kyphosis radiographs - Intraobserver and interobserver variation. J Bone Joint Surg Am72A:328-333.
10. Pruijs J, Hageman M, Keessen W, et al. (1994) Variation in Cobb angle measurements in scoliosis. Skeletal Radiol 23:4.
11. Suzuki S, Yamamuro T, Shikata J, et al. (1989) Ultrasound measurement of vertebral rotation in idiopathic scoliosis. J Bone Joint Surg Br 71:252-255.
12. Cheung CWJ, Zhou GQ, Law SY, et al. (2015) Freehand three-dimensional ultrasound system for assessment of scoliosis. Journal of Orthopaedic Translation 3:123-133.
13. Ungi T, King F, Kempston M, et al. (2014) Spinal Curvature Measurement by Tracked Ultrasound Snapshots. Ultrasound Med Biol 40:447-454.
14. Chen W, Lou EM, Zhang P et al. (2013) Reliability of assessing the coronal curvature of children with scoliosis by using ultrasound images. J Child Orthop7:521-529.
15. Cheung CWJ, Zhou GQ, Law SY et al. (2015) Ultrasound Volume Projection Imaging for Assessment of Scoliosis. IEEE T Med Imaging 34:1760-1768.
16. Gee A, Prager R, Treece G et al. (2002) Narrow-band volume rendering for freehand 3D ultrasound. Comput Graph 26: 463-476.

Author: Yong-ping Zheng
Institute: Interdisciplinary Division of Biomedical Engineering, The Hong Kong Polytechnic University
City: Hong Kong
Country: China
Email: ypzheng@ieee.org

Challenges and Opportunities in Carbon Nanostructured Designing

S.B. Abd Hamid

Nanotechnology & Catalysis Research Centre (NANOCAT), IPS Building, University of Malaya, 50603 Kuala Lumpur, Malaysia

Abstract— **Carbon with possibilities of different microstructures and properties offers a whole range of applications in the area of microelectromechanical systems (MEMS), and nanomechanical systems (NEMS). The behavior of these carbons are governed by their particular dimensionality type, and to a large extend on their spefic physical and chemical characteristics such as structural defects, presence of functional groups, and heteroatoms. Carbon catalysis knowledge is of great relevance for the preparation of novel or improved carbon-based C-NEMS and C-NEMS devices.**

Keywords— **C-MEMs, C-NEMs, Nanocarbon, Carbon chemistry.**

I. INTRODUCTION

Carbon is a very versatile material that, depending on its hybridization and assembly in a one-, two-, or three-dimensional network, exhibits important mechanical, optoelectronic, and chemical properties. Because of its versatility, carbon has found numerous applications, for example, it is found in printer inks, pencils, water purification systems, and thermal isolation and antistatic materials. More elaborate carbon materials such as carbon nanotubes (CNTs), graphene, fullerenes, and nanodiamond are used in sensing, electronic, or field emission applications. In addition, carbon materials with tunable microstructure and surface chemical properties are well suited as catalyst supports and the growth of secondary carbon nanofilaments onto a carbon fiber support is of special interest for composites, electronic, and porous macroshaped carbon materials.

Carbon, because of the various microstructures (with quite different properties) in which it can be found, offers many possibilities for the development of a whole new range of applications in the area of microelectromechanical systems (MEMS) and nanoelectromechanical systems (NEMS). Also, as widely discussed in the rest of this book, such peculiar characteristics allow one to improve the performance of carbon-based devices used in many applications. The behavior of these carbon materials is also governed by their particular dimensionality type, for example, 1D (carbon nanotube or other type of nanofilaments) or 2D (graphene), and depends to a large extent on their specific physical and chemical characteristics, such as structural defects, presence of functional groups, and heteroatoms. Therefore, a broad range of properties and characteristics may be reported for what is apparently the same type of nanocarbon utilized in say an NEMS and NEMS device.

II. CARBON MATERIALS

Carbon is the sixth element in the periodic table and its ground electronic state has a $1s^2$ $2s^2$ $2p^2$ configuration. The carbon atom hybridized orbitals account for the sp^3, sp^2, and sp^1 configurations of tetrahedral, trigonal, and linear symmetries, respectively. The sp^3 hybridization is the basis for diamond and aliphatic compounds, the sp^2 hybridization for graphite and aromatic compounds, and the sp1 for carbyne and two-dimensional ordered linear-chain carbon. The combination of these hybridizations in carbon materials is the underlying focus of interest for the synthesis of carbon–MEMS, as it has a profound effect on their properties. For instance, diamond, being sp^3 hybridized and featuring a tetrahedral geometry, is an isotropic material, transparent, super hard, and with negligible electronic conductivity. Graphite, on the other hand, with sp^2 hybridization is an anisotropic and soft material and a semimetal only within the basal plane.

Three different hybridizations of carbon exist, resulting in a large variety of nanocarbon structures with diamond, graphene, and carbyne (pure nonmixed materials) as limiting cases, respectively. With the term nanocarbon [1], we capture the broad range of carbon materials having a tailored nanoscale dimension and functional properties that significantly depend on their nanoscale features. CNTs and graphene belong to this class of materials and also nanofibers, nanocoils, nanodiamonds, nanohorns, nanoonions, and fullerenes. We thus make a distinction between nanocarbons and the more conventional type of carbon materials, including activated carbon (AC) and other types of porous carbon materials (such as mesoporous carbon materials [2] and carbon aerogels [3]) often utilized as adsorbents and catalysts supports, and the common type of carbon materials utilized in electrodes(such as glassy carbon [GC]) [4], graphite and related materials (pyrolytic and flake graphite and graphitic carbon), and carbon black [5]. Figure 1 shows a simplified classification scheme of carbon materials based on the hybridization of the constituent carbon atoms [6]. Although this figure shows only some of the possible carbon materials, it clearly demonstrates that the range of carbon materials is broad, each with different physicochemical properties and functional behavior.

F. Ibrahim et al. (eds.), *International Conference for Innovation in Biomedical Engineering and Life Sciences*,
IFMBE Proceedings 56,
DOI: 10.1007/978-981-10-0266-3_68

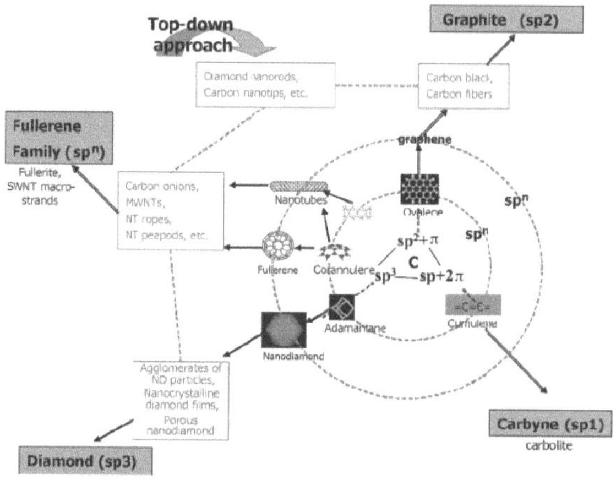

Fig. 1 Classification of carbon nanostructures. The mark 'sp^n' indicates intermediate carbon forms with a noninteger degree of carbon bond hybridization.

III. PROPERTIES OF NANOCARBONS

There are various characteristics that make nanocarbons attractive in MEMS and NEMS applications, such as the unique mechanical properties of graphene and CNTs, which include low mass, high Young's modulus, high thermal conductivity, and high surface area-to-volume ratio. However, as mentioned in the previous sections, these properties not only depend on the structure of the nanocarbon but also greatly depend on the specific characteristics, such as the presence of defects and heteroatoms. Therefore, the specific preparation of these materials may greatly influence their final properties. In addition, the integration of the synthesis of these nanomaterials with that of other materials is very important and introduces some additional constraints as processes might not be compatible, for example, in terms of annealing temperature.

Some nanocarbons (e.g., CNTs, graphene, and nanodiamond, the latter formally diamonds of very small size, 4 to 5 nm, but characterized from a diamond core and graphitic-like surfaces) have typically superior mechanical properties over silicon or conventional-type carbon materials, including those used in MEMS applications such as GC [7]. GC is a hard solid prepared by heat treatment at elevated temperatures (1,000–3,000°C) of polymeric precursors such as copolymer resins of phenol–formaldehyde or furfuryl alcohol–phenol. These polymers are used because of their high carbon yield on pyrolysis (the ratio of carbon present after or before carbonization is ~50 percent). GC is electrically conducting and impermeable to gases and has a low coefficient of thermal expansion. In addition, GC (having a fullerene-related microstructure) has a lower density (1.3 to 1.5 g/cm^3) than that of graphite (2.27 g/cm^3) or diamond (3.52 g/cm^3), because of its porous microstructure (related to bubbles formed during the pyrolysis process). The voids constituting the pores are not connected, and for this reason, GC shows low gas permeability. Moreover, GC has a very high corrosion resistance and inertness under a wide variety of conditions, allowing its use in very corrosive environments. Although some of these properties are better than those of nanocarbon materials, the Young's modulus of GC (10 to 40 GPa) is lower than that of silicon and much lower than that of nanocarbons. On the other hand, this may be useful in some applications, such as in electrostatically driven actuators, where the stiffness of the material is a disadvantage.

CNTs usually exhibit an elastic strain of 5 percent, and therefore, they show plastic deformation before fracture. This property is promising for mechanical energy storage. Defect-free CNTs are the stiffest and strongest materials yet discovered in terms of tensile strength (63 GPa) and elastic modulus (~1 TPa). In Table 1, we compare mechanical properties of CNTs with stainless steel [8]. CNT fibers have the potential to serve as ultrahigh strength (yet lightweight) fibers, electronic interconnects, power conversion devices, motors, transformers, and electro-storage devices.

Table 1 Comparison of mechanical properties of CNTs with stainless steel

Material	Young's modulus (TPa)	Tensile strength (GPa)
SWCNT	~1 (from 1 to 5)	13–53
MWCNT	0.27–0.95	11–150
Stainless Steel	0.186–0.214	0.38–1.55

IV. USE OF CARBON MATERIALS IN MEMS AND RELATED DEVICES

The superior mechanical, electrical, and structural properties of CNTs have resulted in their utilization in a wide range of fields. For example, implementing these materials into flexible or stretchable systems can lead to flexible microelectronics as well as to pressure monitoring devices inside the human body and energy scavenging from lowfrequency motions. The realization of the potential of carbon materials in MEMS and related devices requires the ability to carefully manipulate CNTs, especially on the micro- and nanoscale. The nanodiameter and high surface area of CNTs induce high van der Waals forces between the tubes, resulting in a strong tendency to aggregate with each other in order to minimize the total surface energy. Hence, CNTs have very limited dispersibility in most solvents, but today, it is possible to tune their solvency by controlled functionalization to modify their properties [9].

CNTs, and in general nanofilaments, have been extensively applied in MEMS and NEMS. For example,

CNT-based resonators (where CNTs act as the resonating mass detector element) show the highest mass sensing resolution ever established [10]. A single-clamped DWCNT beam was used instead of an SWCNT to avoid issues related to the chirality of CNTs. Owing to the smaller material density compared to other materials, it is possible to obtain a lower natural resonance frequency, resulting in larger sensor responses at lower resonant frequencies. The optimal sensitivity achieved is of the order of 1.3×10^{-25} kg $Hz^{-1/2}$, but depends on the specific characteristics of the DWCNT, which vary within some degree from filament to filament.

Graphene is another nanocarbon-type material increasingly used in C-MEMS and C-NEMS applications. The low mass and high surface area-to-volume ratio make graphene an excellent candidate for mass and gas sensing applications, respectively. Specifically, graphene-based electromechanical resonators have been demonstrated by using suspended graphene structures over silicon dioxide trenches.

The graphene-based device can be utilized as a gas sensor by monitoring the frequency shift because of mass changes related to the absorption of a gas. Sensitivities of 10–21 g are possible and in theory even higher sensitivities could be reached [10]. In addition to this type of graphene-based mechanical resonators, graphene-based chemical resistors, and field effect transistors have been utilized for gas sensing applications as well, because the resistance of graphene is sensitive to gas molecules [11]. Table 2 summarizes various graphene-based materials in sensing applications.

Table 2 Examples of graphene-based materials in C-MEMS and C-NEMS applications

Sensor	Materials	Detection mode
Physical	Graphene or PDMS	AFM or resistance
Chemical	Monolayer or few layer graphene	FET resistance change
Biosensor	Graphene	Conductance

V. THE SYNTESIS OF NANOSTRUCTURED CARBONS

The synthesis of carbon nanomaterials and the tuning of their peroperties (by posttreatment, functionalization, control of defective state, preparation of hybrid materials, etc) is a scientific area of fast growing interest. The main techniques used to prepare are by CCVD. Attention is focused on how specific properties of the carbon materials prepared are governed by process conditions, carbonaceous source and catalyst used.

VI. CHALLENGES AND OPPORTUNITIES

The knowledge base on nanostructured carbon materials has increased dramatically over the past decade, in terms of better controlled synthesis methods (for different types of materials, nanoarchitectures, and uniformity of the resulting products), posttreatment procedures to tune the carbon materials' properties and functionalities, and in terms of a more advanced understanding of the microstructure and resulting chemical and mechanical properties. These advances have enabled the development of novel nanostructured carbon materials for a large variety of applications from polymer composites to catalysts and adsorbents, materials for advanced energy applications (from energy storage—Li-ion batteries, supercapacitors—to energy conversion—fuel cells, solar devices) and C-MEMS and C-NEMS. In the area of energy, the use of CNTs in battery electrodes enables large electrode–electrolyte contact areas for Li-ion batteries, new storage mechanisms that are not possible in bulk materials, and shorter pathways for both electron and ion transport. For applications in supercapacitors, CNTs offer interesting new possibilities when used in composite materials and to develop electrodes [12]. In dye-sensitized solar cells (DSSCs), nanocarbons can improve charge separation and electron transport and offer novel possibilities for a more efficient architecture, for example, to improve electron collection or develop transparent counter electrodes [13]. In PEM (proton exchange membrane) fuel cells, the use of nanocarbons allow to obtain a more efficient metal dispersions (e.g., lower noble metal loading), improve the stability of the electrodes, and lead to optimal hierarchically organized designs for the electrodes to reduce mass transfer limitations [14].

In C-MEMS and C-NEMS, many novel possibilities exist, as shown in the previous sections. In addition, to graphene and CNTs, other types of nanocarbon materials, including hybrid systems, are available today. In general, these materials offer many new opportunities compared to silicon or even than the more conventional types of carbons such as GC. It should also be mentioned that many methods are available for controlling the intrinsic properties of graphene and CNTs, such as doping, introduction of surface functional groups, and creation of hybrid nanocarbon materials with other types of nanocarbons (e.g., fullerene-type molecules grafted to graphene or CNTs) or with metal or metal oxide nanoparticles. There is thus a plethora of possibilities to tune the properties of nanocarbons and enhance their performance in applications such as sensing and actuating systems as presented earlier.

Integration of nanocarbon materials in commercial C-micro and C-nano devices is still a major challenge, especially in terms of largevolume manufacturing of such devices with reproducible performance. There are, in general, two possible approaches to solve this conundrum: (1) integrated processes to fabricate nanocarbons (e.g., graphene or CNTs) together with other components directly onto the same wafer and (2) transfer processes where we prepare first the high-quality nanocarbons (on a different substrate) and then transfer them to a C-MEMS and C-NEMS device. While the first method is preferable in theory, it often lacks the capability to produce high-quality nanocarbons, which are necessary to achieve good performances. On the other hand, the second approach may lack the capability to attain specific nanoarchitecture that are relevant for the properties and functional behavior of the nanocarbon materials. Therefore, an optimal approach is probably a combination of these two pathways [15].

VII. CONCLUSIONS

As a general concluding remark, and as outlined in the previous sections, not only the nanostructure itself is relevant, but also the specific characteristics of the nanocarbons (type of defects, surface functional groups, heteroatoms, nanoarchitecture, etc.) that determine their performances. A single CNT can go from metallic to insulating behavior, passing through a semiconductor behavior, even if, formally, the nanostructure is the same. The method of preparation should account for these aspects, but a detailed characterization of the carbon nanomaterials remains mandatory. These considerations are valid in general for all types of carbon nanomaterial applications, including the emerging area of energy storage and conversion as well as catalysis. Knowledge in these scientific areas is growing rapidly, with special emphasis on controlled synthesis of nanocarbons, their posttreatment, and understanding of the nature of defects and surface functional groups. This has been achieved by combining novel characterization methodologies with theoretical modeling. This knowledge can now be used for setting a quantum leap forward in C-MEMS and C-NEMS and achieving superior device performance.

ACKNOWLEDGMENT

Financial support from Ministry of Higher Education TRGS TR002B-2014 B and UM Flagship Project FL001B-14 AET are acknowledged.

CONFLICT OF INTEREST

The authors declare that they have no conflict of interest.

REFERENCES

1. Messina, G., and S. Santangelo. 2006. Carbon Topics in Applied Physics. Vol. 100. Berlin, Germany: Springer.
2. Serp, P., and B. Machado. 2015. Nanostructured Carbon Materials for Catalysis. RSC Catalysis Series
3. Gogotsi, Y., and V. Presser. 2014. Carbon Nanomaterials. 2nd ed. Boca Raton, FL: CRC Press.
4. Da, L. (2006. Carbon Nanotechnology. Amsterdam, Netherlands: Elsevier.
5. Tanaka, K., and S. Iijima. 2014. Carbon Nanotubes and Graphene. 2nd ed. Amsterdam, Netherlands: Elsevier.
6. Lu, W., J.-B. Baek, and L. Dai. 2015. Carbon Nanomaterials for Advanced Energy Systems: Advances in Materials Synthesis and Device Applications. Hoboken, NJ: Wiley
7. Centi, G., S. Perathoner, and D.S. Su. 2014. "Nanocarbons: Opening New Possibilities for Nano-Engineered Novel Catalysts and Catalytic Electrodes." Catalysis Surveys from Asia 18, no. 4, pp. 149–63. doi: http://dx.doi.org/ 10.1007/s10563-014-9172-0
8. Hummelgen, I.A., N.J. Coville, I. Cruz-Cruz, and R. Rodrigues. 2014. "Carbon Nanostructures in Organic WORM Memory Devices." Journal of Material Chemistry C: Materials for Optical and Electronic Devices 2, no. 37, pp. 7708–14. doi: http://dx.doi.org/10.1039/c4tc00816b
9. Hanein, Y. 2010. "Carbon Nanotube Integration into MEMS Devices." Physica Status Solidi B: Basic Solid State Physics 247, no. 11–12, pp. 2635– 40. doi: http://dx.doi.org/10.1002/pssb.201000109
10. Podyacheva, O.Yu., and Z.R. Ismagilov. 2015. "Nitrogen-Doped Carbon Nanomaterials: To the Mechanism of Growth, Electrical Conductivity and Application in Catalysis." Catalysis Today 249, pp. 12–22. doi: http://dx.doi. org/10.1016/j.cattod.2014.10.033
11. Gebhardt, P., and D. Eder. 2014. Nanocarbon-Inorganic Hybrids, eds. D. Eder and R. Schlögl, 3–23. Berlin, Germany: De Gruyter Pub.
12. Shi, H., Y. Shen, F. He, Y. Li, A. Liu, S. Liu, and Y. Zhang. 2014. "Recent Advances of Doped Carbon as Non-Precious Catalysts for Oxygen Reduction Reaction." Journal of Material Chemistry A: Material for Energy and Sustainability 2, no. 38, pp. 15704–16. doi: http://dx.doi.org/10.1039/ c4ta02790f
13. Chen, Y., and X. Sun. 2013. Applications of Carbon Nanotubes, ed. A.K. Mishra, 195–210. New York, NY: Nova Pub.
14. Rinaldi, A. 2010. Synthesis of Carbon Nanotubes on Carbon supports and the Purification of Carbon Nanotubes. Jakarta.
15. Liang, C., Z. Li, and S. Dai. 2008. "Mesoporous Carbon Materials: Synthesis and Modification." Angewandte Chemie International Edition 47, no. 20, pp. 3696–717. doi: http://dx.doi.org/10.1002/anie.200702046

Author: Prof Sharifah Bee Abd Hamid
Institute: Nanotechnology and Catalysis Research Centre, Universiti Malaya. 50603 Kuala Lumpur. Malaysia.
Email: sharifahbee@um.edu.my

Author Index

Keyword Index

MIX
Papier aus verantwortungsvollen Quellen
Paper from responsible sources
FSC® C105338

If you have any concerns about our products,
you can contact us on
ProductSafety@springernature.com

In case Publisher is established outside the EU,
the EU authorized representative is:
Springer Nature Customer Service Center GmbH
Europaplatz 3, 69115 Heidelberg, Germany

Printed by Libri Plureos GmbH
in Hamburg, Germany